AF815260

Issues in Business Ethics

Eminent Voices in Business Ethics

Volume 53

Series Editors
Mollie Painter-Morland, Nottingham Trent University Business School
Nottingham, UK
Frank den Hond, Department of Management & Organization,
Hanken School of Economics, Helsinki, Finland

Eminent Voices in Business Ethics is a sub-series of Springer's long-running book series, *Issues in Business Ethics*. Eminent Voices aims, over a period of time, to showcase the lifetime's work of distinguished scholars who have played a leading role in building the field of business ethics; in particular those who have contributed to furthering the philosophical discourse in business ethics and a discussion of its professional implications. These scholars reflect both on how the field of Business Ethics has evolved and how they see it developing, commenting on their contributions to that process and responding to critics work that attracted considerable comment.

Sergiy D. Dmytriyev • R. Edward Freeman
Editors

R. Edward Freeman's Selected Works on Stakeholder Theory and Business Ethics

Springer

Editors
Sergiy D. Dmytriyev
College of Business
James Madison University
Harrisonburg, VA, USA

R. Edward Freeman
Darden School of Business
University of Virginia
Charlottesville, VA, USA

ISSN 0925-6733 ISSN 2215-1680 (electronic)
Issues in Business Ethics
ISSN 2543-0491 (electronic)
Eminent Voices in Business Ethics
ISBN 978-3-031-04563-9 ISBN 978-3-031-04564-6 (eBook)
https://doi.org/10.1007/978-3-031-04564-6

© Springer Nature Switzerland AG 2023
This work is subject to copyright. All rights are reserved by the Publisher, whether the whole or part of the material is concerned, specifically the rights of translation, reprinting, reuse of illustrations, recitation, broadcasting, reproduction on microfilms or in any other physical way, and transmission or information storage and retrieval, electronic adaptation, computer software, or by similar or dissimilar methodology now known or hereafter developed.
The use of general descriptive names, registered names, trademarks, service marks, etc. in this publication does not imply, even in the absence of a specific statement, that such names are exempt from the relevant protective laws and regulations and therefore free for general use.
The publisher, the authors, and the editors are safe to assume that the advice and information in this book are believed to be true and accurate at the date of publication. Neither the publisher nor the authors or the editors give a warranty, expressed or implied, with respect to the material contained herein or for any errors or omissions that may have been made. The publisher remains neutral with regard to jurisdictional claims in published maps and institutional affiliations.

This Springer imprint is published by the registered company Springer Nature Switzerland AG
The registered company address is: Gewerbestrasse 11, 6330 Cham, Switzerland

Paper in this product is recyclable.

About Ed Freeman as a Scholar and a Person: An Introductory Essay to R. Edward Freeman's Selected Works – Including Personal Stories from Some Very Special People, and a Reader's Guide to the Book

Guiding readers through this book is both an honor and a pleasure. I have known R. Edward Freeman for seven years – first, as my academic advisor and then as my mentor. So I thought quite naively that I knew a lot about Ed. However, when I started working with him on this volume, I re-discovered him from many other interesting angles and I am excited about the opportunity to share my discoveries here.

This introductory essay to the book comprises three parts. In the first part, I invite the readers to get to know Ed as a person and as a scholar. The idea here is that before reviewing Freeman's selected works, it would make sense to learn more about the author, his personality, and life. This section of the introduction is based on my personal reflections of interacting with Ed during numerous occasions as well as on a couple of interviews that I conducted with him while preparing the book. Second, I asked several people who are very special for Freeman to briefly share their reflections on what role he played in their lives. I am very grateful that all those I asked have eagerly gotten back to me with their vibrant narratives about Ed. Finally, the last section of the introduction provides an overview of the book's structure and content to guide the readers to those works of Ed Freeman that may be of particular interest to them.

About R. Edward Freeman as a Scholar and a Person

Ed's life has been full of unexpected turns and organically settled paradoxes. However, when people read through Freeman's bio sketches handed out during his talks and other events, they usually encounter neatly organized numeric data, i.e., the number of universities that awarded him with honorary degrees and the number of books and articles he has published. Not so many people know what has brought Ed to these numbers (and big numbers they are!). In this essay, I will try to make up for this information asymmetry by highlighting some interesting events and anecdotes from Freeman's academic career and life.

Education and the First Turning Point (Getting on with the Tragedy) Freeman did his undergraduate studies at Duke University where he originally started in the teaching track intending to become a public-school math teacher. He loved math and excelled at it. In fact, he was so taken by the subject that by his junior year at Duke, he had taken two graduate-level math courses and wondered whether he should perhaps pursue a graduate degree in mathematics. Soon enough, Ed found himself doing what one may call pure mathematics since the only numbers he saw were at the bottom of a page. Though he successfully got through these graduate courses, it became clear for him in the process that he reached his best level in math after stumbling on Lebesgue integration.

At the same time, there was another subject that captivated him to a similar degree – philosophy, and he turned to it wholeheartedly. His first course in philosophy was a Logic class, which resonated with Freeman a lot – this was something he considered himself to be good at because, as he puts it, "logic is just math." He also started playing handball together with his Logic class professor, and during one of these games, the professor introduced Ed to his colleague who was teaching Ethics. Later on, Freeman took his class and got very much fascinated with the subject of morality. With each subsequent philosophy course, he only loved it more and more. Ed recalls that studying Plato was groundbreaking for him as all his previous views on philosophy had been literally swept away. It was especially hard to stay on his feet after studying Plato's renditions of Socrates's dialogues in which Socrates questions how we should live. When he first read the dialogues, Ed was a nineteen-year-old student. Today, fifty years later, Freeman admits that he is still very much into that question.

Ed remembers an episode from one of his Ethics classes with Duke professor Ted Benditt, whose teaching style focused on asking question about ethics. This question-posing class dynamics went on for several seminars until one of the students asked the professor how he understood ethics himself. At first, Benditt got confused but then he replied, "I don't think you understand how this works. This is not about answering questions. This is about asking better questions." Ed recalls that for him that was a moment of pure liberation because he thought he was good at answering questions, but he had never considered before the art of asking better questions. This understanding opened whole new possibilities for Freeman.

Duke was certainly a transformative educational and life experience for Ed, yet it was not a well-paved road. At one point, he considered quitting the school altogether due to financial constraints – he came from a small community in rural Georgia and his family could hardly afford college tuition and expenses, and certainly not an expensive one like Duke. Fortunately, Freeman received a fellowship covering his senior year and one year of a graduate program that helped him continue his studies in philosophy. The fellowship required him to teach and do administration in a junior or community college. The first class he taught was not on math or philosophy but on … chess (today, a chess table stands in the middle of his living room with pieces always ready for a game or two).

While Freeman originally wanted to pursue a teaching career, one of his professors talked him out of the teaching track. At the same time, Ed's interest in philosophy got so strong that he started thinking about applying for a PhD program in philosophy. Soon his plans materialized. After receiving his bachelor's degree in mathematics and philosophy from Duke, Freeman got admitted into the doctoral program in philosophy at Washington University in St. Louis.

Inspired by Rawls's *Theory of Justice*, Freeman in his doctoral work built on the veil of ignorance (a thought experiment used for reasoning about the principles that should govern the societal structure). More specifically, Ed explored decision-making without known probabilities and illustrated it through a formal proof of nihilism, outlining several contradictions in the way we think about decision-making. Yet, his love for math did not dissipate – in his dissertation, Freeman relied heavily on mathematical concepts to support his philosophical argumentation. Given that philosophy is perceived by many as a "soft" discipline where academics do not often deal with advanced scientific formulas, merging math and philosophy made Ed's dissertation quite technical. He has never pursued publication of it.

During his last years of the PhD program, Freeman started looking for opportunities to go to Pennsylvania where his girlfriend was going to do her master's at the University of Pennsylvania (UPenn). Fortunately, one of the professors on Ed's dissertation committee mentioned that there was an open position at the Busch research center at the Wharton School of Business at UPenn and Freeman pursued this opportunity in 1976. Ed's decision was not taken well by his academic advisor at Washington University as he hoped Freeman would stay to teach philosophy, but Ed had already made up his mind and was hired as a researcher by the Busch Center at Wharton in 1976.

In the summer 1980 while continuing to work as a researcher at Wharton, there came another turning point in his life. Ed's older brother was killed in an automobile accident, and it made Freeman really think about what he actually wanted to do with the rest of his life. He needed to find a way to get on with the tragedy. Being in his mid-20s, he knew for sure that he wanted to teach and be a professor, and he also realized more than ever that he wanted to be close to the people he loved. He discussed his future plans with Maureen (the girlfriend he went to Wharton to be with and now wife since 1977), and she was (and still is) very supportive.

Practicing Stakeholder Management and the Second Turning Point (From Consulting to Academia) In addition to scholarship, the research center at Wharton also had a practical orientation as it provided consulting services to companies. Joining the research center at Wharton was for Freeman a major change as he moved from the realm of philosophy to the business domain. Prior to that, he had never even heard about Wharton. Fortunately, Ed found a home in this unfamiliar world. People at Wharton did not care about where one had come from – what mattered to them was what one could actually do. Everyone at the research center was obsessed with helping their clients and this resonated with Freeman. Business appeared to be interesting for Ed, and he also valued the culture of freedom at the research center. About a year after joining, the research center was renamed to Wharton Applied

Research Center (WARC). James Emshoff, ex-McKinsey consultant and a professor at Wharton, led WARC which hosted six scholars working there on a permanent basis. They ran WARC as a management consulting firm – providing advisory services to companies' top management and worked on ideas in their research areas. For Freeman, this research area later materialized into what we know as stakeholder theory and Ed remembers those days with warmness.

Freeman originally thought that his move from philosophy to business would be temporary and he actually planned on going back to philosophy soon. For the first two years at WARC, he kept attending the American Philosophical Association meetings and applying for jobs in philosophy. However, work at WARC kept him busy and, what is more, he really enjoyed it. Little by little, the thought of returning to philosophy disappeared from his horizon. Ed saw that philosophy scholars were not interested much in business ethics since they certainly were not interested in business. Freeman, on the other hand, developed a genuine interest in theoretical and practical aspects of business. Moreover, the firms for which Ed did consulting projects through WARC seemed to value his ideas. In fact, during that time, Freeman received a couple of job offers in the private sector that he seriously considered. In the end, however, he decided to stick with academia.

After several years as a researcher at WARC and a part-time lecturer at Wharton, Freeman decided to leave consulting projects and focus completely on his academic career. The second important transition point came in 1981 when he got an assistant professor position at Wharton. Interestingly enough, it was Ed's consulting work at WARC that actually determined his academic interests. Indeed, WARC was helping companies figure out how to do projects under the ideas that later formed the core philosophy of stakeholder management. Although not yet strictly formulated, this philosophy was so-to-say in the air in both the Busch Center and WARC. Some of the real pioneers of stakeholder theory such as Russel Ackoff, Eric Trist, James Emshoff, Ian Mitroff, and others were scholars that Freeman could learn from. So on one of those days, Jim Emshoff, the Founding Director of WARC, approached Freeman with a question of whether he would like to make something of stakeholder ideas research-wise. Could he figure out what it would be like if executives actually made decisions based on creating value for stakeholders? Originally the idea was used mostly as a way to organize strategic planning. Without much thought, and probably naïvely Ed gladly agreed to give it a try.

There was another event that strengthened Ed's confidence in pursuing this novel research agenda. In the early 1980s, Freeman delivered a presentation at an academic event in Pittsburg, where he mentioned some stakeholder management ideas and his plans on writing a book around those ideas. Bill Frederick, a senior scholar in the field, took Ed aside and told him that "Whatever you do, you need to write this book. You've got something to say that people need to hear." That was just the kind of encouragement that Ed needed. Being young (in his early 30s) and trained as a philosophy scholar, he was very much a novice in the management science field. So the advice from a very senior person in the field meant a great deal to him.

Further down the road, Ed saw even more encouragement coming his way. In 1981, he attended a conference on Corporate Governance at UCLA where he delivered a paper on stakeholder theory and boards[1]. There were many senior scholars in the room who, along with Michael Jensen, could be considered the "Who is Who" of corporate governance. Among the conference guests, there was also Harold Williams, the Chairman of U.S. Security and Exchange Commission under President Jimmy Carter. After Freeman finished, Williams raised his hand. Ed recollects that he expected nothing but criticism from the senior government official for daring to put stakeholders (a relatively new term at that time) on par with stockholders, a group that had unequivocally been revered as the sole beneficiary of firm's financial gains. For a short while, Ed thought that that was it – the end of his young bright academic career. How surprised was he, however, when Harold Williams actually supported his propositions! Ed stood there completely speechless, while Williams went on and on defending Freeman's "brave new" ideas. While Williams' support for stakeholder management did not translate into government policies (Williams left the office soon), it had a strong impact on Ed. It became clear to him that if stakeholder ideas made sense to a senior government official, they were undoubtedly worth fighting for.

Reflecting on Stakeholder Management and the Third Turning Point (Book) In 1982, Freeman wrote the first draft of his seminal book on a stakeholder approach. Even though he had four more years before he had to go up for tenure, he realized it was extremely hard to get tenure at Wharton. He began looking for alternative opportunities to continue his academic career. In the Spring of 1983, the University of Minnesota offered Ed a tenured position, and he decided to join its school of management (now the Carlson School of Management).

Freeman revised his manuscript on stakeholders, and the book got published in the summer of 1983 with a 1984 date under the title *Strategic Management: A Stakeholder Approach*. Later, Ed thought that he should have named his book *Stakeholder Management*. However, considering the entrenched power and indisputability of the shareholder value creation concept at the time, it was probably the right decision to stick to the original title as a compromise – it would have been too risky to focus directly on stakeholders. The word "strategic" in the title defined the book within a familiar and recognizable academic milieu, and as such, it made the book attractable at least for strategic management scholars.

When writing the book, Freeman thought of it as a textbook, though over time he admitted that it did not suit that purpose well. Instead, the ideas proposed in the volume eventually found wide acceptance in the scholarly world. However, the path to recognition was not that straight and easy. At first, there were only a few people in the Social Issues in Management division (SIM) of the Academy of Management (AOM) and the Society for Business Ethics (SBE) who saw promise in Ed's ideas. Within the next 20 years, stakeholder theory gradually travelled to other fields such as strategy, public policy, education, health care, and many others. At one point, one of Freeman's friends decided to track Google searches on the word "stakeholder"

[1]That paper later became one of the chapters in his book, and its shorter version was published in *California Management Review*.

and found there were not that many until the early 1990s when stakeholder ideas really hit the ground running. Today, the book *Strategic Management: A Stakeholder Approach* alone has over 50,000 citations, and beyond the book, there are hundreds of thousands of works on stakeholder theory.

The book became the next major turning point for Freeman. It would not be an overstatement to say that the book changed his life. More than that, it also provided, and continues to provide, ideas and inspiration for many others. With the publication of the book, Ed has become widely known among academics and practitioners as the founder of stakeholder theory, even though he has always modestly insisted that others were the real founders and that he gets far too much credit. Here comes the surprise, however. When working on his book, Ed actually did not regard the stakeholder idea as the most interesting and prominent one in the book. In fact, he originally thought the idea to be rather obvious, and even commonplace. Put simply, he was sure that everyone realized (without his telling them) that stakeholders were important for firms! He had no clue yet that he would have to spend the next several decades engaged in persistent conversations with management scholars and practicing managers trying to persuade them that stakeholders mattered.

What did Freeman see as the most original idea of his book then? Well, he thought that his main contribution would be in what he termed the enterprise strategy – a connection between a firm's activities and its purpose, ethics, and values. He criticized the approach in which the first question that strategy scholars and practitioners were trying to ask was "where are we going?" or "what business are we in?" On the one hand, there was nothing wrong with this logic because a firm would not succeed if its management did not know where they should be heading or what business they were in. On the other hand, Ed believed it was wrong to start developing a strategy plan with that question as the first one in mind. He suggested there should be quite a different starting point for strategy development where a higher-level question precedes anything else – namely, "what do you stand for?" Ed thought that re-grounding strategy in the "what you stand for" idea was the central insight of his book. The idea that what one stands for would determine where one should go or what business one should be in was very much new at that time. Even today, the strategy field has not come to that idea yet as for many the answer to "what you stand for" would be gaining competitive advantage to build shareholder value, which is not what Freeman envisioned in his enterprise strategy.

Business Ethics and the Fourth Turning Point (Darden) At WARC, Wharton, and Minnesota, Freeman always considered himself a business strategy guy who worked, taught, and wrote on strategic matters. At Minnesota, Ed also started teaching business ethics. This teaching experience in tandem with the reluctance of strategy scholars to see beyond shareholder value maximization, pushed Freeman away from strategic management and toward the business ethics field. After connecting to business ethics scholars such as Pat Werhane and Tom Donaldson, Ed learned more about the business ethics movement. He started attending business ethics conferences and taking a more active role in the movement. Over time, Ed served in leadership roles for SBE and SIM, and most recently, Freeman served as a co-editor-in-chief for the *Journal of Business Ethics*.

During his time at Minnesota, Freeman developed a very productive work relationship with his doctoral student Dan Gilbert. Together, they wrote several books and articles. Their co-authored book *Corporate Strategy and the Search for Ethics* (see its excerpt in Chap. 22) can be considered as a sequel to Freeman's 1984 book, even though it did not receive the same traction.

Although overall things were going well for Freeman at Minnesota, he felt not quite himself. It was teaching that was missing from his personal happiness puzzle. Unfortunately, teaching was not prioritized in Minnesota. Ed recollects how one year he was very happy to receive the Outstanding Teacher Award at the School of Management – he thought it to be a great achievement. His happiness quickly dissipated, however, when during the award-granting ceremony a senior professor took Ed aside to tell him that he had to spend more time doing research rather than collecting teaching awards.

Freeman became open to new options where he could enjoy the best of the two worlds – research and teaching. During his project work at a large telecom company, he met two scholars from Darden (the Darden School of Business at the University of Virginia) who told him that Darden took teaching and student learning very seriously. This certainly appealed to Ed, so Alex Horniman (who ran the Ethics Center at Darden) and John Rosenblum (the Darden Dean) invited Freeman to apply for a position at the school. Soon afterward, Ed joined Darden, first as a visiting and then as a full-time professor. Working alongside Horniman and Rosenblum, he soon learned that they were great teachers themselves and Ed was happy with his choice.

Moving to Darden in 1987 was the next major transition point that opened a new era in Freeman's life. He started seeing himself primarily as a business ethicist rather than a strategy scholar. Darden offered Freeman to teach ethics as well as to lead the Olsson Center for Applied Ethics. Ed liked it that Darden did not draw a clear line between strategy and ethics – his department was named Strategy, Ethics, and Entrepreneurship with an intention to integrate ethics in other courses taught throughout the business school.

For Freeman, work at Darden resulted in long and productive relationships with many colleagues. Since joining Darden, he has been awarded with countless teaching, research, and service rewards from the business school, the University of Virginia (UVA), and many other universities and scholarly organizations. In 2010, Ed Freeman was given the University Professor title at UVA, which is a great honor and recognition that allows teaching at any school within the university. During his 35-year teaching career at Darden, Freeman has taught tens of thousands of students at graduate and executive levels. He has also prepared many PhD students with whom he continues to keep close and warm relationships after their graduation. Most notable was in 1993 when Darden hired another philosopher, Patricia Werhane, for its Ruffin Chair in Business Ethics. Freeman and Werhane established a lifelong and productive relationship. Finally, Freeman had another philosopher to talk to, and Werhane had been one of the founders of the academic discipline of business ethics. Together they worked to make Darden a center of excellence for business ethicists around the world.

The Final Turning Point (Pragmatism) Trained as a philosopher, Freeman made several attempts to ground stakeholder theory in different philosophical schools of thought, such as Kantian deontology, feminism, and libertarianism. However, none of these initial attempts really worked to his satisfaction. A turning point came about when he joined Darden and met Richard Rorty who was then a university professor at UVA. When Ed was still teaching at the University of Minnesota, he came across Rorty's book *Consequences of Pragmatism* (1982) in the local bookstore (it was published by the University of Minnesota Press). By the time Ed was to move to Darden, he had already read pretty much everything that Rorty had written. Now Freeman was looking forward to an opportunity to meet Rorty in person. So before coming to Charlottesville, he wrote a letter to Rorty saying that "your description of the typical philosophy department to the question 'who should teach Hegel – the answer is hopefully no one for the reason of mental hygiene' perfectly describes how I was trained." Long story short, after moving to Darden, Freeman, and Rorty became friends. The more they conversed, the more Ed realized that Rorty's ideas could lead him to ask better questions.

Under Rorty's pragmatism, Ed reconsiders and rejects the fact and value, normative and descriptive, and business and ethics dichotomies. What is more, Freeman realizes that, in comparison to other schools of philosophy, the pragmatist vocabulary is much better suited for business scholarship. While Kant's second imperative continues to be applicable – people cannot be treated as mere means to one's ends, the propositions of universal will and generalizing maxims become less so. Thus, in developing stakeholder theory, Freeman has gradually moved away from Kantian and other philosophies toward Dewey's and Rorty's pragmatism (cf. Chaps. 42 and 52).

Rorty's pragmatism was qualitatively different from the pragmatist ideas espoused by Charles Peirce, W.V. Quine, and Nelson Goodman. Freeman was trained in the latter tradition. In fact, Quine's logical empiricist tradition had a strong presence at Washington University at the time Ed was doing his PhD there. For instance, one of Freeman's dissertation committee members, Joe Ullian, co-authored with Quine a well-known volume *The Web of Belief*. Ed, however, found their doctrine to be too rational. It was Rorty who revived Freeman's interest in pragmatism.

Ed started reflecting on Rorty's ideas in his own writings and published an essay in *Academy of Management Review* on the relevance of Richard Rorty to management research (see Chap. 29). Yet another outcome of the reflections on Rorty's ideas was Freeman's criticism of the academia at several scholarly events. When Ed was elected the SIM Division Chair at the AOM in 1988–1989, in his Presidential address called "Let's disband the Academy of Management," he criticized business schools and scholars for not being very serious about seeing themselves as intellectuals (see Chap. 24). The criticism was so pronounced that when the speech was over, Archie Carroll raised his hand and asked "Where do you think you would be able to publish it?" and Freeman replied "That's exactly the point!". He knew it was in fact unpublishable and appears here for the first time. A somewhat similar situation repeated when Freeman delivered his Presidential address to SBE in 1995. His talk was named "The Business Sucks Story," and in it, Ed criticized the way business scholars and practitioners viewed business. Remarkably, Freeman's Presidential address to SBE was published only 25 years later (see Chap. 28).

So what can we surmise about Ed's academic identity? In the 1980–1990s, Ed felt that he was neither fish nor fowl – he was not really a business school professor but also felt like he was hardly a philosopher anymore. At that time, there were only a few people outside of Werhane and their students he could relate to – philosophers simply did not teach management. However, Ed has never given up on his love for philosophy. Rather, he has built on his philosophical knowledge to understand, create and develop his business teachings. Today, Freeman strongly believes that philosophers do have something to say to business people (see his recent book *Humanizing Business: What Humanities Can Say to Business*). In fact, Freeman saw part of his role in academia as getting more philosophical thinking into business schools.

Creativity and Some Other Interesting Facts About Ed Freeman The story about Ed would not be complete if we omitted one phenomenal ability he has developed – creativity. In his classes and research projects, Freeman is often seen generating ideas or prompting others to explore novel and promising paths. I observed Ed on numerous occasions and my take on him is that his personality is full of paradoxes. However, these contrasting dualities coexist quite organically and, in fact, push him toward creativity. I counted at least nineteen paradoxes, though there are probably many more: utilizing space, pursuing diverse interests, coalescing ignorance and knowledge, being a philosopher in business, writing books in the field favoring articles, focusing on some ideas but becoming famous for other ones, both defying the crowd and listening to his audience, reading unrelated literature and making it relevant for management, teaching business students via theater and music, being a great teacher who is wired as an introvert, seeing teaching as performance, making business listen to humanities, being both a public person and a loner, experiencing a different upbringing in his childhood, and having hard work and the unconscious work in unison. It seems that surrounding himself with what others may consider contrasts and paradoxes helps Ed shape for himself the kinds of experiences that allow him to more easily manipulate and combine seemingly opposing and discrete ideas.

Some of the paradoxes are so interesting that it would be a mistake to forego a more detailed account. Take space, for example. Ed's space is not just a backdrop for creative work – it is a meaningful, purposefully produced extension of his mind. I cannot but notice how the spatial organization of Ed's office contrasts with the way space is organized in his sheltered home in Albemarle County where soaring ceilings and large open spaces dominate the design. Compare that to the Darden office space whose every inch on the walls, floor, and tables is taken up by some object – space is in deficit here.

Looking at the cornucopia of books, photos, posters, boxes, and other paraphernalia in his office, one can imagine how someone who does not know Freeman well would have a difficult time figuring out his exact professional affiliation. He can well be an accomplished music professor, a literature professor specializing in poetry, or a philosopher immersed in the study of Freud. Add to that an entire floor-to-ceiling space devoted to taekwondo photos (Ed has a black belt) and you will be amazed at the diversity of Freeman's interests and accomplishments. Ed has passion for music and one may find him performing with his band at different Darden events; he even has his own recording studio – Red Goats Records where he both sings and plays keyboards. Freeman is also a food connoisseur who loves cooking for his guests at home where he often hosts his students, friends, and colleagues. He has a most welcoming heart!

It is only logical to assume that this immense diversity of passions is almost inevitably expected of someone who is open to experiences. Looking at all the turning points throughout Ed's life, it becomes clear that Ed has always been open to trying out something new. Ultimately, all these novel experiences, which he ventured into, have stayed with him and shaped his creative thought.

Yet, among these many passions and talents, Freeman is known to the world as a business ethicist whose stakeholder theory has shaken the then canonical shareholder-centered paradigm of business. By and large, Ed was utilizing the already existing concepts that he was able to adapt to a new context in order to solve problems. Why was Ed Freeman able to creatively use the pre-existing concepts and bring them to a new level of existence? The answer clearly lies in the coincidental action of ignorance and knowledge. It was the utter ignorance of an individual completely new to the field of strategic management combined with the profound knowledge of philosophy that others in the strategic management field were lacking that made it possible for Ed to break through the existing paradigm and see the new applications of the already existing ideas:

> *It was really me making sense of what I was seeing. And I wasn't trained in business. So I didn't know that you are supposed to think about shareholders first and business was just about the money. All those things that are kind of in the old story of business. I just didn't know that because I came from the outside... You were supposed to write journal articles and I had written some of those, but in philosophy no one is going to take you seriously unless you write a book. And so, I wrote a book not realizing in business schools you were not supposed to write books. So again, there is ignorance all the way through this.*

A creative person can and probably should be an active participant in the after-creativity process of evaluation. To get established and sustain his creative potential in the new management field, Freeman had to accept two contrasting principles: defying the crowd and listening to your audience. The first involved sticking to the values and principles brought in from outside the field of management while the latter allowed him learn from the new field.

The same characteristics seem to help Freeman make surprising connections between disparate ideas and transcend unrelated matrices. However, these characteristics do not appear out of nowhere – they are cultivated and sustained through purposefully created and sought-after daily experiences and work. The notion of experience is central in this regard, and creativity is, eventually, bounded by one's own experiences. Thus, the quantity and content of experiences matters. Quantity-wise, some experiences should be consistently integrated into one's daily routines: Ed practices creative writing on a daily basis and tries to start a day by writing a haiku or some lines of a song. There is also a need for occasional bigger-level experiences that can give a stronger burst to creativity: Ed adopted the practice of haiku-writing from Steve Earle, a famed musician whose song-writing seminar he attended. Content-wise, there is a need for experiences directly touching upon Freeman's current work and interests as well as for experiences lying outside your field of expertise. He starts his day by reading. On numerous occasions, I witnessed Ed in his office early in the morning going through a stack of new books and journals lying on his round table.

> *I read a lot. I read a lot of stuff that is unrelated to what I might be working on. But I also spend a fair amount of time talking to people who are working on stuff that I am working on or working on stuff that I am not working on. Bobby (Parmar) and I talk a lot and sometimes it leads to ideas for papers sometimes it doesn't. So I would say there are conversations you would have and then there is the stuff that you are reading.*

Conversations and reading – there is nothing extraordinary in the day-to-day experiences that inspire creativity! These new unrelated-to-work encounters can ultimately turn into passions providing inspiration, energy, and knowledge that can be later linked to one's primary area of expertise. For Freeman, his passion for music and theater spills over to his MBA teaching: every class is a performance where Ed is a musician performing in front of his audience. This teaching approach was chosen purposefully as a creative problem-solving strategy to help Ed overcome his introversion:

> *I am an off-the-chart introvert. And so teaching is a hard thing for me, as much as I've done it. And so the way I have to do it is I have to see it as performance. I've done music and performed. And to be a good performer, you have to be absolutely in the moment and present.*

This brings us to another paradox in Freeman – his love for teaching and his introvertist inclination. Ed loves to teach and has numerous teaching awards. But teaching was a scary proposition for him initially and, to some extent, even today. This may come as a surprise to those who saw Freeman's dynamic and engaging style in a classroom or during his public talks. However, as a creative person, he has learned to overcome his fears of large audience by seeing teaching as performance. Ed has a special relationship with the stage. Even as a child, Ed remembers himself always playing music and often in public – singing in church, doing theater at school, and organizing his own band by the age of 14. For Freeman, performing on stage means creating something together with the people who are there with him. And this is exactly the way he has always thought about teaching. Teaching is performance! For Ed, both are an enormous energy drain, and at the end of any performance or teaching, he is pretty much wiped out.

Similar to performance, Ed can generally tell when the student audience is enjoying what he is doing. As a creative person, he even developed his own vocabulary for the audience reaction. In Ed's words, he enjoys "seeing a light going on" when he says something and then a student adds to it, and then Freeman gets students to interact in a certain way. It is similar to a theatrical approach "Yes, and..." where each next person tries not to criticize the previous speaker but to constructively add to the discussion thread. Eventually, when one of the students says "Ah, I haven't thought about that," it becomes a special delight for Ed who can "see a sort of a light going on" in the classroom.

Freeman has successfully transplanted the idea of "absolute presence" that characterizes theatrical and music performances to his classroom teaching. He also takes to this creative approach when doing presentations in front of large audiences. Ed has taken his theater, music, and literature passion even further by creating in the business school context a number of unprecedented courses that connect business with humanities through an actively creative process. Where else would one expect MBA students to be writing short stories for their final papers, directing and performing a 24-hour prep theatrical production based on the collectively written play, or figuring out how to play an original piece of music together? In a very much the same manner that Freeman is purposefully building his own experiences, he is purposefully organizing experiences saturated with creativity for his students. What allows him to do so is once again his knowledge and experiences from in- and outside the business field that all merge together into one holistic entity:

> *People ask me all the time, "Is music your hobby?" No, I teach, and I write, and travel around talking to people. I write music, and I play music, and I record music, and I work out a lot. I don't have time to have a hobby. It just sort of all fits together in some kind of weird way.*

Another seeming paradox lies in being a public person but having a small circle of friends. Freeman is a prominent figure in academia, who can often be seen giving public talks, participating in scholarly seminars, and attending different events all around the world. He is known by tens of thousands of scholars and practitioners, and he is a legend in the field. However, even though he has over 2,000 friends on Facebook and 7,000 followers on LinkedIn, he does not have a wide circle of close acquaintances. Ed has quite a few very close good friends, and he enjoys being in their company. He once compared himself to his wife who is a much more outgoing person:

> *When we moved to Charlottesville, I knew Maureen would love it because she is very community-oriented, she likes seeing people at a grocery store, she likes going to the restaurants because in a small town there is always somebody there you know. I'm a lot more of a loner. I have a few very close people and I like spending time with them. When they are not there, I miss them a lot. There aren't many of them.*

At the same time, Freeman is an enjoyable interlocutor, and he is always open to anyone who would like to discuss ideas or just simply chat with him. When walking along the faculty hallway at Darden, one can notice that Freeman's office door is always open (unless he is traveling) and anyone is welcome to stop by, exchange greetings, and chat with Ed. From time to time, there can be a line of people waiting to talk to him. He has a small round table in the middle of his office, and everyone is welcome to sit and talk about anything they find interesting.

Patterns of dualities seem to be often surrounding Freeman, and he appears to experience bimodal experiences throughout his life. Ed's childhood can be considered a sort of another paradox from the perspective of how his parents and grandparents raised him:

> *My parents had a lot of issues between them. I got to see some of that stuff and their sort of, well, they did the best they could, but they didn't handle stuff very effectively...On the other hand, I had a grandmother who had lost one of her children in infancy whose name was Edward and the absolute undying love and loyalty of my grandmother to me that was, I was the favorite of 35 grandchildren... Those early childhood things were very influential.*

For Freeman, what was important and what he remembered was exactly this contrast between the two experiences. In other words, he did not consider them separately, but rather as two forces present at the same time. It may well be that as a child experiencing such dualities he learned to understand early on that contrasts indeed can co-exist in the same time and space. Making sense of these early experiences may be a prerequisite for being able to create such purposeful experiences later on in life. There are also other experiences, not necessarily bimodal, that either promote or inhibit creativity in a child, that make a child, so to say. Ed was given an opportunity and enjoyed playing the piano – hence, his passion for music. On the other hand, at one point in middle school, his teacher shamed him in front of the whole class for drawing poorly "like a kindergartener" – Freeman has never drawn pictures since. These experiences may and do have a lasting effect on the creativity trajectory of a child. Therefore, it is only right to seek answers to questions about who creative people are in their childhood experiences. For instance, Ed attributes his gravitation toward business ethics to this influential childhood episode:

> *My father, who was not very well-educated, was the foreman of a crew that painted center lines, lines on the road in Georgia. And I was young, and he took me with him on a trip so that I could ride on the truck that painted lines. And we got back, and he showed me, he said, "Look, I am filling in my expense report and I could just say I spent this much and not tell them that you went with me, but that would be wrong"... so he paid for me... and that, I am sure made a huge impact on me.*

While the duality of experiences is the central theme threading through Freeman's narrative, there are two other important ingredients of creativity that should not be dismissed: hard work and the role of the unconscious. Don't they make yet another perfect dichotomy? Again, Ed is trying not to dichotomize them: both can explain and are present in the creative process. On the one hand, he describes several situations, including writing a song or a stakeholder paper, in which he "plugged in" and had a "flow" where he felt he was "a vehicle" for the creativity to happen. The force of the unconscious seemed to be guiding him in these situations. On the other hand, Ed emphasizes that creativity is inseparable from hard work put into it. It may well be that hard work either precedes or follows the "flow" of the creative process. It may also be that hard work is embedded into the flow, but for an individual engaged in the flow this work portion of the process remains unattended because the focus is delivered completely and utterly on the process of the creation itself.

Freeman believes that creativity does not fall out of the sky but it can be taught. That is exactly why he has taught his classes on literature, theater, and music – he believes in developing creativity in his MBA students (see Chaps. 38 and 39). Ed purposefully organizes learning around creativity for his students and also leaves space for students to organize their own experiences in the process of creation.

During one of the interviews, I asked Ed when he realized that he became a well-known scholar and he replied that he really did not know. After publishing his book in 1984, for the next 10 years he and his wife Maureen were too busy with their three small children, which took up all of his time left after teaching and writing. In those years, there was simply not much time to reflect on how things were going. Ed remembers that suddenly he started getting invitations to receive different awards, with some bearing the title of lifetime achievement. In 2001, World Resources Institute gave Freeman the Faculty Pioneer Award for Lifetime Achievement, and he remembers himself thinking "What, lifetime achievement award? But I'm just 50. Why am I getting a lifetime achievement award?" In 2008, Ed got his first honorary degree from Comillas Pontifical University in Madrid, and it was very special to him. Again, he was quite surprised and he kept thinking that he got too much credit. Soon after, he received honorary degrees at other universities in North America and Europe, and all kinds of awards from the Academy of Management and the Society for Business Ethics. As surprised as he was in the beginning, he also started realizing that people were taking his ideas seriously.

From the very onset, Freeman knew that the companies he had worked with during his time at WARC were taking stakeholders seriously – thus, the roots of stakeholder theory have always been deeply practical. Yet, for academia, it took a much longer time to acknowledge the importance of stakeholders. When summarizing the history of the stakeholder theory development in the book *Stakeholder Theory: The State of the Art* (2010), Ed pointed out that there were a lot of people who wrote about stakeholders in some way or another, and he just tied all those ideas together in a coherent way. Indeed, some people used a stakeholder idea in their strategic planning to organize information and assumptions about the external environment. But Freeman's humbleness should not misguide us from giving him a proper credit. After all, it takes an individual who is a great and creative thinker to coherently tie the ideas together. For instance, there were already many ideas on economic affairs circulating in the Great Britain by the eighteenth century, but it was Adam Smith who organized and structured them in a coherent narrative in his *The Wealth of Nations* that gave a start to what we know today as classical economic theory. And we need to do justice to Freeman as he went far beyond organizing the external environment by stakeholders. In his 1984 book, he asked the question that no one else has dared to ask before and we may say that the management has never been the same once Freeman asked his question:

> *Perhaps the most important area of future research is the issue of whether or not a theory of management can be constructed that uses the stakeholder concept to enrich "managerial capitalism", that is, can the notion that managers bear a fiduciary relationship to stockholders or the owners of the firm, be replaced by a concept of management whereby they must act in the interests of the stakeholders of the organization?"* (p. 249)

The story about Ed would not be complete without the names of individuals who inspire him. On the philosophical side, it is Richard Rorty who had the most influence on Ed and they knew each other personally. It is also Daniel Dennett at Tufts University whom Freeman has not known personally but of whom he has a very high opinion as Freeman found his *Darwin's Dangerous Idea* incredibly inspiring, especially the fact that Dennett in his works was able to cross different fields such as philosophy, psychology, biology, and cognitive science. James Childress at UVA is also very important for Ed as a compelling voice in ethics and biomedical ethics. On the business side, early on, Ram Charan, Jim Emshoff, and Tom Peters were instrumental for Freeman becoming a business school professor. Jim Emshoff was essential in teaching Ed about business. Ram Charan mentored him about business and teaching. And, Tom Peters, especially with his book *In Search of Excellence*, changed Ed's understanding of management. Somewhat later, Freeman got to know Jim Collins and could clearly see why the latter became an eminent voice in management. When it comes to business practitioners, it is John Mackey, co-founder of Whole Foods, and Kip Tindell, co-founder of The Container Store, who have left a strong impression on Freeman's thinking. Looking at scholarly heroes, Ed would probably name Darwin, Freud, De Beauvoir, and Wittgenstein, and he would definitely add Rorty's name to that list. All these great thinkers challenged previous beliefs about human beings and their role in the world. They revolutionized our way of thinking by demolishing seemingly "steady" foundations and opening a path to a better understanding of our selves and our society. Currently, Freeman is writing a book about stakeholder theory in philosophy where he aims to connect some of these philosophical movements and heroes with his vision of stakeholder theory. Though Freeman admits, it is a long project and it will take some time before we see it.

What Freeman does not say, but many of his followers, including myself, can confirm is that Ed has developed his own school of thought – stakeholder school. This is indeed a remarkable lifetime achievement which puts Ed Freeman's name in the group of the greatest thinkers of our time. This book is a tribute to the evolution of his school of thought through Freeman's writings on stakeholder theory, business ethics, and stakeholder capitalism.

Some Personal Stories from Some Very Special People

Many people have supported Ed Freeman throughout his journey. I asked some of them to share their reflections on working with him. Below is what Daniel R. Gilbert, Patricia H. Werhane, Jeffrey S. Harrison, Andrew C. Wicks, Jeanne M. Liedtka, Ellen R. Auster, Robert A. Phillips, and Bidhan (Bobby) Parmar have written.

Daniel R. Gilbert, Jr., Professor Emeritus, Gettysburg College

About lessons from Ed Freeman:

Ed Freeman and I met at the University of Minnesota in 1983 at the launch of a doctoral program called Strategic Management and Organization. We came to know each other at the Metrodome, New French Café, Hungry Mind Bookstore, and handball courts inside Memorial Stadium. Those physical landmarks are gone. What endure are three lessons that Ed taught me.

First, Ed Freeman introduced me to pragmatism and the writings of Richard Rorty. Rorty explained philosophy as a "contest between an entrenched vocabulary that has become a nuisance and a half-formed new vocabulary that vaguely promises great things" (Contingency, Irony, and Solidarity). I have taught and written in such contested, promising spaces ever since.

Second, Ed Freeman taught me that teaching can inspire scholarship. He told me that one impetus for 'Search for Ethics' was something that I said about stakeholders while teaching a class that he was observing. The challenge of explaining a concept to my students has spawned and informed much of my scholarship.

Third, Ed Freeman taught me the discipline of framing a durable question. What is a strategy? was a continuing guide for our writing collaborations. As a corollary, Ed taught me that a question can lead in unexpected directions. In time, I moved away from questions that occupy business ethicists. Applying the discipline that Ed taught me, I know why I have reinvented myself as a writer, and I am thrilled at residing in a new scholarly neighborhood.

Patricia H. Werhane, Professor Emerita, University of Virginia

About Ed Freeman as a colleague and friend:

I was acquainted with Ed Freeman for a very long time, but I got to know him well as a colleague and friend when I came to the Darden School in 1993. I was hired because their only senior faculty woman had retired, and I looked forward to working with Ed to relieve his many teaching and research projects. I was honored to collaborate with such a well-known scholar in the field, and collaborating was a great idea except that instead of reducing Ed's work we both generated even more projects! Our greatest success was to start Darden's Ph.D. program in business ethics for which we are still proud. Our Albatross was the Encyclopedia of Business Ethics, a 700-page tome for which we invited 285 of our best friends to contribute. Although we had an exemplary assistant for the details and a former GE executive to read the submissions, we got the brunt of the personal phone calls. Divorces, illnesses and a plethora of parental deaths plagued our contributors. When the collection was finally completed and on display at the Blackwell booth at the Academy, one of our recalcitrant contributors asked if he could have 2 more weeks to finish his essay.

Despite that challenge, working with Ed was terrific. I learned that once you were his friend, it was a lifetime relationship. His research continues to be incredible as this collection exemplifies, and Ed's kindness and collaborations with graduate students is unsurpassed by anyone else in the field. As co-editor of the Journal of Business Ethics, Ed and Michelle Greenwood have turned a decent journal into the top in the field. What is less well-known is Ed's commitment to teaching, exemplified at Darden where he ordinarily teaches a double load of courses. We are all fortunate to have him as a mentor, colleague, collaborator, and lifelong friend.

Jeffrey S. Harrison, W. David Robbins Chair of Strategic Management, Professor of Management, University of Richmond

About Ed Freeman's contributions:

When I had just begun teaching strategic management at Arizona State University in 1985, a colleague handed me a book called Strategic Management: A Stakeholder Approach, by R. Edward Freeman. I read it and it changed the course of my career. I quickly found out, through journal rejections and tense conversations, that researchers in strategic management were not yet ready for the stakeholder perspective. However, when Caron St. John and I wrote our first strategic management textbook a few years later, we built it on a stakeholder foundation and it was a hit, with adoptions at universities in almost all of the 50 states in the US. It was at this time, while asking for permissions for the book, that I had my first personal contact with Ed. I found him to be as generous as he is brilliant.

Several years later I asked Ed to work with me as co-editor of a special issue of Academy of Management Journal on a stakeholder theme. He graciously agreed, and the special issue had a significant impact on the management field, based on citation count since it was published. Since that time, the field of strategic management has been much more enthusiastic about stakeholder theory, and I have been blessed with the opportunity to work with Ed on many books, conferences, special issues, and articles. He has always been a positive force in my life -- a confidant and cheer leader. I am not alone. He practices what he teaches. He is generous with his time and his wisdom, and has had a huge impact on many students, colleagues, business leaders, societies and even government leaders. For most of us, our work will fade away with time. However, Ed's contributions to the world will shine on through generations.

Andrew C. Wicks, Ruffin Professor of Business Administration, University of Virginia

About Ed Freeman as a mentor and friend:

Ed Freeman changed my life – quite literally, and in a very good way.

I was a graduate student in Religious Studies trying to grow my ability to teach applied ethics, largely with a focus on medical ethics. I found out Ed was teaching at Darden and was encouraged to see if I could persuade him to offer a seminar for me and a few of my peers. So, I called up Ed, and with nothing to offer him except my appreciation. He graciously offered to teach a course in business ethics (along with Tom Donaldson, who happened to be visiting at Darden, and was kind enough to co-teach the course) for me and 3 of my peers. It was an incredible course and a fantastic opportunity for someone like me who wanted to ask big questions – yet do it with my feet on the ground, trying to understand how they make a difference for people living in organizations. From there, Ed agreed to hire me as an RA. I wrote cases for him, helped him revise a textbook, and started writing articles with him – all of which were great experiences. He also served on my dissertation committee. Along the way, he did a lot to mentor me and set me up for success.

From there, Ed had the audacity to recommend I apply for a job in a business school. Given how little I knew about business (or business students) I genuinely thought he was kidding. It took a full 10 seconds of waiting for the laughter to start before I realized he wasn't pulling my leg. Turns out it led to me applying for (and getting) my first academic job, at the University of Washington School of Business. I can honestly say I would never have been able to get that job, or to believe in myself enough to even apply, had it not been for Ed and all he did to make that opportunity possible for me. From there, Ed has continued to be a valued mentor and friend, a frequent co-author, and someone who always finds ways to support me in my work and life at Darden. And, while I think my story about Ed is special (because it happened to me), I know it isn't – mostly because he has done the same thing for so many other graduate students and young scholars. That's the kind of guy he is.

Jeanne M. Liedtka, United Technologies Corporation Professor of Business Administration, University of Virginia

About Ed Freeman as an influencer:

Ed changed my life. Of course, his path-breaking scholarship has influenced the development of scores of us – but I mean much more than that. He has, quite literally, been the driving force in my career. From the moment in the cafeteria line at the Academy of Management Annual Meeting when he said "Come to Darden," through my struggles to emulate his superb performance teaching ethics in the classroom ("This case teaches itself"), through the pleasure of writing with him and seeing his lightening quick mind at work, and right up until today – his has been the not-so-invisible hand that has guided me. It has been a wild and wonderful time. Thank you, Ed!

Ellen R. Auster, Professor of Strategic Management, York University

About working with Ed Freeman:

Ed and I beautifully complement each other as co-authors. For both of us, writing is a dynamic process that emerges and evolves as we banter around ideas and write. We talk, generate an outline, percolate and iterate, then eventually first draft sections and switch, and keep going back and forth until we get to something we like. When we've been in the same location for a conference, we might eat food, drink coffee or wine, listen to music, or wander the streets of whatever city we are in while we work on the ideas for a paper or chapter. When we're not in the same location, it is more of a back and forth over phone calls and email. But it's always an evolving process, that we are curious to see where it takes us and that we both see as fun and interesting. He is wise, counter-intuitive, and has a deep knowledge that is so impressive. He's one of those renaissance thinkers whose breadth of material and disciplines that he draws upon is beyond imaginable. It has been my joy and pleasure to have the opportunity to share and develop thoughts with him.

Rob Phillips, Professor of Strategic Management and Public Policy; George R. Gardiner Professor in Business Ethics, York University

This is a humorous perspective on Ed Freeman which was originally part of a laudation presented in Humboldt University in October 2014:

When Professor Schwalbach asked me to give this laudation, I eagerly accepted the opportunity as both an honor and a privilege. Ed Freeman's accomplishments are legion.

This is all pretty remarkable for a young man from rural Georgia, a math student with a Ph.D. in (ahem) philosophy. What are the chances of such a person becoming one of the most acclaimed, influential and widely-cited of his generation of management scholars? Much of Freeman's scholarly reputation began with the publication, in 1984, of the now classic SM:SA.

The very thought that a philosopher could, over the course of a summer, write such an influential book has long struck many as, and I'm afraid there's no other word for it, supernatural.

It gives me no pleasure today to report that I have uncovered the dark, otherworldly truth about the conjuring of Freeman's classic book. My research for this laudation took me to America's deep south where I met a 102-year old man named "Blind Melon". I was not allowed to take pictures of old Blind Melon because he feared pictures could be used to steal his soul – something he had some personal experience with. Sitting on lawn furniture surrounded by stacks and stacks of dusty old records and (curiously, for an old blind man) hundreds of books, Blind Melon told me a story of a young man's search for scholarly fame. Here's what he said:

Young Ed had searched far and wide for the professor's truth. He had looked to numbers, he had sought truth in logical syllogism, existentialism and categorical imperative, but still he felt no closer to wisdom. After a long night of looking for answers in the bottom of a bottle of Old No. 7 (where so many truth seekers before him had ventured), he fell into a deep sleep. And during this sleep, he dreamed of a sparkling horse who spoke to him. The sparkling horse told him to go to the Cameron Crossroads the next night at midnight. There he met a huge man with eyes the shape of bicycle wheels who called himself Faustoff. Faustoff gave him a strange machine the size of a suitcase that he mysteriously called "S.G".

The machine had a keyboard, but there was no sign of paper or even anywhere to put paper. The idea of a typewriter that used no paper was, back in the 1900s, close enough to witchcraft to raise an eyebrow. But there are two other matters to consider as well.

First, the best-selling electric guitar of all time, is the Gibson SG.

Here's some guy named Derek playing one. But, no matter what some of you are thinking at this point, this is definitely NOT a story about blues guitars or blues players.

The second thing I remembered after hearing Blind Melon's story was that the penultimate line of the preface of Freeman's 1984 book says, "My IBM Personal Computer, 'Dora', and my Osborne 1, 'S.G.' should take credit for the typing of the manuscript." This has always seemed to me to be a lot of credit for a mere machine.

That is, unless the Osborne 1 itself, acquired from Mr. Faustoff at the Cameron Crossroads, is in fact the stakeholder guru we should be honoring today.

Now old Blind Melon was a known exaggerator so it's hard to say how much of this story is true. For example, there's no evidence that photography facilitates the theft of souls - though the global explosion in "selfies" bears obvious marks of demon possession.

If not the work of a word processor from the underworld, how else to explain that there are 6 times more references to Freeman's 1984 book (16,000) than there are copies of the original book in existence (2500). And Freeman's work has more cites (35,000) than the entire journal of Business Ethics Quarterly (just over 25,000 – though there is some double counting as 4 of the top 15 articles in that journal are his work). But to play Dewey's advocate, we should explore other, more earthly, possibilities as well.

It is possible, for example, that his training in philosophy allowed him to see the relevance of ethics and values to management and management scholarship in ways that those specifically trained in economics-derived disciplines were blind to. Though teaching and research on ethics and responsibility has become commonplace in schools of business today, the environment for most of the preceding three decades was characterized by suspicion, disdain, and even hostility. For many in this room, the fact that we have ethics discussions in business schools is due in large part to the efforts of Freeman and a dozen or so others who essentially founded and legitimized the discipline of business ethics in business schools.

And the diversity of Ed's interests – and novelty of his insights – have continued through the years. While remaining deeply enmeshed in a myriad of putatively distinct scholarly areas, the questions he asks have always been driven and anchored by considerations of managerial practice and rooted in pragmatism. And he combines these intellectual pursuits with a full engagement in his own world of practice. His thinking has been influenced over the years by cooking, professional wrestling, martial arts, handball, and theater to name just a few of his passions.

He has most recently tested his ideas against the harsh backdrop of the music business, co-founding a record label, Red Goat Records. Though it should be said that his becoming a music executive and the clear sense of pagan sacrifice in the logo have not helped his defense against Blind Melon's charges of soul mongering.

In addition to the diverse interests that have been key to his originality, Freeman has also been unrelenting in his advocacy of his particular brand of values-based management. It makes me tired just to think about how many speeches, lectures and conversations he's had about stakeholder theory.

Ed managed to stay the course even as business thinking took a turn toward methodolotry and into the agemonic cul-de-sac of the late 1900s and early 2000s that continues in some places to this day. Lesser men would have given up in the face of such hostility – but I am here today as testament to Ed's persistence in defense of a good idea.

In the course of explaining his vision of capitalism and free markets he has heard some objections so frequently in his myriad conversations that he has given them names.

And these objections come from across the political and ideological spectrum. For example, he has objected more than once to what he calls the Business Sucks Story. Versions of the Business Sucks story can be found among socially progressive critics of business and capitalism who believe that business as such is the problem rather than the actions of particular businesses and specific business people. Ed staunchly maintains that free market capitalism is the best system of value creation and trade currently available to us and perhaps the best ever. In the world of business ethicists, holding the line against the business sucks story can itself be a full-time occupation.

Far from here on the political spectrum are those who succumb to what Freeman has called the Separation Fallacy: the belief that there are business decisions on one side, ethics decisions on the other and never the twain shall meet. For Freeman, business is shot through with ethics and values. Ignoring the centrality of values to business decisions – asserting the separation fallacy – is to intentionally make inferior decisions of both sorts. Among Freeman's lasting contributions is to name this fallacy and make us all more vigilant to its insidious effects.

I would be remiss if I didn't briefly mention the limits of persistence. As Mark Twain famously said, "If at first you don't succeed, try, try again… Then quit, no use being a damn fool about it." And Freeman has lessons here as well. He has dashed the hopes of more than a few aspiring scholars through use of the ultimate pragmatist critique, "That's just not a very interesting question." Freeman and I have had versions of this conversation for now over 20 years and it always makes me nervous to persist in a direction he has determined to be fruitless. This is because I suspect in most cases he has been down that road and seen the dead end personally. Ask him sometime to tell you about F_{ij} *and* U_{ij}.

The ultimate testament to Freeman and his ideas is the extent to which others have taken up and advanced the ideas (though not always in directions he would agree with). This speaks to the fecundity of original ideas themselves, but also speaks to Freeman's remarkable generosity. One time-honored tradition for making one's ideas seem more influential is to force graduate students to follow up, test and, most importantly, cite the "foundational" work of the Professor Dr. Dr. This has never been Freeman's approach. Instead he draws both the curious and the critical into his ideas through openness and active engagement with living ideas. He then convinces them through a combination of patience, the previously mentioned persistence, and inspiration that they, themselves can contribute to improving their world. This has proven far more effective than the force-feeding method adopted by many (ahem) "influential" thinkers.

I've already mentioned the 35,000 cites. By my count, Freeman has co-authored with at least 46 different people over the years – only a small percentage of whom were his students directly.

And this brings us back to SG. Irrespective of where he got the original fire of inspiration, Ed's visionary thinking has always been years ahead of his audience. With disconcerting (even preternatural) frequency, I have arrived at various crossroads of my own, along paths that have been (ahem) unpredictable, only to find Ed waiting there.

And I'm not alone. With nothing more (or less) magical than good ideas, generosity and force of will, he continues to lead generations of scholars and practitioners toward finding our own paths to making the world a little better than we found it.

Bidhan (Bobby) L. Parmar, Associate Professor of Business Administration, University of Virginia

About Ed Freeman's inspiration:

R. Edward Freeman is a rare type of scholar who, despite living to see his once marginalized ideas reach widespread influence and acclaim, remains humble and eager to learn.

Ed is most well known for his work on Stakeholder Theory. This view is predicated on the idea that as human beings, we have the ability to enter into free and voluntary relationships with each other, which allows us to collaborate to accomplish things we could not achieve alone. Freeman's scholarship has taken this relational view and redefined both academic conversations and managerial practice in fields such as corporate strategy, management, governance, leadership, and ethics.

Throughout these many topics, Freeman's commitment to philosophical pragmatism has led him to try to help these fields to focus on human practices for coping with an ambiguous and dynamic world rather than on finding unchanging foundations for the truth of our ideas. In this way, Freeman has been more of a problem-dissolver than a problem solver – showing that if we let go of problematic assumptions about how our words connect to the world, we can focus on creating more useful ways of talking.

On a more personal note, Freeman is a legendary teacher, scholar, and friend, and like so many others would attest – my life as a scholar and teacher would not be the same without his influence. His counsel and wisdom have not only benefitted my career in an endless variety of practical ways but more importantly, inspired me (and so many others) to try to do the same for our students and colleagues. In this way, the life that Ed has lived has been his most powerful redescription.

Overview of the Book

This volume comprises over fifty selected works that Freeman wrote at different stages of his academic career. The publications are categorized in Three Parts that reflect major themes in Freeman's body of work: stakeholder theory, business ethics and humanities, and stakeholder capitalism. Some works are single publications, while some others are written together with over thirty co-authors – Ed has productive work relationships with many scholars across the globe.

Part I contains major works that Freeman and his collaborators wrote on stakeholder theory. These publications initiated, nurtured, and shaped what can be called today as the stakeholder school of thought. Though most of the works in the book are organized in the chronological order, **Chap. 1, "The Problems That Stakeholder**

Theory Tries to Solve" is one of the few exceptions. It is an excerpt from the book *Stakeholder Theory: The State of the Art* that has been published rather recently. The choice for the opening chapter of the book was dictated by the need to understand why stakeholder theory is much needed. Freeman, Harrison, Wicks, Parmar, and de Colle argue that stakeholder theory addresses three grand questions: how to manage business in a sustainable way, how to keep ethics and business together, and what to teach students to set them for success.

Chapter 2, "**Stakeholder Management: A Case Study of the U.S. Brewers Association and the Container Issue**" outlines some of the earliest ideas on stakeholder theory. Emshoff and Freeman use brewers to illustrate stakeholder management principles. There are, though, a couple of caveats here: the initial thinking about stakeholders was rather limited to managing the external environment, and this work is abundant with management science formulas which Freeman would have removed from the article as they unnecessarily overcomplicated the story and turned out to be less relevant over time. The next work, **Chap. 3**, "**Stakeholder Management: Framework and Philosophy**", is an excerpt from Freeman's seminal book *Strategic Management: A Stakeholder Approach* (1984). We can see here how stakeholder ideas have progressed since Chap. 2, resulting in organized notions, diagrams, principles, and philosophy of stakeholder management. Not many people know that it took Ed only eight weeks to write the whole book draft – indeed, Freeman can submerge in a flow state and be really productive when he has something to say and is excited about it.

Though **Chap. 4**, "**Theory Building in Strategic Management**", has little to do with stakeholders, we placed it next to the works on stakeholder theory since it shows Ed's perspective on theory development. Freeman and Lorange challenge the traditional "wisdom" in the strategic management field about theory construction, tension between rigor and relevance, and eclectic research methods.

Ten years after the publication of Freeman's seminal book, the scholarship on stakeholder theory started growing. Yet, it came with many critics who claimed that stakeholder theory was built on false premises due to an inherent stakeholder paradox. In **Chap. 5**, "**The Politics of Stakeholder Theory: Some Future Directions**", Ed addresses the criticism and argues against, what he calls, the Separation Thesis. He is doing this by blending business and ethics. Freeman also posits that stakeholder theory can be better understood as a genre rather than a theory. This written piece is among Ed's favorite, and when reading it, one cannot but get impressed with the innovativeness of his ideas.

The early works on stakeholder theory were written in the context of strategic management, and no attempts were made to frame it within any philosophical school of thought for the justification of a stakeholder idea. **Chapter 6**, "**A Feminist Reinterpretation of the Stakeholder Concept**" is the first attempt to provide stakeholder theory with some moral grounding. Wicks, Gilbert, and Freeman position stakeholder theory in Carol Gilligan's feminist tradition and emphasize collaboration and cooperation. Today, Freeman admits that beyond Gilligan's interpretation it would be interesting to explore some other feminist theories that could be applied to stakeholder management. **Chapter 7**, "**Stakeholder Theory: A**

Libertarian Defense" is another attempt to ground stakeholder theory in a moral philosophical foundation. This time, Freeman and Phillips see underpinnings of stakeholder theory in libertarian ideas of freedom and voluntary action. Interesting enough, in his future works Freeman avoids dividing stakeholder theory into normative, instrumental, and descriptive variants, but in this work he keeps the distinction. To this point in time, this work has received some notice among scholars, though it certainly has the high potential to generate more scholarly interest due to so many thought-provoking ideas. The authors outline a number of stakeholder principles that found their way into Ed's future publications on stakeholder capitalism and the new story of business.

By the beginning of 2000s, stakeholder ideas were gaining popularity in other areas beyond business, and **Chap. 8**, "**Business Ethics and Health Care: A Stakeholder Perspective**" contributes to that movement. Gilmartin and Freeman enter the discussion on health-care reform and offer a stakeholder framework as an alternative to ongoing business practices in health care.

In **Chap. 9**, "**A Names-and-Faces Approach to Stakeholder Management**", McVea and Freeman express their concern that as stakeholder theory gained wide scholarly acceptance, it adopted some standard business assumptions which derailed it from its original path. Aiming to reinvigorate the stakeholder scholarship and stay true to the underpinning ideals, the authors suggest we see stakeholders as real people with feelings and aspirations rather than impersonal roles in organizations. **Chapter 10**, "**Enhancing Stakeholder Practice: A Particularized Exploration of Community**" is another step in this direction. Dunham, Freeman, and Liedtka explore the notion of community as a group of stakeholders. The idea comes from what Ed saw as one of the unresolved weaknesses of stakeholder theory: the understanding of community. Due to its delicate nature, the community issue was informally called the soft underbelly of stakeholder theory because no one would criticize stakeholder theory from the standpoint of the ambiguous idea of community – so the authors felt they had to do it themselves. Seeing community as more than just a plateau of land, the authors distinguish among eleven different types of community.

With time, some scholars saw stakeholder theory as overlapping with corporate social responsibility (CSR) in addressing the relationship between business and society. In **Chap. 11**, "**Corporate Social Responsibility: A Critical Approach**", Freeman and Liedtka start a discussion about the revision of traditional ideas in business ethics. The authors criticize the CSR concept on several grounds and call for its abandonment. As an alternative, the authors suggest we begin a new conversation that would embrace the stakeholder proposition, the caring proposition, and the pragmatist proposition. **Chapter 12**, "**A New Approach to CSR: Company Stakeholder Responsibility**" is written 15 years later, but it continues the previous conversation. Claiming that CSR outlived its usefulness, Freeman and Velamuri suggest it be assimilated within stakeholder theory and the abbreviation CSR should instead denote Company Stakeholder Responsibility. **Chapter 13**, "**Corporate Citizenship and Community Stakeholders**" connects stakeholder theory with another business ethics concept – corporate citizenship. Phillips and Freeman continue the conversation on communities as stakeholders and discuss fairness- and reciprocity-based obligations that have become the signature mark of Phillips's research.

In Chapter 14, "**Managing for Stakeholders and the Purpose of Business**", Freeman and Parmar address several widely spread myths about shareholders and offer an alternative way to manage a firm. **Chapter 15**, "**Managing for Stakeholders: Tradeoffs or Value Creation**" explicitly posits the central tenet of stakeholder theory – the jointness of stakeholder interests. Ed also admits that though exploring stakeholder theory as a theory of the firm can be a useful project, he instead aims to bring to the surface a somewhat neglected premise of stakeholder theory – the avoidance of trade-offs in managerial decision-making.

By the end of the 2010s, the volume of stakeholder literature has become so impressive that some scholars started questioning whether it was possible to develop stakeholder theory any further. Envisioning lots of untapped potential and aiming to reinvigorate the stakeholder research, Freeman discusses several unresolved tensions in stakeholder theory in **Chap. 16, "Five Challenges to Stakeholder Theory: A Report on Research in Progress**".

We end Part I of the book with a short essay where Freeman and Sollars give an elegant summary of major developments in stakeholder theory. **Chapter 17**, "**A Puzzle About Business Ethics**" is written like a simple math problem that requires a solution. In fact, it is a business ethics puzzle whose solution, with all the considerations taken, seems straightforward. Interesting enough, the words *stakeholder* and *stakeholder theory* are not mentioned here, but we would need to invent them to solve this simple business ethics puzzle.

In **Part II** of the volume, Ed Freeman shares his views on business ethics and, at times, engages with ideas from a broad spectrum of the humanities. One such example is **Chap. 18**, "**Orwell and Organizations**", where Freeman and Gilbert use Orwell's *1984* as an analogy to explore issues in organizational management. Ed regrets not doing more with this working paper – the time for it had come and gone as it was written when everybody was celebrating Orwell's *1984* book. We are excited to include this paper in our volume – it is certainly a fascinating, even somewhat "hazardous" read as explained in the disclaimer to the essay:

> *A Note to Gentle Readers.*
>
> *This essay is rather different in content and style from the usual academic article on management and organization, and at least part of its message lies in just that difference. Reading it could well turn out to be hazardous. We recommend appropriate caveats and caution.*

In **Chap. 19**, "**The Ethics of Greenmail**", Freeman, Gilbert, and Jacobson explore hostile takeovers in the corporate world from a business ethics perspective. The ideas for this article originated in Ed's job talk that he had given at Darden. The next two works are short essays published in the popular press. **Chapter 20**, "**Airline Horror Stories Indicate an Ethical Problem**" and **Chap. 21**, "**Healthy Tension between Business and News Media**" challenge some ineffective practices in the airline industry and in the news media, respectively. This line of thought continues in **Chap. 22**, "**The Revolution in Management**", where Gilbert and Freeman start with an observation that on any single day *The Wall Street Journal* and other popular press publications are abound with business scandals. The authors explore the root causes of business wrongdoings and suggest that businesses see strategy as purpose and link it to ethics. In **Chap. 23**, "**Values and the Foundations of Strategic Management**", Freeman, Gilbert, and Hartman examine the role of values in strategic management and review possible models of corporate morality. This is one of Ed's favorite works, and by reading it, one can see how far-reaching the ideas presented in the article are.

Chapter 24, "**Let's Disband the Academy of Management**" is Freeman's Presidential address at the SIM Division of the AOM whose background we have previously described in the section on Ed's biography. **Chapter 25**, "**Business Ethics: A Literary View**" examines the separation of business and ethics and argues for having more people educated in the humanities to teach business education. It also argues to value substance over method and appreciate interpretation more than data. Ed's idea of business as a branch of humanities is as relevant today as it was thirty years earlier. **Chapter 26**, "**Business, Ethics and Society: A Critical Agenda**" was one of the very first articles published in *Business & Society* after the International Association for Business and Society took over the journal. It is yet another piece on Freeman's agenda of critical studies in management that he has pursued throughout his career. Freeman and Gilbert start their work by mentioning that

> *In recent papers and presentations we have satirized what the field of business and society has become. We made fun of the fact that ... intellectuals have come to see themselves as scientists with white coats telling us the way the world, even if it is the social world, really is. ...[W]e want ... to give an alternative interpretation of why ... [they] see the question of their own legitimacy as central.*

In their work, the authors re-describe the business and society field, business schools, and business and raise the question of their legitimacy. Another work on business as a humanity is **Chap. 27**, "**Business as a Humanity: Epilogue**". Here, Freeman identifies several important phases in the history of business schools and offers concrete suggestions for business education in the areas of curriculum, faculty, students, and institutions. Finally, the epitome of Freeman's critical approach to the traditional management discourse can be found in **Chap. 28**, "**The "Business Sucks" Story**". The story behind this essay was offered to the reader in the earlier section on Ed's biography.

Chapter 29, "**Book Review Essay: The Relevance of Richard Rorty to Management Research**" is Freeman's tribute to Richard Rorty in a hope to stimulate more reading and discussion of Rorty's work. **Chapter 30**, "**Business As a Human Enterprise: Implications for Education**" takes us back to the discussion about business education. Freeman and Newkirk provide a comprehensive overview of the development of business schools as well as the critiques that business education has accumulated in the process. In their another joint project, **Chap. 31**, "**Business School Research: Some Preliminary Suggestions**", Freeman and Newkirk challenge too much positivism that has overtaken business research, share concerns about the shrinking link between management research and practice, and question the usefulness of MBA in its current form. The authors see the remedy to business education in pluralism. Along the same line, **Chap. 32**, "**Teaching**

Business Ethics in the Age of Madoff" shows that business curriculum is full of analytic techniques, financial methods, and ideology that are deeply hostile to business ethics. Freeman, Stewart and Moriarty address misperceptions common in business education.

Chapter 33, "**The Impossibility of the Separation Thesis**" puts a final blow to a long-standing debate on the co-existence of business and ethics. Harris and Freeman show that their separation is an impossible exercise and any attempts to distinguish facts from values are futile. The same authors in **Chap. 34**, "**Creating Ties That Bind**" develop a theory on integrating normative and descriptive approaches to business ethics by connecting microsocial contracts and hypernorms (values that most of us can agree on).

The next essay was originally written by Freeman as a talk that he gave when receiving his honorary degree from Comillas Pontifical University in Madrid. **Chapter 35**, "**Remoralizing the Debate**" articulates the need to craft conversations on ethical issues in our institutions. In **Chap. 36**, "**Values, Authenticity, and Responsible Leadership**", Freeman and Auster enrich the responsible leadership scholarship by bringing to the discussion the concept of authenticity, i.e., acting on one's values and being true to oneself. In their next work in **Chap. 37**, "**Values and Poetic Organizations: Beyond Value Fit Toward Values Through Conversation**", the authors explore the intersection of leadership, organizational values, and authenticity and develop a typology of interconnected sets of values.

In **Chap. 38**, "**Leveraging the Creative Arts in Business Ethics Teaching**", Freeman, Dunham, Fairchild, and Parmar share their experiences with using literature and theater for teaching ethics to business students. In that, understanding the art of storytelling and narrative plays a key role, as well as attention, creative muscles, and collaboration among the students. A large portion of the study is dedicated to student experiences and reactions. **Chapter 39**, "**Practicing Human Dignity: Ethical Lessons from Commedia dell'Arte and Theater**" contributes to the same topic. S. de Colle, Freeman, Parmar, and L. de Colle demonstrate how experiential exercises from theater can help students develop self-awareness, creativity, ability to listen, non-verbal body language, and public speaking. But most of all, theatrical experience promotes the appreciation of human dignity, leading to more humane organizations.

The next two works are short essays whose audience are practicing managers. **Chapter 40**, "**Ethics and the Algorithm**" explores the pervasiveness of algorithm-based decisions that have been gaining popularity in data-driven organizations. Parmar and Freeman argue that coded decisions are not value neutral as they are programmed based on human judgments. The same authors in **Chap. 41**, "**Which Rules Are Worth Breaking?**" build on Schumpeter's concept of creative disruption and argue that business needs responsible disruption.

Chapter 42, "**A Pragmatist Approach to Business Ethics Research**" is written by strong adherents to pragmatist ideas – Parmar, Phillips, and Freeman – who overview the history of pragmatism, outline key pillars of its doctrine, and argue for the relevance of pragmatism to business scholarship. **Chapter 43**, "**Profit and Other Values: Thick Evaluation in Decision Making**" is among Ed's favorite works. In this article, van der Linden and Freeman find a solution to what previously was considered an unresolved dilemma: managers either focus on a single performance value (whose most common example is profit maximization) or create value for multiple stakeholders. The solution lies in thick evaluation. **Chapter 44, "Unethical, Neurotic, or Both? A Psychoanalytic Account of Ethical Failures Within Organizations**" explores organizational ethical failure from a new perspective for business ethics scholarship – psychoanalysis and clinical psychology. De Colle and Freeman identify five forms of neurotic managerial styles and show how they can lead to ethical fiascos.

By now, the volume has covered Freeman's works on stakeholder theory and business ethics. Though these are already immense contributions to business, his creative scholarly writing goes further. The last section of the volume, **Part III**, is dedicated to Ed's works on what is a logical continuation of all his previous writing – stakeholder capitalism. He knew his works on stakeholder management could improve business practices in many organizations. However, for the effect to be long-lasting, there was a need to shake up the whole economic system. Do not get me wrong – Freeman does support capitalism and believes unreservedly that it is the most efficient way that human beings developed to build their economy. What capitalism needs, though, is reformation that is long overdue. And Ed has become a strong, active voice in a growing movement to make capitalism better.

By the end of the1980s, Freeman has started questioning how the American economy was run. In **Chap. 45, "The Myth of Cowboy Capitalism**", he claims that the real danger facing American business is not its ability to compete, but its ability to cooperate. Ed names the dominating view that business is first and foremost about competition Cowboy Capitalism. This is an excerpt from his work:

> *Executives have to saddle up their horses and ride out to play shoot 'em up with the bad guys – domestic competitors – or band together and duke it out with the savages – foreign competitors.*

Freeman explores this idea further in **Chap. 46, "Understanding Stakeholder Capitalism**" and then along with Liedtka in **Chap. 47, "Stakeholder Capitalism and the Value Chain**". In these works, Ed sees management priority in value creation rather than value capture and outlines key principles of stakeholder capitalism.

In the next works, Freeman starts framing his ideas on stakeholder capitalism in the language of a new story of business. In **Chap. 48, "Toward a Life Centered Ethic for Business**", Freeman and Reichart explore tangible points between business and environment through different mindsets we take on the issue. Though this work has not received scholarly notice to the same degree as many other Freeman's publications, Ed believes that it contains some interesting ideas worthy of further exploration. Some of these ideas where further developed by Freeman in **Chap. 49, "Create a New Story About Business**". Here, he sketches a couple of hypothetical (or quite practical?) situations in business and outlines his recipe for paving the road to a new story of business.

In **Chap. 50**, "**Business Ethics at the Millennium**", Ed laments that most of the work written on business ethics either goes unnoticed or is not taken seriously by the vast majority of the business community. Freeman sees the root cause of the problem in a prevailing view that business and ethical decisions as separate. He offers a solution in seeing stakeholder theory as managerial and addresses four main managerial problems, which altogether leads toward a new understanding of capitalism. Ed composed this 12-page long essay in a Parisian café during a single afternoon.

Chapter 51, "**Poor People and the Politics of Capitalism**" builds further on Freeman's earlier work on Cowboy Capitalism. Freeman, Keevil, and Purnell argue that human beings are not self-interested maximizers as typically portrayed by economists. The authors suggest we use a better language that would do justice to people by stressing their search for meaning, purpose, values, and eagerness to cooperate. The next work is Freeman's tribute to Norman Bowie, a renowned business ethicist well-known for his writing from a Kantian perspective. In **Chap. 52**, "**Bowie's Ethics: A Pragmatist Perspective**", Ed argues that Bowie's Kantian perspective on business ethics has a room for pragmatism as both can align quite neatly.

Chapter 53, "**Short Term vs. Long Term: A Skeptical View … and an Alternative**" argues that flaws in business come not so much from short-termism, but in what we measure to evaluate managers' performance. Freeman posits that managerial compensation should be based not on *stock* options but on *stake* options that represent the level of stakeholder satisfaction. **Chapter 54**, "**Responsible Capitalism: Business for the 21st Century**" has been so rich in ideas about capitalism that it inspired its authors – Freeman, Parmar, and Martin – to follow up with their later book *The Power of And*. Among many interesting ideas, they argue that the unit of business analysis should be not economic transactions but stakeholder relationships where stakeholders are interdependent entities with complex motivations. In the next work, **Chap. 55**, "**The New Story of Business: Towards a More Responsible Capitalism**", Freeman has crystalized all his previous ideas on capitalism by articulating his new story of business.

We end Part III, as well as the whole volume, with **Chap. 56**, "**The Social Responsibility of Business is to Create Value for Stakeholders**", written by Freeman and Elms. The article title is symbolic as it starts the same way as the title for Milton Friedman's doctrine published over 50 years ago. However, it runs in the opposite direction showing that the time for the new narrative of business based on stakeholder ideas has come.

Now, we invite you to proceed to reading Ed Freeman's selected works. We hope that after reading this volume, the audience will have a clear and comprehensive view of Freeman's scholarly narrative. We also hope that the readers will ponder on what has not been said yet and this can generate new creative ideas on stakeholders and beyond.

James Madison University
Harrisonburg, VA, USA

Sergiy D. Dmytriyev

Contents

About the Editors

Sergiy D. Dmytriyev is Assistant Professor of Management at James Madison University. His main areas of research include stakeholder theory, social issues in management, and supererogation in organizations. His latest books are *Humanizing Business: What Humanities Can Say to Business* with Michel Dion and R. Edward Freeman (Springer, 2022); *Research Approaches to Business Ethics and Corporate Responsibility* with Patricia H. Werhane and R. Edward Freeman (Cambridge University Press, 2017); and *The Moral Imagination of Patricia Werhane: A Festschrift* with R. Edward Freeman and Andrew C. Wicks (Springer, 2018). Dmytriyev teaches management consulting, strategic management, and negotiation in Harrisonburg, Virginia. He received his PhD in Business Administration from the University of Virginia's Darden School of Business. Prior to joining academia, Dmytriyev worked for Procter & Gamble, Bain & Company, and Monsanto.

R. Edward Freeman is University Professor, Elis and Signe Olsson Professor, Academic Director of the Institute for Business in Society, and Senior Fellow of the Olsson Center for Applied Ethics at the University of Virginia's Darden School of Business. He is best known for his award-winning book, *Strategic Management: A Stakeholder Approach* (Pitman, 1984; and reprinted by Cambridge University Press in 2010). His latest books are *Models of Leadership in Plato and Beyond* with Dominic Scott (Oxford University Press, 2021); *The Power of And* with Kirsten Martin and Bidhan Parmar (Columbia University Press, 2020); *The Cambridge Handbook of Stakeholder Theory* with Jeffrey Harrison, Jay Barney, and Robert Phillips (Cambridge University Press, 2019); *Research Approaches to Business Ethics and Corporate Responsibility* with Patricia Werhane and Sergiy Dmytriyev (Cambridge University Press, 2017), and *Bridging the Values Gap* with Ellen Auster (Berrett-Koehler Publishers, 2015). He has received six honorary doctorates (Doctor Honoris Causa) from Radboud University in the Netherlands; Universidad Pontificia Comillas in Spain; the Hanken School of Economics and Tampere University in

Finland; Sherbrooke University in Canada; and Leuphana University in Germany for his work on stakeholder theory and business ethics. Freeman is a founding member of Red Goat Records (redgoatrecords.com) bringing the joy of original soul and rhythm and blues music into the twenty-first century. He is also the host of "The Stakeholder Podcast" produced by Stakeholder Media.

Part I
Stakeholder Theory

Chapter 1
The Problems That Stakeholder Theory Tries to Solve

R. Edward Freeman, Jeffrey S. Harrison, Andrew C. Wicks, Bidhan L. Parmar, and Simone de Colle

1 Stakeholder Theory[1]: The Basic Mechanics

Many have argued that the business world of the twenty-first century has undergone dramatic change. The rise of globalization, the dominance of information technology, the liberalization of states, especially the demise of centralized state planning and ownership of industry, and increased societal awareness of the impact of business on communities and nations have all been suggested as reasons to revise our understanding of business.

The dominant way of understanding business and management theory was developed during a time when there was much less concern with turbulence. Weber's ideas about bureaucracy still dominate the managerial landscape, and the economists' idea of an orderly march towards equilibrium still dominates most of the business economics area.[2] Corporations are seen as the property of their owners – shareholders in public corporations – and as limited in their liability for their effects upon

Originally published in: *Stakeholder Theory: The State of the Art*, 3-29, © Cambridge University Press, 2013
Reprint by Springer, Reproduced with permission of The Licensor through PLSclear, https://doi.org/10.1017/CBO9780511815768.002

[1] In Chap. 3 we explain how we ground our approach to "stakeholder theory" in a philosophical pragmatism.

[2] This dependence on a Weberian view of the firm and its underlying ideas about equilibrium may well have been appropriate for understanding a more stable and localized business environment, though Austrians such as von Hayek would still raise many logical questions. However, we believe that in today's world the usefulness of equilibrium-based models is much more limited.

R. E. Freeman (✉) · J. S. Harrison · A. C. Wicks · B. L. Parmar · S. de Colle
University of Virginia, Charlottesville, VA, USA
e-mail: FreemanE@darden.virginia.edu

© Springer Nature Switzerland AG 2023

S. D. Dmytriyev, R. E. Freeman (eds.), *R. Edward Freeman's Selected Works on Stakeholder Theory and Business Ethics*, Issues in Business Ethics 53,
https://doi.org/10.1007/978-3-031-04564-6_1

others. In a world where concerns are primarily domestic, such models may be appropriate, since governments may well be able to abrogate any adverse effects in a way that is fair to all. There is no such world today.

Stakeholder theory has been developed over the last 30 years to counter this dominant mindset. In particular, it has been developed to solve or at least reconceptualize several specific problems. The first might be given the following title:

The Problem of Value Creation and Trade How can we understand business in a world where there is a great deal of change in business relationships, and where these relationships shift depending on the national, industry, and societal context? How is value creation and trade possible in such a world?

As outlined originally in Freeman (1984), stakeholder theory was concerned with the problem of value creation and trade. From its early articulation at the Stanford Research Institute, through the various theorists at Wharton such as Ackoff, Trist, Emshoff, Mitroff and Mason, and Perlmutter, these thinkers were concerned to explain how business could be understood against this backdrop of environmental turbulence to which they saw no end.[3]

It quickly became obvious that trying to solve this problem using the existing fundamental assumptions was fruitless. Most ideas about business assumed the dominance of a kind of economics that assumed that questions of values and ethics were at best "extra-theoretic" if not downright irrelevant. Yet in the real world people were becoming ever more aware of the effects of capitalism on all parts of their lives, so that the second problem may be called:

The Problem of the Ethics of Capitalism As capitalism became the dominant means of organizing value creation and trade, it became clear that restricting attention to its "economic" effects yields a damaging partial view. An increasing number of thinkers have begun to ask questions about the relationship between capitalism and the other institutions in society. Such questions include: (i) How can we understand capitalism so that all its effects can be taken into account by decision makers, rather than externalized on society? (ii) Can we continue to divide the world into the "business realm" and the "ethical realm"? (iii) Is it possible for business executives to "do the right thing," all things considered, no matter how complicated the world is? And (iv) how can we understand both "business" and "ethics" so that we can put them together conceptually and practically?

These questions are relevant to every executive and business thinker today. Given the recent turbulence in financial markets, they have begun to take center stage in the public policy discussions about the reregulation of business. Indeed, the forces of globalization have become stronger, and as information technology has led to more calls for transparency, openness, and responsibility, we have seen an increased

[3] We offer a more comprehensive history of the idea from multiple perspectives in Chap. 2.

interest in understanding how capitalism, ethics, sustainability, and social responsibility can be forged into new ways of thinking about business.

It has become easy to see that solving the problem of value creation and trade only by looking narrowly at the economics of value creation and trade creates the problem of the ethics of capitalism. When the two are combined we find a third, very practical problem. Almost all stakeholder theorists have been engaged in training managers, executives, and MBAs, and have encountered the obvious problem of what to teach.

The problem of managerial mindset also raises a number of questions. (i) How can we utilize and redefine economic theory so that it becomes useful in a turbulent world full of ethical challenges? (ii) How can managers adopt a mindset that puts business and ethics together to make decisions on a routine basis? (iii) How can dealing with turbulence, globalization, and ethics become a routine part of how we understand the manager's job? (iv) What should be taught in business schools?

These questions are relevant to managers who are wondering how to develop their people so that they can be successful in the twenty-first century, and they are relevant to business schools. Business thinkers as diverse as Sumantra Ghoshal and Jeffrey Pfeffer have suggested that current mindsets about business are just not appropriate for the turbulent business environment of today. Since the theories that we teach become "enacted" in the real world, this is much more than an academic issue.[4]

Stakeholder theory suggests that if we adopt as a unit of analysis the relationship between a business and the groups and individuals who can affect or are affected by it, then we have a better chance to deal with these three problems (and surely there are others, or other ways to conceptualize these).[5] Stripped down to its bare essentials, stakeholder theory emerges out of the following four ideas: the separation fallacy, the open question argument, the integration thesis, and the responsibility principle.

Freeman (1994) suggests that most theories of business rely on separating "business" decisions from "ethical" decisions. Indeed, this is the genesis of the problem of the ethics of capitalism, and is seen most clearly in the popular joke about "business ethics as an oxymoron." More formally, we might suggest that we define:

---

[4] In Chap. 3 we attempt to explain why we agree with Sumantra Ghoshal that we enact our theories.

[5] Many will object here that we need to be clearer about whether we are talking about normative stakeholder theory, descriptive stakeholder theory, or instrumental stakeholder theory, as Donaldson and Preston divide the literature (1995). We believe that for some purposes these distinctions are useful, but for others they are not. In particular, we believe that stakeholder theory is inherently managerial. It is about how we do and can understand how we create value and trade with each other. Donaldson and Preston seem to recognize this in their important paper (1995), but it is a point lost on many business school scholars who see research in a narrower vein. See Freeman and Newkirk (2008a, 2008b) for a different view of research in business schools. For a more detailed idea of our pragmatist philosophical views see Wicks and Freeman (1998), as well as Chap. 3.

The separation fallacy

It is useful to believe that sentences such as "x is a business decision" have no ethical content or any implicit ethical point of view. And it is useful to believe that sentences such as "x is an ethical decision, the best thing to do, all things considered" have no content or implicit view about value creation and trade (business).

Wicks (1996) and others have shown how deeply this fallacy runs in our understanding of business as well as in other areas in society.[6] There are two implications of rejecting the separation fallacy. The first is that almost any business decision has some ethical content (Harris and Freeman, 2008). To see that this is true one need only ask whether the following questions make sense for virtually any business decision.[7]

The open question argument

1. **If this decision is made, for whom is value created and destroyed?**
2. **Who is harmed and/or benefited by this decision?**
3. **Whose rights are enabled and whose values are realized by this decision (and whose are not)?**

Since these questions are always open for most business decisions, it is reasonable to give up the separation fallacy. We need a theory about business that builds in answers to the "open question argument."

One such answer would be, "Only value to shareholders counts," but such an answer would have to be enmeshed in the language of ethics as well as business.[8] In short, we need a theory that has as its basis what we might call:

The integration thesis I

Most business decisions or statements about business have some ethical content or an implicit ethical view. Most ethical decisions or statements about ethics have some business content or an implicit view about business.

Yet another way to articulate this idea is:

---

[6] For a recent discussion of the separation thesis, or separation fallacy, see Joakim Sandberg (2008a); the ensuing discussion, Jared D. Harris and R. Edward Freeman (2008); Ben Wempe (2008); John Dienhart (2008); and Joakim Sandberg (2008b). Sandberg's thesis seems to be that by not accepting the fact–value distinction, we somehow make a philosophical mistake. Harris and Freeman argue that the pragmatist view of language that undergirds this distinction is rooted in the philosophical traditions of Wittgenstein, Quine, Dewey, Rorty, and others.

[7] The original open question argument is due to G. E. Moore in Principia Ethica, and was meant to show that "good" and words like that were "supervenient properties." We believe that our current use of the argument is meant to show an openness to most business decisions or, more precisely, theories which may explain business decisions. Thus it is meant to establish that "ethical questions are always there" rather than argue for any particular view of ethics or ethical language.

[8] We shall see later that Friedman, unlike most of his expositors, actually gives such a morally rich answer.

The integration thesis II

1. **It makes no sense to talk about business without talking about ethics.**
2. **It makes no sense to talk about ethics without talking about business.**
3. **It makes no sense to talk about either business or ethics without talking about human beings.**

One of the most pressing challenges facing business scholars is to tell compelling narratives that have the integration thesis at its heart. This is essentially the task that those scholars, called "stakeholder theorists," have begun over the last 30 years. Statement (1) challenges much work done in the name of "value-free economics and science"; (2) challenges much work done by philosophers who have little knowledge of either economics or business; and (3) challenges much work done in all the business disciplines that ignores "the human sciences" or "humanities" or, more concretely, the fact that most human beings are pretty complex (Donaldson and Freeman 1994; Freeman and Newkirk 2008a, 2008b). Stakeholder theory has developed primarily around (1); its future development and usefulness depends largely on how it deals with (2) and (3).[9]

To begin to address (1) we need to go to the very basics of ethics and we suggest that something like the following principle is implicit in most reasonably comprehensive moral views.

The responsibility principle[10]
Most people, most of the time, want to, and do, accept responsibility for the effects of their actions on others.

Clearly the responsibility principle is incompatible with the separation fallacy. If business is separated from ethics, there is no question of moral responsibility for business decisions, hence the joke is that business ethics is an oxymoron. More clearly still, without something like "the responsibility principle" it is difficult to see how ethics gets off the ground. "Responsibility" may well be a difficult and multifaceted idea; there are surely many different ways to understand it. But if we are not willing to accept responsibility for our own actions (as limited as that may be due to complicated issues of causality and the like), then ethics understood as how we reason together so that we can all flourish is likely an exercise in bad faith.[11]

---

[9] We make some suggestions along these lines in Chaps. 9 and 10.

[10] "Responsibility" is a difficult concept. There is a burgeoning philosophical literature on it from the time of Plato. We do not intend that the responsibility principle sets forth any particular view of responsibility. Instead, we intend it as "whatever you think about responsibility, something like this principle is necessary." Of course, portions of it could be modified depending on how responsible you believe people actually are, or on difficult claims about joint causality, institutional roles, and the like.

[11] We might call our overall view on ethics something like "ethics as conversation." It is about trying to work out how we can simultaneously be what Harold Bloom calls "strong poets" and build communities that support human solidarity and flourishing.

One response to the responsibility principle is that some people in fact do not want to be responsible or ethical. They simply want to get away with as much as possible at the expense of others. People sometimes act "opportunistically and with guile." While there is some truth in this view, the question is one of starting points. Start with the responsibility principle and one has to design in how to deal with opportunism. Start with opportunism and one is likely to leave out important ideas such as human dignity, cooperative endeavors, and the creative spirit, all of which, we suggest, are the cornerstones of capitalism. We need a more thorough understanding of the responsibility principle, its origins, and implications, on account of each of these.

It is now easy to see that the genesis of "stakeholder theory" is simply the integration thesis plus the responsibility principle. Give up the separation fallacy, in part because of the open question argument, and there is not much alternative. People engaged in value creation and trade are responsible precisely to "those groups and individuals who can affect or be affected by their actions" – that is, stakeholders. For most businesses, as we currently understand it today, this means paying attention at least to customers, employees, suppliers, communities, and financiers.[12]

"Stakeholder theory" does not mean that representatives of these groups must sit on governing boards of the firm, nor does it mean that shareholders (we prefer "financiers" as a more inclusive term) have no rights. It does imply that the interests of these groups are joint and that to create value, one must focus on how value gets created for each and every stakeholder.[13] How value gets created for stakeholders is just how each is affected by the actions of others as well as managers.

"Stakeholder theory" is fundamentally a theory about how business works at its best, and how it could work. It is descriptive, prescriptive, and instrumental at the same time, and, as Donaldson and Preston (1995) have argued, it is managerial. Stakeholder theory is about value creation and trade and how to manage a business effectively. "Effective" can be seen as "create as much value as possible." If stakeholder theory is to solve the problem of value creation and trade, it must show how business can in fact be described through stakeholder relationships. If it is to solve the problem of the ethics of capitalism, it must show how a business could be managed to take full account of its effects on and responsibilities towards stakeholders. And if it is to solve the problem of managerial mindset, it must adopt

[12] For some purposes it might make sense to pay such attention to others as well. In Chap. 7 we explain why we eschew the line of thought that tries to define all stakeholders for all firms. The business world is simply too diverse.

[13] If there is a jointness, or partial jointness to stakeholder interests, perhaps the insights of Thomas Schelling will be applicable. Schelling (1960) imagined a number of coordination games whereby actors had to coordinate their joint interests. Of course, stakeholder interests may also be in partial conflict, but if the possibility of innovation and the redefinition of interests is always present, then we can more profitably focus on the jointness of interests rather than on the conflict. We believe that the firm could be conceptualized as a "Schelling focal point" that preserves the possibility of innovation through time. We do no more than suggest this idea for others to explore.

a practical way of putting business and ethics together that is implementable in the real world.

For the most part writers on stakeholder theory have taken an approach that looks at reasonably large existing businesses. They have tried to use the idea to address issues such as "corporate social responsibility," "corporate legitimacy," "theory of the firm," and even macro-societal issues such as "building the good society." With rare exceptions little thought has been given to a host of important issues that have concrete practical significance: how are we to understand value creation and trade at the simplest level? How do entrepreneurs create and sustain value? How does value creation and trade take place within and among multiple state regimes? While at first glance these questions may seem intractable, we want to suggest that we can take a stakeholder approach to them to yield some interesting insights, and to highlight some assumptions about business which we may wish to make optional.

There are a number of competing "standard accounts" of value creation and trade. They all revolve around the idea that shareholders or owners or investors are entitled to the residual gains that accrue from value creation and trade. Stakeholder theory suggests that matters are more complicated – that is, that stakeholder relationships are involved, and that human beings are more complex than the standard accounts assume. We shall look, in turn, at the views of four influential theorists, Milton Friedman, Michael Jensen, Michael Porter, and Oliver Williamson. We shall argue that if we see these standard accounts in the proper light they are all compatible with stakeholder theory, but they are not terribly useful for the purposes of solving our three problems. We follow these analyses with a return to the very basics of business – entrepreneurship – and we anchor the basics of stakeholder theory in this realm.

2 The Friedman Problem: Business as Markets and Maximizing Shareholder Value

Since the first formal articulation of stakeholder theory over 25 years ago, there has been much debate about the difference between the views of business that are centered on stockholders and those that are centered on stakeholders. Milton Friedman's article (1970) has long been juxtaposed against stakeholder theory, and the ensuing debates have revealed few new or useful insights. In an attempt to move beyond the narrow, supposed stakeholder/stockholder dichotomy, we spell out our reading of Friedman's controversial article, which we believe to be compatible with stakeholder theory – in fact we see Friedman as an early stakeholder theorist.

Friedman writes, "It may be in the long-run interest of a corporation that is a major employer in a small community to devote resources to providing amenities to that community or to improving its government"; he goes on to say that it is wrong to call this social responsibility because "they [the actions] are entirely justified in its [the corporation's] self interest" (Friedman 1962: 132).

For Friedman, supporting stakeholder interests is not about social responsibility; it is about capitalism. According to Friedman the purpose of business is to "use its resources and engage in activities designed to increase its profits so long as it stays within the rules of the game, which is to say, engages in open and free competition, without deception or fraud" (Friedman 1962: 133).

All this sounds well and good to us. A key difference between our view and Friedman's is what makes business successful. Friedman believes that it is maximizing profits. We believe that in order to maximize profits, companies need great products and services that customers want, solid relationships with suppliers that keep operations on the cutting edge, inspired employees who stand for the company mission and push the company to become better, and supportive communities that allow businesses to flourish. So in our view Friedman could have written the above quotation as:

Business is about making sure that products and services actually do what you say they are going to do, doing business with suppliers who want to make you better, having employees who are engaged in their work, and being good citizens in the community, all of which may well be in the long-run (or even possibly the short-run) interest of a corporation. Stakeholder management is just good management and will lead to maximizing profits.

Under this reading Friedman is at least an instrumental stakeholder theorist.[14] He may also believe that individuals have a responsibility not to destroy the basis of capitalism – freedom, in his view. In his book *Capitalism and Freedom* he spells out that one of the virtues of the market economy is that it protects individuals from social conformity and abuse of political power. For Friedman, power must be checked and used responsibly. Since in his view economic freedom is a large subset of political freedom, we may deduce that he would agree that economic powers are also subject to responsible use. Friedman may come to something like stakeholder theory out of more than just instrumentalism; he could see it, as we do, as the very basis of capitalism.[15]

There may also be a difference in the theories about the way the world works. Friedman may actually believe that if you try to maximize profits you will do so. We believe that trying to maximize profits is counterproductive, because it takes attention away from the fundamental drivers of value – stakeholder relationships. There has been considerable research that shows that profitable firms have a purpose and values beyond profit maximization.[16] Profit maximization may be better thought of as a result or outcome.

Both we and Friedman agree that business and capitalism are not about social responsibility. We contend that stakeholder theory is about business and value

[14] Instrumental stakeholder theorists believe that creating value for nonshareholder stakeholders actually creates the most value for shareholders. See Jones (1995), Donaldson and Preston (1995), and Jones and Wicks (1999a, b).

[15] We explain how stakeholder theory can be the basis for a new capitalism in Chap. 9.

[16] We review this literature in Chaps. 4 and 8. Of particular note is the work of Collins and Porras (1997) and Graves and Waddock (1994).

creation and, as we said above, it is managerial. Economics may not fundamentally be about value creation in real business. At its best it may be an idealized and abstract view of markets built around the goals of prediction, not around the way that actual business works. It is clearly useful for many purposes, but perhaps not for solving the problems of understanding business in the twenty-first century.

Despite these differences, we believe that Friedman's maximizing shareholder value view is compatible with stakeholder theory. After all, the only way to maximize value sustainably is to satisfy stakeholder interests.

3 The Jensen Move: Business as Agency

Michael Jensen, in a paper titled "Maximization, stakeholder theory, and the corporate objective," argues that stakeholder theory needs an objective function, namely value maximization. He says,

Value maximization states that managers should make all decisions so as to increase the total long-run market value of the firm. Total value is the sum of the values of all financial claims on the firm – including equity, debt, preferred stock, and warrants. (Jensen 2002: 236).

Jensen argues that stakeholder theory is incomplete because it does not offer answers to the questions, how do we keep score? and how do we want the firms in our economy to measure better versus worse? His argument is built on two major premises.

First, Jensen states that purposeful corporate behavior requires a single value objective function. He gives the example of a manager who is forced to choose between maximizing profit or market share – given that every incremental increase in market share comes at higher cost. Here he believes that managers are forced to choose between the two goals and that value maximization offers them an objective principle for making the trade-off. He continues,

A firm can resolve this ambiguity by specifying the tradeoffs among the various dimensions, and doing so amounts to specifying an overall objective function such as $V = f(x,y, \ldots)$ that explicitly incorporates the effects of decisions on all the goods or bads (denoted by $(x,y, \ldots)$) affecting the firm (such as cash flow, risk, and so on). (Jensen 2002: 238).

We do not believe that the complexity of management can be made so simple. Primarily, the variety of metrics used in a firm can not be folded so easily into a single overall objective function. Firms and people do not simply arrange values and preferences in hierarchical and easily understandable decision trees. Jensen's view ignores lexicographical orderings, or dictionary orderings. Additionally, to create a final score or objective measure of the kind that Jensen wants, different metrics must be weighted. The process of choosing weights for these metrics requires some other notion of purpose or mission – it requires firms to answer the questions, who are we? and who do we want to be? These questions go beyond objective value maximization.

Second, Jenson claims that total firm value maximization makes society better off. He also admits that for this to be true some special conditions must be in place. He says,

When monopolies or externalities exist, the value maximizing criterion does not maximize social welfare. By externalities I mean situations in which the decision-maker does not bear the full cost or benefit consequences of his or her choices; water and air pollution are classic examples. (Jensen 2002: 239).

For Jensen, Ronald Coase provides the solution to these issues by reassigning property rights to avoid a second-best solution. But Coase's ideas of rights assignment come from a utilitarian perspective for which there are few arguments. Jensen, and Coase, simply ride roughshod over the idea of rights, assuming as Charles Fried (1981) has argued, that everything is alienable, even our right to bargain at all. Fried suggests, and we agree, that such a view is at best incoherent.[17] So, Jensen's faith that total firm value maximization makes society better off is dependent on a number of further arguments. While these arguments may be interesting to economists and philosophers, they do not serve much purpose in understanding how value gets created.

Jensen as much as acknowledges this point as he comes to see stakeholder theory as the primary vehicle for understanding how value creation and trade takes place. He says,

We can learn from the stakeholder theorists how to lead managers and participants in an organization to think more generally and creatively about how the organization's policies treat all important constituencies of the firm. This includes not just financial markets, but employees, customers, suppliers, the community in which the organization exists, and so on. (Jensen 2002: 245).

Jensen refers to the coupling of the objective function and stakeholder theory as enlightened value maximization. Like Friedman, Jensen can be seen as an instrumental stakeholder theorist. He believes that managers need to make trade-offs and that they should be guided by the principle of enlightened value maximization. For a second time we see that if we interpret stakeholder theory as a theory about how value gets created, we have little difference with economists like Friedman and Jensen.[18]

4 Porter's Strategy: Business as Competitive Strategy

Since 1980 Michael Porter has explicated a way to think about business that takes the metaphor of "competitive strategy" as a central one. Porter (1980) situates the theory of business squarely in the "structure–conduct–performance" paradigm of

[17] This issue is addressed in more detail in Freeman and Evan (1990).

[18] Similarly we argue in Chap. 4 that the recent theory of incomplete contracting in economics lays down yet another economic keystone for a stakeholder theory of business. See especially the recent work of Joe Mahoney (2005), Russ Coff (1999), and other strategic management theorists who ground stakeholder theory in this approach. More fundamentally see Oliver Hart (1995).

industrial economics. The general idea is that effective strategy (conduct) is a function of the structure of an industry, and that a particular performance results. Porter's second major book (1985) added a description of five forces that determine the nature and level of competition in an industry, as well as suggestions for how to use this information to develop competitive advantage. He provided a value chain that described a firm's primary resource transformation processes and the activities that support those processes. By comparing a firm's value chain to that of competitors, managers could devise ways to develop competitive advantage. More recently Porter and Kramer (2006) have suggested that companies can add thinking about corporate social responsibility and sustainability to their strategic arsenals to gain advantage.

Business is, according to this view, a struggle for advantage. Companies compete with each other to find an advantage that will last – that is, sustainable competitive advantage. The search for advantage is to be found primarily in industry structure and the company's contribution to the value that gets created in the industry. External issues such as corporate responsibility and environmental sustainability can lead to advantage, especially if companies find innovative ways to approach these challenges that are better than those of industry rivals.

While Porter puts more emphasis on "industry" and "competitive strategy" than he does on stakeholder theory, there is much compatibility between the two approaches. To begin with, if we take a somewhat broader view of "value chain," we can easily see that it is just the stakeholders who are a part of this chain. Porter recognizes this fact in making "bargaining power of customers and suppliers" a critical force. But, just as clearly, the bargaining power of employees, the ability of a community to approve regulations or legislation that affects the value chain, and the emergence of other value chain actors such as nongovernmental organizations (NGOs) that call for responsibility and sustainability, are all sources of "advantage." Indeed, the latest Porter work, on corporate social responsibility (CSR), seems to acknowledge this fact.

One might easily develop a Porter-like stakeholder theory along the following lines of a "stakeholder – conduct – performance" paradigm. At any point in time a company exists in a network of stakeholder relationships, a subset of which we might designate as "industry" if we are so inclined. Businesses then try to craft a value proposition that meets the needs or expectations of a certain group of these stakeholders.[19]

Implicit in the Porter account is the idea that the interests of stakeholders can conflict. While this is indeed true, such an account underestimates the extent to which the interests of customers, suppliers, employees, financiers, and communities go together. Without a jointness to their interests, there will be no deal among them. If these interests can be kept "going in the same direction," then the deal can be sustainable. We could describe such a deal as "competitive advantage," or

[19] In Freeman (1984) we find several ways to connect stakeholder theory with Porter's view.

alternatively we could describe it as a "system of cooperation."[20] In the view of most economists like Porter, business and capitalism are to be seen as systems of competition for resources. It only takes a very slight twist to see this same system as one of social cooperation and value creation. Stakeholder theory focuses on the jointness of stakeholder interests rather than solely on the trade-offs that sometimes have to be made. It does not deny that such trade-offs are necessary, but suggests that they also represent opportunities to think beyond trade-offs to a question of value creation. Stakeholder theory solves the value creation question by asking how we could redefine, redescribe, or reinterpret stakeholder interests so that we can figure out a way to satisfy both, or to create more value for both. Porter's view, like that of Friedman, Jensen, and others, offers a good building block from which to ask the value creation question.

5 The Williamson Result: Business as Transaction Cost Economizing

In a pathbreaking paper, Ronald Coase (1937) questioned the economic orthodoxy of the time, and wondered why some transactions seem to be organized by markets as economic theory demands, while others seem to be organized by hierarchical arrangements, such as firms. Coase's answer was that most of the time there is a cost to using the pricing mechanism, and that when these "transactions costs" are sufficiently high, someone will organize the transaction via a hierarchy or firm, as opposed to a market. The literature on "transactions costs" or "markets and hierarchies" is now a well-established area of social science.

Indeed, Oliver Williamson (1984), one of Coase's main modern disciples, has suggested that we can understand transactions cost theory in terms of contracts, and that the standard account of firms as a nexus of contracts follows. Shareholders still bear the residual risk, while other stakeholders have arranged bilateral contracts with built-in safeguards, so that shareholders are entitled to the returns. There is no need to give a "stakeholder account" of transactions cost theory on this interpretation of Williamson's view.[21]

The first point to make here is that, like the standard account of Friedman, this view does not offer much practical insight into how to create value and trade. The best it can do is to exhort us to "understand the structure of transaction costs." While this may not seem much, recent work on e-business, supply chain management, and other issues resulting from the application of information technology offer much insight into the actual practice of value creation and trade. However, on closer

[20] This seems to be the major intuition behind the so-called "resource-based view" of the firm. See Barney and Arikan (2001).

[21] There is a burgeoning literature on transaction cost economics in the economics and strategic management disciplines.

examination of these issues, each appears to be simply a detailed analysis of particular stakeholder relationships. After all, how can one see supply chain management as anything other than integrating the supplier–firm–customer chain of interests? So it may be that to turn transactions cost theory towards the practical understanding of value creation and trade, one needs to overlay a stakeholder network.

Second, Freeman and Evan (1990) have questioned Williamson's analysis here by introducing the idea that if contracts have safeguards, then the question of who pays the cost of the safeguards is relevant. For instance, if management and labor contract against a backdrop of the liberal state, complete with safeguards for labor such as labor boards, processes that must be followed under penalty of law, and so on, then both parties have successfully exported the costs of the safeguards of their contract to society as a whole. Indeed, we suggest that a distinction between exogenous safeguards (where the costs are externalized to society or other stakeholders) and endogenous safeguards (where the parties to the contract pay the cost of contracting including safeguards) is crucial for seeing the necessity for a stakeholder approach to markets and hierarchies.

In a more recent paper, Williamson and Bercovitz (1996) seem partly to accept this idea. They suggest that shareholder boards be seen as endogenous safeguards. They even suggest that stakeholder-oriented "boards of overseers" may well be a good idea to get more stakeholder input into the value creation process, of which stakeholders are clearly a part. But they fail to deal adequately with the criticism that safeguards have costs. The implication of such a view is that the contractual arrangements that we observe will be a function of how parties to the contract have been able either to accept or to offload the costs of safeguards. This process is not necessarily a transaction cost economizing process, but rather a political one. If the parties to the contract can externalize the costs of safeguards to others, such as taxpayers, then we would expect to see them use their own power in the political process to realize such gains. In fact, we are appealing to nothing more than the strict "opportunism" assumption in transactions cost theory.[22] In Williamson's well-known diagram, slightly revised, it would be difficult to tell whether a particular governance mechanism appeared at node B or at node D (see Fig. 1.1).

In summary the argument is this. Assume a version of the modern state, the rule of law, and a set of institutions that makes contracting viable. One can then understand the creation of value and trade against this backdrop of background institutions. In a world in which these institutions emerge so that financiers have the right to the residuals of the firm, something like the standard story emerges. Absent these institutions we are left wondering who pays or should pay for whose safeguards. If this is in fact an open question, then a series of other questions is relevant. Could it be interesting to imagine a world where there are only endogenous

[22] The only way to explain voluntary interactions with stakeholders or endogenous safeguards would be to appeal to either a lack of political power, or something like the responsibility principle and subsequent stakeholder theory. The recent financial crisis is a textbook case study of how destructive exogenous safeguards can be.

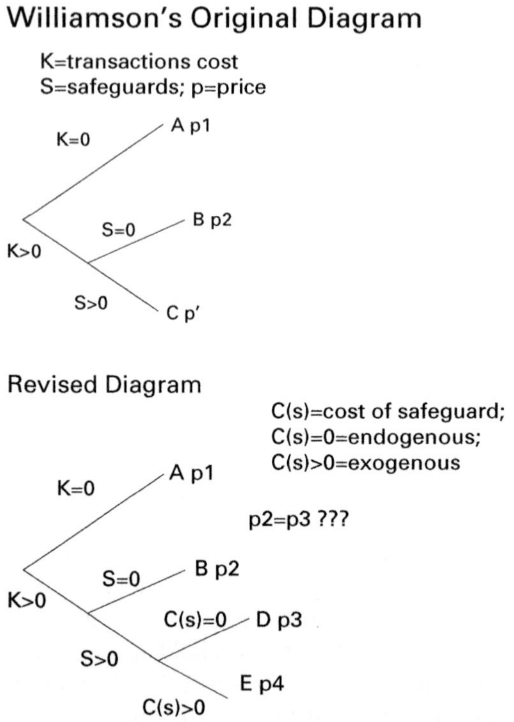

Fig. 1.1 Transactions cost theory

safeguards? A world in which there are no background institutions, or where there is only the presumption that value creation and trade will continue over time? A world in which there are many conflicting and competing background institutions and there is the desire for value creation and trade to continue over time?

We want to suggest that these last questions must take us substantially beyond what has been done so far in the transactions cost literature, and must put us firmly in the middle of stakeholder theory. Transaction costs economics (TCE) simply focuses too heavily on one sort of governance mechanism – traditional boards of directors. And TCE is too concerned with yielding the traditional view of economics. However, we have suggested that one can use TCE reasoning to see that if the cost of safeguards were assigned differently, then other arrangements may well be possible. We do not see those arrangements because of the way in which we currently think about safeguards as "primarily the government's job."

However, TCE's idea of stakeholder boards of overseers is actually quite interesting. Suppose such a board's task were (i) to reduce information asymmetry among key stakeholders so that management could more easily create even more value; (ii) to view the interest of financiers, customers, suppliers, communities, and employees as joint; and (iii) assume the continuation of the corporation through time. It may well turn out that such a board becomes a very effective "governance mechanism" to help managers create as much value as possible for stakeholders.

6 Business as Entrepreneurial Opportunity: Basic Ideas of Stakeholder Theory

In a path-breaking set of articles that both summarizes and extends the entrepreneurship literature, Sankaran Venkataraman has suggested that understanding entrepreneurship can fill the gap left by the standard accounts of business activity. He suggests that.

> In most societies, most markets are inefficient most of the time, thus providing opportunities for enterprising individuals to enhance wealth by exploiting these inefficiencies (the Weak Premise of Entrepreneurship).

and, alternatively, that

> Even if some markets approach a state of equilibrium, the human condition of enterprise, combined with the lure of profits and advancing knowledge and technology, will destroy the equilibrium sooner or later (the Strong Premise of Entrepreneurship). (Venkataraman 1997: 121)

Venkataraman explicitly connects entrepreneurship with the stakeholder literature by claiming that

> The essence of the corporation is the competitive claims made on it by diverse stakeholders. It is a fact of business life that different stakeholders have different and often conflicting expectations of a corporation. Indeed, the firm itself can be said to be an invention to allow such conflict to be discovered, surfaced, and resolved, because conflicting claims have to be discovered and methods for resolution executed … This inherent conflict is a feature not only of the established giant corporation, but also of the very act of creation of the productive enterprise. Entrepreneurship involves joint production where several different stakeholders have to be brought together to create the new product or service. (Venkataraman 2002: 46)

According to this view, the existence of entrepreneurial activity in a society acts as an equilibrating force. It offers an alternative to stakeholders whose needs are not being met by the current arrangement.[23] There is both a weak and strong equilibrating process.

> The weak equilibrating process holds that whenever a stakeholder justifiably believes that the value supplied by him or her to a firm is more than the value received, the entrepreneurial process will redeploy the resources of the "victimized" stakeholder to a use where value supplied and received will be equilibrated. The strong equilibrating process holds that if the redeployment of individual stakeholders does not work freely and efficiently and serious value anomalies accumulate within firms and societies, the entrepreneurial process will destroy the value anomalies by fundamental rearrangements in how resources and stakeholders are combined.[24]

The very processes of entrepreneurial activity, whereby entrepreneurs find or create opportunities because they have knowledge or experience that others do not, depend on understanding how stakeholder interests have been or cannot be satisfied. In the

---

[23] This is the reason behind the insight that "behind every disgruntled stakeholder and critic of a company lies a business opportunity."

[24] Venkataraman (2002), at 50.

following paragraphs we want to unpack these processes in more practical terms to see how value creation and trade can actually come about.

Suppose that Smith has a particularly good recipe for bread. He finds that friends and relatives are always taking second and third helpings of bread at dinner, asking for the recipe, and cornering Smith for tips on how to bake such good bread. Smith reasons that if the bread is so good, there must be people who are willing to pay for the bread, and after all who can't use the extra cash? So, Smith starts to sell his bread to others. Perhaps he delivers it to steady customers or even "contracts" with the local grocer to sell the bread in her store. Smith has become an entrepreneur. And perhaps the standard account can explain Smith's success or failure.

On the standard account we would expect the growth and development of Smith Bread Company to be a function of the market for bread. We would try to understand the structure of that market, for instance the number of buyers and sellers, the product ranges of each, the price points of the offerings, and so on. If Smith Bread Company succeeds, it would be because Smith is able to offer a similar product at perhaps a lower price, or perhaps with another feature that buyers of bread want. If Smith Bread Company fails, it would be because others offered the same product at a lower price. In fact, the strict neoclassical view of the standard account would suggest that all the information regarding features and product performance is reflected in the price of the product. A "second best" version of this view, à la Michael Porter's view of strategy, argues that in most real markets it would be slightly more complex, and Smith could position the company to take advantage of those complexities or not. In short, Smith's success or failure is a matter of the market for bread. Understand that market and you'll understand all you need to know about Smith Bread Company.

None of this gives much advice to Smith or explains how Smith Bread Company really came about. This view of markets – as consisting of buyers and sellers – is interesting only to the extent that the question, "how does this market work?" is an interesting question. Understanding the Dutch flower market, the Chicago futures market, the coffee exchange in Uganda, and others is a set of interesting questions but ultimately they are questions about the distribution of value in very specialized situations, rather than its creation.[25]

Let us go further and suppose further that Smith's bread is a big hit with all who try it. Soon Smith must quit his full-time job (perhaps Smith is professor of economics and moral philosophy at Edinburgh) and devote all day to baking bread. He quickly realizes that the kitchen oven is being monopolized by the bread baking, so he invests money in another oven and fixes up the spare room to do nothing but bake bread. But even this is not enough. The demand for Smith's bread is so great that he decides to invest his savings and perhaps talk to his local banker about a loan. Smith builds a bread factory, hires workers to bake the bread in the ovens. Smith spends his time directing the baking and selling of bread.

[25] Amartya Sen' s On Ethics and Economics (1987) diagnoses what has gone wrong with economics as a discipline. Also see Julie Nelson (2006) for an insightful critique.

The transaction cost view is relevant here. It would suggest that Smith Bread Company is successful simply because Smith is correct that he can organize some of the transactions internally via the authority relationship, such as hiring workers to bake the bread, rather than buying bread in the market for bread and reselling it. Indeed, this view might tell us that if it could be done more cheaply, it may well be in Smith's interest to begin to grow his own wheat. The success of Smith's venture will not only be a function of the market for bread, but also a function of the markets for the factors of production, such as the labor market and the market for ingredients such as wheat and yeast.

While this view is a more fine-grained analysis of what is happening to Smith Bread Company, it gives little practical advice, for how is Smith to know whether the costs of organizing transactions inside the firm are actually lower than using the market mechanism?[26]

Now let us take a more fine-grained view of Smith's enterprise. What must Smith do to be successful? He must buy raw materials from suppliers such that he can be assured they are of good quality. He must have employees who will make the bread as Smith would, and this is easier when these employees come to want to make the bread as Smith would. He must find customers who want and enjoy his bread so much that they want to buy it again and again.[27] To the extent that he has extended his financial resources to include the bank, relatives, or even shareholders, Smith must make a return for these other financiers, as well as profits (in some form) for himself. And, perhaps more subtly, Smith must be a good citizen in the community. At a minimum Smith must not use his property to harm others. Suppose, for instance, that Smith's new bakery emitted noxious fumes, smelled by other members in the community. Smith would come under pressure to do something about it, and if Smith lived in a relatively free society, community members could claim that Smith has damaged them, and sue for relief, either through the courts, or via legislation.

Venkataraman has suggested that the conflicts that exist between actors in the factor markets will ultimately be sorted out by the entrepreneurial process (the strong or weak force). But, alternatively, we can look at these conflicts from the standpoint of Smith and the stakeholders in Smith Bread Company. From Smith's point of view, his job is to try and resolve these conflicts in a way that is good for the "joint enterprise" that is Smith Bread Company. When customers have complaints, he wants to resolve these complaints so that they don't stop buying bread. Now, there will be limits to what Smith is able to do, and against these limits the entrepreneurial forces will operate. When employees become disgruntled so that they do not put forth their best effort or even think about leaving, Smith wants to find a way to keep creating a sufficient level of value for them to stay. Smith always asks the value creation question. Again, there will be limits, but, practically speaking,

[26] Paradoxically, this knowledge itself is a transaction cost.

[27] Strictly speaking, this premise is not necessary. There are some businesses that rely on one-time purchases. But, in the real world, managers and entrepreneurs think of customers as wanting to buy again and again. This is the whole reason for brands.

Smith always seeks to test these limits by creating as much value as possible for all stakeholders.

As a practical solution to this problem, Smith needs to see the interests of stakeholders as moving roughly in the same direction. And he needs to see that the interests of one stakeholder may well be enhanced in the presence of others.[28] Many stakeholder theorists have focused on the inherent conflict between stakeholder interests and, in doing so, they have forgotten that stakeholder interests are also joint. Many other theorists have asserted that stakeholder theory claims that all stakeholders are equally important. Again, they have forgotten the real world beginnings of the theory.[29] All are not equally important at all points of time, but all have the equal right to bargain over whatever their interests are. (We take this to be a simple statement of some notion of classical liberalism.) And all interests have roughly to go together over time, otherwise, in a relatively free society, stakeholders will turn to the state for restitution.

In short, when Smith successfully, over time, satisfies customers, financiers, suppliers, employees, and the community, then Smith Bread Company flourishes. Notice that the success of Smith Bread Company is still dependent on the market for bread, and the various factor markets, but Smith now has some tangible advice about how to create value and sustain it. Figure 1.2 gives the standard picture of value creation and trade among stakeholder theorists.[30] Smith creates value for stakeholders.

7 The Stakeholder Mindset

The basic idea of creating value for stakeholders is quite simple. Business can be understood as a set of relationships among groups which have a stake in the activities that make up the business. Business is about how customers, suppliers, employees, financiers (stockholders, bondholders, banks, etc.), communities, and managers interact and create value. To understand a business is to know how these relationships work. And the executive's or entrepreneur's job is to manage and shape these relationships.

---

[28] A simple example of what we have in mind here is the fact that our interests are better served when we are on the same faculty as Venkataraman where we can easily work together.

[29] These are especially problematic interpretations for the so-called "normative" disciplines. See Chaps. 7 and 8.

[30] Of course, this is only one of many possible pictures. Clearly, the corporation is not the center of the universe, though for some purposes it is useful to see the world from the corporation at the center point of view. For other purposes it is useful to diagram stakeholders as being at the center. The Danish company, Novo Nordisk puts "people with diabetes" at the center of its stakeholder map. See Freeman, Harrison, and Wicks (2007) for more suggestions about stakeholder maps. Fassin (2008) contains several useful depictions of the stakeholder idea. Figure 1.2 is due in large part to conversations with Robert Phillips. See in particular Phillips (2003).

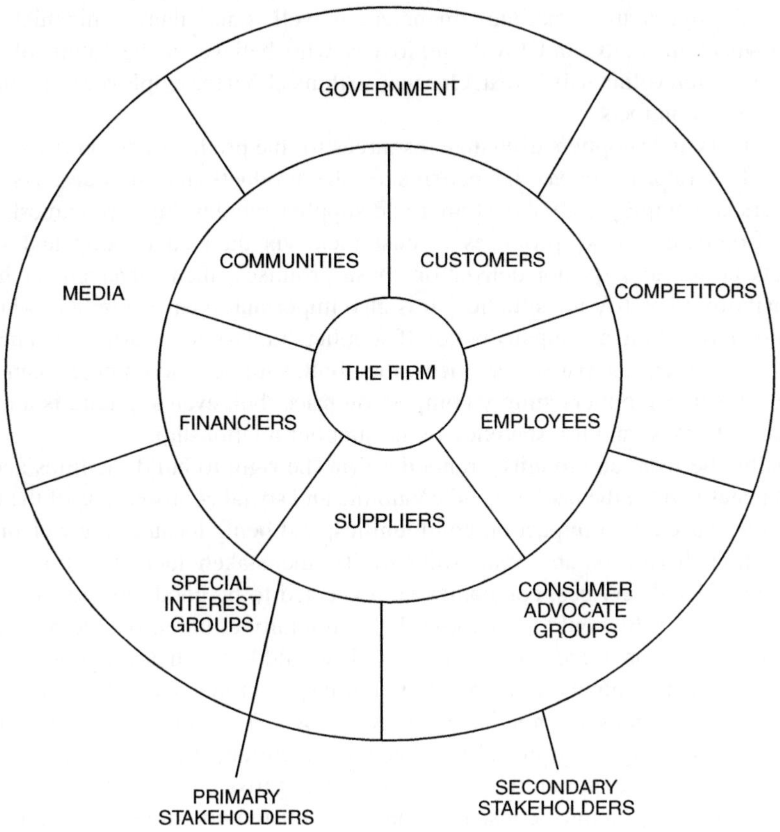

Fig. 1.2 Creating value for stakeholders. (Source: R. Edward Freeman, Jeffrey S. Harrison, and Andrew C. Wicks 2007. Managing for Stakeholders: Survival, Reputation, and Success. New Haven: Yale University Press. Originally from a conversation with Robert Phillips)

Owners or financiers (a better term) clearly have a financial stake in the business in the form of stocks, bonds, and so on, and they expect some kind of financial return from them. Of course, the stakes of financiers will differ by type of owner, preferences for money, moral preferences, and so on, as well as by type of firm. The shareholders of Google may well want returns as well as be supportive of Google's articulated purpose of "do no evil." To the extent that it makes sense to talk about the financiers "owning the firm," they have a concomitant responsibility for the uses of their property.

Employees have their jobs and usually their livelihood at stake; they often have specialized skills for which there is usually no perfectly elastic market. In return for their labor, they expect security, wages, benefits, and meaningful work. Often, employees are expected to participate in the decision making of the organization, and if the employees are management or senior executives, we see them as shouldering a great deal of responsibility for the conduct of the organization as a whole. Employees are sometimes financiers as well, since many companies have stock ownership plans, and loyal employees who believe in the future of their companies often voluntarily invest. One way to think about the employee relationship is in terms of contracts.

Customers and suppliers exchange resources for the products and services of the firm, and in return receive the benefits of the products and services. As with financiers and employees, the customer and supplier relationships are enmeshed in ethics. Companies make promises to customers via their advertising and when products or services do not deliver on these promises, then management has a responsibility to rectify the situation. It is also important to have suppliers who are committed to making a company better. If suppliers find a better, faster, and cheaper way of making critical parts or services, then both supplier and company can win. Of course, some suppliers simply compete on price, but, even so, there is a moral element of fairness and transparency to the supplier relationship.

Finally, the local community grants the firm the right to build facilities and, in turn, it benefits from the tax base and economic and social contributions of the firm. Companies have a real impact on communities, and being located in a welcoming community helps a company create value for its other stakeholders. In return for the provision of local services, companies are expected to be good citizens, as is any individual person. It should not expose the community to unreasonable hazards in the form of pollution, toxic waste, and so on. It should keep whatever commitments it makes to the community, and operate in a transparent manner as far as possible. Of course, companies do not have perfect knowledge, but when management discovers some danger or runs afoul of new competition, it is expected to inform and work with local communities to mitigate any negative effects as far as possible.

While any business must consist of financiers, customers, suppliers, employees, and communities, it is possible to think about other stakeholders as well. We can define "stakeholder" in a number of ways. First of all, we could define the term fairly narrowly to capture the idea that any business, large or small, is about creating value for "those groups without whose support, the business would cease to be viable." The inner circle of Fig. 1.2 depicts this view. Almost every business is concerned at some level with relationships between financiers, customers, suppliers, employees, and communities. We might call these groups "primary" or "definitional." However, it should be noted that as a business starts up, sometimes one particular stakeholder is more important than another. In a new business start-up, there are sometimes no suppliers, and paying a lot of attention to one or two key customers, as well as to the venture capitalist (financier), is the right approach.

There is also a somewhat broader definition that captures the idea that if a group or individual can affect a business, then the executives must take that group into consideration in thinking about how to create value. Or, a stakeholder is any group or individual that can affect or be affected by the realization of an organization's purpose. At a minimum some groups affect primary stakeholders, and we might see these as stakeholders in the outer ring of Fig. 1.2 and call them "secondary" or "instrumental."

There are other definitions that have emerged during the last 30 years, some based on risks and rewards, some based on mutuality of interests. And the debate over finding the one "true definition" of "stakeholder" is not likely to end.[31] We prefer a more pragmatic approach of being clear of the purpose of using any of the proposed definitions. Business is a fascinating field of study. There are very few principles and definitions that apply to all businesses all over the world. Furthermore, there are many different ways to run a successful business or, if you like, many different flavors of creating value for stakeholders. We see limited usefulness in trying to define one model of business, based on either the shareholder or stakeholder view, which works for all businesses everywhere. We see much value to be gained in examining how the stakes work in the value creation process, and the role of the executive.

Executives play a special role in the activity of the business enterprise. On the one hand, they have a stake like every other employee in terms of an actual or implied employment contract. And that stake is linked to the stakes of financiers, customers, suppliers, communities, and other employees. In addition, executives are expected to look after the health of the overall enterprise, to keep the varied stakes moving in roughly the same direction, and to keep them in harmony.[32]

No stakeholder stands alone in the process of value creation. The stakes of each stakeholder group are multifaceted, and inherently connected to each other. How could a bondholder recognize any returns without management paying attention to the stakes of customers or employees? How could customers get the products and services they need without employees and suppliers? How could employees have a decent place to live without communities? Many thinkers see the dominant problem of stakeholder theory as how to solve the priority problem, or "which stakeholders are more important?" or "how do we make trade-offs among stakeholders?" We see this as a secondary issue.[33]

First and foremost we need to see stakeholder interests as joint, as inherently tied together. Seeing stakeholder interests as "joint" rather than opposed is difficult. It is not always easy to find a way to accommodate all stakeholder interests. It is easier to trade off one against another. Why not delay spending on new products for customers in order to keep earnings a bit higher? Why not cut employee medical benefits in order to invest in a new inventory control system?

---

[31] We examine this debate in Chaps. 7 and 8.

[32] In earlier versions of this argument we suggested that the notion of a fiduciary duty to stockholders be extended to "fiduciary duty to stakeholders." We believe that such a move cannot be defended without doing damage to the notion of "fiduciary." The idea of having a special duty to either one or a few stakeholders is not helpful. See Phillips, Freeman, and Wicks (2003) for the arguments. We have also put the point in other places as the need to "balance" the interests of stakeholders. We prefer the metaphor of thinking about keeping stakeholder interests in "harmony." "Harmony" depicts a "jointness" to the interests that is perhaps the major contribution of a stakeholder approach to business. The notes are different but they must blend together.

[33] See especially Mitchell, Agle, and Wood (1997) for an example. We discuss this paper in Chap. 8.

Stakeholder theory suggests that executives try to reframe the questions. How can we invest in new products and create higher earnings? How can we be sure that our employees are healthy and happy and are able to work creatively so that we can capture the benefits of new information technology such as inventory control systems? In a recent book reflecting on his experience as CEO of Medtronic, Bill George (2003) summarized the stakeholder mindset:

> Serving all your stakeholders is the best way to produce long term results and create a growing, prosperous company ... Let me be very clear about this: there is no conflict between serving all your stakeholders and providing excellent returns for shareholders. In the long term it is impossible to have one without the other. However, serving all these stakeholder groups requires discipline, vision, and committed leadership.

Even the well-known takeover artist Henry Kravis has climbed aboard the stakeholder approach to business. Speaking at the Super Return Conference in Dubai in 2008, the KKR chief said:

> All of us need to accept responsibility for the damage done to the free-market system ... We've moved too slowly to replace management in some situations ... We wait too long hoping they'll improve, but they never do ... You have to focus on all the stakeholders. It's a new thing for us and something we're really hammering. Long-term value is only achieved if growth benefits all stakeholders in a company, from owners to employees, communities and even governments. We are also conscious we are fiduciaries to millions of hard-working men and women and university endowments... Trust must be earned over the long haul and maintained constantly. We have not always adequately explained what we do to the man on the street. Even some of our investors, although happy with the returns we deliver, don't fully understand what we do and why they should invest with us.[34]

The primary responsibility of the executive is to create as much value as possible for stakeholders. Where stakeholder interests conflict, the executive must find a way to rethink the problems so that these interests can go together, so that even more value can be created for each. If trade-offs have to be made, as often happens in the real world, then the executive must figure out how to make the trade-offs, and immediately begin improving the trade-offs for all sides. **A stakeholder approach to business is about creating as much value as possible for stakeholders, without resorting to trade-offs.**

We believe that this task is more easily accomplished when a business has a sense of purpose. Furthermore, there are few limits on the kinds of purpose that can drive a business. Wal-Mart may stand for "everyday low price." Merck can stand for "alleviating human suffering." The point is that if an entrepreneur or an executive can find a purpose that speaks to the hearts and minds of key stakeholders, it is more likely that there will be sustained success.

Purpose is complex and inspirational. The Grameen Bank wants to eliminate poverty. Tastings (a local restaurant) wants to bring the taste of really good food and wine to lots of people in the community. And all of these organizations have to generate profits, or else they cannot pursue their purposes. Capitalism works because

[34] We are grateful to Michael O'Brien and several other students for bringing this quotation to our attention. The source is Dan Primack (2008).

we can pursue our purpose with others. When we coalesce around a big idea, or a joint purpose evolves from our day-to-day activities with each other, then great things can happen.

To create value for stakeholders, executives must understand that business is fully situated in the realm of humanity. Businesses are human institutions populated by real live complex human beings. Stakeholders have names and faces and children. They are not mere placeholders for social roles. As such, matters of ethics are routine when one takes a "managing for stakeholders" approach. In the words of one CEO, "The only assets I manage go up and down the elevators every day."

8 Conclusion

Stakeholder theory has evolved to address the problems of (i) understanding and managing a business in the world of the twenty-first century (the problem of value creation and trade); (ii) putting together thinking about questions of ethics, responsibility, and sustainability with the usual economic view of capitalism (the problem of the ethics of capitalism); and (iii) understanding what to teach managers and students about what it takes to be successful in the current business world (the problem of managerial mindset).

By focusing on the basic mechanics of stakeholder theory we argue that we can understand capitalism as a set of relationships between customers, suppliers, communities, employees, and financiers (and possibly others), all of whom consist of human beings fully situated in the realm of both business and ethics. This approach is consistent with the main ways in which we understand capitalism. In particular we have argued that it is consistent with the market-based approach of Milton Friedman, the agency theory approach of Michael Jensen, the strategic management approach of Michael Porter, and the transactions cost theory of Oliver Williamson. However, we believe that a more useful theory can be built by understanding the "distinctive domain of entrepreneurship" and entrepreneurial theory. By focusing on the jointness of the stakes of stakeholders, and the entrepreneurial dictum to create as much value as possible, we have suggested several features of the stakeholder mindset.

References

Barney, J.B., and A.M. Arikan. 2001. The resource-based view: Origins and implications. In *Handbook of Strategic Management*, ed. M.A. Hitt, R.E. Freeman, and J.S. Harrison, 124–188. Oxford: Blackwell.

Coase, R.H. 1937. The nature of the firm. *Economica* 4 (16): 386–405.

Coff, R.W. 1999. When competitive advantage doesn't lead to performance: The resource-based view and stakeholder bargaining power. *Organization Science* 10 (2): 119–133.

Collins, J., and J.I. Porras. 1997. *Built to Last: Successful Habits of Visionary Companies*. New York: HarperCollins.
Dienhart, J. 2008. The separation thesis: Perhaps nine lives are enough. *Business Ethics Quarterly* 18 (4): 555–559.
Donaldson, T.J., and R.E. Freeman, eds. 1994. *Business as a Humanity*. New York: Oxford University Press.
Donaldson, T., and L.E. Preston. 1995. The stakeholder theory of the corporation: Concepts, evidence, and implications. *Academy of Management Review* 20 (1): 65–91.
Fassin, Y. 2008. Imperfections and shortcomings of the stakeholder model graphical representation. *Journal of Business Ethics* 80 (4): 879–888.
Freeman, R.E. 1984. *Strategic Management: A Stakeholder Approach*. Boston: Pitman.
———. 1994. The politics of stakeholder theory: Some future directions. *Business Ethics Quarterly* 4 (4): 409–421.
Freeman, R.E., and W. Evan. 1990. Corporate governance: a stakeholder interpretation. *Journal of Behavioral Economics* 19 (4): 337–359.
Freeman, R.E., and D. Newkirk. 2008a. Business school research: Some preliminary suggestions. In *1st IESE Conference, Humanizing the Firm and Management Profession*. Barcelona: IESE Business School.
———. 2008b. Business as a human enterprise: Implications for education? In *Rethinking Business Management*, ed. Samuel Gregg and James R. Stoner Jr., 131–148. Princeton: Witherspoon Institute.
Fried, C. 1981. *Contract as Promise: A Theory of Contractual Obligation*. Cambridge, MA: Harvard University Press.
Friedman, M. 1962. *Capitalism and Freedom*. Chicago: University of Chicago Press and Phoenix Books.
Freeman, R.E., J.S. Harrison, and A.C. Wicks. 2007. *Managing for Stakeholders: Business in the 21st Century. Managing for Stakeholders: Survival, Reputation, and Success*. New Haven: Yale University Press.
George, B. 2003. *Authentic Leadership: Rediscovering the Secrets to Creating Lasting Value*. Hoboken: Jossey-Bass.
Graves, S.B., and S.A. Waddock. 1994. Institutional owners and corporate social performance. *Academy of Management Journal* 37 (4): 75–83.
Harris, J., and R.E. Freeman. 2008. The impossibility of the separation thesis. *Business Ethics Quarterly* 18 (4): 541–548.
Hart, O. 1995. *Firms, Contracts, and Financial Structure*. Oxford: Oxford University Press.
Jensen, M.C. 2002. Maximization, stakeholder theory, and the corporate objective. *Business Ethics Quarterly* 12 (2): 235–256.
Jones, T. 1995. Instrumental stakeholder theory: A synthesis of ethics and economics. *Academy of Management Review* 20 (2): 404–437.
Jones, T.M., and A.C. Wicks. 1999a. Convergent stakeholder theory in management research. *Academy of Management Review* 24 (2): 206–221.
———. 1999b. Letter to AMR regarding "Convergent Stakeholder Theory". *Academy of Management Review* 26 (4): 621–623.
Mahoney, J.T. 2005. *Economic Foundations of Strategy*. Thousand Oaks: Sage.
Mitchell, R., B.R. Agle, and D.J. Wood. 1997. Toward a theory of stakeholder identification and salience: Defining the principles of who and what really counts. *Academy of Management Review 22* (4): 853–886.
Nelson, J.A. 2006. *Economics for Humans*. Chicago: University of Chicago Press.
Phillips, R., R.E. Freeman, and A.C. Wicks. 2003. What stakeholder theory is not. *Business Ethics Quarterly* 13 (4): 479–502.
Porter, M. 1980. *Competitive Strategy*. New York: Free Press.
Porter, M., and M.K. Kramer. 2006. Strategy and society: The link between competitive advantage and corporate social responsibility. *Harvard Business Review* 84 (12): 78–92.

Primack, D. 2008. *PE Week Wire*. October 17, 2008. http://hosting.mansellgroup.net/enable–mail/ThomsonNewLetter/HostedWires/NewsLetters/octl7–08.htm
Sandberg, J. 2008a. Understanding the separation thesis. *Business Ethics Quarterly* 18 (2): 213–232.
———. 2008b. The tide is turning on the separation thesis? *Business Ethics Quarterly* 18 (4): 561–565.
Schelling, T. 1960. *The strategy of conflict*. Cambridge: Harvard University Press.
Sen, A. 1987. *On Ethics and Economics*. Oxford: Blackwell.
Venkataraman, S. 1997. The distinctive domain of entrepreneurship research: An editor's perspective. In *Advances in Entrepreneurship, Firm Emergence, and Growth*, ed. J. Katz and R. Brokhous. Greenwich: JAI Press.
———. 2002. Stakeholder value equilibration and the entrepreneurial process. In *Ethics and Entrepreneurship*. Ruffin Series 3., ed. R.E. Freeman and S. Venkataraman, 45–57. Charlottesville: Philosophy Documentation Center.
Wempe, B. 2008. Understanding the separation thesis: Precision after the decimal point. *Business Ethics Quarterly* 18 (4): 549–553.
Wicks, A. 1996. Overcoming the separation thesis: The need for a reconsideration of SIM research. *Business and Society* 35 (1): 89–118.
Wicks, A.C., and R.E. Freeman. 1998. Organization studies and the new pragmatism: Positivism, anti-positivism, and the search for ethics. *Organization Science* 9 (2): 123–140.
Williamson, O.E. 1984. Corporate governance. *Yale Law Journal* 93 (7): 1197–1230.
Williamson, O.E., and J. Bercovitz. 1996. The modern corporation as an efficiency instrument: The comparative contracting perspective. In *The American Corporation Today*, ed. C. Kaysen, 327–359. New York: Oxford University Press.

R. Edward Freeman is University Professor, Elis and Signe Olsson Professor, Academic Director of the Institute for Business in Society, and Senior Fellow of the Olsson Center for Applied Ethics at the University of Virginia's Darden School of Business.

Jeffrey S. Harrison is a University Distinguished Educator and the W. David Robbins Chair of Strategic Management in the Robins School of Business, University of Richmond.

Andrew C. Wicks is Ruffin Professor of Business Ethics in the Darden School of Business, University of Virginia.

Bidhan L. Parmar is Associate Professor and the Shannon Smith Emerging Scholar in Business Administration in the Darden School of Business, University of Virginia.

Simone de Colle is Associate Professor of Business Ethics & Strategy, IESEG School of Management, Paris, and Visiting Associate Professor of Business, Organizations and Society at New York University Abu Dhabi.

Chapter 2
Stakeholder Management: A Case Study of the U.S. Brewers Association and the Container Issue

James R. Emshoff and R. Edward Freeman

Today's executive faces an increasing amount of turbulence from the external environment. As managers and management scientists seek to reduce and explain this environmental turbulence, external pressures seem to become even more complicated. Debates on topics range from corporate governance, social responsibility, and business ethics to raw materials management, the global economic situation, and governmental regulation. The management literature is replete with calls for the need to take the external environment into account when formulating strategies.[1] Some have even gone so far as to suggest that the support of external groups is important to the success of certain corporate policies.[2]

Originally published in: *Applications of Management Science*, 1, 57–90 © Emerald Publishing Limited, 1981
Reprint by Springer, Reproduced with permission of The Licensor through PLSclear, https://doi.org/10.1108/mansc

[1] See K. Andrews, *The Concept of Corporate Strategy* (Homewood, III.: Dow Jones-Irwin, 1971), Chap. 3; C. R. Christensen et al., *Business Policy.*(Homewood, III.: Irwin, 1973) pp. 229–431; Arnold Cooper et al., "Strategic Responses to Technological Threats." In *Business Policy,* W. Glueck, ed. (New York: McGraw-Hill. 1976); pp. 75–85: 1. Ansoff. *Corporate Strategy* (New York; McGraw-Hill, 1965), Chaps. 6–7, pp. 103–138; and C. W, Hofer and D. Schendel *Strategy Formulation: Analytical Concepts* (St. Paul, Minn.: West Pub., 1978), Chaps. 4–5, pp. 69–157.

[2] See R. Ackoff, *A Concept of Corporate Planning* (New York; Wiley, 1970); and F. Paine and W. Naumes, *Organizational Strategy and Policy* (Philadelphia; Saunders, 1978). Most of the literature referred to in note I treats stakeholder concerns as static, against which the corporation must respond. We believe that stakeholders views may be susceptible to change, i.e., that there are "win-win" strategies for stakeholders and the firm. For a summary of another approach to stakeholder strategy, see I. MacMillan, *Strategy Formulation: Political Concepts* (St. Paul, Minn.: West Pub., 1978).

J. R. Emshoff · R. E. Freeman (✉)
University of Virginia, Charlottesville, VA, USA
e-mail: FreemanE@darden.virginia.edu

© Springer Nature Switzerland AG 2023

S. D. Dmytriyev, R. E. Freeman (eds.), *R. Edward Freeman's Selected Works on Stakeholder Theory and Business Ethics*, Issues in Business Ethics 53,
https://doi.org/10.1007/978-3-031-04564-6_2

External pressures are indeed a critical fact of life for today's executive. No longer can these forces be ignored in the hope that they simply will go away. If executives are to begin to manage the external environment effectively, and hence regain some measure of control over the future of their organizations, three important aspects of management must be strengthened. First, a framework must be established to explicitly blend the management of the external environment into the overall management of the business. Second, managerial tools and techniques must be developed that facilitate the definition and implementation of effective corporate strategies for dealing with the external environment. Finally, philosophical principles of strategic management must be articulated that are proactive, i.e., principles which do not take the external environment as given, but rather as opportunistic.

This chapter describes one approach to these three aspects of the management of the external environment. The framework that we present—stakeholder management—has roots in the management literature.[3] We shall attempt to explicate the concepts systematically, and to provide a description of a concrete managerial process for the implementation of "stakeholder thinking." However, an important caveat is necessary at the outset; namely, the process described will not help managers to manipulate and control external groups. Rather, the process is based on philosophical principles of cooperation and negotiation, and on the belief that stakeholder positions can be mutually supportive of corporate objectives. Our analysis is divided into three major sections: a description of the stakeholder concept and framework; an application, through a case study, of the stakeholder management process; and an analysis of the underlying philosophy of stakeholder management.

1 A Stakeholder Framework for Management

Traditionally, executives have viewed the stockholders of a company as having ultimate external control over managerial policies and practices. Such positional power is based on the stockholder's financial stake in the company. Although this group continues to be an important factor in managerial decision making, many other and often highly visible external groups exist as well. And while each group's degree of control over a corporation's behavior varies, each believes that it has a legitimate *stake* in the corporation. Moreover, such groups often evaluate a corporation's actions in vastly different ways. It is precisely these diverse and often conflicting external pressures that make the achievement of managerial control over external groups so elusive.

[3]Ackoff, in *Redesigning the Future* (New York: Wiley, 1974), credits Ansoff, op. cit., with the concept. A referee has informed us that the concept was coined in a Stanford Research Institute internal memorandum. Other purveyors of the term include F. E. Emery and E. L. Trist, *Towards a Social Ecology* (New York: Plenum, 1973), and their more recent work on sociotechnical systems; W. Dill, "Strategic Management in a Kibitzer's World," in I, Ansoff, R. Declerck. and R. Hayes, eds., *From Strategic Planning to Strategic Management* (New York: Wiley, 1976; and W. Rothschild, *Putting It All Together* (New York: AMACOM, 1976),

We use the term *stakeholder* in a generic sense to identify any group whose collective behavior can directly affect the organization's future, but which is not under the organization's direct control.[4] Examples of typical stakeholders are suppliers, customers, owners, employees, competitors, trade associations, special interest groups, and regulatory agencies. The exact nature and number of stakeholders will depend on the particular organization.

We view organizational health, in a stakeholder sense, as being analogous to human health in a biological sense: unless all organs of the body are functioning satisfactory (i.e., all stakeholder relations are in cooperative balance), the overall system cannot function effectively. Similarly, as is the case with the human body, we believe that the organization can take actions equivalent to diet control and exercise to maintain proper balance on a number of fronts.

Though most organizations have some way of dealing with many of these stakeholder groups, such methods usually consist of a fragmented, though concerted, attempt at dealing with one or two stakeholders at the expense of ignoring or applying minimal resources against the rest. Pressure to address a multitude of externally influenced issues commands more and more of an executive's time: "We must insure that raw material shortages do not reoccur." "How shall we position ourselves for dealing with proposed government energy policies?" "This marketing campaign needs revision." "It's wage and benefit time again." Such issues crowd managers agendas, and ate normally handled on a case-by-case basis. Management's attention to each tends to be proportional to the momentary leverage the stakeholder has over the organization. The failure of most corporations to develop systemic strategies for dealing with external pressure is indicative of the fact that a stakeholder framework of managing the external environment is not normally utilized.

The stakeholder structure forces managers to explicitly recognize all the diverse groups which have a stake in the organization. By drawing stakeholder maps, such as the one illustrated in Fig. 2.1, managers become conscious of *all* the external pressures they face, not just those which happen to be highly visible at one particular point in time.

Two basic principles underlie the stakeholder management framework:

1. The central goal is to achieve maximum overall cooperation between the entire system of stakeholder groups and the objectives of the corporation (i.e., the behavior of all groups is maximally supportive of the objectives).
2. The most efficient and effective strategies for managing stakeholder relations involve efforts which simultaneously deal with issues affecting multiple stakeholders.

[4]See J, R. Emshoff and R. E. Freeman, "Managing the External Environment," *New Jersey Bell Journal* 2(1): 12–18 (Spring 1979): and R. E. Freeman, A. Finnel, and J: R. Emshoff, "The USBA and the Container Issue," Wharton ARC Working Paper No. 23–78 (Philadelphia: 1978); and Emshoff and Freeman, "Stakeholder Management," *The Wharton Magazine* (Fall 1979). For an early view of the concept, see A. Berle and G. Means, *The Modern Corporation and Private Property* (New York: Harcourt Brace Jovanovich, 1932; rev. ed., 1968.

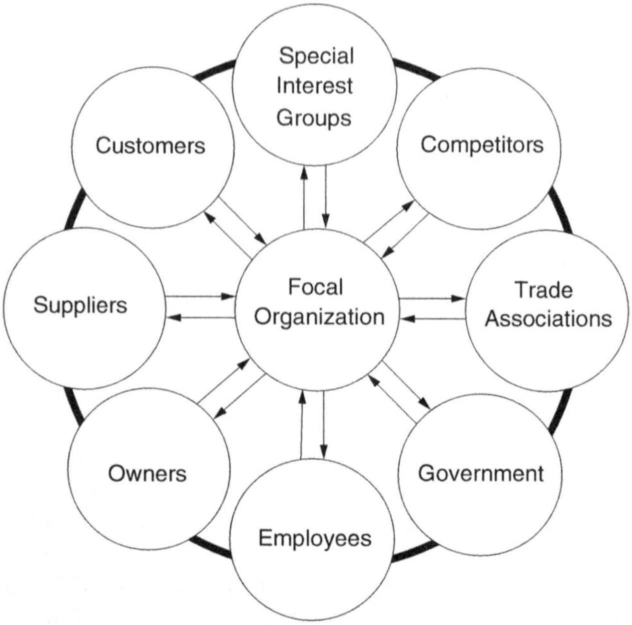

Fig. 2.1 Typical stakeholder map

A few examples of the pitfalls of dealing with stakeholders on a one-by-one basis may better illustrate these points. The 15 years since the 1960s have witnessed philosophical debate over whether a conflict exists between a corporation's social and profit responsibilities.[5] A number of organizations are experimenting with a strategy that attempts to integrate these divergent perspectives. Some have pledged a portion of profits from branded product sales to support specific social causes. In return, social advocates can signal their support of a cause through their purchasing behaviors. This strategy works best for products with functionally equivalent brands, such as soap, coffee, or razor blades, but it has also been used by big-ticket items in modified forms. A Japanese car manufacturer offered to plant a tree for any person who took a test drive—an obvious appeal to environmentalists. A direct purchase strategy was utilized by a consumer goods company during the New York City fiscal crisis of the late 1970s when it contributed some of its profits from selected products to the city's high school athletic program. A regional beer is quietly supporting gay liberation in California and is reported to dominate other brand sales in gay bars. These examples of the simultaneous support of profit and social responsibilities require that the issues be analyzed *together*, as we propose in stakeholder theory.

[5] For a summary of this debate see Fred Sturdivant, *Business and Society* (Homewood, Ill.: Irwin, 1977). For a stakeholder approach to the problem see R. E. Freeman, "Corporate Social Responsibility: A Stakeholder Approach," Wharton ARC. Working paper No. 26–79 (Philadelphia: 1979).

Industry takeover of government services which can be handled more efficiently through the private sector should result in substantial financial benefits, even if the specific returns don't match those of the rest of the corporation. For example, some states use the banking system as a distribution point for food stamps, welfare checks, and other social payments. Due to the sophisticated control procedures used by banks, thefts and related losses have been limited to a fraction of those in a government-run system. Hence the systemwide cost-benefit analysis of banking participation should enable the negotiation of an adequate fee. But often both banks and government take a narrow view of the problem, and in many cases the distribution is done through more expensive means. Thus individuals see a small gain as owners, and a big loss as taxpayers.

2 The Stakeholder Management Process

While the stakeholder concept provides a theoretical framework for analysis, concrete, innovative managerial processes must be developed for actually dealing with the external environment.[6] It is not sufficient to simply define a conceptual framework for managing the external environment. Nor is it enough simply to state that managers should assume an integrated perspective which deals with all stakeholders. Effective stakeholder management calls for concrete managerial processes which turn philosophy and concepts into action. Such a process must satisfy three requirements.

1. It must be capable of identifying strategic options for the company to pursue.
2. It must provide decision makers the inputs from a wide range of key executives who have diverse backgrounds and experiences with various stakeholders.
3. It must encourage executive development of fresh insights into stakeholder behaviors—insights which can be translated into day-to-day strategies for managing stakeholder relationships.

Figure 2.2 depicts a conceptual flow diagram of one such process.

To illustrate the stakeholder management process, we shall use a case study of the problems encountered by the United States Brewers Association (USBA) in trying to address the issue of beverage container legislation. This evaluation will suggest some alternative strategies that the USBA could have taken to manage this problem.[7]

---

[6] Processes for dealing with the external environment have their roots in the modern theories of organization and social psychology, viz., the collection edited by W. M. Evan, *interorganizational Relations* (Philadelphia: Univ. of Pennsylvania Press; 1978). Our own approach has some roots in the theory of games and the belief that there are few, "real-world" zero-sum games. The interested reader may see especially D. Luce and H. *Raiffa, Games and Decisions* (New York: Wiley, 1957); and J. F. Nash, "Two Person Cooperative Games," *Econometrica* 21(1953): 128–140.

[7] The analysis performed here is by the authors and not members or representatives of the USBA. The analysis is only illustrative of stakeholder management techniques and does not reflect either effective or ineffective handling of an administrative situation.

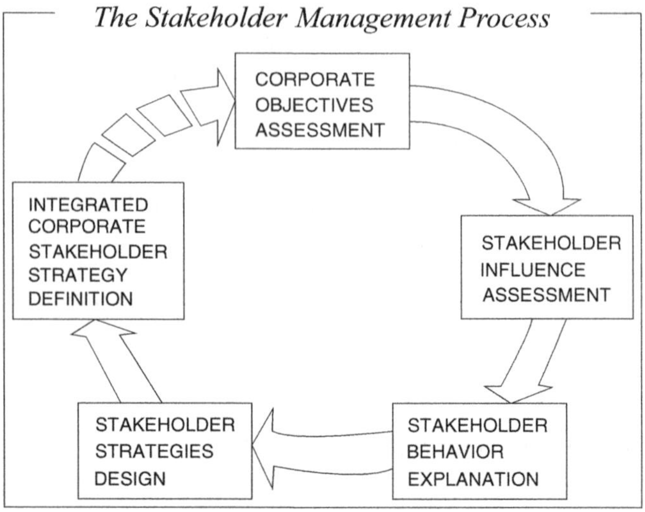

Fig. 2.2 The stakeholder management process

2.1 The USBA and the Container Issue: A Short History[8]

In the mid-1960s, beer and soft drink containers emerged as a significant environmental problem. Public concern prompted controversy over whether the beverage industry should restrict the use of convenience-oriented one-way beverage containers, reinstitute returnable bottles as the sole source of beverage packaging, or continue with the present system that contains both one-way and returnable bottles and cans.

Proponents of beverage container legislation have proposed four types of solutions to solve "the beverage container problem:"

1. Enactment of antilittering educational and law enforcement campaigns
2. Taxes on beverage containers or on other contributors to litter, solid waste, and energy and material depletion problems (these tax revenues would be used to help solve such problems)

[8]The material in this and subsequent sections is adopted from Freeman, Finnel, and Emshoff. op. cit.; and from: W. Adams, *The Structure of American Industry* (New York; Macmillan, 1977); *A History of Packaged Beer and Its Market in the U.S.* (American Cap Co., 1966); *The National Impact of a Ban on Non-refillable Beverage Containers* (Midwest Research Institute, 1971); *Energy and Economic Impacts of Mandatory Deposits* (Research Triangle Institute, 1976); J. Steinman, *Beer Marketer's Insights* (West Nyack, N. Y., Feb. 2. 1977); *Brewers' Almanac* (Washington, D.C.: United States Brewers Association, 1975); R. S. Weinberg, *The Effects-of Convenience Packaging on the Malt Beverage Industry, 1947–1969* (1972). *A Study of the Impacts on the USBA of a Ban on One-Way Beverage Containers* (Philadelphia: Wharton School, University of Pennsylvania. 1976).

3. Mandatory deposits on all beverage containers (without prescribing the containers to be used by the relevant industries) so that consumers would have an incentive to return them
4. Ban on all one-way beverage containers, and return to an all-refillable-bottle system

In 1972 Oregon and Vermont passed laws that required minimum deposits on all beer and soft drink containers. In 1973 Washington State passed a model litter law which placed a tax of 0.0005 s per container on all beer and soft drink bottles and cans. And in 1976 four states—Marne, Michigan, Massachusetts, and Colorado—placed proposed beverage container legislation on the ballot as referenda.

Until the referenda in 1976, challenges to the beverage industry had been fought on a well-known battleground—the state legislatures and the Congress, whose Non-Returnable Beverage Container Prohibition Act of 1973 failed to pass the test of Senate hearings. But when four state legislatures placed legislation on the ballot in a general election, the United States Brewers Association found itself on unfamiliar turf. As the principal voice for the industry, the USBA was faced with the task of defining a strategy with which to attack the issue of beverage container legislation.

The USBA is the oldest trade association in the United States. Formed in 1862 to assist the government in enforcing a beer tax, the association has gradually assumed as its principle task the management of the industry's relationship with government on restrictive legislation issues as well as taxes.

Throughout the 1950s, 1960s, and early 1970s, the structure of the brewing industry evolved from a large number of local and regional breweries to a highly concentrated national industry. In 1974, the top five brewers held a 67% market share. Due to this concentration, the industry came under intensified government scrutiny. Additionally, public concern abounded over alcoholism and the container issue. Combined, these factors effected an adverse public image of the industry.

To counter these negative forces, the USBA became the focal point of an attempt to balance the interests of brewers and packagers. Its strategy was to assemble a strong coalition of beverage-related industries and labor and to propound an economic argument that the status quo, with respect to the container issue, should be maintained.

By 1975, glass accounted for about 9% of total packaging sales in the United States. Over half of all glass container shipments went to the beer and soft drink industries, making those industries the glass producers' Iargest customers. Despite technological developments in plastic containers, glass manufacturers began to invest in new glass plants and equipment in order to hold their product's price advantage over metal cans and the new plastic bottles.

Beverages accounted for 49% of total can shipments in 1975. While the market share of cans had stabilized by then, the aluminum industry's introduction of the two piece aluminum can had been gaining at the expense of steel. Historically, the leading manufacturers of metal beverage cans have not been highly diversified in their operations, and consequently their revenues and profits have been significantly affected by any changes in the market demand for their products. And while

American Can and Continental Can are now finally realizing the benefits of diversification, the metal can remains a major product in their near-term futures. In general, the industry is making little or no new investment in can-manufacturing facilities but is sinking considerable R&D (research and development) funds into recycling capabilities.

Soft drink manufacturers are not as heavily dependent on one-way containers as is the brewing industry. Nevertheless, they would be affected by either a ban, a deposit, or a tax on containers. Hence they gave public support to the USBA position of maintaining the status quo on the container issue.

The retail side of the industry is an interesting case. There are three major outlets for beer and soft drinks: (a) off-premises consumption outlets (liquor stores and grocery stores); (b) on-premises consumption outlets (bars, restaurants, and hotels); and (c) vending machines.

Off-premises outlets accounted for 71% of beer sales and 81% of soft drink sales in 1974. While supermarkets feel that they must carry a full line of beer and soft drinks to attract customers, these items are for the most part low-profit items (around 2.5%). However, retailers generally took no public position on the container issue for fear of alienating segments of their customers.

Labor unions argued that legislation which makes the returnable bottle the dominant container would decrease the demand for skilled workers. Labor was especially effective in supporting the USBA position at the state and national levels. Except in Michigan, where no position was taken, the unions supported the USBA position of no change in the status quo.

Opposition to the USBA position appeared in a variety of forms. Officials of several states advocated container legislation, and where they were successful—as in Oregon, Vermont, and Washington—they publicly promoted the effectiveness of their laws to influence officials in other states. On the other hand, some state legislatures were wary of passing container legislation directly, as is evidenced by their relegation of the question to referenda. Environmental groups, such as the Sierra Club and Friends of the Earth, organized grass roots support for the referenda as well as national support for container legislation.

One major opponent of the USBA's position was the federal government's Environmental Protection Agency (EPA). The EPA held that since beverage containers are estimated to contribute between 15 and 30% of all roadside litter and over 10% of all urban litter, the beverage industry should be singled out for regulation. EPA studies of the environmental impact of container restrictions indicated only marginal negative and numerous positive effects of mandatory deposit laws.

The story of the USBA's position in 1976 would be incomplete without some words about the voters who were charged with deciding the issue in the referendum states. In general, voter reaction was mixed. Some were strongly in favor, some strongly opposed, and some neutral. Many were confused about the many issues: jobs, energy, the economy, and litter. It is against this backdrop that the USBA assumed its own mandate of managing the turbulent external environment.

2.1.1 Phase 1: Assess Corporate Objectives

The first phase in the stakeholder management process is to carefully *state* and *prioritize* the corporation's—i.e., the focal organization's—objectives. Unless an organization understands what it hopes to accomplish. It will be ineffective in analyzing the external environment. An assessment of external forces surrounding an organization must be focused, and a clearly stated set of objectives is the best means of focusing.

In the case of the USBA, which we shall treat as the focal organization, its four core objectives were the following:

O_1: Maintain the status quo with respect to the container issue
O_2: Remove all government activities from the brewing industry
O_3: "Solve" the alcoholism problem as best the USBA could
O_4: Balance the interests of the brewing industry

Naturally, certain objectives are more important than others. Since the USBA had a finite amount of resources to allocate to its objectives, the next step is to prioritize or develop weights on the four objectives. We have developed an interactive technique for doing so.[9] Briefly, the technique involves making pairwise comparisons of each set of two objectives on a scale from 1 to 9. The resulting matrix of comparisons possesses mathematical properties such that a set of weights, W_1, W_2, …, W_n, can be induced on the objectives, O_1, O_2, …, O_n, such that

$$\sum_{i=1}^{n} W_1 = 1.0.$$

Applying this technique to the USBA's objectives, the weights of those objectives are:

$$\begin{aligned} O_1 &= .38 \\ O_2 &= .10 \\ O_3 &= .20 \\ O_4 &= .32 \\ & \overline{1.00} \end{aligned}$$

Thus O_1 (maintaining the status quo with respect to the container issue) is the most important objective, while O_4 (balancing the interests of the brewing industry) is almost as important. Neither O_2 (removing all government activities from the brewing industry) nor O_3 ("solving" the alcoholism problem) is nearly as important as O_1 or O_4.

[9] J. R. Emshoff and T. L. Saaty. "Prioritized Hierarchies as a Vehicle for Long Range Planning," Wharton ARC Working Paper No. 11–78 (Philadelphia: 1978).

2.1.2 Phase 2: Assess Stakeholder Influence

Once priorities for objectives have been established, the next phase of the stakeholder management process is to assess the influence of the organization's stakeholders on these objectives. This phase of the analysis contains several steps:

a. Identify the stakeholders
b. Specify stakeholder behaviors
c. Assess cooperative potential
d. Assess competitive threat
e. Choose target stakeholders

a. *Identify the Stakeholders.* Our recent historical perspective of the USBA indicates the presence of a number of stakeholders. By 1976, the year of the first state container referendum, the rendering of its stakeholder map had evolved to the inclusion at least eleven external groups:

S_1	Can Manufacturers
S_2	Glass Bottle Manufacturers
S_3	Retailers
S_4	Soft Drink Manufacturers
S_5	Brewers
S_6	The EPA
S_7	State Official Advocating Restrictive Container Legislation
S_8	Voters in Referendum States
S_9	Organized Labor
S_{10}	Environmentalists
S_{11}	State Officials in Referendum States

The USBA's stakeholder map in 1976 is depicted in Fig. 2.3.

b. Specify Stakeholder Behaviors. Since an organization's ultimate concern is in achieving its objectives—in this case, maintaining the status quo on the container issue and balancing the interests of the brewing industry—it must analyze how each stakeholder can influence its objectives. This calls for the specification of three sets of behaviors for each stakeholder: (1) *its actual behavior*—what a stakeholder is currently doing to affect the corporation's objectives; (2) *its cooperative potential*—what a stakeholder could potentially do to help the corporation achieve its objectives; and (3) *the competitive threat*—what a stakeholder could potentially do to prevent the corporation from achieving its objectives.

For the USBA we can list the actual behavior of each of its eleven stakeholders. In order to simplify our analysis, we shall be concerned only with objective O_1, "maintain the status quo with respect to the container issue." Including O_4, "balance the interests of the brewing industry," would probably add even more stakeholders to the picture and make our task of illustrating stakeholder management techniques overly cumbersome. Table 2.1 provides the actual behaviors of the USBA's stakeholders in 1976 for objective O_1.

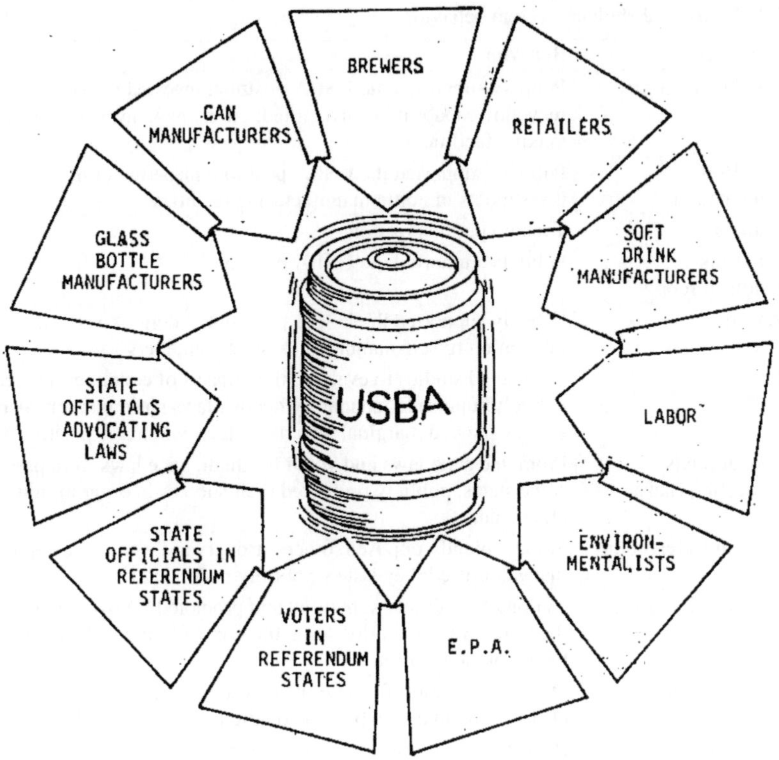

Fig. 2.3 USBA stakeholder map for 1976

The purpose of the stakeholder management process is to devise strategies which yield more cooperation among the stakeholders of an organization. Consequently, we need to look not only at what each stakeholder's actual behavior is, but also how each could behave more cooperatively. To do so we must draw on our knowledge of each stakeholder and its "stakeholder profile." which is created by listing these actual behaviors, as in Table 2.1. Creativity is called for in developing cooperative behaviors, since they are activities we would observe under ideal circumstances.

For example, while the Retailers took no public position on container legislation for fear of alienating some segments of their markets, they could have been more cooperative by publicly supporting the USB A positions. Indeed, they could have threatened to eliminate the sale of beverages—which, after all, were low-margin items—if container restrictions were passed. This action would have been feasible since the Retailers would bear greater labor costs as a result of restrictions on one-way containers, and would have to settle for either lower volume or still lower margins. In creating cooperative behaviors for each stakeholder, overly optimistic aspirations are discouraged. "Pie in the sky" behaviors, such as "The Retailers could have contributed $100 million to the campaign against legislation," are impractical. On the other hand, we don't want to restrict cooperative behaviors so that we miss

Table 2.1 USBA stakeholders' actual behavior

	Stakeholder	Behavior
S_1	Can Manufacturers	Publicly supported the USBA position: invested R&D into recycling capabilities; diversified; put no new investment into can making facilities
S_2	Glass Bottle Manufacturers	Publicly supported the USBA position; made major new investments in bottle manufacturing facilities
S_3	Retailers	Took no public position
S_4	Soft Drink Manufacturers	Publicly supported the USBA position
S_5	Brewers	Publicly supported the USBA position; undertook vertical integration ID self-manufacture some containers
S_6	EPA	Conducted studies to evaluate the impacts of container restrictions; publicly supported mandatory deposit laws since publicly available studies showed marginal negative effects and many positive ones
S_7	State Officials Advocating Laws	Promoted state laws and voted for them; once laws were passed in some states, publicly promoted their success in order to influence other states
S_8	Voters in Referendum States	Some violently opposed, others strongly in favor, some neutral, many confused: key issues not fully understood
S_9	Organized Labor	Nationally, publicly supported and promoted USBA position; on the state level, generally doing the same except in Michigan where they took no position
S_{10}	Environmentalists	Organized support for national and state container laws; made efforts to build and expand a coalition to support the law
S_{11}	State Officials in Referendum States	Passed the responsibility to make the decision on to voters

feasible strategies at the end of our process. Sound business judgment and adequate knowledge about a stakeholder's actual behavior will usually serve to balance the analysis. Table 2.2 provides an analysis of the cooperative behaviors for each of the USBA's stakeholders. Note that some stakeholders may already be as cooperative as possible. In these cases, cooperative behavior will be the same as actual behavior.

In addition to looking at more cooperative stakeholder behaviors, we must examine how their actual behaviors could deteriorate. That is, how could each stakeholder behave more competitively toward the USBA? By assessing both cooperative and competitive behaviors, executives can draw the boundaries of both positive and negative effects on the organization. For example, though the USBA had the solid backing of the brewing industry in 1976, a competitive behavior would be for each brewer to go its own way. The concerns we mentioned with respect to cooperative behavior also apply to competitive behaviors. Table 2.3 indicates the competitive behavior for each stakeholder.

c. Assess Cooperative Potential. Once we have specified each stakeholder Cooperative behavior, we need to assess how important it is to get each stakeholder to be more cooperative. Since resources are limited, it may not be justifiable to try to induce every stakeholder to become more cooperative. We must therefore examine the most, important stakeholders and use them as pressure points to make the total

Table 2.2 USBA stakeholders' cooperative behavior

	Stakeholder	Behavior
S_1	Can Manufacturers	Commit more heavily to can manufacturing: reduce margins to make cans more attractive; increase R&D in new technologies
S_2	Glass Bottle Manufacturers	Shift cost allocations so that there are higher margins on nonreturnables; make the cost of a shift to all-returnable systems much higher
S_3	Retailers	Publicly support USBA position; threaten to eliminate the sale of all beverages if restrictions are placed on containers
S_4	Soft Drink Manufacturers	Same as actual behavior
S_5	Brewers	Same as actual behavior
S_6	EPA	Withdraw support of present public position or at least decrease the constant attention which they focus on the container issue
S_7	State Officials Advocating Laws	Withdraw support of container legislation
S_8	Voters in Referendum States	Vote to defeat laws that make efforts to get other voters to do the same
S_9	Labor	Assume support.in all states
S_{10}	Environmentalists	Withdraw support of container laws
S_{11}	State Officials in Referendum States	Force vote in state legislatures to defeat referenda and take a negative public position if referenda are put on the ballot

Table 2.3 USBA stakeholders' competitive behavior

	Stakeholder	Behavior
S_1	Can Manufacturers	Diversify more rapidly; use cans as a "cash cow" through aggressive price increases
S_2	Glass Bottle Manufacturers	Develop and publicize advances in returnable technology
S_3	Retailers	Publicly support tax legislation
S_4	Soft Drink Manufacturers	Move toward a standard returnable bottle
S_5	Brewers	Each brewer go its own way
S_6	EPA	Same as actual behavior
S_7	State Officials Advocating Laws	Same as actual behavior
S_8	Voters in Referendum States	Vote in favor of referenda and make efforts to get other voters to do the same
S_9	Labor	Withdraw support of USBA position
S_{10}	Environmentalists	Same as actual behavior
S_{11}	State Officials in Referendum States	Take pro position on referenda

stakeholder environment more cooperative. Thus we need to ask the question, "What would be the benefit to the organization, in terms of attaining its objective, of getting stakeholder A rather than stakeholder B to be cooperative?" Again, we can use the interactive judgment technique[10] to facilitate a solution to this question.

[10] Ibid.

The outcome of this technique is a set of weights, similar to those of Phase 1, which indicate the relative importance of inducing each stakeholder to shift from its actual behavior to its cooperative behavior.

For each objective j and each stakeholder S_l, these weights produce a set of favorable movements F_{ij}, such that

$$\sum_{l=1}^{m} F_{ij} = 1.0.$$

Hence F_{ij} represents the relative worth of getting stakeholder S_l to move from his actual behavior to his cooperative behavior.

For example, it is more important to get voters to be cooperative—i.e., to vote against the referenda—than it is to get the glass bottle manufacturers to shift their cost allocations, thus making the move to an all-returnable system more expensive. The *cooperative potential,* as we call these weights, for the USBA stakeholders is given in Table 2.4. Note that stakeholders who are already very cooperative have low cooperative potential.

d. Assess Competitive Threat. Likewise, in assessing each stakeholder's competitive threat we can ask, "What would be the negative impact on the organization's attainment of its objectives if stakeholder A rather than stakeholder B became more competitive?" Using the same judgment technique, we can derive a set of weights which measure the competitive threat of each stakeholder. For each objective j and each stakeholder S_l, these weights produce a set of unfavorable movements U_{ij}, such that

$$\sum_{l=1}^{m} U_{ij} = 1.0.$$

Hence U_{ij} represents the relative harm done by inducing stakeholder S_l to move from its actual behavior to its competitive behavior. Table 2.4 gives a sample set of weights for the competitive threats of the USBA stakeholders. For instance, the

Table 2.4 USBA stakeholders' cooperative potentials and competitive threats to objective 1

	Stakeholder	Potential (F_{ij})	Threat (U_{ij})
S_1	Can Manufacturers	.13	.15
S_2	Glass Bottle Manufacturers	.04	.16
S_3	Retailers	.13	.06
S_4	Soft Drink Manufacturers	0.00	.07
S_5	Brewers	0.00	.18
S_6	EPA	.15	0.00
S_7	State Officials Advocating Laws	.02	0.00
S_8	Voters in Referendum States	.20	.20
S_9	Labor	.10	.16
S_{10}	Environmentalists	.15	0.00
S_{11}	State Officials in Referendum States	.08	.02

USBA would be worse off, in terms of maintaining the status quo on container legislation, if the brewers became competitive and each went his own way than if the retailers supported tax legislation. Note that since competitive threat measures relative *changes* in behavior stakeholders who are already fairly hostile to the USBA—i.e., the EPA, Environmentalists, etc.—have low competitive threats.[11,12]

e. Choose Target Stakeholders. The results of the previous four steps are the identification of each stakeholder's actual behavior, measures of the cooperative and competitive potentials of each, and a more precise mapping of the stakeholder environment. Managers are now prepared to choose the most important stakeholders for further analysis. There are a number of ways to choose target stakeholders; the final analysis will depend on the circumstances surrounding each case. One can look at stakeholders who have the highest cooperative potential—for instance, Voters and Environmentalists. In this case, one would choose according to

$$\max\left\{F_{ij}\right\} \qquad \text{for largest } O_j.$$

One can look at the stakeholders who have the highest competitive threat—for instance. Voters and Brewers—choosing according to

$$\max\left\{U_{ij}\right\} \qquad \text{for largest } O_j.$$

One might focus on the stakeholders that offer the most potential improvement for the overall set of objectives. In this case, we would choose the largest

$$\sum_{j=1}^{n} O_j F_{ij}.$$

One might also try to get some measure of the overall strength of each stakeholder, taking into account both cooperative potential and competitive threat. However, at present there are some technical scaling problems with obtaining this measure. The quantity

$$\sum_{j=1}^{n} O_j\left(F_{ij} + U_{ij}\right)$$

is not guaranteed to always be meaningful, since the units of measurement of F_{ij} and U_{ij} need not have the same magnitude.

---

[11] One possible addition is to build in a set of likelihoods or probabilities of stakeholders playing their cooperative or competitive strategies. Yet the flavor of stakeholder management would make these probabilities *conditional* on the final strategy adopted by the firm, since the stakeholder environment is not assumed to be static: Such a conditional probability step in the process is omitted here for the sake of simplicity.

[12] J. Steinman, *Beer Marketer's Insights* (West Nyack, N.Y., Feb. 2. 1977).

For our analysis of the USBA's stakeholders in 1976, we shall choose the following stakeholders with a high cooperative potential: Can Manufacturers, Voters, Environmentalists, the EPA, and Retailers. In addition, we shall look at the following stakeholders who have high competitive threats: Voters, Brewers, Labor, Glass Bottle Manufacturers, and Can Manufacturers.

Combining these choices, our list of target stakeholders is:

S_1	Can Manufacturers
S_8	Voters
S_{10}	Environmentalists
S_6	EPA
S_5	Brewers
S_7	Labor
S_2	Glass Bottle Manufacturers
S_3	Retailers

2.1.3 Phase 3: Explain Stakeholder Behavior

By the end of Phase 2, we have a fairly clear picture of the impact of each stakeholder's behavior on our objectives. We have also made decisions as to which stakeholders are most important. In order to lay the groundwork for the development of strategies for each stakeholder group, we need to go beyond just listing behaviors. We need to explain *why* a stakeholder might behave in a certain way. Why, is a stakeholder behaving as he actually is? Why would he exhibit cooperative behavior? Competitive behavior? Phase 3 enables us to explain each stakeholder's repertoire of possible behaviors. This assessment enables us (a) to understand each stakeholder's, actual behavior, its needs, and beliefs about other stakeholders; and (b) to understand what the world would have to be like in order to induce a stakeholder to become either cooperative or competitive. Both elements set the stage for strategy design.

The explanation of a stakeholder's behavior involves two critical factors; understanding (a) the stakeholder's objectives, and (b) the stakeholder's beliefs about its own environment. For example, in order to explain the Can Manufacturers' actual behavior—i.e., publicly supporting the USBA position while diversifying, making no new investment in can manufacturing facilities, and investing R&D into recycling capabilities—we might put forth as their objectives (1) to protect their core business, and (2) to minimize risk, This explanation alone, however, would not fully explain the Can Manufacturers' behavior, Why? Because if this group believed that there was little or no chance of container legislation being enacted, then for them to diversify from a highly profitable business, as can manufacturing is, would be irrational. Hence we must list the environmental factors for the Can Manufacturers which link, their objectives to their behaviors. We organize these factors around the Can Manufacturers' beliefs about other relevant stakeholders, which include:

- *USBA*: Coalition with USBA generates maximum protection of business.
- *Voters and State Legislatures*: Good chance of enactment of container legislation which gives advantages to refillable bottles.
- *Glass Bottle Manufacturers*: Cans are losing their competitive edge against glass.

Once the Can Manufacturers' objectives and beliefs about other stakeholders have been identified, we should have a complete, logical explanation for their behavior, that can be phrased in the farm:

> In order to achieve Objective O, Stakeholder S has chosen Behavior B, for Reasons R.

For example, the explanation of the Can Manufactures actual behavior can be phrased as:

> In order to protect their core business and to minimize future risks, the Can Manufacturers publicly supported the USBA position while diversifying, invested R&D into recycling capabilities, and made no new investment into can-manufacturing facilities—because they believed that a coalition with the USBA would maximize the protection of their business, that the enactment of unfavorable container legislation was probable, and that, because of such legislation, cans would lose their competitive advantage against glass bottles.

If the explanation is not logically valid, then more analysis is required in order to specify more carefully the relevant objectives and factors.

Similar explanations are then derived for cooperative and competitive behaviors. For example, in order to explain the Can Manufacturers' cooperative behavior—i.e., committing more heavily to can manufacturing, reducing profit margins to make cans more attractive, and putting R&D into new can technologies—we may put forth as their objectives: (1) to spend resources, to support their can business; and (2) to promote the business. In addition to these objectives, a number of reasons or environmental factors are necessary to explain their cooperative behavior:

- *Brewers*: Will stop vertical integration, leaving can-making business to can manufacturers.
- *USBA*: The USBA's economic argument will be important in deterring legislation.
- *Glass Bottle Manufacturers*: Cans have long-run technological advantages over glass.

By blending objectives and reasons, we obtain the following explanation:

> In order to spend resources to support their basic business and to try to achieve success in that business, the Can Manufacturers committed themselves more heavily to can manufacturing, reduced profit margins to make cans more attractive, and put R&D resources into new can technologies—because they believed that the brewers would stop vertically integrating, leaving the can making business to them; that cans have long-term technological advantages over glass, making it a profitable business down the road; and that the USBA's economic argument would be an important deterrent to possible container legislation.

A similar analysis produces the following explanation for the Can Manufacturers' competitive behavior:

> In order to move into long-run lower-risk businesses, the Can Manufacturers diversified more rapidly and used cans as a "cash cow" through aggressive price increases—because they believed that there was a good chance of passage of unfavorable legislation, that the USBA's economic argument either doesn't work or won't work, that nobody is willing to buy the can business from them, and that they can make aggressive price increases stick and end up with a good return on investment even with a decline in volume.

Tables 2.5, 2.6, 2.7, 2.8, 2.9, 2.10, 2.11, and 2.12 outline similar explanations for each of our remaining target stakeholders. It must be emphasized that the analysis of a particular stakeholder rests on the validity of the organization's beliefs about that stakeholder, rather than the "facts" about the stakeholder. As such, the certainty of these beliefs should be assessed. To do so, a number of questions need to be asked: Is the organization prepared to act, based on the strength of its beliefs? Where necessary, can research be initiated to clarify positions or to increase the level of certainty? What would be the time and effort required to improve the certainty? Is the wait justified? This part of the process requires considerable staff analysis in order to insure that executives' beliefs are based in fact.

That such questions are critical to the strategy development in subsequent phases of the analysis is clear. Suppose, for example, that while the Can Manufacturers are bent on minimizing risks, they would like to do so in the context of staying in the can business. Hence, their true objectives are somewhere between those necessary for the explanation of their actual behavior and their cooperative behavior. If the USBA, as focal organization, were uncertain of its assessment of the Can Manufacturers behavior, it would initiate research to try to uncover that stakeholder's

Table 2.5 Explanation of can manufacturers' behaviors

Objectives	Reasons
Actual behavior	
Protect core business	*USBA:* Coalition position gives maximum protection of business
Minimize future risks	*Voters and State Legislators:* Good chance of enactment of container legislation which gives advantages to refillable bottles *Glass Bottle Manufacturers:* Cans are losing their competitive advantage against glass
Cooperative behavior	
Spend resources on basic business	*USBA:* The USBA's economic argument will be important in determining legislation
Promote the business	*Glass Bottle Manufacturers:* Cans have a long run technological advantage over glass *Brewers:* Will stop vertical integration, leaving can-making business to can manufacturers
Competitive behavior	
Move into long-run lower-risk business	*Legislation and voters:* Good chance for legislation *USBA:* Economic argument doesn't or won't work *Customer:* Will accept price increase yielding good ROI even will decline in volume *Industry:* No one willing to buy the can business

Table 2.6 Explanation of voters' behaviors

Objectives	Reasons
Actual behavior	
Reduce Utter	*Industry:* Voters are confused as to what position is on litter
Obtain beverages at reasonable prices	*Government:* Voters are confused as to how deposits will help reduce litter
Cooperative behavior	
Reduce litter	*Industry:* Can effectively solve problem if left to themselves
Obtain beverages at reasonable prices	*Government*: There is too much government interference in the marketplace and proposed solutions will raise prices
Competitive behavior	
Reduce litter	*Industry:* Arguments are self-serving. Industry doesn't want to change because they will lose money; prices really won't change
Obtain beverages at reasonable prices	*EPA and Environmentalists:* Their proposed solution will help reduce litter

Table 2.7 Explanation of environmentalists' behaviors

Objectives	Reasons
Actual behavior	
Get a national law enacted (ultimately to ban nonremmables)	*Voters and Legislators:* By acting state-by-state, pressure will build for national law; Environmentalists believe that they can assemble a coalition of voters and legislators of Sufficient influence to get measures passed
Gain similar environmental reforms in other areas	*Media:* Will focus attention on container issue which will be helpful to other environmental issues *Public:* Benefits to public outweight costs
Cooperative behavior	
Gain environmental reforms across industries	*Media:* Publicity on this issue will undermine other more important environment issues *Voters and Legislatures:* Not enough support; defeat would weaken more important causes

true objectives. The USBA would then be ready to design a strategy that would induce the Can Manufacturers to be more cooperative.

2.1.4 Phase 4: Design Stakeholder Strategies

The first three stages of the stakeholder management process constitute the organization's initial attempt at understanding the external environment: The final two phases are an endeavor to formulate action-oriented responses to the threats and opportunities that have been found to exist. At this point in the analysis, executives in the organization have identified their own objectives and assessed how each stakeholder can influence them; they have also developed an analysis of the reasons for the behaviors of each target stakeholder. In addition, the executives have developed some level of confidence that their analysis is correct, or have initiated

Table 2.8 Explanation of EPA's behaviors

Objectives	Reasons
Actual behavior	
Get a national container law enactment (preferably a mandatory deposit law)	*Public:* Would reap significant environmental benefits in the form of reduced litter; will be exposed positively to the EPA through bottle or container legislation
Use the container issue as a step toward achieving major environmental action	*Consumer:* Long-term benefits through lower beverage prices *Government:* Must push returnable system or it won't come about *Brewing Industry:* Could without harm, absorb economic effects of legislation *Other Industry:* Will cooperate on reform if EPA wins container battle
Cooperative behavior	
Concentrate government and public attention on other areas of the environment when EPA positions are stronger and changes are easier to enact	*Government and Public:* Becoming disenchanted with present EPA focus; other issues are easier *Industry:* Has assembled too powerful a coalition against legislation for EPA to fight *Third-Party Researchers:* Arguments for legislation, which EPA is promoting cannot be substantiated

Table 2.9 Explanation of Brewers' behaviors

Objectives	Reasons
Actual behavior	
Continue to fight to defeat legislation as long as it is possible and profitable to so do	*Legislators and Voters:* Proposals can be defeated in most localities, states, and the Congress; delaying legislation is profitable since the payback period on can lines is 9 months; if laws are passed on state-by-state basis, they will not cause much chaos—thus it would be better to push for national legislation *Can Manufacturers:* Will escalate prices in the future; the Brewers can manufacture cans profitably *Consumers:* Will not be alienated by the Brewers' position *Industry:* Current system of containers is more profitable than all-returnable system
Extended business in a vertical manner to reap maximum profits themselves from price escalation of inputs	
Competitive behavior.	
Obtain a law that provides maximum security and profits associated with each brewer's system	*USBA:* Can no longer act as coalition spokesman *Voters and Legislators.* Some legislation is inevitable in most states and cannot be delayed much longer *State and Local Governments:* Different state and local laws would cause difficulties for national brewers *Consumer:* Price-sensitive
Reduce container costs by using cheaper alternatives	

Table 2.10 Explanation of labor's behaviors

Objectives	Reasons
Actual behavior	
Protect jobs and salaries of present union members	*Voters and Legislators:* Enactment of legislation would cause more job losses than gains for union members
Promote positions consistent with those of its membership	*USBA:* USBA position is the best way to protect union members
	Union Membership: More concerned about loss of jobs than other issues that would be improved by legislation
Cooperative behavior	
Prevent state legislation in states, even where it has little direct impact on jobs	*Local Unions:* Can be influenced by national or state unions; this issue is important enough to risk influencial position on other issues
	Voters and Legislators: Legislation in some states will have other spinoffs
Competitive behavior	
Get legislation passed Protect jobs and salaries of members	*USBA:* Do not accept analyses that USBA has made to show that there would be a loss of jobs from legislation

Table 2.11 Explanation of glass bottle manufactures' behaviors

Objectives	Reasons
Actual behavior	
Protect and expand core business lines	*Voters and Legislators:* Doesn't matter which way legislation goes—protected either way *Can Manufacturers:* Losing competitive advantage vis-a-vis glass *USBA:* USBA and Glass Bottle Manufacturers are philosophically in agreement *Government and Voters:* Glass Bottle Manufacturers are concerned about intervention in other areas
Cooperative behavior	
Maintain as many product alternatives as possible	*Customers:* Total sales revenues will fall under all returnables, tax, or deposit system (i.e., return rate will be high)
Protect historically potent product markets	*Brewers:* Will manufacture to protect themselves if only one container alternative exists
Competitive behavior	
Expand core business lines	*Customers:* Total sales revenues and/or profits will increase under the new system
Become sole source of supply	*Brewers:* Will not manufacture glass bottles

research to clarify or solidify uncertainties The strategy design phase builds on the knowledge and understanding base of these earlier phases and centers on trying to increase the chances of cooperative behavior for each target stakeholder or, alternatively, to decrease the chances of competitive behavior. For each target stakeholder we can proceed along two roads.

Table 2.12 Explanation of retailers' behaviors

Objectives	Reasons
Actual behavior	
Maintain as low a public profile as possible on this issue	*Consumer:* May be alienated if retailers take one side in the container issue *Competitors:* Each retailer only carries beverages so as not to be competitively disadvantaged *Legislators and Voters:* Legislation may not hurt retailers and may provide a positive way to do away with beverages
Cooperative behavior	
Maintain the status quo	*Consumers:* Decrease in sales by legislation which will hurt revenues and/or profits; consumers want the convenience of buying beverages at supermarkets *Beverage Industry:* Support of retailers on this issue will make a difference in final outcome
Competitive behavior	
Achieve best possible compromise position	*Public and Industry:* Antagonism minimized by support of tax legislation *Government:* Regulation which leads toward returnables would be economically disadvantageous

The first technique of strategy design is to identify direct strategy options—i.e., tactics designed to induce more cooperation and/or reduce the competitive behavior of the stakeholder whose behavior the organization is trying to change. One example of a direct strategy is to simply provide beneficial information to the stakeholder. Sometimes "win-win" strategies, from which both organization and stakeholder emerge as winners, are the results of one side becoming aware of other options or of some unforeseen consequences of its present behavior. Additionally, the organization can increase the attractiveness of more cooperative actions by introducing new reasons for choosing the alternative. Whichever type of direct strategy is considered appropriate, the analysis must be built on a foundation of an extensive understanding of each stakeholder; otherwise, the consequences can be disastrous to the organization.

In the case of the USBA, the advertising campaign sponsored by the industry in the four states that faced referenda is an example of a direct strategy designed to induce more cooperative voter behavior. The campaign argued from an economic platform—stressing jobs, energy, and costs—and pointed to the consequences of moving to an all-returnable container system. In stakeholder terms the USBA endeavored to directly provide Voters additional justification for acting more cooperatively, i.e., to vote on container legislation. Unfortunately, the industry was acting on its misperception of the Voters' true concerns and objectives, as revealed by a statement by Henry King, president of the USBA:

> The two devastating defeats in referenda led to a review of positions, past events, and future problems. The net result of the studies [done by the USBA] is that we've reconfirmed that litter is the problem—that the public really cares little about other issues. While some brewers question the validity of the study most don't. One study found that our advertising

> may have been counterproductive, particularly with themes of jobs and cost. The theme that deposit legislation won't do the job is the theme that works best. Our advertising was perceived to be self-serving.

The message in King's argument is clear: Had the USBA initially been aware that the voters' real concern was ecologically based, then the advertising campaign would have stressed the existence of other options to deposit legislation.

The second technique in strategy design is to use the explanatory analysis from Phase 3 to generate indirect-strategy options. Indirect strategies use third parties as pressure points on the target stakeholders. Consider, for example, the recent action by the National Organization of Women (NOW), a major proponent of the Equal Rights Amendment (ERA). Rather than dealing directly with the legislatures in states which had failed to pass the ERA, NOW developed an indirect strategy. Its members organized a boycott of potential convention cities in those states, hoping that affected businessmen would pressure state legislators to pass the amendment. Such indirect strategy options are often subtle and frequently require working through a stakeholder who was initially thought.to be unimportant. The selection of additional target stakeholders, through whom the organization feels it can exert influence over other stakeholders, may be necessary.

USBA Stakeholder Strategies: Can Manufactuers In order to increase the cooperative behavior of the Can Manufacturers, e.g., to get them to commit more heavily to can manufacturing, the USBA needs to convince them that the can business has long-term growth potential. One possible strategy to reach this end would be for the USBA to try to get the Brewers to stop their vertical integration by arguing that the Can Manufacturers are better equipped to make cans. Tricky antitrust issues complicate this strategy. But if it worked, the Can Manufacturers would have an incentive to stop diversifying. Another strategy might be to persuade the Can Manufacturers to invest new R&D in cans, by arguing that the Brewers are strongly committed to metal cans as a source of packaging. In addition, the USBA should have reinforced its economic argument to the Can Manufacturers.

Voters The prime concern of the Voters is litter. Hence an advertising campaign which gives recognition to the problem and promotes the USBA's (or its represented industries') endeavors to rectify the litter mess would be effective. Such a campaign would capitalize on the fact that deposits alone do not solve the problem. But a totally negative campaign would appear to the electorate to be self-serving to the USBA.

Unfortunately, such a strategy is precisely what the USBA now argues it should have pursued, as indicated by Henry King's statement. This tactic, however, presents a risk of alienating Labor without USBA assurances that it was also truly concerned about the issue of jobs. Another possibility is to work through the state legislatures to pressure them to decide the issue, rather than putting it on the ballot in a general election. Such a strategy would require understanding the attitudes and behaviors of the state legislatures as well.

Environmentalists The probability is small of making the Environmentalists, a somewhat hostile group, more cooperative. Hence the USBA finds itself in a virtual no-risk situation, since the Environmentalists are already as competitive as possible. Therefore, a radical strategy may be in order. Since the Environmentalists are ultimately concerned with ecological quality, the USBA could play a leading role in pollution control across a number of industries. Such a strategy could weaken the hostility of the Environmentalists' platform and reinforce the USBA's negotiative position for the future.

Environmental Protection Agency It is difficult to envision a worse relationship than the USBA's association with the EPA. Perhaps an anti-EPA campaign by the USBA could induce the agency to back off; but this too would probably be seen as self-serving to the industry. Again, a strategy worthy of consideration is to take on leading-edge work in pollution control. Such a step would mean, essentially, doing part of the EPA's job for them. The USBA could argue that the private sector is better equipped to control pollution than a government agency. Such an argument would have to be backed up by a concrete program comprehensive enough to convince the government of the USBA's sense of urgency.

Brewers The strategy for ensuring continued cooperation from the Brewers must be designed around the Brewers* commitment to strengthening the coalition built by the USBA. The driving force behind this strategy may well involve the now-familiar economic argument on the potentially disastrous effects of any container legislation. Tied to this argument must be a statement that the Brewers can only work effectively through a coalition, and that short-term losses in localities and states can be recovered by delaying legislation as long as possible.

Labor The USBA should be more concerned with the possible deterioration of Labor's support than with increasing pressure on national unions to twist the arms of their locals to more fully support the USBA in Michigan and other states. Again, the economic argument—that container legislation would cost jobs—will be important in inducing Labor to remain relatively cooperative. If Labor doesn't accept this argument, then the USBA could face difficulties with this stakeholder.

Glass Bottle Manufacturers The Glass Bottle Manufacturers are in philosophical agreement with the USBA and are actually publicly supportive of the industry's position; nevertheless, they sit in an extremely favorable position regardless of how the container issue is decided. One strategy to increase this stakeholder's cooperation would be for the USBA to convince the Brewers (a) that some contingency planning is necessary to prepare for the possible ban on metal containers, and (b) that the Brewers could be restricted to the Glass Bottle Manufacturers as a sole supplier. The brewing industry would then commission a study, or perhaps several large brewers would do so independently, to examine the feasibility of vertical integration into glass bottle manufacturing.

Retailers Since the Retailers have taken no public position for fear of alienating their customers, they offer some hope of cooperative behavior, thus cementing a coalition against legislation along the entire product cycle from suppliers to retailers. The Retailers also have the potential to mount a massive public educational campaign by distributing antilegislation leaflets along the soft drink shelves of their stores. A consumer survey which indicates that consumers want the convenience of being able to buy beverages in supermarkets could be beneficial. If the results of such a survey were positive, it would show that the Retailers could take a public position without fear of alienating a large segment of their customers. Hence their cooperative behavior of threatening to eliminate beverage sales if container legislation were passed would have enormous leverage and could make the difference at referendum time. If, on the other hand, the results of the survey were negative, the Retailers could still be cooperative if they were in philosophical agreement with the USBA. And if they carried out their threat of eliminating beverage sales, they would rid themselves of a low-profit business which their customers really didn't care about. Conversely, negative results could provide the Retailers the incentive to act competitively by supporting tax legislation.

Each of these stakeholder strategies rests on the intent of the USBA to induce a particular stakeholder to be more cooperative or less competitive. There is no guarantee that the strategies blend into one consistent package. The accomplishment of this integration is the task for the final phase of analysis.

2.1.5 Phase 5: Define Integrated Corporate Stakeholder Strategy

Stakeholder management is based on corporate implementation of one or multiple integrated strategies which deal with the entire stakeholder environment. Sometimes, however, the organization will have the luxury of few options besides dealing with specific stakeholders independently. In some cases, such a predicament can lead to an even more turbulent environment.

For example, the USBA could consider dealing independently with the EPA. Brewers, and Can Manufacturers. But by promising to promote pollution control in the brewing industry, the USBA might realize a degradation of the relatively cooperative behavior presently exhibited by the Can Manufacturers. Yet to the extent that resources can be used to simultaneously address multiple stakeholder issues, the benefits of these efforts become increasingly attractive.

Integrated stakeholder strategies can be formulated through two methods. One is to pool all of the strategies that were designed for individual stakeholders and try to synthesize the common threads into a metastrategy for global management of the environment. In our example, such an analysis might require a number of iterations to assure consistency across stakeholders. But we can generate a tentative strategy.

Critical to the strategies of dealing with the Can Manufacturers. Brewers, Glass Bottle Manufacturers. Retailers, and Labor is the *validity* of the economic argument. Essential to government officials who are considering regulatory or legislative options is the *believability* of the industry's economic arguments. Picture the integrated strategy that might emerge from these crucial variables. The USBA could commission an unbiased study conducted by an independent third party to assess the true aggregate systems cost of a mandated change to an all-returnable system. This would be the *forced cost* of dealing with the problem. The study could then assess the costs of dealing with the problems through the optimal free market mechanisms, e.g., through municipal solid waste recovery systems, cost-efficient litter cleanup, etc. The second cost, the so-called *free market cost*, ought to be lower than the forced cost; otherwise the industry should be willing to convert to the mandated system voluntarily.

Assuming the free market cost is considerably lower than the forced cost, the USBA could initiate a voluntary "set-aside" program equal to the free market cost and use it to support the start-up of businesses to efficiently deal with the problems. Members of the industry would contribute to the fund in proportion to the damage they would suffer through the less efficient forced-cost solution.

This systemic solution has many advantages. First, the industry is seen as making a direct, positive contribution to what is publicly considered a critical problem. Second, it doesn't face confrontation with its adversaries over distorted economics because it is acting an research-based information. Third, everyone in the industry should gain by this action relative to the alternative. The critical hurdle the USBA must now overcome to implement this strategy is to convince all affected parties that if no initiatives are taken by the industry, the forced costs will in fact be imposed by government legislation or regulation. Hence timing proposal appropriately and tracking industry sentiment become critical to success.

The second technique to creating integrated strategies is to work directly from the explanatory analysis of individual stakeholder behaviors. This approach attempts to identify common threads in these behaviors that can be exploited through redesigned corporate stakeholder strategies. For example, critical to the success of the USBA's position is that resources be invested in new can technologies. Otherwise an erosion of the economic position of these manufactures will occur. Moreover, legislation restricting their use will be more likely.

Understanding each stakeholder's rationale for or against supporting new can technologies can also facilitate new strategy designs by the USBA. For example, the EPA and Environmentalists would want to encourage such technology if they thought it would be directed at improving litter, pollution, solid waste, or resource usage conditions. Can Manufacturers would be willing to devote resources to such developmental efforts if they were sure that a profitable market for new products would emerge. Brewers would support such efforts if they facilitated overall market growth and didn't undermine their own competitive position.

As a third-party networker, the USBA may be able to use this input to create an integrated proposal that gives each party assurances of achieving their objective without the incidence of conflict. One possibility is special economic credits from the government for ecologically superior containers, with guaranteed market access for any product that meets such standards.

3 Stakeholder Management Philosophy

Clearly, some relatively attractive strategy options were available to the USBA. The stakeholder management process systematical drives the thinking necessary to define and implement those options and subsequently, to effectively manage the external environment. However, a number of general and philosophical issues surface from a study of the USBA case.

The stakeholder approach to managing external forces is based on the belief that those forces can be *positively* managed. Corporations need not assume a defensive, reactive posture. Normally, however, successful implementation of the methodology requires a reevaluation of the organization's management system. Organizations committed to strengthening their interactions with the external environment should begin to modify managerial behavior in four areas so as to adopt appropriate managerial postures for effective stakeholder management. Organizations should:

1. Develop a generalized marketing approach
2. Define a stakeholder negotiation process
3. Establish a proactive, risk-taking decision philosophy
4. Redesign the resource allocation process

3.1 Develop a Generalized Marketing Approach

Effective stakeholder management requires an understanding of each stakeholder's behaviors. What circumstances motivate a stakeholder to act in a certain way? What secondary environmental factors affect such behaviors? Naturally, these questions must be answered with the values of the stakeholder in mind. The organization's marketing philosophy can and should be generalized to achieve this. Marketing principles should be applied to each stakeholder, as opposed to restricting them to customers.

Intraorganizational conflicts, such as territorial perspectives, often prevent the application of sound marketing principles to the external environment. Market research is usually channeled only toward the needs of the customer, and "marketing" to other stakeholders is confined to advertising special programs or promoting corporate image. Significant decisions are made on other groups with little understanding of what they want and need. Product marketing strategies, for example, would never be developed in the absence of significant research to establish and explain consumer reactions, yet when the government is a stakeholder, research is seldom even considered. This imbalance must be corrected.

The USBA case is a case in point. The association argued from an economic base because it knew those issues best. It showed that container legislation would raise prices, cost skilled jobs, and disrupt distribution systems. Voters, however, were most concerned about litter, a problem not touched by the industry's campaigns. Clearly, the USBA failed to recognize the perspectives of this pivotal stakeholder.

3.2 Define a Stakeholder Negotiation Process

Stakeholder strategies must be designed to provide balanced cooperation among stakeholders. As such stakeholder management becomes a problem of conflict management, to which the principles of conflict resolution can be applied.

Most organizations see issues in the external environment as threats to the status quo; if the consumer wins, then the company must lose. However, this reasoning is normally invalid; there are often win-win situations as well as win-lose and lose-lose situations. General Motors' management saw Ralph Nader's "Campaign GM" as a win-lose situation. Their actions—or better, their reactions—allowed Campaign GM to attract even more publicity, and did more to undermine GM's image than negotiation could have. In fact, over the years General Motors implemented some portion of every one of Campaign GM's original proxy statements.

Negotiation includes at least five crucial activities:

1. Collect information
2. Disseminate information
3. Bargain on issues
4. Form coalitions
5. Initiate unilateral actions

Information collection is crucial because it improves the understanding of each stakeholder's position. Without the benefit of accurate information, misperceptions can drive ineffective strategies, as happened to the USBA. Information dissemination is equally important, just as it is essential for the corporation to understand its stakeholders' positions, it is equally crucial for stakeholders to understand the corporation's position. This does not mean that business must disclose trade secrets or sensitive strategic decisions, but that it must understand the principles of negotiation.

The importance of information collection and dissemination was validated in a study done by The Wharton Applied Research Center some years ago for the Arms Control and Disarmament Agency.[13] Research was initiated to define the causes of conflict situations in which people and organizations act on a variation of the Golden Rule, namely "Do unto others as you *believe* they would have done unto you," Valid information about both parties, the Wharton study confirmed, is critical to the creation and implementation of mutually acceptable solutions. If there are misperceptions or ambiguities, each side will tend to slightly bias its beliefs by assuming less cooperative motivations on the part of the other parties. This situation mutually reinforces the positions that the conflict is real. The net result is often a spiral into a serious conflict dilemma, all because each party's original positions were not well understood by its counterparts.

The third activity is negotiation itself. Each organization must make proposals and counterproposals to identify feasible solutions. By negotiating, each party can

---

[13] J. R. Emshoff, *The Analysis of Behavioral Systems* (New York; Macmillan, 1974).

try to create win-win situations instead of win-lose or, worse yet, lose-lose situations, which can occur if the conflict escalates beyond control.

The fourth activity is the understanding and forming of coalitions. An organization or stakeholder can often join forces with another Stakeholder to strengthen each party's position in relation to some third party. Since coalitions are prohibited by law regarding some business practices—such as setting prices—business tends to be paranoid even on issues over which there are no legal prohibitions. The focal company need not act in isolation or solely in conjunction with other businesses.

Finally, we need to consider the effect of unilateral action by which the focal organization adopts an independent strategy with some expectations as to each stakeholder's response. The Wharton conflict study showed that such action has a high potential for escalating the conflict, and thus should be used cautiously. Yet most organizations manage their stakeholder relations almost exclusively through such actions. They collect and disseminate little *useful* information, negotiate only when forced, and treat each stakeholder and issue separately. As a result, strategic responses are often haphazard. Procedures must be established so that the full spectrum of negotiation options is considered and appropriately used.

The USBA was accustomed to "negotiating" in the legislative arena. Their apparent failure to negotiate with other stakeholders can be seen as part of the reason that the results of the 1976 referenda were so poor. More precisely, their failure to negotiate with these stakeholders may be seen as a key to the very existence of those referenda.

3.3 Establish a Proactive, Risk-Taking Decision Philosophy

Many successful organizations take the initiative with their customers; proactive and risk-taking behavior with this stakeholder is a natural part of business. But for many of the other stakeholders, the organization is defensive; it only reacts when coerced through explicit initiatives from the stakeholder. When action is required in these situations, the objective is usually to return the environment to the state before the "disturbance" arose by neutralizing the antagonistic stakeholder's initiatives.

The reasons for adopting a reactive posture for stakeholders other than customers are easy to understand. The risks of an inappropriate proactive posture with one of these stakeholders are highly visible and can be costly, but because successful proactive behavior *prevents* a problem from occurring, it never becomes visible. Inappropriate reactive postures cannot be so clearly established because action is only taken when it is clearly required. Thus, faced with a choice of either, initiating action in an environment that is superficially nonthreatening in order to maintain the status quo, or waiting until the environment demands a response, most managers adopt the safer, risk-avoidance course.

A posture of risk avoidance results in a short-term, perspective which ignores the early warning signals of future conflicts. This increases the likelihood that such conflicts will materialize. Moreover, the organization greatly reduces the chance that it can control the eventual outcome of the conflict and thereby achieve jointly satisfactory results with its stakeholders. Thus a short-term, risk-avoidance, reactive posture—characteristic of most organizations today—should be replaced by a long-term, risk-taking, proactive posture.

To successfully implement this philosophy, managerial performance must be measured on all the real costs, benefits, and risks—not just the visible ones—of actions and strategies. If actions are taken based only on visible costs, then a reactive and short-term focus will develop. Such a myopic perspective drives unilaterial actions in a state of crisis. Without previous communications, the crisis escalates. To prevent such crises and to successfully achieve this "success factor." senior management must acknowledge all the expected consequences of action *and* inaction in dealing with stakeholders. They must learn to evaluate all the real costs and benefits, not just the visible ones.

3.4 Redesign the Resource Allocation Process

Corporate management processes typically reflect the traditional perspective of the organization's function, i.e., the optimization of the supply-conversion-selling activities. Most companies are organized on the basis of a product/business line, a function, or a matrix. Product management is separated from external relationships, and each area is in turn divided into discrete factors. But because of these factors, the information needed to build integrated stakeholder management strategies does not surface at a central place in the organization; therefore, appropriate resources are never committed among stakeholders.

Typically, the functional oriented organization places too much resource emphasis on highly visible stakeholders such as customers, and too little on stakeholders that are nobody's direct responsibility, such as special interest groups. Without a systemic perspective on the stakeholder environment, the resources needed to maintain balance within the stakeholder system will probably not be properly allocated. As a result, crises in various parts of the system will continue to occur. The focus on firefighting in specific areas must be redirected if the stakeholder process is to be realized in practice. To achieve this, the resource allocation process and sometimes even the organization itself must be redesigned.

For example, when the stakeholder process was applied in a large international packaged goods company, we asked the senior managers in the company to assess the influence of its stakeholders on the company's future. Government ranked first and the consumer ranked last, yet expenditures were such that 95% of the available resources were spent trying to influence the consumer. Working closely with these managers, we have begun to reverse this trend. The company is experiencing a much less turbulent environment and is building bridges with those stakeholders who will influence the future directions of the company.

4 Conclusions

The United States Brewers Association faced four potentially devasting referenda in 1976. The association's strategies for addressing the issues represented by those referenda provide a vehicle by which a process for managing the external environment can be examined. The process does not guarantee a favorable solution to externally driven problems. It does, however, generate at least a comprehensive understanding of the environment and a framework for strategy analysis and enrichment that endeavors to provide executives some measure of control over the turbulence they are now facing. The problems which stakeholder management addresses are too serious to ignore; they must be tackled if the corporation is to maintain control over its own future.

Acknowledgments James R. Emshoff is Vice President of Marketing and Development—Swanson Division, Campbell Soup Company, Camden, N.J. R. Edward Freeman is Senior Project Manager, The Wharton Applied Research Center, and Lecturer, Department of Management; The Wharton School, Philadelphia. Pa. We wish to acknowledge the support, comments, and ideas of the members of the Wharton ARC "Stakeholder Project," especially Arthur Finnel, Gordon Sollars, and David Reed.

James R. Emshoff is a retired executive in health care, finance and other sectors, and a co-founder of the Wharton Applied Research Center, now Center for Applied Research.

R. Edward Freeman is University Professor, Elis and Signe Olsson Professor, Academic Director of the Institute for Business in Society, and Senior Fellow of the Olsson Center for Applied Ethics at the University of Virginia's Darden School of Business.

Chapter 3
Stakeholder Management: Framework and Philosophy

R. Edward Freeman

1 Introduction

Organizations have stakeholders. That is, there are groups and individuals who can affect, or are affected by, the achievement of an organization's mission. I have shown that if business organizations are to be successful in the current and future environment then executives must take multiple stakeholder groups into account. The purpose of this chapter is to discuss how the stakeholder management framework can be used to better understand and manage both internal and external change, and how the management philosophy which accompanies this framework fits into our more customary way of thinking about organizations.[1]

[1]The ideas presented in this chapter form part of a paper, "Managing Stakeholders: One Key to Successful Adaptation" presented to the Academy of Management National Meeting in August 1982. I wish to thank the participants in the Symposium on managing adaptation, and its chairperson, Professor Bala Chakravarthy, for many helpful comments. In addition, several Faculty members at the University of Pittsburgh's Graduate School of Management and Rutgers University's Department of Management have made helpful comments. In particular Barry Mitnick and Aubrey Mendelow have been encouraging over the past year.

Originally published in: *Strategic Management: A Stakeholder Approach*, 52–82, © Cambridge University Press, 2010
Reprint by Springer, Reproduced with permission of The Licensor through PLSclear, https://doi.org/10.1017/CBO9781139192675.006

R. E. Freeman (✉)
University of Virginia, Charlottesville, VA, USA
e-mail: FreemanE@darden.virginia.edu

© Springer Nature Switzerland AG 2023

S. D. Dmytriyev, R. E. Freeman (eds.), *R. Edward Freeman's Selected Works on Stakeholder Theory and Business Ethics*, Issues in Business Ethics 53,
https://doi.org/10.1007/978-3-031-04564-6_3

2 The Stakeholder Framework

The literature discussed in Chap. 2 yields a broad range of definitions of the stakeholder concept. From the standpoint of strategic management, or the achievement of organizational purpose, we need an inclusive definition. We must not leave out any group or individual who can affect or is affected by organizational purpose, because that group may prevent our accomplishments. Theoretically, therefore, "stakeholder" must be able to capture a broad range of groups and individuals, even though when we put the concept to practical tests we must be willing to ignore certain groups who will have little or no impact on the corporation at this point in time. Such a broad notion of "stakeholders" will include a number of groups who may not be "legitimate" in the sense that they will have vastly different values and agendas for action from our own. Some groups may have as an objective simply to interfere with the smooth operations of our business. For instance, some corporations must count "terrorist groups" as stakeholders. As unsavory as it is to admit that such "illegitimate" groups have a stake in our business, from the standpoint of strategic management, it must be done. Strategies must be put in place to deal with terrorists if they can substantially affect the operations of the business.

The stakeholder concept must capture specific groups and individuals as "stakeholders." As we move from a theory of strategic planning to a theory of strategic management, we must adopt an action orientation.[2] Therefore, if the stakeholder concept is to have practical significance, it must be capable of yielding concrete actions with specific groups and individuals. "Stakeholder Management" as a concept, refers to the necessity for an organization to manage the relationships with its specific stakeholder groups in an action-oriented way.

The very definition of "stakeholder" as "any group or individual who can affect or is affected by the achievement of an organization's purpose" gives rise to the need for processes and techniques to enhance the strategic management capability of the organization. There are at least three levels at which we must understand the processes which an organization uses to manage the relationships with its stakeholders.[3]

First of all, we must understand from a rational perspective, who are the stakeholders in the organization and what are the perceived stakes. Second, we must

[2] See Schendel and Hofer (1979) for a collection of essays that catalog the development of strategic management. Freeman (1983) is an overview of how the stakeholder concept fits into the development of strategic management theory, as well as a conceptual history of the term, "stakeholder."

[3] My use of "rational," "process" and "transactional" parallels Graham Allison's (1971) three levels of organizational analysis. However, the three levels are not mutually exclusive as is often interpreted from Allison's account. Each level of analysis offers a different "lens" for viewing the organization and offers different kinds of explanation for some underlying phenomena broadly called "organization behavior." While the explanations at each level need not be identical, they do need to be consistent. Hence, the concept of "fit" among the three levels. The application of this three-leveled conceptual scheme is not unique to the stakeholder concept, as it is conceivable that we could define the process and transactional levels to complement a "portfolio approach" to strategic management.

understand the organizational processes used to either implicitly or explicitly manage the organization's relationships with its stakeholders, and whether these processes "fit" with the rational "stakeholder map" of the organization. Finally, we must understand the set of transactions or bargains among the organization and its stakeholders and deduce whether these negotiations "fit" with the stakeholder map and the organizational processes for stakeholders.

We might define an organization's "Stakeholder Management Capability" in terms of its ability to put these three levels of analysis together.[4] For instance, an organization which understands its stakeholder map and the stakes of each group, which has organizational processes to take these groups and their stakes into account routinely as part of the Standard operating procedures of the organization and which implements a set of transactions or bargains to balance the interests of these stakeholders to achieve the organization's purpose, would be said to have high (or superior) stakeholder management capability. On the other hand, an organization which does not understand who its stakeholders are, has no processes for dealing with their concerns and has no set of transactions for negotiating with stakeholders would be said to have low (or inferior) stakeholder management capability. Each of these levels of analysis needs to be discussed in more detail, if the stakeholder management framework is to become a useful managerial tool.

3 The "Rational" Level: Stakeholder Maps

Any framework which seeks to enhance an organization's stakeholder management capability must begin with an application of the basic definition. Who are those groups and individuals who can affect and are affected by the achievement of an organization's purpose? How can we construct a "stakeholder map" of an organization? What are the problems in constructing such a map?

In Chap. 1 we saw that the traditional picture of the firm consisting of customers, suppliers, employees and owners had to change to encompass the emergence of environmentalists, consumer advocates, media, governments, global competitors, etc. I based this argument on an analysis of the changes in the business environment of the last twenty years. The resulting generic stakeholder map can serve as a starting point for the construction of a stakeholder map of a typical firm. Ideally the starting point for constructing a map for a particular business is an historical analysis of the environment of that particular firm.[5] In the absence of such an historical document, Exhibit 1.5 can serve as a Checkpoint for an initial generic stakeholder map.

---

[4] Chakravarthy (1981) defines a similar concept of the adaptive capabilities of an organization using "management capability" and "organization capability."

[5] For instance, as in a clinical case study, viz., Emshoff and Freeman's (1981) analysis of the brewing industry around the issue of beverage Container legislation or an in-depth historical study as per Miles (1982) of tobacco companies.

Exhibit 3.1 depicts a stakeholder map around one major strategic issue for one very large organization, the XYZ Company, based primarily in the U.S. The executives in this organization, however, believed that Exhibit 3.1 could be used as a starting point for almost any issue of importance to the company. Unfortunately, most attempts at "stakeholder analysis" end with the construction of Exhibit 3.1. As the literature of the last chapter suggests, the primary use of the stakeholder concept has been as a tool for gathering information about generic stakeholders. "Generic stakeholders" refers to "those categories of groups who can affect...." While "Government" is a category, it is EPA, OSHA, FTC, Congress, etc. who can take actions to affect the achievement of an organization's purpose. Therefore, for stakeholder analysis to be meaningful Exhibit 3.1 must be taken one step further. Specific stakeholder groups must be identified. Exhibit 3.2 is a chart of specific stakeholders to accompany Exhibit 3.1 for the XYZ Company. Even in Exhibit 3.2 some groups are aggregated, in order to disguise the identity of the Company. Thus, "Investment Banks" would be replaced by the names of those investment banks actually used by XYZ.

Most very large organizations have a stakeholder map and accompanying stakeholder chart which is relatively similar to the above exhibits. There will be

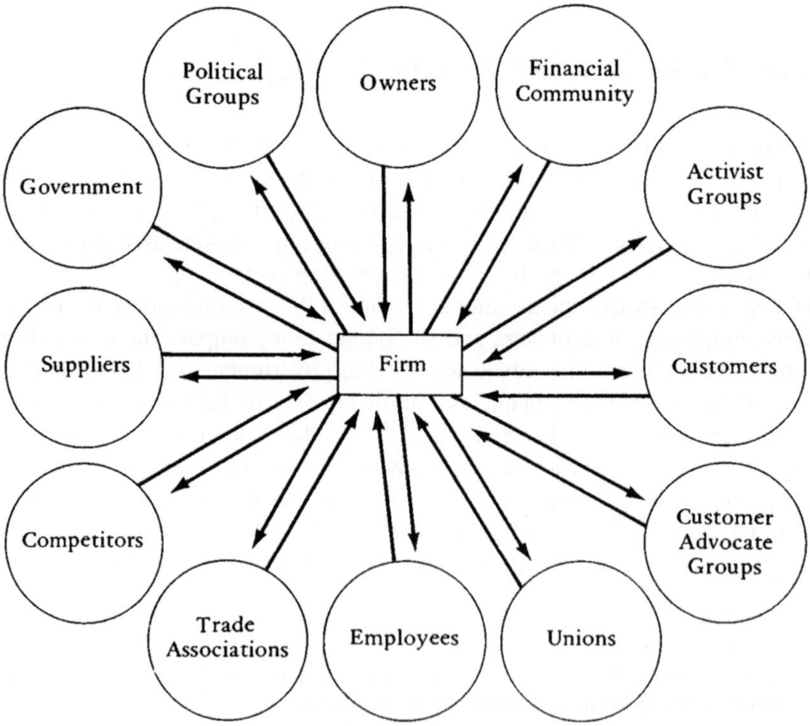

Exhibit 3.1 Stakeholder Map of a Very Large Organization

Owners	**Financial Community**	**Activist Groups**
Shareowners Bondholders Employees	Analysts Investment Banks Commercial Banks Federal Reserve	Safety and Health Groups Environmental Groups "Big Business" Groups Single Issue Groups
Suppliers	**Government**	**Political Groups**
Firm #1 Firm #2 Firm #3 etc.	Congress Courts Cabinet Departments Agency #1 Agency #2	Political Party #1 Political Party #2 National League of Cities National Council of Mayors etc.
Customers	**Customer Advocate Groups**	**Unions**
Customer Segment #1 Customer Segment #2 etc.	Consumer Federation of America Consumer's Union Council of Consumers etc.	Union of Workers #1 Union of Workers #2 etc. Political Action Committees of Unions
Employees	**Trade Associations**	**Competitors**
Employee Segment #1 Employee Segment #2 etc.	Business Roundtable NAM Customer Trade Org. #1 Customer Trade Org. #2 etc.	Domestic Competitor #1 Domestic Competitor #2 etc. Foreign Competitor #1 etc.

*The actual names of most stakeholder groups are disguised.

Exhibit 3.2 Specific Stakeholders in a Very Large Organization*

variations among industries, companies, and geographies at the specific stakeholder level, but the two exhibits can be used as a checklist of stakeholder groups. In the several industries analyzed in subsequent chapters there is little variation at the generic level.

Exhibit 3.3 is an analysis of the stakes of some of those specific stakeholder groups listed in the stakeholder chart (Exhibit 3.2). Thus the stake of Political Parties #1 and #2 is as a heavy user of XYZ's product, as being able to influence the regulatory process to mandate change in XYZ's operations and as being able to

Customer Segment #1	Political Parties #1 and #2
High Users of Product Improvement of Product	High Users of Product Able to Influence Regulatory Process Able to get Media Attention on a National Scale
Customer Segment #2	**Consumer Advocate #1**
Low Users of Product No Available Substitute	Effects of XYZ on the Elderly
Employees	**Consumer Advocate #2**
Jobs and Job Security Pension Benefits	Safety of XYZ's Products
Owners	
Growth and Income Stability of Stock Price and Dividend	

Exhibit 3.3 "Stakes" of selected Stakeholders in XYZ Company

elevate XYZ to national attention via the political process. The stake of XYZ's owners varied among specific stakeholder groups. Those employees of XYZ, and the pension funds of XYZ's unions are concerned with long term growth of XYZ's stock, as their retirement income will depend on the ability of XYZ to earn returns during their retirement years. Other shareowner groups want current income, as XYZ has been known for steady though modest growth over time. Customer Segment #1 used a lot of XYZ's product and was interested in how the product could be improved over time for a small incremental cost. Customer Segment #2 used only a small amount of XYZ's product, but that small amount was a critical ingredient for Customer Segment #2, and there were no readily available substitutes. Thus, the stakes of the different customer segment stakeholders differed. One consumer advocate group was concerned about the effects of XYZ's product decisions on the elderly, who were for the most part highly dependent on XYZ's products. Another consumer advocate group was worried about other XYZ products in terms of safety.

As these three exhibits from the XYZ Company show, the construction of a rational "stakeholder map" is not an easy task in terms of identifying specific groups and the stakes of each. The exhibits are enormously oversimplified, for they depict the stakeholders of XYZ as static, whereas in reality, they change over time, and their stakes change depending on the strategic issue under consideration. Similarly, the construction of an accurate portfolio is no easy task as the problems with measuring market share have shown.[6] The task becomes even harder when we consider several implications of these three exhibits.

The first implication is that just as Merton (1957) identified the role set for individuals in society, and Evan (1966) generalized this notion for organizations to the organization set, we might combine these notions into a "stakeholder role set," or the set of roles which an individual or group may play qua being a stakeholder in an organization. For example, an employee may be a customer for XYZ's products, may belong to a union of XYZ, may be an owner of XYZ, may be a member of Political Party #1 and may even be a member of a consumer advocate group. Many members of certain stakeholder groups are also members of other stakeholder groups, and *qua stakeholder in an organization* may have to balance (or not balance) conflicting and competing roles. Conflict within each person and among group members may result. The role set of a particular stakeholder may well generate different and conflicting expectations of corporate action. For certain organizations and stakeholder groups, a "stakeholder role set" analysis may be appropriate. Exhibit 3.4 is an example of the stakeholder role set of employees and a government official.

The second implication of Exhibits 3.1, 3.2, and 3.3 is the interconnection of stakeholder groups, or the interorganizational relationships which exist, a phenomenon well studied in organization theory.[7] XYZ Company found that one of their Unions was also a large contributor to an adversarial consumer advocate group who was pressuring a key government agency to more closely regulate XYZ. Networks of stakeholder groups easily emerge on a particular issue and endure over time. Coalitions of groups form to help or oppose a company on a particular issue. Also, some firms are quite adept at working indirectly, i.e. at influencing Stakeholder A to influence Stakeholder B, to influence Stakeholder C.[8]

More traditional examples include the emergence of the courts as a key stakeholder in takeover bids. Marathon Oil successfully used the courts and the agencies involved in anti-trust to fend off a takeover bid from Mobil, while finding U.S. Steel to come to the rescue. AT&T recently marshalled the support of employees and stockholders to try and influence the Congress through a letter writing campaign. While there is some research on power and influence networks, little is known in the

[6] The point here is that any theory must explicitly define the range of entities over which the propositions in the theory range. Sometimes it is convenient to speak of "stakeholders" as referring to categories, or sets, of specific groups. But, I insist that, strictly speaking, it is specific groups and individuals which are real, and hence, which can be strictly said to "hold stakes." For a philosophical treatment of the rather nominalistic position taken here see Nelson Goodman (1955).

[7] The literature on interorganizational relations is quite enormous and is rich in insights for strategic management. Evan (1976), Negandhi (1975), Nystrom and Starbuck (1981) are excellent collections of articles, each of which contains review articles which summarize the State of the art.

[8] See Miles (1982) analysis of the tobacco industry, and Wilson (1981) for analyses of coalitions among interest groups.

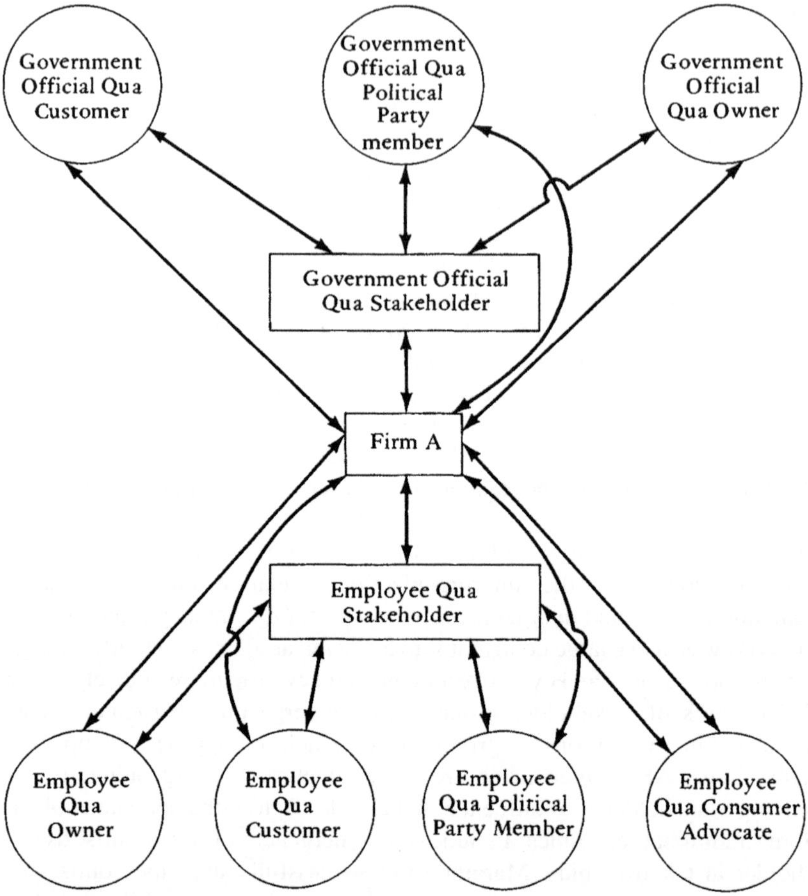

Exhibit 3.4 Possible Stakeholder role set of Employees and Government Officials

way of formulating strategies for utilizing such networks in a positive and proactive fashion. Little is known, prescriptively, about what range of alternatives is open to managers who want to utilize such an indirect approach to dealing with stakeholders. Exhibit 3.5 depicts several networks, and illustrates the necessity of thinking through the possible networks that can emerge or be created to accomplish organizational purposes. We will return to the question of how to analyze networks and coalitions in Chap. 5.

The courts and some government agencies play a Special role as part of the process by which groups interact. They have a Special kind of "stake," one of formal power. While they usually do not initiate action, they can serve as resolver of conflicts, or as guarantor of due process. If we generalize this notion we see that another implication of Exhibits 3.1, 3.2, and 3.3 is the phenomenon of the differing kinds of stakes and the differing perceptions of stakes that various groups have.

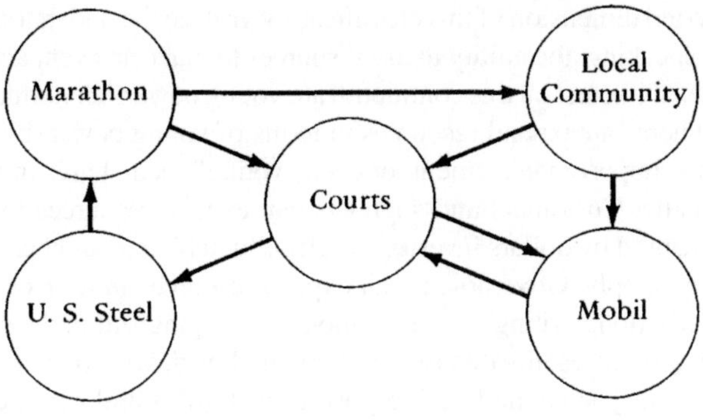

Example #2: AT&T and House Bill 5158

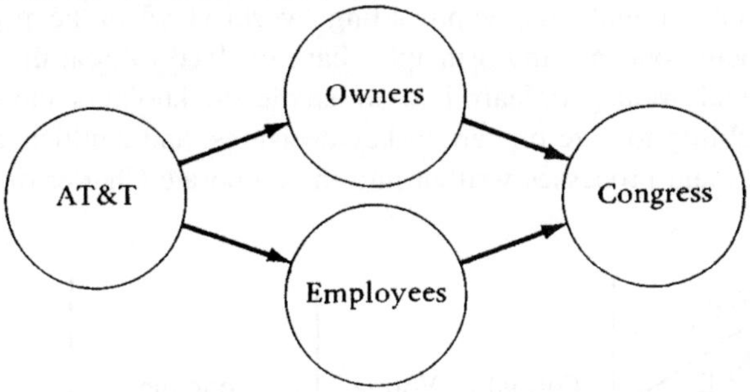

Exhibit 3.5 Typical Indirect or Coalition Strategies

"Stake" is obviously multi-dimensional, and not measured solely in dollar terms. However, exactly what the dimensions are of "stake" is a more difficult question. Exhibit 3.3 ranges across a broad spectrum of phenomena from more traditional dollar returns to stockholders to a call for "voice" in running the affairs of XYZ (Hirschman 1970). Clearly we need to understand "stake" in more detail.

One analytical device depicts an organization's stakeholders on a two dimensional grid.[9] The first dimension categorizes stakeholders by "interest" or "stake." The idea is to look at the range of perceived stakes of multiple stakeholders. While there are no hard and fast criteria to apply here, one typical categorization is to classify "stake" from "having an equity interest in the firm" to "being an influencer" or in Dill's (1975) terms, "being a kibbitzer, or someone who has an interest in what the firm does because it affects them in some way, even if not directly in marketplace terms." We might place a middle category between equity and kibbitzer and call it having a "market" stake. These three catagories of a continuum are meant to represent the more traditional theory of the firm's differing stakes of owners (equity stake), customers and suppliers (market stake) and government (kibbitzer).

[9] For a discussion of this grid in the context of corporate governance see Freeman and Reed (1983).

The second dimension of this classificatory grid can be understood in terms of power, or loosely speaking, the ability to use resources to make an event actually happen.[10] The three points of interest on this continuum are voting power, economic power and political power. Owners can expend resources in terms of voting power, by voting for directors or voting to support management, or even "voting" their shares in the marketplace in a takeover battle. Customers and suppliers can expend resources in terms of economic power, measured by dollars invested in R&D, switching to another firm, raising price or withholding supply. Government can expend resources in terms of political power by passing legislation, writing new regulations or bringing suit in the courts.

Exhibit 3.6 represents this two dimensional grid, with owners being the textbook case of an equity stake and voting power; customers and suppliers having a market stake and economic power; and government having an influencer stake and political power. The diagonal of Exhibit 3.6 represents the development of classical management thought, and the prevailing "world-view" of the modern business firm. Management concepts and principles have evolved to treat the stakeholders along this diagonal. Managers learn how to handle stockholders and boards of directors via their ability to vote on certain key decisions, and conflicts are resolved by the procedures and processes written into the corporate Charter or by methods which

Power / Stake	Formal or Voting	Economic	Political
Equity	Stockholders Directors Minority Interests		
Economic		Customers Competitors Suppliers Debt Holders Unions	Foreign Governments
Influencers			Consumer Advocates Government Nader's Raiders Sierra Club Trade Association

Exhibit 3.6 Classical Stakeholder Grid

[10] The approach to "power" outlined here is quite simplistic, and should be viewed as illustrative rather than definitive. Pfeffer (1981) is suggestive of a more comprehensive analysis of the concept which could be applied to the "power and stakes" grid.

involve formal legal parameters. Strategic planners, marketing managers, financial analysts and operations executives base their decisions on marketplace variables and a long tradition of wisdom and research based on an economic analysis of marketplace forces. Public relations and public affairs managers and lobbyists learn to deal in the political arena, to curry the favor of politicians and to learn to strategically use PACs, "perks" and the regulatory process.

As long as the "real world" approximately fits this diagonal case of Exhibit 3.6, there are few problems. Each set of managerial problems and issues has an established body of knowledge upon which to draw in times of change. Another way of further supporting the argument of Chap. 1 is to say that the world can no longer be seen in terms of the diagonal of Exhibit 3.6.

For instance, in the auto industry one part of government has acquired formal power, the Chrysler Loan Guarantee Board, while in the steel industry some agencies have acquired economic power in terms of the imposition of import quotas or the trigger-price mechanism. The SEC might be viewed as a kibbitzer with formal power in terms of disclosure and accounting rules. Outside directors, now, do not necessarily have an equity stake. This is especially true of women, minority group members and academics who are becoming more normal for the boards of large corporations, even though it is far from certain that such directors are really effective and not merely symbolic. Some traditional kibbitzer groups are buying stock and acquiring an equity stake. While they also acquire formal power, the yearly demonstration at the stockholders meeting or the proxy fight over social issues is built on their political power base. Witness the marshalling of the political process by church groups in bringing up issues such as selling infant formula in the third world or investing in South Africa at the annual stockholders meeting. Unions are using political power as well as their equity stake in terms of pension fund investing, to influence management decisions. Customers are being organized by consumer advocates to exercise the voice option and to politicize the marketplace.

In short the nice neat orderly world of Exhibit 3.6 is no longer realistic. The real world looks more like Exhibit 3.7 which catalogs some of the differing stakes mentioned above. Of course, each individual organization will have its own separate grid, and given the complexity of the stakeholder role set, there may be groups which fall into more than one box on the grid. The "messiness" of Exhibit 3.7 lends credence to the search for alternative applications of more traditional management knowledge and processes. Getting the last two degrees of knowledge out of the diagonal of Exhibit 3.6 is simply no longer good enough. We must find innovative ways of understanding both the power and stakes of a variety of influential and interconnecting stakeholder groups. Thus, MacMillan (1978) has argued that elements of strategic planning, traditionally reserved for market stakeholders with economic power, can be applied to the pure political case. While there is a long tradition of applying economic analysis to public policy questions, we are beginning to see the application of political concepts to economic questions, via recent discussions of co-determination and quality of work life.[11]

[11] For an interesting distinction between economic and political explanations see the work of Hirschman (1970, 1981).

Power / Stake	Formal or Voting	Economic	Political
Equity	Stockholders Directors Minority Interests		Dissident Stockholders
Economic		Suppliers Debt Holders Customers Unions	Local Governments Foreign Governments Consumer Groups Unions
Influencers	Government SEC Outside Directors	EPA OSHA	Nader's Raiders Government Trade Associations

Exhibit 3.7 "Real World" Stakeholder Grid

The second issue which a "power and stakes" analysis surfaces is the issue of congruent perceptions among organization and its stakeholders. There may be differing perceptions of both power and stake depending on one's point of view. An organization may not understand that a particular union has political power, and may treat the union as a "purely economic entity," only to be surprised when the union gets a bill introduced in the legislature to prevent a proposed plant closing. The ABC Company completely misread the power and stake of a group of realtors who were upset over a proposed change in ABC's product. The legislature in the State where ABC operates was composed of a number of realtors, who easily introduced a bill to prevent the proposed product changes. It was only by some tough eleventh hour negotiations that ABC escaped some completely devastating legislation. The DEF Utility could not understand why a consumer advocate group was opposing them on a certain issue which had no economic effect on the group. Finally they spoke to a consumer leader who told them that the only reason that the group was opposing them was that they had not informed the group of the proposed rate change before the case was filed. In short the consumer group perceived that they had a different stake than that perceived by the management of DEF. DEF managers naturally believed that so long as the proposed rate change was in the economic interest of the consumer group and its constituency there would be no problem. The consumer group perceived things differently, that they had a vital role to play as influencer or kibbitzer.

Analyzing stakeholders in terms of the organization's perceptions of their power and stake is not enough. When these perceptions are out of line with the perceptions of the stakeholders, all the brilliant strategic thinking in the world will not work. The congruence problem is a real one in most companies for there are few organizational processes to check the assumptions that managers make every day about their stakeholders. The rational analysis proposed here in terms of stakeholder maps must be tempered by a thorough understanding of the workings of the organization through an analysis of its strategic and operational processes.

4 The "Process" Level: Environmental Scanning and the Like

Large complex organizations have many processes for accomplishing tasks. From routine applications of procedures and policies to the use of more sophisticated analytical tools, managers invent processes to accomplish routine tasks and to routinize complex tasks. To understand organizations and how they manage stakeholder relationships it is necessary to look at the "Standard Operating Procedures," "the way we do things around here," or the organizational processes that are used to achieve some kind of "fit" with the external environment. While there are many such processes, I shall concentrate on three well known and often used ones which purport to assist managers in the strategic management of corporations: Portfolio Analysis Processes, strategic Review Processes and Environmental Scanning Processes. Variations of each of these strategic management processes are used in many large complex organizations. Each is usually inadequate in terms of taking complex stakeholder relationships into account and can be enriched by the stakeholder concept.

As mentioned in Chap. 2, a good deal of research during the past twenty years has gone into understanding how a Corporation can be seen as a set or portfolio of businesses.[12] Discrete business units are easier to manage and factors for success may well be easier to discern at the business level, than at the aggregated level of the corporation as a whole. The idea is to look at this set of businesses as Stocks in a portfolio, with selection and nourishment given to winners and the door given to losers. Corporate planners and division managers (or strategic Business Unit managers) plot the firm's set of businesses on a matrix which arrays an external against an internal dimension. The external dimension is usually labeled "Industry Attractiveness" and is usually measured by the growth rate of the industry under consideration. The internal dimension is usually labeled "Business Strengths" and is usually measured by market share. The corporate managers, after plotting the portfolio of businesses, seek to arrive at a balanced portfolio which maximizes returns (measured by Return on Equity or Earnings per Share or Return on Investment, etc.) and minimizes risks. Managers of particular businesses are then given a strategic mission based on their place in the portfolio and the potential of the business in question.

As an analytical tool and a management process, Portfolio Analysis can easily be out of touch with the stakeholder maps of most firms, as depicted in earlier exhibits. It simply looks at too narrow a range of stakeholders, and measures business Performance on too narrow a dimension. While industry growth rate may be influenced by a number of non-marketplace stakeholders, to rely on it solely is to forego opportunities to influence stakeholders which may determine the future growth rate of the industry. For example, in the auto industry foreign competitors

[12] For a more complete discussion of portfolio theory see Abell (1980), Rothschild (1976), Lorange (1980), and the literature referenced in these works.

and governments, U.S. government agencies, the Congress, the courts, Ralph Nader and the Center for Auto Safety, environmental groups, the United Auto Workers, etc. all have an influence on future growth rates in the industry. However, if market share is relied upon as the sole criterion to measure competitive strength, we will not necessarily invest resources to deal with all of the groups who can influence future market position. Market share is too broad a measure and an overreliance on it can be detrimental.

To illustrate, consider the fate of JKL Company after spending several million dollars in R&D to develop a new product which would serve as a substitute to a large established market. JKL believed that the product offered high growth potential, and in accordance with accepted theory, introduced the new product before getting approval from a key government agency which closely regulates the industry in which JKL would be competing. The product was later found to be carcinogenic and JKL took a large loss. Market share was not the sole indicator of success for JKL.

Or, consider Proctor and Gamble's experience with Rely tampons. P&G had entered a mature market with a new product and spent heavily to gain market share. When reports linking Rely with toxic shock syndrome surfaced, P&G voluntarily removed Rely from the market rather than jeopardize future products and its corporate reputation. Industry attractiveness was not the sole criterion for the success of Rely. The future attractiveness of the market together with the possibility of tarnishing P&G's excellent reputation, caused them to make a decision that was quite expensive. Even though it cannot be shown that use of Rely caused the disease, the mere possibility of a linkage was enough for P&G to recall the product.

Similarly, Johnson and Johnson acted quickly to recall the entire stock of Extra-Strength Tylenol after several deaths were reported as a result of criminal tampering with bottles of the product. Someone allegedly put cyanide capsules in bottles of the product after it was on retail store shelves. Johnson and Johnson's actions were lauded on "60 Minutes," a show sometimes critical of the actions of large corporations. They have reintroduced the product in "tamper proof" packages, and advertised heavily. Portfolio analysis simply cannot prepare the corporation to deal with issues such as those faced by these companies. Industry or market attractiveness analysis is not sophisticated enough to yield practical conclusions in areas where economics, social and political forces and new technologies combine.

The point of this critique of portfolio analysis is not that managers must be certain of success before taking action, nor that since market share and industry attractiveness do not yield certainty they must be rejected. But, rather, that *the strategic processes that we use must, as a minimum, raise the right questions.* Portfolio analysis processes are enormously useful in helping managers understand some of the factors for success in a business, yet for the most part they ignore non-marketplace stakeholders who can often, though not always, determine the success or failure of a business.

A related issue is that to view the corporation as a portfolio of businesses to be managed like stocks in an investment portfolio runs the risk that managerial processes will become overly concerned with the financial Performance of the corporation.[13] While financial Performance is vital to the health of a business, it is but one criterion used by external stakeholders to judge the viability of the corporation over time. When interpreted too narrowly, portfolio processes are asking for more regulation of "externalities," more social critics and ultimately less productive work.

A second strategic management process, made famous by Harold Geneen at ITT is the strategic review process (Pascale and Athos 1981; Charan 1982). The idea of this process is for the top executives in a corporation to periodically meet with division or strategic Business Unit (SBU) managers in a formal review session. Progress towards the planned goal is reviewed and new strategies are sometimes formulated. Top executives are usually accompanied by staff experts who have unearthed hard questions for the reviewee to answer. These reviews are usually built into the strategic planning cycle and are used as methods of communicating expectations and evaluating both personal and business Performance.

The major problem with strategic reviews, in terms of being in synch with the stakeholder map of an organization, is that they do not encourage and reward an external orientation or stakeholder thinking. The emphasis from the point of view of the divisional manager under review is to "look good" to the senior executives who are reviewing performance. The formality of most strategic review processes and the mixing of personal and business evaluation make it difficult for the division manager to pay attention to multiple stakeholder concerns, which may contradict established corporate wisdom about the factors for success in a particular business. The nature of the organizational beast is such that it doesn't like and doesn't reward bad news and can hardly tolerate innovation. (How else can we explain the State of U.S. business?) It is much easier to play "Blame the Stakeholder" after the fact. "What senior executive in his right mind can hold a division manager accountable for a regulation which accounts for lost profits?" While responsibility for profits has been decentralized in most large multi-business firms, the responsibility for managing non-marketplace stakeholders (and some marketplace stakeholders) has not. Corporate Public Relations and Public Affairs are for the most part responsible for insuring a stable business climate for all the corporation's businesses. Division managers naturally perceive that they have a lack of control over critical stakeholder variables. During one seminar on stakeholder analysis with division managers the predominant response was "Great stuff, too bad my boss isn't here to hear it." Upon giving the same seminar to the top levels in the corporation the predominant response was "Great stuff, too bad our people (the division managers) weren't here to hear it." While too much should not be made of an isolated case, processes like the strategic review process can exacerbate the inability of the organization to ask the right questions.

[13] The critique of portfolio theory surfaced here is quite general in that it applies equally well to "misuses" of other processes. The point is that the processes must be capable of "fitting" with the other levels of analysis. They must describe the world as it is, and must prescribe transactions that are consistent with such a description.

A third strategic management process which explicitly tries to focus the organization externally is Environmental Scanning.[14] Adopting a metaphor of the radar technology, the idea is for corporate managers to "put up their antennae" and to scan the business horizon for key events, trends, etc. which will affect the business in the future. There are several versions of environmental scanning, each of which has strengths and weakness. Scenario building, whereby several key events and trends are linked together to form a possible future for the organization, is a favorite technique of some corporate planners and a product of several Consulting firms. Another technique is trend analysis, whereby key variables, usually demographic and economic are monitored for change. And, futures research, which predicts the future, is yet another technique for helping managers scan the external environment.

While all of these processes are useful, most of them do not yield concrete action Steps. It is hard to see how a 10 year forecast can help the SBU manager worried about how to overcome the latest regulation. Consequently, most corporate plans have an environmental scan in the front section of the plan, which states the environmental assumptions on which the plan is based. These assumptions are usually stated in terms of an econometric forecast of macro-economic variables such as inflation, unemployment, interest rates, etc. If the assumptions have not been forgotten by the time the plan produces concrete strategic programs, they surely will be by the time the results are reviewed. Then, no one is held accountable for using the wrong assumptions.

Focusing the strategic management processes in a corporation is a necessary condition for success in the current business environment. However, this external focus must be pervasive, from "front-end" analysis to control processes. Our portfolio analysis, strategic review and environmental scanning processes must get better and more sophisticated, yet this is not the whole story.

Organizational processes serve multiple purposes. One purpose is as a vehicle for communication, and as Symbols for what the Corporation Stands for[15,16] "The way we do things around here" depicts what activities are necessary for success in the organization. And, the activities necessary for success inside the organization must bear some relationship to the tasks that the external environment requires of the organization if it is to be a successful and ongoing concern. Therefore, if the external environment is a rich multi-stakeholder one, the strategic processes of the organization must reflect this complexity. These processes need not be baroque 25-step rigid analytical devices, but rather existing strategic processes which work reasonably well must be enriched with a concern for multiple stakeholders.

---

[14] See Schendel and Hofer (1979) for several review articles on the State of the art in environmental scanning.

[15] Lorange (1980) explores the Communications aspects of strategic management, and recommends a 3 × 3 matrix to diagram such processes.

[16] See Freeman (1983), and Chap. 4 below for an analysis of "what do we stand for" and the relationship of enterprise level strategy to the stakeholder concept and managerial values.

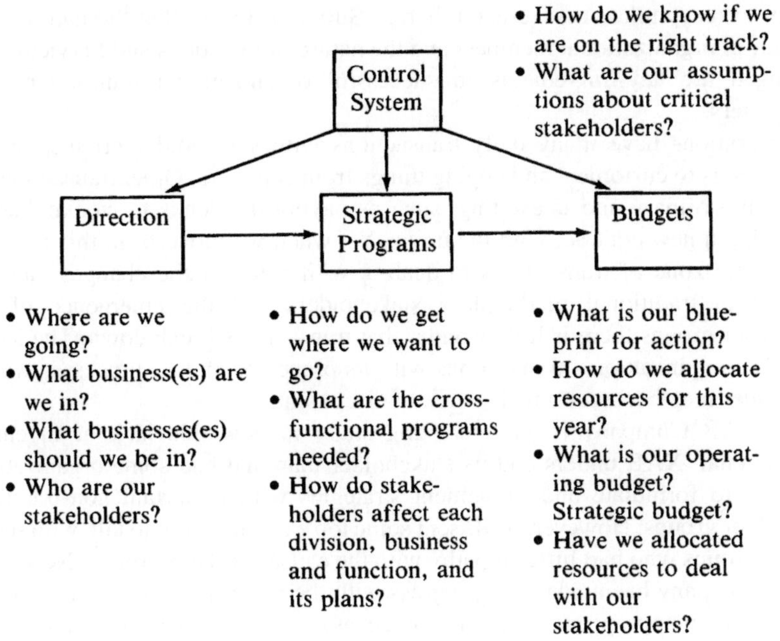

Exhibit 3.8 Typical strategic Planning Process Schematic (Lorange 1980)

For instance, strategic management processes such as Exhibit 2.2 can easily be enriched by adding "who are our stakeholders" to a concern with corporate mission; "how do stakeholders affect each division, business and function, and its plans" to the formulation of strategic programs; "have we allocated resources to deal with our stakeholders" in the budget cycle; and "what are our critical assumptions about key stakeholders" to the control process. Exhibit 3.8 depicts a revised version of Lorange's schema for strategic management processes. Each of these questions which are added will be discussed in more detail in subsequent chapters. The point is that relatively simple ideas can be used to encourage managers to think through the external environments of their businesses, and that such ideas must be added to organizational processes if they are to continue to be useful and to "fit" the stakeholder picture of the firm that is emerging.

5 The "Transactional" Level: Interacting with Stakeholders

The bottom line for stakeholder management has to be the set of transactions that managers in organizations have with stakeholders. How do the organization and its managers interact with stakeholders? What resources are allocated to interact with which groups? There has been a lot of research in social psychology about the so called "transactional environment" of individuals and organizations, and I shall not attempt to recapitulate that research here.[17] Suffice it to say that the nature of the behavior of organizational members and the nature of the goods and Services being exchanged are key ingredients in successful organizational transactions with stakeholders.

Corporations have many daily transactions with stakeholder groups, such as selling things to customers and buying things from suppliers. Other transactions are also fairly ordinary and unexciting, such as paying dividends to stockholders or negotiating a new contract with the union. Yet when we move from this relatively comfortable zone of transactions to dealing with some of the changes that have occurred in traditional marketplace stakeholders and the emergence of new stakeholder groups there is little wonder that transactions break down. The lack of "fit" of an organization's transactions with its processes and its processes with its stakeholder map becomes a real source of discontent.

The XAB Company is an interesting study in how this lack of fit can be dysfunctional. XAB understood its stakeholder map and had some organizational processes to formulate and implement strategies with important non-traditional stakeholder groups. However, XAB sent some top executives out to talk with several of these groups who had little empathy with the causes of these groups. Needless to say the Company has made little progress with them. Perhaps the strategy and the processes are inappropriate given the objectives of the Company. However, another interpretation is that the transactions between Company and stakeholders have not given the strategy and processes a fair test.

New England Telephone adopted a stakeholder approach to implementing a plan for charging for Directory Assistance in Massachusetts (Emshoff and Freeman 1979). The rational analysis of the stakeholder environment was sound and the planning process used to chart out an implementation scenario was successful. However, its transactions with several key stakeholders, most notably and ironically, its own union, as well as the State Legislature, were not successful. The union got a piece of legislation prohibiting the company's plan passed in the State legislature, and even though the Company was successful in persuading the Governor of Massachusetts to veto the legislation, as there was no public support, the State legislature overrode the Governor's veto, at the cost of $20 million to the customers of New England Telephone.

Consumer complaints are an area where there is usually a noticeable breakdown in the organization's Stakeholder Management Capability. Many large corporations simply ignore consumer complaints and dismiss them as that 5% of the market which they had rather someone else serve. Not only are there few successful processes for dealing with consumer complaints, but the transactions involved are material for every stand-up comic who ever walked. Nothing is more frustrating to the consumer than being told "sorry, I wish I could help you, but it's company policy to do things this way." One consumer leader commented that being told it was

[17] Emery and Trist (1965), Pfeffer and Salancik (1978) and many others have looked at the transactional level of organizations. Van de Ven et al. (1975) describe several different models of transactions.

Company policy may well finish the incident for the manager, but it begins the incident for the consumer advocate.[18] Several successful companies seem to "overspend" on handling consumer complaints. IBM's commitment to Service, P&G's consumer complaint department and the Sears philosophy of taking merchandise back with no questions asked, yield valuable lessons in understanding the nature of transactions with customers. These companies act as if consumer complaints yield an opportunity for understanding customer needs which ultimately translates into a good bottom line and satisfied stakeholders.

Other sets of transactions, which often get out of line with process and rational analysis, include the firm's relationships with the media, shareholder meetings, meetings with financial analysts, encounters with government officials and day to day interactions with employees and unions.

Many managers actively perspire during "60 Minutes" in fear of being before the sharp tongues of the reporters and the skillful editing of the news show producers. Some organizations have become proactive and given their senior executives special training on "How to Meet the Press."

Shareholder meetings have become rituals for most corporations, except for the occasional meaningful proxy fight à la Rockwell-SCM. Rather than carry out meaningful transactions with shareholders in accordance with a clearly thought out strategy and process, executives now treat stockholders to lunch and Speeches (with the stockholders' money) and a round of abuse from corporate critics who have bought one share of stock in order to be heard.

Meetings with financial analysts are another opportunity for transactions which can be made consistent with a firm's strategy and processes. Many executives understand that U.S. firms have underinvested in modern plant and equipment relative to foreign competition, and that they have lost sight of the marketing prowess of some of their competitors. How U.S. corporations can regain their competitive edge is a source of much debate in managerial and academic circles. Yet to regain competitive position will be neither easy nor inexpensive. Many U.S. firms will have to "take a hit on earnings" for several years in a row to be truly competitive. Most financial analysts are by their nature short-term focused. If executives use meetings with analysts to tout earnings per share, which may be inflated in real terms, then analysts will continue to expect short-term results. Talk of an investment strategy to regain competitive edge will be just talk. The transactions which executives make with analysts must square with the strategy of the organization regardless of the pain. By taking a leadership position in this area perhaps the thoughtful Company can change the expectations of financial analysts. Of course, there is a vicious "chicken-egg" cycle here, that illustrates the dilemma of attempting to change stakeholder expectations. If we are measured on short-term Performance results, and such a system is reinforced by expectations from the financial community, then to break the cycle involves additional pain. If strategic investments

[18] Interview with Professor Currin Shields, University of Arizona, and past President of the Conference of Consumer Organizations, a national consortium of local consumer advocate organizations.

really are necessary then we must bite the bullet, and work to change the expectations of analysts, stockholders, and even board members, even at substantial personal risk.

Transactions with government officials often take place under adversarial conditions. Because government is a source of trouble for many companies, their transactions with government show their discontent. One company is reported to have rented a truck and dumped the requested documentation on the doorstep of the government agency which requested it. When stakeholder relationships are viewed on both sides as adversarial it is a small wonder that anyone ever changes. The Business Roundtable, as a transactional organization for large businesses with the government, published a study decrying the cost of regulation and calling for regulatory reform. While it is clear that the regulatory process has gotten out of control in some areas, a more helpful transaction would have been to try and gain some formal input into the regulatory process. To gain such input would mean that a firm's transactions with the government could be made congruent with its organizational processes, and the firm could formulate strategies for influencing government in a positive way, breaking down the adversarial barriers of so many years and so many hard-fought battles.

Perhaps the most fruitful area for transactional analysis is with the employee stakeholder group. One large Company announced that it was committing to "Quality of Work Life," and set up national and local committees to form a partnership with its employees for the long term. However, shortly thereafter the Company announced that many employees were in fact "surplus," and offered incentive programs for early retirement. Its transactions were simply inconsistent with its stated future direction for this stakeholder group. Much has been written lately about Japan and Theory Z (Ouchi 1981), and co-determination in Europe. However, before U.S. managers launch into different directions with employees, perhaps we should understand whether our current managerial principles can work. When processes are set up to treat employees one way, no matter how well-meaning or "humanistic" they may be, and day-to-day transactions treat them another, it is not lack of theory that is the problem. The real importance of the suggestion box in Japan, and quality circles that work, is the consistent message that they send to employees, that their ideas have some impact on the firm.

If corporate managers ignore certain stakeholder groups at the rational and process level, then there is little to be done at the transactional level. Encounters between corporation and stakeholder will be on the one hand brief, episodic and hostile, and on the other hand non-existent, if another firm can supply their needs. Successful transactions with stakeholders are built on understanding the "legitimacy" of the stakeholder and having processes to routinely surface their concerns. However the transactions themselves, must be executed by managers who understand the "currencies" in which the stakeholders are paid. There is simply no substitute for thinking through how a particular individual can "win" and how the organization can "win" at the same time.

Clearly, there must be some "fit" among the elements of an organization's Stakeholder Management Capability-defined as its understanding or conceptual map of its stakeholders, the processes for dealing with these stakeholders, and the transactions which it uses to carry out the achievement of organization purpose with stakeholders. Exhibit 3.9 illustrates how some criteria might be used to measure the Stakeholder Management Capability of an organization. Whether an organization falls into the "Understands Correct Stakeholder Map" or "Does Not Understand Correct Stakeholder Map" is a relatively easy test. If, over time, an organization is continually surprised and continually plays "Blame the Stakeholder" then something is amiss. Whether an organization's processes and transactions are in line with that stakeholder map is a more difficult problem, for as I have shown, we do not have an adequate understanding as to what processes are appropriate for the multitude of stakeholders which firms now have. I shall return to the issue of defining a firm's Stakeholder Management Capability in Part II, Chaps. 4, 5, and 6, by way of suggesting several processes to be used to understand and manage stakeholder relationships. However, before attempting such a task it is necessary to be more explicit about the underlying philosophy which accompanies the Stakeholder Management model. How can the multitude of Charts, graphs, and maps be integrated into the current managerial wisdom of running a successful business?

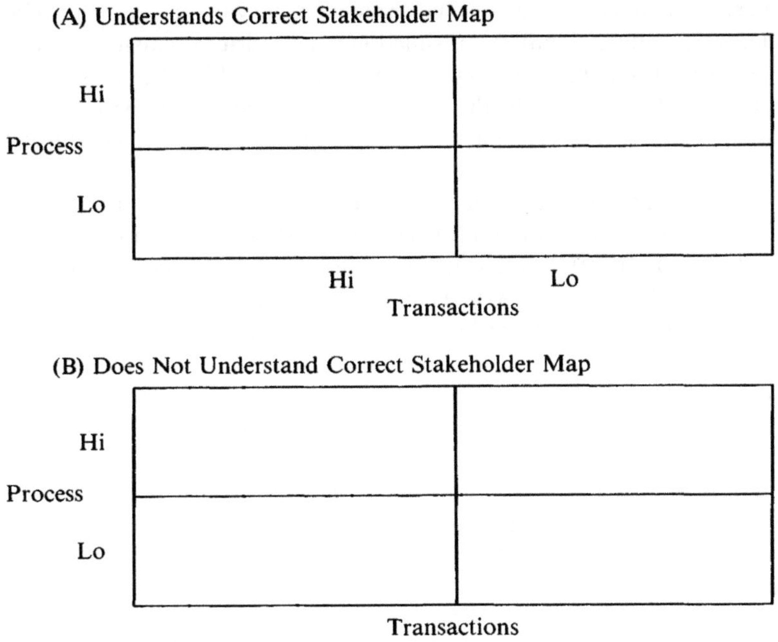

Exhibit 3.9 Stakeholder Management Capability = f (Stakeholder Map, Organizational Process, Stakeholder Transactions

6 The Stakeholder Philosophy: A Plea for Voluntarism

While the temptation to play "Blame the Stakeholder" is a strong one, the major problem facing U.S. managers is really not an external one, but an internal one. Pogo's saying is once again applicable, "we have met the enemy, and he is us." The challenge for us is to reorient our thinking and our managerial processes externally, in order to be responsive to stakeholders. There are three levels of analysis which must be consistent, rational, process and transactional. However, there are several common themes, or philosophical propositions, which can serve as "intellectual glue" to hold these ideas together. Such a philosophy of management is necessary if we are to undertake the rather considerable task of regaining managerial competence in the new business environment, without losing even more of our competitive position in the marketplace. We must learn to use our current knowledge and skill-base to respond quickly to the "stakeholder challenge" and to create some initial "win-win" situations, if meaningful change is to occur.

Such a philosophy of management must be based on the idea of voluntarism, if it is to be implemented in U.S. based companies. Not only is voluntarism the only philosophy which is consistent with our social fabric, but the costs of other approaches are simply too high. Voluntarism means that an organization must on its own undertake to satisfy its key stakeholders. A Situation where a solution to a stakeholder problem is imposed by a government agency or the courts must be seen as a managerial failure. Similarly, a situation where Firm A satisfies the needs of consumer advocates, government agency, etc. better than Firm B, must be seen as a competitive loss by Firm B. The driving force of an organization becomes, under a voluntarism philosophy of management, to satisfy the needs of as many stakeholders as possible.

Consider the current "Stakeholder Dilemma" in which many firms find themselves. The following story is a simplified illustration based on several real situations.

An Activist Group (AG) is worried about some aspect of the ABC Company's Product Y. AG believes that if ABC is allowed to continue to produce and sell the product as it now exists, harm will be done to the public and to some of AG's constituents. AG is a credible group in some circles, especially with a key government agency and the national media. While it has not always been successful in getting large corporations to be responsive to its claims, it has had some successes. AG does not have a large reservoir of resources, nevertheless, it can devote adequate resources to the pursual of this current case. ABC believes that there is nothing wrong with its product, and that they should be allowed to continue to sell it. ABC is a veteran of several campaigns against its products, and it has won some and lost some in the past, but each has been expensive to wage.

Let us assume that ABC has two major strategic responses. It can *Negotiate* with AG to reach a mutually agreeable solution with respect to Product Y by listening to the concerns of the leaders of AG, explaining the Position of ABC on Product Y, exploring Solutions to AG's concerns, voluntarily agreeing how AG and ABC are to proceed on this and future areas of mutual concern, involving other interested parties in the discussions, etc. Or, it can *Play Hard Ball,* by ignoring AG's concern, perhaps disparaging AG and the cause that it Stands for, respond when AG files a formal complaint, try to delay AG through countersuits, etc.

Of course, AG also has two very similar strategies. It can *Negotiate* with ABC by attending meetings with ABC managers and presenting the concerns of AG, attempting to understand the needs of other interested parties in Product Y, working with ABC to find a mutually acceptable solution, or a mutually acceptable process for finding a solution to the issue in a timely fashion, etc. AG can also *Play Hard Ball* by trying to make a splash in the media, bringing formal action in the Courts against ABC, complaining to government agencies which regulate ABC, tying ABC up on other issues unrelated to Product Y, introducing legislation prohibiting the sale of Product Y, etc.

Clearly if both parties negotiate, then an agreement which both find mutually satisfactory is the result.[19] Both parties may have to compromise, or at least be willing to compromise, if negotiation is to proceed in good faith, else Negotiate is an identical strategy to Play Hard Ball. If ABC negotiates and AG decides to play hard ball (perhaps after the first Session, AG decides to double cross ABC), then ABC will be embarrassed and vulnerable to AG's formal challenges by having admitted that there may be some legitimacy to AG's Claims about Product Y. AG members will have the feeling of having "beaten" ABC and may well be successful in their challenge to Product Y. Managers in ABC will not trust AG, and will respond in a "win-lose" way to AG's initiatives. On the other hand if AG tries to negotiate and ABC responds by playing hard ball, the same sorts of feelings arise for AG members as in the case where AG double-crossed ABC. If both parties Play Hard Ball, then the outcome is a long drawn out process with a solution imposed by the courts, government agencies and legislation—plus the cost of doing battle.

The most preferred outcome for ABC, in a cold and calculating sense, is for AG to negotiate and for ABC to double cross, since ABC then "beats" AG. The most preferred outcome for AG is for ABC to negotiate and for AG to double cross, thereby "beating" ABC. Yet when each plays its preferred strategy, Play Hard Ball, the result is far inferior to the result of playing the Negotiate strategy. In a real sense, by following the dictates of self-interest, both lose.

Exhibit 3.10 sets out the form of this stakeholder problem. Of course, it is identical in form with the so called "Prisoners' Dilemma Game" which illustrates the difficulty of achieving cooperative Solutions under communication constraints.[20] In the classical form of the game, two suspects to a crime (which they actually

[19] The structure of the payoffs of the game outlined here presupposes that the issue is vague enough for there not to be a "clearly optimal" solution, but that a solution which is mutually acceptable is possible, and further that this mutually acceptable solution is preferable by both parties to a solution which is imposed by external parties, such as government.

[20] There is a vast literature on the Prisoners' Dilemma, however, a clear discussion of the game can be found in Luce and Raiffa (1957). The game described here is similar to the plight of wheat farmers that is taught in every introductory economics class and chronicled by Garrett Hardin in the "Tragedy of the Commons."

outcomes	ABC's Strategies	
AG's Strategies	Negotiate	Hardball
Negotiate	Mutually acceptable solution	ABC "wins", does not give in
	Each must compromise	AG "loses"
Hard-Ball	AG "wins", does not give in	ABC & AG have solution imposed from outside
	ABC Loses	Cost of Adversarial proceedings

Payoff Table

AG \ ABC	N	HB
N	(2,2)	(4,1)
HB	(1,4)	(3,3)

1 = Most Preferred Outcome

4 = Least Preferred Outcome

Exhibit 3.10 The "Stakeholder Dilemma" Game

committed) are caught, and interrogated separately. Each is told that if they confess a light sentence will be passed depending on whether or not the other confesses. If one prisoner turns state's evidence and the other does not, the book is thrown at the non-confessing prisoner. If both confess, then each gets a medium-length sentence, while if neither confess they are convicted of a much lesser Charge. Neither confessing yields a preferred outcome to both confessing, but self-interest dictates confessing. The payoff structure is identical to Exhibit 3.10 with "Negotiate" replacing "Don't Confess" and "Play Hard Ball" replacing "Confess." If the prisoners could communicate they would form an agreement not to confess, or agree to get revenge if the other double-crossed. The lack of communication and the ability to form binding agreements dooms the prisoners to a heavy sentence.

The striking fact about the Stakeholder Dilemma version of this game is that there are absolutely no such communication constraints upon ABC and AG, and there are no constraints which prevent binding agreements. The managerial processes of both groups simply do not include considering communication and responsiveness as normal managerial activities. The Status quo imposes similar Prisoners' Dilemma-like constraints on ABC and AG.

The "Stakeholder Dilemma" game is one which is played out in some form in many organizations.[21] The only way out is to *voluntarily* adopt a posture of

[21] I am not claiming that every game that a Corporation plays with stakeholders is a Prisoners' Dilemma game, but only that some interactions are Prisoners' Dilemmas. The use of game theory

negotiation with stakeholder groups. Why negotiate voluntarily? Because, there is no other way to keep from having a solution imposed upon the organization from outside. And, to accept such an imposition of a solution to a problem is to give up the managerial role. Additionally, there seems to be no reason to pay the enforcement costs of adversarial proceedings. How many managers, lawyers and other professionals in large organizations spend most of their time in some sort of adversarial proceedings with stakeholders? Could not these resources be put to work more productively?

Our managerial processes must make managers "Free to Cooperate," rather than forcing them to play the Stakeholder Dilemma Game. Negotiation must become accepted practice, rather than conflict escalation through formal Channels. The "try it, fix it, do it" mentality (Peters and Waterman 1982) which many companies have used successfully with customers, must be applied to other stakeholder groups. This implies that voluntarism as a basic managerial value must permeate the organization which is successful in managing its relationships with multiple stakeholders.

This philosophy of voluntarism can be summarized in several prescriptive propositions which build on successful managerial theories and techniques. These propositions should be taken as tentative Statements of a theory which needs much more elaboration, but which are hopefully practical suggestions.

Organizations with high Stakeholder Management Capability design and implement communication processes with multiple stakeholders.

An example of a communication process is the recent formation by some Utilities, of Consumer Advisory Panels, whereby the company brings issues which are usually settled in the formal regulatory process to the attention of leaders of consumer advocate groups well in advance of actually filing the rate case. Company executives and consumer leaders can negotiate on issues of mutual concern and avoid the costly adversarial proceedings of the rate case on a number of issues.

Organizations with high Stakeholder Management Capability explicitly negotiate with stakeholders on critical issues and seek voluntary agreements.

An example of explicit negotiation is AT&T's convening an industry-wide conference of telecommunications executives, academics and consumer leaders over the issue of how to reprice local telephone Service to bring it in line with its true costs. The outcomes of such a meeting are multiple and not all have been successful. However, the tenor of negotiation was set, and at least some of the local telephone companies have begun to explicitly follow up and negotiate on issues before the rate case proceedings.

Organizations with high Stakeholder Management Capability generalize the marketing approach to serve multiple stakeholders. Specifically, they overspend on understanding

---

in strategic management, as an explanatory tool, is a long-neglected research issue. McDonald (1977) is one source. Recent work in applying game theory at the conceptual level can be found in Brams (1981) and Muzzio (1982). Both of these works by political scientists yield interesting insights into the workings of individuals in organizations.

stakeholder needs, use marketing techniques to segment stakeholders to provide a better understanding of their individual needs and use marketing research tools to understand, viz., the multi-attribute nature of most stakeholder groups.

We might define "overspending" as paying extra attention, beyond that warranted by considerations of efficiency, to those groups who are critical for the long term success of the firm. Overspending on stakeholders without whose support the Company would fail can make sense in a number of instances. For instance P&G overspends on customers, interviewing several thousand customers a year. AT&T overspends on the attention it pays to the regulatory process, which was for a long time, its major source of revenue. Oil companies should, likewise, consider adopting a conscious policy of overspending on OPEC and government and stakeholders who can convey a positive image to the public. Chemical companies have not overspent on environmentalists, for the most part, with the results being onerous regulations and reputations as "spoilers of the environment."

Organizations with high Stakeholder Management Capability integrate boundary Spanners into the strategy formulation processes in the organization.

Many organizations have public relations and public affairs managers who have a good working knowledge of stakeholder concerns, and marketing and production managers who have expertise in the needs of customers and suppliers. However, these managers are not always a part of the strategic planning process. Hence, their expertise is lost. The assumption is that those managers who are rewarded to be sensitive to stakeholder needs are in the best position to represent their interests inside the organization. For this representation to occur successfully, those boundary Spanners must have some credibility and some meaningful role to play in the organizational processes.

Organizations with high Stakeholder Management Capability are proactive. They anticipate stakeholder concerns and try to influence the stakeholder environment.

The micro-computer industry is full of firms who practice anticipation as a way of life. These firms, some of them quite small, spend resources trying to "guess" what will best serve the customer in the future and where the market will be. Similarly, larger computer manufacturers, should be "guessing" that issues such as "privacy" and "individual freedom" and "Computer literacy" will be major concerns as we move to technologies where "1984" is a distinct possibility. Several Utilities try to anticipate the concerns of intervenors in their rate cases, and actively seek out those groups which will be critical to try and influence their views.

Organizations with high Stakeholder Management Capability allocate resources in a manner consistent with stakeholder concerns.

Emshoff (1980) tells of analyzing the stakeholders in a large international firm and ranking the stakeholders in order of importance. A rough check was also made of how the firm's resources were allocated to deal with those groups who would be most important in the future. The results of his investigation were that almost no resources were being allocated to deal with those groups felt to be absolutely critical to the future success of the Company. Many executives are not reticent to play "Blame the Stakeholder," yet are not willing to devote resources to changing a particular stakeholder's point of view.

Managers in organizations with high Stakeholder Management Capability think in "stakeholder-serving" terms.

Just as many successful companies think in terms of "how to serve the customer" or "how to serve the employees," it is possible to generalize this philosophy to "how to serve my stakeholders." The "reason for being" for most organizations is that they serve some need in their external environment. When an organization loses its sense of purpose and mission, when it focuses itself internally on the needs of its managers, it is in danger of becoming irrelevant. Someone else (if competition is possible) will serve the environmental need better. The more we can begin to think in terms of how to better serve stakeholders, the more likely we will be to survive and prosper over time.

7 Summary

The purpose of this chapter has been to explicate the stakeholder management framework and philosophy in general terms. I have shown that the three levels of analysis, rational, process and transactional must be consistent if the stakeholder concept is to make a difference in the way that organizations are managed. I have offered a brief sketch of the principles of voluntarism which I believe must go hand in hand with the application of the stakeholder concept to strategic management processes. In the following section I shall try to elaborate on how this is to be accomplished.

References

Abell, D. 1980. *Defining the Business*. Englewood Cliffs: Prentice Hall, Inc.

Allison, G. 1971. *Essence of Decision*. Boston: Little Brown.

Brams, S. 1981. *Biblical Games*. Cambridge: MIT Press.

Chakravarthy, B. 1981. *Managing Coal*. Albany: SUNY Press.

Charan, R. 1982. The Strategic Review Process. *The Journal of Business Strategy* 2 (4): 50–60.

Dill, W. 1975. Public Participation in Corporate Planning: Strategic Management in a Kibitzer's World. *Long Range Planning* 8 (1): 57–63.

Emery, F., and E. Trist. 1965. The Causal Texture of Organizational Environments. *Human Relations* 18: 21–31.

Emshoff, J. 1980. *Managerial Breakthroughs*. New York: Amacom.

Emshoff, J., and E. Freeman. 1979. Who's Butting Into Your Business. *The Wharton Magazine* 1 (44–48): 58–59.

———. 1981. Stakeholder Management: A Case Study of the U.S. Brewers and the Container Issue. *Applications of Management Science* 1: 57–90.

Evan, W. 1966. The Organization Set: Toward a Theory of Inter-Organizational Relations. In *Approaches to Organizational Design*, ed. J. Thompson, 175–190. Pittsburgh: University of Pittsburgh Press; also in W. Evan. 1976. *Organization Theory: Structures, Systems, and Environments*. New York: John Wiley and Sons.

———. 1976. *Organization Theory: Structures, Systems, and Environments*. New York: Wiley.

Freeman, E. 1983. Strategic Management: A Stakeholder Approach. In *Advances in Strategic Management*, ed. R. Lamb, vol. 1, 31–60. Greenwich: JAI Press.

Freeman, E., and D. Reed. 1983. Stockholders and Stakeholders: A New Perspective on Corporate Governance. In *Corporate Governance: A Definitive Exploration of the Issues*, ed. C. Huizinga. Los Angeles: University Press.

Goodman, N. 1955. *Fact, Fiction, and Forecast*. New York: Bobbs Merrill Co.

Hirschman, A. 1970. *Exit, Voice and Loyalty*. Cambridge: Harvard University Press.

———. 1981. *Essays in Trespassing*. Cambridge: Cambridge University Press.

Lorange, P. 1980. *Corporate Planning: An Executive Viewpoint*. Englewood Cliffs: Prentice Hall, Inc.

Luce, D., and H. Raiffa. 1957. *Games and Decisions*. New York: Wiley.

MacMillan, I. 1978. *Strategy Formulation: Political Concepts*. St. Paul: West Publishing Co.

McDonald, J. 1977. *The Game of Business*. New York: Anchor Press.

Merton, R. 1957. *Social Theory and Social Structure*. Glencoe: The Free Press.

Miles, R. 1982. *Coffin Nails and Corporate Strategies*. Englewood Cliffs: Prentice Hall Inc.

Muzzio, D. 1982. *Watergate Games*. New York: New York University Press.

Negandhi, A., ed. 1975. *Interorganization Theory*. Canton: The Kent State University Press.

Nystrom, P., and W. Starbuck, eds. 1981. *Handbook of Organizational Design, Volumes 1 and 2*. New York: Oxford University Press.

Ouchi, W. 1981. *Theory Z*. Reading: Addison Wesley.

Pascale, R., and A. Athos. 1981. *The Art of Japanese Management*. New York: Simon and Schuster.

Peters, T., and R. Waterman. 1982. *In Search of Excellence*. New York: Harper and Row.

Pfeffer, J. 1981. *Power in Organizations*. Marshfield: Pitman Publishing Inc.

Pfeffer, J., and G. Salancik. 1978. *The External Control of Organizations*. New York: Harper and Row.

Rothschild, W. 1976. *Putting It All Together*. New York: AMACOM.

Schendel, D., and C. Hofer, eds. 1979. *Strategic Management: A New View of Business Policy and Planning*. Boston: Little, Brown and Co.

Van de Ven, A., D. Emmett, and R. Koenig. 1975. Frameworks for Interorganizational Analysis. In *Interorganization Theory*, ed. A. Negandhi, 19–38. Canton: The Kent State University Press.

Wilson, G. 1981. *Interest Groups in the United States*. New York: Oxford University Press.

R. Edward Freeman is University Professor, Elis and Signe Olsson Professor, Academic Director of the Institute for Business in Society, and Senior Fellow of the Olsson Center for Applied Ethics at the University of Virginia's Darden School of Business.

Chapter 4
Theory Building in Strategic Management

R. Edward Freeman and Peter Lorange

1 Introduction

Strategic management research is undergoing a period of change and proliferation. For instance, Ansoff (1977), Ansoff et al. (1976), Schendel and Hofer (1979), Gluck et al. (1980), Quinn (1981), and Lorange (1980a) have each argued that a shift in paradigm, or framework, is occurring, and the concept of "strategic management" has emerged to give direction to a diverse body of literature. Also, within the last several years new journals devoted to strategy have appeared. Similarly, there has been a renewed emphasis in the more popular press on strategy. Schools of management have seen a boom in courses in strategic management. Strategy-based consulting firms have proliferated. In short, there seems to be an abundance of examples underscoring the rich activity level presently enjoyed by strategic management.

We believe that the time is ripe to reexamine several foundational issues in strategic management research, particularly questions pertaining to the following: research method and technique; the unit of analysis for research; the focus, interpretations, and uses of research; the role of researchers; and the development of

---

Originally published in: *Advances in Strategic Management*, 3, 9–38, © Elsevier Books Limited, 1985

Reprint by Springer, Reproduced with permission of The Licensor through PLSclear, DOI 10.1108/astm

R. E. Freeman (✉)
University of Virginia, Charlottesville, VA, USA
e-mail: FreemanE@darden.virginia.edu

P. Lorange
IMD, Lausanne, Switzerland

© Springer Nature Switzerland AG 2023

S. D. Dmytriyev, R. E. Freeman (eds.), *R. Edward Freeman's Selected Works on Stakeholder Theory and Business Ethics*, Issues in Business Ethics 53,
https://doi.org/10.1007/978-3-031-04564-6_4

synthesizing frameworks for better understanding the myriad pieces of research that are now being conducted and in order to build more powerful theories of strategic management.

The purpose of this paper is to articulate the terrain of foundational issues in strategic management research as we see it and to suggest a framework for understanding and justifying the generation of knowledge in this Held, i.e., a framework for theory building.

The chapter will be divided into five main sections. First we shall argue that it will be necessary to delineate the theoretical base or context within which one intends to do research before addressing specific substantive research issues. Next, we shall suggest that the complicated nature of strategic management research requires a conceptual framework which can serve as a guiding device or heuristic for researchers. Third, we propose a framework with four dimensions: specification of the strategic decision, identification of decision level, delineation of the purpose of the decision, and specification of the decision context. We shall argue for this framework as a four-dimensional heuristic to classify and understand strategic planning research. Fourth, we shall show how some research issues can be better understood using our framework and, more importantly, how linkages across streams of research can be made. The illustrative research topics shall focus around "strategic business unit definition" and "strategic control and stakeholder management." We shall argue that our framework points toward a building block approach for reaching a full-blown theory of strategy and that the current state of the art is not that there is too much theory but rather that there exists too little theory to enable researchers to integrate the bits and pieces into a coherent whole. A consequence of our argument is that the controversy of "rigor vs. relevance" is a non sequitur. Finally, we shall examine some implications of our argument for further research in strategic management.

2 Fundamental Contexts for Substantive Issues

Before examining each of the four elements of our proposed framework, we need to add a further complication. Careful research in any field requires a distinction between "questions in a theory" and "questions about a theory" (Rudner 1966). Exhibit 4.1 lists some questions about strategic management research which deal with the process of research as well as the application of the research. "What is the unit of analysis in strategic management research" is a different kind of question from "How do firms in the steel industry set direction?" The questions in Exhibit 4.1 are meant to illustrate these "meta-level" concerns about theory building, i.e., they are about *theories of strategic management* or *research in* strategic management, rather than the *substance of* strategic management. We shall call these questions *fundamental questions.*

Exhibit 4.1 Fundamental Questions in Strategic Management

1. What is "strategic planning?"
2. What is the *unit of analysis* in strategic planning research?
3. What is the *scope* of strategic planning research?
4. What are the best techniques for discovering knowledge in strategic planning? (Logic of discovery?)
5. What is the appropriate methodology, or criterion, for justification of research in strategic planning? (Logic of justification?)
6. What is the role of strategic planning research in the larger context of management theory?
7. What are the overlaps with OB, OT, OD, OR, DSS, etc.?
8. What are the pitfalls in strategic planning research? What are some heuristic meta-level principles to guide researchers?
9. What place does *replicability* play in strategic planning *research?—Predictability—Explanation?*
10. What are the value issues inherent in strategic planning research, in its methods of discovery and justification?

Exhibit 4.2 illustrates the more familiar content questions in strategic management research. Most of the research of the past 50 years has been an attempt to answer one or more of the questions in Exhibit 4.2, though our list is far from exhaustive. These questions are about the phenomena being studied, i.e., the behavior of firms or industries or individuals or some combination of these entities. We shall call these questions *substantive questions.*

Before proceeding let us in parenthesis note that there are a number of ways to classify substantive questions in strategic management. For our present purpose let us assume that any substantive question in either a question about the *direction* of a firm or set of firms (i.e., the definition of its businesses, relationships with the external environment, analysis of strengths-weaknesses-opportunities-threats, values of management, behavior in an industry, etc.) or about the *implementation* of direction (i.e., control and evaluation, management systems and structures, etc.), where we define "direction" and "implementation" broadly. Nothing turns on this assumption, and other classifications of this dimension may be substituted—and will in fact be discussed below.

We shall claim that fundamental questions bear an important logical relationship to substantive questions. Answers to fundamental questions will set the context for substantive questions, as well as yield methods and processes for how substantive questions are to be answered. Unfortunately, most research studies pay little if any explicit attention to delineating such a theoretical context. Jemison (1981), an exception, has argued that the answer to question 2 in Exhibit 4.1, the unit of analysis question, can help to clarify the relative contributions of fields related to strategic management, e.g., industrial organization, or marketing and administrative behavior.

Exhibit 4.2 Familiar Content Questions in Strategic Management

Direction:

1. How do firms set direction for the organization as a whole and for subsets or divisions and businesses of the organization?
2. What direction has been set for firms in Industry A, or in firms with characteristics C_1 ..., C_n?
3. What are the appropriate processes for setting direction?
4. What are appropriate directions for firms with characteristics C_1, ..., C_n?
5. How do firms manage their relationships with important stakeholders?
6. How should firms manage their relationships with important stakeholders?
7. To what extent should (do) firms participate in joint ventures?—Coalition strategies?—Collective actions?
8. How do (should) firms generate growth internally?
9. How can firms exit from certain industries?

Implementation:

10. What can strategic planners do to achieve strategic direction?
11. How can the planning process be tailor-made?
12. What is the difference between "strategic" and "operating" systems and structures?
13. What can strategic managers do to incorporate direction into day-to-day activities?
14. What is the "strategy, systems, structure" connection in firm A?
15. What are appropriate control processes?
16. How many SBUs should (does) a firm have?
17. What is the connection between reward systems and strategic behavior?

For example, one researcher might address the question "How do organizations set direction for themselves as a whole" with an analysis of the strategic planning systems in place and a careful look at the internal perceived validity of these systems. The unit of analysis (either explicit or implicit) of such a study may be the behavior of certain individuals in the organization (Lorange 1980; Quinn 1981). Another researcher may argue that in a large complex organization overall direction is a function of the cumulative effects of the directions of its divisions or departments, and so understanding of overall direction hinges on the analysis of the direction of these subunits and the dimensions along which their performance is measured (Lawrence and Lorsch 1969). Thus, the latter researcher sees a different unit of analysis, the actions of subunits, as critical. Both of these approaches are perfectly valid. However, each one would have to specify the fundamental theoretical contexts it is assuming in order for the specific substantive issues to be meaningfully interpretable.

We shall argue throughout the following sections that research programs (Lakatos 1981) in strategic planning must attempt to sort out these issues and to deal with them in a systematic fashion. Answers to fundamental questions in isolation from substantive issues is empty philosophy. And research in rich content areas in isolation from fundamental constraints and opportunities leads to fragmented, unjustifiable pseudo knowledge.

3 The Need for a Framework

There has been much talk about the "shift in paradigm" in strategic management. Ansoff et al. (1976), Ansoff (1977), Schendel and Hofer (1979), and Saunders and Thompson (1980) as well as others have written about the stages of development of the field and the "shift in paradigm" from "business policy" to "strategic planning" to—'strategic management." Attempting to apply the ideas of Thomas Kuhn (1970) on the history and philosophy of science, these scholars have pointed out that the way that researchers conceptualize "strategic management" has undergone significant changes. However, "paradigm" is a misnomer, in our context, in that it assumes that there exists a well-accepted practice, field, or discipline where the conditions for knowledge are clear-cut and the methods for justification are without question. While questions of research method and technique do occasionally arise in more mature disciplines such as physics and biology, there is hardly the amount of hoopla which we see in strategic management. Indeed even in these more mature disciplines, many have argued that the notion of "paradigm" does not adequately describe how knowledge is created and justified. See, for instance, the essays in Gutting (1980) and Lakatos and Musgrave (1970) for a beginning to the burgeoning literature on this topic. Kuhn himself (Kuhn 1970), in the postscript to the second edition of *The Structure of Scientific Revolutions,* recants the notion of "paradigm," suggesting "disciplinary matrix" as a way of denoting the concepts, norms, and practices of science. See Barnes (1982), Feyerabend (1975), and Laudan (1977) for some of the resulting complications. Without such a mature disciplinary matrix, large-scale "conceptual shifts" are virtually impossible. We believe that the issue is precisely that the "field" of strategic management is "preparadigmatic" and that we need to turn our attention to the process of adding to the disciplinary matrix at work to build a theory that is both comprehensive and testable. Research in strategic management increasingly points toward the necessity for a framework for research which elaborates the basic building blocks of a theory of strategic management, thereby seeing the justification of new knowledge as largely one of putting it into an overall context. We believe that there are three closely related arguments which can be marshaled in favor of such a framework. We shall designate these arguments or challenges as (a) the "world-view" syndrome, (b) the reconciliation problem, and (c) the argument from scholarship.

(a) The World-View Syndrome

Many have fallen prey to the trap of elaborating a way of thinking about a problem that may be overly technique oriented or dogmatic and thus becomes "the gospel." In its extreme form there might be waves of applications of the latest technique or "knowledge" to virtually every strategic problem, whether it needs it or not. For instance, managers might rush to plot growth/share matrices, to draw experience curves, to design a three-cycle planning process, to map stakeholders, to learn from "excellent companies," or to analyze the elements of a firm with tools borrowed from the efficient market portfolio theory of finance. Predictably, implementation of the latest technique bogs down after the initial enthusiasm wears off, and then books are burned, consulting contracts tom up, curse words uttered, etc., and the field moves on to embrace the next technique that has come over the horizon and is now in vogue.

The lesson for researchers from this is that, given that strategic management is by its nature an applied as opposed to a pure research field, the researcher should be doubly obligated to more carefully state his or her assumptions on which "theory and results" depend, being precise as to how they can and cannot be applied and how they are similar to and different from those of other studies. It is impossible to fulfill this obligation without a framework which sets forth the dimensions of building blocks along which other research may be judged, differentiated, or reconciled. Thus coping with the world-view syndrome calls for a research framework and is one argument in support of our attempt here.

(b) The Reconciliation Problem

A second major problem facing researchers in strategic management is the current piecemeal approach to research. Several streams of research or research programs are emerging in the literature. For example Porter (1980) has set forth a theory of competitive strategy, and others (e.g., Harrigan 1981), have attempted to work out answers to questions arising within this theory. Vancil (1979) and earlier Lorange and Vancil (1976) have systematically analyzed the role of strategic planning processes. The PIMS program attempts to delineate empirical performance effects from various strategic choices. Industry Market structure issues and their impacts have been studied by Montgomery (1982) and others. Emshoff and Mitroff (1978) and Mason and Mitroff (1982) have analyzed one piece of the planning process, the critical role of assumption making.

However, virtually no one has attempted to address how aspects of these streams of research might be reconciled, in the sense of linking these various research questions together, but without, of course, going so far as to attempt to find a "unified science of management" or indubitable "first principles." Even though many researchers try to cite the relevant literature for their research, there is a curious absence of attempts to link research programs.

There might be at least two main reasons for such a lack of reconciliation. The first is one of understanding. Consider, for instance, how Porter's (Porter 1980) work on generic competitive strategy yields insight into strategy formulation process for stakeholder groups (Freeman 1984). By attempting to intellectually bridge seemingly radically different research questions new insights can be gained. The second reason for reconciliation is related: if we could systematically link differing research programs, we might identify new research questions that might lead to expanding the frontiers of knowledge. Once more, however, we find that the key to our reconciliation argument is the establishment of a framework which can delineate such linkages in a systematic way.

(c) The Argument From Scholarship

A third major argument for the need for a framework for strategic planning research stems from a plea for adhering to careful principles of scholarship. We believe that group and institutional norms may well frustrate the generation of new and exciting insights, failing to keep a perspective on the eclectic nature of research in strategic management. Rather than substitute a set of norms, we believe that researchers should begin to systematically probe the nature of various streams of research to uncover nuggets of knowledge which may be used. To illustrate, consider the following quote from the excellent article by Saunders and Thompson (1980):

> On the other hand, the conceptual papers were long on compilation and assimilation and short on genuinely new flashes of insight and understanding. An original, stimulating conceptual paper is probably difficult to come by anywhere, anytime, but in the policy field the advice of critics and commentators may be much to the point: a turn away from feeble attempts at the insight type and toward hard examination of applicable data in an empirical framework is what is needed now.

On the one hand, the authors provide more evidence for our claim that a framework for research is needed, i.e., we need more insight and linkage. However, the inference that these will necessarily come through "hard examination of applicable data" is not warranted. There are too few concepts which yield hypotheses which are genuinely testable, much less replicable. "Testability" and "replicability" are troublesome concepts. Neither is very useful if the underlying logic of the concepts being tested is unclear. And, indeed, they are only part of the logic of justification, where concepts such as "simplicity," "conservatism," "modesty," and "refutability" also play important roles in determining the truth of theories (Quine and Ullian 1978). Perhaps in a hundred years, theorists will look upon the state of the art in strategic management in the 1980s in a way similar to how physicists view those early scientists who believed in the "ether." It may be that the very concepts of "strategy" and "structure" are quite primitive. Our theory-building activity must not rule out the thought experiments necessary to make "breakthrough" conceptual changes. Of course, such conceptual breakthroughs must yield a set of experiments which in principle and in actual fact allow the concept to take its place in the background theory. The danger of strategic management traveling a narrow path of empiricism is too great to ignore. Empirical validity is but one criterion for research. Other disciplines, in particular recent advances in biology and physics, should point to the necessity for strong conceptual frameworks, calling for logical as well as empirical analysis.

In what follows we shall briefly set forth a framework for research in strategic planning in response to the challenges just discussed. We shall illustrate how this can be used as a heuristic for placing a particular work of research on the conceptual map. There exist two basic approaches for the scientific elaboration of such a framework. The first is to count and classify studies in one or another category, thus yielding some "frequency" validity for a set of categories (Saunders and Thompson 1980). The second is to propose, a priori, a framework which can then be used, once it is refined and applied, to yield a number of empirical hypotheses. However, such hypotheses are not testable within a theory of strategic management. They are contained in a theory which is about "theories of strategic management." We shall take the second approach, arguing that the task which presently faces us is more one of unifying framework construction than of framework validation.

4 A Classification Framework

In this section we shall propose and argue for a simple a priori framework for sorting out research questions in strategic planning. In doing so, we shall introduce in a stepwise mode four dimensions, argue for the relative importance of each dimension, and illustrate the strengths and weaknesses of each dimension in the framework, indicating where some benchmark studies can be placed. In order to construct an argument for an a priori framework, we need to show why the dimensions of such a framework are "natural," i.e., why they have some intuitive appeal and explanatory power. We shall assume that the question before us is to explain "strategic planning research" to a novice in the field or, even more difficult still, to a person from Mars.

Suppose that we go back to basics and try to understand this concept called "strategic management." In order to articulate the building blocks for a theory of strategic management we need to understand the concept in nontechnical terms, i.e., in terms which are not a part of the theory of strategic management itself. Ultimately, if we are to be precise and logically correct, we need to literally "point out" the objects of the theory (Rudner 1966; Hempel 1965; Quine 1960; Quine and Ullian 1978). Ostentation (as discussed below) anchors our theory in reality and makes certain that a theory is grounded in real cases. We propose to show how such a philosophy of science approach can be applied to strategic management by trying to explain (or give an account of) "strategic management" to someone who has no idea what it is.

Suppose that a creature from Mars, let us call him Mike, lands in the midst of a meeting of the Strategic Management Society. After the initial shock has worn off, we find that the creature does understand our language but is unclear about what "strategic management" means. What kinds of questions could we reasonably expect this creature to ask? And how would we go about making our research understandable? In short, what dimensions of our work would we need to explain in order for Mike the Martian to understand us?

The reason for assuming the existence of Mike the Martian is twofold. First, it forces us to give an explanation of strategic management which does not rely on technical terms; that is, since Mike does not understand what strategic management is, he ipso facto does not understand the technical terms and their referent objects, such as "strategy," "structure," "corporation," "market," "hierarchy," etc. within the context of strategic management. Hence, we must make use of ostentation, pointing out, to show Mike what we mean. Secondly, by assuming that Mike has a rudimentary knowledge of our language, our sociology, psychology, perhaps even our philosophy and economics, we can clarify the central metaphors which are embedded in any interdisciplinary activity such as strategic management (Quine 1960). The use of non-native speakers of a language is a standard technique in philosophy even though it has had little use in the social sciences [but see Nozick 1969 and Brams 1983 for a discussion of Newcomb's problem and how the assumption of such an extraterrestrial being helps to clarify some basic principles of decision theory and economics].

4.1 The Task Dimension: Strategic Decisions

We shall argue that one of the first questions that Mike the Martian would ask us, trying to elicit the referent objects of the term, is "What is strategic planning about?" Just as we would explain that physics is *about* "the behavior of complex wave equations in a Hilbert space" and that biology is *about* "the origins, random mutations, and survival of species of life," so do we need to explain what strategic planning is about. Indeed a fundamental question for any discipline is "What is the discipline about?" The logic of "about" is far from noncontroversial (Goodman 1961), but suffice it to say that research in strategic planning, just as research in physics or biology or even history, is about some phenomenon or other.

We would tell Mike the Martian that strategic planning is about a set of decisions that organizations make. Of course, we would have to explain "decision" and "organization," but we shall assume that Mike has some rudimentary knowledge of our psychology and sociology. We might call this dimension of strategic planning the "task dimension" (or "decision dimension") to connote that strategic planning is about certain tasks (or decisions). Various words such as "planning," "evaluating," "controlling," etc. have evolved to give us a taxonomy of the kinds of tasks (or decisions) which organizations undertake. However, we shall define these tasks quite loosely as "direction-setting tasks" or "direction-implementing" tasks. For example, direction-setting tasks involve deciding what business to enter, what mix of businesses to invest in, what the strengths and weaknesses of the organization are, and what opportunities and threats in its external environment are important. To illustrate we might tell Mike about John Connally's decision to abandon the packer can business at Crown, Cork and Seal or about George Romney's decision to produce the Rambler at American Motors (Christensen et al. 1973). Direction-implementing tasks refer to a narrower set of decisions designed to achieve the desired "strategic direction." For example, "How should Tom Pringle allocate resources to achieve a certain profit figure at Texas Instruments," or "What system of control and measurement should be used at Dansk Designs to insure that the new products stay on track" (Christensen et al. 1973). We would then take Mike to a couple of organizations and show him around, pointing out which decisions were strategic and which were not.

Schendel and Hofer (1979) have proposed a 6 fold classification of "strategic tasks" and is a candidate for possible refinement of our twofold schema. We adopt the two-dimensional classification for the sake of simplicity. Nothing critical turns on our adoption of the terms *direction-setting tasks* and *direction-implementing tasks* vs. a more complicated taxonomy. Our argument is that some classification of the task dimension is necessary and of primary importance. If we cannot explain what strategic management is about, then further questions regarding how the research is to be understood are meaningless.

For our purposes it is important to note that we are taking the set of "strategic decisions" as the basic set of entities (or ontology) or primitives of any piece of research in strategic planning. We have described these entities as direction-setting decisions and direction-implementing decisions, and we have given examples of each (see Exhibit 4.2). Can this primitive concept of "strategic decision" be broken down further? The answer is surely affirmative, and it is here that fundamental questions become more important. By analyzing the "unit of analysis" of a particular research study we might decide that there may be a "more primitive" typology which would yield fruitful research hypotheses. For example, in classifying strategic decisions we might note that some researchers use "strategic decision" to refer to "actions of an organization as a Whole" such as "U.S. Steel diversified into oil by acquiring Marathon." Other researchers might take as a unit of analysis the discrete actions of certain managers, e.g., "The chief executive officer of U.S. Steel petitioned the Board for approval to acquire Marathon." Even other researchers might take as a unit of analysis the entire system of organizations and people involved in a strategic decision: "To understand U.S. Steel's decision to acquire Marathon, one must analyze the actions of government agencies, Mobil, unions, key officials, and others."

Whether or not these different units of analysis are reducible to some more fundamental unit is a question which we need not answer here, though we would have to tell Mike the Martian that there is little agreement about what constitutes the basic interpretation of the terms "organization" and "decision," just as the physicists would tell Mike that whether light is a wave or a particle is not a question that can be answered within the current state of the theory. We would caution Mike that in the strategic management literature some authors may use "strategic decision" to refer to discrete actions of managers while others may use it to refer to the behavior of an organization or set of organizations, showing that this ontological issue is far from settled. We would also caution Mike to guard against "the fallacy of composition" whereby researchers simply assume, without reason or argument, that properties of individuals can be applied to groups of individuals, i.e., that organizations have goals and objectives and can act rationally. [A simpleminded example of the fallacy of composition is the "logical" step from "All horses have tails" to "Herds of horses have tails." While it seems obviously stupid in this context, consider the inference from "Individuals make choices" to "Corporations make choices." The issue is far from clear. One way of parsing much of the literature is on the basis of who believes that the fallacy of composition is involved in the above inference. For an enlightening discussion relevant to strategic management, see Hardin 1982.] We would also tell Mike that direction-setting (DS) and direction-implementing (DI) decisions differ in degree rather than in kind and that he should expect that better (more refined) taxonomies will replace the DS–DI distinction. Nonetheless, we will have answered Mike's most basic question—"What is strategic management about?"—and, in doing so, set the stage for the second question.

4.2 The Decision Level Dimension

The second question that Mike the Martian might naturally ask is whether or not there exist some other meaningful dimensions of "strategic decisions." From his rudimentary knowledge of our psychology and sociology and after we have shown him around several organizations, Mike might discern that organizations are fairly complex phenomena and that the point of belonging to an organization is that the division of labor and hierarchy makes it easier to accomplish complex tasks. Thus, there might well be very different levels of analysis for strategic decisions. Mike need not settle the metaphysical questions of which level, if any, is primary. However, it is important that he see that strategic decisions, both DS and DI decisions, can be made at a number of organizational levels and units, and by a number of organizational actors. The literature is replete with distinctions among subunits in organizations and with an explicit acknowledgment that DS and DI decisions are dependent on organizational level. Thus, resolving the question "What should our portfolio of businesses be" is a corporate or organization-wide task, while finding an answer to "How can we compete successfully in this business" is a task with a narrower organizational scope. This element, "level of analysis," recognizes the fact that organizations are complex and that "value added" gets created in different forms at various levels of large organizations.

Again, we propose only one of several possible taxonomies for the issue of "level of analysis." That is, we propose to divide strategic decisions into four levels, suggested by several authors (e.g., Schendel and Hofer 1979; Lorange and Vancil 1976: (1) enterprise; (2) corporate; (3) business family; and (4) business element (or functional, in some cases). Enterprise-level strategic decisions (sometimes called "collective") seek to position an organization in the broader social context (Astley and Fombrun 1983; Freeman 1984). Typical areas for research include antitrust, "deregulation," and other issues of social policy as it affects organizations, as well as decisions to form coalitions and collectives through vehicles such as trade organizations or with other members of the organization set (Evan 1976). Issues such as corporate governance and ethical norms can also be envisioned as enterprise-level decisions, as they sometimes act as constraints on other levels of strategic decisions. Corporate level strategic decisions deal with the issues of generating and allocating strategic resources among a corporation's businesses, as well as assessing the economic and political risk exposure of the firm in total. Familiar aspects of these decisions include business portfolio direction decisions, risk analysis, and capital budgeting and other resource allocation models, as well as corporate-wide planning process issues. Business family-level strategic decisions are only beginning to be researched as more and more firms seek some rational principle other than financial criteria for grouping a set of businesses together into "groups," "sectors," "strategic business units" (SBUs), and "divisions." Research issues of cross-business synergies, niche options, and entry/exit barriers are beginning to be defined and better understood. Business element-level strategic decisions (resolving questions like "How do we compete successfully in a given business?") have been the most vigorously studied in the past few years, with concepts such as market share or return on investment beginning to be integrated into theories of business competition (Porter 1980) and testable hypotheses being generated (Abell and Hammond 1979; Henderson 1979). Just as there is an obvious continuum from DS decisions to DI decisions, so is there a continuum from business element-level to enterprise-level decisions depending on the size and degree of organizational complexity of the firm(s) being studied.

In order to explain to Mike the Martian what strategic management research is, we have argued that first we must be clear regarding what strategic management is about, namely, strategic decisions. We have argued that the DS–DI continuum is one way (not the only way) of parsing the set of strategic decisions that are made. Next, we have argued that both DS and DI decisions are made at various levels in an organization, and that "level of decision" or "level of analysis" is a useful second dimension for Mike to understand. We have proposed a taxonomy of this dimension which involves four categories standardly found in the literature. Some might argue that there are better second dimensions than "level of analysis." In particular, size or industry are favorites for research studies. While these may be useful variables with which to study organizations, we shall argue that "level of analysis" or "level of decision" is more general than these variables and in reality can capture both of them. Many firms compete in more than one industry, thereby necessitating a division of the firm into some meaningful parts to capture industry-wide effects. Level of analysis is such a division. DS and DI decisions would be expected to differ in firms of varying size, and again "level of analysis" captures the possibility of this difference, as it should be possible to compare the corporate level decisions of small firms with business element decisions of larger firms. In short "level of analysis" begs no questions of size or industry, yet is compatible with studies of both variables.

Hence, we believe that it can usefully serve as a second dimension to explain strategic planning research.

4.3 The Decision Purpose Dimension: Content and Process

Armed with the knowledge of different kinds of strategic decisions and the various levels in the organization at which these decisions can emerge, Mike the Martian may well wonder about the purpose of a particular DS or DI decision at a particular organizational level. Is it a decision to do something or a decision to decide or to facilitate doing something? Does the research in question study the decision itself or how the decision is made, that is, the content of a decision or set of decisions or the process of decision making? Hence, just as in biology it is legitimate to worry about the behavior of a creature as well as the processes which give rise to that behavior, so in strategic management we need to distinguish between the questions "What is to be done" and "How do we decide what is to be done?" For example a content research study would focus on questions such as "Shall we acquire a company in the steel business" or "Shall we launch an intensive competitive campaign against company x," whereas process research would focus on "How can we decide what direction to go in" or "What processes can we use to encourage long-range thinking on the part of our division managers?" Lorange (1980) has argued that a necessary ingredient for strategic planning to be successful is the involvement and commitment of the chief executive officer (CEO). Research on process can cover one particular decision or a connected series of decisions (a system of decision making). Critical aspects of such process issues might deal with how an organization reached consensus for pursuing a particular decision and why key people are motivated accordingly to pursue that decision.

It is here that a controversy usually arises in strategic management—for there is a good deal of research on fundamentally "content" questions as well as on fundamentally "process" questions, and there is a great deal of controversy on which kind of research is more "important." Quinn (1981) has stirred the controversy by questioning the very nature of content research as being too wedded to the rational model of organizations and not sensitive enough to the political aspects of organizational life. For many years the strategic planning literature was solely concerned with the process of strategy and policy, and with the associated concepts of control, evaluation, implementation, etc., as processes. More recently more analytical models have been developed which focus on the content of strategy. In the earliest stages these models have been *prescriptive,* i.e., "build market share," but lately have become *descriptive* as in the PIMS studies. Hofer (1975) summarized research on the content of strategy and compiled a list of 54 variables that the literature indicated were of major importance. We need not solve this controversy here, but we shall again point out to Mike the Martian the need to beware of the biases which a particular body of research may have toward answering this question. We shall take the position that, given the present state of the art, both content and process research present viable alternatives, and so the question is less one of which is more fruitful and more one of how can these two seemingly different kinds of research be linked together.

As in the previous two dimensions, we have drawn on dichotomous distinctions in the literature to contrast the ends of each dimension. However, the process of making strategic decisions will obviously affect the content of those decisions. At the extremes one can argue that the process of decision-making is merely discovering the appropriate decision from a rational content model or, conversely, that the content of a decision is whatever emerges from whatever process is used. Content research often ignores the importance of process, while process research often ignores the fact that rational models can describe and prescribe choice.

We are now armed with three sets of distinctions which serve as the answer to three key questions: What is strategic management about? Are there differences among levels of decision making? Is there a difference between what is decided and how it is to be decided? The specific answers which we have proposed to these questions yield three separate dimensions along which strategic management research can be classified and hence raise the need for one final dimension, discussed next.

4.4 The Decision Context Dimension

A careful look through Exhibit 4.2, as well as an examination of several meanings of the word "strategy," points to a notion that has not yet been captured in the three dimensions proposed so far. Consistently the definition of "strategy" used in the literature relies on a notion of fit between an organization's internal capabilities and its external environment. This internal–external distinction is implicit in many of the questions in Exhibit 4.2, as it is in each of the three preceding dimensions to strategic management research. However, we believe that it should be made explicit simply because it is so pervasive. Of the 54 variables analyzed by Hofer (1975), not all of which are independent and unrelated, 44 are focused externally, i.e., they measure characteristics of. industry structure, competitors, customers, and suppliers as well as broader economic and political phenomena. Vancil's (1979) work on decentralization tries to measure the internal ability of firms to use structure as a key component of their strategic planning system. Lorange (1980) discussed how the strategic situational setting of a particular internal organizational unit calls for tailor-making of the planning process to meet these particular situational needs.

These three studies give us examples of the final element of our framework: the setting or the context of a particular decision. On the one hand, some research is focused toward understanding how a firm relates to its external environment or how models of adaptation can be used to explain such interaction (Chakravarthy 1981). On the other hand, another body of research focuses on the internal side, the processes and systems, the strengths and weaknesses that make up the organization.

Research on industry analysis focuses on the external fit or tailor-making between the firm and its environment, while research on particular business strategies focuses on the internal tailor-making of the management processes and decisions to the particular situational needs. It is important, we believe, to understand the particular setting of a decision or set of decisions which form the basis for a research study, again (above all) to be able to link it to other work in the field.

4.5 *Putting the Pieces Together*

There are other candidates for questions to which Mike the Martian must find the answers if he is truly to understand strategic management research. Perhaps different research methodologies and techniques would help Mike to better understand the field. Perhaps a dimension of the relevant disciplines which can be brought to bear on a particular question may also help. For the sake of simplicity, we shall stop with four dimensions: (1) strategic taks; (2) decision level; (3) decision purpose; an (4) decision context. Exhibit 4.3 depicts our framework, holding strategic decisions constant at DS for reasons of graphic simplicity. Exhibits 4.4, 4.5, 4.6 and 4.7 are illustrative of research topics which appear in each cell. As already stressed, our purpose is less to put studies in the "right" unambiguous box than to illustrate how these four elements work together. It is also possible to suggest some benchmark

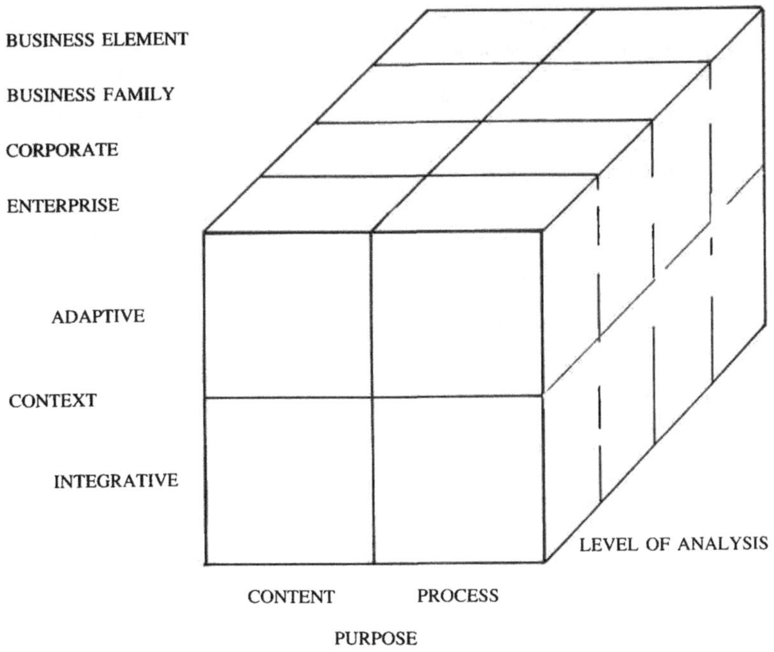

Exhibit 4.3 A heuristic for theory building in strategic management

CONTEXT	CONTENT	PROCESS
ADAPTIVE	Social Policy Anti-trust Policy "Deregulation" Macro-economic Policy	Collective Action Trade Orgs. Coalition Strategy "Executive as Public Servant"
INTEGRATIVE	Business Ethics	Corporate Governance
	PURPOSE	

Exhibit 4.4 Enterprise level, direction-setting research topics

Context	Content	Process
Adaptive	Portfolio of Ind. Corporate Growth Entry/Exit Barriers Acquisition/Mergers	Strategy for Corp. Stakeholders
Integrative	Risk Exposure Models --BCG --CAPM SBU Definition	Planning Process
	Purpose	

Exhibit 4.5 Corporate level, direction-setting research topics

Context	Content	Process
Adaptive	Industry Growth Entry/Exit Barriers Competitiveness Niche Options	Industry-specific Stakeholders --Union --Consumers --Gov't.
Integrative	Industry Know-how Business Synergies	Cross-business Integration
	Purpose	

Exhibit 4.6 Business family, direction-setting research topics

		Purpose	
		Content	Process
Context	Adaptive	Growth Rate Niche Exit/Entry Barriers Competitiveness	Business/Plant Stakeholder Relations QWL Local Govt.
	Integrative	Market Share Experience Curve Business Definition Distinctive Competence	Business Processes Cross-functional Plans

Exhibit 4.7 Business element, direction-setting research topics

studies as examples in each category. Thus, Hofer's (1975) study mentioned earlier found that most of the content research was at the business level and focused externally. Lorange's (1980) research on the planning process has been primarily process research focused internally. Freeman's (1984) research on stakeholders has been process research focused externally. Porter's (1980) work has been content research focused externally, though at the business level. The PIMS studies are business element studies that focus externally.

Research on such issues as antitrust matters, ethics, collective and coalition processes, and corporate governance and the role of boards are at the enterprise level. See Weidenbaum (1980), Astley and Fombrun (1983), Lindbloom (1977), French (1979, 1982), Ewing (1978), Bowie (1981), De George (1981), Dill (1978), Jones (1978), and Freeman and Reed (1983) for a sample of relevant enterprise level research, which has been for the most part ignored by strategic management theorists. Research on more traditional policy content areas such as growth, niche, entry barriers, and competitiveness, focused externally, and market share, experience curve, business definition, and distinctive competence, focused internally, is found at the business level.

5 Some Examples of the Use of the Framework

In order to illustrate how this framework can be used as a heuristic for theory building in strategic management we propose to take three issues and show how linkages among research areas can be built, using the four elements of our framework. The issues which we shall choose form the basis for our own research, and it is our hope that other researchers can apply our framework to their research areas to assess their research in terms of potential "world-view" syndrome fallacies, explicitness of linkages or reconciliation with other areas, and nondogmatism in the choice of research methodologies.

5.1 *Strategic Business Unit Definition*

Let us begin by considering some definitional aspects of what might be seen as an SBU. a common definition is one of identifying the core unit around which to carry our competitive business strategy formulation and implementation: a particular business segment is an SBU if and only if it can be delineated as a business segment which is associated with some specific notion of attractiveness as well as relevant competitive strengths to attract the target customers with one's own products or services and thereby beat a set of specific competitors in question. Thus, along our first dimension, defining an SBU is in part a DS decision and, at another level, a DI decision. For a discussion of how to define the business along these lines see Abell (1980).

The original use of the SBU concept seems to have been somewhat different than as just defined, however, at least if we examine the use of the term at General Electric, the company which seems to have been the first to use SBU. Here the SBU concept relates to the delineation of a number of competitive product/market business strategies which should be seen as interrelated in the marketplace or in the use of internal resources, such as research and development (R&D), manufacturing, or sales, whereby considerable synergies might accrue.

We thus see that the SBU concept can have different meanings depending on how we specify at least one of the four dimensions in our classification framework, namely, the level of analysis or decision level. Upon more in-depth discussion we shall in fact find that we must be extremely careful in stating our dimensions or premises for research on the SBU concept.

5.1.1 Decision Dimension

a. *Direction Setting*. This is a central issue for most aspects of the research on the SBU concept, in that it established the need to identify the strategic entities around which to formulate strategies: what might be the entities or "building blocks" that the firm should break itself into for the purpose of developing successful strategies, i.e., for being in a position of creating a distinctive concept of what is to be the strategic "value added"? Thus, the question of "in what strategic direction are we going" can only be dealt with by assuming a clear delineation of the organizational units we are dealing with. Our point here is a logical extension of the answer that we gave to Mike the Martian. We must get clearer about the referent objects for "SBU."
b. *Direction Implementing*. The question of implementing direction (i.e., "How are we going to get to where we want to go?") will bring on such issues as who is responsible for carrying out the various aspects of a particular strategy, how many resources of various kinds are needed, and how does the organization control its progress in implementing these plans. The definition of the

organizational units for our strategic analysis, often denoted by "strategic structure," is important here, too, particularly when we are delineating to what extent, if any, the resulting strategic structure might be different from the formal operating organizational structure. If there is an absence of congruence between what we might see as the strategic and the operating structures, then a key requirement for implementation will be to make it explicit how to link the two structures. Thus, while the SBU definition has an important potential effect on the implementation of strategic direction, it has a direct impact on the setting of strategic direction.

5.1.2 Decision Level Dimension

a. *Business Element.* The issue to be delineated here is how a competitive strategy might be formulated around an organizational unit so as to succeed in reaching a particular target customer group with the firm's own product or service in competition with those of explicitly considered competitors. The business element level of analysis is thus directly linked to the task of setting strategic direction, in that it focuses on how to delineate realistic, successful strategic direction at the competitive strategy level. From a direction implementation point of view there is also a link, in that the question is "how to win the competitive battle" for a particular business element.
b. *Business Family.* The issue here is to consider potential interrelationships between several business elements, in terms of achieving synergies internally or in the marketplace. Again, however, we see that it is next to impossible to discuss this without having a relatively explicit view of how the various business elements in question have been defined. From a DS point of view the issue thus is one of identifying the relevant synergies; from a DI viewpoint, however, the issue is one of delineating how the synergies can be achieved. As mentioned above, while the original conception of an SBU as it was developed by General Electric (see Vancil 1979) is analogous to what we have denoted "business families," another common use of the SBU label today seems to be to use it synonymously with what we have denoted "business elements." While there is no right or wrong definition, it seems clear that considerable confusion might arise due to the lack of uniformity of usage.
c. *Corporate level.* At this level the focus is one of delineating the overall portfolio strategy of the firm, to be built up as a pattern of business elements grouped into business families and leading to some corporate-wide expected consequences, in terms of strategic resource generation and allocation, as well as economic and political risk exposure. The definition of SBUs has thus an important, but indirect, bearing on this level of analysis, in that the way the unit of analysis gets determined might have a profound impact on the way the firm's top management "looks at itself." This might in turn affect the resulting portfolio strategy as follows:

1. From a purist point of view one might argue that the preferred units of analysis should be clearly delineated business elements since it will be the competitive exposures the firm faces which create the resulting resource generation/utilization balance as well as the risk exposure balance.
2. From a synergy point of view the business families may be more appropriate building blocks, in that they allow for seeing business elements as interdependent when appropriate. This also allows for at least some resources to be allocated to the families, allowing a family to allocate resources to its own elements. This delegation of strategic focus might increase organizational commitment and quality of judgment, even though the portfolio might be seen as "suboptimal" from a rational perspective.
3. From a cognitive point of view it might be difficult to balance literally hundreds of business elements into a corporate-wide pattern; to balance a relatively smaller set of business families, however, might seem easier. However, how can families be balanced, given that they are themselves "balances" of business elements?

There is no universally right answer to how to delineate the strategic task at the corporate level; however, to be precise about task and level seems key in order to be explicit about the nature of what magnitudes of trade-offs and biases one is dealing with.

d. *Enterprise level.* This level considers the firm within a larger societal context. The way the firm defines its SBUs might have impact at this level too. First, each SBU's environment will be a function of how it is defined. Also, however, several SBUs may be facing the same or in parts overlapping environments, thus calling for a reconciliation of the SBU strategies. For instance, General Electric is attempting to develop the so-called arena concept in which several SBUs (i.e., business families) are coordinated in terms of their overlapping environments, particularly when it comes to overlapping technology, or in joint foreign country markets.

5.1.3 Purpose Dimension

a. *Content.* Let us consider an SBU which has been focused on the DS task and at the business element level (or, alternatively, a piece of research which focuses on describing such a group of SBUs or even prescribing a course of action for them). The content dimension will deal with applying familiar models of business attractiveness/competitive strength analysis. Business attractiveness can be linked to growth rate (Henderson 1979), product life cycle (Abell and Hammond 1979), and the competitive structure of the business (Porter 1980), as well as other factors. The competitive strength dimension can be associated with relative cost advantages due to exploitation of experience curve benefits from larger market share or scale economies. Quality leadership may also lead to a competitive advantage. Various models have been suggested for pursuing strategic direction of an SBU's strategic position, both normative and empirical.

Shifting now to the analytical issue associated with the DS task at the business family level, we may also apply analytical schemes for creating synergies—either internally, by joint utilizations of functional skills, or in the marketplace, by coordinated positioning of products for several niches within the overall business. Chandler (1962) provides a good example of this analytical task when he describes how the various lines of business of General Motors were delineated to achieve overall market synergy.

b. *Process.* Consider the same SBU focus as above, i.e., emphasis on setting direction at the business element level. However, let us now shift our emphasis to the managerial process. The issue here is how the SBU organization goes about analyzing its attractiveness and its strengths. What interactive or iterative mechanisms are useful in reaching a consensus view on this? What formats might be useful for the organization to follow in order to reach its focus?

At the business family level the issue similarly is one of delineating a working process for identifying and agreeing upon the various synergies. While the analytical issues are clear, the challenge now is to delineate a managerial process for building organizational commitment.

At the corporate level the process issue is partly one of developing a mode of interaction with the business family or business element levels, to create a bottom–up/top–down consensus. There is also a process issue at the portfolio level: how in an organizational context is management to receive information from the businesses, as well as how to collect information about potential outside candidates for takeover; how is it to consolidate this into a tentative overall portfolio picture; and how is it to give feedback and initiate modifications in some of the business strategies when necessary, consistent with the overall portfolio strategy.

5.1.4 Context Dimension

a. *External.* The definition of the SBU can also be seen in an external context as helping the firm to adapt to the external environment. At the business element level the issue of comparing several members of a population of SBUs to determine more effective business strategies would be an external focus. At the firm level the comparison of portfolio strategies might be the issue. Industrial economics and industry analysis provide a prominent example of a research stream that takes an external context.
b. *Internal.* The internal focus is represented by the management of the organization itself attempting to look at its own opportunities and to adapt the internal capabilities of the firm to the environmental realties. The particular decision-making characteristics of such a focus makes it different than that of external observation of a population of SBUs. Again, the issue is to make explicit the viewpoint for the setting of the research.

5.2 *Strategic Control and Stakeholder Management*

Let us now give a somewhat briefer discussion of how research on strategic control and stakeholder analysis may be enriched and reconciled with other streams of research by using our framework.

In terms of the task dimension one would normally see strategic control within the implementation sphere as a DI decision. It should be noted, however, that strategic control may also be seen as a vehicle for incrementally resetting direction (Lorange 1983; Quinn 1981). If such a "logical incrementalism" viewpoint is taken, then our first classification dimension should help us make this distinction by recognizing the contrast with more classical conceptions of control. The level of analysis will provide another set of distinctions regarding the classification of who is to be involved with the control tasks and what they are controlling. At the corporate level, for instance, this would be different than at the business element level, especially if we view the control task as a way of resetting direction. Content issues deal with understanding sources for strategic decisions and explaining deviations as well as suggesting corrective actions, while process issues deal with the methods and procedures that organizations use to control, again varying by level of analysis. Traditionally control has had an internal context, that is, measurements are taken and corrective action initiated within the firm. Recent work on understanding the external environment would argue, however, that control processes may need an external flavor. If the assumptions that underlie a given direction for a firm are discovered to be invalid, perhaps action needs to be taken to try and make the assumptions true. For example, recent initiatives by executives in the political arena can be viewed as seeking to get the external environment of a firm or set of firms "back in control."

Traditionally stakeholder analysis has been viewed as a DS task which helps managers at the corporate level understand the external environment in order to formulate strategies (Emshoff and Freeman 1979, 1981). However, the other dimensions of our framework yield other research questions and other uses of the concept for managers. Trivially, stakeholder analysis, or more precisely "stakeholder management," can be used in implementation. Emshoff and Finnel (1979), Mason and Mitroff (1982), and Freeman (1984) all discuss the notion of "internal stakeholders," i.e., those groups and individuals within a corporation who must be influenced if strategy is to be implemented. As organizational forms become more complex, and as hybrids of the traditional structures become more normal (Mintzberg 1979), the concept of internal stakeholders may be more useful. On the decision-level dimension, it has become clearer in recent years that responsibility for dealing with the external stakeholders in a firm needs to spread out at a number of levels. Again, General Electric has been the leader by making sector executives responsible for political issues that affect their individual sectors. There are stakeholder issues that must be handled at the business level as well, as the pervasiveness of agencies such as the Federal Trade Commission (FTC), Food and Drug Administration (FDA), Environmental Protection Agency (EPA), Occupational Safety and Health Administration (OSHA), etc. and critics of particular product lines and issues has emerged. Originally stakeholder analysis was formulated along the process dimension to deal with the standard operating procedures that organizations could develop to deal with multiple stakeholder groups. However, Freeman (1984) has formulated a content approach whereby an analysis of behavior can yield generic prescriptions for strategies in a manner similar to the business attractiveness/ competive strength matrix, even though more research is called for in this area. The context of stakeholder analysis has always been external; however, again using the concept of internal stakeholders, we can focus the issue internally to better understand the internal environment that managers face.

6 Putting the Framework Together

Our examples of the use of this framework have indicated that by paying attention to its four structuring dimensions new research issues and hypotheses can be generated. The above examples are not meant to be analyzed in a complete and comprehensive way, but rather to be suggestive and to serve as illustrations as to how the framework can be used by researchers. Let us now turn to a larger issue of how research streams can be linked together and why an eclectic approach to research is necessary if our framework is to be useful.

We not only can use the four dimensions to enrich and fill out a particular line of research or a particular set of managerial problems, but we can also make connections among the issues themselves. We have made several such connections and, while that is not evidence of all connections being fruitful, it is evidence that connections can be made. Additionally, several boxes in the grid depicted in Exhibit 4.3 are for the most part absent. Let us explain in more detail.

Originally Freeman and Reed (1983) argued that stakeholder analysis was to be as a DS process issue at the corporate level with an external focus. By moving diagonally in the grid and understanding the issue as an implementation issue, research on other implementation issues became available. Indeed Freeman (1984) proposes a method for formulating strategic programs, budgets, and control which depends heavily on previous research in these areas. Alternatively, by understanding the original "box" for stakeholder analysis, even if it represents no more than the researcher's own narrow "world view," Freeman (1984) was able to draw on other research in that box. For example, Porter (1980) deals with DS questions, even though the level is the business element level, which have an external focus. It is relatively easy to see that if it makes any sense to describe generic strategies for competitors, customers, suppliers, new entrants, or providers of substitutes, then one can generalize the notion that there are five forces that shape competitive strategy to include a sixth force, the relative power of other stakeholders. Porter's generic strategies can then be interpreted as prescriptions for other stakeholders as well. Freeman (1984) explores a number of hypotheses which come from just such a move. Such diagonal moves, and explicit linking with other research in the same "box," we shall argue, can only serve to strengthen the scholarly basis for strategic management.

Once such linkages occur, however, problems arise as to interpretation of research results, for such results are only valid within a particular framework. Hence, we shall argue, the methodology for a particular study is relatively less important than the conceptual value of that study. Our framework presupposes no primacy of one research method over another. It concentrates on *conceptual linkages*. Questions of research technique are questions to be answered given that certain conceptual linkages exist or can be made to exist.

Thus, our framework must be understood as calling for more theory construction. We believe that the absence of explicit linkages across streams of research in the literature points to the need for more theory rather than less. And it is to these conclusions and implications that we now turn.

7 Conclusions

We have argued that a four-dimensional framework for understanding research in strategic management can be useful in formulating new research questions and in interpreting extant literature. The proof of our claim is in the pudding, that is, whether or not such a framework can actually lead to such theory-building activity that is judged fruitful. We have tried to show, from our experience and research, how this framework has been useful. However, our investigations into theory construction in strategic management have led us to four tenative conclusions which we will now state in a form that seems to contradict the currently accepted "wisdom" in the field. We hope that these four claims will be understood as working hypotheses which, again, we have found useful in our own research.

Claim No.1 *There Is Too Little Theory In Strategic Management.*

The direction of current research in strategic management as it has evolved from the case method of business policy is that it is too "squishy." There are too many concepts, and there is too little empirical research that supports the concepts.

We want to argue that precisely the opposite is nearer to the truth. Our simple framework indicates that it is possible to go a long way toward constructing theory out of currently existing literature and that to test "concepts" or piecemeal relationships is to further fragment a rather meager quantity of real knowledge. Theories hang together. Particular claims, hypotheses, and sentences simply "do not confront the tribunal of experience alone" (Quine 1960; Quine and Ullian 1978). Without adequate attention to the context of concepts and general claims, empirical support of these claims is virtually meaningless. Thus, while there may be too many generalizations, too many unfounded ideas, and too much sloppy conceptual logic, there can never be too much theory. At least there can never be too much "rigorous theory." This leads us to the second claim.

Claim No.2 *There Is No Trade-Off Between Rigor And Relevance.*

Recent conferences have all addressed themselves to the supposed trade-off between rigor and relevance. The argument is that, since strategic management is an applied discipline, we must sacrifice rigor in favor of relevance to practitioners.

We believe that our framework indicates that such an argument is a non sequitur. There is simply no trade-off when rigor is interpreted as "logical rigor." Theories which are not logically rigorous, i.e., which are not tied to previous research in the field by some framework or other, are simply inadequate. Practitioners can not be assured of valid results when such "theories" are applied. There are insufficient grounds for accepting such claims as "knowledge." There is simply no substitute for a good theory, and there is no trade-off to be made in terms of the underlying logic of the concepts. However, the debate is usually cast in the form of "rigorous methodology" vs. "relevant results." Here again it is a false trade-off—for, as research in the sciences shows, research methodology must be broad. Conceptual and empirical techniques are relevant.

Claim No.3 *Strategic Management Requires Eclectic Research Methods.*

The trend in strategic management is toward more and more empirical research methods, where "empirical" is to be interpreted as referring to a set of statistical techniques that are "accepted practice" in other areas of the social sciences. Such a trend is a natural outgrowth of institutional norms and the quest for certainty which goes hand in hand with intellectual curiosity.

Large-scale data base analysis, quantitative models, and survey research all form a part of the modem social scientist's tool kit. However, these methods are only part of the tool kit. While they may make for exciting and impressive journal reading, they are not the valued end result. Such an end result, we believe, is a theory which explains and predicts. The use of such methods, as well as the techniques of pure mathematics and logic, have led to enormous advances in physics and biology. But it is not the empirical method itself which is so impressive; it is the theoretical structures which have emerged. Such structures have emerged only through conceptualizations which can be linked together and then subjected to rigorous empirical tests. It is theory formulation which is critical. Without the insight, no matter its source, theory validation is "an academic exercise" in the most perjorative sense of that phrase. Narrow empiricism leads to theories which have neither the conceptual richness nor the full range of empirical evidence that should be garnered in their support. This leads us to a final claim with respect to an overall assessment of strategic management theory.

Claim No.4 *Strategic Management Is Preparadigmatic. More Logical Analysis And Theory Construction Is Needed.*

Much has been written about paradigms, and in particular about shifts in the paradigm of strategic management. It has been argued that with the maturing of the strategic management paradigm, researchers can now get on with the business of applying the paradigm and testing its results.

The very existence of paradigms raises a whole thicket of philosophical problems. However, suffice it to say that we believe that our crude attempt at a framework shows that much more work needs to be done before a "paradigm" can even be formulated. In shifting the nature of physics, Albert Einstein stood on the shoulders of giants. He used available tools to work out his insights, and he invented those tools which were not available but were necessary if the insight that was general relativity was to be grounded. Strategic management is light years away from such *conceptual* sophistication. To accept current "theory" as a paradigm to be cleaned and polished by further validation of its hypotheses is to run the risk of turning an emerging and fruitful area of research into mere dogma.

This paper has been an attempt to continue the dialogue on foundational issues in strategic management, in the spirit of Schendel and Hofer (1979). By attempting to formulate a crude framework and to state the claims which are necessary if such a formulation is to be useful, we have argued that much more thinking in this area is necessary. If knowledge in strategic management is to be conceptually sound and of benefit to practicing managers, such critical thinking must occur while "strategic management" is being formulated.

References

Abell, D. 1980. *Defining the Business*. Englewood Cliffs: Prentice-Hall.
Abell, D., and J. Hammond. 1979. *Strategic Market Planning*. Englewood Cliffs: Prentice-Hall.
Ackoff, R. 1974. *Redesigning the Future*. New York: Wiley.
———. 1981. *Creating the Corporate Future*. New York: Wiley.
Aldrich, H. 1979. *Organizations and Environments*. Englewood Cliffs: Prentice-Hall.
Allison, G. 1971. *Essence of Decision*. Boston: Little Brown.
Andrews, K. 1965. *The Concept of Corporate Strategy*. Homewood: Irwin.
———. 1980. *The Concept of Corporate Strategy*. Rev. ed. Homewood: Irwin.
Ansoff, I. 1965. *Corporate Strategy*. New York: McGraw-Hill.
———. 1976. Managing Strategic Surprise by Response to Weak Signals. *California Management Review (Winter)*. 18: 21–33.
———. 1977. The State of Practice in Planning Systems. *Sloan Management Review*. (Winter) 18 (3): 1–24.
———. 1979. *Strategic Management*. New York: Wiley.
———. 1979a *The Changing Shape of the Strategic Problem*. In Schendel and Hofer (1979), 30–44, below.
Ansoff, I., R. Declerk, and R. Hayes, eds. 1976. *From Strategic Planning to Strategic Management*. New York: Wiley.
Astley, G., and C. Fombrun. 1982. The Telecommunications Community: An Institutional Overview. *Journal of Communication 32* (4): 56–68.
———. 1983. Collective Strategy: The Social Ecology of Organizational Environments. *Academy of Management Review 8* (4): 576–587.
Athos, A., and R. Pascale. 1981. *The Art of Japanese Management*. New York: Simon & Schuster.
Barnes, B. 1982. *T. S. Kuhn and Social Science*. New York: Columbia University Press.
Bowie, N. 1981. *Business Ethics*. Englewood Cliffs: Prentice-Hall.
Brams, S. 1983. *Superior Beings*. New York: Springer.
Chakravarthy, B. 1981. *Managing Coal*. Albany: State University of New York Press.

Chandler, A. 1962. *Strategy and Structure*. Cambridge, MA: MIT Press.
———. 1977. *The Visible Hand*. Cambridge, MA: Harvard University Press.
Christensen, C., K. Andrews, and J. Bower. 1973. *Business Policy: Text and Cases*. 3rd ed. Homewood: Irwin.
De George, R. 1981. *Business Ethics*. New York: McGraw-Hill.
De George, R., and J. Pichler, eds. 1978. *Ethics, Free Enterprise, and Public Policy*. New York: Oxford University Press.
Dill, W. 1958. Environment as an Influence on Managerial Autonomy. *Administrative Science Quarterly* 3: 409–443.
———. 1975. Public Participation in Corporate Planning: Strategic Management in a Kibitzer's World. *Long Range Planning* 8 (1): 57–63.
———. 1976. *Strategic Management in a Kibitzer's World*. In *From Strategic Planning to Strategic Management*, ed. H.I. Ansoff et al., 125–136. New York: John Wiley.
———. 1978. *Running the American Corporation*. Englewood Cliffs: Prentice-Hall.
Duncan, R. 1979 *Qualitative Research Methods in Strategic Management*. In Schendel and Hofer (1979), below.
Emshoff, J. 1980. *Managerial Breakthroughs*. New York: AMACON.
Emshoff, J., and A. Finnel. 1979. Designing Corporate Strategy. *Sloan Management Review 21* (3): 41–52.
Emshoff, J., and E. Freeman. 1979. Who's Butting Into Your Business? *The Wharton Magazine (Fall)* 44–48: 58–59.
———. 1981. Stakeholder Management: A Case Study of the U.S. Brewers and the Container Issue. *Applications of Management Science 1*: 57–90.
Emshoff, J., and I. Mitroff. 1978. Improving the Effectiveness of Corporate Planning. *Business Horizons (Oct.)* 21 (4): 49–60.
Evan, W. 1976. *Organization Theory: Structure, System and Environment*. New York: Wiley.
Ewing, D. 1978. *Freedom Inside the Organization*. New York: McGraw-Hill.
Feyerabend, P. 1975. *Against Method*. London: Vesso Editions.
Freeman, E. 1984. *Strategic Management: A Stakeholder Approach*. Marshfield: Pitman.
Freeman, E., and D. Reed. 1983. Stockholders and Stakeholders: A New Perspective on Corporate Governance. In *Corporate Governance: A Definitive Exploration of the Issues*, ed. C. Huizinga. Los Angeles: University of California at Los Angeles Extension.
French, P. 1979. The Corporation As A Moral Person. *American Philosophical Quarterly 16*: 207–217.
———. 1982. Crowds and Corporations. *American Philosophical Quarterly 19*: 271–277.
Gluck, F., S. Kaufman, and A. Walleck. 1980. Strategic Management for Competitive Advantage. *Harvard Business Review 58* (4): 154–161.
Goodman, N. 1961. About. *Mind 70*: 1–24.
Grant, J., and W. King. 1982. *The Logic of Strategic Planning*. Boston: Little, Brown.
Gutting, G., ed. 1980. *Paradigms and Revolutions*. Notre Dame, Ind: University of Notre Dame Press.
Hardin, R. 1982. *Collective Action*. Baltimore: John Hopkins University Press.
Harrigan, K. 1981. Barriers to Entry and Competitive Strategies. *Strategic Management Journal 2* (4): 395–412.
Hatten, K. 1979 *Quantitative Research Methods in Strategic Management*. In Sehendel and Hofer (1979), pp. 448–467, Below.
Hayes, R., and W. Abernathy. 1980. Managing Our Way to Economic Decline. *Harvard Business Review 58* (4): 67–77.
Hempel, C. 1965. *Aspects of Scientific Explanation*. New York: Free Press.
Henderson, B. 1979. *Henderson on Corporate Strategy*. Cambridge, MA: Abt Books.
Hofer, C. 1975. Towards a Contingency Theory of Business Strategy. *Academy of Management Journal* 18: 784–810.

Hofer, C., and D. Schendel. 1978. *Strategy Formulation: Analytical Concepts*. St. Paul: West Publishing Co.

Hofer, C., E. Murray, R. Charan, and R. Pitts. 1980. *Strategic Management*. St. Paul: West Publishing Co.

Huizinga, C., ed. 1983. *Corporate Governance: A Definitive Exploration of the Issues*. Los Angeles: UCLA Extension.

Jemison, D. 1981. The Importance of an Integrative Approach to Strategic Management Research. *The Academy of Management Review* 6 (4): 601–608.

———. 1981a. The Contributions of Administrative Behavior to Strategic Management. *The Academy of Management Review* 6 (4): 633–642.

Jones, R. 1978 *The Relations between the Board of Directors and Operating Management*. In Dill (1978), above.

Kuhn, T. 1970. *The Structure of Scientific Revolutions*. 2nd ed. Chicago: University of Chicago Press.

Lakatos, I. 1981. History of Science and Its Rational Reconstructions. In *Scientific Revolutions*, ed. I. Hacking, 107–127. New York: Oxford University Press.

Lakatos, I., and A. Musgrave, eds. 1970. *Criticism and the Growth of Knowledge*. New York: Cambridge University Press.

Laudan, L. 1977. *Progress and Its Problems*. Berkeley: University of California Press.

Lawrence, P., and J. Lorsch. 1969. *Organization and Environment*. Homewood: Irwin.

Lindbloom, C. 1977. *Politics and Markets*. New York: Basic Books.

Lorange, P. 1979 *Formal Planning Systems: Their Role in Strategy Formulation and Implementation*. In Schendel and Hofer (1979), pp. 226–241, below.

———. 1980. *Corporate Planning: An Executive Viewpoint*. Englewood Cliffs: Prentice-Hall.

———. 1980a. Evolution des méthodes de planification. *Revue'economique et sociale* 38 (3/4): 134–166.

———. 1983. Strategie Control. In *Latest Advances in Strategie Management*, ed. R. Lamb. Englewood Cliffs: Prentice-Hall.

Lorange, P., and R. Vancil. 1976. *Strategic Planning Systems*. Englewood Cliffs: Prentice-Hall.

Mason, R., and I. Mitroff. 1982. *Challenging Strategie Planning Assumptions*. New York: Wiley.

Mintzberg, H. 1979. *The Structuring of Organizations*. Englewood Cliffs: Prentice-Hall.

Mitroff, I., and J. Emshoff. 1979. On Strategic Assumption-Making. *Academy of Management Review 4* (1): 1–12.

Montgomery, C. 1982. The Measurement of Firm Diversification: Some New Empirical Evidence. *Academy of Management Journal* 25 (2): 299–307.

Nozick, R. 1969. Newcomb's Problem and Two Principles of Choice. In *Essays in Honor of C. G. Hempel*, ed. N. Rescher et al., 114–146. Dordrecht: Reidei.

Peters, T. 1981. *The Excellent Company Study*. New York: McKinsey.

Peters, T., and R. Waterman. 1982. *Searching for Excellence*. New York: Harper & Row.

Pfeffer, J., and G. Salancik. 1978. *The External Control of Organizations*. New York: Harper & Row.

Porter, M. 1980. *Competitive Strategy*. New York: Macmillan.

Quine, W. 1960. *Word and Object*. Cambridge, MA: Harvard University Press.

Quine, W., and J. Ullian. 1978. *The Web of Belief*. New York: Random House.

Quinn, J. 1981. *Strategies for Change*. Homewood: Dow Jones–Irwin.

Rothschild, W. 1976. *Putting It All Together*. New York: AMACOM.

Rowe, A., R. Mason, and K. Dickel. 1982. *Strategic Management and Business Policy*. Reading: Addison-Wesley.

Rudner, R. 1966. *Philosophy of Social Science*. Englewood Cliffs: Prentice-Hall.

Rumelt, R. 1979. *Evaluation of Strategy: Theory and Models*. In Schendel and Hofer (1979), pp. 196–212, below.

Saunders, C., and J. Thompson. 1980. A Survey of the Current State of Business Policy Research. *Strategic Management Journal 1* (2): 119–130.

Schendel, D., and C. Hofer, eds. 1979. *Strategic Management: A New View of Business Policy and Planning*. Boston: Little Brown.

——— 1979a. *Introduction*. In Schendel and Hofer (1979), pp. 1–22, above.
Vancil, R. 1979. *Decentralization*. Homewood: Dow Jones–Irwin.
Weidenbaum, M. 1980. Public Policy No Longer A Spectator Sport For Business. *The Journal of Business Strategy 1* (1): 46–53.
Wind, Y., and V. Mahajan. 1981. Designing Product and Business Portfolios. *Harvard Business Review 59* (1): 155–165.

R. Edward Freeman is University Professor, Elis and Signe Olsson Professor, Academic Director of the Institute for Business in Society, and Senior Fellow of the Olsson Center for Applied Ethics at the University of Virginia's Darden School of Business.

Peter Lorange is Emeritus Professor of Strategy and Honorary President of IMD.

Chapter 5
The Politics of Stakeholder Theory: Some Future Directions

R. Edward Freeman

1 Introduction

Several recent papers have raised important conceptual questions about the idea of "stakeholder management" or "stakeholder theory." Kenneth Goodpaster has sounded a significant challenge by diagnosing a "stakeholder paradox" at the heart of stakeholder theory.[1] James Kuhn and Donald Shriver have attacked the very idea

This paper has had a long gestation period. I want to acknowledge the comments and conversations of a number of people. The main ideas in this paper have been refined over a number of years via seminars at Georgetown University, the Society for Business Ethics, Dartmouth College, Loyola University, the Wharton School and others. I am especially indebted to Max Clarkson, Michael Deck and the Canada Council for Social Sciences for sponsoring a workshop on Stakeholder Theory in May of 1993, and to Peter Pruzan, Ole Thuyson, and Werner Peterson and the Danish Council for Social Science for sponsoring a conference on ethics and stakeholders in Copenhagen in the Fall of 1993. Participants at both of these conferences produced many more refinements and suggestions than I can individually attribute. Also, I have benefited from conversations with Kendall D'Andrade, John Boatright, Steve Brenner, Archie Carroll, Phil Cochran, Robbin Derry, Tom Donaldson, Ken Goodpaster, Ron Green, and Terry Halbert, as well as my colleagues at The Darden School, Rosalyn Berne, Andrea Larson, Jeanne Liedtka, Tara Radin, and Patricia Werhane. It is now difficult to separate out how much of these ideas I owe to Daniel R. Gilbert, Jr.

Originally published in: ***Business Ethics Quarterly***, 4(4), 409–421 © Cambridge University Press, 1994
Reprint by Springer, https://doi.org/10.2307/3857340

[1] Kenneth Goodpaster, "Business Ethics and Stakeholder Analysis," *Business Ethics Quarterly*, Vol. 1, No. 1, 1991, pp. 53–73.

R. E. Freeman (✉)
University of Virginia, Charlottesville, VA, USA
e-mail: FreemanE@darden.virginia.edu

© Springer Nature Switzerland AG 2023

S. D. Dmytriyev, R. E. Freeman (eds.), *R. Edward Freeman's Selected Works on Stakeholder Theory and Business Ethics*, Issues in Business Ethics 53,
https://doi.org/10.1007/978-3-031-04564-6_5

that stakeholder relationships are to be managed at all, proposing instead a "constituency view" that sees the corporation and its stakeholders as a voluntary community.[2] Martin Meznar, James Chrisman and Archie Carroll have straddled this controversy by explicitly connecting stakeholder management to business strategy and adopting a utilitarian ethic for its defense.[3] And, John Boatright has argued that while the special nature of stockholder claims can't be justified—there is no argument for the special nature of stakeholder claims.[4]

The purpose of this paper is to enter this conversation with the goal of clarifying certain foundational issues. I want to show in Part II, along with Boatright, that there is no stakeholder paradox, and that the principle on which such a paradox is built, the Separation Thesis, is nicely self-serving to business and ethics academics. In Part III I suggest that if we give up such a thesis we find there is no stakeholder theory but that stakeholder theory becomes a genre that is quite rich. It becomes one of many ways to blend together the central concepts of business with those of ethics. Ratvher than take each concept of business singly or the whole of "business" together and hold it to the light of ethical standards, we can use the stakeholder concept to create more fine-grained analyses that combine business and ethics; or more simply, we can tell many more, and more interesting, stories about business. In Part IV I shall sketch one such story based on some recent papers on stakeholder theory. And, finally in Part V I want to suggest that the way that the debate on the stakeholder concept has proceeded is illustrative of how we can recast the role of business ethics along more pragmatist lines.

2 Goodpaster's Argument

In "Business Ethics and Stakeholder Analysis" Kenneth Goodpaster suggests that "stakeholder analysis" (as he calls it) has two competing interpretations. The Strategic Interpretation says that managing stakeholders is a means, perhaps a mere means, towards the achievement of both stockholder or managerial ends. So, managing stakeholder relationships makes good business sense, in that it allows the firm and its managers to achieve its objectives understood, according to Goodpaster, in narrow economic (profit-maximizing) terms. On the other hand, the Multi-Fiduciary Interpretation says that managers and directors have fiduciary obligations to stakeholders, one of which is stockholders, and that managing stakeholder relationships is non-optional, it is morally required.

---

[2] James Kuhn and Donald Shriver, *Beyond Success*, New York: Oxford University Press, 1992.

[3] Martin B. Meznar, James J. Chrisman, and Archie B. Carroll, "Social Responsibility and Strategic Management," *Business & Professional Ethics Journal*, Vol. 10, No. 1, Spring 1991, pp. 47–66.

[4] John Boatright, "What's So Special About Stakeholders?" *Business Ethics Quarterly*, Vol. 4, No. 4, October 1994 (this issue).

From these two competing interpretations Goodpaster deduces the Stakeholder Paradox:[5]

> It seems essential, yet in some ways illegitimate, to orient corporate decisions by ethical values that go beyond strategic stakeholder considerations to multi-fiduciary ones."

While it may be difficult to see this sentence as paradoxical, Goodpaster does so presumably because:[6]

> It can be argued that multi-fiduciary stakeholder analysis is simply incompatible with widely-held moral convictions about the special fiduciary obligations owed by management to stockholders. At the center of the objection is the belief that the obligations of agents to principals are stronger or different in kind from those of agents to third parties.

Plainly, unless this "argument" has some merit, there is no paradox, for the Multi Fiduciary Interpretation need only hold that managers bear fiduciary, or "special fiduciary" obligations to stakeholders. What John Boatright has rightly shown is that such "widely held" moral convictions cannot be sustained. So this interpretation just denies what the Strategic Interpretation claims. There is no paradox, just difference, to be resolved by way of justification. That Goodpaster thinks that the above argument has some merit, at least as far as discrediting the multi-fiduciary view, is clear from his analysis. He cleverly delivers the punch line as "strategic stakeholder synthesis (business without ethics) or the effective loss of the private sector with a multi-fiduciary stakeholder synthesis (ethics without business)."[7]

Put in these terms there is clearly a paradox, one that Goodpaster resolves by creating "nonfiduciary obligations to third parties surrounding any fiduciary relationship."[8] These obligations are moral ones, but not fiduciary. These are made real by a kind of projection principle, called the "Nemo Dat Principle" that says that stockholders can't expect managers to disobey reasonable community standards of ethics. Goodpaster concludes that:[9]

> the foundation of ethics in management—and the way out of the stakeholder paradox—lies in understanding that the conscience of the corporation is a logical and moral extension of the consciences of its principals. It is not an expansion of the list of principals, but a gloss on the principal-agent relationship itself. Whatever the structure of the principal-agent relationship, neither principal nor agent can ever claim that an agent has 'moral immunity' from the basic obligations that would apply to any human being toward other members of the community.

While this last sentence is surely correct, what Goodpaster fails to see is that moral immunity is in fact what gets claimed and justified in an invisible utilitarian sleight of hand.[10]

---

[5] Goodpaster, *op. cit.*, at 63.

[6] *Ibid.*

[7] *Ibid.* at 67.

[8] *Ibid.*

[9] *Ibid*, at 68.

[10] We need look no further than any introductory finance book to find such claims of moral immunity.

None of this will do—and I want to recommend that the stakeholder concept can provide a more than adequate dissolution of the so-called paradox. At least part of the difficulty is in the way that Goodpaster has set up the dichotomy between the Strategic Interpretation and the Multi-Fiduciary Interpretation. It is surely true that some, myself included, have written as if managing stakeholder relationships is just good business. It is common sense to spend time and attention worrying about those groups and individuals who can affect you or whom you can affect. The problem lies in the question "What is a stakeholder?"

In much earlier work, there was very little philosophical sophistication. Stakeholders were just generic groups, unindividuated without clear membership conditions. A number of people began to suggest that stakeholders had to be seen in a nominalist fashion. They were individuals who stood in a certain relationship, via membership in some group, or via some role-related activity, to the corporation. Once they are seen as individuals, human beings, they must be seen as moral beings, without moral immunity, initially on the same rather level playing field. The question now becomes what principles should govern their interaction.[11]

One such principle, which I will call "The Principle of Who and What Really Count,"[12] says that the primary function of the corporation is to enhance the economic well-being, or serve as a vehicle for the free choices of, the owners of the corporation. And, owners are defined as those who hold legal title to shares of stock in the corporation. This principle is embodied in the law of corporations which has historically directed managers and directors to "manage the affairs of the corporation in the interests of stockholders, using sound business judgement." So, the only possible interpretation of "stakeholders" given the Principle of Who and What Really Count, is Goodpaster's Strategic Interpretation. Given this principle we have to find some way to build ethics into the discussion or we have Goodpaster's "business without ethics."

But, the law of corporations is not the only law, and as I have argued elsewhere[13], managers and directors have ignored the rights of stakeholders on peril of legislation in a free political system. So, we have protections in the form of the Wagner Act, the Consumer Product Safety Commission, the Uniform Commercial Code, the Foreign Corrupt Practices Act, the Securities Exchange Commission, the Environmental Protection Agency, and many more. These laws and agencies have arisen chiefly because the Principle of Who and What Really Count has not been called into

---

[11] In *Strategic Management: A Stakeholder Approach* I made a rather limited attempt to articulate these principles. Daniel R. Gilbert, Jr. showed me that once stakeholders were treated as moral agents, nothing less than a thoroughgoing redescription of corporate life would suffice. He and I have been engaged in various aspects of this project for the last 10 years.

[12] I owe the formulation of these principles to a long-standing conversation of many years with Professor Jesse Taylor of Appalachian State University.

[13] Cf., R. Edward Freeman and William M. Evan, "Corporate Governance: A Stakeholder Interpretation," *The Journal of Behavioral Economics,* Vol. 19, No. 4, 1990, pp. 337–59; and, William M. Evan and R. Edward Freeman, "A Stakeholder Theory of the Modern Corporation: Kantian Capitalism," in T. Beauchamp and N. Bowie, (eds.) *Ethical Theory and Business*, 4th Edition, Englewood Cliffs: Prentice Hall, 1993, pp. 75–93.

question. The mythology of laissez-faire capitalism is unfortunately easily propped up by arguments that claim, of course managers have the same moral obligations that you and I have, even if they often disregard them in favor of invisible hand explanations. Economists, business theorists and evidently some business ethicists think that it is enough to see morality as a side-constraint on the maximization of stockholder wealth, justifiable only if a greater good or other moral end is served.

Goodpaster's version of this error is to claim that morality governs any principal-agent relationship, and while that is surely true, he does not go on to ask the obvious question: What really is the nature of the principal-agent relationship in business? Now John Boatright has argued that Goodpaster is wrong in using principal-agent language to unpack the moral nature of business. But, the use of agency theory is not the main issue. One can be careful in articulating the obligations and duties to both principals and agents to obtain a model of the firm that is a multiple principal/multiple stakeholder model. Or, one can subject the agency model to a careful reinterpretation that makes ethical issues quite explicit.[14]

In fact, the very idea of the Multi-Fiduciary Interpretation is to call the normal way of thinking about business into question. It suggests that we reject a principle which is implicit in both Goodpaster's and Boatright's analysis, and is indeed prevalent in most of the business literature, including the business ethics literature. Let me call this principle, "The Separation Thesis" and describe it as follows:

The Separation Thesis

> The discourse of business and the discourse of ethics can be separated so that sentences like, "x is a business decision" have no moral content, and "x is a moral decision" have no business content.

Obviously, I am using "moral" and "business" as descriptive terms picking out decisions for analysis. But, it is ingrained in all that we do in business schools to separate the discourse of business from the discourse of ethics. The Separation Thesis pervades business ethics. As long as we can separate the discourse of business and the discourse of ethics there will be room for people to connect them—to hold particular business concepts up to the light of ethical discourse and to do criticism on a large scale either condemning or glorifying business as a practice. We in essence subscribe to a cousin of the Principle of Who and What Really Count, which we may call "The Principle of Make It Up As You Go Along." As long as the discourse distinguishes "business" and "ethics" we will need business ethicists to make it up as we go along—holding business, piece by piece, to the light of reason. And, as long as business ethics is separate, business theorists are free to make up supposedly morally neutral theories such as agency theory which can be used to justify a great deal of harm.

---

[14] See for instance the essays in N. Bowie and E. Freeman (eds.) *Ethics and Agency Theory*, New York: Oxford University Press, 1993.

Now the whole point of the stakeholder approach is to deny the Separation Thesis, reject the Principle of Who and What Really Counts, and to set aside the Principle of Make It Up As You Go Along. This approach borrows from Quine and other pragmatists, "Sentences (particularly sentences about business) do not confront the tribunal of experience alone." There is always a context to business theory, and that context is moral in nature. It is only by recognizing the moral presuppositions of business theory, refining them, testing them by living differently, and revising them that we can invent and reinvent better ways to live. All of this is just Dewey updated for business, but the Separation Thesis runs the gamut from managerial decision making to Nobel decorated economists. And, we should make it optional.

So, the argument among us goes like this: Goodpaster wants the Separation Thesis to save business from ethics, and a version of moral theory to save ethics from business.[15] Boatright suggests that there is no argument for treating stockholders (or stakeholders) in a special way. And I want to suggest that if you give up the Separation Principle and its subsidiary principles, then the argument between Goodpaster and Boatright is beside the point.

3 Normative Cores

I want to suggest how things would look if we dropped the idea that we can meaningfully talk about business and ethics by keeping the concepts, ideas and theories of each autonomous. In other words I want to suggest, it is not meaningful to talk about either stockholders or stakeholders without engaging in discourse that is at once normative, descriptive, instrumental and metaphorical.

In an important recent paper, Tom Donaldson and Lee Preston[16] have argued, like Boatright, that the normative ground of any adequate theory of business is inconsistent with a stockholder theory. They turn to an analysis of the various uses of "stakeholders" and conclude that the stakeholder theory is consistent with a modern view of property. They rightly suggest that the "stockholder theory" is an idea whose time has come and gone, and that we should get on with the task of connecting the stakeholder theory with traditional normative conceptual schemes. Given Boatright's analysis, and Donaldson and Preston's summary of the arguments against the stockholder theory I believe we can safely say that the stockholder theory is or at least should be intellectually dead. If so, I want to suggest something that seems paradoxical: There is no such thing as the stakeholder theory.

---

[15] My colleague Professor Patricia Werhane has rightly point out, that on his own positive account of the moral nature of the corporation, Goodpaster does not need to accept the Separation Thesis. That he does, is evidenced by his formulation of the so-called "paradox."

[16] Thomas J. Donaldson and Lee Preston, "The Stakeholder Theory of the Corporation: Concepts, Evidence, Implications," College Park: CIBER Occasional Paper #37, January 1994, forthcoming in *Academy of Management Review.*

The temptation has been for a long time to depict the stakeholder concept as a kind of rallying cry against the stockholder theory. Armed with stakeholder maps on our shields and banners, we have marched forth to browbeat the infidels, mostly economists and finance theorists, and others who want to be like them, such as marketing theorists and accountants, and show them that the stakeholder theory is "better" than the stockholder theory. This certainly is the tone if not the intent of Brenner and Hosseini, as well as Meznar, Chrisman, and Carroll, as well as the intent of some of my earlier writing.[17] But, if the Separation Thesis cannot be maintained, then the issue is what kind of moral content a theory has, not whether it has any moral content or not.

"The stakeholder theory" can be unpacked into a number of stakeholder theories, each of which has a "normative core," inextricably linked to the way that corporations should be governed and the way that managers should act. So, attempts to more fully define, or more carefully define a stakeholder theory are misguided. Following Donaldson and Preston, I want to insist that the normative, descriptive, instrumental, and metaphorical (my addition to their framework) uses of 'stakeholder' are tied together in particular political constructions to yield a number of possible "stakeholder theories." "Stakeholder theory" is thus a genre of stories about how we could live. Let me be more specific.

A "normative core" of a theory is a set of sentences that includes among others, sentences like:

(1) Corporations ought to be governed…
(2) Managers ought to act to…

where we need arguments or further narratives which include business and moral terms to fill in the blanks. This normative core is not always reducible to a fundamental ground like the theory of property, but certain normative cores are consistent with modern understandings of property. Certain elaborations of the theory of private property plus the other institutions of political liberalism give rise to particular normative cores. But there are other institutions, other political conceptions of how society ought to be structured, so that there are different possible normative cores. Such a "reasonable pluralism" is what I had in mind with the idea of "enterprise strategy," but even that concept was filled with the language of the Separation Thesis, and too much in the instrumental/descriptive mode.[18]

So, one normative core of a stakeholder theory might be a feminist standpoint one, rethinking how we would restructure "value-creating activity" along principles

[17] See Steven N. Brenner and Jamshid C. Hosseini, "The Stakeholder Theory of the Firm: A Methodology to Generate Value Matrix Weights," *Business Ethics Quarterly*, Vol. 2, No. 1, April 1992, pp. 99–120.

[18] Creating a "reasonable pluralism" is a straightforward Rawlsian project whose whole point is to ensure that there exist multiple worldviews. Rawls' argument is that only a liberal society, such as "justice as fairness" describes, makes such a reasonable pluralism possible.

	A. Corporations ought to be governed…	B. Managers ought to act…	*C.* The background disciplines of "value creation" are…
Doctrine of Fair Contracts	…in accordance with the six principles.	…in the interests of stakeholders.	—business theories —theories that explain stakeholder behavior
Feminist Standpoint Theory	…in accordance with the principles of caring/connection and relationships.	…to maintain and care for relationships and networks of stakeholders.	—business theories —feminist theory —social science understanding of networks
Ecological Principles	…in accordance with the principle of caring for the earth.	…to care for the earth.	—business theories —ecology —other
⋮	⋮	⋮	⋮

Exhibit 5.1 A reasonable pluralism

of caring and connection.[19] Another would be an ecological (or several ecological) normative cores. Mark Starik has argued that the very idea of a stakeholder theory of the *firm* ignores certain ecological necessities.[20] Exhibit 5.1 is suggestive of how these theories could be developed.

In the next section I shall sketch the normative core based on pragmatic liberalism. But, any normative core must address the questions in columns A or B, or explain why these questions may be irrelevant, as in the ecological view. In addition each "theory," and I use the word hesitantly, must place the normative core within a more full-fledged account of how we could understand value-creating activity differently (column C). The only way to get on with this task is to see the stakeholder idea as a metaphor, in addition to the normative, descriptive, and instrumental uses as per Donaldson and Preston. The attempt to prescribe one and only one "normative core" and construct "a stakeholder theory" is at best a disguised attempt to

---

[19] See for instance A. Wicks, D. Gilbert and E. Freeman, "A Feminist Reinterpretation of the Stakeholder Concept," *Business Ethics Quarterly*, Vol. 4, No. 4, October 1994 (this issue); and E. Freeman and J. Liedtka, "Corporate Social Responsibility: A Critical Approach," *Business Horizons*, Vol. 34, No. 4, July–August 1991, pp. 92–98.

[20] At the Toronto workshop Mark Starik sketched how a theory would look if we took the environment to be a stakeholder. This fruitful line of work is one example of my main point about pluralism.

smuggle a normative core past the unsophisticated noses of other unsuspecting academics who are just happy to see the end of the stockholder orthodoxy.

4 The Doctrine of Fair Contracts

In several recent papers William Evan and I have struggled with articulating a more explicitly moral basis for the stakeholder idea.[21] In one paper we suggested a kind of neo-Kantian principle for treating stakeholders as ends rather than mere means. And, in the second we attempted to show how such a Kantian conception is consistent with a reinterpretation of Oliver Williamson's transaction-cost economics. Each attempt was problematic precisely because it juxtaposed a single stakeholder theory with the orthodox stockholder theory. The intuitions behind these attempts were sound, namely, that a new narrative for how we understand the human process of joint value creation was necessary. The Kantian framework, however, ought to impose an order on this process which did not admit the pluralism in which we are so obviously enmeshed.

The central insight of these ideas is that the usual story of value-creation is suspect. Stories which depict the firm as either (1) the private property of owners; (2) the necessary arrangements if we are to maximize the greatest good for the greatest number; or, (3) the result of a voluntary contracting process, all appeal to the Separation Thesis to rule out certain effects of the firm on other stakeholders.[22] We need new narratives which recognize these effects from the beginning and which do not appeal to the Separation Thesis.

If we begin with the view that we can understand value-creation activity as a contractual process among those parties affected, and if for simplicity's sake we initially designate those parties as financiers,[23] customers, suppliers, employees, and communities, then we can construct a normative core that reflects the liberal notions of autonomy, solidarity, and fairness as articulated by John Rawls, Richard Rorty and others.[24] Notice that building these moral notions into the foundations of how we understand value creation and contracting requires that we eschew separating

---

[21] Infra note 14.

[22] To show this rather than just claim it, one would have to examine how each of these three normative grounds of the stockholder theory themselves are used to justify the Separation Thesis. Such an analytic task is important, but beyond the scope of my present purposes. I claim here that each of these three reasons are used to justify treating stockholder claims as different from stakeholder claims, and used to separate the affairs of business from the affairs of ethics. I examine these issues in *Managing for Stakeholders*, forthcoming.

[23] In private correspondence, David Schrader has recently argued correctly that "owners" is a misnomer, and that its use subtly cuts against the stakeholder idea. I have adopted "financiers" to call to mind Theodore Dreiser's novel as well as recent events on Wall Street.

[24] J. Rawls, *Political Liberalism*, New York: Columbia University Press, 1993; and R. Rorty, "The Priority of Democracy to Philosophy" in *Reading Rorty: Critical Responses to Philosophy and the Mirror of Nature (and Beyond)*, ed. Alan R. Malachowski, Cambridge, MA: Blackwell, 1990.

the "business" part of the process from the "ethical" part, and that we start with the presumption of equality among the contractors, rather than the presumption in favor of financier rights.

The normative core for this redesigned contractual theory will capture the liberal idea of fairness if it ensures a basic equality among stakeholders in terms of their moral rights as these are realized in the firm, and if it recognizes that inequalities among stakeholders are justified if they raise the level of the least well-off stakeholder. The liberal ideal of autonomy is captured by the realization that each stakeholder must be free to enter agreements that create value for themselves, and solidarity is realized by the recognition of the mutuality of stakeholder interests.

One way to understand fairness in this context is to claim *a la* Rawls that a contract is fair if parties to the contract would agree to it in ignorance of their actual stakes. Thus, a contract is like a fair bet, if each party is willing to turn the tables and accept the other side. What would a fair contract among corporate stakeholders look like? If we can articulate this ideal, a sort of corporate constitution, we could then ask whether actual corporations measure up to this standard, and we also begin to design corporate structures which are consistent with this Doctrine of Fair Contracts.[25]

Imagine if you will, representative stakeholders trying to decide on "the rules of the game." Each is rational in a straightforward sense, looking out for its own self-interest. At least *ex ante*, stakeholders are the relevant parties since they will be materially affected. Stakeholders know how economic activity is organized and could be organized. They know general facts about the way the corporate world works. They know that in the real world there are or could be transaction costs, externalities and positive costs of contracting. Suppose they are uncertain about what other social institutions exist, but they know the range of those institutions. They do not know if government exists to pick up the tab for any externalities, or if they will exist in the nightwatchman state of libertarian theory. They know success and failure stories of businesses around the world. In short they are behind a Rawls-like veil of ignorance, and they do not know what stake each will have when the veil is lifted. What groundrules would they choose to guide them?[26]

The first groundrule is "The Principle of Entry and Exit." Any contract that is the corporation must have clearly defined entry, exit, and renegotiation conditions, or at least it must have methods or processes for so defining these conditions. The logic is straightforward: each stakeholder must be able to determine when an agreement exists and has a chance of fulfillment. This is not to imply that contracts cannot

[25] Notice that one could equally give a conservative notion of fairness, that a contract is fair if it is in fact agreed upon. On such an argument, the real world would assume a large element of fairness. While there would be a disagreement over which normative core is better, the argument around the legitimacy of the corporation would be moot. The question would be what constitutes fairness, not what constitutes the corporation. The stakeholder genre in fact makes moot this large question of corporate legitimacy. It is a genre that is inherently liberal, rather than radical.

[26] I am especially grateful to Professor John Hasnas for critical comments on these principles.

contain contingent claims or other methods for resolving uncertainty, but rather that it must contain methods for determining whether or not it is valid.

The second groundrule I shall call "The Principle of Governance" and it says that the procedure for changing the rules of the game must be agreed upon by unanimous consent. Think about the consequences of a majority of stakeholders systematically "selling out" a minority. Each stakeholder, in ignorance of its actual role, would seek to avoid such a situation. In reality this principle translates into each stakeholder never giving up its right to participate in the governance of the corporation, or perhaps into the existence of stakeholder governing boards.[27]

The third groundrule I shall call "The Principle of Externalities," and it says that if a contract between A and B imposes a cost on C, then C has the option to become a party to the contract, and the terms are renegotiated. Once again the rationality of this condition is clear. Each stakeholder will want insurance that it does not become C.[28]

The fourth groundrule is "The Principle of Contracting Costs," and it says that all parties to the contract must share in the cost of contracting. Once again the logic is straightforward. Any one stakeholder can get stuck.

A fifth groundrule is "The Agency Principle" that says that any agent must serve the interests of all stakeholders. It must adjudicate conflicts within the bounds of the other principals. Once again the logic is clear. Agents for any one group would have a privileged place.

A sixth and final groundrule we might call, "The Principle of Limited Immortality." The corporation shall be managed as if it can continue to serve the interests of stakeholders through time. Stakeholders are uncertain about the future but, subject to exit conditions, they realize that the continued existence of the corporation is in their interest. Therefore, it would be rational to hire managers who are fiduciaries to their interest and the interest of the collective. If it turns out the "collective interest" is the empty set, then this principle simply collapses into the Agency Principle.[29]

Thus, the Doctrine of Fair Contracts consists of these six groundrules or principles:

[27] This principle has been much criticized in a number of seminars, but I remain convinced that it is necessary. It ensures the liberal ideal of citizenship for all. Each stakeholder is sovereign in the matter of their own interests and they are jointly sovereign in matters of their mutual interests. Think of this principle as ensuring that no stakeholder can sell another into virtual slavery. It is also possible to invent actual mechanisms that rely on unanimous consent of representatives rather than individual stakeholders.

[28] Again, this principle has been criticized as failing to understand that opportunity costs are pervasive. My reply is that we should build our understanding of economics on a reasonable moral order, rather than *vice-versa*; or, so much the worse for the idea of "opportunity cost."

[29] The presumption of a reasonable future is an important yet overlooked area of democratic theory. The idea of "race relations" depends on a shared understanding of "reasonable future" or more poignantly a "philosophy of hope."

(1) The Principle of Entry and Exit
(2) The Principle of Governance
(3) The Principle of Externalities
(4) The Principle of Contracting Costs
(5) The Agency Principle
(6) The Principle of Limited Immortality

Think of these groundrules as a doctrine which would guide actual stakeholders in devising a corporate constitution or charter. Think of management as having the duty to act in accordance with some specific constitution or charter.

Obviously, if the Doctrine of Fair Contracts and its accompanying background narratives are to effect real change, there must be requisite changes in the enabling laws of the land. I propose the following three principles to serve as constitutive elements of attempts to reform the law of corporations.[30]

The Stakeholder Enabling Principle

Corporations shall be managed in the interests of its stakeholders, defined as employees, financiers, customers, employees, and communities.

The Principle of Director Responsibility

Directors of the corporation shall have a duty of care to use reasonable judgment to define and direct the affairs of the corporation in accordance with the Stakeholder Enabling Principle.

The Principle of Stakeholder Recourse

Stakeholders may bring an action against the directors for failure to perform the required duty of care.

Obviously, there is more work to be done to spell out these principles in terms of model legislation. As they stand, they try to capture the intuitions that drive the liberal ideals. It is equally plain that corporate constitutions which meet a test like the doctrine of fair contracts are meant to enable directors and executives to manage the corporation in conjunction with these same liberal ideals.

I have only been able to sketch how one normative core could be used to construct a new narrative about how we create value. In the following section I want to suggest that how this construction has proceeded, from the early work on

[30] For a recent and thorough analysis of the current state of so-called "constituency statutes" see Eric W. Orts, "Beyond Shareholders: Interpreting Corporate Constituency Statutes," *George Washington Law Review*, Vol. 61, November 1992, p. 14.

stakeholders in the 1960s to the rich ground we currently till, offers some rather different ideas about how business ethics ought to proceed.

5 Conclusion

Much of business ethics has proceeded under the influence of the Separation Thesis. Philosophers have been hesitant to get their hands dirty by understanding the day to day life of value-creation activity. Business theorists have been loathe to enter what they see as the intellectual morass of moral theory. Happily this scenario is changing as there is more dialogue among philosophers, business academics and managers.[31] Yet the shadow of the Separation Thesis looms large. Unless we can reinvent business from the ground up, we remain open to the charge of reserving a special place for business ethicists ready to "make it up as we go along." Thus, our task is to take metaphors like the stakeholder concept and embed it in a story about how human beings create and exchange value.

To see the role of the business ethicists as reinventing the corporation and describing and redescribing the complex human beings who work in the corporation is to become a pragmatist. For the pragmatist, the question is less, "what is true," than, "how should we live," or better still, "how does this narrative allow us to live," or "what does this way of talking allow us to do." So, for instance, on pragmatist's grounds the stakeholder idea is part of a narrative about how we do and could live, how we could experiment with different institutional arrangements, and how we do and could organize a sphere of our lives built mostly around something we have come to call "work."

Seeing the stakeholder idea as replacing some shopworn metaphors of business with new ones—such as stakeholders for stockholders, humans as moral beings for humans as economic beings, and the Doctrine of Fair Contracts for the current panoply of corporate chartering laws—is to give up the role of finding some moral bedrock for business. Finding such bedrock, required by the Separation Thesis, is especially fruitless on pragmatist grounds for there are no foundations for either business or ethics. All we have is our own history, culture, institutions, and our imaginations. For the pragmatist it is "just us" rather than "justice" or "justification" in any sense of foundational bedrock. The cash value of our metaphors and narratives just is how they enable us to live, and the proof is in the living.

There is a great deal of important intellectual work to be done in business ethics. For too long philosophers, ethicists, liberal theorists and others have looked down their noses at business as not worthy of serious intellectual attention. Liberals carp about business, conservatives carp about liberals, and radicals bemoan the good old

[31] So, Andrew Stark's recent lament "What's the Matter with Business Ethics," *Harvard Business Review*, Vol. 71, No. 3, May–June 1993, pp. 38–48, misrepresents and misunderstands the current landscape of both business and ethics, though his main points were true enough 10 years ago.

days when in appeared as if the work of redescribing the good society and the good life had been done by Marx and others.[32]

It is time to get on with it, and get on with it in the pragmatist vein. Redescribing corporations means redescribing ourselves and our communities. We cannot divorce the idea of a moral community or of a moral discourse from the ideas of the value-creation activity of business. To do so, entails the acceptance of a principle, the Separation Thesis, which has for too long been used to close off discussion and to silence conversation. Likewise, if business as an institution is immune to such redescriptions as have been redescribed in this paper, we have only the Separation Thesis to blame. Unless we get on with this work of redescription that crosses the bright lines between discourses that have been established by theory and practice, talk of human progress, as difficult as it is, is likely to come to a rather immodest and abrupt halt.

R. Edward Freeman is University Professor, Elis and Signe Olsson Professor, Academic Director of the Institute for Business in Society, and Senior Fellow of the Olsson Center for Applied Ethics at the University of Virginia's Darden School of Business.

[32] I explore the difficulties inherent in these dynamics in an unpublished paper, "The Business Sucks Story."

Chapter 6
A Feminist Reinterpretation of the Stakeholder Concept

Andrew C. Wicks, Daniel R. Gilbert, Jr., and R. Edward Freeman

1 Introduction

Language matters. Language profoundly shapes who we are, how we think and what we do. It is essential that we remain continually aware of the dominant metaphors we use to describe ourselves and our practices because they often shape us in ways that we don't initially recognize or desire. By examining old metaphors and considering them alongside new and enriching possibilities, we can lay the groundwork for constructive change. This type of change need not entail a wholesale rejection of current understandings and practices, but can function to revise or extend the values we share and the possibilities they generate in how we live. Such a way of thinking about language has been articulated best by postmodemists who suggest that attention to metaphors, grammar, and ways of speaking is not an obscure intellectual exercise, but a central task to developing alternative vocabularies that open up new possibilities and ways of living.[1]

[1] See Richard Rorty's writings on pragmatism for the best and most detailed version of this approach, particularly his discussion of the ironist in *Contingency, Irony and Solidarity* (New York: Cambridge University Press, 1989). Rorty maintains that language and metaphors are far more powerful in shaping our sense of "rationality" than any analytic conceptions of "reason" or logic. Indeed, he suggest that it is through new and innovative ways of speaking that we can alter the "logical space of reasons" to include the descriptions and metaphors of oppressed groups in a way

Originally published in: *Business Ethics Quarterly*, 4(4), 475–497 © Cambridge University Press, 1994
Reprint by Springer, DOI 10.2307/3857345

A. C. Wicks · D. R. Gilbert, Jr.·R. E. Freeman (✉)
University of Virginia, Charlottesville, VA, USA
e-mail: FreemanE@darden.virginia.edu

© Springer Nature Switzerland AG 2023

S. D. Dmytriyev, R. E. Freeman (eds.), *R. Edward Freeman's Selected Works on Stakeholder Theory and Business Ethics*, Issues in Business Ethics 53,
https://doi.org/10.1007/978-3-031-04564-6_6

The stakeholder concept is a relatively new metaphor for both describing how a business operates and defining what its most basic purposes are. The idea has developed and changed as one traces its evolution in the business and society literature. The emergence of the modem corporation created difficulties for theorists, as they struggled to articulate its nature and the sorts of moral responsibilities appropriate to it. Work on corporate social responsiveness and corporate social responsibility helped to expand the horizons of our moral imagination and articulate the relationship between corporations and ethics. In this context of change, as authors groped to describe how business and ethics fit together, several thinkers drew upon, but pressed beyond, more settled metaphors until the idea of the "stakeholder" achieved a level of development rich enough to encompass a whole new way of understanding the firm. It captured the attention of different thinkers as a powerful tool to push our view of corporations beyond what it had been perceived to be and articulated a new image of the firm which transcended the boundaries of previous conceptions. From the beginning, the stakeholder concept has been about not only opening questions of who the firm serves and how it operates, but also posing the more basic question of "what are we doing with business?" In so doing, stakeholder theorists have stripped away the facade of such constructs as agency relationships and efficiency calculations, to remind us that the most fundamental questions about economic activity are unavoidably and profoundly moral.

Yet, while stakeholder management has already become widely recognized and discussed in academic circles, and practiced within numerous corporations, it requires ongoing re-examination, critique and development. Freeman's work on the stakeholder concept underscores the extent to which this idea has evolved and continues to undergo fundamental change. Many credit him with the first systematic discussion of the stakeholder idea (Freeman 1984), however, the intellectual framework that is used to articulate the idea is dramatically different from that in his later book with Gilbert (Freeman and Gilbert 1989), and the views presented in a recent essay (Freeman and Gilbert 1992). In the first work (Freeman 1984), stakeholders were heralded as important players who were to be dealt with if a particular firm was to be successful. Rather than being tied to the identity of the corporation, or individuals who the firm is accountable to for creating value, they were seen more as means through which the firm achieved its own pre-ordained ends (e.g., maximizing profits). In Freeman's first book with Gilbert (1989), stakeholders assumed a more central and morally significant connection to the firm. Indeed, they were seen as ends—individuals with "personal projects" and interests the corporation (now a means) was constructed to serve. Finally, their most recent essay (1992) suggests some important shortcomings of this version, principally that it relies too much on a individualistic-autonomous-masculinist mode of thought to make it intelligible and discounts many of the feminist insights which we will

---

that make "sense"—he cites the work of feminist thinkers who were able to sensitize us to problems like marital rape, spousal abuse, abortion rights, and the problems of pornography through stretching the boundaries of language and offering potent new metaphors. For this latter argument, see his essay "Feminism and Pragmatism" in *Michigan Quarterly Review* 30 (Spring 1991): 231–58.

discuss more fully in this paper (such as the capacity to care and the importance of relationships). Thus, far from having achieved a fundamental "paradigm shift" when the stakeholder idea was first presented, its articulation and revision are a part of a process of change which is ongoing, dynamic and ad hoc. We suspect that future discourse will similarly transform what is valuable about stakeholder analysis through the use of a variety of forms of critical analysis, creative innovation, and the development of new vocabularies for thinking about the meaning of business.

The impetus for this paper is to explore many of the concepts and language used to discuss business activity and structure stakeholder relations. We think that many of the metaphors which have been used to understand and explain the stakeholder concept, and business activity in general, are couched in language which is distinctly "masculine" and that a re-interpretation of the stakeholder idea using "feminine" or "feminist" descriptions could create important new possibilities. There are two underlying reasons that make a re-interpretation of the stakeholder concept necessary.

The first reason is a moral concern. The stakeholder concept was not simply some new strategy or mechanism for business to increase profits. Rather, we see it as an attempt to articulate the meaning of the corporation and the sense of responsibility that businesses feel to those both inside and outside the 'walls' of the firm in a more useful and compelling manner. In this vein, these moral insights have been stunted by remaining couched in terms which limit the extent to which the moral commitments of businesses can be expressed. The stakeholder concept can be enriched by replacing some of the masculine metaphors with feminist alternatives and, in so doing, more fully express its latent possibilities.

The second, and subsidiary, motivation for this paper is that transforming these older "masculine" metaphors is critically relevant for businesses that want to improve their responsiveness and adaptability in a complex and rapidly changing global economy. As the sections below will make clear, there are a number of related trends emerging that have already shaped—and will continue to shape—the business world, which have a strong affinity with the feminist alternatives we will propose. The emergence of forms of collaboration, including the explosion of joint ventures, highlights a significant and growing trend in business. As Barbara Gray notes.

> Over the past fifteen years examples of collaboration have been emerging in virtually every sector of society—business, government, labor, and communities. The incidence of collaboration among businesses is becoming increasingly common. Especially in the telecommunications, computer, automobile, aerospace, robotics, and biotechnology industries, international joint ventures represent critical additions to firms' portfolios. Evidence of labor-management cooperation is increasing in the labor arena, and some scholars believe a major transformation of industrial relations is under way (Kochan et al. 1987; Katz 1985; Schuster 1985). Since the early 1970's, numerous public-sector and environmental disputes have been successfully resolved using collaborative processes, often with the assistance of a mediator (Carpenter and Kennedy 1988; Susskind and Cruikshank 1987).... The judicial system is also turning to collaboration to settle complex multiparty cases. (Gray 1989: 26–7)

Gray's (and to a lesser extent Waddock's) work on collaboration, and Peters' work on the importance of "thriving on chaos," along with the work of organizational theorists and others, will help us paint a picture of the business world where our proposals can help firms make practical changes for the better. Thus, our argument will operate simultaneously on two levels. First, we will maintain that a feminist re-interpretation of the stakeholder concept helps us better express the meaning and purposes of corporations. Second, we will also defend the view that this model yields important insights for corporations that want to improve their adaptability and responsiveness. While we want to suggest that stakeholder firms can achieve success defined in the traditional financial sense, our larger argument is about changing how we understand "success" and other core metaphors that shape our understanding of organizations.[2] Changing the language and metaphors we use doesn't preclude us talking about "business" or wealth, but it does alter the discourse and logic which drive how we think about these matters and the corporation. One result of embracing this alternative conception of corporations is to get past the idea of business ethics as an oxymoron and to better understand how corporations can be value-driven and still thrive in a modern free market economy.[3]

To understand the gender differences we have been alluding to, it is important to set out what we mean by "masculinist" and "feminist" metaphors and to demonstrate how they are manifest in practice. First of all, we use the term "feminist" in a highly qualified manner. We don't wish to downplay the significance and the diversity of feminist scholarship today—there are numerous authors who have made distinguished contributions and their work represents a broad spectrum of views. Necessarily, we will be drawing from only a few of these authors, primarily Carol Gilligan's work on the ethic of care, who represent a small but significant segment of the scholarship which attends to gender issues—thus, any time we use the term "feminist" in the paper, we refer only to these authors. In addition, the strict connection between sexual difference and the term "feminist" raises another important qualification in how we use the term "feminist" (or "masculinist"). While the terms help contrast a set of orientations to the world, there is no such sharp dividing line when we refer to these terms and their associated metaphors. Instead, both modes of thought are present in varying degrees in both males and females—all people share the sensibilities which give voice to these issues.[4] Thus, to speak of

[2] We will not provide evidence or sustained argument about whether stakeholder theory, and this particular formulation of it, is consistent with financial success. Others, including Peters in his collaborative projects on excellence, have made this type of argument. We can also look to particular companies which adopt forms of stakeholder management—either implicitly or explicitly—as an indication that this is a reasonable conclusion [e.g., The Body Shop, Ben & Jerry's, Coming, Johnson & Johnson, Merck, etc.] For the purpose of this paper, we want to leave open the possibility that stakeholder companies can be financially successful in the traditional sense.

[3] In the process of offering a new set of metaphors and an alternative discourse we will create a frame of reference within which the terms business and ethics can constructively interact and reinforce each other. Stakeholder theory gives us a means of reinterpreting traditional business discourse so that ethics can have a viable place. Our feminist reinterpretation takes this project a step further and advances the attempt to make ethics and business more interconnected and pleonasmic.

[4] Many have argued that the kinds of discourse we call "feminist" are more prevalent in most women (what Gilligan calls the "care perspective") while "masculine" discourse is more prominent in men (what Gilligan calls the "justice perspective").

the "care perspective" is not to speak only—or even primarily—to women, but to basic moral sentiments that we all share. At the same time it reminds us that while we shall criticize some of the language, ideas, and structures associated with the "justice perspective," we are not arguing against all masculinist concepts nor are we suggesting that the ones we criticized be abolished.[5] They too shape an important part of our moral perceptions. Rather, we strive to bring to the surface a mode of thinking which can help us revise outworn and overly rigid metaphors which have unquestioningly dominated the business world for too long and offer in their place new concepts, language, and metaphors which sustain a distinctive account of stakeholder management.

2 A Survey of Some Masculine Metaphors Behind the Stakeholder Concept

It would be an absurdly complex and reductive task to try and isolate the myriad of images and structures which operate in our everyday environment that are distinctly masculine. We seek to scratch the surface by exposing what we think are some of the most egregious examples of masculine metaphors which have come to shape our thoughts about business and the stakeholder concept. We will look at five specific cases: (1) the notion that corporations should be thought of primarily as an "autonomous" entity, bounded off from its external environment; (2) that corporations can and should enact or control their external environment; (3) that the language of competition and conflict best describes the character of managing a firm; (4) that the mode of thinking we employ in generating strategy should be "objective"; and (5) that corporations should structure power and authority within strict hierarchies.

2.1 Corporations are Autonomous Entities

The first assumption we want to explore is the idea that the corporation is an autonomous entity which is fundamentally separate from its environment. Part of what this approach exemplifies is the celebration of the individual and the respect

[5] The care orientation makes the human capacity to extend care the touchstone of moral activity. Justice places notions of rights and the capacity to logically reason with abstract moral concepts as the key to moral reflection. Care emphasizes the importance of human interaction and the ability to flourish in a network of relationships. Justice makes the individual primary and focuses on a system of rules and responsibilities that enable people to live together and respect each others' projects. For a more complete description of the terms care and justice, see Gilligan's work. While these terms provide an important backdrop for our argument, we will focus on the terms "masculinist" and "feminist" in the paper.

for personal freedom which the post-Enlightenment West, and particularly the U.S., has embraced. One theme particularly characteristic of the American experience has been the image of the pioneer, an image which describes both a heroic individual and a context in which that individual is able to act in and dominate a formidable environment. Such an outlook makes sense on the frontier—a setting where one's actions appear to have little direct influence on, or connection with, those of outside one's clan (or firm). One of the assumptions embedded in this world view is that the "self" is fundamentally isolatable from other selves and from its larger context. Persons exist as discrete beings who are captured independent of the relationships they have with others. While language, community, and relationships all affect the self, they are seen as external to and bounded off from the individual who is both autonomous from and ontologically prior to these elements of context. The parallel in business is that the corporation is best seen as an autonomous agent, separate from its suppliers, consumers, external environment, etc. Here too, while the larger market forces and business environment have a large impact on a given firm, it is nonetheless the individual corporation which has prominence in discussions about strategy and preeminence in where we locate agency.[6]

In terms of how we understand stakeholders, they are people who effect or are affected by the corporation, but they are not integral to its basic identity. This is reflected in the understanding of stakeholders offered by a number of authors including: Mason and Mitroff, "Stakeholders are all those claimants inside and outside the firm who have a vested interest in the problem and its solution" (Mason and Mitroff 1981: 43); Thompson, "A stakeholder in an organization is (by definition) any group or individual who can affect or is affected by the achievement of the organization's objectives" (Freeman 1984: 46); and O'Toole, stakeholders are "all the parties who have a stake" in the firm (O'Toole 1987: 42). These definitions all share the implicit premise that the basic identity of the firm is defined independent of, and separate from, its stakeholders.[7] The macro level view of the world of business is seen as a collection of atoms, each of which is colliding with other atoms in a mechanistic process representative of the interactions and transactions of various firms.

---

[6]See Peter French, "The Corporation as a Moral Person," *American Philosophical Quarterly* 3 (1979): 207–15.

[7]This view finds a classic articulation in Alfred P. Sloan, Jr. as related in his autobiography. *My Years With GM* (NY: Doubleday, 1963). While Sloan's views about the autonomous nature of the corporation are characteristic of his era, they would fit in well with much of the strategy literature in the 1990s.

2.2 *Companies Should Enact and Control Their Environment*

The second assumption is that firms confront an external environment which they should seek to enact and/or control. The premise here is that in a world of complexity and change, individuals need order and stability so that their basic needs can be met. Part of the psychology behind this desire for control is the sense of fear that is often generated when people confront uncertainty, dynamic action, and disorder. Things that disrupt order or promise change and complexity are seen as a threat to the self, and the reaction is to seek control; to conquer that which is other or disruptive; to dominate as a way to restore order. As the work of Gilligan suggests, this sort of response is more associated with the "justice" or masculine orientation. In the world of philosophy we can see this impulse expressed in Plato's search for Forms which composed the simplistic and harmonic essence of reality, hidden by the misleading complexities and illusions of our sense perceptions. Similarly other philosophers have carried on this tradition through the search for an "Archimedean point"—a point where the world can literally be stopped and carefully analyzed to grasp fundamental truths which are not undermined or deconstructed by dynamic action.

In the business world, strategists assume an outlook where the individual firm confronts an external world of market forces and competitors which put in peril its prospects for success. It is only by seeking and finding a measure of control over these forces that a given firm can flourish. Failure to exercise control most likely translates into lost opportunity, unfavorable market conditions, government restrictions, threats of decreased profits from hostile consumer and labor groups which cumulatively can threaten the very survival of the corporation. Rather than be caught failing to act, or simply reacting to the initiatives of others, the smart corporation will anticipate the activity within its environment and work to try to control and change these actions so that it can dictate the dynamics of what happens.[8] Mason and Mitroff embrace this approach throughout their work, but most noticeably in an example of the snaildarter (a three inch fish) and a planned dam project by the Tennessee Valley Authority. As they note,

> Snaildarters are stakeholders that have a low probability of being involved, but they *threaten* the success of the policy, plan or strategy if they are. The purpose of generating 'snaildarters' is to force the proponents of a perspective to think of deeper assumptions they will need to *overcome* if their plan is to be successful. They need to explain how they will meet their snaildarters and *conquer* them (Mason and Mitroff 1981: 99; emphasis added).

[8] See also Michael Porter's writings, especially Chaps. 4 & 5 of his book *Competitive Strategy* (Free Press, 1985), for use of military metaphors of strategy that are in this vein. A similar sense of hostility towards the external environment and the need to control it are characteristic of his views.

2.3 *The Metaphors of Conflict and Competition Best Describe How Firms Should Be Managed*

The third assumption we wish to explore is the idea that the language of conflict and competition best describe how firms ought to be managed. The insight behind this approach is that competition makes for stronger, healthier companies that are self-reliant and better able to introduce innovations which serve customer needs. Competition brings out the best in the people in the firm and helps to trim away waste—either one can respond to the challenge by "getting lean and fit" or one loses out and gives way to a healthier business which can. At the more specific level of stakeholder management within a given firm, this impulse assumes the divergence and conflict among various stake holder interests. As the snaildarter example in the last section demonstrates, there are numerous stakeholders in any given policy making context and many find themselves directly at odds with other stakeholders. Their interests conflict and the resulting process used to mediate the conflict is a competition to determine the weightier or more compelling interest. To further illustrate this type of thinking, should a corporation appease a group of environmentalists by closing down a plant which has consistently failed to meet their standards of pollution control or should it protect the jobs of the thousands of workers who depend on the continued operation of the plant? How are strategic planners to approach this problem?

James O'Toole tells us that good management involves weighing or balancing the merits of various competing interests, pursuing those which are most compelling or beneficial to the firm, and working to resolve conflict (O'Toole 1987: 42, 77). Managers discuss how the interests of stakeholders are affected, weigh their options, and then make a decision. Together with the first two assumptions we have explored, this impetus for competition and conflict makes sense and emerges as a logical extension of a picture of the world of individual actors all striving to control external forces.

2.4 *Strategy Formulation Should Be Objective*

The fourth assumption is that strategy formulation should be done in an objective manner. As firms strive to get beyond the messiness and chaos of the world of experience, as well as the biases of their own perceptions, good companies look to factual data and disinterested calculation to construct a strategy. The turn to science, and the suspicion of human perceptions, reflects a deeper assumption that empirical inquiry helps us to make contact with "hard" facts and what's really going on in the world (as opposed to what our senses and interpretations tell us is happening). Science helps strip away the artifices of language, culture and context to reveal *real* reality in its untutored and pure form. It is through the careful collection of facts and empirical investigation that managers can piece together what is going on in the world around it and formulate an appropriate strategy.

While the facts play an important role in plotting the strategic direction of a firm, objectivity also serves another purpose. Objectivity pushes decision makers to distance themselves from their leanings, biases, and perceptions—to try to be as rationally detached and analytical as possible (as in Mason and Mitroff's "dialectics" which seeks to "systematize doubt," p. 15). This preference for rationality, science, and detachment reflects a deep suspicion of emotions, experience and attachment which has characterized much of Western thought since Descartes. It expresses itself throughout the business strategy literature, as Mason and Mitroff's program exemplifies: they relentlessly challenge perceptions as a way of getting at more objectively derived and truer views of the world.

One of the implications of this approach is that stakeholders are silenced. That is, as we move toward a form of thinking which detaches us from the identities, emotions, needs, and perceptions of particular individuals, we simultaneously distance ourselves from stakeholders. The resort to quantification and abstraction glosses over the reality of stakeholders as particular persons and it implicitly erodes their legitimacy as agents who deserve a "voice" in the corporation. The quest for objectivity has the insidious effect of closing off a wealth of knowledge and experience which would appear to be essential—indeed, the most basic source for strategy formulation—if we are committed to the core ideas of stakeholder theory.

2.5 Power and Authority Should Be Embedded in Strict Hierarchies

The final assumption we wish to isolate is the view that corporations should structure power and authority within strict hierarchies. Perhaps the most clear example of organizations which invested power and authority in a system of strict hierarchies is the military, though it is probably no less true of the medical profession, the Catholic Church and numerous other organizations which embrace the ethos of hierarchy. This conception of power and authority in the organization establishes a clear chain of command all the way from people on the assembly line up the CEO and specifies the tasks that are appropriate to each role. In this type of system, people are primarily responsible for carrying out assigned duties rather than acting as a creative agent set free to make decisions and exercise judgment over important matters.

The desire and logical impetus for hierarchy stems from the four assumptions explored above. Hierarchy helps to simplify, organize and structure the organization by defining roles, duties, and functions clearly. By creating carefully detailed schematics of organizational life and elaborate structures to harness its forces, hierarchy helps to satisfy the rational desire for order, simplicity, and symmetry. It also helps to simplify issues of blame and responsibility for the individual in the firm by assigning them specific roles appropriate to their position in the hierarchy.

It also makes for a more quick and efficient exercise of control, since the chain of command makes clear who is in charge and who has power—thus, there are fewer decision makers and a greater sense of respect for authority from above. While it tends to harmonize and order the world of the corporation, it does reduce the capacity for creative and innovative action being introduced from below.

Though what we have argued is undoubtedly familiar, it is worth pointing out that the language of hierarchy is prevalent in the business literature. Russell Ackoff refers to the organization as a series of systems and subsystems, just as the organization itself is involved in a hierarchy of systems which must be controlled and organized to be more effective in realizing their respective tasks. Each layer shapes and controls the layer below it (Ackoff 1974: 18). Mason and Mitroff utilize their image of the policy-maker as a 'steerer' of forces which are lower on the totem pole of authority. O'Toole lauds the decision-making process of Vanguard companies for its structure, formality, continuity, and consistency in generating decisions (O'Toole 1987: 266–61).[9]

Despite these compelling virtues, adopting hierarchies exacerbates the problem of silencing stakeholders which we exposed in the previous section on objectivity. Not only do hierarchies tend to exclude the voices of stakeholders outside the organization, it narrows and ranks the range of input from those inside the firm. The resort to hierarchies creates a organizational structure, and accompanying ethos, which erodes the effort to be inclusive, to operate in terms of the needs and interests of stakeholders, and to recognize the validity of stakeholder concerns. Thus, between objectivity and hierarchies, stakeholders are silenced twice over.

3 A Feminist Reading of the Stakeholder Concept

These five different metaphors are some of the more powerful images shaping our understanding of the business world today. They also happen to have a profound influence on how we see the stakeholder concept.[10] From the standpoint of several feminist authors this kind of thinking is far from 'natural,' necessary, or even healthy, but is instead an over-played theme from one part of our moral imagination which needs to be re-considered in light of other powerful metaphors. Just as the assumptions in the first part appeared complimentary and conceptually related, so too are the revisions we will offer connected and mutually reinforcing. Indeed, as

[9] Mason and Mitroff as well as O'Toole can be seen as clear descendants of James March and Herbert Simon in terms of their views of the organization. For them, at bottom, the study of the organization revolves around the problem of information processing which, in turn, dictates the need for hierarchical control mechanisms.

[10] While we argue that these metaphors are rampant in the literature on business and in our culture and we present evidence to support this claim, we do not maintain that they are universally held or that there are not serious challenges to these views on grounds similar to the arguments we make in this paper.

we suggested at the outset of the paper, such metaphors are more than mere words—they help create an entire new vocabulary and framework within which we can think about the organization and its purposes.

3.1 Corporations as Webs of Relations Among Stakeholders

Rather than taking the individual as primary and independent of its environment, the feminist reading we offer suggests that persons are inextricably embedded in context.[11] While people retain a distinct personal identity, they are not isolated from their connections with others or the language and culture which pervades their existence. Persons are fundamentally connected with each other in a web of relationships which are themselves integral to any proper understanding of "the self," such that any talk of autonomy or search for personal identity must be qualified and located within this more organic and relational sense of the world. The stakeholder concept, understood in feminist terms, makes explicit how the boundaries of the self extend into areas far beyond what we can easily recognize and into areas clearly 'outside' the corporation. The us/them and internal/external distinctions fade into a sense of communal solidarity in which one sees the corporate identity as manifest within an entire network of stakeholders and a broader social context. Thus, our proposed new stakeholder interpretation of the firm is: *the corporation is constituted by the network of relationships which it is involved in with the employees, customers, suppliers, communities, businesses and other groups who interact with and give meaning and definition to the corporation.*

This approach does not go to the opposite end of an individualism by collapsing personal identity into some homogenous collective; rather it seeks to acknowledge that the individual and the community, the self and the other are two sides of the same coin and must be understood in terms of each other. If our redescription of the stakeholder concept can be thought of as providing the resources for an interesting story to be told through the workings of particular corporations, it is a story which would continue to celebrate the individual and the capacity for self-creation. We simply wish to overtly connect that with collective activity and the need for the creation of community in a manner which integrates the projects of the individual and the collective. Part of this stems from the moral significance of relationships and the capacity for care, both of which are taken as hallmarks of healthy private activity (in the family, voluntary associations, church congregations, etc.), but which have been either systematically devalued and largely excluded from the public world (business and politics especially) or viewed as anti-thetical to it.

We suggest, as feminist scholars have been doing for some time, that the capacity to care has tremendous moral importance as well as potential to enrich human

[11] A number of other authors take a similar view from a non-feminist perspective. See Alisdair MacIntyre, Michael Sandel, Stanley Hauerwas, Charles Taylor and others who have been grouped under the broad category of "communitarians."

activity in both public and private spheres. An additional moral concern which connects with this emphasis of self as embedded in community is the growing sense that the world is becoming too complex and fragile for us to embrace a rampant competitive individualism. Companies and persons cannot ignore the responsibility they bear for all of their actions which affect others, even when there is no legal duty to refrain from such behavior. Replacing the identity of persons as isolated and autonomous with a model which fundamentally ties them together not only makes these sensibilities intelligible, but provides an impetus for acting on them. It also makes it easier for us to talk about our responsibilities to future generations, as well as our connection with the environment and the need to preserve and sustain it.

While there are moral reasons to look to this feminist reading of identity, there are also practical reasons which make it an appealing alternative, and perhaps a necessary corrective. Everyone from diplomats and politicians to the popular media is echoing the same refrain: the world is becoming more interdependent and the actions of one person, company, or country are increasingly seen to have profound implications for the prospects of others. In a world where the actions of one directly affects everyone and the fate of each depends on the actions of all, the model of the pioneer or heroic individual acting in isolation has outlived its usefulness (Gray 1989, p. 269). A number of population ecologists and organizational theorists have also explored this theme of interdependence and emphasized its significance for how business will have to operate to survive (Gray 1989; Astley 1984; Astley and Fombrun 1983; Aldrich 1979; Hannan and Freeman 1977).[12] Not only does their work suggest that the business world is becoming increasingly interdependent, but that the image of firms as autonomous agents striving to plot out a course of action is misguided or largely misleading in the current business environment. As Astley and Fombrun note,

> …[P]opulation ecologists criticize the notion of strategic choice (Aldrich 1979). They argue that, at a macro level, historical, political, economic, and social factors determine the fate of whole populations of organizations, so that the actions of single organizations count for little in the long run course of events. The field of business policy has failed to respond to this criticism in a way that adequately preserves the notion of strategic planning as a proactive process. This paper argues that such a response is possible if one takes into account the role played by organizations as constituent members of an overarching interorganizational collectivity… [R]esponse to the environmental determinism of population ecology can be made by recasting the concept of strategy in terms of collective mobilization of action and resources oriented toward the achievement of ends shared by the members of interorganizational networks (Astley and Fombrun 1979: 577).

Managers who seek to retain the individualism and the sense of isolation from their stakeholders and the "external environment" will find themselves at a serious disadvantage and perhaps even completely ineffective.

---

[12] Astley and Fombrun, like many of the other thinkers we draw from in this paper, are not "care" oriented thinkers. There are important affinities between their work and the account we are offering of stakeholder theory, but Tom Peters, Jack Welch, the Club of Rome, and other figures we cite are not necessarily "caring" individuals nor do they write from a "care" perspective. However, their insights are valuable and lend support to our view as we construct the larger narrative of the paper.

Tom Peters' work also highlights this theme. His emphasis on teams of employees working on the development and production of new products as a way to improve quality, innovation, customer responsiveness, and worker satisfaction leads us in the same direction. The "American functionalism" which he so rightly criticizes for its rigidity and segmentation, tends to isolate and fragment the process of work and the relationships between important stakeholders (especially customers, suppliers, and other intraorganizational departments). Teams not only improve ones' ability to succeed and respond in today's dynamic business environment, they embrace a picture of the firm as a world of interdependence which exists within the "walls" of the company and extends to all other stakeholders. Beyond teams, he emphasizes the importance of the ethos of cooperation and collective action throughout the firm (Peters 1987).

3.2 Corporations Should "Thrive on Chaos" and Environmental Change

While the masculine narrative of the world depicted in the first part of the paper viewed change and uncertainty as threatening, the feminist rereading we propose would embrace these dynamic forces as enriching and part of how we experience the world. The forces which were seen as messing things up and posing threats to the firm are now viewed as important vehicles for creating diversity, offering new opportunities, and reflecting the natural processes of change. In addition, there is less fear in a dynamic context because the self is not as clearly bounded off from the "external environment." The sense of interdependence arising from feminist thought makes the world appear less alien and artificial and more of an outgrowth of oneself—a fellow traveler whose fate is bounded up with one's own. Thus, the impulse is to create harmonious relationships with one's environment, to nurture and sustain it as well as oneself, rather than to try to conquer or control it.

Here again, trends in the business world suggest that this view yields important and timely insights. The world of organizational theorists and population ecologists convey that not only is interdependence a fact of life, but that it is increasing as "turbulence" continues to rise and the necessity for cooperation grows (Gray 1989: 1, 27, 231; Astley and Fombrun 1983: 577). Part of accepting change is an acute awareness that, no matter how successful one is and how unshakable one's assumptions about how to manage are, what works today will need to be re-examined and revised tomorrow. Because of this, firms which want to heed this wisdom will approach their own future and their success with a management philosophy which takes change and adaptation as a guiding principle.

Another aspect of this approach is the emergence of patterns of cooperation to respond to the fluctuating business environment. The explosion of joint ventures and global alliances reflects the growing realization that we must face change and uncertainty with shared risks, joint action, and embrace further interdependence rather than moving away from it. Competitors as entrenched and hostile as Japanese and American auto makers have launched extensive joint ventures and cooperative agreements to enrich their prospects and to better serve their customers. Examples abound, particularly in high-technology and communications industries, but have found their way into a myriad of different markets and industries as well as labor-management relations (Gray 1989, p. 26). All of these firms have come to acknowledge the best way to respond to turbulence is to accept change and to join forces with others who have a stake in their future to help improve the prospects of all. Jack Welsh of General Electric has maintained that "cooperative alliances, designed to achieve global scale and strong positions for both companies involved in the transaction," would become an extremely popular way for firms to manage change and be successful (Van Bever 1987: 5). The Third Club of Rome has offered similar sentiments.

> What is required is a new international order in which all benefit from change. What is required are fundamental institutional reforms, based upon a recognition of a common interest and mutual concerns, in an increasingly interdependent world (Tinbergen 1976: 7-8).

Perhaps the most ardent advocate of this view is Tom Peters. His recent book, *Thriving On Chaos,* is devoted to elaborating on this fundamental insight. His book makes extensive references and examines numerous case studies which demonstrate the critical need for companies to internalize an attitude which both accepts and thrives on change as a part of its proper function. In one of his more pointed and passionate statements on this subject, Peters claims:

> Success will come to those who love chaos—constant change—not those who attempt to eliminate it. The fleet-of-foot, value-adding, niche-market creator thrives on the very uncertainty that drives others to distraction. Stability and predictability are gone for good—and therefore must not be the implicit or explicit goals of organizational design, the layout of factory or operations center, pay schemes, strategic planning, objective-setting exercises, accounting systems, or job evaluations. Victory will go to those who master instability by constantly working on responsiveness-enhancing capabilities (Peters 1987: 394).

In terms of understanding the new version of the stakeholder theory, the alternative metaphor highlights the need for stakeholder groups to recognize their interdependence, to embrace it, and to work together to meet the changing needs and expectations of each group. Where the masculinist formulation put the corporation as an autonomous entity, bounded off from its environment and the key stakeholder groups who bring change—customers, communities, suppliers—our reading places these groups squarely within the domain of the organization. As such, with communication and mutual support the firm is able to internalize them and make their changing needs, wants and expectations part of the organization (i.e., there is a built-in tendency to be dynamic and to both anticipate and respond to change).

3.3 Replacing Conflict and Competition with Communication and Collective Action

The metaphors of conflict and competition run rampant in business thought. They describe a world where managers must make "tough choices," "kill or be killed," "outgun the opposition," and "look out for number one." Where the masculinist thought celebrates these competitive and violent metaphors, feminist thought strives to replace them with more cooperative alternatives. Feminists recognize that adversarial approaches are often harmful to human relationships, undermining trust and the potential for cooperation. One can see, particularly in the area of divorce law and child custody cases, how this approach has systematically engendered serious and lasting harms to persons and family relations. The conclusion that many feminists would have us draw is not just that there are places where an adversarial approach is simply inappropriate, but that there are harmful consequences to any set of human relationships when one assumes this type of posture. Peters maintains that competition and adversarial relations among stakeholders, particularly groups like suppliers, distributors, and customers, spells doom for the firm precisely because cooperation and trust are such a vital ingredient for success in the current business environment (Peters 1987: 108).

At a more practical level, an adversarial outlook tends to see unavoidable conflicts where there are winners and losers and nothing in between. Feminists see this as a blindness deriving from a world view which makes competition and these "tough" choices the only options. They would instead seek to find win-win situations where what initially appear to be conflicts of interest among stakeholders can be turned into forms of collaboration which work for the benefit of all. Barbara Gray's book. *Collaboration,* is devoted, in part, to illustrating how conflict and either/or choices have been turned into mutually advantageous collective action, among the most impressive of which is the emergence of the European Economic Community (Gray 1989: 280).

When their best efforts to avoid putting stakeholder interests in direct conflict seem doomed to failure, managers and executives should be able to generate effective communication to resolve conflicts. Indeed, such skills are key to establishing successful collective action, or the forms of collaboration discussed by Gray. The emphasis on caring and relationships makes communication a capacity which becomes vital to any vibrant community or human interaction that would realize these moral values. Particularly when a given relationship entails cooperation for mutual enrichment, or coordinated action, effective communication is a premium. Communication provides the mechanism for persons to interact with and learn from one another, to build trust, to find points of agreement and disagreement, to discover how a relationship can enrich each party involved, and to sculpt a form of interaction that fits them.

As more and more companies embrace forms of cooperation, collective action, decentralized authority, and teams, managers are echoing this stress on communication. In an article with executives and managers from companies such as

Heinz, Corning, Sara Lee, Chesapeake Corporation, and Colgate, a Fortune reporter concludes that "communication is the most important source of personal power" and that "smart CEO's of decentralized companies realize that they have to actively encourage communication among the different parts of the organization" (Stewart 1989: 58–62).

By facilitating collective action and constructive responses to interdependence, communication helps realize a number of important goods. Not only does collective action allow both individuals and groups to be a part of decision-making and implementation—forms of participation which are fulfilling and help build a sense of community—it brings together greater resources to address a given problem. Gray (1989) and Waddock (1989, 1991) have documented a number of different partnerships of public and private organizations to deal with problems which neither were capable of dealing with effectively on their own. The urgency of this form of action is underscored when one looks at the increase in turbulence in numerous markets and the growth of interdependence across the globe. As Astley and Fombrun note,

> …[S]trategic choice is radically inhibited in highly competitive markets in which opportunistic behavior on the part of any single organization is overwhelmed by long term economic trends that determine the availability and distribution of environmental resources. From this point of view, whole populations survive or fail regardless of the actions taken by particular organizations within them. However, what population ecologists fail to recognize is that, although such considerations diminish the importance of strategic action at the level of the single organization, they do not diminish the importance of strategic action at a collective level. Indeed, the very impotence of organizations' acting in isolation merely elevates the importance of collective action (Astley and Fombrun 1983: 582).

Encouraging participation and collective action also helps validate decision-making. By giving stakeholders a voice and a measure of control with other stakeholders, it helps participants more willing to accept a "second best" result, to acknowledge the differences which both divide and hold the company together, and to make them feel as though their concerns matter. Where adversarial approaches can make certain stakeholders feel as though they are being dumped on or excluded, and thereby harming their relationship to the company, fostering participation, collective action, and empowerment builds precisely the sort of environment firms need to be successful.

4 Strategy as Solidarity

Where the assumption in the first part of the paper was that we need to mistrust human perceptions and distance ourselves from our natural judgments, feminists take precisely the opposite approach. They reject the idea that we need to stand outside problems to understand them and instead draw on the human capacity to engage situations with empathy and communication.[13] There is a fundamental skepticism of science contained within feminist thought which is expressed in three insights: first, the frank recognition that science and empirical inquiry are meaningless—they cannot tell us what to do or how to live; second, that the resort to detachment and other supposedly "objective" methods are merely masculine approaches which reflect a bias toward abstraction and calculating rationality; and third, relationships and particularized forms of cooperation cannot be easily translated into abstract formulations. Indeed, an objective approach appears directly counter (and counter-intuitive) to the basic goals of the feminist understanding of business activity—namely, creating value for specific persons, advancing personal projects, and nurturing the relationships among various stakeholder groups. For these reasons, the scientific approach of objectivity appears uniquely ill-suited for the formulation of strategy.[14]

Where strategy as objectivity seeks to describe decisions as "dictated by the numbers" or as "business decisions," the feminist view of solidarity requires that managers make choices based on the responsibilities and relationships it has with specific stakeholders. Feminist thought challenges the idea that these "objective" decisions are what they purport to be—that instead of being disinterested and neutral, they merely distance us from the responsibility of choice and mask the reality that the actual motivation is the pursuit of profit for those who have power in the organization. Starting with solidarity requires that the use of scientific data be done in the service of specific wants, that it not impede the ability to talk in terms of particular persons and their needs, and that the "numbers" or facts it produces can never dictate decisions.

This is not to say that feminists reject the idea that empirical study can offer helpful insights. As Sandra Harding notes,

> To repeat the metaphor I borrowed from behaviorism, science functions primarily as a "black box": whatever the moral and political values and interests responsible for selecting problems, theories, methods, and interpretations of research, they reappear at the other end of inquiry as the moral and political universe that science projects as natural and thereby helps to legitimate. In this respect, science is no different from the proverbial description of computers: "junk in; junk out." It is within moral and political discourses that we should expect to find paradigms of rational discourse, not in scientific discourses claiming to have disavowed morals and politics. This assertion of the priority of moral and political over scientific and epistemological theory and activity makes science and epistemology less important, less central, than they are within the Enlightenment world view. Here again, feminism makes its own important contribution to postmodernism—in this cast, to our understanding that epistemology-centered philosophy—and, we may add science-centered rationality—are only a three-century episode in the history of Western thinking (Harding 1986:250-1).

[13] The position we take here on strategy has important affinities with the work of Gilbert—particularly his account of strategy through convention—in his book *The Twilight of Corporate Strategy* (New York: Oxford University Press, 1992).

[14] See also Marilyn Friedman, "The Impracticality of Impartiality," *Journal of Philosophy* LXXVI (1989): 645–56.

We should invert the traditional order of priority. For feminists we should search through the rich tapestry of experience, language, and impression to construct a picture of both problems and solutions that is complex and reflects the variety of perceptions of the stakeholders involved. Rather than looking to science, statistics, or "facts" to create an image of a problem which all can accept, a feminist reading would have us begin by making various stakeholders work through the painful process of piecing together their different impressions of the situation, what is at stake, and how it can best be dealt with. While this process may appear to "cloud" the picture, it also expands the breadth of possibilities for envisioning a situation or designing a response (see also Gray 1989: 5, 12, 14, 228).

Not only does this process acknowledge the idea that our relationship to the world is human and therefore a construction built upon language, culture, and custom, it helps to build solidarity among stakeholders. As we have already argued, going about strategy this way achieves important moral goods—trust, commitment, participation—which enrich the lives of stakeholders. Yet, it also helps achieve practical goals which are becoming increasingly significant: improving communication and coordination, building a clear consensus about both identifying problems and defining solutions, getting people to legitimate the process and outcome through their involvement. Each of these help unite the firm, allow it to be more responsive and decentralized, and increase the likelihood that the actions of employees will be in cinch with the direction set by the company. The success of Japanese management, particularly their stress on consensus and communication, testifies to the pragmatic value of this approach. As Barbara Gray emphasizes,

> Arriving at a negotiated order requires joint appreciation. An essential notion of appreciation is the power of collective thought to transform existing circumstances. Boje (1982), Smircich (1983), and Bartunek (1984) have described how leaders induce change by introducing new meanings for existing circumstances. This "myth-making" process occurs during collaboration as stakeholders negotiate an image of their desired future. Myth making involves abandoning existing interpretations and creating new or expanded interpretations for existing problem domains. Individually, stakeholders are often constrained from envisioning new possibilities. Their own visions are only partially formed or are blocked by a sense of individual powerlessness. Part of the process of myth making involves introducing new language (Boje 1982; Smircich 1983) and new symbols of what is important. Fundamental to this process of transformation is the idea that "society is made and imagined and that it can therefore be remade and reimagined" (Unger 1987, p. 8). A new negotiated order emerges, and stakeholders begin to invent collective symbols and language to describe the agreements they reach about the domain. When people share the fundamental assumptions that undergird their thinking, they can forge a collective appreciation (Gray 1988: 271–2).

We advocate that managers drop the quest for objectivity and embrace the quest for solidarity and communicatively shared understandings. Strategic direction is never determined by numbers, abstract surveys, or utility functions, although these may play a role in terms of how stakeholders see their interests and how they may be realistically pursued. The strategic direction of the firm should always be thought of and developed in terms of "us"—the interests, desires, and needs of all stakeholder groups—rather than a firm charting its path as a lone actor. Indeed, if one comprehends the sort of shift entailed in our new formulation of the stakeholder concept, one cannot comprehend what formulating strategy "alone" looks like. The firm has become, in terms of its purposes, decision-making structure, and identity, a vehicle for creating value for all stakeholders. As such strategy is merely the direction set out by the interests, desires, and needs of stakeholders, as forged through negotiations and agreements of these groups.[15]

4.1 Replace Hierarchy with Radical Decentralization and Empowerment

The resort to hierarchy as a means of structuring power and authority made sense in the set of assumptions described in the first part of this paper. Yet, after tracing our reinterpretation of the first four of these premises, it becomes clear that a feminist reading would downplay or collapse much of the hierarchy. One reason for this shift lies in the realization that decentralization, team structures, worker empowerment all make for a more meaningful work experience. Frederick Taylor's management revolution began with a rational and mechanistic approach that broke tasks down to their simplest components and then capitalized on repetition and mass production. Yet precisely what made it efficient and structured was what made it alienating for workers. On the alternative reading we are suggesting, one of the primary goals in not the search for structure, order, and simplicity, but the need to connect workers with their jobs and unleash their creative potential. Getting the most from workers means giving them responsibility and offering them jobs where they have a direct connection with the end product and sustained involvement with the main aspects of what is being done. Where the first view would celebrate the U.S. automobile assembly line of the 1970's where workers perfected one small task and did it repeatedly, our view would look to the Swedish Volvo plants and their use of teams of workers who assemble entire sections of cars, rotate tasks, and become involved in the schedules and structure of their work. The disbursement of authority and the sharing of responsibility that emerges in team-type work environments shifts the role of managers closer to that of teachers and coordinators and away from the tasks of dictating and supervising.

Another virtue of the movement away from hierarchy is the reinforcement of attitudes which promote worker involvement, increase trust, and bring out added commitment and productivity. As common-sense suggests, trust is built by those who are willing to extend it, and responsibility is borne only when we are prepared to delegate it. While the equation doesn't work every time, or for all workers, the business environment would indicate that firms can ill afford to not risk taking these

[15] See Ed Freeman's *Strategic Management: A Stakeholder Approach* (Boston: Pittman, 1984) for a discussion for a model of negotiation among stakeholders that is similar to what we advocate here, particularly his level three negotiation.

chances. James R. Houghton, CEO of Corning, puts it simply: "The age of hierarchy is over" (Stewart 1989: 53). Precisely because the workplace and the global economy has changed so drastically, firms must increasingly move away from centralized authority and hierarchy. The idea is not to completely abolish company structure and division of labor, but to put it in the service of humanizing work practices and increasing employee involvement and responsibility. Strategic solidarity (described in the last section), thorough communication, and loose organic structures all help to sustain and nurture the complex relationships that allow the corporation to be productive as well as responsive to change. Managers and executives are no longer those who formulate "the plan" and then get their drones to follow orders and then spend their time supervising others and making sure people follow orders. Instead, managers become motivators, coordinators, and creators who help workers share information and get them involved in the development, planning, and revision of key ideas as well as their implementation.

Part of the pragmatic advantage of this power-sharing and hierarchy flattening is an increase in total power. This makes intuitive sense, given that extending responsibility to employees allows them to take advantage of opportunities, to free management up from having to act like parents supervising their children, and brings them into the loop as contributors. A recent article in *Training* notes:

> [Ricardo] Semler points out that managing without managers is based on the assumption that employees are trustworthy adults. "Think about that," he wrote in *HBR*. "Outside the factory, workers are men and women who elect governments, serve in the army, lead community projects, raise and educate families, and make decisions every day about the future…But the moment they walk into the factory, the company transforms them into adolescents." His radical notion: "We hire adults, and then we treat them like adults. To make teams work, management must learn to give employees responsibility, just as parents must learn to allow their children to make decisions," says Coming's O'Brien. "When I see high-performance teams working, how enthusiastic they are, how goo they are at their jobs, I think, shame on us for not empowering employees years ago" (Lee 1990: 32).

Ralph Stayer, CEO of Johnsonville Foods, claims:

> Flattening pyramids doesn't work if you don't transfer power too. Before, I didn't have power because I had people wandering around not giving a damn. Real power is getting people committed. Real power comes from giving it up to other who are in a better position to do things than you are. Control is an illusion. The only control you can possibly have comes when people are controlling themselves (Stewart 1989: 53).

The wisdom of these statements is reflected in ongoing changes in the business world towards decentralization and power-sharing. A recent survey indicates that more than half (55%) of the Fortune 500 and Service 500 companies that were polled indicated they were reducing management layers and only 7% said they had added new layers. The survey also finds that 74% of the CEO's of these companies, "…[A]re more participatory, more consensus oriented, and rely more on communication than on command: only 5% say the reverse" (Stewart 1989: 66).

Peter's work also makes the issue of flattening the hierarchy and increasing employee involvement a top priority for firms in the emerging global economy: "The ability to take action close to the market—and fast—is the first requirement for competing. Therefore both information availability and the authority to move forward at the front line are musts" (Peters 1987: 504). He is equally insistent about battling that stepchild of hierarchy and centralized authority, bureaucracy: "Moaning about bureaucracy is a time-honored management prerogative. Now, however, bureaucracy is beyond moaning about; it is a block to survival. *The campaigns against bureaucracy must become strategic priorities of the first order*" (Peters 1987: 377, no emphasis added).

Shifting to decentralized authority signals a change in managerial attitudes especially towards workers, but to all stakeholder groups as well. The reading we are offering on the shift of authority entails that management acknowledge: the legitimacy of the different viewpoints of stakeholders within the organization; the stakes that each of these groups have in the future of the firm, and more fundamentally; that each of them have value to contribute to the organization. By pushing for more decentralization, the firm not only nurtures communication and interdependence among stakeholders, it is able to draw more fully on the latent abilities and creativity of all. The worker benefits from more control and involvement in work, while the firm as a whole is better able to serve the interests of all stakeholders (via new ideas, greater efficiency, increased responsiveness, etc.).

5 Conclusions

Feminist thought brings a fresh perspective and important insights to our understanding of the stakeholder concept. Its potential contributions are magnified by the fact that it offers a richer appreciation of the stakeholder idea and the resources to articulate many of the most pressing moral problems of the day, as well as many practical insights which will make firms better able to create value and thrive. Our efforts here are a part of an ongoing project to sustain a more thorough and comprehensive reexamination of how we think about ourselves and about business.

We have argued that stakeholder management, understood in feminist terms, is about creating value for an entire network of stakeholders by working to develop effective forms of cooperation, decentralizing power and authority, and building consensus among stakeholders through communication to generate strategic direction. In this context, it is imperative that managers are able to sustain networks of communication, share decision making authority with workers, and help tap into the creativity of workers and suppliers.

Rather than argue that this picture requires that we drop or excise the metaphors we isolated in the first part of the paper, we advocate a shift towards the feminist insights while retaining these other tools to help shape and moderate management practices. The feminist reading we offered was not meant to be a foil or symmetric opposition to the framework we saw in place. Even though the assumptions we looked at in each of the two groups tended to be logically or intuitively connected to represent a coherent worldview, favoring one does not require that we exclude the other. For instance, to move towards a feminist understanding of managing for stakeholders puts collaboration and cooperation as the central directive of management—firms should strive to create value for stakeholders by developing networks of cooperation which maximize their resources. Competition doesn't become irrelevant in this picture, it merely becomes a secondary virtue—a firm becomes competitive as an effect of successful collaboration and teamwork.

Yet we do insist that this reading breaks new ground and transforms how we think about managing for stakeholders and about corporations. It offers a new vocabulary, with its own distinct set of concepts and metaphors, that enables us to envision an innovative and enriching context within which to think about the firm. Another virtue of this reading is that it returns the focus in conversations about stakeholder theory to the central moral questions inherent in the meaning of organizations and makes those issues the crux of the matter—both in terms of why it is compelling and how it is implemented. We believe this paper represents a progressive step in our attempt to construct a picture of business ethics which makes intellectual sense, expresses our most important aspirations and values, and enables firms to thrive in the global economy. We trust that ongoing conversation about these matters will help us wrestle with the most urgent problems of the day as we confront new economic, moral, and political challenges of unprecedented magnitude. While our situation might appear dire, we should remember that such occasions provide the greatest opportunities for change and reconstruction.

References

Ackoff, Russell. 1974. *Redesigning the Future*. New York: Wiley.

Aldrich, H. 1979. *Organizations and Environments*. Englewood Cliffs: Prentice-Hall.

Astley, W.G., and C.J. Fombrun. 1983. Collective Strategy: Social Ecology of Organizational Environments. *Academy of Management Review* 8: 576–587.

Astley, W.G. 1984. Toward an Appreciation of Collective Strategy. *Academy of Management Review* 9: 526–535.

Bartunek, J. 1984. Changing Interpretive Schemes and Organizational Restructuring: The Example of a Religious Order. *Administrative Science Quarterly* 29 (3): 355–372.

Boje, D. 1982. Towards a Theory and Praxis of Transorganizational Development: Stakeholder Networks and Their Habitats. Working paper 79-6. *Behavioral and Organizational Science Study Center, Graduate School of Management*, Los Angeles: University of California.

Carpenter, S.L., and W.J.D. Kennedy. 1988. *Managing Public Disputes: A Practical Guide to Handling Conflict and Reaching Agreements*. San Francisco: Jossey-Bass.

Freeman, R.E. 1984. *Strategic Management: A Stakeholder Approach*. Boston: Pittman.

Freeman, R.E., and Daniel R. Gilbert. 1989. *Corporate Strategy and The Search for Ethics*. Englewood Cliffs: Prentice-Hall.

———. 1992. Business Ethics and Society: A Critical Agenda. *Business & Society* 31: 9–17.

French, Peter. 1979. The Corporation as a Moral Person. *American Philosophical Quarterly* 3: 207–215.

Friedman, Marilyn. 1989. The Impracticality of Impartiality. *Journal of Philosophy* LXXVI: 645–656.

Gilligan, Carol. 1982. *In A Different Voice*. Cambridge: Harvard University Press.

Gray, Barbara. 1985. Organizations as Constructions and Destructions of Meaning. *Journal of Management* 11: 83–98.

———. 1989. *Collaborating: Finding Common Ground for Multiparty Problems*. San Francisco: Jossey-Bass.

Hannan, M., and J. Freeman. 1977. The Population Ecology of Organizations. *American Journal of Sociology* 82: 929–964.

Harding, Sandra. 1986. *The Science Question in Feminism*. Ithaca: Cornell University Press.

Katz, H. 1985. *Shifting Gears*. Cambridge, MA: MIT Press.

Kochan, J., H. Katz, and R. McKersie. 1987. *The Transformation of American Industrial Relations*. New York: Basic Books.

Lee, C. 1990. Beyond Teamwork. *Training* 27 (June 1990): 25–33.

Mason, R.O., and I.I. Mitroff. 1981. *Challenging Strategic Planning Assumptions: Theory, Cases and Techniques*. New York: Wiley.

O'Toole, J. 1987. *Vanguard Management: Redesigning the Corporate Future*. Garden City: Doubleday.

Peters, T. 1987. *Thriving on Chaos: Handbook for a Management Revolution*. New York: Knopf.

Schuster, M. 1985. Models of Cooperation and Change in Union Settings. *Industrial Relations*. 24 (3): 382–394.

Smircich, L. 1983. Organizations as Shared Meaning. In *Organizational Symbolism*, ed. L.R. Pondy, P. Frost, G. Morgan, and T. Dandridge. Greenwich: JAI Press.

Stewart, T. 1989. New Ways to Exercise Power. *Fortune* (November 6) 52-66.

Susskind, L.E., and J. Cruikshank. 1987. *Breaking the Impasse*. New York: Basic Books.

Tinbergen, J. 1976. *Reshaping the International Order: A Report to the Club of Rome*. New York: Dutton.

Unger, R. 1987. *Social Theory: Its Situation and Its Task*. Cambridge: Cambridge University Press.

Van Bever, D. 1987. HBS Focuses Microscope on General Electric CEO. *Harbor News* 51 (November 2): 1–5.

Waddock, Sandra. 1986. Public-Private Partnerships as Social Problem Solving: Product and Process. In *Research in Corporate Social Performance and Policy*, ed. J.E. Post, vol. 8, 273–300. Greenwich: JAI.

———. 1989. Understanding Social Partnerships: An Evolutionary Model of Partnership Organizations. *Administration & Society* 21: 78–100.

———. 1991. A Typology of Social Partnership Organizations. *Administration & Society* 22: 480–515.

Andrew C. Wicks is Ruffin Professor of Business Administration at the Darden School of Business, University of Virginia.

Daniel R. Gilbert, Jr. is Professor Emeritus at Gettysburg College.

R. Edward Freeman is University Professor, Elis and Signe Olsson Professor, Academic Director of the Institute for Business in Society, and Senior Fellow of the Olsson Center for Applied Ethics at the University of Virginia's Darden School of Business.

Chapter 7
Stakeholder Theory: A Libertarian Defense

R. Edward Freeman and Robert A. Phillips

1 Introduction

Quine (1960) once wrote that "sentences do not confront the tribunal of experience alone." And, he might have added, "nor do arguments and theories?" Sentences, as well as arguments and theories, always import their background conditions and related theories with them. Context plays a crucial role in social phenomena, and sometimes we need to resist the efforts of greedy reductionists (Dennett 1995)[1] to set context aside and focus on the precision of hypotheses, the collection of data, and the rituals of method. Sometimes it is important to point out a feature of the everyday landscape which is often taken for granted, because in doing so, we can come to see new relationships and new features that previously were hidden (Wisdom 1970).[2] Such is the role of what has come to be called "stakeholder theory," and its relationship to our understanding of business activity.

The interpretation of business activity can be approached in several ways. One way is to take for granted the usefulness (others might say "truth" or "moral

Originally published in: *Business Ethics Quarterly*, 12(3), 331–349 © Cambridge University Press, 2002
Reprint by Springer, https://doi.org/10.2307/3858020

[1] Dennett (1995) distinguishes between "reductionism" and "greedy reductionism"

[2] Wisdom (1970) argues that philosophy's unique contribution to intellectual life is to illuminate such hidden features.

R. E. Freeman (✉)
University of Virginia, Charlottesville, VA, USA
e-mail: FreemanE@darden.virginia.edu

R. A. Phillips
York University, Toronto, ON, Canada

© Springer Nature Switzerland AG 2023

S. D. Dmytriyev, R. E. Freeman (eds.), *R. Edward Freeman's Selected Works on Stakeholder Theory and Business Ethics*, Issues in Business Ethics 53,
https://doi.org/10.1007/978-3-031-04564-6_7

legitimacy" here) of a way of understanding value creation and trade (or business activity), whereby individuals are presumed to be nakedly greedy, responsible to others for the effects of their action only in so far as they are caught doing harm, and the legitimacy of a state that pervasively regulates all aspects of value-creation and trade, from rules governing the height of ladders to thousands of pages of the tax code. Let us call this way of understanding value creation and trade, or business activity, "the Standard Story" or "Shareholder Capitalism," or "Cowboy Capitalism," and let us make the requisite academic changes and disclaimers to include sophisticated agency relationships, transaction costs, and other assorted modifications of the standard story so as not to be accused of the Straw Man Fallacy.

Within business ethics it is fairly well accepted that the standard story is the main way that we understand business and capitalism. A great deal of interesting work has emerged that tries to point out various places where the standard story fails a variety of ethical tests.[3] These ethical tests are, for the most part, developed outside of the standard story, and indeed, outside of much consideration for value creation and trade, or business activity, at all.

Yet another group of scholars has developed a critique of the standard story that goes like this. The standard story is fine as far as it goes, but it doesn't go far enough. We need to add the idea that the collections of individuals that we call "corporations" need to understand the *social* effects of their actions. And, we need to link the social effects of corporate action to the economic effects. The background conditions and theories of the standard story simply need to be broadened to include a set of ideas about "social" and its link to "economic."[4]

Yet another way to understand business activity would be to change the background conditions of the standard story itself. Such a method might ask how value creation and trade takes place in a world in which individuals have a complex psychology, where individuals and groups of individuals desire to be, and mostly are, responsible for the effects of their actions on others (good and bad), and where many are, or certainly ought to be, deeply skeptical of the view that the state looks out for their interests. Rather, individuals see the state as mostly pervasive and intrusive without particular insights into either the best height for ladders or the most effective means for building a just society. The standard story would have to be reformed to take account of these questions, and such reform would have to go well beyond the external critique of business ethicists, and the broadening of "economic" proposed by business and society scholars. It would have to include specific appeals to practice and a notion of "best practice" from the real world of value creation and trade. It would necessarily include examples of businesses that voluntarily manage key relationships and thrive. In short such a revisionist account of the background conditions of the standard story should be firmly enmeshed in business, i.e., in the practice of value creation and trade.

---

[3] See for instance the work of Bowie. Brenkert, Donaldson and Dunfee. Solomon. Werhane to name but a few.

[4] See for instance the work of Carroll. Cochran, Post, Wood. Waddock and others.

However haltingly it has been presented, this latter narrative has been at work in the background (some might say "deep background") in the development of what has come to be called "stakeholder theory." In the following sections we want to refocus "stakeholder theory" toward its "Libertarian" background conditions. In Sect. 2 we briefly summarize what has come to be called "stakeholder theory." In Sect. 3 we set out what makes a particular theory "libertarian," and what kinds of arguments could roughly be called "libertarian arguments." In Sect. 4 we construct some libertarian arguments for a variety of forms of stakeholder theory. These arguments seek to change the background conditions of the standard story and they lead to a revised understanding of capitalism called "stakeholder capitalism." Finally, in Sect. 5 we generalize the ideas in the previous sections and construct an argument for understanding capitalism in both stakeholder and libertarian terms, a view which we call "stakeholder capitalism."

2 Stakeholder Theory

Stakeholder theory is a managerial conception of organizational strategy and ethics (Donaldson and Preston 1995; Evan and Freeman 1993; Freeman 1984, 1994, 1996; Freeman and Evan 1990; Hill and Jones 1992; Jones 1995; Mitchell et al. 1997; Orts 1992, 1997; Phillips 1997; Rowley 1997). The central idea is that an organization's success is dependent on how well it manages the relationships with key groups such as customers, employees, suppliers, communities, financiers, and others that can affect the realization of its purpose. The manager's job is to keep the support of all of these groups, balancing their interests, while making the organization a place where stakeholder interests can be maximized over time. The identification of stakeholder groups is currently among the central debates in the scholarly and popular literature (Mitchell et al. 1997; Phillips 1997). but most scholars would include employees, customers, suppliers, financiers, and local communities, at a minimum.

Contributions to stakeholder theory have come from, among others, such disciplines as:

- Ethics (Boatright 1994; Burton and Dunn 1996; Donaldson and Dunfee 1999; Goodpaster 1991; Phillips 1997; Phillips and Reichart 2000; Starik 1995; Wicks et al. 1994; Van Buren 2001);
- Strategy (Berman et al. 1999; Carroll 1993; Clarkson 1994, 1995; Freeman 1984; Frooman 1999; Mitchell et al. 1997):
- Law (Lampe 2001; Orts 1992, 1997);
- Economics (Alkhafaji 1989; Barton et al. 1989; Freeman and Evan 1990); and

- Organization theory (Donaldson and Preston 1995; Freeman 1994, 1996; Evan and Freeman 1993; Hill and Jones 1992; Jones 1995; Rowley 1997; Williamson and Bercovitz 1996).[5]

Stakeholder theory has also received significant attention in the discourse of political economy, particularly in the U.K. (e.g., Hutton 1995; Kelly et al. 1997; Plender 1997). These authors propose a "stakeholder economy" that features a large-scale role for government in the process of value creation and trade. They argue that, while the stakeholder concept was originally applied to the private sector as a theory of organizational ethics (Phillips and Margolis 1999), expanding the concept to include public institutions and the entire national or world economy is a conceptual advance (Rustin 1997; Barnett 1997). This has led some to claim that the stakeholder approach comes from a socialist worldview. One goal of this paper is to clearly differentiate this brand of stakeholder theory from stakeholder theory in general, and our own conception in particular.

We do note, however, that even though we disagree with this "stakeholder economy" approach, the strategy of these thinkers is a central one to the stakeholder project similar to that employed by Freeman and others. Change the background conditions of the standard story. Revise our understanding of capitalism accordingly. See what works. On the "stakeholding" account, these background conditions revolve around the articulation of "liberal" principles instead of libertarian ones, and therefore encourage a large role of the state. The disagreement between liberal and libertarian versions of stakeholder capitalism may turn out to be less over fundamental ethical principles and more over what actually works in the world of value creation and trade.

Finally, it is a mistake to tar all stakeholder theory with the same broad brush. Freeman (1994) suggests that the theory is better understood not as a monolithic theory, but rather as a genre of stakeholder "theories." While the very nature and definition of "stakeholder theory" is itself a contentious issue, the idea is quite simple. A "stakeholder theory" is one that puts as a primary managerial task the charge to influence, or manage, or balance the set of relationships that can affect the achievement of an organization's purpose.

3 Libertarian Arguments

What makes an argument a "libertarian" one?[6] This is a question that is quite controversial, even among libertarians. While they come in many colors and shapes, most libertarians agree that "liberty" or "personal freedom" or "freedom" has

---

[5] Many of the works exemplifying the various categories are cross-disciplinary, so neither the works cited nor the categories themselves should be construed as mutually exclusive.

[6] We owe a great debt to Gordon Sollars in the following paragraphs, though this interpretation of what counts as libertarian is our own.

intrinsic value. Some might say that such an idea is definitional of humans, or shows us at our best. Others might say that developing "an attitude of liberty" is the only hope for human society.[7] Other libertarians put the matter in terms of "rights" or "natural rights" and concentrate on "negative rights." They claim that one human being has the right not to be interfered with by others, where "interfered with" is parsed in terms of physical harm. Most libertarians argue from freedom and liberty as primary to suggest that a strong system of individual property rights is the best means to preserve freedom and liberty. Property rights become an extension of the right to one's own physical body and its movements, so long as these movements don't interfere with others (Locke 1690/1952). After all, the argument goes, if one can't have the right to use one's body the way one wants to, it is difficult to see how one can be free, in any meaningful sense of "free."

Typically, libertarians do not want to trade off "equality" with "freedom," especially when "equality" is understood in terms of the distribution of wealth in society. And, they believe that the existence of a "more than minimal nightwatchman state" that redistributes wealth cannot be justified (Nozick 1974).

Rawls's first principle of justice is a paradigm case of a libertarian principle. He claims that "each person is to have an equal right to the most extensive basic liberty compatible with a similar liberty for others" (Rawls 1971: 60). "Basic liberty" here refers to "political liberty, freedom of speech and assembly; liberty of conscience and freedom of thought; freedom of the person along with the right to hold personal property, and the freedom from arbitrary arrest and seizure as defined by the concept of the rule of law" (Rawls 1971: 61).

However, Rawls is no libertarian, given his extensive arguments for a redistributive state or his second principle of justice.[8] (Some libertarians like John Hospers would quarrel with Rawls's list of basic liberties and argue that most are actually reducible to property rights.) Indeed Robert Nozick takes Rawls's first principle as a starting point for a libertarian theory. Jean Hampton (1997: 145) argues that:

> In a sense he [Nozick] agrees with the importance and priority of Rawls's first principle of justice, but in Nozick's view that first principle, requiring equal liberty for all, has implications for the kind of conception of justice that can be endorsed by a liberty-loving society. It must be one that allows people maximal (and equal) freedom to do with their property what they choose to do, without being subjected to interference by the state.

---

[7] See the work of James Buchanan Buchanan argued that developing an "attitude of liberty" was the central hope for democratic societies at a Liberty Fund Conference at George Mason University in 1986.

[8] There have been many thousands of journal pages devoted to Rawls's second principle, The Difference Principle, but relatively little attention paid to the Equal Liberty Principle, the first principle. This is quite odd given that the first principle is supposed to be lexically ordered with the second, i.e., there is an absolute priority of equal liberty over the difference principle. It is interesting to speculate that if one fully works out the best mechanisms for assuring equal liberty, those same mechanisms may also promote equality as far as possible. Get the first principle right, in the details, and one may not need the second principle. This is the spirit of the argument of Lomasky (1987).

Nozick's argument is roughly: pick any redistribution that you want. No matter how perfect, people would undertake trade to improve their lot, thereby necessitating more redistribution, negating the effects of the voluntary trades. It is a hallmark of libertarian views that voluntary acts among consenting adults ought to count as morally permissible, provided that they impose no substantial costs on any third party.

So, libertarian theories are often contrasted with liberal theories.[9] According to Alan Ryan (1995: 296) the line is not an easy one to draw.

> Both are committed to the promotion of individual liberty; both rest most happily on a theory of human rights according to which individuals enter the world with a right to the free disposal of themselves and their resources. The line of cleavage lies between the libertarian view that government is not a necessary evil but a largely (and for so called "anarchocapitalists" a wholly) unnecessary evil and the liberal view that government power is to be treated with caution, but like any other instrument may be used to achieve good ends.

It is interesting to note here that the difference between libertarian theories and liberal theories may well be a difference in what actually happens in the real world. Libertarians may believe that the best means to promote liberty, etc., are markets and limited, at most, governments, while liberals may believe that the best means to promote liberty include the welfare state. While the differences in theory are not always so obvious, political philosophers have done a good job of actually ignoring what we might call "differences of fact."[10]

Alternatively, libertarianism can be contrasted with utilitarianism. Jean Hampton has put the point quite contentiously: "Whereas utilitarianism might be said to allow individuals to be held hostage to the well-being of the community, libertarianism might be said to allow the community's well being to be held hostage to the rights, and in particular the property rights, of individuals" (Hampton 1997).

While the focus of libertarian theory has been on some form of freedom/liberty, some analysis of rights, property rights in particular, and some arguments for the limited role of government, there is also another less obvious tenet of most libertarian thought. Most libertarians must have strong beliefs about personal responsibility.

First of all, on any libertarian account, persons are responsible for themselves. The equal liberty principle makes no sense if A may do whatever she likes to B. The "compatible with a like liberty for all" clause is crucial. The libertarian must assume that people are capable of controlling their actions so that they do not harm others. Second, when such boundary crossings or harms occur, the offending party must make some attempt at reparation. The alternative to this strong notion of individual responsibility is making some collective like the state responsible for repairing any

[9] Some libertarians take issue with the distinction between liberalism and libertarianism claiming that the term "liberal" originally applied to what is today termed libertarianism and the advocates of current "liberalism" co-opted the term from "classical liberals" (i.e., libertarians) like Locke. See Lomasky (1987). For the sake of clarity we will employ the more common contemporary usage.

[10] To say there are differences of "fact" is not to countenance the fact-value distinction. More carefully we might say that libertarians and liberals tell different stories about our relationship to our history, and indeed, even write quite different histories.

damage that is done, but the whole point of the minimal state is that such "collective responsibility" carries severe "freedom denying" penalties. So, libertarians do or ought to accept some variant of a principle about responsibility. The Responsibility Thesis says:

> The basis for ethics or the moral point of view is that most people, most of the time, take or want to take responsibility for the effects of their actions on others. And, if they did not, then what we call "ethics" and "morality" would be meaningless.[11]

Now there may well be a particular libertarian version of such a principle which specifies the necessary and sufficient conditions under which a person must make someone that she has harmed "whole," But it is hard to conceive of a libertarian theory that doesn't have some version of the responsibility principle.

We claim that an argument is a libertarian argument if it (1) relies on freedom, liberty, the equal liberty principle, or some kindred notion; (2) relies on basic negative rights, like those defined by Rawls's first principle, and including individual property rights; (3) allows for the creation of positive obligations through various voluntary actions (e.g., contracting and promising); (4) countenances at most a minimal state, as defined by Nozick and others; and, (5) assumes that human beings are largely responsible for the effects of their actions on others.

4 Some Libertarian Arguments for Stakeholder Theory

Now, we want to suggest that in the background of stakeholder theory are libertarian arguments that draw on these five ideas. There are at least two branches of "stakeholder theory" or "managing for stakeholders" that we need to take into account. The first branch is based on what Freeman (1999) has called "the Instrumental Thesis." This thesis suggests:

> To maximize shareholder value over an uncertain time frame, managers ought to pay attention to key stakeholder relationships.

It is easy to see libertarian assumptions running in the background here. We might assume that the corporation is the private property of the shareholders, and that managers have been hired by the shareholders to do what's best for the shareholders. Shareholders must be responsible for the uses of their property, even by their agents, according to the responsibility part of libertarian theory (*respondeat superior*). So, managers who are boundedly rational and acting under real uncertainty, must take the interests of stakeholders into account, else they might misuse shareholders' property to harm others and violate their right to freedom. This argument says

[11] We need appeal to nothing more here than a standard Darwinian account, *a la* Daniel Dennet (1995) that the "responsibility meme" is evolutionarily stable. However there is a deeper connection worth exploring among libertarians, existentialists, and pragmatists.

nothing about treating all stakeholders equally[12] nor does it suggest, even remotely, that managers should take from one stakeholder group and give to another. Rather, the argument recognizes that the stakeholder framework is largely managerial, in the sense that Donaldson and Preston (1995) have pointed out (cf. Freeman 2000).

Why is it important to point out the libertarian roots of the argument for the instrumental thesis? In *Strategic Management: A Stakeholder Approach.* Freeman suggested, in a section entitled "A Plea for Voluntarism," that a growing tendency to look outside the firm for the causes of decline masked the fact that managers had it within their power to influence the external environment (cf. Pfeffer and Salancik 1978). Only by seeing stakeholder relationships subject to managerial influence could managers actually begin to do their jobs of leading the corporation toward its purpose, whatever that purpose may be. He suggests:

> Such a philosophy of management [the stakeholder one, as interpreted as the instrumental thesis] must be based on the idea of voluntarism, if it is to be implemented in U.S. based companies. Not only is voluntarism the only philosophy which is consistent with our social fabric, but the costs of other approaches are simply too high. Voluntarism means that an organization must on its own will undertake to satisfy its key stakeholders. A situation where a solution to a stakeholder problem is imposed by a government agency or the courts must be seen as a managerial failure (Freeman 1984: 74).

Today we might add that it is because of bounded rationality and uncertainty that we cannot trust that a governmental solution will continue to be optimal, even though it might look more favorable than one that deals directly with key stakeholders. Give the state power over a particular area, and it rarely relinquishes control. Again, we appeal to libertarian principles to get this voluntarism argument off the ground. The prominence of responsibility in libertarian thought is reflected in this early discussion of the motivational structure of managing for stakeholders. So, we conclude that the Instrumental Thesis can have a decidedly libertarian flavor, even though it need not, i.e., there are certainly non-libertarian arguments for the instrumental thesis, for it does not depend on the idea of exclusive private property rights or the limited state.

The second branch of stakeholder theory is based on what we might call the "Normative Thesis." It claims:

> Managers ought to pay attention to key stakeholder relationships.

There is no starting point of property rights of shareholders here. In fact some have claimed that the normative thesis must be defended from the pure starting point of ethical theory. While we are skeptical of such claims, we do believe that such a defense can be offered on libertarian grounds.

The first defense argues that in fact we do live in a world of property rights. If shareholders are in fact the owners of the corporation in some sense, then managers must respect the property rights of shareholders, unless they are trumped by more important liberty rights of other stakeholders (or shareholders themselves). Most libertarians would argue that property rights are fairly basic, and rarely trumped by other considerations, indeed that most other considerations can be reduced to

[12] Cf. Marcoux (2000) and Sternberg (2000).

property rights. Therefore, if we add the ingredients of bounded rationality, uncertainty and responsibility, the argument for the normative thesis looks a lot like the argument for the instrumental thesis even where we have not assumed that managers are the explicit agents for the shareholders.

A variant of this argument suggests that each stakeholder has property rights. Consumers have the property right to their wealth. Suppliers have the property right to the supplies that they sell to the corporation. Employees have a property right to their labor. Communities have a property right to public goods.[13] In order to respect these property rights, managers must pay attention to stakeholders.[14]

Yet another variant of this argument relies on the notion of voluntary action. On one interpretation, the firm is a nexus of contracts or the centerpiece of an ongoing multilateral agreement, based on voluntary consent. As indicated above, libertarian theory is not antithetical to the notion of positive obligations among actors. Rather, the argument is that no such duties exist apart from the voluntary actions of actors. There is no natural right to a job, for example, though actors can create reciprocal obligations based on the voluntary action of entering into an employment contract. If there is a weak presumption that the agreement is ongoing, managers must take the interests of all parties to the contract, or the nexus, into account.

In addition to consent- or contract-based obligations, Phillips (1997) has argued for obligations of stakeholder fairness derived from the voluntary acceptance of the benefits of a mutually beneficial cooperative scheme. Such obligations are also consistent with libertarianism inasmuch as they countenance neither a natural nor a hypothetical basis for organizational obligations to stakeholders, but rather one based on voluntary' actions of the organization (or its management). Relying, then, on the responsibility principle, we would expect managers to try to keep the joint interests of stakeholders in balance in any instance where positive obligations have been created through consent, contract, voluntary acceptance of benefits, or any other voluntary action.

In a series of papers, Evan and Freeman made such a libertarian argument in which they went further and tried to set forth conditions under which it would be in the interests of each stakeholder group to conceptualize the firm as such a voluntary agreement. A little-noticed condition in that paper was that such an agreement ought to be enforceable without depending on a specific state regime. Hence, the agreements were to be self-enforcing in order to limit the role of the state.

The classical role of the state to solve coordination problems (*a la* Prisoner's Dilemma), provide or regulate public goods, and serve as a court of last resort, was to be usurped by these voluntary agreements, precisely to limit the involvement of the state in the affairs of the corporation. Evan and Freeman relied on a Rawlsian-like veil of ignorance to make the argument, but it was one with strong libertarian, instead of liberal, background assumptions, Evan and Freeman stipulated (by tacit

[13] Now any libertarian would severely constrain the interpretations of this statement, but however it turns out, even if it is analyzed away as public goods are turned into private goods the point still stands, as some system of property rights will hold sway.

[14] This seems to be the general approach of Donaldson and Preston 1995.

appeal to libertarian background conditions) that the agreements which came out of the veil must work in any set of conditions, and could not depend on particular kinds of state regimes. Indeed, if Evan and Freeman had perhaps more carefully set out who these hypothetical stakeholders were who were trying to set forth ideal conditions for fair agreements, perhaps they would have explicitly made them (1) holders of the equal liberty principle; (2) committed to negative rights, especially property rights; (3) able to create and accrue positive commitments and obligations among members of private associations; (4) distrustful of the state in any guise; and (5) committed to being responsible for the effects of their actions on others. In short they would have been libertarians.

Despite the libertarian arguments that can be adduced for both the instrumental and normative branches of stakeholder theory, ultimately we have to come to see stakeholder theory as managerial. According to Donaldson and Preston (1995), stakeholder theory is managerial in the sense that it recommends courses of action for managers and deals *at once* with normative, instrumental, and descriptive claims. They write.

> The stakeholder theory is managerial in the broad sense of that term. It does not simply describe existing situations or predict cause-effect relationships: it also recommends attitudes, structures, and practices that, taken together, constitute stakeholder management. Stakeholder management requires, as its key attribute, simultaneous attention to the legitimate interests of all appropriate stakeholders, both in the establishment of organizational structures and general policies and in case-by-case decision making. This requirement holds for anyone managing or affecting corporate policies, including not only professional managers, but shareowners, the government, and others. Stakeholder theory does not necessarily presume that managers are the only rightful locus of corporate control and governance. Nor does the requirement of simultaneous attention to stakeholder interests resolve the longstanding problem of identifying stakeholders and evaluating their legitimate "stakes" in the corporation. The theory does not imply that all stakeholders (however they may be identified) should be equally involved in all processes and decisions (1995: 75–76).

To further fill out the conception of stakeholder theory as managerial, we would add that to see stakeholder theory as managerial is to see it as intimately connected with the practice of business. First and foremost, stakeholder theory is about business and capitalism. And it is here that the libertarian background assumptions of stakeholder theory come out the strongest.

Business is that human institution that is about value creation and trade. Libertarian theorists point out that value creation and trade are older than the idea of governments, and that both have gone on independently and in spite of specific state regimes. Note that this is not the claim that the state has no impact on value creation and trade, but the claim that it need not. Value creation and trade take place across state regimes, and among actors who live in a multiplicity of different state regimes. Indeed, as trade becomes increasingly global in scope, the regulatory power of any single government is diminished.

Like Follett (1994) and Barnard (1938) before. Freeman (2000) has argued that the engine of value creation and trade is the human desire to create. It comes from the many values that we hold that drive us to take a stand and do something with our lives. It comes from what Harold Bloom calls the "strong poet," the person who shows us how to live differently. Freeman also argues that the desire for solidarity fuels capitalism, the desire to come together and build something which no single person can accomplish. Now both of these desires emerge in theory when we admit a complex psychology of human beings. There is no reason for libertarians to deny such a complex psychology, though often one finds libertarian analyses which assume that most people are narrowly economic, or self-interested, etc.

If we come to see stakeholder theory as concerned with the practice of value creation and trade, we are led back to "the plea for voluntarism." Regardless of the purpose of the firm, since managers are boundedly rational, and since the world is uncertain, they must pay attention to the consequences of their actions on others. To ignore these others is to put oneself and one's company beyond the pale of morality and ethics, conceived of as liberal, Kantian, Utilitarian, or libertarian.[15]

5 From "Stakeholder Theory" to a Libertarian Stakeholder Capitalism

The arguments of the previous two sections give rise to the question of whether there can be a systematic way to understand business activity that is both libertarian in spirit and attends to the managerial interests inherent in the stakeholder approach. We want to argue that there are a number of principles that any such stakeholder capitalism would have to satisfy.

The hallmark of libertarian theory is one of consent and agreement. Free people have the right to make agreements with others, even if some of these agreements limit their own freedom.[16] The first three criteria for what makes a theory a libertarian one are freedom, rights, and the creation by consent of positive obligations. Our underlying notion of why capitalism works is ultimately due to these three ideas and how they interact. Business is founded (and businesses are created) on this idea of making agreements with each other. And we are free to make these agreements because others are not permitted to interfere (so long as they are not substantially affected).[17] Entrepreneurs see the possibility of creating value where others do not. They contract with suppliers, employees, suppliers of finance, and customers, as well as others, to start and build firms. In other words they create a set of positive obligations among those parties that are affected. We might capture these intuitions in the following principle:

[15] That business is in fact "beyond the pale" at least in its perception is precisely the claim that the Separation Thesis is at the heart of this debate about the proper role of stakeholder theory.

[16] The limiting case of whether or not a libertarian theory countenances "selling oneself into slavery" is beyond our scope here, and is probably uninteresting.

[17] We use "substantially" here because we note that counterexamples are easy to come by, but as Rawls. cautions us, we shouldn't always give in to them.

> *The Principle of Stakeholder Cooperation* says that value is created because stakeholders can jointly satisfy their needs and desires by making voluntary agreements with each other.[18]

Value creation and trade is not a zero-sum game. Capitalism works because entrepreneurs and managers put together and sustain deals or relationships among customers, suppliers, employees, financiers, and communities. The support of each group is vital to the success of the endeavor and the outcomes are synergistic. This is the cooperative common-sense part of business that every executive knows. It is deeply libertarian since it is rooted in the notion that voluntary action is the well-spring of capitalism. When stakeholders pool their resources to create something, no one has the right to prevent their actions, provided they do not impose substantial harms on innocent third parties.

Having the freedom to make agreements is as important for customers who purchase products as it is for employees who agree to take direction and work for some corporate objectives in return for money, satisfaction, knowledge, or whatever the particular agreement authorizes. More subtly, communities are also a part of the agreement structure of business, since they provide air, water, schools, roads, protection from harm, and other so-called "public goods."[19]

Whether these agreements among stakeholders are to be understood as bilateral agreements or multilateral agreements is an interesting question. However, if these agreements are to be sustainable over time, they must include some element of fairness, as Freeman (1994) has argued. There are many different conceptions of fairness that may be applied here. And there are many kinds of agreements that can be termed "fair." Freeman (1994) suggests one method based on Rawls's theory of justice (cf. Phillips and Margolis 1999), called "the doctrine of fair contracts." but there are others that would work equally well (Deutsch 1975). In a relatively free and open society, if agreements are not fair, parties to the agreement will seek alternatives, and may well seek state intervention and recompense.

In addition to basic fairness, and prior to it from a libertarian perspective is the question of whether parties to an agreement must be responsible for the outcomes of that agreement. We have suggested above that a strong notion of responsibility is necessary for libertarian theories to be tractable, and if parties to an agreement are responsible, the question of fairness may well become subsidiary. Again, the failure of a notion of responsibility to find a place at the center of our understanding of business is perhaps the reason for the large regulatory role played by government and the courts in our current economy. If liberty is to be preserved, then stakeholder capitalism must include a principle that ensures that parties to the agreements will be responsible for the effects of their actions.

---

[18] The following sections contain some paragraphs from R. Edward Freeman, "Stakeholder Capitalism." *Financial Times,* July 26, 1996. We are grateful to the editors and publisher for permission to use this material here.

[19] Whether these goods could be provided by private means, as some libertarians argue, is not at stake here.

> *The Principle of Stakeholder Responsibility* claims that parties to an agreement must accept responsibility for the consequences of their action. When third parties are harmed, they must be compensated, or a new agreement must be negotiated with all of those parties who are affected.[20]

Responsibility is a tricky concept. It is easy to find simple counterexamples using vague terms like harm (suppose I am "harmed" when you wear pink shirts, for example.). Such examples (see Marcoux and Child, for instance) are a function of the vagueness of "harm" or "responsibility" and count no more against our use of these central moral ideas than against anyone else's use of them. So, regardless of the weaknesses of our best current theory of "responsibility," our claim is that some notion of responsibility is central to the development of stakeholder capitalism.

A second feature of this principle is that it applies reciprocally to all stakeholders. If an entrepreneur, manager, or firm has responsibility for the effects of its actions, so too, do customers, communities, suppliers, financiers, and employees. Firms are not the sole carriers of responsibility in today's world. Customers have a duty to use products as they were intended, or else take reasonable care, including the burden of responsibility, when they do not. Employees have a responsibility to support their employers within reason. Suppliers have the duty to do their best to make the supply chain work properly and be efficient.[21] And shareholders have a responsibility to elect responsible directors who will take seriously their "duty of care" to manage "the affairs of the corporation."

The first two principles of stakeholder capitalism ground business in the freedom of individuals to make agreements, and in the concomitant responsibility that comes with exercising such freedom. The alternative to these two principles is a view that capitalism rests on the idea that "anything goes" and "let the buyer beware," Long a part of the common parlance of capitalism, it is high time to put these ideas to rest. The only possible outcome of these principles is a belief by people that "in business what you do is what you can get away with" and that we need the state to protect citizens from the naked self-interest of business. The resulting "bad public relations" for capitalism and the growth of the modern state are both a testament for our argument that we need to return to the basics of how business can and should operate.[22]

---

[20] We are grateful to an anonymous referee for the suggestion of this principle.

[21] Recent attention to "sweat-shop" working conditions in the developing world has raised questions regarding the responsibility of corporations for the actions of their sub-contractors as well those farther removed in the supply chain (sub-sub-contractors) both "upstream" and "downstream." This is an interesting stakeholder question, but one outside the scope of the present inquiry.

[22] Some might argue that we have committed the naturalistic fallacy here—that we have mistaken how businesses do act with how they should. To begin, we pragmatists will have none of the prescriptive-descriptive, or fact-value distinction. Wicks and Freeman (1998) tries to outline how such a pragmatist view of management theory would work. The story of business that is deeply embedded in our society is not the story of stakeholder capitalism. Thus, it is important to clearly set forth the underlying principles on which this new story, the story of stakeholder capitalism, rests.

Out of this bad reputation for capitalism comes the idea that in business people act solely in their own self interest, and often they go further and act narrowly in a selfish manner. Thus, the popular and scholarly press is filled with "horror stories" about yet another egregious infelicity involving some large multinational. Recently, Werhane (1999) has suggested that most of these cases look like the story of Rashomon, where the description of what happened depends on the point of view that you have. While there are some undoubtedly real "horror stories," it is the assumption that business people must be narrowly self-interested that is really doing the damage. After all, one response to the "horror story" scenario is that in fairness, if business is tarred with the bad, it must at least be feathered with the good consequences. But even this response misses the point. What is missing is a complex psychology of the actors in business. Surely, some are self-interested, and, just as surely, some are other-regarding. Managers, entrepreneurs, and stakeholders must have as complex a psychology as ordinary human beings, for the simple fact that they are ordinary human beings. Once we remove the narrow view of business required by the non-stakeholder story, we can open up the complex entities that are involved.

People have many and varied values. These values cause them to pursue their hopes and dreams in a number of ways. At least some of these pursuits (what Freeman and Gilbert [1988] called "personal projects") involve collaborating and contracting with others to create something of value. Stakeholder capitalism must rest on something like the following principle:

> *The Principle of Complexity* claims that human beings are complex psychological creatures capable of acting from many different values and points of view.

This principle may well seem trivial. But, we argue that the current backdrop of business makes it necessary to spell it out. We are not just self-interested narrowly economic maximizers. The discussions of self-interest vs. altruism, which seem embedded in literatures such as corporate social responsibility, as well as finance, simply miss the mark. Sometimes we are selfish and sometimes we act for others. Many of our values are jointly determined and shared. Capitalism works because of this complexity, rather than in spite of it. A central task of managers and entrepreneurs is to determine an answer to fundamental values questions that may bind together a business entity. There are no obvious "right" answers here. There are many different ways to engage in value creation and trade and also be "an ethical person."[23] Such a principle might rely on the background assumption that free and responsible people who respect each other's freedom will develop complex interests and values, which will lead to many options for themselves and those with whom they form

[23] That there are many ways to run a business is the insight behind the often ignored idea of "enterprise strategy" and its theoretical analog "normative core" It is a separate story whether or not "being an ethical person" makes any sense in isolation from the ideas of value creation and trade. If value creation and trade are fundamental to the human experience, then separating out "ethical person" as the above sentence does, is also illegitimate. Another way to say this is that our analysis points out the need for a political philosophy or a conception of ethics where value creation and trade, rather than the state, play a central role.

relationships. Capitalism, on this view, becomes the voluntary associations of free, responsible, cooperating, consenting, and complex adults.

The final two principles are subsidiary to the first three. They are necessary to correct the mistaken impressions left by the standard story of capitalism. Perhaps it is possible to construct a stakeholder capitalism without these principles, but let them serve as explanatory of the kinds of behaviors that will result from such a new story.

> *The Principle of Continuous Creation* says that business as an institution is a source of the creation of value. Cooperating with stakeholders and motivated by values, businesspeople continuously create new sources of value.

Schumpeter's old saw is "the principle of creative destruction." It claims that at the center of capitalism is the emergence of new value that necessarily destroys old value. We believe that this principle focuses too much attention on capitalism as a closed system. While value destruction does in fact take place, the genius of the corporate form, combined with the genius of entrepreneurship is that such destruction need not destroy businesses. Corporations can (some would say have the duty to) continuously create value. Obviously, continuous creation is possible because of the three principles outlined above, sharing the libertarian roots of these ideas.

In short, the principle of continuous creation claims that the creative force of humans is the real engine of capitalism. One creation doesn't have to destroy another, rather there is a continuous cycle of value creation which raises the well-being of everyone. People come together to create something, be it a new computer program, a new level of service, a way to heal the sick, or simply to work together. It is the creative spirit that results from freedom-loving people that makes capitalism work.

> Finally, *The Principle of Emergent Competition* says that competition emerges from a relatively free society so that stakeholders have options. Competition emerges out of the cooperation among stakeholders, rather than being based on the primal urge to "get the other guy."

This principle seeks a corrective measure in the now largely dominant idea that capitalism is first and foremost about "anything-goes competition." In a relatively free and open society, people are "free to compete," or "free to offer alternatives." The entrepreneurial process ensures that there is an outside "equilibrating force" that adds pressure for managers to manage for stakeholders. When they don't, yet another stakeholder network is capable of forming (Venkataraman 2002). Competition is important in Stakeholder Capitalism, but it is an emergent property. The psychological connection between the desire to create and achieve and the desire to compete—to beat others—is an interesting question. Perhaps in the end, both are necessary, or connected in important ways. Suffice it for our purposes to call attention, once again, to the creative force and the complex human psychology that is necessary to combine libertarian ideas of freedom with managerial ideas of stakeholder theory.

Stakeholder capitalism invokes a focus at the level of how value is created, rather than at the societal level of value distribution, or the accrual of large amounts of capital and control over it. Capitalism in this sense has little to do with nineteenth-century robber barons, the emergence of institutions for trading stocks, or the ability to accumulate and make available to entrepreneurs large supplies of capital. All of these issues are important, but they are subsidiary to the value-creation process. Our five principles form a basis for assessing the most basic nature of business, or value creation and trade. How far the enactment of our current story is from these principles is a matter for further debate for academics and policymakers at both the corporate and societal levels. Our suggestion is that, by focusing on these principles, we can reorient capitalism toward an ethics of freedom and responsibility—one that inherently marries business and ethics.

Stakeholder Capitalism requires that freedom-loving human beings be at the center of any process of value creation and trade. It underscores the responsibility thesis that common decency and fairness are not to be set aside in the name of playing the game of business. It suggests that we should demand the best behavior of business, and that we should enact a story about business that celebrates its triumphs, admonishes its failures, and fully partakes of the moral discourse in society as a routine matter. Yet, Stakeholder Capitalism is no panacea. It simply allows the possibility that business becomes a fully human institution. There will always be businesspeople who try to take advantage of others, just as there are corrupt government officials, clergy, and professors. Stakeholder Capitalism bases our understanding and expectations of business not on the worst that we can do, but on the best. It sets a high standard, recognizes the common sense practical world of global business today, and asks managers to get on with the task of creating value for all stakeholders.

References

Alkhafaji, A.F. 1989. *A Stakeholder Approach to Corporate Governance: Managing in a Dynamic Environment*. New York: Quorum Books.

Barnard, Chester I. 1938. *The Functions of the Executive*. Cambridge, MA: Harvard University Press.

Barnett, Anthony. 1997. Towards a Stakeholder Democracy. In *Stakeholder Capitalism*, ed. Gavin Kelly, Dominic Kelly, and Andrew Gamble, 82–98. London: MacMillan Press.

Barton, S.L., S.C. Hill, and S. Sundaram. 1989. An Empirical Test of Stakeholder Theory Predictions of Capital Structure. *Financial Management* 18 (1): 36–44.

Berman, Shawn L., Andrew C. Wicks, Suresh Kotha, and Thomas M. Jones. 1999. Does Stakeholder Orientation Matter? The Relationship Between Stakeholder Management Models and Firm Financial Performance. *Academy of Management Journal* 42 (5): 488–506.

Boatright, John R. 1994. What's So Special About Shareholders. *Business Ethics Quarterly* 4 (4): 393–408.

Burton, B.K., and C.P. Dunn. 1996. Feminist Ethics as Moral Grounding for Stakeholder Theory. *Business Ethics Quarterly* 6 (2): 133–148.

Carroll, A.B. 1993. *Business and Society: Ethics and Stakeholder Management*. 2nd Edition. Cincinnati, OH: South-Western Publishing.

Carroll, A.B., and Ann K. Buchholtz. 2000. *Business and Society: Ethics and Stakeholder Management*. 4th ed. Cincinnati: South-Western.

Clarkson, M.B.E. 1994. *A Risk Based Model of Stakeholder Theory*. Toronto: University of Toronto Working Paper.

———. 1995. A Stakeholder Framework for Analyzing and Evaluating Corporate Social Performance. *Academy of Management Review* 20 (1): 92–117.

Dennett, Daniel. 1995. *Darwin's Dangerous Idea*. New York: Simon and Schuster.

Deutsch, M. 1975. Equity, Equality, and Need: What Determines Which Value Will Be Used As the Basis for Distributive Justice? *Journal of Social Issues* 31 (3): 137–149.

Donaldson, T., and T.W. Dunfee. 1999. *Ties That Bind*. Boston: Harvard Business School Press.

Donaldson, Thomas, and L.E. Preston. 1995. The Stakeholder Theory of the Corporation Concepts. Evidence, and Implications. *Academy of Management Review* 20 (1): 65–91.

Evan, William M., and R. Edward Freeman. 1993. A Stakeholder Theory of the Modern Corporation: Kantian Capitalism. In *Ethical Theory and Business*, ed. Tom L. Beauchamp and Norman E. Bowie, 4th ed. Englewood Cliffs: Prentice-Hall.

Follett, Mary Parker. 1994. *Mary Parker Follett—Prophet of Management: A Celebration of Writings from the 1920s*, ed. Pauline Graham. Boston: Harvard Business School Press.

Freeman, R. Edward. 1984. *Strategic Management: A Stakeholder Approach*. Boston: Pitman Publishing Inc.

———. 1994. The Politics of Stakeholder Theory: Some Future Directions. *Business Ethics Quarterly* 4 (4): 409–422.

———. 1996. Stakeholder Capitalism. *Financial Times*, July 26.

———. 1999. Divergent Stakeholder Theory. *Academy of Management Review* 24 (2): 233–236.

———. 2000. Business Ethics at the Millenium. *Business Ethics Quarterly* 10 (1): 169–180.

Freeman, R. Edward, and William Evan. 1990. Corporate Governance: A Stakeholder Interpretation. *The Journal of Behavioral Economics* 19 (4): 337–359.

Freeman, R. Edward, and Daniel R. Gilbert Jr. 1988. *Corporate Strategy and the Search for Ethics*. Englewood Cliffs: Prentice-Hall.

Frooman, J. 1999. Stakeholder Influence Strategies. *Academy of Management Review* 24 (2): 191–205.

Goodpaster, K.E. 1991. Business Ethics and Stakeholder Analysis. *Business Ethics Quarterly* 1 (1): 53–73.

Hampton, Jean. 1997. *Political Philosophy*. Boulder: Westview Press.

Hill, C.W.L., and T.M. Jones. 1992. Stakeholder-Agency Theory. *Journal of Management Studies* 29: 131–154.

Hutton, Will. 1995. *The State We're In*. London: Jonathan Cape.

Jones, Thomas M. 1995. Instrumental Stakeholder Theory: A Synthesis of Ethics and Economics. *Academy of Management Review* 20 (2): 404–437.

Kelly, Gavin, Dominic Kelly, and Andrew Gamble, eds. 1997. *Stakeholder Capitalism*. London: MacMillan Press.

Lampe, Marc. 2001. Mediation as an Ethical Adjunct of Stakeholder Theory. *Journal of Business Ethics* 31: 165–173.

Locke, John. 1690/1952. *The Second Treatise of Government*. New York: Macmillan Publishing Company.

Lomasky, Loren. 1987. *Persons, Rights, and the Moral Community*. Oxford: Oxford University Press.

Marcoux, Alexei M. 2000. Balancing Act. In *Contemporary Issues in Business Ethics*, ed. J.R. DesJardins and J.J. McCall, 4th ed., 92–100. Wadsworth.

Mitchell, R.K., B.R. Agle, and D.J. Wood. 1997. Toward a Theory of Stakeholder Identification and Salience: Defining the Principle of Who and What Really Counts. *Academy of Management Review* 22 (4): 853–886.

Nozick, Robert. 1974. *Anarchy, State, and Utopia*. New York: Basic Books.

Orts, Eric. 1992. Beyond Shareholders: Interpreting Corporate Constituency Statutes. *George Washington Law Review* 61: 14–135.a.

———. 1997. A North American Legal Perspective on Stakeholder Management Theory. *Perspectives on Company Law* 2: 165–179.

Pfeffer, Jeffrey, and Gerald R. Salancik. 1978. *The External Control of Organizations: A Resource Dependence Perspective*. New York: Harper and Row.

Phillips, Robert A. 1997. Stakeholder Theory and A Principle of Fairness. *Business Ethics Quarterly* 7 (1): 51–66.

Phillips, R.A., and J.M. Margolis. 1999. Toward an Ethics of Organizations. *Business Ethics Quarterly* 9 (4): 619–638.

Phillips, R.A., and J. Reichart. 2000. The Environment as a Stakeholder? A Fairness Based Approach. *Journal of Business Ethics* 23: 2.

Plender, John. 1997. *A Stake in the Future: The Stakeholding Solution*. London: Nicholas Brealey Publishing.

Quine, W.V.O. 1960. *Word and Object*. Cambridge, MA: MIT Press.

Rawls, John. 1971. *A Theory of Justice*. Cambridge, MA: Harvard University Press.

Rowley, Timothy. 1997. Moving Beyond Dyadic Ties: A Network Theory of Stakeholder Influences. *Academy of Management Review* 22 (4): 887–910.

Rustin, Mike. 1997. Stakeholding and the Public Sector. In *Stakeholder Capitalism*, ed. Gavin Kelly, Dominic Kelly, and Andrew Gamble, 72–81. London: MacMillan Press.

Ryan, Alan. 1995. Liberalism. In *A Companion to Political Philosophy*, ed. Robert Goodin and Philip Pettit, 291–311. Oxford Blackwell's Publishing.

Starik, Mark. 1995. Should Trees Have Managerial Standing? Toward Stakeholder Status for Non-Human Nature. *Journal of Business Ethics* 14: 207–217.

Sternberg, Elaine. 2000. *Just Business*. New York: Oxford University Press.

Van Buren, Harry J., III. 2001. If Fairness is the Problem, Is Consent the Solution? Integrating ISCT and Stakeholder Theory. *Business Ethics Quarterly* 11 (3): 481–499.

Venkataraman, S. 2002. Stakeholder Value Equilibration and the Entrepreneurial Process. *Business Ethics Quarterly, Ruffin Series* 3: 163–173.

Wang, J., and H.D. Dewhirst. 1992. Boards of Directors and Stakeholder Orientation. *Journal of Business Ethics* 11: 115–123.

Werhane, P.H. 1999. *Moral Imagination and Management Decision Making*. New York: Oxford University Press.

Wicks, A.C., and R.E. Freeman. 1998. Organizaation Studies and the New Pragmatism: Positivism, anti-positivism and the search for ethics. *Organization Science* 9 (2): 123–140.

Wicks, Andrew C., Daniel R. Gilbert Jr., and R. Edward Freeman. 1994. A Feminist Reinterpretation of the Stakeholder Concept. *Business Ethics Quarterly* 4 (4): 475–498.

Williamson, O.E., and J. Bercovitz. 1996. The Modern Corporation as an Efficiency Instrument: The Comparative Contracting Perspective. In *The American Corporation Today*, ed. C. Kaysen, 327–359. New York: Oxford University Press.

Wisdom, John. 1970. *Paradox and Discovery*. Berkeley: University of California Press.

R. Edward Freeman is University Professor, Elis and Signe Olsson Professor, Academic Director of the Institute for Business in Society, and Senior Fellow of the Olsson Center for Applied Ethics at the University of Virginia's Darden School of Business.

Robert A. Phillips is George R. Gardiner Professor in Business Ethics, Professor of Sustainability, and Director of the Centre of Excellence in Responsible Business at the York University's Schulich School of Business.

Chapter 8
Business Ethics and Health Care: A Stakeholder Perspective

Mattia J. Gilmartin and R. Edward Freeman

1 Introduction

This article suggests that we need more dialogue on the appropriate model of understanding health care as a business. We argue that the existing dialogue focuses on the standard model of business that we call "cowboy capitalism," which does not do justice to the noble cause of health care organizations. We outline an alternative model of "stakeholder capitalism" that has been developed in recent years by scholars working in business ethics, and suggest that it is a more appropriate model for thinking about health care reform. In doing so we attempt to provide the ethical foundations for much of the work on stakeholder management and health care that has emerged in the last few years. In Sect. 2 we outline the problem of understanding health care in terms of cowboy capitalism. In Sect. 3 we explain how stakeholder capitalism offers a better alternative for understanding business. In Sect. 4 we suggest six areas for applying the ideas of stakeholder capitalism, all of which require further research and dialogue.

Originally published in: *Health Care Management Review*, 27(2), 52–65 © Wolters Kluwer Health, Inc., 2002
Reprint by Springer, https://doi.org/10.1097/00004010-200204000-00006

M. J. Gilmartin · R. E. Freeman (✉)
University of Virginia, Charlottesville, VA, USA
e-mail: FreemanE@darden.virginia.edu

© Springer Nature Switzerland AG 2023

S. D. Dmytriyev, R. E. Freeman (eds.), *R. Edward Freeman's Selected Works on Stakeholder Theory and Business Ethics*, Issues in Business Ethics 53,
https://doi.org/10.1007/978-3-031-04564-6_8

2 The Changing Nature of the Health Care Business

Managers, health care professionals, and service workers are faced with significant internal and external changes in what has been termed the economic era of health service delivery (Brown 1994). External forces shaping the strategic management environment include prospective payment mechanisms to reduce the cost and improve the quality and efficiency of service; growing concern over the 43 million uninsured people in a nation where health care is viewed as a right; and an increasing consumer movement to improve the nature of the health care service interaction between the patient and provider.

Internally, the conceptual model of health care service is changing from a fragmented disease-focused, hospital-based model of health care service to a community-focused system of disease prevention, health promotion, and primary care (Goldsmith 1989; Shortell et al. 1995). Strategically, organizations are faced with creating a fundamentally different system of health care service delivery.

In the United States, the prevailing social value system of health care embraces individualism, sees provider (physician) autonomy as the preeminent value, and neglects a community orientation (Preister 1992). These ideals have shaped the identity of the delivery organization to enact an altruistic role in the service to the uninsured and poor, while achieving exponential economic growth, wealth, power creation, and self-interest over a greater social good (Stevens 1989; Starr 1982). Managed care financing was offered in part as a means to correct the excess and deprivation in health care service delivery (Enthoven 1988; Enthoven and Kronick 1989). Theoretically, the free market forces of managed care were intended to create an equilibrium so that the 40 million uninsured in the United States could gain access to a basic set of health care services. A secondary purpose of introducing free-market forces into the health care sector was to improve the overall quality of health care service.

As a result of these changes, academics, policy pundits, and average citizens have become increasingly skeptical and critical of our health care system in the United States. The major targets seem to be profit-seeking corporations, such as pharmaceuticals, for-profit hospitals, health maintenance organizations (HMOs), preferred provider organizations (PPOs), and other "greedy caregivers." There is a growing concern that "markets" do not work in health care, because critics claim that we can't view health care as just another business.

In particular, Medicare and managed care financing have been singled out as anathemas to the conduct of health care service. The scale of the "moral" crisis in health care service delivery has permeated the national discourse in both the professional and popular media. At the heart of the discourse is the perceived amoral nature of health care's "corporate" transformation in juxtaposition with the social function of health care service. Opponents of the "corporatization of health care" argue that the entry of profit-seeking enterprises has undermined the very core of ethical and just health care service delivery. We believe that these discussions reflect some common misperceptions about both business and health care that we have in our society.[1]

Freeman (Freeman 1994, 1996, 2000) has suggested that our current understanding of business rests on "the Separation Thesis" which is as follows:

> The discourse of business and the discourse of ethics can be separated so that sentences like "x is a business decision" have no moral content, and "x is a moral decision" have no business content.

Business theorists and executives too often appeal to the separation thesis to justify their concepts and their actions. The Separation Thesis leads to the idea that business is amoral at best, and at worst is probably immoral and in need of zealous oversight by regulators and philosophers. We might call this view of business "cowboy capitalism" to represent the free-wheeling-anything-goes idea that comes to mind. (One popular version of cowboy capitalism applies to publicly traded corporations and suggests that the sole responsibility of managers is to the shareholders of those corporations. There is much literature on this view. Suffice it to say that cowboy capitalism is broader than its application to publicly traded corporations. It reaches into the depths of American culture.) According to this view, capitalism works primarily because people are self-interested, even greedy, and seek to win in the competitive struggle that is business.

Obviously, when cowboy capitalism is applied to health care, its noble cause of healing and caring for the sick becomes much less noble. Patient care takes on the motives of a crass materialism, and the values that cause many practitioners to choose the health care field are cheapened. It is easy to see how the view emerges that "markets and capitalism can't be applied to health care." We shall argue that the problem lies with understanding capitalism as cowboy capitalism.[2]

Within health care, there are at least four misconceptions that depend on the Separation Thesis and cowboy capitalism: (a). Health care is not a business; (b). health care operates at a higher moral standard than most sectors of the economy; (c). good business practices have caused the crisis in health care; and (d). free markets must not be allowed to operate in health care. These misconceptions are supported by a set of stories that focus on narrow, bottom-line decision making, and that highlight the failure of the market mechanism within the health care delivery sector.

A dominant idea is that health care is not a business, never has been a business, and should not be operated as a business. A closer examination of the evolution of the health care sector provides a convincing argument that health care is solidly a

---

[1] There is a vast literature here. Almost any issue of popular and managerial magazines contain articles decrying the state of health care. For instance, see a recent issue of *California Management Review* 43, no. 1 (2000), edited by Sara Beckman and Michael Katz. "The Business of Health Care Concerns Us All" by the editors, and James C. Robinson's "Deregulation and Regulatory Backlash in Health Care" are paradigmatic of the kinds of discussions that take place.

[2] Thomas Rice's recent book, *The Economics of Health Reconsidered*, ASHR, 1998, also contains a number of arguments about the form of market competition that is available in health care. He suggests, and we agree, that the idea of "unfettered, anything goes, capitalism "is a mistake."

business endeavor (Salmon 1995). In 1998 health care spending in the United States topped $1.1 trillion dollars, accounting for approximately 13.5% of the gross domestic product (Levit et al. 2000). Although the health care service sector is dominated by not-for-profit organizations the designation is a convention for organizational missions and tax structures and does not negate the need for revenue generation for sustained operations.

Historian Rose Mary Stevens traces the development of the post World War II health care industry in her book *In Sickness and in Wealth* (Stevens 1989). The modern health care system is built on a value system of volunteerism used to define the role of the hospital and health care in service to society, with the simultaneous growth of physician power, cost-free reimbursement, and technology development. Similarly, Salmon (1995) traced what he terms the corporatization of the American health care sector. The corporatization of health care is conceived as the dual entry of for-profit firms into the sector and the development of corporate purchasing arrangement for service delivery. Salmon identifies four interrelated factors that have fueled the development of the American industrial-medical complex: national health care policy decisions; insurance reimbursement patterns; the funding of medical education, research and practice; and the role of employers and federal government as the primary purchasers of care.

The second issue is that health care operates at a higher ethical standard than other service industries in the national economy. Relying on their own version of the separation thesis, health care delivery organizations have created the perception that they are unique from other enterprises because the output of service relates to life, death, health, and wellness. Professionals in the domains of health care management, and medical and nursing practice have carried on a discourse regarding the schism in the values of professional providers and the amoral nature of the efficiency-maximizing, profit-oriented business enterprise called managed health care. A focus of the professional discourse rests on the mismatch of biomedical clinical ethics and a narrow conception of business ethics in fulfilling the goals of patient care service. Biomedical ethics are based on the principles of respect, beneficence, nonmalfeasance, and justice (Beauchamp and Childress 1994; Friedman and Savage 1998) used to guide clinical decision making within a system that values individuality and autonomy of the provider and the patient. The entry of a narrow set of corporate values based on profit generation, cost minimization, and the efficient allocation of economic resources for health care delivery represented by managed care are viewed as in opposition to the shared value of do no harm in patient care service.

Heater (1996) considers the identity of the health care consumer under the various modes of financing. She argues that over time the identity of the primary customer of health care service delivery has shifted from patients or those in need, to the physicians, to the third party payer/benefits provider. For Heater the health care delivery organization focuses its actions on satisfying the expectations of the benefits providers at the expense of the patient, professional groups, and society. Managed care complicates this analysis by shifting the focus of the third party payment form patient to the employer. David (1999) found that nurses encountered three tensions in their ability to fulfill professional ethics within the corporate health care environment. These tensions revolve around the ability of the practitioner to preserve the essence of nursing, while ensuring patient safety within defined productivity and cost-efficiency guidelines.

Physicians continue the discourse about the amoral nature of managed care in the wake of the changing power balance between providers and third party payers. As Thomas Bodenheimer (1996) puts it, physicians are faced with a reactionary effort to return to the "golden era" of medicine where insurers paid for whatever fees physicians charged, or a righteous movement to correct the excesses of "*unfettered market competition*" (emphasis ours). In the era of managed care the physicians' professional autonomy to prescribe a treatment plan in pursuit of their noble cause of care of the sick, irrespective of the cost efficiency or effectiveness of a given treatment, is at stake.

Given the values of health care service one would expect that health care organizations would be expert at creating an environment where the patient is treated with empathy, civility, integrity, and generosity—namely the attributes of exemplary customer service (Barry et al. 1990). Collins and Porras (1994) show many instances of more traditional for-profit businesses that have such values as their corporate purpose. For example 3 M's core values include "a respect for the individual initiative and personal growth." Merck's core values include "profits, but profit from work that benefits humanity." Hewlett-Packard's core values include "respect and opportunity for HP people" and "affordable quality for HP customers" and "profit and growth as a means to make all else possible." Marriott's core values include "people are #1—treat them well, expect a lot, and the rest will follow." Disney's core values include "to bring happiness to millions and to celebrate, nurture, and promulgate wholesome American values." There is little analytical difference between these companies and most health care delivery institutions. But, because we see "health care as a business" in the sense of cowboy capitalism, we never make the transition to seeing how great companies have been managed for many years. The unfortunate irony is that health care delivery organizations have adopted an ethic of customer-oriented service only in recent years (Eisenberg 1997).

The third issue is the apparently common wisdom of the popular discourse suggests that the application of good or at least standard business practices have caused the health care crisis. Examples of cowboy capitalism are used to bemoan the extravagant profits of the insurance/health maintenance organizations/pharmaceutical companies and their executives while care is rationed, professional autonomy and working conditions erode, physician salaries dwindle, and delivery organizations must do more with less (Chamberlain 1999).

Etheredge et al. (1996) examine the effects the "socially amoral economic forces" of purchasing power of buyers; price competition; the practices of the managed care industry; the drive for market share; assumption of insurance risk; investment capital; and new roles for employers and patients in the managed care marketplace. These market-based dynamics frame an analysis of system change effecting four primary groups in the health care sector—employers, providers, health plans, and consumers. The authors suggest that the introduction of market forces into the health care system have resulted in the following amoral actions of the sector's major actors. The employers, who purchase health care benefits, focus on the price of health care benefits without good measures of quality and value. Health plans focus on increasing market clout by competitive pricing strategies. Providers focus on declining professional prospects and fears about their future in terms of financial and clinical rewards and autonomy. And, consumers worry about less choice in providers and eligibility for various types of health care services.[3]

Finally, there is the argument that the application of free markets should never take place in health care. Managed care and its corollary financing mechanisms have created budget horizons and economic constraints that have intensified the scarcity of resources available to produce services.[4] The potentially immoral nature of managed care is highlighted when its processes focus on economics, cost containment, and profit at the expense of care of appropriate quality, the integrity of professionals, and the primacy and autonomy of patients (Camunas 1998). Cost containment and profit maximization for the institution is used to discuss reimbursement policies that micromanage the clinical judgment of health care professionals, the restructuring and de-skilling of hospital-based nursing staff, and the exclusion of the vulnerable and at-risk from insurance coverage under managed care arrangements (Chamberlain 1999). These discussions highlight the many ethical challenges necessary to balance the noble cause of health care service in an environment that is focused on short-term economic objectives indicative of actions more aligned with the separation thesis.

The era of managed care is not the first time that health care organizations have coped with resource scarcity and market mechanisms. Based in part on the third party payer system, health care delivery organizations have not been required to attend to enterprise issues such as the balance of production cost and revenue, exemplary customer service, and process innovation to reduce cost and improve quality required in typical markets. A market mechanism of buyers, sellers, and other stakeholders has existed in the health care sector since the inception of the

[3] Recently Bigelow and Arndt studied the application of business techniques such as product line strategy, cost accounting systems, diversification, restructuring, and reengineering to see whether the results of the adoption of such "businesslike" approaches are positive. They conclude that the claims are largely unsubstantiated. We suggest that it is difficult to untangle these approaches from their underlying reliance on ST and cowboy capitalism. So we would take this work as more evidence that the wrong model of business has been applied to health care. See Bigelow, B., and Arndt, M. "Great Expectations: An Analysis of Four Strategies." *Medical Care Review* 51, no. 2 (1994): 205–32; Arndt, M., and Bigelow, B. "The Adoption of Corporate Restructuring by Hospitals." *Hospitals and Health Services Administration* 40, no. 3 (1995): 332–47, and Arndt, M., and Bigelow, B. "Reengineering: Déjà vu All Over Again." *Health Care Management Review* 23, no. 3 (1998): 58–66.

[4] Nobel Laureate Milton Friedman has recently argued that the dissatisfaction with managed care is a natural outcome of Medicare, which essentially made health care free for a large part of the population, driving demand to infinity. Friedman argues that rationing has to occur in such circumstances. His solution, a reform of insurance markets to put the onus back on the patient to choose and share risks, is controversial and beyond our present purposes. It merely points out that the application of "markets" to health care is more complicated than it might first appear. See Friedman, M. "How to Cure Health Care." *The Public Interest* [Winter 2001]: 3–30.

modern health care system. The narrow view of markets based solely on costs and profits neglects many positive attributes of the free market such as the power of invention, innovation, and entrepreneurship. Another positive feature of the free market mechanism absent in the discourse is the role of the client/consumer in shaping new or better services, or the development of new relationships between stakeholder groups to serve the broad interests of a particular community.

The entry of managed care and the rise of cowboy capitalism within the health care sector is used to highlight the failure of the market mechanism in the sector. The failure of various organizations to satisfy stakeholder interest over time has led to a raft of legislation that seeks to protect the interests of professionals and consumers alike. The most dramatic stories of the managed care backlash focus on legislation to ensure consumer rights and safety for services such as mandatory 24-hour lengths of stay following child-birth, nursing staff-patient ratios, or basic level of services provided by health plans (Anonymous 1999). In the medicopolitical arena, the American Medical Association's desire to be represented by a collective bargaining unit to regain power over clinical autonomy and economic interests highlights the shifting power balance within the health care sector (Scheffler 1999).

The issues described above rely on a set of common misconceptions about health care and its institutions. They are anchored in the separation thesis, which becomes a self-serving mechanism to focus on the "worst" aspects of business, rather that learning from the "best" aspects of business to bring about fundamental change in health care service. These critiques based on cowboy capitalism are used to stress the shifting power balance between the key stakeholders in the health care sector and serve two distinct purposes. First, they are used to reinforce our society's beliefs, values, and mental images about what health care service should be like and how consumers, professionals, and organizations should behave. Second, they reinforce and point out the negative beliefs that we hold about the business enterprise as a heartless, greedy behemoth. Ironically, these critiques of applying markets to health care rightly point at the need for reform. However, the necessary reform must be informed by the best of what business can be, namely a model of noble cause service based on the principles of stakeholder capitalism.

3 Stakeholder Capitalism

A group of scholars working in a number of disciplines for the last 15 years have begun to challenge this orthodoxy. Their ideas have coalesced around the notion that organizations can be conceived of as interactions of groups who have a stake in the organization. Such a view of "stakeholder theory" can serve as an antidote to cowboy capitalism. By focusing on stakeholders, these theorists, and the many executives who have begun to "manage for stakeholders," have refocused the role of both managers and organizations toward understanding the consequences of their actions on all the parties that are affected.

The key insight of this group of scholars is that ethics is relevant to business, either because business actions harm or benefit various stakeholders, or because stakeholders have rights, or because stakeholders should be part of the processes which meaningfully affect their lives. Regardless of the particular ethical orientation we can see the foundation of what has come to be called "stakeholder theory" resting on a fundamental ethical principle: The Responsibility Thesis (RT).[5] Or:

> The basis for ethics, or the moral point of view, is that most people, most of the time, take or want to take, responsibility for the effects of their actions on others. And, if they did not, then what we call "ethics" and "morality" would be meaningless.

Now it is easy to see the resulting problem, when the Separation Thesis and the Responsibility Thesis are juxtaposed. They are simply incompatible. If the discourse of business is separate from ethics, then there can be no question of responsibility, hence, the idea emerges that business is at best amoral. Given the nature of health care delivery, which wears its ethics on its sleeve, the combination of a market view based on the Separation Thesis and a values-driven view, based on Responsibility Thesis and stronger ethical principles, just doesn't seem to work. Conflicts will inevitably be diagnosed as stemming from the amorality of the market mechanisms, or the lack of responsibility of for profit firms, or alternatively by "business people" as stemming from the "fact" that caregivers don't understand business.

Many of the scholars who have developed the view that the Responsibility Thesis is relevant to business have developed sophisticated models of "corporate social responsibility," but many of these approaches still separate out the "economic" from "political," "social" and "ethical." And, there are countless studies that try to show that if a company is socially responsible then it also is profitable. The problem with this literature is that there is no clear consensus on what is economic and what is social, and there is a temptation to therefore marginalize the "social" as "not really part of the business." (Freeman and Liedtka 1991)

We want to suggest that if we replace the Separation Thesis and the resulting cowboy capitalism with a fully morally loaded notion of capitalism called stakeholder capitalism, we will be better able to address some of the fundamental problems of health care delivery .[6,7] These problems exist not because of the use of

---

[5]The Responsibility Thesis is a restatement of a number of principles found throughout ethical theory. Its clearest form is in Hazel Barnes, *An Existentialist Ethic,* Chicago, 1967, where she outlines the idea that choosing to adopt the ethical point of view, and hence the relevance of ethical discourse, involves seeing oneself as responsible for one's actions. Freeman, R.E. and Phillips, R.A. (in press) "Stakeholder Theory: A Libertarian Defense." *Business Ethics Quarterly* trace the idea to the foundations of liberal democratic theory.

[6]For a related use of stakeholder theory in health care see Kenman L. Wong, *Medicine and the Marketplace: The Moral Dimensions of Managed Care.* Notre Dame: University of Notre Dame Press, 1998, at 130ff; and, Edward Spencer, Ann Mills, Mary Rorty, and Patricia Werhane, *Organization Ethics in Health Care.* New York: Oxford University Press, 2000, at 61ff.

[7]This section contains passages from Freeman, R. Edward, "Business Ethics at the Millennium." *Business Ethics Quarterly* 10, no. 1 (2000): 169–80. We are thankful to the editor of *Business Ethics Quarterly* for permission to use this material here.

markets, but because we use the wrong understanding of markets, one imbued with the Separation Thesis rather than the Responsibility Thesis.

Stakeholder capitalism can be seen to rest on a few fundamental principles (Enthoven and Kronick 1989). First of all **The Principle of Stakeholder Cooperation** says that value is created because stakeholders can jointly satisfy their needs and desires. Value creation and trade is not a zero sum game. Capitalism works because entrepreneurs and managers put together and sustain deals or relationships among customers, suppliers, employees, financiers, and communities. The support of each group is vital to the success of the endeavor. This is the cooperative common sense part of business that every executive knows, but the separation thesis leads us to believe that the shareholder is always more important than others. Try building a great company without the support of all stakeholders. It simply cannot be sustained.[8]

Because this principle is rooted in the interests of stakeholders, the corporation becomes a clearinghouse or nexus of activity where stakeholders satisfy their desires. Far from being meaningless, the corporation becomes an institution imbued with meaning from many different perspectives. Freeman and Gilbert Jr. (1987) have suggested that employees need not pursue only careers in such corporations, but the corporation becomes a vehicle, a mere means, if you like, to employee and other stakeholder ends. Note however, that such meaningful work has to satisfy the desires and interests of other parties to the agreement. All corporations are not managed for the benefit of any one group, though some may in fact be so managed.

Of course this is plainly obvious in the sphere of health care delivery. Blair, Fottler, Savage, and colleagues (Blair and Fottler 1990; Blair et al. 1996; Savage et al. 1997; Blair and Buesseler 1998) have documented the interactions among multiple stakeholders in health care. These stakeholders have multiple interests, and we need to come to see health care delivery mechanisms as clearing-houses for the satisfaction of these interests. Of course sometimes these interests do conflict, but the job of the manager or entrepreneur is to design the institution so that over the long run, the interests of all are satisfied (Quinn and Rohrbaugh 1983). When the interests of one particular group are not satisfied over time (as opposed to on a particular issue) then that group will use the political process to try to get regulatory relief. Hence, the raft of impossible regulations with which health care professionals must cope. Alternatively, Venkataraman (Venkataraman, 2002) suggested that entrepreneurs represent an equilibrating force, whereby new institutional forms could be created to better satisfy stakeholder needs. To do so, however, requires that we come to see capitalism along the lines set forth here, rather than with the shopworn metaphors of cowboy capitalism.

Second, **The Principle of Complexity** claims that human beings are complex creatures capable of acting from many different values. We are not just economic

[8] In addition there has been an explosion of the use of alliances and partnerships of all types that add emphasis to this argument that business is based on the idea of cooperation. See Spekman, R., and Isabella, L. *Alliance Competence: Maximizing the Value of Your Partnerships*. Wiley, 2000, and the literature referenced there for an introduction to these issues.

maximizers. Sometimes we are selfish and sometimes we act for others. Many of our values are jointly determined and shared. Capitalism works because of this complexity, rather than in spite of it. It is the wellspring that yields value creation. Given the multiplicity of values, we would expect to see groups of people (some of whom share like values) working together to create what they cannot do alone. All businesses are not alike.

If human beings are complex and multifaceted, it becomes a central task to determine an answer to fundamental values questions that may bind together a particular health care institution. There are no obvious "right" answers here. There are many different ways to engage in value creation and trade and also be "an ethical person."[9] Equivalently there are many possible organizational forms that will deliver health care.

Third, **The Principle of Continuous Creation** says that business as an institution is a source of the creation of value. Cooperating with stakeholders and motivated by values, businesspeople continuously create new sources of value. This creative force of humans is the engine of capitalism. The beauty of the modern corporate form is that it can be made to be continuous, rather than destructive. One creation doesn't have to destroy another, rather there is a continuous cycle of value creation that raises the well being of everyone. People come together to create something, be it a new computer program, a new level of service, a way to heal the sick, or simply to work together.

It is here that there is the most to be gained by health care institutions. Failing to see that markets are a source of innovation, as well as the distribution of scare resources, is perhaps the largest gap in the health care delivery literature. Markets do distribute via the pricing mechanisms, however, entrepreneurs and managers create and innovate. Indeed modern medicine is a testimony to the power of the creative force of markets when we examine the device industry or the pharmaceutical industry. Markets do more than distribute benefits and burdens. They are a source of innovation and new technology and new organizational forms. Health care delivery has not been as robust.

Finally, **The Principle of Emergent Competition** says that competition emerges from a relatively free society so that stakeholders have options. Competition emerges out of the cooperation among stakeholders, rather than being based on the primal urge to "get the other guy." Some forms of cooperation may well better satisfy some stakeholders' needs, so that in a free society stakeholders are free to form many different cooperative schemes. Competition is important in stakeholder capitalism, but it is not the primary force. It is in its ability to manage the tension

[9] That there are many ways to run a business is the insight behind the often-ignored idea of "enterprise strategy" and its theoretical analog "normative core." It is a separate story whether or not "being an ethical person" makes any sense in isolation from the ideas of value creation and trade. If value creation and trade are fundamental to the human experience, then separating out "ethical person" as the above sentence does, is also illegitimate. Another way to say this is that our analysis points out the need for a political philosophy or a conception of ethics where value creation and trade, rather than the state, play a central role.

created by simultaneous cooperation and competition that stakeholder capitalism distinguishes itself.

The obsession with applying the metaphors of competitiveness, and the resulting narrow view of competitiveness that is "cost reduction" has become a dominant feature of health care delivery. While value can be created via "cost reduction," little innovation occurs. And, the resulting concern with "efficiency" does little to motivate employees who have real moral values in their hearts. Again the problem is with how this "cost reduction" comes about, through a concern with values of stewardship, as at Park Nicollet in Minneapolis (Liedtka, 1995) or the actions of a large player in the marketplace, like Medicare.

Stakeholder capitalism implies that human beings are required to be at the center of any process of value creation and trade. It underscores the responsibility thesis that common decency and fairness are not to be set aside in the name of playing the game of business. It is a far more robust way of understanding health care delivery in all of its complexity. It suggests that we should demand the best behavior of health care delivery institutions, and that we should enact a story about them that celebrates their triumphs, admonishes their failures, and fully partakes of the moral discourse in society as a routine matter. Yet, stakeholder capitalism is no panacea. It simply allows the possibility that health care delivery institutions become fully human institutions, based on collaboration and the noble cause of health care, as well as being profitable. But, it recognizes that there will always be people who try to take advantage of others, just as there are corrupt government officials, clergy, and professors. Stakeholder management and the underlying principles of stakeholder capitalism provide a new story of organizational action, a foundation for a moral discourse about the business of health care, and the potential for greater levels of human and organizational achievement within the health care context. We need to be more specific about how this can work.

4 Stakeholder Capitalism as a Framework for Reform

If we come to see health care organizations as businesses in the sense of the responsibility thesis and stakeholder capitalism, we want to suggest that it is possible to revitalize the dialogue about health care reform that will lead to more innovation and value creation. Drawing on the principles of stakeholder capitalism described above we attempt to reframe the debate about the vital and necessary role of business in creating a better health care system in the United States. The stakeholder capitalism framework we develop here is based on the six areas of: (1) seeing the caregiver as potential entrepreneur; (2) seeing businesses such as Wal-Mart, GE, and Hewlett Packard as role models; (3) formulating integrative strategies for multiple stakeholder groups; (4) unleashing the power of innovation; (5) seeing human beings as the creators of value; and, (6) committing to regulatory reform to improve performance.

(1) We must see "caregivers" as potential entrepreneurs.

The most basic market mechanism in health care service is the physician as entrepreneur to build medical practices, attain economic compensation, and fulfill their noble cause. Drucker (1985) describes the successful entrepreneur as a person who "… tries to create value and make a contribution. Successful entrepreneurs aim high. They are not content simply to improve on what already exists, or to modify it. They try to create new and different values and new and different satisfactions" (Drucker 1985) (pp. 28–29). The entrepreneur identifies an opportunity to bring a good or service to people that satisfies a spoken or unspoken need, and ultimately makes life better or more fulfilling. New market opportunities are based the ability of the entrepreneur to identify a gap in services based on a change in demographics, understand an incongruity between how things are and how they should be, or combine new scientific and nonscientific knowledge to create a good or service.

American physicians have had a strong entrepreneurial tradition throughout their professional development in the form of independent practice (Stone 1997). Easily identifiable examples of entrepreneurial physician activities include the development of ambulatory surgery, primary care, or radiology centers to provide patients with more convenient access to physician services than the traditional hospital setting. Physician entrepreneurs in academic medicine have created new clinical and surgical procedures or instruments, clinical software, or web-based services that improve the overall process of diagnosis, treatment, or recovery speed for clients.

The integration of clinical innovation, patient-focused services, and the most effective use of health care resources have been extended in the development of specialist clinics or services in the health care focused factories described by Herzlinger (1998). In the manufacturing sector, the focused factory model has led to the development of unique organizational core competencies, customer focus, and a process of continuous improvement and innovation. Applied to health care service the focused factory model addresses the fragmentation of current care delivery by focusing on the needs of one particular patient/consumer group. The early stage focused factories provide physicians with an avenue to sustain their noble cause of providing patient care services while restoring financial and clinical autonomy many feel is lost under managed care. Two early physician-led healthcare-focused factories applied to high-volume procedures or chronic care services are exemplified in the Texas Heart Institute founded by cardiac surgeon Dr. Denton Cooley, or the California-based cancer care provider Salick Health Care, founded by Dr. Bernard Salick (Drucker 1985).

Of course this model need not be restricted to physicians. Other clinicians such as physician's assistants, nurses, and nurse practitioners can also be seen as entrepreneurs capable of building their own businesses. For example, nurse practitioners seem particularly able to focus on keeping people healthy. It is surely plausible that there is a value creation opportunity in signing up individuals and families and working with them to prevent disease or manage chronic diseases. We have seen the growth of home health care businesses, rehab clinics, and other specialties, each of which necessitates seeing the caregiver as entrepreneur.

An increasing emphasis on primary care, home-based, or health and wellness services along with changing working conditions in the acute care sector have created opportunities for nursing entrepreneurship (Roggenkamp and White 1989). Examples of services developed by nurse entrepreneurs include home health care agencies, legal or organizational development consulting practices, midwifery-led childbirth centers, or primary care clinics. University based enterprises such as the University of Pennsylvania's Penn Nursing Network, or Columbia University's Advanced Practice Nurse Associates offer a range nurse-managed elder care, primary care, and health and wellness services. These nurse-led services fulfill a community health care need; provide educational and research sites for students and faculty; and generate revenue to sustain these activities (University of Pennsylvania, School of Nursing, Penn Nursing Network 2000; Columbia University, School of Nursing, Advanced Practice Nurse Associates (CAPNA) 2000).

The *principle of stakeholder cooperation* provides a basis for the next generation of entrepreneurial activities in health care service. Within a dynamic market environment, the entrepreneur acts as a stabilizing force to balance and coalesce stakeholder interests by providing a new product, service, or invention (Savage et al. 1997). The changes in population demographics, advances in information and durable medical technologies, and the growing consumer dissatisfaction with existing health care services provide a host of entrepreneurial opportunities. The above examples of consumer-focused health care service demonstrate the wide range of provider-led innovation and entrepreneurship base on unique knowledge, skills, and abilities.

Health care professionals practice their art and science with the noble goal of achieving improvements in health or the elevation of suffering from disease in the patients that they serve. However, the noble goal of patient care and the value of this service to society have not been linked in a manner that promotes continuous change and development. Within health care, a traditional resistance to entrepreneurship has been identified as one barrier to the development of more consumer/patient-service-oriented health care delivery institutions (Bodenheimer 1996). A managing-for-stakeholders framework recasts the professional role as one of entrepreneur possessing a wellspring of knowledge, skill, and expertise necessary for innovative solutions for the underlying problems in health care service.

Extending the image of the health care provider as entrepreneur with a professional responsibility to contribute to service change activities presents a challenge for those who manage and educate health care professionals. New skills in the areas of human resource management, leadership, and enterprise management are necessary. Professional socialization to develop an ethos of entrepreneurship, opportunity identification, change, and innovation into practice routines is required in the organizational setting. Finally, research to gain an understanding of managerial and leadership characteristics of entrepreneurs and organizations to support these activities are also needed.

(2) We need to take Wal-Mart, GE, and Hewlett Packard, for example, as instructive models for health care delivery.

Much of the literature on health care delivery focuses on the "cost" of health care, lamenting the steady tide of rising costs. Surely this is a mistake for two reasons. First, the cost of health care is related to the value produced, and the value produced is surely in the eyes of the consumer. Rarely are there arguments that try to quantify the costs and benefits of a particular health care delivery system, or component. Of course such quantification is complex, but without it, these cost estimates are the equivalent of telling people that we spend $X billion on flour without telling them how much bread we bake. Second, the cost of health care is relative to a particular system of accounting for those costs, which is relative to a particular way of delivering health care services. Just as Wal-Mart forever changed retailing by questioning the costs of retailing, and delivering goods and services in a different manner, so too is innovation possible in health care. Just as Wal-Mart is committed to everyday low prices there is room for some entrepreneur to create how to deliver great health care at everyday low prices. The financing mechanisms will follow such innovation, even though the current third party payer-government insurance system may not survive.

Coming to see Wal-Mart, and other great companies such as GE and Hewlett Packard as instructive examples in health care will ensure that the *Principle of Emergent Competition* can function. When a particular delivery system fails to meet customer needs, the customer will have the opportunity for exit, affiliating with another network or organization that is more in tune with their particular case. Of course, the idea of "customer" in health care is complex and multilayered. But, ultimately health care delivery organizations must be responsive to the end user who pays. And, it doesn't matter if that end user has bought insurance, has benefits from work (paid for by his/her labor), or pays out of pocket. Health care organizations that want to grow and continue to gain customers will have to be innovative in managing their supply chain (similar to Wal-Mart's emphasis on distribution, GE's emphasis on supply chain management, and even Hewlett Packard's spin-off of Agilent). They will have to invest in infrastructure and capabilities, just as Wal-Mart invested in a state-of-the-art information system with a satellite network as the backbone. And, more to the point, they will have to use their noble cause to determine what counts as a cost and what counts as an investment. Costs have to be minimized while investments have to be cared for and nourished. A noble cause of great health care at everyday low prices is an idea whose time has come. And, there are other noble causes for health care organizations. Just as GE has a different purpose from Merck, so too can this be true for health care organizations. But, such a reconceptualization is possible only if we come to see health care as a business in the sense that we described earlier, enmeshed in a network of stakeholders and capable of pursuing its own noble cause.

(3) We need to build integrative strategies that appeal to multiple stakeholders.

The principle of *stakeholder cooperation* says that value is created because stakeholders can jointly satisfy their needs and desires. Capitalism works because entrepreneurs and managers put together and sustain relationships among customers, suppliers, employees, financiers, and communities. The power of this management principle relies on the ability of the managers to satisfy the needs of various stakeholder groups over time. A variety of collaborative relationships among internal and external stakeholders are emerging to improve the quality of patient care delivery within the inpatient setting or serve a greater public health need of local communities.

Increasing pressures to focus on community and population based health needs within a framework of financial risk sharing has led to the development of strategic alliances and collaborative organizational relationships between external stakeholder groups (Kaluzny and Zuckerman 1992; Bums et al. 2000).

Collaborative relationships among delivery organizations, which traditionally compete in a local market, are used to combine resources and provide services to meet an overarching community health need. The efforts of Mercy Health Services in Detroit, Michigan provides one of a growing number of examples of the power of stakeholder cooperation in achieving new levels of community focused health care service. In collaboration with Henry Ford Health System and the University of Michigan Health System the three organizations have developed targeted community health programs. Programs developed under this strategic alliance include an immunization registry and rate improvement program for school-aged children, a care delivery agency, and a health and human services center located in the Ann Arbor Public Schools (Mercy Health Services 1997).

Internal stakeholder relationships can be used to forge a patient-centered service culture based on collegial relationships among health care professionals. The failure of collaborative working relationships between nurses, nurses and physicians, and other health care professionals unfortunately characterized many organizational settings (Aroskar 1998). A growing body of health services research documents the power of collaborative working relationships between health care providers and the effects of these relationships on the quality of patient care (Knaus et al. 1986; Shortell et al. 1994; Aiken et al. 1994) The *principle of stakeholder cooperation* provides a powerful management tool to open a dialogue between professional groups to build new relationships based on a shared understanding of the knowledge, expertise, and challenges that each group brings to the production of patient care services.

For example, Jon Meliones, Chief Medical Director, describes the development and implementation of the balanced scorecard method to support the noble mission of high quality and compassionate care within an efficient organization at Duke Children's Hospital. The adaptation of the approach aligned the goals of financial health, customer satisfaction, employee satisfaction, and so forth. By focusing the internal processes on serving multiple stakeholders simultaneously, the organization was able to execute a turnaround (Meliones 2000).

(4) We need to unleash the power of innovation in health care delivery.

The clearest examples of the creative powers of the managing for stakeholders approach come from the durable medical and pharmaceutical segment of the health care market. Within these sectors, free-market mechanisms have created an environment of innovation and learning to get new products to market, and support investments in research and development. Organizations in these health care segments have effectively paid attention to the internal and external factors affecting the overall success of the enterprise. Investments in technological innovation have given us equipment to peer into the inner recesses of the human body, pharmaceuticals that cure or keep disease at bay, and a multitude of technological and service innovations to improve health and wellness. The newest wave of research in this sector blends advances in biotechnology to unlock the mysteries of the human body and ultimately eradicate disease at the cellular level.

In the last 50 years the advances in durable medical and pharmaceutical technologies have exceeded all expectations in treating or curing a host of complex diseases. The advance of these technological innovations have shaped professional education and practice, clinical care routines, and the organization of health care services to focus on complex care needs of the very sick. However, the complexity of care that these technological innovations support have contributed to the current crisis in health care service that retard the positive and creative effects of market-forces within the sector.

Christensen et al. (2000) discuss the role of disruptive innovations as a possible cure for the resistance to a market based model of health care service. Disruptive innovations are typically developed by upstart companies that introduce a cheaper, simpler, more convenient product or service that starts by meeting the needs of less-demanding customers. Over time, the disruptive innovations take over the market because they enable a larger population of less-skilled people to do things in a more convenient, less expensive setting that historically could be performed only by expensive specialists in centralized, inconvenient settings. What seems exotic and expensive at an earlier time, actually becomes simpler and less expensive, as the disruption takes hold. Generally, the losers here are the organizations that cannot manage the disruption.

The stakeholder capitalism principle of *continuous creation* provides a frame to examine the power of innovation within a market setting. The continuous cycle of value creation, human creativity, and the cooperation of various stakeholders to create something new are illustrated in the innovation dynamic. Changing patterns in the prevalence of chronic disease and greater access to health information by consumers creates the opportunity to develop less complex and costly services. Cooperation, rather than competition, among stakeholders is a key element in the development of disruptive health care innovations. Christensen and colleagues suggest that the emergence of disruptive health care professionals and disruptive delivery organizations hold substantial hope for improving the quality and availability of health care services. The market-based model of disruptive health care innovation relies on not only the development of provider roles and services that align practitioner skills with health care needs, but also delivery sites that either maximize consumer convenience or the intended therapeutic effect of the service. The development of either type of disruptive innovation threatens many existing patterns, relationships, and power structures of the current health care system.

A managing-for-stakeholders approach provides a useful model to emphasize the potential gains in well being for all the participants involved in developing new service innovations. For example, consumers would have access to services that allow them to manage their chronic condition in a more effective manner or single working parents would be able to obtain antibiotics and care for their sick child during business hours with minimum disruption to their work responsibilities. Health care professionals have the potential to deliver patient care services that use their knowledge, skill and clinical expertise to its full potential. Payers would benefit through reductions in the cost or improvements in the quality of services provided to their beneficiaries. Additionally, a broader range of health care services, delivered in alternative settings has the potential to meet the health care needs of a larger population. By extension, society would benefit through the development of services that meet a wider array of health needs and the gains in productivity or more effective resource consumption within the health care sector could be used for other national investments.

(5) We need to understand that human beings are the creators of value.

The stakeholder capitalism *principle of complexity* states that human beings are complex creatures capable of acting from many different values. Sometimes we act selfishly and at other times we act on behalf of others. Within this view, organizations are seen as enterprises that celebrate the human condition and strive to fulfill the aspirations of all participants in achieving the organization's objectives. In response to changing revenue patterns during the 1980s, health care delivery organizations, like their corporate counterparts were searching for new management methods and organizational practices to improve service delivery and financial performance. Looking to the humanistic management practices advocated by Peters and Waterman (1982) health care delivery organizations began to understand and leverage the power of values-based strategic planning and the role of employee knowledge and skill in improving organizational functioning. During this period the Magnet Hospitals became the exemplar for humanistic management practice leading to improved nursing job satisfaction and retention, and patient care outcomes.

Stories about exemplary health care delivery organizations illustrate the power of understanding organizational culture and the management of human resources in terms of the stakeholder principle of complexity. Some organizations have created almost mythical reputations for their excellence in health care service throughout the industry such as the Mayo Clinic in Rochester, Minnesota, or the John Hopkins Hospital in Baltimore, Maryland. Designated Centers of Excellence have emerged across the country to recognize institutions known for their expertise in treating specific diseases with impressive patient and organizational outcomes.

The power of humanistic management practices relies on seeing human aspirations and achievements as a central part of organizational success. Selfish and altruistic actions of organizational actors are necessary to achieve a range of personal, professional, and organizational goals. An understanding of the complexity and creativity of human actions in the value creation process provides a foundation to improve the nature of health care service delivery. A managing-for-stakeholders approach provides one model to develop practice environments that support health care professionals in their ability to practice their art and science in a manner that supports clinical excellence and expertise to achieve the noble cause of providing services to those in need of health care.

(6) We need to commit to ongoing regulatory reform to make the market work better.

Health care delivery is a business where one customer, the U.S. government through Medicare, exerts enormous influence on the way that the market operates. While the details of necessary Medicare reform would fill volumes, suffice it to say here, that Medicare prevents the establishment of a full range of the benefits of capitalism. Since the sole object of Medicare is to keep prices as low as possible, many innovations that may initially cost more, but eventually lead to greater benefits at lower prices, are simply never tried. In any other industry we would say that the customers had "conspired" to fix prices with the suppliers. Why should a health care delivery organization innovate if the result is not reimbursable by Medicare, or if the initial results skew Medicare performance data, on which prices or diagnosis related groups (DRGs) are based? The existence of such a customer with high bargaining power is a known barrier to innovation, as Porter and his colleagues have shown over the years.

Secondly, one standard way to achieve lower prices in an industry is to increase the number of suppliers. Suppliers then try to innovate to achieve lasting collaborative relationships with stakeholders. However, when the entry of new participants in an industry is highly regulated or controlled by some third party, we should again expect a dampening of the entrepreneurial force that gives stakeholders options via the *Principle of Emergent Competition.* Such is the case in the highly regulated arena of health care delivery, from the control of the number of physicians by the American Medical Association, to the establishment of the system of Certificates of Need, each serving as vestigial limbs of an outdated worldview of planning at a centralized level. Centralized planning doesn't work in the sense that it stifles innovation and drives up prices. We have almost learned this lesson in health care delivery and we need to be more vigilant in implementing it.

Finally we need to end the assumption, often implicit in regulatory regimes, that businesses can't or won't seek both profits and the noble cause of health care. This mindset has done enough damage. We need to have high expectations of our health care organizations: (a) that they be healthy themselves, which means that they earn profits; and (b) that they are committed to the noble cause of health care. We are not arguing that government does not have a proper role to play in health care and its reform. (That is the subject of a project of far larger scope than the current one). Rather we are suggesting the way that we understand its role must be consonant with the very best that business has to offer: stakeholder capitalism, and that government needs to apply these principles to its own actions in the health care arena.

In summary we have argued that far from being the cause of the current malaise in health care, capitalism, understood as stakeholder capitalism, has not been given a real chance. We need to see health care as a business, searching for profits and searching to fulfill the noble cause of health care. We need more capitalism not less. We need more role models from business not fewer ones. And, we need to do so urgently, because the noble cause of health care requires us to do no less.

References

Aiken, L.H., H.L. Smith, and E.T. Lake. 1994. Lower Medicare Mortality Among a Set of Hospitals Known for Good Nursing Care. *Medical Care* 32 (8): 771–787.

Anonymous. *Congressional Budget Office Cost Estimate: S.6 Patients' Bill of Rights Act of 1999*. Available at http://www.cbo.gov, Accessed 13 July 1999.

Aroskar, M.A. 1998. Ethical Working Relationships in Patient Care: Challenges and Possibilities. *Nursing Clinics of North America* 33 (2): 313–324.

Barry, L.L., V.A. Zeithaml, and A. Parasuraman. 1990. Five Imperatives for Improving Service Quality. *Sloan Management Review* 31 (4): 29–38.

Beauchamp, T., and J. Childress. 1994. *Principles of Biomedical Ethics*. 4th ed. New York: Oxford University Press.

Blair, J.D., and J.A. Buesseler. 1998. Competitive Forces in the Medical Group Industry: A Stakeholder Perspective. *Health Care Management Review* 23 (2): 7–27.

Blair, J.D., and M.D. Fottler. 1990. *Challenges in Health Care Management: Strategic Perspectives for Managing Key Stakeholders*. New York: Jossey-Bass.

Blair, J.D., T.T. Rock, T.M. Rotarius, M.D. Fottler, et al. 1996. The Problematic Fit of Diagnosis and Strategy for Medical Group Stakeholders—Including IDS/Ns. *Health Care Management Review* 21 (1): 7–22.

Bodenheimer, T. 1996. The HMO Backlash—Righteous or Reactionary? *The New England Journal of Medicine* 332 (21): 1601–1604.

Brown, M. 1994. Commentary: The Economic Era: Now For Real Change. *Health Care Management Review* 19 (4): 73–81.

Bums, L.R., G.J. Bazzoli, L. Dynan, and D.R. Wholey. 2000. Impact of HMO Market Structure on Physician-Hospital Strategic Alliances. *Health Services Research* 35 (1): 101–132.

Camunas, C. 1998. Managed Care, Professional Integrity and Ethics. *Journal of Nursing Administration* 28 (3): 7–9.

Chamberlain, C. *HMO Horror Stories: How Managed Care Got Its Bad Rap*. (January 22, 1999). ABCNEWS.com available at http://more.abcnews.go.com/sections/living/DailyNews/hmo-horror.html. Accessed 18 Aug 1999.

Christensen, C.M., R. Bohmer, and J. Kenagy. 2000. Will Disruptive Innovations Cure Health Care? *Harvard Business Review* 78: 102–112.

Collins, J., and J. Porras. 1994. *Built to Last*. New York: Harper Books.

Columbia University, School of Nursing, Advanced Practice Nurse Associates (CAPNA), website. Available at http://www.capna.com. Accessed 16 Dec 2000.

David, B.A. 1999. Nurses' Conflicting Values in Competitively Managed Health Care. *Image: Journal of Nursing Scholarship* 31 (2): 188.

Drucker, P.F. 1985. *Innovation and Entrepreneurship: Practice and Principles*, 1–278. New York: Harper and Row.

Eisenberg, B. 1997. Customer Service in Health Care: A New Era. *Hospitals & Health Services Administration* 42 (1): 17–31.

Enthoven, A. 1988. Managed Competition of Alternative Delivery Systems. *Journal of Health Politics, Policy and Law* 13 (2): 305–359.

Enthoven, A., and R. Kronick. 1989. A Consumer Choice Health Plan for the 1990s: Universal Health Insurance in a System Design to Promote Quality and Economy. *New England Journal of Medicine* 320 (2): 94–101.

Etheredge, L., S.B. Jones, and L. Lewin. 1996. What is Driving Health System Change? *Health Affairs* 15 (4): 93–104.

Freeman, R.E. 1994. The Politics of Stakeholder Theory: Some Future Directions. *Business Ethics Quarterly 4*: 409–422.

———. (1996). Understanding Stakeholder Capitalism. *The Financial Times*.

———. 2000. Business Ethics at the Millennium. *Business Ethics Quarterly* 10 (1): 169–180.

Freeman, R.E., and D.R. Gilbert Jr. 1987. *Corporate Strategy and the Search for Ethics, Englewood Cliffs*. NJ: Prentice Hall.

Freeman, R.E., and J. Liedtka. 1991. Corporate Social Responsibility: A Critical Approach. *Business Horizons* 34 (4): 92–98.

Freidman, L.H., and G.T. Savage. 1998. Can Ethical Management and Managed Care Coexist? *Health Care Management Review* 23 (2): 56–62.

Goldsmith, J. 1989. A Radical Prescription for Hospitals. *Harvard Business Review* 67: 104–111.

Heater, B.S. 1996. The Current Health Care Environment: Who is the Customer? *Nursing Forum* 31 (3): 16–21.

Herzlinger, R.E. 1998. The Managerial Revolution in the U.S. Health Care Sector: Lessons from the U.S. Economy. *Health Care Management Review* 23 (3): 19–29.

Kaluzny, A.D., and H.S. Zuckerman. 1992. Strategic Alliances: Two Perspectives for Understanding their Effects on Health Services. *Hospitals and Health Services Administration* 37 (4): 477–490.

Knaus, W.A., E.A. Draper, D.P. Wagner, and J.E. Zimmerman. 1986. An Evaluation of Outcomes from Intensive Care in Major Medical Centers. *Annals of Internal Medicine* 104: 410–418.

Levit, K., C. Cowan, H. Lazenby, A. Sensenig, et al. 2000. Health Spending in 1998: Signals of Change. *Health Affairs* 19 (1): 124–132.

Liedtka, J. 1995. *The Park Nicollet Medical Center*. Video-tape Series. Charlottesville: The Darden School Case Bibliography.

Meliones, J. 2000. Saving Money, Saving Lives. *Harvard Business Review* 78: 57–65.

Mercy Health Services. 1997. *Mission a Guiding Force: President's Report*. Fiscal Year.

Peters, T.J., and R.H. Waterman. 1982. *In Search of Excellence: Lessons from America's Best Run Companies*. New York: Warner Books.

Preister, R. 1992. A Values Framework for Health System Reform. *Health Affairs* 11 (1): 84–107.

Quinn, R., and J. Rohrbaugh. 1983. A Spatial Model of Effectiveness Criteria: Towards a Competing Values Approach to Organizational Analysis. *Management Science* 29 (3): 363–377.

Roggenkamp, S.D., and K.R. White. 1989. Four Nurse Entrepreneurs: What Motivated Them to Start Their Own Business. *Health Care Management Review* 23 (3): 67–75.

Salmon, J.W. 1995. A Perspective on the Corporate Transformation of Health Care. *International Journal of Health Services* 25 (1): 11–42.

Savage, G.T., R.L. Taylor, T.M. Rotarius, and J.A. Buesseler. 1997. Governance of Integrated Delivery Systems/Net-works: A Stakeholder Approach. *Health Care Management Review* 23 (1): 7–20.

Scheffler, R.M. 1999. Physician Collective Bargaining: A Turning Point in U.S. Medicine. *Journal of Health Politics, Policy and Law* 24 (5): 1071–1076.

Shortell, S., J. Zimmerman, D. Rousseau, R.R. Gillies, D.P. Wagner, E.A. Draper, et al. 1994. The Performance of Intensive Care Units: Does Good Management Make a Difference? *Medical Care* 32 (5): 508–525.

Shortell, S.M., R.R. Gillies, and K.J. Devers. 1995. Reinventing the American Hospital. *Milbank Quarterly* 73 (2): 131–160.

Starr, P. 1982. *The Social Transformation of American Medicine: The Rise of a Sovereign Profession and the Making of a Vast Industry*. New York: Basic Books.

Stevens, R. 1989. *In Sickness and In Wealth: American Hospitals in the Twentieth Century*. New York: Basic Books.

Stone, D.A. 1997. The Doctor as Businessman: The Changing Politics of a Cultural Icon. *Journal of Health Politics, Policy, and Law* 22 (2): 533–556.

University of Pennsylvania, School of Nursing, Penn Nursing Network website. Available at http://www.upenn.edu. Accessed 16 Dec 2000.

Venkataraman, S. (2002). Stakeholder Value Equilibration and the Entrepreneurial Process. *Business Ethics Quarterly, Ruffin Series No. 3,* Charlottesville, VA.: Philosophy Documentation Center. 45–57.

Mattia J. Gilmartin is the executive director of NICHE and a senior research scientist at NYU Rory Meyers College of Nursing.

R. Edward Freeman is University Professor, Elis and Signe Olsson Professor, Academic Director of the Institute for Business in Society, and Senior Fellow of the Olsson Center for Applied Ethics at the University of Virginia's Darden School of Business.

Chapter 9
A Names-and-Faces Approach to Stakeholder Management: How Focusing on Stakeholders as Individuals Can Bring Ethics and Entrepreneurial Strategy Together

John F. McVea and R. Edward Freeman

Just as the ideas of stakeholder theory, stakeholder management, or stakeholder capitalism approach acceptance as a mainstream core idea in management theory, we want to suggest a somewhat radical rethinking of it. In almost all of its in-carnations, stakeholder theory merely recapitulates some rather standard business assumptions. First, it rests on the idea that value is created when entrepreneurs or managers put together a deal that simultaneously, and over time, satisfies the groups of stakeholders who play a critical role in the ongoing success of the business: customers, suppliers, employees, communities, and shareholders. Of course, any entrepreneur knows this as second nature. Business just is creating value for stakeholders. That it has taken nearly 40 years from its inception to be accepted into the mainstream is a testament to either the wide gap between theory and practice or the stranglehold that the dogma of shareholder value has on business theory and within business schools. The second assumption that stakeholder theory has done little to question is the separation thesis, that is, the notion that business and ethics can be meaningfully separated. Freeman (1984, 2000) and Wicks (1996) argued that this assumption underlies most of our current thinking about business. Although

Authors' Note: We would like to thank Jean Liedtka, Ed Hartman, Sankaran Venkataraman, Laura Dunham, Betsy Lofgren, S. Ramakrishna Velamuri, and three anonymous reviewers at *JMI* for their helpful comments in preparing this article.
Originally published in: ***Journal of Management Inquiry***, 14(1), 57–69 © Sage Publications, 2005
Reprint by Springer, DOI 10.1177/1056492604270799

J. F. McVea
University of St. Thomas, Saint Paul, MN, USA

R. E. Freeman (✉)
University of Virginia, Charlottesville, VA, USA
e-mail: FreemanE@darden.virginia.edu

© Springer Nature Switzerland AG 2023

S. D. Dmytriyev, R. E. Freeman (eds.), *R. Edward Freeman's Selected Works on Stakeholder Theory and Business Ethics*, Issues in Business Ethics 53,
https://doi.org/10.1007/978-3-031-04564-6_9

Freeman (1984) argued that the stakeholder management approach lent itself to moral and normative analysis, it was only later, in the hands of business ethicists, that the normative branch of stakeholder theory began to emerge. And, when it did, it was almost in isolation from the real world of business.

The separation thesis and the assumption of the usual stakeholder roles of customer, supplier, and so on, have been recapitulated and reified in the very structure of business knowledge and business schools. Marketing worries about the customer. Production and operations management worry about suppliers (and sometimes customers). Organization behavior worries about employees. Finance and accounting worry about shareholders. And, social issues, business ethics, corporate responsibility, or corporate citizenship worry about communities. There are at least three important points here.

The first is that increasingly, in the real world of value creation and trade, these role assumptions are too simplistic. Customers may play a significant role in designing, not simply just purchasing, innovative products and services, such as Linux software. Employees who are direct shareholders may act differently from employees who are not shareholders and differently again from those who may hold shares only through their pension funds. Enron stands as a testament to how employee and shareholder roles can easily become confused such that the current theory that defines stakeholders solely through traditional roles may be not very useful.

Second, these role assumptions fail to capture opportunities for integration, that is, designing innovative entrepreneurial strategies and products that satisfy multiple stakeholders simultaneously. This can be done only if managers or entrepreneurs have localized, even idiosyncratic, knowledge of particular stakeholders and their potential. Business policy or strategic management was originally formulated to fulfill this imaginative and integrative role; however, it seems to have fallen headlong off the shareholder-value cliff.

Finally, when combined with the separation thesis and its application in disciplines such as marketing and finance, where theorists eschew moral concerns, these generic assumptions are more likely to lead to the kind of horror stories to which we have been treated in the business press during the past 2 years. After all, if shareholders and customers are just roles, not real people with names, faces, and families, then one may easily rationalize taking from one group and giving to another (perhaps to which one belongs) as just business.

So, let's start over. Let's understand the key insight of stakeholder theory as being that managers or entrepreneurs have to put together a deal that simultaneously satisfies those who can affect or be affected by what the company does, but let us also be open about how to define stakeholder groups or if they are groups at all. In particular, we want to suggest that if we come to see stakeholders as individuals with names and faces, we have a better chance of putting business and ethics together. Let's move on from theories that revolve around the legitimacy of abstract faceless roles and the division of the same old theoretical pie, between the same old generic groups. Let's instead focus attention on the task of understanding and illuminating the challenges faced by real entrepreneurial practitioners—the challenge of better understanding how collections of idiosyncratic individuals can work together to create new value in entrepreneurial firms in innovative ways that benefit all those who are affected.

1 The State of Stakeholder Theory: An Impetus for a New Direction

During the past 15 years, stakeholder theory has become more than a simple strategic management theory and has developed into what has been called a broad "research tradition" (Trevino and Weaver 1999). However, that so-called tradition has departed sharply from the direction set by the early work on stakeholder management. The original stakeholder framework was explicitly managerial and grew out of a series of clinical studies of management practitioners that were carried out during 10 years (Freeman 1984). However, in its contemporary guise, stakeholder theory has had its greatest influence on theorists and academics rather than practitioners.

Stakeholder theory today seems to be primarily of concern to business ethicists and management scholars in the area of business and society. Despite the differences in approaches between stakeholder theory as social science and stakeholder theory as normative ethics (Jones and Wicks 1999), the stakeholder concept has been tremendously successful in gaining acceptance as a core management idea. However, it seems to us that throughout its development, the content of stakeholder research has been moving further and further away from the real challenges faced by practicing entrepreneurs.

It is ironic that despite this dwindling practitioner focus, the challenges of the contemporary business environment are making the stakeholder perspective more relevant than ever for the practicing entrepreneur. In an economy that is increasingly influenced by the role of networks, it is becoming ever more important to view firms as networks of relationships that extend well beyond the traditional boundaries of the organization (Kelly 1998; Larson 1992; Rowley 1997). At the same time, technology is dramatically reducing the costs of building and operating complex stakeholder networks, and these networks are becoming easier to use, richer in content, more interactive, and instantaneous in operation (Kelly 1998). Given these circumstances, stakeholder theory would appear to offer a unique and neglected contribution to decision-making processes, particularly in innovative and entrepreneurial fields. Other theoretical frameworks emphasize the central role of market structure (Porter 1985), or the role of internal competencies (Penrose 1959; Prahalad and Hamel 1990), or the role of learning (Senge 1990) in the strategy process. However, the stakeholder approach is unique in its emphasis on the relationships that make up the networks from which new value creation ideas are generated. Thus, it should be emphasized that although we believe that the approach we are advocating has important implications to strategy in general, in this article, we are focusing our thoughts on the challenges facing entrepreneurial businesses attempting to create new markets for novel and innovative products and services (Venkataraman 1997). We will leave it for another occasion to address the specific stakeholder challenges of large, mature, traditional, and commodity-type industries.

2 A Names-and-Faces Approach

Before defining more fully the characteristics of what we have called "a names-and-faces approach," we must first address a common intuitive response. Some might say that while adopting a more personal names-and-faces approach might be an interesting idea, it is too complex and too costly to be of any practical use. We will initially respond to this criticism not through theory or logic but rather by telling the story of the development of a business strategy that successfully implemented a counterintuitive approach in the face of similar criticisms. It should be noted that we are not proposing that this example of Hertz Gold represents a fully-fledged version of an ethical and entrepreneurial names-and-faces strategy. Furthermore, as we are proposing what we believe is a radical approach, we recognize that this and other examples that we use in this article give no more than a flavor of the possibility of a names-and-faces approach and may indeed have been driven by entirely different motivations. Nevertheless, we believe that partial examples and similar approaches can provide us with an inspiring counterpoint to the conventional wisdom that deepening stakeholder relationships is a cost-creation, rather than a value creation, exercise.

2.1 Hertz Gold: The High Cost of Individualized Relationships?

Until very recently, the rental car industry did not place great emphasis on highly individualized relationships with any of its stakeholders, including customers. The business was complex, and it was assumed that what customers most valued was a cheap rental contract. Thus, the dominant business strategy was to pursue automobile purchasing scale and to leverage headquarter administrative costs through geographic expansion. Soon, nearly all the rental businesses were occupying the same market space, with consequent downward pressure on business margins. In the 1990s, Hertz decided to move beyond this generic strategy and develop closer personal ties with its customers. This entrepreneurial vision eventually became what we today know as Hertz Gold. Customers could sign up for a program that bypassed the anonymous counter lines and provided personalized service. Using digital screens that showed the customers' name and the location of a car of their preference, customers were directed to a covered collection point where an employee had already opened the trunk, started the car, and set the heating or air-conditioning a few minutes before the customer arrived. Checkout and check-in were handled by employees in the rental lots, equipped with mobile printers and data-processing devices. Employees, freed from lines and infuriating paperwork, could greet customers by name—prompted by handheld devices. Customer preferences could be recorded and applied to future bookings.

The new system was enabled by cheap portable communications technology but also required the complete redesign of all the supporting management systems. By treating its customers more like "names and faces," Hertz provided them with what felt like a personalized and luxurious service. However, astonishingly, "by doing only and exactly what each customer required, Hertz discovered that its Gold service was actually less costly to provide than its standard service" (Gilmore and Pine 1997, p. 91). As we said earlier, we are not trying to suggest that the Hertz Gold approach represents a fully fledged attempt to treat all stakeholders as names and faces. Clearly Hertz focused all of its efforts on individualizing its relationships with certain customers, as opposed to other stakeholders. However, what we believe it does illustrate is the possibility of technology and entrepreneurial strategy working together to make such individualized approaches feasible and cost-effective.

The situation faced by Hertz is typical of the types of context faced by many contemporary businesses. To deal with the complexity of making decisions within large organizations, problems tend to become defined in terms of generic stakeholder groups. This approach has two effects. First, it results in generic strategies that are similar to those of competitors—at a certain generic level most firms' customers look the same. Stakeholder management, in this incarnation, fails to confer any particular advantages in terms of value creation for the firm and thus fails to generate strong instrumental claims about stakeholder management leading to superior performance. Furthermore, it fails to explain the source of new entrepreneurial opportunities. Rather, it focuses on how current returns are distributed among stakeholders. This is what we will call the entrepreneurial problem of generic stakeholder theory.

Second, generic analyses disguise the real human relationships that exist between the firm and its stakeholders. Stakeholders are treated not as morally important individuals, but as abstractions, characterized by the roles that they play. This approach makes it easy to separate decision making from those stakeholders who are affected by decisions. When this separation has occurred, it becomes difficult to reattach human values to business decisions again. This is the source of what we will call the *normative problem* of generic stakeholder theory. We believe that this separation is at the root of many of the contemporary ethical problems that have plagued certain businesses pursuing innovative strategies. When stakeholders are thought of in terms of their generic roles rather than as individuals of moral worth, it is much easier to formulate strategies that are not only inhuman and unethical but also counter to the long-term interests of the firm. However, it is our proposition that to the extent that firms can analyze business problems by recognizing stakeholders as individuals, they will be able to develop idiosyncratic strategies that are novel, entrepreneurial, and hard to replicate. Furthermore, such strategies would more easily recognize the moral worth of the individual.

2.2 *Defining a Names-and-Faces Approach*

The names-and-faces approach that we are proposing is an attempt to describe a managerial approach to stakeholder theory. A truly managerial theory "does not simply describe situations or predict cause-effect relationships; it also recommends attitudes, structures and practices" (Donaldson and Preston 1995). As such, a managerial approach does not attempt to precisely separate out descriptive, normative, and instrumental claims. Rather it focuses on concrete business problems, addressing simultaneously why stakeholder management might result in better outcomes and how it might incorporate stakeholder interests into business strategies. To achieve these ends, any framework must be intimately connected with the process of value creation and trade (Freeman 2000). Thus, a managerial approach to stakeholder theory would focus on a fundamentally different set of questions from much of the existing literature. A managerial approach would not attempt to justify the existence of the firm, similar to many normative theories. Rather, it would focus on how managers make decisions that create value while taking account of the interests of all the stakeholders affected by the decision. We now consider each of the three critical characteristics, outlined above, of our proposed names-and-faces approach in turn: a value creation focus, a decision-making focus, and individual relationship focus.

A Focus of Entrepreneurial Value Creation In focusing on the value creation process, we are adopting a perspective that has been developed by the emerging field of entrepreneurship. Traditional business strategy has focused on explaining competitive advantage and relative firm performance. Indeed, Jones (1995) used this approach to build an instrumental stakeholder theory that explains why stakeholder management should produce superior results. However, although competitive advantage is an important strategic issue, Shane and Venkataraman (2000) pointed out that it offers only an incomplete explanation of value creation. For a start, such an approach presupposes the existence of products, firms, and markets and thus precludes an understanding of the creation of fundamentally new opportunities. Furthermore, we believe the fixation with competitive advantage has resulted in, or at least coincided with, the strategy literature abandoning its original integrative and purpose-defining function and instead promoting the primacy of shareholder wealth at the expense of all other considerations. In contrast, entrepreneurial theories of value creation try to explain the creation and exploitation of new goods and services. We believe that the entrepreneurial perspective is particularly important to developing a contemporary managerial stakeholder theory because of the importance of change and new technological development in today's economy. Furthermore, the entrepreneurial perspective allows us to focus on how collections of stakeholders come together and create mutually beneficial projects.

So far we have established that a managerial stakeholder theory should explain the discovery and exploitation of new entrepreneurial opportunities from the perspective of a broad range of stakeholders. We now make a critical proposition. We propose that paying attention to individual stakeholders, rather than analyzing them as generic groups, is required to explain at least one important driver of this entrepreneurial process from opportunity creation or discovery to exploitation.

In supporting this proposition, we first consider the process of opportunity discovery. According to the entrepreneurial account, value creation opportunities exist because of asymmetries of information and beliefs (Schumpeter 1934). There are two sources of these opportunities—erroneous guesses that others have made about the future and new information that has been created (Shane and Venkataraman 2000). Penrose (1959) demonstrated that new information and learning are inevitable parts of doing business. New information is routinely created by all stakeholders as part of their regular tasks either because of the inevitability of organization slack or because of the idiosyncratic nature of the way people work. However, this new information is not instantaneously spread to everyone in society. Thus, it remains for the perspicacious individual to make use of his or her own unique "information corridor" (Shane and Venkataraman 2000) to identify the opportunities that others might overlook. Thus, as Hayek (1945) observed,

> Practically every individual has some advantage over all others in that he possesses unique information of which beneficial use might be made, but which use can be made only if the decisions depending on it are left to him or are made with his active cooperation. (p. 521)

If we accept that a significant driver of new value creation is rooted in the idiosyncratic knowledge of individual stakeholders, then it seems that the role of stakeholder management should be to harness as many of these new value creation opportunities as would be fruitful for the benefit of the stakeholders. Doing so requires access to the fragmented bits of information diffused across numerous individual stakeholders. Thus, to the degree that managers make decisions based on generic analyses of stakeholder groups, they are blinding themselves from much of the richness of new value creation opportunities. Analysis at the level of generic stakeholder groups may encourage management to believe that various stakeholders' interests are in irreducible conflict and so must be offset by each other. Whereas, when stakeholders are considered individually, there may be opportunities for creative and entrepreneurial solutions that allow stakeholders to transcend what seems, prima facia, to be an inherent conflict. As a result, generic analyses may discourage the evaluation and creation of new opportunities. These unrecognized opportunities may then either lie fallow or migrate to benefit competing entrepreneurial networks.

A Focus on Strategic Decision Making As stakeholder theory developed during the past two decades, researchers have concentrated their analyses on the firm, to the detriment of analyzing the decision process. This focus has led normative stakeholder theorists to such issues as the purpose of the firm, the role of the firm, the structure of the firm, and the governance of the firm. Likewise, it has led social scientists studying stakeholder interests to explore questions such as "does the firm that practices stakeholder management outperform the firm that does not" and "what are the broader measures of performance of the firm that stakeholder theory might imply." Although much valuable work has been produced, the results have also been problematic for both streams of stakeholder research. Ethical theory has become mired in a philosophical debate about the normative foundations of the firm, producing a multitude of competing philosophies. At the same time, the social scientific approach has focused on developing ever more complex ways of measuring corporate social performance to prove either the existence or the superiority of stakeholder firms—a case that has yet to be persuasively demonstrated (Wood and Jones 1995). Although the issues of corporate governance, legal status, and social performance indices are important in their own right, they are rarely the central challenges of entrepreneurial practitioners. Thus, the analytical focus on the institution of the firm has led to the neglect, by stakeholder theorists, of issues that are central concerns of real entrepreneurs; and this, in turn, has led to a widening gap between theory and practice in the field.

We believe that it is time to change the level of our analysis and focus our efforts on illuminating the actual process of making real entrepreneurial decisions. Such a change will refocus our research on issues that are more relevant to the practitioner. Furthermore, it will also help us sidestep a number of issues that have become significant distractions to progress. If we focus our analysis on the challenges of specific decisions, then a number of theoretical problems that plague generic stakeholder theory disappear such as the question of management fiduciary duty. Whether managers' duties are to stakeholders in general or stockholders in particular becomes moot at the level of strategic decision making. From the perspective of a particular decision, it becomes clear that if management tries to make decisions by focusing attention exclusively on their shareholder relationships, they will never discover all the information they need to make the best entrepreneurial decisions. From an entrepreneurial perspective, the value creation ability of the firm is embedded within the network of hundreds of idiosyncratic stakeholder relationships—an ability that remains unliberated by a theoretical commitment to focus on serving the shareholder. Furthermore, such an approach would not even be in the long-term interests of shareholders. From a decision-focused perspective, it becomes clear that maximizing shareholder wealth is a theoretical end rather than a practical means to value creation.

A Focus on Individual Relationships The problem of identification, that is, who is and who is not a stakeholder, is an important issue among stakeholder theorists. However, over the years, stakeholder theory has offered "a maddening variety of signals on how questions of stakeholder identification might be answered" (Mitchell et al. 1997). On one hand, some theorists have claimed that the lack of specificity in how stakeholder theory deals with this problem illustrates the need for a fleshing out of the stakeholder approach through the development of more detailed theories

(Clarkson 1998; Phillips 1997;Wood and Jones 1995). On the other hand, others have used the problem of stakeholder identification as a demonstration of the hopelessly intractable nature of the stakeholder approach per se (Marcoux 2000).

Historically, most theorists have defined stakeholders according to a set of simple generic roles with the firm: customers, suppliers, employees, shareholders, and community. Few have questioned the value of analyzing stakeholders as generic groups based on their roles. As a result, much discussion has taken place about the normative claims of stakeholder groups in general (Burton and Dunn 1996; Clarkson 1998; Wicks et al. 1994), the legitimacy of various stakeholder groups (Blair 1998; Boatright 1994; Phillips 1997), and the priority that management should give to these groups (Mitchell et al. 1997).

In contrast, we propose an approach that is founded on the real set of individual stakeholder relationships of real firms in the contemporary business environment. We believe that current stakeholder trends are making general group designations ever more inadequate. Stakeholder migration is a critical issue in today's economy. Stakeholders have begun to challenge the borders between traditional stakeholder groups. Numerous examples across all stakeholder boundaries suggest a fundamental shift in stake-holder dynamics is occurring. A simple and accurate description of stakeholders by neat categories is becoming more and more difficult, particularly in the most entrepreneurial sectors of the economy such as high technology and medical sciences. For example, consider the following conflation of roles:

1. Management and employees: Organizations have become flatter as front-line employees have become better informed than their managers. The role of the manager has changed as he or she focuses on empowerment and coaching. The growth of team-based structures, collegial relationships, and campuslike work environments has blurred the distinction between management and employees.
2. Customers and employees: Customers are frequently completing tasks formerly carried out by employees. For example, "When you pump gas at the filling station, are you working for the gas station or for yourself" (Kelly 1998, p. 120). In other areas of contemporary business, this trend has gone even further. Open system software, such as Linux, depends on customers to design and continuously update the product. Amazon.com relies on customers to write book reviews and to generate purchase suggestions for other like-minded consumers. The result is the rise of "prosumers," a combination of the producer and consumer roles (Toffler 1980) that are fast becoming part of our daily lives.
3. Shareholders and employees: Employees and management have become significant holders of equity particularly in the high-growth service industries. In many of today's leading firms it is becoming hard to distinguish between owners and employees.

In each of these examples the boundaries between stakeholder groups are becoming blurred, and stakeholder relationships are becoming more complex. Obviously, any generic classification of stakeholder groups is a simplification. However, we propose that the traditional simplifications are no longer adequate. As pointed out earlier, to manage stakeholder relations according to the traditional groupings (customers, employees, suppliers, shareholders, community) would be to blind the entrepreneur to some of the critical characteristics of the contemporary business environment. However, this generic approach has a second important side effect. Generic stakeholder analysis is an important driver of the normative problem of stakeholder theory.

The normative problem is a result of what Freeman (1988) referred to as the separation thesis. According to this view, business discourse has been divided into separate business-is-business decisions with no moral content and moral decisions with no business content. To try and resolve this separation, philosophers have developed a series of competing moral foundations for stakeholder theory. However, these theorists have failed to clearly demonstrate the superiority of one moral foundation over another. As a result, practitioners are left with a confusing array of frameworks through which to incorporate ethics into the business world.

We believe that the articulation of stakeholder theory from the perspective of generic stakeholder groups has contributed to this theoretical separation in several ways. First, the generic perspective can encourage practitioners to feel distant from their stakeholders as individuals and thus to discount their effects on them. Thinking about stakeholders as faceless groups, defined according to abstract roles, can result in individual moral responsibility becoming easier to ignore. Much the same point has been made by Jones (1991) in identifying the role of proximity as a driver of ethical behavior. According to this account, managers will restrict their highest level of ethical consideration for situations that they feel to be morally intense, and perceived moral intensity increases with the degree of proximity the manager feels toward the objects of his or her actions. This feeling of proximity to others can be strongly influenced not only by physical distance but also by how we think of those people. We are suggesting that generic thinking dulls our sense of proximity to those stakeholders. Thus, we believe that the conventional language of generic stakeholder groupings can facilitate a neglect of individual ethical consideration.

A second tendency encouraged by the generic perspective is what we might refer to as a just-business attitude to decision making. It is true that all theorists regard the interests of stakeholder groups as having intrinsic value. However, it is often difficult for practitioners to incorporate the normative dimension within the everyday language of business decision making. The moral basis for the consideration of stakeholder groups is generally justified with reference to complex, abstract, and philosophical foundations within the ethics literature. The language and the content of these foundational approaches are alien to mainstream business practitioners. Thus, when time-pressed decision makers run up against the difficulties of incorporating Kantian, rule utilitarian, or virtue ethics into the language of business, there is a tendency to leave out the difficult abstractions and leave in the familiar business considerations, resulting in a just-business approach.[1]

[1]This amoral approach to business is sometimes defended by the observation that business must still operate within the bounds of the law. However, the law is too blunt an instrument to fully complete this task (Stone 1975). As numerous recent examples have demonstrated, many of the

In contrast to the generic perspective, from a names-and-faces perspective it is much more straightforward to see decision making as an ethical process from the ground up. First, a names-and-faces approach can increase the feeling of proximity that the decision maker has for his or her stakeholders. Second, if we view firms as a network of individual personal relationships, then we have less need to import abstract moral foundational arguments to justify the nature of our consideration. We believe that we can make more progress in integrating ethical responsibility in business if it is founded in the physical reality of the personal relationships that make up the business rather than in the ability of managers to operationalize philosophical abstractions. We believe that if we start from the position that stakeholders are simply individuals with a stake in the decision at hand, then the relationships themselves become sufficient motivation for moral consideration. We need to dig no deeper. This does not of course imply that just by considering stakeholders as people with names and faces, ethical problems are resolved. Managers will always be free to be ethically lax or to make bad ethical decisions. They will be free to be imaginative in creating expansive solutions or thoughtless in neglecting or reinforcing enduring conflicts. Furthermore, stakeholder management will remain difficult, complex, and perhaps at times, intractable. Nevertheless, adopting a names-and-faces approach frees us to consider the practice of business as an inherently ethical process where ethical consideration is integral to normal decision making. Thus, we believe that ethical problems involved in making business decisions can become practical considerations for entrepreneurial value creation rather than theoretical or metaphysical abstractions.

In sum, we want to begin the task of radically rethinking the role of the stakeholder framework based on three critical characteristics: a focus on value creation, a focus on decision making, and a focus on individual stakeholder relationships. This names-and-faces approach to stakeholder management states that firms that treat their stakeholders as individuals with names and faces will develop more value-creating strategies and will also incorporate ethics as an inherent part of the decision-making process. Reframing the stakeholder enterprise along these lines will obviously be an extensive undertaking. At this stage, we can provide no more than an agenda and a call to arms. However, having outlined this initial sketch, we end by turning to a parallel field of business theory: mass customization. A brief review of this literature provides, we believe, not only another inspiring story of how researchers can successfully break through the apparent contradictions and constraints of academic conventions and make a tangible contribution to the practice

---

most damaging examples of stakeholder negligence or exploitation occur in that gray area between where the law ends and where personal ethics begin. Furthermore, expanding and overspecifying the law beyond its effective scope seems only to create the opportunity for new loop holes, or worse, leaves the impression that there is no need for ethics—that we just need to wait for the law to catch up. We believe that such faith in the law is naïve at best. In an entrepreneurial world where the boundaries of possibility are always in transition, there will always be opportunities for unethical, but legal, entrepreneurial decisions. This is precisely the reason that the separation thesis is so dangerous and must be challenged.

of value creation and trade but also the identification of a number of principles that can be applied to a names-and-faces approach. By identifying and applying these principles, we hope to begin the task of making more tangible and accessible the approach we have advocated in the first part of this article.

3 Names and Faces: Lessons from Mass Customization

Mass customization was a response to severe market turbulence. In the late 1980s manufacturers were stuck in a paradox. Consumers seemed to exhibit an insatiable demand for variety and personalized products. However, variety and complexity of product range was a major driver of manufacturing cost. Cost and customization seemed, intractably, to be on conflicting sides of the challenge facing practitioners. Thus, the more firms succeeded in satisfying their customers, the more they seemed destined to raise costs and destroy their own financial performance. Mass customization was a way of thinking that helped manufacturers break out of this paradox.

Mass customization is a collection of techniques that allows manufacturers to treat customers as individuals by ensuring that they are each supplied with exactly what they want. However, it is not only customers who gain from this individualized approach. Amore individualized service leads, in turn, to greater customer satisfaction, and this, in turn, leads to higher prices for the manufacturer.[2] Furthermore, there is often less wasted effort if a manufacturer makes each product with only the attributes each customer requires. Thus, the intellectual ideas and techniques developed through this research helped demonstrate how practitioners could bring opposing constraints into alignment. However, developing this win-win strategy for all stakeholders did not come without considerable effort.

First, mass customization required a change of strategic mind-set. Just like a names-and-faces approach to stakeholder management, the concept of mass customization in manufacturing was counterintuitive to the existing way of thinking. “Mass customization is an oxymoron, which is the putting together of seemingly contradictory notions, like jumbo shrimp and artificial intelligence” (Davis 1993, p. ix). Second, it required the creation of innovative management systems and technologies that were capable of supplying “variety and customization through flexibility and quick response” (Pine 1993, p. 44).

What principles can be gleaned from the development of the mass customization approach that might shed some light on our ambitions for stakeholder relationships as a whole? Three principles seem to be shared by successful mass customizing firms that might point a way forward for our own project—first, the importance of

---

[2] In a working paper, Ebben and Johnson (2003) found evidence that “meaningful differences exist between small firms in terms of standard and made to order products” (p. 22). They concluded that firms that have assumed a flexible or made-to-order strategy tend to outperform firms that try to pursue efficiency strategies as well as flexibility.

intense individual and lasting relationships; second, modular design; and third, flexible delivery systems.

3.1 *Intense, Individual, and Lasting Relationships*

Mass customization requires the development of more intimate relationships with customers than simple exchange-transactions. "(It) encourages more customer interaction and information, faster cycle times, and greater variety and customization" (Pine 1993, p. 194). For example, in the drug industry a new approach called pharmacogenomics is being developed. Individuals can use devices, such as those made by Affymetrix, to encode part of their DNA profile on chips. Drug companies can then use these chips to custom design treatments for each individual, taking into account their age, family history, and genetic make up (Cox and Alm 1998). The information generated by one treatment can be used to develop the next customized treatment. Intimate information exchange and deep relationships of trust are necessary to enable such transactions to succeed. Through mass customization lifelong customer relationships become the norm; simple exchange transactions become the exception.

When this operating principle is applied to stakeholder theory, in general, it suggests two critical advantages of a names-and-faces approach: an increase in strategic options and greater stakeholder support. First, by developing intense, lasting, and individual stakeholder relationships, management will uncover a much broader range of strategic options. Just as mass customization enables an almost endless range of product options, so a names-and-faces approach to stakeholders, in general, will produce an almost endless number of entrepreneurial options in finance, human resources, purchasing, environmental performance, and community relations. For example, instead of assuming permanently intractable relationships with environmental concerns, pioneering energy groups can begin dialogues that try to craft innovative strategies that take account of the environmental concerns of individuals—whether they are also employees, customers, or members of environmental organizations.

Second, as mass customization has shown, individualized relationships can produce much higher levels of stakeholder satisfaction. Stakeholder theory is built on the assertion that the success of a firm depends on its ability to satisfy stakeholders over the long run. Thus, to the degree that a names-and-faces approach can raise the satisfaction of stakeholders, it should eventually raise the long-term performance of the firm. Stakeholder satisfaction can be an extremely tangible aspect of corporate performance (Jones 1995). Lower employee turnover will reduce hiring and training costs, loyal suppliers will reduce quality qualification costs, supportive communities will reduce legal and public relations overhead, and stable shareholders will reduce stock market volatility and lengthen planning horizons.

3.2 Modular Design

The second operating principle developed by successful mass customizers is that of using modular design to provide customers with flexibility at low cost. Modular design allows customers flexibility over the aspects of the product that they value most, while allowing the firm to standardize those aspects that customers value least. For example, Hewlett Packard printers use a standardized printer engine for all customers everywhere. However, power supply components are designed as modules operating at 110 volts or 220 volts. These modules can then be slotted in depending on the customer needs in the eventual market of sale (Feitzinger and Lee 1997, p. 116).

When the principles of modular design are applied to stakeholders in general, they suggest three critical management tasks. First, management must develop an understanding of the nature of the stake each stakeholder has in the business. Second, management needs to identify which aspects of each stake are most important to each stakeholder. Thus, a critical priority for management should be to develop a deep and personal understanding of what really matters to each stakeholder. Third, a critical management task is to determine which aspects of stakeholder relationships should be standardized and which can be left flexible through modular design.

For example, let us consider a computer manufacturer where different employees may hold slightly different stakes in the business. One employee may be a fanatical computer geek, another might be a significant owner of company stock, and a third might be politically active in the local community. The firm might consider offering a range of employee benefit modules such as fixed salary pay, overtime, performance incentive pay, product discounts, priority access to new product launches, matching gifts to local charities, and/or paid leave for community service. Each employee could then select from these modules, within restrictions set by management, to design her or his own compensation package that satisfies each better than a standard compromise employment package. By treating each employee stakeholder as a name and face with an idiosyncratic stake in the firm, management is able to provide customized solutions that create value for the employees and reduce waste, thereby providing benefits to all other stakeholders in the firm.

3.3 Flexible Delivery Systems

The third principle of mass customization is the development of flexible delivery systems. "The key to mass customizing effectively is postponing differentiating a product for a specific customer until the latest possible point in the supply network" (Feitzinger and Lee 1997, p. 116). The classic example of postponed differentiation is the Benetton innovation of knitting sweaters in greige (undyed) yarn and then dying the garments at the last minute depending on the latest color trends. Flexible delivery systems decentralize critical management decisions and move the decision point much closer to the consumer. The critical change is that manufacturing emphasis switches from a push to a pull system. In a push system, customer demand is forecasted centrally, and then product is designed, manufactured, and pushed through the supply pipeline. In a pull system, the customer selects from a broad pallet, and then the assembly and production is pulled through the supply pipeline. Moving from push to pull system required more than just a change of attitude.

> Of equal if not greater importance in achieving low costs and customization are advances in management. The development of just-in-time delivery, lean production techniques, time-based competition, cross-functional teams, and a host of other advances has increased flexibility and responsiveness and therefore the ability to increase variety and customization without parallel increases in cost. (Pine 1993, p. 49)

If this operating principle is applied to stakeholder management, in general, it suggests that a names-and-faces approach could invert the process of making entrepreneurial strategies. Many critical decisions would be decentralized and moved much closer to the point of direct interaction with stakeholders. Strategy itself would become a pull rather than a push process. Stakeholders themselves would become the origin of much of what is today considered corporate strategy. However, these changes will only be possible after considerable advances in management practices and systems have taken place. These decentralized processes would allow stakeholders to make trade-offs between different aspects of their stakes in the business. Where this is not possible, and conflict persists, management priority should be to try to creatively develop new modules and alternatives that allow stakeholders to transcend the conflict.

Although the idea of decentralizing strategy might itself seem oxymoronic, an example from the area of retirement planning might help illustrate the nature of the point. As a tradition, management appointed an independent board to manage investments in the pensions for all employees; it was much too costly and complex to take into account the individual circumstances and preference of each employee. Eventually, however, the development of the 401(k) account allowed firms to let employees choose between a number of investment vehicles. Moving a step further, with the advent of network technology, it is now common for firms to provide employees with Internet-based trading systems where they can finely tune their retirement plans to their own needs and do so on a daily basis. This customized service has resulted in more satisfied employees, employees treated more like individuals about whom the organization cares individually, and lower operating costs; it has also moved the decision point on retirement planning strategy out toward the stakeholders. Retirement planning strategy has been turned on its head. Make no mistake, the firm still has a retirement benefits strategy; however, the strategy is mostly crafted by decentralized employees. The only centralized strategy required is the design of the system that facilitates stakeholder dialogue and that determines the level and the types of trade-off that will be offered.[3]

In summary, the current study of the mass customization literature has yielded a number of interesting insights for our proposal of a names-and-faces approach to stakeholder management. We have identified three major operating principles for successful implementation of mass customization, relating to relationships, design, and delivery.

4 What Are the Limitations of This Approach?

Is it too tall an order to propose that managers and entrepreneurs develop this level of understanding with their stakeholders? Of course, it could be argued that the process of stakeholder management is intractable because complete analysis would require the consideration of an infinite number of "knock-on effects" on an almost infinite number of stakeholders. We would respond to this in two ways. First, we are proposing a names-and-faces approach as a moral and strategic aspiration that should drive the entrepreneurial enterprise. That it is challenging, difficult, and demanding is not in dispute. However, in our personal lives we would not dismiss the importance of, for instance, being considerate to others on the grounds that it would be too challenging to expect anyone to even try. Second, our proposal is underpinned by a belief in human, entrepreneurial, and moral progress. One of the distinguishing features of human culture is the ability to develop and amend ways of dealing with complex, open-ended, imperfectly defined problems through experimentation and learning. The level of complexity in dealing with individualized stakeholder relationships is no different from that of many other management problems. In the traveling salesperson problem, the complexity of determining the exact shortest route between a large number of customers defies precise calculation. In inventory management, it is impossible to fully optimize stock levels in the face of uncertain demand. Good management involves the development of heuristics to find reasonable solutions to infinitely complex problems. Traveling salespeople do not remain analytically paralyzed; rather, they get to their customers in a reasonable manner. Warehouse managers juggle their stock, meet demand, and control their costs; and stakeholder managers can create strategies that take account of individual stakeholders, within reasonable time constraints, in a morally acceptable way. Furthermore, the Hertz Gold story reminds us that technology can help us develop better and better solutions to exactly these sorts of intractable problems. In sum, we are simply trying to challenge the assumption that entrepreneurial decision making

[3] It should be noted that the purpose of this example is to illustrate how technology has been successfully used to ease the tension between individualized service and cost for stakeholders other than customers. We believe that the illustration holds notwithstanding any wider criticisms about the knock-on economic effects of the general switch to individual pension plans.

cannot take account of individual relationships. Moral and entrepreneurial progress can go hand in hand.

5 Summary and Conclusion

We proposed that it is time for a radical rethinking of the stakeholder approach to business. If we are correct in our assessment, stakeholder theory stands at something of a crossroads. One direction leads toward a more robust theory of the status quo, that is, marketing understood in stakeholder terms, operations, stakeholder terms, and so on. A second direction leads toward a more pronounced split between theory and practice, between normative and descriptive (if this is even thinkable), whereby when all is said and done, stakeholder theory represents the wild ravings of ethicists who don't know anything about the real world of business. A third direction leads toward the production of more studies like most of those in the special issue of *Academy of Management Journal* (Harrison and Freeman 1999)—good solid empirical studies that describe the way managers and others think the world is. A final direction is perhaps the deal with the devil that we propose: come to see stakeholders as human beings with names, faces, and families. The opportunities for satisfying their individual needs, exploring their multiple role relationships and determining how these can be better understood and fulfilled would open up a vast new space for value creation and trade. Ultimately, this space would require that the modern corporation can be remade more in line with the moral ideas of classical liberalism and pragmatism (Freeman and Phillips 2002), and ultimately, the disciplines of business and business schools would have to be rethought or continue down their current pathway to oblivion.

References

Blair, M. 1998. Whose interests should be served? In *The corporation and its stakeholders*, ed. M. Clarkson, 202–234. Toronto: University of Toronto Press.

Boatright, J. 1994. Fiduciary duties and the shareholder-management relation: Or, what's so special about shareholders? *Business Ethics Quarterly 4*: 393–407.

Burton, B., and C. Dunn. 1996. Collaborative control and the commons: Safeguarding employee rights. *Business Ethics Quarterly 6*: 277–288.

Clarkson, M. 1998. A stakeholder framework for analyzing and evaluating corporate social performance. *Academy of Management Review 20*: 92–117.

Cox, M., and R.G. Alm 1998. Falling prices: good values in the American economy. *Current*, 14–19.

Davis, S. 1993. Foreword. In *Mass customization: The new frontier in business competition*, ed. J. Pine, ix–xii. Boston: Harvard Business School Press.

Donaldson, T., and L. Preston. 1995. The stakeholder theory of the corporation: Concepts, evidence, and implications. *Academy of Management Review 20*: 65–91.

Ebben, J., and A. Johnson 2003. *Efficiency, flexibility, or both? Evidence linking strategy to performance in small firms*. Manuscript submitted for publication.

Feitzinger, E., and H.L. Lee. 1997. Mass customization at Hewlett Packard: The power of postponement. *Harvard Business Review 75*: 116–122.
Freeman, R.E. 1984. *Strategic management: A stakeholder approach*. Boston: Pitman.
———. 1988. The politics of stakeholder theory. *Business Ethics Quarterly 10*: 169–180.
———. 2000. Business ethics at the millenium. *Business Ethics Quarterly 10*: 169–181.
Freeman, R.E., and R.A. Phillips. 2002. Stakeholder theory: Alibertarian defense. *Business Ethics Quarterly 12* (3): 331–350.
Gilmore, J.H., and B.J. Pine. 1997. The four faces of mass customization. *Harvard Business Review 75*: 91–102.
Harrison, J., and R. E. Freeman, eds. 1999. *Academy of Management Journal* [Special edition], 42.
Hayek, F.A. 1945. The use of knowledge in society. *American Economic Review 35* (4): 519–530.
Jones, T.M. 1991. Ethical decision making by individuals in organizations: An issue-contingent model. *Academy of Management Review 16* (2): 366–395.
Jones, T. 1995. Instrumental stakeholder theory: Asynthesis of ethics and economics. *Academy of Management Review 20*: 92–117.
Jones, T., and A. Wicks. 1999. Convergent stakeholder theory. *Academy of Management Review 24*: 206–221.
Kelly, K. 1998. *New rules for the new economy: 10 radical strategies for a connected world*. New York: Viking.
Larson, A. 1992. Network dyads in entrepreneurial settings: A study of the governance of exchange processes. *Administrative Science Quarterly 37*: 76–104.
Marcoux, A. M. 2000, July 24. Business ethics gone wrong. *CATO Policy Report*. Washington, DC: CATO Institute.
Mitchell, R., B. Agle, and D. Wood. 1997. Toward a theory of stakeholder identification and salience: Defining the principle of who and what really counts. *Academy of Management Review 22*: 853–886.
Penrose, E.T. 1959. *The theory of the growth of the firm*. New York: John Wiley.
Phillips, R. 1997. Stakeholder theory and a principle of fairness. *Business Ethics Quarterly 7*: 51–66.
Pine, B.J. 1993. *Mass customization: The new frontier in business competition*. Boston: Harvard Business School Press.
Porter, M. 1985. *Competitive advantage*. New York: Free Press.
Prahalad, C.K., and G. Hamel. 1990. The core competence of the corporation. *Harvard Business Review 68* (3): 79–91.
Rowley, T.J. 1997. Moving beyond dyadic ties: A network theory of stakeholder influences. *Academy of Management Review 22*: 887–911.
Schumpeter, J.A. 1934. *The theory of economic development: An inquiry into profits, capital, credit, interest, and the business cycle*. Cambridge: Harvard University Press.
Senge, P. 1990. *The fifth discipline*. New York: Doubleday.
Shane, S., and S. Venkataraman. 2000. The promise of entrepreneurship as a field of research. *Academy of Management Review 25* (1): 217–226.
Stone, C. 1975. *Where the law ends: The social control of corporate behavior*. New York: Harper & Row.
Toffler, A. 1980. *The third wave*. New York: Morrow.
Trevino, L., and G. Weaver. 1999. The stakeholder research tradition: Converging theorists—not convergent theory. *Academy of Management Review 24* (2): 222–251.
Venkataraman, S. 1997. The distinctive domain of entrepreneurship research. In *Advances in entrepreneurship, firm emergence and growth*, ed. J. Katz, vol. 3, 119–138. Greenwich: JAI.
Wicks, A. 1996. Overcoming the separation thesis: The need for a reconsideration of SIM research. *Business and Society 35* (1): 89–118.
Wicks, A., D. Gilbert, and E. Freeman. 1994. A feminist reinterpretation of the stakeholder concept. *Business Ethics Quarterly 4*: 475–497.
Wood, D.J., and R.E. Jones. 1995. Stakeholder mismatching: A theoretical problem in empirical research in corporate social performance. *International Journal of Organizational Analysis 3*: 229–267.

John F. McVea is Professor of Entrepreneurship at the University of St. Thomas Opus College of Business.

R. Edward Freeman is University Professor, Elis and Signe Olsson Professor, Academic Director of the Institute for Business in Society, and Senior Fellow of the Olsson Center for Applied Ethics at the University of Virginia's Darden School of Business.

Chapter 10
Enhancing Stakeholder Practice: A Particularized Exploration of Community

Laura Dunham, R. Edward Freeman, and Jeanne M. Liedtka

1 Introduction

The development of stakeholder theory during the last decade has led to a number of interesting theoretical and practical questions. Areas of focus have ranged from establishing a normative justification for stakeholder management (Donaldson and Preston 1995; Evan and Freeman 1993; Phillips 1997), to examining the theory's implications for existing models of corporate governance (Goodpaster 1991; Boatright 1994), to exploring its link to corporate social responsibility and performance (Wood and Jones 1995; Waddock and Graves 1997), to integrating stakeholder with other strategic management approaches (Harrison and John 1994).

One recurring question of both theoretical and practical significance that has garnered significant scholarly attention has centered upon the identification of legitimate stakeholders—the question of "who and what really counts" (Freeman 1994). While many scholars and practitioners have simply used a catchall stakeholder map that includes general references to customers, employees, suppliers, investors and community, other theorists have explored the subject in detail. Mitchell et al. (1997) review exhaustively the vast literature on the subject and conclude that stakeholder theory "offers a maddening variety of signals on how questions of stakeholder identification might be answered.... Among the various ways of identifying stakeholders, as well as in the agency, behavioral, ecological,

Originally published in: *Business Ethics Quarterly*, 16(1), 23–42 © Cambridge University Press, 2006
Reprint by Springer, https://doi.org/10.5840/beq20061611

L. Dunham · R. E. Freeman (✉) · J. M. Liedtka
University of Virginia, Charlottesville, VA, USA
e-mail: FreemanE@darden.virginia.edu

© Springer Nature Switzerland AG 2023

S. D. Dmytriyev, R. E. Freeman (eds.), *R. Edward Freeman's Selected Works on Stakeholder Theory and Business Ethics*, Issues in Business Ethics 53,
https://doi.org/10.1007/978-3-031-04564-6_10

institutional, resource dependence, and transaction cost theories of the firm, we have found no single attribute within a given theory that can guide us reliably on theses issues." (Mitchell et al. 1997: 853–54). Their work, like that of the other scholars they cite, focuses on the development of a set of broad theoretical criteria for categorizing and prioritizing stakeholders. They choose power, legitimacy, and urgency as most key; others have focused on the holding of moral claims (Carroll 1989; Donaldson and Preston 1995; Hill and Jones 1992); or even on the bearing of risk (Clarkson 1995). While this ongoing line of scholarly inquiry has been an important contributor to the refinement of stakeholder theory, we have concerns about its ability to have a similar impact on stakeholder practice. Conversely, other business disciplines have formulated highly detailed definitions of key constituencies and strategies for managing them (e.g., marketing analyses of customers), whereas stakeholder theory's articulation of key constituencies remains vague and incomplete.

In this paper, we want to propose an alternative approach that, we feel, has special promise for impacting management practice. Nowhere is this shortcoming more evident than in stakeholder theory's treatment of the constituency known as "community," and so we have elected to focus the remainder of our paper on this category, in an attempt to demonstrate what we believe to be the shortcomings of current approaches and the promise of an alternative, *particularizing* focus.

While essential to any definition of corporate stakeholders, the concept of community has been left largely undefined and subject to broad interpretation. In fact, one might say that community as a stakeholder has come to represent something of a default, a sort of error term containing all sorts of interests and externalities that fail to find homes within customer, supplier, employee, or shareholder groups. Such a broad-brushed approach to describing this (or, we would argue, any) stakeholder group can only result in superficial and broad-brush ethical consideration and will inevitably ignore or fail to take account of important and marginalized interests. In other words, while we may agree that "community" represents a legitimate and, perhaps, high priority stakeholder group, we are left with no real guidance as to the specific ethical stance the corporation should take toward any particular community constituency. And in a day and age when various interest groups can form with unprecedented speed and can exert a powerful influence, as recent events at world trade summits in Seattle; Washington, D.C.; Toronto; and Genoa demonstrate, developing a finer grained understanding of community has become increasingly important.

In this paper we set forth what we call "The Problem of Community" as indicative of the definitional problems of stakeholder theory. We then begin the process of gaining greater specificity around our notions of community and the role of community in stakeholder theory and management. We identify and describe four distinct sub-categories of community. In doing so, we identify two important developments in the areas of community that we believe are particularly relevant to the stakeholder theorist and practicing manager, and that is the emergence of two fairly new forms of community. These two new variants of community—the virtual advocacy group and the community of practice—represent two very different notions of community, with strikingly different implications for stakeholder theory and practice. By breaking down our conceptualization of community into these finer grained and more cohesive sub-categories, new understandings emerge regarding the very different constellations of moral issues, interests, and claims that were previously aggregated under the more amorphous heading of "community." This, in turn, suggests several different stakeholder management approaches that are more tailored to the distinctive needs and issues of the different community segments.

In undertaking this project, we suggest how the Problem of Community is symptomatic of particularized work that needs to be done, work that would allow the linking of stakeholder theory with other background theories such as marketing, accounting and finance, production and operations management, organization studies, and the law. We believe such efforts are essential in moving stakeholder theory forward.

2 The Problem of Community

While Freeman (1984) explicitly traces the origins of the stakeholder idea to others, his book is often taken as a starting point for scholars in the Business, Ethics, and Society area. In it Freeman proposes stakeholder theory as a strategic management approach aimed at enabling the firm to survive in turbulent times by becoming more responsive to the many constituencies that could play a role in the firm's success. He defines stakeholders as "any group or individual who can affect or is affected by the achievement of a corporation's objectives" (Freeman 1984: 46). This wide or inclusive definition is meant to signal that managers must pay attention to external groups regardless of whether or not they happen to like those groups, or are disposed to communicate or not with them. They must gain the support of those who directly affect their achievements, but also understand how the firm will affect others, as these may take longer-term retaliatory action.

However, in order to bound the discussion for practical purposes. Freeman categorizes stakeholders into the following groups: customers, supphers, employees, financiers, and communities. In creating this narrow definition of stakeholders. Freeman returns to an older definition by Ackoff (1974), focusing on those groups that the firm relies upon—"those groups without whose support, the firm would fail to exist."

Both the wide and narrow definitions have served as starting points for subsequent theorists, but have generated a tension within the stakeholder literature. On the one hand are theorists who say that, in order to analytically and descriptively use the stakeholder theory, and to avoid the charge of managerial irrelevance, the wide or inclusive definition should be used. On the other hand, there are those who say we must start a theory of legitimacy, or a normative stakeholder theory with rather more careful claims. Surely the claims of a customer outweigh in a moral sense the claims of a terrorist group, no matter what effects are brought about by each. And so, to avoid the proliferation of stakeholders and hence devalue the idea of a group's importance qua stakeholder, the narrow or exclusive definition is put to use.

On both accounts, however, there is rather a large theoretical silence. On the wide or inclusive view, particular communities or interest groups may or may not be included, so we are left wondering, if we even get to that level of detail, just what communities should be included as stakeholders. On the narrow or exclusive view, we are simply told that "community" is a stakeholder without regard to what a community actually is. This leaves a potentially large hole in our narrow definition of stakeholders, enabling any number of entities to enter the picture under that nebulous heading. And the narrow definition then risks collapsing into the wide definition.

Thus, stakeholder theory suffers with the "problem of community": wanting 'community' to do some conceptual work, and wanting corporations to be held accountable for their actions vis-à-vis actual communities; but not specifying exactly who or what they are, so that we could know whether or not we are successful.

To see the magnitude of this issue reflect for a moment on the communities of which you are a member. We list the following as candidates: (1) the place, town, where we live; (2) a subdivision of this place; (3) the school where we teach; (4) a subdivision of this school, such as the people who are interested in business, ethics, and society; (5) the set of people who are involved in our children's schools; (6) the state or region or country that we live in; (7) the sets of people who play chess, golf, participate in martial arts, etc.; (8) the people that attended the same graduate school that we did; (9) the set of scholars in the country or world who teach business, ethics and society; (10) the electronic communities that we are a part of; (11) at least ten or more other uses of "community" that you have thought of that we have not.

So, what exactly is a community? It is reasonable to say that the ordinary usage of this term applies to at least some of the "communities" referred to above. However, we are still left casting about for those defining attributes that enable us to characterize some of these entities as communities and to exclude others. If we turn to the academic literature for the answer, we don't fare much better. In its long history as a subject of academic study, the term "community" has been made to encompass numerous, often conflicting, conceptions. One well known study of the literature, almost fifty years old, found over ninety competing definitions of "community" and the only element they all held in common was that they dealt with people (Hillery 1955).

Given its ambiguity, one is tempted to give up the quest for greater clarity and precision in our understanding of the term. To be told, however, that "community" is just systematically ambiguous is akin to being told that "customer" is too broad a term and little understood, but please agree that it is important to act in the interests of customers. Can we remedy this situation with a "Theory of Community" that can serve to buttress stakeholder theory?

3 Defining Community

The multiplicity of meanings attached to the term "community" presents a daunting challenge to the scholar in search of precision. If stakeholder, as a concept, suffers from a relative lack of detailed exploration, community suffers from the opposite—philosophers, psychologists, sociologists, anthropologists, political scientists, even urban planners, have explored the terrain of "community" with enthusiasm since Aristotle's time. Somewhere, we suspect, are yet to be interpreted cave drawings defining the term.

Complicating the matter are the eulogistic overtones often carried with the term. "Community" has both an evaluative usage as well as a descriptive/empirical one. It is frequently linked with conceptions of the "good life" and is thus often approached less as an empirically observable phenomenon than a normative prescription for social interaction, an approach that accommodates any number of both complementary and conflicting notions.

This normative use has its roots in the work of the first modem scholars of community sociology (Bell and Newby 1971; Gilligan and Harris 1989). These theorists, writing during the social upheaval of the nineteenth century, focused upon what they perceived as a disintegration of social cohesion and interpersonal ties in the wake of industrialization. This concern was probably best captured by a work that came to characterize most subsequent work in the field. The book was Ferdinand Tönnies's *Gemeinschaft und Gesellschaft* (translated as *Community and Civil Society*), first published in 1887. In it, Tönnies established the perspective that has shaped much work on community since. On the one hand, according to Tönnies's view, there was communal society—Gemeinschaft—defined as small, close knit, and unified by shared experience, values and accepted norms. It was traditional, slow to change and characterized by a sense of solidarity among its members. On the other hand was the associative community, or Gesellschaft, which was the polar opposite. It was defined as large scale, impersonal, and linked only by the transactions involved in the pursuit of individual self-interests.

Tönnies's approach has remained embedded in much subsequent work on community, with theorists often plotting various kinds of social interrelationship on a continuum between these two end points. And while theorists in related fields of study have often pointed to the tone underlying Tönnies's notions of Gesellschaft, and found it critical to the advancement of such other socio-political-economic phenomena as democracy and capitalism, Gesellschaft remains in community studies the less normatively attractive of the two. Furthermore, Gemeinschaft is intertwined with notions of homogeneous, place-based communities, and this has, until recent years, hindered theorists from fully exploring the notion of community as it might apply outside such circumstances.

Despite this initial bias toward smaller, place-based entities, community scholars have been forced to grapple with the changes in social interaction brought about by industrialization, urbanization, and technological change. Webber (1963) was one of the first to articulate a notion of community unbounded by geographic restrictions.

For him, "community without propinquity" was becoming increasingly feasible, as a growing middle class with greater access to information was finding it possible to achieve connection with others outside their immediate locality. Thus, the community to which we could belong was "no longer the community of place … but an interest community which within a freely communicating society need not be spacially concentrated for we are increasingly able to interact with each other wherever we may be located." (Webber 1963: 29). Webber's work set the stage for a broadening of the notion of community away from purely place-based definitions.

This brings us to our current project. Given the tensions between the descriptive and prescriptive uses of the term, and the multiplicity of definitions in operation, can we define community in such a way that we can begin understanding its implications for stakeholder theory and facilitating its use by practitioners? Can we develop some definitions of community that have enough descriptive power to allow us to identify legitimate community stakeholders and to develop some guidelines on how corporations should interact with them? In order to do so, we will avoid getting involved in debates about the "good life" and focus on the work of contemporary sociologists and political scientists who have sought to develop more empirically based descriptions of community.

3.1 Geography, Interaction, and Identity

In developing their definitions of community, most scholars have generally agreed that communities can be characterized by three factors: *geography, interaction,* and *identity* (Lee and Newby 1983). Communities primarily characterized by *geography* represent people residing within the same geographic region but with no reference to the interaction among them. Communities primarily identified by regular *interaction* represent a set of social relationships that may or may not be place based. Communities characterized primarily by *identity* represent a group who share a sense of belonging, generally built upon a shared set of beliefs, values, or experiences. In this case, the individuals need not live within the same physical locality.

These three factors serve as the basis for our identification of four more or less distinct subcategories of "community" relevant to stakeholder theory. These include community of place, community of interest, community as virtual advocacy group, and community of practice. While community scholars have traditionally emphasized communities based on geography, or place-based communities, we believe that communities built on interaction and identity are playing an increasingly important role in the business world. Furthermore, recent cultural and technological changes have begun altering the nature of the interactions and the kinds of identities involved in community activity, and these changes are of particular relevance to the stakeholder theorist and practicing manager. In recent years, for instance, community interactions have moved from primarily face-to-face to a mix of face-to-face and electronically mediated. At the same time, the types of identities shaping various community groups has expanded from more traditional proactive/agenda-driven identities to include groups built upon more negative, or "oppositional," identities.

These factors help us to bring a finer lens to the broad concept of community and to differentiate different kinds of communities that most business organizations will need to deal with. Figure 10.1 identifies the four distinctive subcategories of community that will be explored in this paper. The first two—communities of place and of interest—are well recognized in the stakeholder literature. The other two are not; yet, we feel that each of these—virtual advocacy groups and communities of practice—are emerging as increasingly significant.

In analyzing each in turn, we hope to demonstrate that the kind of particularizing work we advocate adds a level of insight beyond that offered by the broad conceptual categories offered by scholars to date.

For example, as we apply Mitchell et al.'s (1997) analysis to these subcategories of community, and seek to determine the degree of power, legitimacy, and urgency to ascribe to each, we often find that the answer is "it depends." In the community of place, for instance, while clearly a legitimate stakeholder, its power and urgency would likely depend upon a variety of specific factors—the issue at hand, the history of the relationship, the politics at play, among other factors. A similar concern would be raised for the community of interest. Communities of practice might or might not be legitimate, powerful, or characterized by a sense of urgency, depending upon the corporate culture and competitive environment in which they operate. Perhaps the Mitchell, Agle, and Wood framework is most useful in examining a virtual advocacy group such as the world trade activists, labeling these as "dangerous" stakeholders due to the combination of probably low legitimacy with significant power and urgency.

We now turn to a more in-depth exploration of each of these four communities.

		Communities of Place	*Communities of Interest*	*Virtual Advocacy Groups*	*Communities of Practice*
Geography		X			
Interaction	*Face-to-Face*	X	X		X
	Electronic		X	X	
Identity	*Proactive/ Agenda-Driven*		X		X
	"Opposi-tional"			X	

Fig. 10.1 The categories of community

4 Communities of Place

In the community of place, it is the physical proximity of its members that defines the community. Interaction can run the gamut from non-existent to the more regular interaction of the smaller neighborhood. This is perhaps the most common conceptualization of what is meant by "community" by stakeholder theorists—firms must take account of the effects of their behavior upon those who live in close physical proximity to their operations. Many of the most catastrophic failures of business firms to meet their obligations to community have occurred within this realm. Love Canal, Bhopal, and Chernobyl all provide devastating testimony to the consequences, albeit unintended, of corporate behavior on nearby locales.

In recent years, corporations have become increasingly active within the communities in which they are located (Altman 1998). This emphasis on place-based communities within close proximity to company operations has driven increased activity by corporate community relations departments. These departments are charged with getting to know their local communities and becoming involved in various civic activities. The activities range from making donations to encouraging employee volunteerism to supporting various community-based programs (Altman 1998). Thus, corporations are increasingly acknowledging the significance of such constituencies and taking steps toward both identifying the issues of importance to these stakeholders and acting upon that knowledge.

However, challenges remain. When relationships between the corporation and the community of place become contentious, the task of evaluating moral claims and developing appropriate responses to community stakeholder groups can be complex and confusing. A case in point illustrates the difficulties—think, for instance, of the controversy that erupted over GM's 1981 decision to build a state-of-the-art auto plant in Detroit that required the razing of one of the area's older, ethnic neighborhoods, Poletown. Given the economic decline the area had experienced, and GM's threat to move production to cheaper sunbelt states, the cities of Detroit and Hamtramck threw their support behind the project, and sought to use their power of eminent domain to transfer to GM 465 acres of land, which was home to 1300 families, 140 businesses, six churches and one hospital (Nolan 2004). Opposition to the plan developed among the local residents, resulting in numerous protests and, ultimately, lawsuits. In the end, although many families agreed to the buyout agreement, a core group of Poletown residents, led by the local parish priest, continued to resist the plan. The standoff between GM and these Poletown residents was ended by a decision by the Michigan Supreme Court which allowed the transfer of the property to GM, but it took police action to remove the last group of protesters from a twenty-nine-day sit-in at a local church.

The case demonstrates the complexities involved in navigating the many subsets that can exist within the subcategory we have called "community of place." For GM, those subsets included both the cities of Detroit and Hamtramck, as well as the smaller neighborhood groups in the Poletown area. Each had their own agenda and presented competing moral claims to GM, to the courts, and to the media.

Thus, although in some areas corporations appear to be making great strides in both identifying and addressing issues of importance to these local stakeholders, corporations continue to face challenges developing a finer grained understanding of the many different, and often competing, communities of place they can encounter within a single geographic region. While the traditional generic prescription of stakeholder theory to acknowledge the legitimacy of these communities' claims and to interact in good faith remains appropriate, it is incumbent upon both business scholars and practitioners alike to develop a better set of tools for identifying and understanding and interacting with these subgroups. In the final section of this paper, we offer a framework that sets out to begin this task.

5 Communities of Interest

Although Webber's early claims about the advent of the community without propinquity were initially greeted with skepticism, time has proven his essential insight to be profoundly correct. Communications technologies have removed the barriers of distance from the sort of social interactions that form the basis of community. With the arrival of the Internet, the speed and ease with which communities of interest can form and grow has accelerated well beyond what Webber could have imagined. And given their rapidly growing numbers as well as the increasing amount of influence they wield, communities of interest have become increasingly important constituencies for the corporate manager.

Communities of interest run the gamut in terms of purpose and area of focus. They range from hobbyists, to various membership groups centered on religious affiliation, civic activities, and charitable causes to political action groups (Walker 1983). In this paper, we will focus on those interest groups most likely to be involved in advocacy efforts targeting business firms, and therefore most likely to be pertinent to stakeholder theory.

Advocacy groups include both those built on voluntary membership and those in which membership is not entirely voluntary. The latter includes trade unions and business corporations involved in advocacy effort. Voluntary advocacy groups include both those tied to various occupational or professional specialties as well as those open to all individuals regardless of their qualifications (Walker 1983). The latter—the citizen action group—includes those advocacy groups organized around issues or causes, such as the Sierra Club, or Common Cause. All of these advocacy groups tend to spend a substantial portion of their efforts focused on issues of public policy. However, the citizen action group has become increasingly involved in advocacy efforts targeting specific firms.

Citizen action groups have become more important to business, not only through their increasing focus on business issues but also due to their rapidly increasing numbers and influence. Recent decades have seen an explosion in the number of citizen action groups (Walker 1983; Petracca 1992). Certainly, rapid social and economic changes have given rise to new interests and issues of concern. Greater affluence, higher levels of education, improved access to information have all contributed to the growing pool of potential group members as well as those entrepreneurs or patrons (Salisbury 1969; Walker 1983) responsible for spearheading the formation of such groups. Furthermore, these groups are forming in a more emotionally charged environment as "postindustrial changes have altered the pattern of conflict in society and created an intensely emotional setting in which groups rise or fall" (Loomis and Cigler 1998: 20–21). Thus business managers are faced with an increasing number of groups focused passionately on a narrow set of issues.

While confrontations between such groups and the firm can become heated, the traditional citizen action group is shaped by a set of attributes that make it possible for the manager to build a productive relationship with these particular stakeholders. Members of such groups have traditionally been linked by the common ideological goal that they share. These goals shape an agenda-driven identity—these groups identify a set of objectives, a core set of issues, some desired outcomes and a focused set of strategies for achieving those outcomes. Given their purpose, aimed at achieving a specific set of ends, these interest groups recognize the need to negotiate with, as well as challenge, their targets. This in tum makes possible the realization of win-win solutions between the interest group and its target organizations. Such win-win solutions are made feasible, if not mandated, by the interdependence that often exists between the interest group and its target.

A case in point would be the behavior and outcomes achieved by AIDS activists in the late eighties and early nineties. Members of ADDS activist groups shared internal alignment around a specific set of issues that they saw as key and a set of outcomes they wanted to achieve. These specific issues—focused around drug testing, availability, and pricing—and their desired outcomes were of a level of complexity that lent themselves to a clear and workable set of strategies. In addition, AIDS activists and their targets, the pharmaceutical companies, had a clear interdependence upon one another. Activists were influential customers and pharmaceutical firms were producers of an urgently needed product. This spurred the pursuit of win-win solutions that increased the likelihood of mutually acceptable problem resolution. During the period between 1987 and 1993, AIDS activists successfully drove the reshaping of the US drug industry's testing protocols, quadrupled AIDS research funding, and reduced the cost of the main treatment, AZT, to one quarter of its original price (Liedtka 2000).

Much stakeholder activity will continue to stem from the work of these types of interest groups. These groups, like communities of place, have been explored in the stakeholder literature—they are clearly seen as key stakeholders whose ends business managers must understand and be prepared to respond to, even helping to shape the specific issues and outcomes which guide the groups' actions. In order to do so, business managers must be prepared to acknowledge the areas of interdependence between the activists and the firm and to seek solutions acceptable to both parties.

Of course, communities of interest run the gamut in terms of the degree to which members are united by a shared agenda and are willing to negotiate win-win solutions with business. However, in seeking to add clarity to our understanding of community stakeholders, we suggest reserving the term "communities of interest" for those groups that do exhibit a proactive, agenda-driven identity that recognizes an interdependence with business and is willing to engage in productive dialogue, however fraught with tension. This allows us to draw a sharp distinction between such groups and a newer form of interest group, whose membership is less driven by a sense of identity than by specific issues and whose orientation toward business is more adversarial and less conducive to mutually satisfying solutions. We have termed these latter entities "virtual advocacy groups."

6 Virtual Advocacy Groups

In recent years, a new breed of citizen action group has sprung up, a kind of pressure group that we have termed "virtual advocacy groups." These groups are marked less by an "agenda-driven" sense of group identity—a feature of more traditional citizen action groups—and more by an "oppositional" sense of identity. The most public of this new breed of community emerged in 1999 at the Seattle World Trade Organization meeting (Bernard 1999) where various virtual advocacy groups opposed to "globalization" appeared, whose purpose appeared to be the short-term goal of disruption, rather than any longer term goal around problem resolution or shared win-win with the organizations being targeted.

The emergence and maintenance of such groups has been made possible by technological advances facilitating electronic communication among group members. The purpose, tone and tactics of such groups have also been shaped by technology. Godwin (1992), for instance, has posited that the advent of "direct marketing" appeals via mail, telephone and television, and their role in recruiting new members and spurring old members into action, has resulted in an increasingly confrontational tone between interest groups and their targets. According to Godwin, in order to galvanize existing and potential members, direct marketing appeals by interest groups emphasize highly emotional issues, use fear as the primary appeal and stress "the more negative and conflictual aspects of politics" (Godwin 1992: 324).

Furthermore, direct marketing may change the kind of people involved in advocacy efforts. Godwin suggests that outside of direct marketing, most people join groups based on the variety of business and social contacts they maintain on a regular basis. These individuals tend to hold memberships in various organizations with different areas of emphasis, and this "crosscutting" pattern of membership activity helps moderate their views, build tolerance, and encourage an ongoing engagement in the political process. Groups that rely heavily upon direct marketing, on the other hand, tend to attract more extreme, intolerant and politically alienated members. As Godwin puts it, "persons motivated by direct marketing are less committed to democratic norms and to the political system itself." This in tum translates into a lesser degree of effectiveness in addressing the issues of concern than that achieved by groups formed via social and business contacts. While good at generating protest, "direct-marketing-based groups are less able to lobby behind the scenes, concentrate on narrow issues, or pay close attention to implementation" (Godwin 1992: 317).

It should be noted that Godwin also found that direct-marketing recruits became more tolerant and less combative the longer they stayed involved with the group. Nonetheless, he says, the widespread use of direct marketing techniques has created an environment in which interests groups and their members have become more combative, aggressive, and negative in their outlook, and less capable of generating solutions.

If Godwin is right, then surely the Internet represents the culmination of these trends. For while most direct marketing still takes place via mail, telephone and television, the Internet has become a powerful new tool for advancing the same sorts of messages to far larger numbers of individuals at a fraction of the cost. Events in Seattle; Washington, D.C; Toronto; and most recently, Kyoto, underscore the speed with which large numbers of people can be mobilized through Internet appeals as well as the highly charged nature of the confrontations taking place between certain interest groups and their business and government targets. These virtual advocacy groups differ markedly along every attribute identified as features of traditional interest groups.

First, the "virtual" advocacy groups that descended upon Seattle, for instance, had no specific shared agenda. Unlike the AIDS activists, the summit protesters were not in agreement on a core set of issues, but in fact varied significantly in their areas of focus—"the UAW, tree huggers, and turtle lovers marching together" was how the *New York Times* described the unlikely partisans. Their temporary unity and identity rested entirely upon what they were against, rather than what they were for. They were further distinguished from the AIDS activists by the levels of complexity inherent in the issues that motivated them. The questions surrounding globalization, the ultimate target of their concern, were numerous and difficult. Environmentalism, employment, wage rates—these are inherently "wicked" problems, to borrow Rittel's (1972) nomenclature, and thus more difficult to resolve. Seattle was less about the failure of organizations to live up to a widely held set of community expectations, than it was a clash of view about what the future ought to look like. Thus, these virtual advocacy groups were unable and/or apparently uninterested in articulating a core set of issues, a clear vision of desired outcomes, or workable strategies beyond simple disruption. Finally, returning to our previous example, while AIDS activists and the pharmaceutical companies were mutually reliant, the summit activists and their targets had little overt interdependence. They were thus neither motivated towards nor capable of generating win-win solutions.

Such groups pose far thornier questions for business organizations than do traditional communities of place or interest—questions too important for scholars or businesses to ignore. If the ability to build collaborative relationships rests on a perceived interdependence that brings interest groups to the table interested in the possibility of creating "win/win" solutions to resolvable issues (Freeman and Liedtka 1997), how can the stakeholder manager effectively address the virtual advocacy group? Given their amorphous nature, can one even characterize the virtual advocacy group as an actual stakeholder constituency? And yet given their increasing visibility and power to disrupt, can the stakeholder manager afford to ignore them?

These questions concerning the legitimacy of moral claims and related questions of the appropriate stakeholder management approaches can only be meaningfully addressed at the level of the particular. In the case of virtual advocacy groups with characteristics similar to the WTO activists an approach that focuses more on *issue,* rather than *relationship* management, may be most appropriate. We will explore this idea more fully in the last section of this paper.

7 Communities of Practice

Our final definition of community—community of practice—is a relatively new one and has come to denote professional work groups united by a sense of shared interests, values and purpose. These groups are marked by a strong sense of identity, mutual obligation and an openness that facilitates learning and change within organizations.

Lave and Wenger (1993), who originated the term, have defined the community of practice as "an activity system about which participants share understandings concerning what they are doing and what that means in their lives and for their community. Thus, they are united in both action and in the meaning that that action has, both for themselves and for the larger collective" (98). In the business enterprise, communities of practice encompass those individuals involved in regular, close interactions. Brown and Duguid (2000) elaborate: communities of practice, they say, are "relatively tight knit groups of people who know each other and work together directly. They are usually face-to-face communities that continually negotiate with, communicate with and coordinate with each other directly in the course of work. And this negotiation, communication and coordination is highly implicit, part of work practice.... These groups allow for highly productive and creative work to develop collaboratively" (Brown and Duguid 2000: 143).

Communities of practice balance the needs of the group against the needs of the individuals within the group, seeking the achievement of the collective's goals while maintaining a concern for the developmental needs of the individual. The two are seen as related. It is possible to envision a business organization as a community of practice—to see it as held together by a shared concern for both the outcomes it achieves for stakeholders (be they customers or shareholders) and the personal development and learning of its members. In fact, it would see these two as inseparable, in that increased capabilities at the organizational level flow from development at the individual level. It is this notion of community that appears to come closest to approximating Aristotelian notions of community in the world of business practice.

The challenge for the stakeholder managers here is to identify the various communities of practice that operate within their organizations, or that link their organizations with outside constituencies, and then support those communities. This requires that the stakeholder manager seek to support those actions and policies committed to preserving, supporting, and enriching the development of each individual's uniqueness within the context of community, and linking this with community purpose, rather than subordinating the individual in the name of community building. As we explore the literatures related to this view of community, we might find linkages to areas such as learning theory and relational ethics most useful (Liedtka 1999).

8 Coming to Grips with "Community": Implications for Stakeholder Theory

In this paper, we have examined four different subcategories of community, each of which suggests potentially different discussions of both moral claims and management approaches. In this final section of this paper, we examine the ramifications for stakeholder theory and present a framework for defining and determining appropriate stakeholder management approaches to those groups.

As we have attempted to bring greater specificity to our understanding of community, several overarching themes emerge: First, despite the neatness implied in our identification of these four types of community, it is clear that each of these categories, in and of themselves, yields countless variations. Whether we are talking about communities of place or of interest or of practice, the challenge for the stakeholder theorist and practitioner remains to develop a more granular understanding of the many individual neighborhoods, interest groups, and communities of practice that concern them. We have endeavored here to begin to outline the key attributes and major trends affecting these different groups, and, in doing so, to highlight the great differences between them in terms of moral issues, interests, and claims. This, in tum, should confirm the importance of efforts aimed at unpacking our notions of community and forming more particularized views of the various subcategories comprised by that title. However, the specifics associated with each particular community group as they affect or are affected by the corporation, and the evaluation of their claims and selection of appropriate strategies for managing the interaction, remains to be explored within an individual firm's unique circumstances.

While it may seem an overwhelming task for the manager to identify, understand, and prioritize these groups, it may actually be more manageable that it appears. McVea and Freeman (2005) have suggested that, in order to remain viable as a strategic management framework, stakeholder theory and application need to move away from a view of stakeholders as generic groups. Instead, stakeholder theorists and practitioners must move toward a "names and faces" orientation, seeking a highly specific understanding of and communication with each stakeholder. They point to the successes of manufacturers and marketers in the area of mass customization, that is, the tailoring of unique products and services for each customer, and suggest that learnings from these fields could be applied to stakeholder theory. And certainly, given the proliferation of community groups and the very particular nature of their concerns that we have only touched upon here, this approach seems well worth further exploration. Ryan and Schneider (2002) have already demonstrated the potential utility of this particularized approach to stakeholder analysis in the investor category. In their work, they posit twelve differentiating criteria to assist managers in evaluating their relationships with their institutional investors.

A second major theme emerges around the degree of symbiosis implied between the corporation and its various communities. The community of place pursues much of its existence outside any particular influence of the corporation. Their interests certainly coincide and, at times, collide, but short a major disruption or other significant event, these communities and the corporation can generally pursue their objectives without extensive dialogue or active engagement with the other. Communities of interest, on the other hand, can often find their objectives strongly affected by the actions of the corporation, thus necessitating greater involvement between the two. And at the furthest extreme, communities of practice and the corporation find themselves fully intertwined, each an essential element of the other.

Hence, although much work remains for the practitioner, in terms of developing a finer-grained understanding of community stakeholder groups, our classification scheme here provides some insights into the relationship between the corporation and these different groups that can serve as the basis for different approaches to stakeholder management. These differences are captured in Fig. 10.2.

Traditionally, stakeholder theory has prescribed a generic willingness to acknowledge the legitimacy of the claims of various "communities" by interacting with them in good faith and by considering the consequences of the firm's actions on them in making its decisions. Our discussions here suggest at least two additional possibilities. First, instead of merely *interacting* with different communities as stakeholders, the firm might opt to enhance the degree of cooperation and work to be *in* community with some of them. Communities of practice offer a case in point here, where committing to build a spirit of true collaboration that moves beyond the arms-length stance of traditional stakeholder management, based on shared values and aspirations, offers new possibilities. This cycles us back to our earlier debate

Potential Strategies for Stakeholder Management

Low	*Degree of Cooperation*	High
Control/Containment	**Traditional Stakeholder Prescription**	**Collaboration**
recognize that claims may not be legitimate	recognize legitimacy of claims	deepened commitment to partnership built on shared values
focus on issue vs. relationship	interact in good faith	

Fig. 10.2 Potential strategies for stakeholder management

concerning the adoption of the more inclusive definition of stakeholders (those whom the firm's actions effect) versus a more exclusive definition (those upon whom the firm relies). We hypothesize that a firm ought to interact *with* other communities that it *affects or is affected by*, seeking to understand their perspectives, listen to their preferences, and evaluate the impact of its actions on them. Such interaction is best characterized as being marked by a spirit of *cooperation*. On the other hand, it ought to be *in* closer community with those upon whom it *relies* for support—employees, suppliers, and customers. Such interaction requires a deeper commitment than that necessary for the first set of communities. It requires a more active pursuit and support of the kind of spirit imbued in the community of practice—sharing interests, actions, and values. The firm's interaction with these groups must be marked by a spirit of *collaboration*.

The growth of "virtual" interest groups, such as the WTO activist community, suggests a second alternative and distinctly opposite approach to that advocated here with respect to relied-upon communities. While traditional stakeholder management remains an important contributor to thinking about how corporations can build effective relationships with particular stakeholder groups, those groups for whom it is most suited appear to be characterized by an internally aligned identity and purpose, formed around a specific set of actionable issues, sharing a commitment to progress through dialogue and in a context of interdependence with the firm in question. However, as we have suggested here, some of the new virtual advocacy groups may lack most—or all—of these characteristics. Rather than seeking constructive engagement with their corporate or governmental targets, they are more likely to engage in the kinds of "guerilla" public relations tactics that are difficult, if not impossible, for the corporation to effectively address through traditional stakeholder approaches. The growing number and influence of these groups suggest a refocusing of traditional stakeholder theory may be in order here, shifting from an emphasis on building either cooperative or collaborative relationships to an emphasis on using control mechanisms aimed at *containment*. An emphasis on containment would (1) focus on the *issues* raised by the activists, rather than on creating a relationship with them; (2) create mechanisms to allow such groups to voice their concerns and would ensure that their perspective was understood, but not necessarily that their membership would be involved, in corporate decision-making processes; and (3) recognize and accept the likelihood of ongoing adversarial relationships.

In sum, we suggest that the business practitioner can draw upon three distinct strategies, when approaching community stakeholders—(1) collaboration, (2) cooperation, and (3) containment. These approaches are marked by different objectives, processes, and interactions and are summed up in Fig. 10.3.

Although they do not track perfectly, these strategies can be broadly mapped against the four categories of community which have been developed in this paper, as shown in Fig. 10.4. Of course, each firm's and stakeholders' situations are distinct, and it remains the responsibility of the practitioner to develop a more specific understanding of each stakeholder group and then determine the appropriate strategy. For instance, in the previously cited case of GM and Poletown, it has been

	Collaboration	Cooperation	Containment
Objectives	Support stakeholder development	Negotiate win-win solutions	Minimize potential damages by SH
Nature of interactions	Open, trust-based	Cordial, reciprocal	Adversarial
Frequency of Interactions	High	Mixed	Low
Duration of Interactions	Ongoing, long-term	Intermittent	Short-term
Process Focus	Building/supporting shared identity	Developing mutual understanding and constructive solutions	Identifying and monitoring
Key Actions	o Building shared vision o Information sharing o Training/professional development	o Selective information sharing o Ongoing dialogue	o Information collection o Pre-emptive public relations

Fig. 10.3 Strategies for approaching community stakeholders

	Collaboration	Cooperation	Containment
Communities of Place		XXX	
Communities of Interest		XXX	
Virtual Advocacy Groups			XXX
Communities of Practice	XXX		

Conclusion

Fig. 10.4 Mapping strategies against the categories of community

suggested by some observers that the neighborhood group most resistant to and vocal about GM's plans represented a small minority of local residents who were ultimately co-opted by outside agitators, in this case Ralph Nader and his organization. If such characterizations were true, a business practitioner might do well to adopt strategies better suited for the kinds of adversarial virtual advocacy groups described in this paper than those most effective with more mainstream community of place stakeholders.

9 Conclusion

We set out, in writing this paper, to elaborate upon the limitations engendered by stakeholder theory's current lack of specificity and what we see as the shortcomings of current scholarly attempts to address it. We recognize that, to a certain degree, there will always be problems with stakeholder categories in terms of the detail and precision with which we can approach them. All stakeholder groups must ultimately be characterized with a certain degree of generality in order to remain tractable for both the theorist and practitioner. However, our conceptualizations must generate enough specificity to enable both scholars and practicing managers to acquire the kinds of meaningful insights that lead to strategic and ethical analysis with real teeth.

Nowhere is this tension between breadth and specificity, or the general and the particular, more problematic than in the area of community. As currently conceived in the stakeholder literature, community is a particularly amorphous and centerless stakeholder group. It can almost be seen as something of a default, a sort of error term containing a variety of interests and externalities that fail to find homes within customer, supplier, employee, or shareholder groups. Given this state of affairs, we have two choices for dealing with this particular stakeholder category. One way is to try to abstract and generalize our commitments to community by attempting to identify the consensus, or center of gravity, of this amorphous whole. This path leads towards difficult philosophical abstractions and away from practical managerial strategies. We have argued that this would be a mistake which can only result in superficial and broad-brush ethical considerations that will inevitably ignore or fail to take account of important and marginalized interests.

We have proposed that a better way to progress would be to abandon the complex philosophical challenges of developing general prescriptions towards community and to apply ourselves to finding more useful and coherent sub-categories of community as a set of stakeholders around which managers can develop meaningful and practical strategies. In attempting to begin this process of particularizing the community as a stakeholder, we have identified four important subcategories of community:

1. Communities of place
2. Communities of interest
3. Virtual advocacy groups
4. Communities of practice

The second proposal of this paper is that rather than focusing on the dividing of stakeholders into legitimate and illegitimate ones and then giving them due, or no, consideration, we should be open to accepting all interests as being legitimate, in that they are at least candidates for our consideration, but that we should take different ethical stances towards these groups depending on the nature of our relationship with them. We believe that if stakeholder theory is to rise above a crude form of utilitarianism, it must take account of the unique natures of the various relationships. We have suggested that three levels of ethical stance—collaboration, cooperation, and containment—will allow us to develop a more nuanced approached to the community as a stakeholder without resorting to broad brushed approaches that risk marginalizing important and potentially dangerous groups.

Of course simply elaborating upon the categories does not make tensions or conflicts go away. Nor does it eliminate the need for much further work and deeper analysis upon the part of practitioners dealing with their individual firm's own unique circumstances. However, the complexity implied does not to us seem to be an argument against attempting to make particularized progress any more than an argument about the impossibility of truly marketing to "markets of one" suggests that we should abandon all attempts to identify and serve customer segments.

Gaining a more particularized view of this stakeholder group does suggest to us that the traditional prescription that stakeholder theory offers can be elaborated upon in multiple directions, some increasing and other decreasing the degree of cooperation, in ways that would make its prescriptions more useful for management practice. While we have offered some preliminary thoughts here, much work, both empirical and theoretical, remains to be done before those possibilities can be fully articulated. Nonetheless, the category of community, in its many forms and guises, represents an increasingly important set of constituencies for both the theorist and the practitioner, and one to which we must apply greater rigor in our analysis. In doing so, it could also provide a model from which future analyses of other stakeholder categories could benefit.

One way to accomplish this task is to take seriously the notion that we need more finely grained narratives about actual companies and actual communities. We need to get past the genre of corporate horror story, and community "feel good" approaches to consider other notions of community as we have outlined here. Taking a closer and detailed look at companies that see themselves as "in community with stakeholders" has the potential to lend insight into what actually is good practice. Alternatively, examining companies that have a geographical or an electronic notion of community will equally yield insight. Only by taking this next step and constructing real stories of companies and communities can we more fully explore the potential contribution of stakeholder theory to improving business practice, in both a moral and a bottom line sense.

References

Ackoff, R. 1974. *Redesigning the Future*. New York: John Wiley and Sons.

Altman, B.W. 1998. Transformed Corporate Community Relations: A Management Tool for Achieving Corporate Citizenship. *Business and Society Review* 102 (103): 43–51.

Bell, C., and H. Newby. 1971. *Community Studies: An Introduction to the Sociology of the Local Community*. London: George Allen and Unwin Ltd.

Bernard, E. 1999. The Battle in Seattle: What Was That All About? *New York Times* (December 5): B1, B5.

Blair, M. 1995. *Ownership and Control: Rethinking Corporate Governance for the Twenty-First Century*. Washington: Brookings Institute.

Boatright, J. 1994. Fiduciary Duties and the Shareholder-Management Relation: Or, What's So Special about Shareholders? *Business Ethics Quarterly* 3: 393–407.

Brown, J.S., and R. Duguid. 2000. *The Social Life of Information*. Boston: Harvard Business School Press.

Carroll, A. 1989. *Business and Society: Ethics and Stakeholder Management*. Cincinnati: South-Western.

Clarkson, M. 1995. A Stakeholder Framework for Analyzing and Evaluating Corporate Social Performance. *Academy of Management Review* 20: 92–117.

Donaldson, T., and L. Preston. 1995. The Stakeholder Theory of the Corporation: Concepts, Evidence, and Implications. *Academy of Management Review* 20: 65–91.

Evan, W., and E. Freeman. 1993. A Stakeholder Theory of the Modern Corporation: Kantian Capitalism. In *Ethical Theory And Business*, ed. T. Beauchamp and N. Bowie, 5th ed. Englewood Cliffs: Prentice Hall.

Freeman, R.E. 1984. *Strategic Management: A Stakeholder Approach*. Boston: Pitman Publishing Inc.

———. 1994. The Politics of Stakeholder Theory. *Business Ethics Quarterly* 4: 409–421.

Freeman, R.E., and J. Liedtka. 1997. Stakeholder Capitalism and the Value Chain. *European Management Journal* 15 (3): 286–296.

Gilligan, J.H., and C.C. Harris. 1989. Community and Community Studies: Studying Local Social Life. In *Investigating Society*, ed. R. Burgess. London: Longman Group UK Ltd.

Godwin, R.K. 1992. Money, Technology and Political Interests: The Direct Marketing of Politics. In *The Politics of Interests: Interest Groups Transformed*, ed. M. Petracca. Boulder: Westview Press.

Goodpaster, K. 1991. Business Ethics and Stakeholder Analysis. *Business Ethics Quarterly* 1: 53–73.

Harrison, J., and C.H.St. John. 1994. *Strategic Management of Organizations and Stakeholders: Theory and Cases*. Minneapolis/St. Paul: West Publishing Co.

Hill, C.W.L., and T.M. Jones. 1992. Stakeholder-Agency Theory. *Journal of Management Studies* 29 (2): 131–155.

Hillery, G.A. 1955. Definitions of Community: Areas of Agreement. *Rural Sociology* 20: 194–204.

Lave, J., and E. Wenger. 1993. *Situated Learning: Legitimate Peripheral Participation*. New York: Cambridge University Press.

Lee, D., and H. Newby. 1983. *The Problem of Sociology: An Introduction to the Discipline*. London: Hutchinson.

Liedtka, J. 1999. Competitive Advantage and Communities of Practice. *The Journal of Management Inquiry* 8 (1): 5–16.

———. 2000. *Stakeholders in Seattle*. Presented at the Academy of Management Meeting, Social Issues in Management Division, Toronto, Canada.

Loomis, B.A., and A.J. Cigler, eds. 1998. *Interest Group Politics*. 5th ed. Washington: Congressional Quarterly Press.

McVea, J., and R.E. Freeman. 2005. A 'Names and Faces' Approach to Stakeholder Management: How Focusing on Stakeholders as Individuals Can Bring Ethics and Entrepreneurial Strategy Together. *Journal of Management Inquiry* 14 (1): 57–69.

Mitchell, R., B. Agle, and D. Wood. 1997. Toward a Theory of Stakeholder Identification and Salience: Defining the Principle of Who and What Really Counts. *Academy of Management Review* 22: 853–886.

Nolan, J. 2004. Auto Plant vs. Neighborhood: The Poletown Battle. *Detroit News* website: http://info.detnews.com/history/story/index.cfm?id=18&category=business.

Petracca, E. 1992. *The Politics of Interests: Interest Groups Transformed*. Boulder: Westview Press.

Phillips, R. 1997. Stakeholder Theory and a Principle of Fairness. *Business Ethics Quarterly* 7: 51–66.

Rittel, H. 1972. On the Planning Crisis: Systems Analysis of the First and Second Generations. *Bedrift Sokonomen* 8: 309–396.

Ryan, L., and M. Schneider. 2002. The Antecedents of Institutional Investor Activism. *Academy of Management Review* 27 (4): 554–573.

Salisbury, R.H. 1969. An Exchange Theory of Interest Groups. *Midwest Journal of Political Science* 8: 1–32.

Tönnies, F. 2001. *Community and Civil Society*, ed. Jose Harris; trans. Jose Harris and Margaret Hollis. Cambridge: Cambridge University Press.

Waddock, S.A., and S. Graves. 1997. The Corporate Social Performance-Financial Performance Link. *Strategic Management Journal* 18: 303–317.

Walker, J.L. 1983. The Origins and Maintenance of Interest Groups in America. *The American Political Science Review* 77: 390–406.

Webber, M. 1963. Order in Diversity: Community Without Propinquity. In *Cities and Space*, ed. L. Wingo. Baltimore: John Hopkins Press.

Wood, D.J., and R.E. Jones. 1995. Stakeholder Mismatching: A Theoretical Problem in Empirical Research in Corporate Social Performance. *International Journal of Organizational Analysis* 3: 229–267.

Laura Dunham is the Dean of the Opus College of Business at the University of St. Thomas.

R. Edward Freeman is University Professor, Elis and Signe Olsson Professor, Academic Director of the Institute for Business in Society, and Senior Fellow of the Olsson Center for Applied Ethics at the University of Virginia's Darden School of Business.

Jeanne M. Liedtka is United Technologies Corporation Professor of Business Administration at the University of Virginia's Darden School of Business and former chief learning officer at United Technologies Corporation.

Chapter 11
Corporate Social Responsibility: A Critical Approach

R. Edward Freeman and Jeanne M. Liedtka

Our conversations about corporations need to change for us to truly create the good society.

The idea of corporate social responsibility has failed to help create the good society. Long seen by academics and managers alike as the missing link in capitalism, the concept of corporate social responsibility has not delivered on its promise. Furthermore, it has become a barrier to meaningful conversations about corporations and the good life. Corporate social responsibility, in all of its many masks, has outlived its rather limited useful life, and we call for its immediate demise.

The purpose of this article is to suggest some reasons as to why the concept of corporate social responsibility should be abandoned. And, we offer an alternative: an ongoing conversation about corporations and the good life.

Such an alternative conversation needs to be centrally connected to other conversations about how we human beings can live. In particular, we suggest that reconceptualizing the corporation as a network of relationships makes possible a social world in which "caring" has primary significance. In addition, we suggest a whole host of different metaphors that can be developed once we drop the search for corporate social responsibility.

Originally published in: *Business Horizons*, 34(4), 92–98 © Elsevier, 1991
Reprint by Springer, DOI 10.1016/0007-6813(91)90012-K

R. E. Freeman (✉) · J. M. Liedtka
University of Virginia, Charlottesville, VA, USA
e-mail: FreemanE@darden.virginia.edu

© Springer Nature Switzerland AG 2023

S. D. Dmytriyev, R. E. Freeman (eds.), *R. Edward Freeman's Selected Works on Stakeholder Theory and Business Ethics*, Issues in Business Ethics 53,
https://doi.org/10.1007/978-3-031-04564-6_11

Understanding our "responsibilities" does not enable us to create new forms of economic and social life. In fact, we suggest that it gets in the way. We believe that managers and business school professors need to return to some more basic questions about the nature of the human self and the communities we are capable of creating.

1 Why Corporate Social Responsibility Is Not a Useful Idea

The purpose of this section is to give seven reasons (summarized in Fig. 11.1) why we should give up the idea of corporate social responsibility. We draw on our experience of some years in helping managers think about these ideas, as well as our reading of the management literature on this and other subjects. We have to confess that after years of trying to make sense of this idea we have come full circle to agree with Milton Friedman (1970) that the concept of corporate social responsibility is a dangerous idea. We have slightly different reasons for our conclusions.

1. The origins of the concept are suspect, as they derive primarily from the field of economics, and fail to include, among others, history, religion, and culture.
2. The different models of corporate social responsibility all accept the terms of the debate as set forth by Milton Friedman's argument that sees corporations only as profit maximizers.
3. Corporate social responsibility accepts the prevailing business rhetoric of "capitalism: love it or leave it."
4. Corporate social responsibility is inherently conservative—it starts with the standard received wisdom and then attempts to "fix" its unintended consequences.
5. Corporate social responsibility promotes incompetence by leading managers to involve themselves in areas beyond their expertise—that is, repairing society's ills.
6. Corporate social responsibility accepts a view of business and society as separable from each other, each with a distinct ethic, linked by a set of responsibilities.
7. The language of rights and responsibilities is, itself, both limiting and often irrelevant to the world of the practicing manager.

Fig. 11.1 Seven reasons to abandon the concept of corporate social responsibility

1.1 The History of Corporate Social Responsibility Is a History of Economics[1]

The idea of corporate social responsibility has its roots in the writings of Andrew Carnegie and others. Carnegie, founder of U.S. Steel, articulated two principles he believed were necessary for capitalism to work. First, the charity principle required more fortunate members of society to assist its less fortunate members, including the unemployed, the disabled, the sick, and the elderly. These "have nots" could be assisted either directly or indirectly, through such institutions as churches, settlement houses, and other community groups. Second, the stewardship principle required businesses and wealthy individuals to see themselves as the stewards, or caretakers, of their property. Carnegie's view was that the rich hold their money "in trust" for the rest of society. Holding it in trust for society as a whole, they can use it for any purpose society deems legitimate. However, it is also a function of business to multiply society's wealth by increasing its own through prudent investments of the resources that it is caretaking.

These ideas gained wide acceptance over the years. Coupled with the threat of government intervention and regulation, they helped form the expectation that corporations add social needs and concerns to their economic purpose. But fulfilling the economic purpose was seen as the primary task of the corporation, and was so recognized by the law of corporations for many years.

1.2 Milton Friedman's Argument Led to More Theories About the Essentially Economic Role of the Corporation

Milton Friedman's now-famous argument is that corporations should pursue their economic self interest, and that any attempt to promote corporate social responsibility, however it might be defined, amounts to moral wrong. Friedman questioned the logic of corporate social responsibility as it had developed. He insisted that in a democratic society, government was the only legitimate vehicle for addressing social concerns. (He also argued that government was the best vehicle for meeting such concerns, but it is difficult to believe, given Friedman's libertarianism, that such an argument was in good faith.)

The response of management thinkers was to develop more sophisticated models of corporate social responsibility, variously called corporate social responsiveness, the social policy process, social issues in management, business and public policy, corporate social performance, and so forth. While there are real and relevant

---

[1] This section is based on Freeman (1991). For a complete discussion of the history of the concept of corporate social responsibility, see Frederick (1987) and Preston (1988). Our understanding of the history of these ideas is due to Professors Frederick and Preston, as well as James Post and Ed Epstein.

differences among these models, they share an important common ground. They seem to accept the terms of the debate on Friedman's ground: that business can (or cannot) or should (or should not) address social issues in addition to economic ones. We want to suggest that the very question of corporate social responsibility, the very terms of the debate itself, should be put to one side and forgotten.

1.3 Our Business Rhetoric: "Capitalism: Love It or Leave It"

Friedman's rhetoric runs strong throughout business, public policy, and academic circles. The fall of socialist regimes in Eastern Europe, the opening of the Soviet Union, and even unrelated events such as the Allied victory in the Persian Gulf War are all taken as evidence of the triumph of democracy and capitalism. One pundit has even called for "the end of history," claiming that the battle between the West and other systems is over: intellectually, there are no real competing ideas.

Underneath this rhetoric is a belief that capitalism can be seen as an immutable system with socialism positioned as its alternative (see Selden 1990). Because socialism appears so clearly to have failed in its ability to produce any semblance of a "good society" no matter how defined, capitalism—in all its unfettered glory—is seen as our best hope, to be accepted with all its warts as clearly the lesser of two evils. Adam Smith is seen as having the last laugh after all.

However, after examining the "good society" that capitalism has created—the damage to the environment, the hunger and homelessness that exist even in wealthy areas of the world—can anyone today (even Milton Friedman) really believe that the pursuit of self interest has culminated in the common good? The alternative to capitalism as we know it is not socialism, but a better form of capitalism—one that recognizes the existence of the commons and acts to prevent the single-minded individualism capable of destroying it.

At the heart of capitalism is the problem of the commons, so aptly described by Garrett Hardin in his article "The Tragedy of the Commons" (1968). Hardin suggests that if there is a common grazing area to the community, and if everyone pursues self interest and grazes cattle, then the commons will be destroyed. This argument is a specific form of the "prisoner's dilemma," which can be found in another form in Plato's *Republic.* This problem, in whatever guise, points out a fundamental tension in our notions of "self vs. community," "public vs. private," and other cherished, usually unquestioned, third-level assumptions about human beings. A willingness to enter into the conversation about what a "commons- sensitive" vision of capitalism might look like is a pre-condition for linking corporations and the good society—that is, if we believe that a good society is concerned with more than the accumulation of wealth by a privileged group.

1.4 Corporate Social Responsibility Is Inherently Conservative

Corporate social responsibility starts with the standard received idea of the corporation as essentially and primarily an engine of economic production and distribution. Models of corporate social responsibility, from arcane academic ones to real programs like those at Dayton Hudson, Control Data, and Chemical Bank, try to fix this idea by adding social responsibility and stirring. For instance, if our economic system, based on seeing the corporation as essentially economic, leads to a neglect of communities, then we need corporate social responsibility programs to address community needs. If we see the corporation as an economic system based on hierarchy and command and control, we will develop a view of human beings as being obedient (for the most part) and willing participants in the resulting social milieu. Since the human consequences of living in command and control systems are unsavory, say corporate social responsibility theorists, we need to treat employees as capable of going beyond command and control, so they are not exploited. Further, we need to do so because we have a social responsibility to our employees—or a moral responsibility, as current jargon would have it.

Why not just abandon the idea of the corporation as a command and control system, and therefore abandon its social responsibility to treat employees as more human? We could start with the idea that employees are complex human beings who have hopes and dreams they want to realize. And corporations can be used by employees to realize those dreams in a number of dimensions. A special "social responsibility" to treat employees better would simply not be relevant, and we would have many more possibilities open in terms of creating organizational forms that help us realize complex human purposes.[2]

1.5 Corporate Social Responsibility Promotes Incompetence

The idea of corporate social responsibility leads managers to make decisions about issues beyond their expertise; further, it reinforces what managers and business school professors take to be the real and legitimate competence of business executives to deal with social ills.

Philanthropy represents acts of charity, of benevolence taken on by choice. Ethical conduct, on the other hand, is obligatory and expected for maintenance of the relationships between the organization and its constituents. Adding social goals directly to the corporation's obligations brings with it an obvious right to involvement in the process of achieving those goals. Not surprisingly, the perspective brought by direct business involvement is often one that focuses on efficiency and short-term outcomes.

---

[2] We take this point to be a large part of the substance of Tom Peters' call for chaos in organizations as well as the current movement toward employee empowerment, "flat organizations," and the decentralization of command and control organizations.

Because managers have little training in understanding the complexity of human beings and their communities, and because they are encouraged to believe that humans are economic and economical beings, approaches to social problems are at best well meaning and simplistic. We are constantly bombarded with the notion that welfare recipients need better "incentives," or that teachers need to be held "accountable" in the same way division presidents are accountable. Business managers, and their politicians, constantly try to privatize the public sector, if not by assigning actual property rights, then by assigning "management expertise" understood as "economic logic."

Well-meaning managers turn social programs into corporate strategies, subordinating the needs of external constituencies to the needs of corporations. Corporate social responsibility programs are often the province of a special committee, an executive assistant, a foundation, or public relations/affairs. Several foundations with which we are familiar routinely wrestle with their mission as furthering the economic ends of the corporation or doing something else. Giving to particular community groups either furthers the ends of the corporation or it doesn't. Separating these ends into economic ones and social ones, and letting the latter serve a legitimating function for the former, leads us astray. It causes us to not question the impacts of economic ends on the very groups to whom we claim to owe social responsibilities.

1.6 Business and Society Are Not Separable

We need to closely examine the idea that the ethics of business, couched in the language of the "social responsibility of business," can usefully be thought of as separable from the ethics of all of society. American business values are inevitably the product of American culture. (We take this point to be true of business set in any culture, not just American business). For the most part business values have represented a small subset, rather than a wide spectrum, of these values. The relationship between business and the larger cultural context has been one of mutual influence, yet with few exceptions managers and management thinkers have not carefully explored these linkages.[3] In reviewing the legacy of the 1980s, what we find most troubling is not the extent to which respect for business has sunk to new depths—it is the extent to which it has risen to new heights. Sociologist Robert Bellah and his colleagues argue (1985) that the archetypal American Character is no longer the religious or civic leader, it is the businessman. Why the recent concern with business ethics—is it because the ethics of American business are deteriorating alarmingly? More likely, we believe, is the hypothesis that it is the ethics of the

[3] Notable among these exceptions are Cavanagh (1976) and Scott and Hart (1990). More commonly, management thinkers are often heard to decry the Japanese success in business with explanations of the homogeneous nature of Japanese society and the reliance on communal versus individual values. If our management thinkers are to be believed, Japanese business is closely connected with the underlying cultural values, but American business is not—a strange theory indeed.

community at large that are changing—and the values of business are increasingly becoming the values of America. If business values have historically been, for society, its moral floor, what are the implications of them becoming the norm? If, in fact, business values now drive the American consciousness, they take on a new importance—an importance that can only be recognized if we drop the charade that privileges business with a discrete set of ethics, immune from the standards applied in the rest of our lives.

Arguments within the business community for the self-interested pursuit of profit maximization gain as much support from Americans' traditional focus on individualism as they do from Adam Smith. An emerging group of communitarians such as Amitai Etzioni have begun to question the implications of this for the quality of life that we lead. The strain of individualism runs deep in American ideology—it was one of what de Tocqueville called our "habits of the heart." But, at the time de Tocqueville was writing, in the 1830s, he saw other habits of the heart as well—the concept of the ethical community as represented by Massachusetts' John Winthrop, a leader of the Puritans. Winthrop's focus was on "moral freedom," which consisted not of individuals free to pursue self interest at will (which reflects more closely our current notions of freedom) but rather the liberty to pursue only that which is good, just, and honest. De Tocqueville also noted America's republican tradition of commitment to civic responsibility, as exemplified by individuals like Thomas Jefferson.

Robert Bellah and his colleagues argue that our problems as a country today relate, in part, to the loss of the tempering forces of civic responsibility and religion on the individualism so much a part of our heritage. They argue that our focus on the liberation and fulfillment of the individual has become a cancer, because it has allowed us "to think of commitments—from marriage and work to political and religious involvement— as *enhancements* of the sense of individual well-being rather than as *moral imperatives.*" Creating the good society will require that the business community accept a new set of moral imperatives. This is not to suggest that business take on the traditional role of government— quite the contrary. Instead, it suggests that business acknowledge its complicity in the creation of our current society—and abandon the defense of "business as usual," shielded and decoupled from the larger questions raised by the good society. The business community must, indeed, have its separate conversation, but the fundamental values that determine the nature of that conversation are irrevocably shared with the society in which business functions.

1.7 Rights and Responsibilities Are Only Part of the Issue

Although the language of rights and responsibilities is a central portion of our moral discourse, we want to join the group of thinkers like Carol Gilligan, Alasdair MacIntyre, Nell Noddings, John Rawls, Richard Rorty, Sara Ruddick, and Michael Sandel who suggest, each differently, that rights and responsibilities are not the whole story. The language of rights and responsibilities dooms our discussion to remain an essentially intellectual exercise with little relevance to the practicing manager. A focus on the allocation of rights and responsibilities as an end point in the discussion is helpful only in a world where we can reach some level of consensus around that allocation.

Such a view envisions rights as fixed and knowable and responsibility as a pie, to be sliced into discrete pieces—some of which are laid upon the plate of government, others to producers, retailers, consumers, and so on. The curbside appeal of such an approach is undeniable, provided, of course, agreement can be reached on how big a slice each party receives. The complexities evoked by the differing natures of the situations faced spawns a series of debates, some of which occur at the abstract level of stockholders versus stakeholders, others of which filter down into a seemingly endless array of topic discussions (bribery, product misuse, and so on). Throughout, the knife remains poised but unable to cut through the verbiage to affix responsibility in a way that would let us actually deal with the issue at hand.

The difficulties inherent in this framework do not end with the problem of consensus, however. In fact, even if we assume that such an elusive consensus about who owns particular rights and responsibilities can be achieved, our search for a useful framework remains unsatisfied—for we must also believe that the recipients of the responsibilities actually live up to them. Thus, we may ascribe to agricultural chemical firms the responsibility to produce a "good" product in a safe manner, and to government the role of educating the populace to ensure safe use of pesticides. When, as is often the case, one party fails to meet its responsibilities, the question of interest becomes, "How much of the responsibilities ceded to others am I, the other party, willing to shoulder to meet legitimate needs?" If government, in the above case, is irresponsible and fails to educate its people, the chemical firm has not solved the dilemma at hand—preventing misuse of its product; it has merely traded in one perplexing question for another.

2 Toward a New Conversation

We need a new conversation that brings with it new possibilities for corporate life. We suggest that there are at least three avenues we can pursue to replace the worn-out language of corporate social responsibility with some new, rich, and more useful conversations. We shall put these proposals into the form of three propositions summarized in Fig. 11.2 to crystallize the metaphors. They are:

- The Stakeholder Proposition—Corporations are connected networks of stakeholder interests;
- The Caring Proposition—Corporations are places where both individual human beings and human communities engage in caring activities that are aimed at mutual support and unparalleled human achievement; and
- The Pragmatist Proposition—Corporations are mere means through which human beings are able to create and recreate, describe and redescribe, their visions for self and community.

PROPOSITION 1: Corporations are connected networks of stakeholder interests.

This proposition expands the conversation to include suppliers, employees, and customers, among others, making them legitimate partners in the dialogue.

PROPOSITION 2: Corporations are places in which both individual human beings and human communities engage in caring activities that are aimed at mutual support and unparalleled human achievement.

This proposition pushes us beyond the language of rights and responsibilities to a focus on the ethic of care, which recognizes needs and affirms the self and its linkage with others.

PROPOSITION 3: Corporations are mere means through which human beings are able to create and recreate, describe and redescribe, their visions for self and community.

This proposition urges us to see the projects of "self-creation" and "community creation" as two sides of the same coin, and see in institutions many possibilities for different ways of living together to pursue the joint ends of individual and collective good.

Fig. 11.2 Three propositions for new conversations

2.1 The Stakeholder Proposition

Because stakeholders are broadly defined to include suppliers, community, employees, customers, and financiers, if the stakeholder proposition is accepted into the normal way we talk about corporations, the question of social responsibility just doesn't come up. Once we come to see each of these groups, and the individuals within them, as legitimate partners in the dialogue about "what is this corporation going to be," the social responsibility of the resulting entity is moot. Social responsibility to whom? All of the affected parties are a part of the conversation. Since social responsibility depends on the external effects on those not a party to the corporation, we have simply made the issue disappear.

Such a move is not just intellectual trickery. Language is important; in fact, it is the only tool we have to deal with a rather difficult real world. If we come to see corporations as connected sets of stakeholders, all of whom are "in it together," then we are able to live in a way that doesn't carve up the world into "economic, social, political, and technological" parts.

2.2 The Caring Proposition

The caring proposition offers a nice link to a body of work that has gone unnoticed by most management thinkers: feminist theory. Of course, there are many "feminisms," and many conflicts and tensions within each, but we want to illustrate the kinds of conversations we can have if we take this work seriously. Scholars working in what has become known as "liberal feminism" or "feminist standpoint theory" or "different voice feminism" have suggested that we need to explore an ethics of caring and connection. Caring that is personal and supportive is juxtaposed against an ethics of rights that is deliberately and logically impersonal. Rights are abstract, formal, immutable. They reflect the reality of those who define them, independent of the perspective of those who own them. The language of the ethic of care sees humans as enmeshed in communal institutions as well as striving for individuality. It recognizes human needs as concrete, personal, changing, defined from within by those who possess them. The ethic of care would acknowledge the differing perspectives of the many voices that exist in the corporate context, and makes such an acknowledgement primary.

Where does this talk of employee rights lead us? Does the employee, acting as agent for the institution, have a "right" to voice—and, beyond that, to autonomy, to freedom of choice? Does the organization have the "right" to demand such abdication of individual rights in exchange for the daily wage? Is this the place to begin? What is clear is that humans have the *need* to be heard, to be respected, the need to live in accordance with the values they hold. Organizations, as well, have needs: needs that must be met today to keep the organization a viable vehicle for meeting both its own and its constituents' needs in the future. To focus exclusively on rights is a zero-sum game. And in business, which has inherently to do with relationships among individuals and groups, zero-sum games are self-defeating, as Chester Barnard recognized over 50 years ago.

The language of care begins with the premise that we *do* care, not that we have a responsibility to care. It builds on a view of human nature as connected with others, concerned with maintaining and nurturing relationships. Drawing on our years of interaction with practicing managers, and with management students, we find this view of human nature both more accurate than views of humans as calculated, self-interested economic beings, as well as more compelling in the possibilities it offers. Rather than distancing managers from their altruistic selves, as the focus on rights does, the ethic of care reaffirms the self and its link with others. Rather than creating a false dualism that asks individuals to be managers from 9 to 5 and humans for the duration of their day, it leaves us whole.

2.3 The Pragmatist Proposition

The ethics of caring is only one of many metaphors we can use to describe and redescribe corporate life. Embodied in the pragmatist proposition is the suggestion that if we come to see the projects of "self creation" and "community creation" as two sides of the same coin, we will stop juxtaposing "self interest" with "altruism," "public self with private self," even "rights and responsibilities with caring and connection." If we come to see our institutions, like the corporation, as just one more way we can live together, we can easily accept the pragmatist advice to "be experimental." There are lots of ways to reconfigure corporations, many methods of governance, many ways that humans can thrive and flourish in the joint pursuit of their individual and collective good. What holds us back is our language, our view of ourselves, and our worn out metaphors for corporate life.

The pragmatist project for self-creation and community creation is less a matter of economics and politics (the fundamental grist for the business and society mill) and more a matter of history, culture, poetry, and literature. Roland Barthes (1975) suggests that we learn to "read without restraint" or come to find "jouissance" in reading. We must see corporations in the same way, as places in which we can be fully unrestrained human beings, places of "jouissance" rather than grey flannel, places of liberation and achievement rather than oppression and denial, and finally, places where we can have real and relevant conversations about our differing visions of the good life.

References

Barthes, Roland. 1975. *The Pleasure of the Text.* Trans. Richard Miller. New York: Hill and Wang.

Bellah, Robert, et al. 1985. *Habits of the Heart.* New York: Harper & Row.

Cavanagh, Gerald. 1976. *American Business Values in Transition.* Englewood Cliffs: Prentice-Hall.

Frederick, William C. 1987. Corporate Social Responsibility and Business Ethics. In *Business and Society*, ed. S. Prakash Sethi and Cecilia M. Falbe, 142–161. Lexington: Lexington Books.

Freeman, R. Edward. 1991. *A Note on Ethics and Business.* UVA-E-071, The Darden School, University of Virginia.

Friedman, Milton. 1962. *Capitalism and Freedom.* Chicago: University of Chicago Press.

Friedman, Milton. 1970. The Social Responsibility of Business is to Increase Its Profits. *New York Times Magazine,* September 13, 122–126.

Hardin, Garrett. 1968. The Tragedy of the Commons. *Science,* December 13, 1243–1248.

Preston, Lee E. 1988. *Social Issues and Public Policy in Business and Management: Retrospect and Prospect.* College Park, Md.: University of Maryland.

Scott, William, and David Hart. 1990. *Organizational Values in America.* New Brunswick: Transaction Publishers.

Selden, Arthur. 1990. *Capitalism.* Oxford: Basil Blackwell.

R. Edward Freeman is University Professor, Elis and Signe Olsson Professor, Academic Director of the Institute for Business in Society, and Senior Fellow of the Olsson Center for Applied Ethics at the University of Virginia's Darden School of Business.

Jeanne M. Liedtka is United Technologies Corporation Professor of Business Administration at the University of Virginia's Darden School of Business and former chief learning officer at United Technologies Corporation.

Chapter 12
A New Approach to CSR: Company Stakeholder Responsibility

R. Edward Freeman and S. Ramakrishna Velamuri

1 The Problem: Has the Idea of Corporate Social Responsibility Outlived Its Usefulness?

Assume that the CEO of Firm A is asked the following: 'Well, I know that your company makes products that consumers like, and that those products make their lives better. And I know that suppliers want to do business with your company because they benefit from this business relationship. I also know that employees really want to work for your company, and are satisfied with their remuneration and professional development. And, let's not forget that you're a good citizen in the communities where you are located;[1] among other things, you pay taxes on the profits you make. You compete hard but fairly.

[1] We admit that there are many ways of being a good corporate citizen.

Originally published in: *Corporate Social Responsibility*, 9–23 © Palgrave Macmillan, 2006
Reprint by Springer, https://doi.org/10.1057/9780230599574_2

The ideas in this paper have been developed with a number of co-authors over the years in several places. In particular see Wicks et al. (2005), Freeman and McVea (2001), Freeman et al. (2005a, b). We are grateful to a number of people for helpful conversations, in particular Professors Gianfranco Rusconi, Dr Valeria Fazio, Dr Mette Morsing, doctoral students at the Copenhagen Business School doctoral consortium on Corporate Responsibility, numerous participants in the EABIS conference in Gent, Professors Jeff Harrison, Robert Phillips and Andrew Wicks.

R. E. Freeman (✉)
University of Virginia, Charlottesville, VA, USA
e-mail: FreemanE@darden.virginia.edu

S. R. Velamuri
Mahindra University, Hyderabad, Telangana, India

© Springer Nature Switzerland AG 2023

S. D. Dmytriyev, R. E. Freeman (eds.), *R. Edward Freeman's Selected Works on Stakeholder Theory and Business Ethics*, Issues in Business Ethics 53,
https://doi.org/10.1007/978-3-031-04564-6_12

You also make an attractive return on capital for shareholders and other financiers. However, are you socially responsible?'

We confess to having absolutely no idea what 'socially responsible' could mean here. If a firm is doing all the things that Firm A does, then it deserves to be applauded and offered as an example for other firms, large and small, to emulate. If it is not doing them as satisfactorily as we think it ought to, then we could perhaps offer to help it do them better, rather than appeal to actions and responsibilities that might lie outside the domain of its day-to-day activities. In summary, by talking of business and social responsibility as if they were two separate things, we might unintentionally be promoting the idea that they involve discrete thought processes and activities. In our opinion, the challenge is to promote a different way of doing business that integrates considerations of business, ethics and society.

Herein lies the problem with Corporate Social Responsibility. Corporate social responsibility reinforces the 'the separation thesis', or the idea that we should separate 'business' from 'ethics or society'. This separation is an idea that reaches very deeply into western culture. It is reinforced by the disciplines of business, by our major theoretical frameworks in management, and by executives and business thinkers themselves. At its worst it generates an absolutely destructive idea of capitalism, i.e. that capitalism is about 'anything goes'. After all, the theory says, 'its just business'. Viewed in this way, corporate social responsibility becomes an 'add-on' to ameliorate the supposedly harsh consequences of this view of capitalism.

Let us go back to the example of our Firm A, and examine its decision to hire employees. Has it done something that is 'for the business'? We believe that the answer to that question is a resounding and unqualified 'Yes.' Has it done something that is 'for society'? We believe that the answer to that question is also a resounding and unqualified 'Yes.' So, how do matters of employment count – in the social ledger or the business ledger? A similar argument can be made for customers and communities, and for suppliers and financiers as well. All these individuals and organisations are full-fledged members of society, as well as being stakeholders in Firm A. If they benefit in their dealings with Firm A, then society benefits too, both directly and in a number of indirect ways.

Corporate social responsibility is often about seeming to 'do good works'. And, while there is certainly nothing wrong with doing more good, there can be an implication that companies need to do good works because the underlying structure of business is not good, or morally neutral. We believe that this is a destructive idea, because it fails to recognise the central role business has played in improving the well-being and prosperity of hundreds of millions of people around the world. And, it often causes companies to act in bad faith and get involved in matters where they have little expertise.

This is not Milton Friedman's argument that the only social responsibility is to increase profits; rather it is a practical matter, that giving money to the opera doesn't make up (in any moral sense) for short-changing customers or communities. The focus needs to be on how value is created in the basic business proposition. How does this company make customers, suppliers, communities, employees and financiers better off? Capitalism is a system of social cooperation – a system of how we work together to create value for each other. Seeing it any other way can lead to dangerous social policies, and to the tarnishing of the one institution – business – that still has to play a central role in lifting hundreds of millions of people out of poverty in Asia, Africa and Latin America.

The second problem with corporate social responsibility is that it is focused on 'Corporate' social responsibility. Why is it not called Business Social Responsibility? The focus on 'Corporate' implies that corporations, due to their size and success and perhaps their shareholding pattern, have to shoulder responsibilities that smaller and more closely held businesses do not. Why? It could be argued that large and successful corporations have a greater responsibility to society than small and less successful ones, because they have greater resources to shoulder society's burdens, and 'can' implies 'ought'. However, we believe that talking of responsibilities that are contingent on size and success is highly problematic.

In short, our argument is that if you take a 'creating value for stakeholders' approach to business, and if you acknowledge that ethics and values are as important in these relationships as they are in our other relationships with our fellow human beings, then the idea of 'corporate social responsibility' is just superfluous. There is nothing natural about categories such as 'economic, political, social, etc.' and we want to suggest that such a conceptual scheme – that separates the social responsibilities of a corporation from its business responsibilities – has long outlived its usefulness.

We propose to replace 'corporate social responsibility' with an idea we call 'company stakeholder responsibility'. This is not just semantics, but a new interpretation of the very purpose of CSR. 'Company' signals that all forms of value creation and trade, all businesses, need to be involved. 'Stakeholder' goes back to the very first paragraph of this essay and suggests that the main goal of CSR is to create value for key stakeholders and fulfill our responsibilities to them. And 'Responsibility' implies that we cannot separate business from ethics.[2] We will argue that taking a stakeholder approach to business is ideally suited to integrate business, ethics and societal considerations. Stakeholder theory is about value creation and trade – it is a managerial theory about how business works. It does not subscribe to the separation thesis, so it asks at once business and ethics questions about each stakeholder relationship. The remainder of this paper is structured as follows. Section 2 provides a brief history of the how the stakeholder approach to management developed. Section 3 outlines four successive levels of commitment to the stakeholder approach to CSR – the basic value proposition, sustained stakeholder cooperation, understanding of the broader societal issues, and ethical leadership. These four levels can be considered steps that firms can take as they progressively increase their commitment to the stakeholder approach. In Sect. 4, we discuss ten principles of the stakeholder mindset that a firm must follow for it to reach the highest level of stakeholder commitment – that of ethical leadership. In Sect. 5, we summarise the argument.

---

[2] Note that we are using 'ethics' in its broadest sense to encompass obligations to employees, and other stakeholders. This is sometimes referred to as an 'American' usage, whereby the 'European' usage is much narrower. CSR is our broad term here, and we think it is more specific and more useful than distinguishing 'ethical' from 'social'. We are grateful to Dr Valeria Fazio for many conversations on this issue.

2 A Brief History of the Stakeholder Idea[3]

A stakeholder approach to business emerged in the mid-1980s. One focal point in this movement was the publication of R. Edward Freeman's *Strategic Management: a Stakeholder Approach* (1984), building on the process work of Russell Ackoff, Eric Trist, Ian Mitroff, Richard Mason and James Emshoff. The impetus behind stakeholder management was to try and build a framework that was responsive to the concerns of managers who were being buffeted by unprecedented levels of environmental turbulence and change. Traditional business frameworks were neither helping managers develop new strategic directions nor were they helping them understand how to create new opportunities in the midst of so much change. As Freeman observed '[O]ur current theories are inconsistent with both the quantity and kinds of change that are occurring in the business environment of the 1980s … A new conceptual framework is needed' (Freeman 1984: 5). A stakeholder approach was a response to this challenge. An obvious play on the word 'stockholder', the approach sought to broaden the concept of business beyond its traditional economic roots, by defining stakeholders as 'any group or individual who is affected by or can affect the achievement of an organization's objectives'. The purpose of stakeholder management was to devise a framework to manage strategically the myriad groups that influenced, directly and indirectly, the ability of a firm to achieve its objectives. While the stakeholder framework had roots in a number of academic fields, its heart lay in the clinical studies of management practitioners that were carried out over ten years through the Busch Center, the Wharton Applied Research Center, and the Managerial and Behavioral Science Center, all at the Wharton School, University of Pennsylvania, by a host of researchers.

While the 1980s provided an environment that demonstrated the power of a stakeholder approach, the idea was not entirely new. The use of the term stakeholder grew out of the pioneering work at Stanford Research Institute (now SRI International) in the 1960s. SRI's work, in turn, was heavily influenced by concepts that were developed in the planning department of Lockheed and these ideas were further developed through the work of Igor Ansoff and Robert Stewart. Thus, the stakeholder approach is firmly rooted in the practice of management.

Recently, Giles Slinger has revisited the early history of the idea of stakeholders. Through more extensive interviews, and the examination of a number of historical documents, Slinger rewrites the history as told in Freeman (1984). The essential difference is that the early use of the stakeholder idea was not particularly oriented towards the survival of the firm.[4]

---

[3] For a fuller discussion of the history of the stakeholder idea see Freeman (2005, in press).

[4] Slinger's argument can be found in his doctoral dissertation, 'Stakeholding and Takeovers: Three Essays', University of Cambridge, 2001. An abridged version is in 'Spanning the Gap: The Theoretical Principles Connecting Stakeholder Policies to Business Performance', Centre for Business Research, Department of Applied Economics, Working Paper, University of Cambridge, 1998.

SRI argued that managers needed to understand the concerns of shareholders, employees, customers, suppliers, lenders and society, in order to develop objectives that stakeholders would support. This support was necessary for long term success. Therefore, management should actively explore its relationships with all stakeholders in order to develop business strategies.

The stakeholder approach has been used extensively by business ethicists to explore the ethical consequences on stakeholders of managerial action. Donaldson and Preston (1995)[5] proposed four different ways in which scholars had applied the stakeholder approach to business ethics: as a normative theory, which posits that managers ought to take into consideration the interests of all stakeholders; as a descriptive theory, which limits itself to describing how managers in fact treat stakeholders; as an instrumental theory, which takes the position that managers who take into consideration stakeholders' interests will enjoy better firm performance; and finally, as a managerial theory, that is, as a guide to managerial action. We believe that this fourth managerial perspective on the stakeholder approach has received the least attention in recent times, in spite of having been at the roots of the stakeholder concept at SRI.

According to Freeman and McVea (2001), the stakeholder approach has seven distinguishing characteristics. First, it offers a single strategic framework that allows a manager to deal with changes in the external environment without the need for new strategic paradigms. In this way, it is particularly suited for the dynamic environments that are so prevalent today. Second, the stakeholder approach is a strategic management process rather than a strategic planning process. The focus is less on predicting the future and more on proactively plotting the direction of the firm. Third, a central concern of the stakeholder approach is the achievement of the organisation's objectives through the harnessing of support of all those who are affected by the firm's actions, as well as all those who can affect the progress of the firm. Fourth, the stakeholder approach emphasises the critical role of values-based management, by recognising that a diverse collection of stakeholders will cooperate with the firm over the long term only if they share a core set of values. Fifth, it is at once a prescriptive and a descriptive framework. It advocates a holistic approach to management, integrating economic, social, political and ethical considerations. Sixth, rather than take a stylised view of stakeholders based on very general roles-based groupings (such as shareholders, suppliers, etc.), the stakeholder approach places great importance in acquiring a fine grained understanding of the particular stakeholders of each firm. This deep understanding enables the management to develop tailored solutions for particular stakeholders, as with mass customisation. Finally, it starts off with the premise that a firm can exist and sustain itself only if it offers solutions that balance the interests of multiple stakeholders over time. Taking a stakeholder approach to CSR means we have to focus on integration across stakeholders and on practical managerial solutions that create value for customers, employees, suppliers, communities and financiers.

[5] An important paper on the stakeholder approach in the business ethics literature is Donaldson, T. and Preston, L. E. 1995. 'The Stakeholder Theory of the Corporation: Concepts, Evidence, and Implications'. *Academy of Management Review*: 20(1), 65–91.

In the section that follows, we outline four levels of commitment to Company Stakeholder Responsibility.

3 Four Levels of Commitment to the Stakeholder Approach[6]

3.1 Level 1: Basic Value Proposition

At this most basic level, the entrepreneur or manager needs to understand how the firm can make the customer better off, while at the same time offering an attractive value proposition to employees, suppliers, communities and financiers. It is important to note that it is not possible to sustain making customers better off, without at the same time making the other stakeholders better off. For example, Naturhouse, a Spanish dietary supplements retail chain, has grown at an annual rate of between 40% and 50% for seven years in a row from 1997 to 2004, in a mature and highly competitive market. It took the company five long years, from 1992 to 1997, to develop and refine a new business model with strong value propositions to its stakeholders. Once the company had finally come up with a model that offered its suppliers a decent price, its franchisees an attractive return on investment and effort, and its customers a unique combination of products and nutritional advisory services, it was able to grow at an explosive rate. What this example highlights is so obvious that we too often take it for granted: a business model that simultaneously satisfies the different stakeholders is a prerequisite for any company to start doing business profitably. Business failure and mediocre performance are often attributable to the firm's inability to articulate strong enough value propositions simultaneously to all its stakeholders.

3.2 Level 2: Sustained Stakeholder Cooperation

Once the most basic level of stakeholder awareness has been achieved, the entrepreneur or manager must understand that the continued survival and profitability of the company depend on effectively sustaining the cooperation amongst the stakeholders over time. The competitive, macroeconomic, regulatory and political environments are so dynamic that they make it necessary for the initial stakeholder arrangements to be revised on a constant basis. Each revision will invariably upset the delicate balance struck in the value propositions of the company to the different stakeholder groups. For example, the entry of a new competitor from a low cost country such as

---

[6]The first three levels of commitment are explored in greater detail in Wicks et al. (2005). The origins of these ideas can be found in part in Freeman (1984) in the idea of 'enterprise strategy'. We are grateful to our coauthors for permission to develop these ideas in the context of Company Stakeholder Responsibility.

China might mean the company has to reduce its price to its customers. This reduction might well involve a short term reduction in the prices paid to suppliers, a reduction in the wages of employees relative to output, or a reduction in the return to financiers. It is important for the manager to have a deep understanding of how these trade-offs affect each stakeholder, the limits to the sacrifice a given stakeholder will accept, and how these current sacrifices can be compensated in the future. Indeed, management according to the stakeholder approach is the effective balancing over time of multiple stakeholder interests.

3.3 Level 3: An Understanding of Broader Societal Issues

According to Haaland-Matlary (2005), the manager today is asked to be aware of and responsive to more and more international issues, without the moral compass of the nation state or religion to guide her any more. The insecurity caused by the increase in terrorism further compounds matters. Often, companies are caught flat-footed in the face of unexpected developments.

> [I]t was only after the fact in Nigeria that Shell took a major interest in human rights … [W]hen Amnesty International accused Telenor, a Norwegian telecommunications company, of racist policies in Malaysia, the company's management froze and responded more than a week late. Even though Amnesty International's case was poor, the damage was done. (Haaland-Matlary 2005)

What this means is that managers can no longer decline to take positions on issues that apparently are not purely business related. Shell paid a heavy price through loss of reputation by its refusal to use its considerable leverage with the Nigerian government to try to halt the execution of political activist Ken Saro-Wiwa. A proactive attitude is necessary towards all stakeholder groups, both primary, i.e. those that have direct business dealings with the company, and secondary, such as NGOs and political activists who can affect the operations of the company.

3.4 Level 4: Ethical Leadership

Recent research points to a strong connection between ethical values and positive firm outcomes such as sustained profitable growth and high innovativeness. The Good Work Project, started in 1995 by three teams of investigators led by Howard Gardner, Mihaly Czikszentmihalyi and William Damon, examined the relationship between ethics and performance. This project involved intensive, face-to-face interviews with professionals in a variety of domains – 'journalism, genetics, business, jazz music, theater, philanthropy, and higher education' (Gardner et al. 2000, 'The Good Work Project: a Description', Unpublished Document). Based on forty interviews with business leaders in the US, Damon (2002) concludes:

> We found that a strong sense of moral purpose not only promotes a business career but also provides a telling advantage in the quest to build a thriving enterprise. In fact, a sense of moral purpose stands at the center of all successful business innovations. Far from being a constraining force that merely keeps people honest and out of trouble, morality creates a fertile source of business motivation, inspiration, and innovation. (Damon 2002)

Damon (2002) proposes the three faces of morality: restrictive, philanthropic and generative. Restrictive morality mainly consists of controlling behaviours that can be destructive, such as cheating, lying, or committing sexual harassment; it provides us with guidance of what not to do. Philanthropic morality promotes contributions to worthy social causes. Damon points out that this form of morality is triggered after the individual or organisation has achieved success and profits. It is often practised out of enlightened self-interest. The third type of morality is of the generative kind, which is based on a:

> proactive promotion of positive moral initiatives, as opposed to embracing ethical postures as a way to avoid ethical breaches. A positive moral initiative may be as simple as the urge to serve customers better by bringing them a superior or less-expensive product, or as complex as making the news available to everyone on earth 24 hours a day. But whether the impetus is large or small, a sense of moral purpose beats at the heart of every great business success, because it's that pulse that creates and sustains the innovation that achieves success.
>
> Generative morality arises from a deep inner belief that sparks imagination and gives birth to a new business concept. It also fosters the sense of commitment that sustains the concept during inevitable periods of doubt, stress, and temporary reversals. It provides a reason to go to the mat for an idea, a steel foundation for the persistence needed to implement any innovation. And it's the key to giving your company the moral advantage. (Damon 2002)

We believe that this form of proactive ethical leadership is possible only if there exists a deep understanding of the interests, priorities and concerns of the stakeholders.

We believe that there are some general principles which make up a 'mindset' or 'worldview' that is necessary to understand and practise all four levels of Company Stakeholder Responsibility.

4 Ten Principles of Company Stakeholder Responsibility

1. ***We see stakeholder interests as going together over time.*** The very idea of managing for stakeholders is that the process of value creation is a joint process. Let us take the case of a typical CEO of a large international company. We'll make the CEO an amalgam of a number of real CEOs, and call him Bob Collingwood, and his company Woodland International.[7] Collingwood's company's products and services must create value for customers, first and fore-

---

[7] Some readers may recognise Bob Collingwood as the harried hero of Freeman (1984). In reality he is a composite of the thousands of executives who have been kind enough to have conversations with us about the ideas here.

most, so that they are willing to pay for them. Suppliers must be willing to do business with Woodland International, so that products and services can be created in the first place, and if the suppliers are committed to making Woodland even more effective and productive, then both will be winners. Woodland must offer employees jobs (wages and benefits) that are acceptable, and if Bob and his colleagues can get employees to share the purpose of Woodland, to come to work engaged and ready to create value, then all will be winners. Woodland needs to be a good citizen in the communities in which it operates, if for no other reason than in a relatively free and open society citizens can use the political process to force Woodland to be a better citizen. If on the other hand, Woodland acts as a responsible citizen it may well generate very positive goodwill, and be able to operate more freely. Finally, Wood- land needs to show returns to its shareholders, meet obligations to debt holders, banks and others. Profits don't conflict with other stakeholders, rather they are the scorecard which tells us how well we are managing the whole set of stakeholder relationships. Bob and his colleagues must keep these stakeholder interests in balance, hopefully mutually reinforcing each other.

2. ***We see stakeholders as real people with names and faces and children****. They are complex.* Of course people are complex, and that should go without saying. However, much of the popular thinking about business people assumes just the opposite. We often make assumptions that business people are only in it for their own narrowly defined self-interest. One main assumption of the traditional shareholder-centred view is that shareholders only care about returns, and therefore their agents, managers, should only care about returns. Most human beings are more complicated. Most of us do what we do because we are self-interested and interested in others. Business works in part because of our urge to create things with others and for others. Working on a team, or creating a new product or delivery mechanism that makes customers lives better or happier or more pleasurable all can be contributing factors to why we go to work each day. And, this is not to deny the economic incentive of getting a pay cheque. The assumption of narrow self-interest is extremely limiting, and can be self-reinforcing – people can begin to act in a narrow self-interested way if they believe that is what is expected of them, as some of the scandals such as Enron have shown. We need to be open to a more complex psychology – one any parent finds familiar as they have shepherded the growth and development of their children. We have encountered story after story where managers 'discovered' that their 'adversaries' were a lot more like them than they had originally thought. In short they discovered that these 'adversaries' shared a great deal of their own humanity: a lesson which we should all remember.
3. ***We seek solutions to issues that satisfy multiple stakeholders simultaneously.*** Bob Collingwood's problem is that his world is fragmented. Issues and problems come at him and his team from lots of places, in lots of forms. He could spend his entire job just talking to customers, or employees. He needs to find a way to develop programmes, policies, strategies, even products and services that satisfy multiple stakeholders simultaneously. Now the first step in that pro-

cess is to actually recognise that he needs to look for simultaneous solutions. For instance, suppose that he is under pressure to make a particular service more affordable to low-income citizens. Under the traditional shareholder-centred view, he might see this as an illegitimate 'tax on shareholders'. Such a view would not lead to much innovation and to a constant friction with critics and regulators. He might take this criticism as a call for innovation and productivity, so that if he can figure something out, he can develop a new market (lower-income customers), satisfy some critics, and become a good citizen in the community. The difference in mindsets is fairly substantial, and so will be the search for a solution.

4. ***We engage in intensive communication and dialogue with stakeholders – not just those who are friendly***. Obviously we need intensive dialogue through multiple methods with customers, suppliers, employees and shareholders, but communities, critics and other secondary stakeholders count as well. Critics are especially important dialogue members. Critics are trying to give Collingwood and his team another point of view about Woodland International. One way to see critics is to view them as representing unmet market needs, since the critic wants the company to act differently. It is the job of the executives to see if there is some underlying business model, so that this unmet need can be turned into an entrepreneurial opportunity creating wins for all stakeholders. Not every critic can be satisfied, not every critic has a legitimate point of view, and not every need can be met. But, too often, executives don't meet with their critics enough to determine whether or not there is an opportunity to create value. Dialogue is the foundation of a free society, and the foundation of capitalism itself. Despite fictional stories about 'spot market transactions where every player just knows the prices', real business is built on a foundation of solid, honest and open communication. Indeed, most management meetings we have been a part of for the last twenty-five years have all, at some point, reinforced the need for 'better communication'. There is no difference in adopting a managing for stakeholders view – it is just more difficult and even more intense.
5. ***We commit to a philosophy of voluntarism – to manage stakeholder relationships ourselves, rather than leaving it to government***. When executives and pundits are committed to the traditional shareholder view, there is a temptation to look at the myriad stakeholder pressures and play 'Blame the Stakeholder'. But, the real problem here is our mindset. In short 'we have met the enemy, and he is us'. The challenge for us is to reorient our thinking and our managerial processes to be responsive to stakeholders. We believe that such a stakeholder mindset must be based on the idea of voluntarism. Voluntarism means that an organisation must, of its own will, undertake to satisfy its key stakeholders. A situation where a solution to a stakeholder problem is imposed by a government agency or the courts must be seen as a managerial failure. Similarly, a situation where Firm A satisfies the needs of consumer advocates, government agency, etc. better than Firm B, must be seen as a competitive loss by Firm B. The driving force of an organisation becomes, under a voluntarism mindset, to create as much for stakeholders as possible.

6. ***We generalise the marketing approach***. We need to 'overspend' on understanding stakeholder needs, using marketing techniques to segment stakeholders to provide a better understanding of their individual needs and using marketing research tools to understand the multi-attribute nature of most stakeholder groups.

   We might define 'overspending' as paying extra attention, beyond that warranted by considerations of efficiency, to those groups who are critical for the long term success of the firm. Overspending on stakeholders without whose support the company would fail can make sense in a number of instances. For instance, many fast moving consumer goods (FMCG) companies overspend on customers, interviewing several thousand a year. Traditionally telecom companies overspent on the attention they paid to the regulatory process, which was for a long time their major source of revenue. Oil companies should, likewise, consider adopting a conscious policy of overspending on OPEC and government and stakeholders who can convey a positive image to the public. Chemical companies have recently begun to overspend on environmentalists, trying to clean up their image as 'dirty companies' and 'spoilers of the environment'. 'Overspending' is not necessarily measured in monetary terms. 'Spending' may be in terms of more time or more energy or whatever the relevant resource required by a given stakeholder group.
7. ***Everything that we do serves our stakeholders. We never trade off the interests of one versus the other continuously over time***. Just as many successful companies think in terms of 'how to serve the customer' or 'how to serve the employees', it is possible to generalise this philosophy to 'how to serve our stakeholders'. The 'reason for being' for most organisations is that they serve some need in their external environment. When an organisation loses its sense of purpose and mission, when it focuses itself internally on the needs of its managers, it is in danger of becoming irrelevant. Someone else (if competition is possible) will serve the environmental need better. The more we can begin to think in terms of how to better serve stakeholders, the more likely we will be to survive and prosper over time. A managing for stakeholders approach asks the company to clearly articulate how its basic business proposition makes its stakeholders better off.
8. ***We negotiate with primary and secondary stakeholders***. The basic idea behind the stakeholder approach is that if a group or individual can affect a company or be affected by a company then there needs to be some interaction and some strategic thinking. Many executives get caught up in whether or not a particular stakeholder group, especially critics, is 'legitimate' or not. And, while this is an important issue for some purposes, the stakeholder mindset encourages executives to meet, interact and negotiate with both 'legitimate' stakeholders, and those whose legitimacy may be questioned from an overall point of view.

   In very practical terms, groups which have some power must be taken into account, regardless of whether or not in a 'pure capitalism' system, they should be there at all. In our relatively free and open society, the consequences of not negotiating with a broad range of stakeholders is that they use the political process to 'negotiate' indirectly by pressuring government to enact a set of rules that is not likely to be optimal to company interests. You can think of this idea as 'managerial legitimacy', i.e., if a group has some power to affect the company then it is legitimate to spend managerial time worrying about that group. Often, because these interactions start off with stereotypes of both business and critic behaviour, careful attention to process can turn the relationship into one positive for both sides.

9. ***We constantly monitor and redesign processes to make them better serve our stakeholders.*** A hallmark of the stakeholder mindset is that in today's world no one 'gets it right' all the time. Whatever your interactions and strategies are with stakeholders, they can always be improved. The classic case for such improvement comes from thinking about the environment. By paying attention to the environment and environmentalists, companies from McDonalds to 3M have radically redefined their production processes to turn waste streams into new products, realise millions of dollars in cost savings, and gain a reputation as companies that are environmentally friendly and willing to work with environmental groups.
10. ***We act with purpose that fulfills our commitment to stakeholders. We act with aspiration towards fulfilling our dreams and theirs.*** We believe that the key idea which holds this stakeholder mindset together is the idea that businesses can have a purpose. And, there are few limits on the kinds of purpose that can drive a business. Wal-Mart may stand for 'everyday low price'. Merck can stand for 'alleviating human suffering'. The point is that if an entrepreneur or an executive can find a purpose that speaks to the hearts and minds of key stakeholders, it is more likely that there will be sustained success.

    Purpose is complex. Running a purposeful business is even more complicated. Once we give up the traditional shareholder view as the only possible purpose for a business, the field is wide open. Perhaps 'maximising shareholder value' is a good purpose for a business, but surely it is not the only one. Purpose is inspirational. The Grameen Bank wants to eliminate poverty. Fannie Mae wants to make housing affordable to every income level in society. Tastings, a restaurant in Virginia, wants to bring the taste of really good food and wine to lots of people in the community. All of these organisations have to generate profits, or else they cannot pursue their purposes. And, they cannot generate profits or fulfill purpose without intense engagement with their stakeholders.

5 A New CSR: Company Stakeholder Responsibility

We have argued that Corporate Social Responsibility has outlived its usefulness, because it is flawed in two respects. First, it promotes the 'separation thesis', the idea that business issues and social issues can be dealt with separately. This flaw promotes the potentially destructive idea that the underlying structure of business is either not good or is morally neutral. We have proposed a stakeholder approach that takes into consideration the intertwined nature of economic, political, social and ethical issues. It is centered in the practice of management, and provides the manager with a pragmatic framework for action. The second flaw with Corporate Social Responsibility is its focus on corporations. We do not see why social responsibility only applies to corporations, rather than to all organisational forms. The stakeholder approach that we have proposed applies as much to an entrepreneurial start-up and to a mid-sized closely held firm as it does to a corporation with diffuse ownership.

Based on the stakeholder approach we have outlined, we have proposed a new CSR – Company Stakeholder Responsibility – as a new capability for organisations to develop. We have outlined four levels of commitment to this new CSR, and we have suggested ten principles that can help executives and business thinkers begin to apply this approach. (Boxes 12.1 and 12.2 are a summary of Company Stakeholder Responsibility.)

Box 12.1: Four Levels of Commitment to Company Stakeholder Responsibility

The Basic Value Proposition

How do we make our stakeholders better off?
What do we stand for?

Principles for Sustained Stakeholder Cooperation

What are our principles or values on which we base our everyday engagement with stakeholders?

Broader Societal Issues

Do we understand how our basic value proposition and principles fit or contradict key trends and opinions in society?

Ethical Leadership

What are the values and principles that inform my leadership?
What is my sense of purpose? What do I stand for as a leader?

Box 12.2: Ten Principles for Company Stakeholder Responsibility

1. We see stakeholder interests as going together over time.
2. We see stakeholders as real people with names and faces and children. They are complex.
3. We seek solutions to issues that satisfy multiple stakeholders simultaneously.
4. We engage in intensive communication and dialogue with stakeholders, not just those who are 'friendly'.

(continued)

Box 12.2 (continued)

5. We commit to a philosophy of voluntarism – to manage stakeholder relationships ourselves, rather than leaving it to government.
6. We generalise the marketing approach.
7. Everything that we do serves our stakeholders. We never trade off the interests of one versus the other continuously over time.
8. We negotiate with primary and secondary stakeholders.
9. We constantly monitor and redesign processes so that we can better serve our stakeholders.
10. We act with purpose that fulfills our commitment to stakeholders. We act with aspiration toward our dreams and theirs.

References

Damon, W. 2002. The Moral Advantage. *Optimize*. Available on the Internet at: http://www.optimizemag.com/issue/003/ethics.htm.

Donaldson, T., and L. Preston. 1995. The Stakeholder Theory of the Corporation: Concepts, Evidence, and Implications. *Academy of Management Review* 20: 65–91.

Freeman, R.E. 1984. *Strategic Management: A Stakeholder Approach*. Boston: Pitman Publishing Inc.

———. 2005. The Development of Stakeholder Theory: An Idiosyncratic Approach. In *The Oxford Handbook of Management Theory: The Process of Theory Development*, ed. Ken G. Smith and Michael A. Hitt. Oxford: Oxford University Press.

Freeman, R.E., and J. McVea. 2001. A Stakeholder Approach to Strategic Management. In *Handbook of Strategic Management*, ed. M. Hitt, R.E. Freeman, and J. Harrison, 189–207. Oxford: Blackwell.

Freeman, R.E., K. Martin, B. Parmar, M. Cording, and P. Werhane. 2005a. Leading through Values and Ethical Principles. In *Inspiring Leaders*, ed. C. Cooper and R. Burke. Oxford: Oxford University Press.

Freeman, R.E., R. Phillips, J. Harrison, and A. Wicks. 2005b. *Managing For Stakeholders*, Olsson Center Working Papers. The Darden School, University of Virginia.

Gardner, H., M. Czikszentmihalyi, and W. Damon. 2000. *The Good Work Project: A Description*. unpublished document.

Haaland-Matlary, J. 2005. Kjernekar: Ethical Integrity in a Chaotic World. *IESE Business School Alumni Magazine* 96 (Jan–March): 12–15.

Slinger, G. 1998. *Spanning the Gap: The Theoretical Principles Connecting Stakeholder Policies to Business Performance*, Centre for Business Research, Department of Applied Economics, Working Paper. University of Cambridge.

Wicks, A.C., R.E. Freeman, and B. Parmar. 2005. *Business Ethics in an Era of Corporate Crisis*. Darden School Working Paper.

R. Edward Freeman is University Professor, Elis and Signe Olsson Professor, Academic Director of the Institute for Business in Society, and Senior Fellow of the Olsson Center for Applied Ethics at the University of Virginia's Darden School of Business.

S. Ramakrishna Velamuri is Professor of Entrepreneurship and Dean at the School of Management, Mahindra University.

Chapter 13
Corporate Citizenship and Community Stakeholders

Robert A. Phillips and R. Edward Freeman

1 Introduction

It is fair to say that the concept of corporate citizenship has witnessed a meteoric rise in terms of scholarly attention since Logsdon and Wood presented their original paper (Wood and Logsdon 2001, 2002; Logsdon and Wood 2002). Some elaborations and extensions of corporate citizenship make reference to earlier scholarship, others do not. Stipulating that work on corporate citizenship is intended to add to the conversation around the role of business in society, it is reasonable to assume that scholars adopting (and adapting) the language of corporate citizenship find something there that allows for better description, analysis and synthesis of this role. Though what 'better' may mean here remains an open question, a sensible place to begin considering the question is to compare and contrast corporate citizenship with more established ways of conceiving business's role in society such as, in the case of this chapter, stakeholder theory.

Among the challenges of comparing corporate citizenship and stakeholder theory is the fact that neither theory can currently claim a defining consensus regarding the content and limits of their respective domains. To differing degrees, both are less monolithic concepts than ways of conceiving of and arranging the complex

Originally published in: *Handbook of Research on Global Corporate Citizenship*, 99–115, © Edward Elgar Publishing Limited, 2008
Reprint by Springer, https://doi.org/10.4337/9781848442924.00002

R. A. Phillips
York University, Toronto, ON, Canada

R. E. Freeman (✉)
University of Virginia, Charlottesville, VA, USA
e-mail: FreemanE@darden.virginia.edu

© Springer Nature Switzerland AG 2023

S. D. Dmytriyev, R. E. Freeman (eds.), *R. Edward Freeman's Selected Works on Stakeholder Theory and Business Ethics*, Issues in Business Ethics 53,
https://doi.org/10.1007/978-3-031-04564-6_13

relationships between business organizations and other social actors. A further difficulty shared by both corporate citizenship and stakeholder theory is that summarizing the literature may, or may not, include writings that unreflectively use the terminology of citizenship or stakeholders without reference to any particularly deep thoughtful or coherent conceptual foundations. Much of the heavy lifting in numerous treatises on corporate citizenship is done by merely inserting the word 'citizenship' and assuming that author and reader have the same understanding of what this denotes and implies. The identity of the citizens, the basis of this status, and the rights, duties and obligations of such status are only a few of the matters on which such agreement is assumed, but often not made explicit. Logsdon and Wood (2002) – pioneers in the area of corporate citizenship – recognize this ambiguity in the use of 'citizen' and briefly address it; though Moon et al. (2005) have criticized the sufficiency of these efforts and offer their own elaboration.

Similar to the unreflective use of citizenship terminology, discussions of corporate citizenship also employ the term 'stakeholder' with little or no reference to the now extensive literature on these relationships. To be fair, corporate citizenship is hardly alone in this failure to consider the deeper foundations and challenges of extant stakeholder scholarship, though perhaps we should expect more of a literature stream so closely related in subject matter to stakeholder theory. In many cases, studies of corporate citizenship employ the terminology of 'stakeholders' suggesting a conflation of the two theories. Our reading of the literature on corporate citizenship leads us to believe that the language of stakeholders is adopted nearly universally. This makes the failure to consider prior theorizing about stakeholder relationships still more troubling.

When used in this somewhat superficial fashion, corporate citizenship may be fairly described as a recapitulation of well-rehearsed corporate social responsibility (CSR) concepts – with the added benefits and *dangers* of importing central and implied, but underexamined concepts of citizenship from political theory. Others, both within this volume and elsewhere (Waddock 2001; Windsor 2001; Wood and Logsdon 2001; Moon et al. 2005) have examined the relationship between corporate citizenship and historical notions of both CSR and citizenship itself. We shall not repeat these efforts. Nor shall we undertake here to remedy the dearth of conceptual depth concerning the 'citizenship' within the corporate citizenship literature – others in this volume take on this question with greater expertise and focus.

Instead we shall focus on stakeholder theory. In particular, we shall focus on 'community' stakeholders. We discuss the source and limits of organizational obligations to communities and describe a typology of potential relationships that organizations may have with communities. While this discussion of community stakeholders may well be interpreted as a description of corporate citizenship obligations and how they can or should be discharged, we do not see this interpretation as necessary nor shall we attempt to fully render such an interpretation. The concept of citizenship is too complex to convincingly make such a case here. Our more humble goal is to discuss firm-community relationships as one potential lever in beginning this more complicated endeavor of reconciling and distinguishing corporate citizenship and stakeholder theories. And in the process, we hope to highlight challenges involved with the unreflective use of stakeholder terminology in discussions of corporate citizenship.

2 Corporate Citizenship and Community Stakeholders

When considered in terms of stakeholder theory, corporate citizenship focuses particular attention on the community as a stakeholder. It focuses managerial and scholarly attention toward consideration of the role of the organization in its community, with a concomitant de-emphasizing of other core stakeholders. Although there are myriad ongoing debates about the scope and content of stakeholder theory – together with employees, financiers, suppliers and customers – local communities are universally counted among the archetypal five stakeholders (Phillips 2003). But this apparent agreement may itself mask ambiguities similar to those associated with the concepts of citizenship and stakeholder alluded to above. That is, stakeholder theorists may reasonably be accused of replacing 'citizen' with 'community member' and running foul of a similar imprecision as corporate citizenship scholars.

As with corporate citizenship, there has been some effort to better specify the content of 'community' within stakeholder theory (Dunham et al. 2001, 2006). According to Dunham et al. (2006, p. 24), 'community as a stakeholder has come to represent something of a default, a sort of error term containing all sorts of interests and externalities that fail to find home within customer, supplier, employee, or shareholder groups'. They suggest that communities can be subdivided into what they term 'communities of place', 'communities of interest', 'communities of practice' and 'virtual advocacy groups'. Communities of place are the shared geographic locations which most people associate, most of the time, with 'community'. Actors living in, more or less, close proximity to one another are, of necessity, mutually interdependent. At the very least these actors share the same natural environment (for example, water, air) and infrastructure (for example, roads, schools, police, retail establishments). And, typically, these actors share even deeper interdependencies such as local norms of behavior, dress, language and other necessary aspects of consistent, ongoing social interaction. These mutual interdependencies appear sufficient in the eyes of many stakeholder theorists to establish at least 'local' communities as organizational stakeholders. But, for any particular organizational issue, the intensity of the local community members' interest will vary.

This is not the case for what Dunham et al. (2006) call 'communities of interest'. Although they may be geographically local, the more relevant – indeed defining – characteristic of these groups is their interest in a specific topic. The stakeholder relevance of these groups is contingent upon the organization's actual or potential ability to affect – for better or worse – this interest. The coordination capacity of communities of interest has increased dramatically with the advent of new innovations in communication and information technology. These innovations have also improved the focal organization's ability to communicate and coordinate with these communities. Communities of interest are able to significantly aid or harm the focal organization's ability to achieve its goals. This ability to affect the achievement of the firm's objectives puts communities of interest squarely within Freeman's (1984) original definition of a stakeholder (that is, those who can affect or are affected by the achievement of the firm's objectives).

While communities of interest are often willing and able to help the firm advance its goals, 'virtual advocacy groups' – according to Dunham et al.'s definition – are consistently hostile. Relying even more extensively on information technology, the ties among the members of these groups may be entirely confined to cyberspace. But, of course, this does not lessen their ability to harm the achievement of the focal firm's objectives. Dunham et al.'s identification of such virtual communities adds another layer of complexity to the already challenging elaboration of community stakeholders.

Finally, Dunham et al. invite readers to consider communities of practice as a new variant of community that has risen to prominence among practitioners. They write that conceiving of a business organization as a community of practice is:

> to see it as held together by a shared concern for both the outcomes it achieves for stakeholders (be they customers or shareholders) and the personal development and learning of its members. In fact, it sees these two as inseparable, in that increased capabilities at the organizational level flow from development at the individual level. (Dunham et al. 2006, p. 35)

A mirror image of virtual advocacy groups, Dunham et al.'s elaboration of communities of practice presents additional positive opportunities for managers to reciprocally advance firm/community objectives.

Of particular relevance to this chapter, Dunham et al. suggest the importance of the 'degree of symbiosis' between the various community types and the focal organization. The intensity and nature (that is, cooperative or hostile) of involvement between the organization and the community group in question are among the attributes a scholar or practitioner must consider in determining what manner and intensity of attention is due to a particular community stakeholder. That is, the manner and intensity of a community's interactions with an organization will tend to define and delimit the extent of an organization's stakeholder-based obligations toward, and responsibility for, that community.

And here we may re-engage, more specifically, the question of the relationship between stakeholder theory and corporate citizenship. Stakeholder theory, as we conceive it, is more concerned with the 'core' functions of a, firm than many elaborations of corporate citizenship with which we are familiar. That is, the threshold of relevance for inclusion in a stakeholder-theoretic analysis involves a greater degree of symbiosis and reciprocal impact between the community in question and the organization than appears necessary for consideration on many accounts of corporate citizenship. We take this up in greater detail in the next section.

3 Stakeholders and Global Citizenship

Elsewhere (Phillips et al. 2003), we have attempted to describe some of the limits to stakeholder theory that we believe render it more theoretically rigorous and managerially useful. Among the delimitations we claim that stakeholder theory does not apply to entire economies. We argue that, '"Stakeholder" is not synonymous with "citizen" or "moral agent" as some wish to interpret it. Rather, a particular and much closer relationship between an organization and a constituency group is required for stakeholder status. The theory is delimited and non-stakeholder should remain a meaningful category' (Phillips et al. 2003, p. 491). We believe that this delimitation of the stakeholder domain as organizational ethics rather than political philosophy is one point of distinction between stakeholder theory and much – though not all – elaboration of corporate citizenship.

We would further assert that stakeholder theory is, or should be, importantly distinct from broader conceptions of CSR. While there are myriad justifications for ascribing social responsibilities to business firms, such responsibilities often extend beyond core firm/stakeholder relationships. While recognizing that many (perhaps most) writers on the subject of business, ethics and society employ the language of stakeholders and CSR interchangeably, we maintain that stakeholder theory's focus on core business relationships entails significant differences between the two frameworks – theoretically and practically. The implications of this distinction are illustrated below using corporate philanthropy as an example.

As with CSR, the use of stakeholder terminology by the majority of corporate citizenship scholars suggests an implied synonymy between stakeholder theory and corporate citizenship. For example, Waddock (2001, p. 27) writes, 'Relationships with stakeholders constitute the essence of corporate citizenship'. Further, the similarities between CSR and corporate citizenship may lead one to believe that some see all three as merely a different language for discussing many of the same ideas. We believe that there are important and useful distinctions to be made. These distinctions return us to the topic of community stakeholders.

Figure 13.1 suggests a possible source of confusion and ambiguity both in stakeholder theory itself as well as in corporate citizenship's use of stakeholder concepts. As the figure indicates, it is difficult if not impossible to draw a bright line distinguishing stakeholders from non-stakeholders. More typically there are gradations. That is, there are groups who obviously contribute to or threaten the value-creation activities of the firm. These most intensive relationships are represented by the darkest areas of the ovals. There are also groups whose relevance and connection to the firm is tenuous or even nonexistent. The white space in Fig. 13.1 is filled with such non-stakeholder groups. The challenge and difficulty arises in the darker and lighter gray areas within the ovals. Within each stakeholder group, there are 'names and faces' whose relevance, power and bases of moral obligation are vague, perhaps even indeterminate. 'Potential' stakeholders – potential employees, potential suppliers, potential financiers and so on – may reside in this gray area.

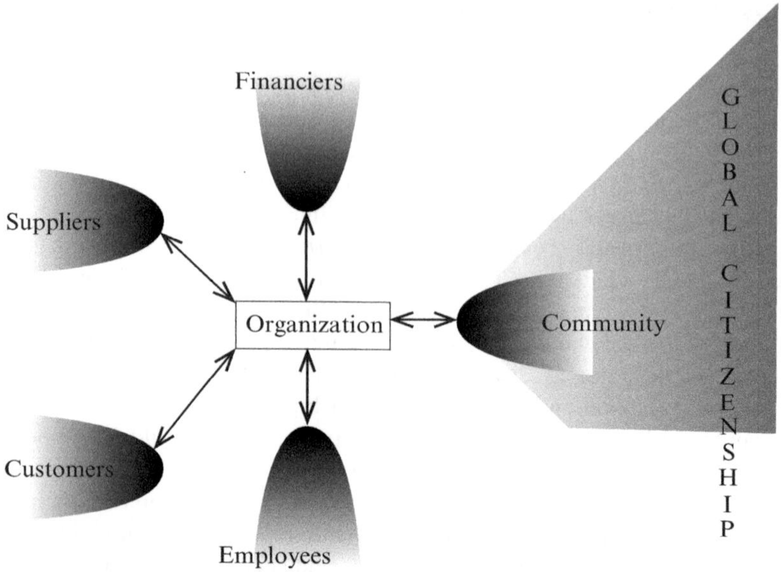

Fig. 13.1 Community stakeholders as transition between stakeholder theory and corporate citizenship

While a source of some vexation in delimiting stakeholders of all sorts, the gradation of stakeholder obligation and responsibility presents particular difficulties in sorting out prospective community stakeholders. As indicated in Fig. 13.1, many writers have a tendency to use 'community' far too expansively – a catch-all category for groups, individuals and causes that do not fit in one of the other categories, but which the author ardently desires to be a stakeholder. Such treatment is represented by the conical shape extending well outside the community oval. This way of engaging the language of stakeholder theory subjects it to precisely the sort of criticism from which we have tried elsewhere to rescue it. Specifically, this conception of community makes stakeholder theory once more susceptible to accusations that it is little more than warmed-over CSR and that such excessive breadth threatens theoretical rigor ('if everyone's a stakeholder, what value is added by use of the term?').

This difficulty is more pronounced still when stakeholder concepts are unreflectively employed in the context of corporate citizenship. It is our contention here that corporate citizenship begins to diverge from stakeholder theory as it begins to move into the conical extension of community. Indeed, much of the writing on corporate citizenship we would place well into the space labeled 'global citizenship'. At this point, we argue that corporate citizenship bears little overlap with stakeholder theory and begins to more closely resemble its CSR predecessors. Employing the language of stakeholder theory to defend the idea of 'global citizenship' runs contrary to our understanding of stakeholder theory as limited to the core value-adding functions of the organization.

4 Corporate Citizenship, Stakeholder Theory and Philanthropy

As with the predecessor literature on CSR, corporate citizenship scholarship makes extensive reference to a firm's charitable activities – often as a gauge of the quality of a firm's citizenship behaviors. While in many cases a laudable endeavor, we believe that philanthropy not related to the firm's core value-adding function lies outside the sphere of stakeholder theory's concern. We should point out that philanthropy related to the firm's core business is more amenable to stakeholder analysis. When a pharmaceutical company faces a decision regarding the provision of free or discounted medicines which such companies may be specifically or uniquely qualified to provide (Hsieh 2004; Dunfee 2006), or if a micro-processor company conducts research on how to make a $100 laptop computer, stakeholder theory may yield insight; or when a business working in a lesser developed area provides social services for employees and local citizens (Matten and Crane 2005, p. 166); or when philanthropy relates directly to core stakeholder interests (for example, donation matching programs, time off for employee charitable work, and so on), stakeholder theory may be of some assistance. However, in the normal course of events, corporate donations to groups lying outside of such core firm-stakeholder obligations are supererogatory – that is, potentially praiseworthy, but not obligatory.

That said, such donations are not always particularly praiseworthy. Often such donations also have little impact on the nature and quality of relationships between the firm and its stakeholders. Philanthropy unrelated to the core business often has a limited, even negligible, impact on the culture and character of the firm making such donations (we discuss in more detail below the issue of who and what an organization would like to be). Here we are talking about what we shall call 'checkbook citizenship'. Cash donations to some or other unrelated philanthropy have little effect on the firm or its core stakeholders – such donations generally have only a tangential impact. It is tantamount to organizational values as an after-thought – an appendage tacked on to what the firm *really* does.

A second danger associated with philanthropy unrelated to the firm's core value-adding functions is that, in all too many cases, it amounts to an attempt to whitewash (or greenwash in the case of the natural environment or bluewash in the case of efforts related to the United Nations) otherwise harmful or dangerous firm activities. Such a company, 'thinks that singin' on Sunday's gonna' save his soul, now that Saturday's gone' (Johnson 2003). These efforts appear to little influence the firm, its stakeholders or its core activities.

When unrelated charity and philanthropy are central to discussions of corporate citizenship, such discussions diverge from our understanding of stakeholder theory. Checkbook citizenship is not stakeholder theory; such donations are outside the purview of core stakeholder responsibilities and obligations. Although a complete discussion of what these stakeholder responsibilities and obligations are is beyond the scope of this chapter (see, for example, Phillips 2003), we shall spend the remainder of our space elaborating on the obligations and responsibilities to *community* stakeholders. In so doing, we hope to show how a stakeholder-theoretic consideration of communities – delimited to discussion of core value-adding activities – can address a limited set of questions pertaining to corporate citizenship. We shall consider a firm relationship with community stakeholders from two overlapping perspectives. We shall first discuss Phillips's obligations of stakeholder fairness as a means of ascertaining the (necessarily imprecise) limits of firm obligations to community stakeholders. We follow this discussion with an elaboration of possible postures a firm might take toward community stakeholders thus delimited.

5 Obligations of Stakeholder Fairness

We have argued here that the concept of corporate citizenship – as we understand it – bears a much larger footprint than stakeholder theory. Figure 13.1 suggests that corporate citizenship (and the notion of 'global corporate citizenship' in particular) extends well beyond a firm's obligatory stakeholder concerns. Obligations extending indefinitely and universally are of limited use in managerial (or personal or political) decision making. This leads us to inquire as to the source of such stakeholder obligations and their limits. This is the question taken up by Phillips (2003) in his discussion of obligations of stakeholder fairness.

Obligations of stakeholder fairness make stakeholder status largely a matter of reciprocity. Reciprocity is a widely – perhaps universally – recognized moral norm. Leaders since Odysseus[1] (Donlan 1998) ignore the demands of reciprocity at their own peril. Reciprocity is not only normatively ubiquitous, it is also a deeply and powerfully embedded feature of human psychology (Cialdini 1984). One means, therefore, of establishing obligations to stakeholders is through application of a principle of stake-holder fairness. That is,

> whenever persons or groups of persons voluntarily accept the benefits of a mutually beneficial scheme of co-operation requiring sacrifice or contribution on the parts of the participants and there exists the possibility of free-riding, obligations of fairness are created among the participants in the co-operative scheme in proportion to the benefits accepted. (Phillips 1997, p. 57)

By voluntarily joining with groups of suppliers, customers, employees, shareholders, communities and others for mutual social and economic benefit, reciprocity-based obligations of fairness are created within such networks. Those particular groups and individuals to whom such obligations are owed are stakeholders.

[1] 'The rule of reciprocity, that one gives of one's own accord, with the expectation that a suitable return will follow, was a powerful regulator of social behaviour at every stage of Greece's history. The Homeric epics provide our earliest observation of its operation' (Donlan 1998, p. 51).

This understanding of reciprocal stakeholder obligations allows scholars and practitioners to place a limit on the extension of stakeholder-specific responsibility of firms to communities. As a matter of stakeholder theory, firms bear no *additional* responsibility for groups outside the cooperative endeavor – subject to the caveat below. One implication of this is that using stakeholder terminology to describe responsibilities of 'global' citizenship is spurious. Such usage renders stakeholder theory a morass of 'everyone is a stakeholder of everyone else'. Such ambiguity is neither helpful nor useful to corporate citizenship, stakeholder theory or managerial practice.

There are two important caveats that bear mention regarding the limits of stakeholder-based obligations. The first is that a group or individual may be a stakeholder even without being the subject of a direct reciprocal stakeholder obligation. In addition to this *direct* moral stakeholder legitimacy, a group may also bear an *indirect* legitimacy derived from a relationship with the focal organization or its stakeholders. For example, if a group has the power to aid or hinder the achievement of the firm's core activities – even though not engaged in a reciprocal relationship with the firm – that group may be a 'derivative' stakeholder. Dunham et al.'s 'virtual advocacy' groups discussed above fit this description. The group's power makes it a legitimate object of managerial attention, even if the firm owes it no particular obligation beyond mitigating (in the case of hostile power) or embracing (in the case of productive power) these potential effects.

In cases relevant to corporate citizenship, for example, philanthropy related to the core value-adding activities of the firm may positively influence employee or community stakeholder relations. Or, such charity may allay the harmful criticism of a particularly powerful activist group. Donations are made, in these cases, due to the actual or potential effects on other stakeholders, not due to any stakeholder obligation owed to the charity itself or any obligation to the powerful, but hostile activist group. Reciprocity between the firm and charity recipients or between the firm and hostile activists is, at best, attenuated.

The second caveat to the limits on stakeholder obligation that bears mention here is that stakeholder-based obligations are not the sole source of moral obligation a firm may have. That is, actors may have non-reciprocal obligations, duties and responsibilities of many sorts. These could be based on, for example, familial relations, human rights, and even direct duties of charity and citizenship. The language of stakeholders implies a particular sort of relationship characterized by an obligation arising from an intentional exchange between actors as opposed to duties arising from such unintentional characteristics as being human, a member of a clan or a citizen of a state. The absence of a reciprocal stakeholder relationship does not give a firm license, for example, to violate the human rights of a group or individual. It merely means that this particular form of actor-generated moral relation is absent.

Thus, there may exist more general duties of philanthropy that a firm must meet as a powerful social actor. But this does not make the donations obligatory from a stakeholder perspective, nor does it make those in need stakeholders. As an analogy, let us say that I agreed to write a book chapter for you and thereby incurred an obligation to you. It may also be the case that I have a generalized duty of charity that may be (partially) fulfilled by helping rebuild post-hurricane Katrina New Orleans or going to Africa to help with HIV-AIDS missions. While all would be excellent uses of my time, only one of these counts as an obligation I have voluntarily undertaken. Similarly, we would reserve the term 'stakeholder' for those core actors to whom the organization has a certain sort of reciprocal obligation and those whose relevance can be derived from these specific stakeholder relationships.

Relevant to corporate citizenship, this implies that there may be any number of other moral relations between a firm and its fellow social actors. For example, there may be a general duty of charity that may apply to firms as repositories of enormous resources. There may also be more specific duties of charity or assistance (Hsieh 2004; Dunfee 2006) between firms with particular – even unique – capacity to help. Corporate citizenship scholars may well appeal to these sources of duty, obligation and responsibility in advancing their claims. But these are distinct from reciprocity-based stakeholder obligations.

While we hope that the distinctions and clarifications above add a measure of increased precision to discussions of community stakeholders and corporate citizenship, difficult cases remain. The case of second- and third-tier suppliers in a network of nested supplier relationships continue to defy easy placement within a stakeholder framework (Phillips and Caldwell 2005). One widely discussed variant of this challenge is a firm that contracts a portion – perhaps all – of its manufacturing to another, ostensibly independent firm. This subcontractor, in turn, may further outsource some or all of this work to still another firm, and so on (for example, Nike, Mattel, Wal-Mart, YUM! Brands, and so on *ad infinitum).* What sort of stakeholder obligations does the first firm have *vis-à-vis* the second-, third- and fourth-tier firm? Or the fourth-order firm's employees and local community? Although we shall not attempt to systematically resolve this ambiguity at the nexus between community stakeholder obligations and duties of corporate citizenship, its relevance as an increasingly prominent borderline case merits a few comments.

One reasonable question to ask when assessing the stakeholder-based moral obligations of firms to second- and third-level suppliers is the actual level of independence between the organizations. Independence within such nested and networked value chains is a function of – among other criteria perhaps – the exclusivity of interaction. Briefly, claims of independence between firm and supplier are attenuated to the degree that the firm is the sole (or even super-majority) customer of that supplier. And so on down the value chain. If a firm buys all, or nearly all, of a supplier's output and this supplier itself buys all, or nearly all, of the output of its own suppliers, this value chain begins to more closely resemble a hierarchy – or at a minimum, a network organization (Powell 1990). Arguments for stakeholder obligations to such value chain members are strengthened as the degree of transactional exclusivity rises. The frequency of firm/supplier transactions and the duration and history of the relationship may also contribute to a general notion of the 'intensity' of the value chain relationships and hence the power of fairness-based stakeholder obligations.

Beyond claims to the normative stakeholder legitimacy of putatively independent members of a firm's value chain, there are also increasingly prevalent and prominent attributions of value chain responsibility from powerful stakeholders. When arising from powerful or otherwise morally legitimate stakeholders, these attributions of value chain responsibility can give rise to derivative stakeholder legitimacy. Issues relating to outsourced manufacturing by Nike and Mattel are among the more prominent examples of companies that were compelled by stakeholder pressure to take greater responsibility for actions occurring deep within their value chains. But they are hardly alone.

There is, we have claimed, a point at which relationships with other value chain members become so tenuous and distant from a business's core value-adding activities that the framework of 'global citizenship' becomes more applicable than that of stakeholder theory. While exclusivity and duration of relationships as well as other sources of derivative stakeholder obligations are all suggestive of where the line between stakeholder obligation and global corporate citizenship may be drawn in networked value chains, the line is neither a bright nor a static one. The standards applied by local communities and other stakeholders create a line that moves around in the gray area, often dramatically. Work remains for scholars and managers in better defining these obligations and community expectations. With these points in mind, we turn now to a discussion of the possible ways an organization might engage with its relevant stakeholder communities.

6 A Typology of Community Engagement

If the above is correct, there are many ways for companies to engage their communities. We propose five strategic postures that a company may take with respect to a community. A 'strategic posture' is like Porter's (1980) notion of 'generic strategy'. It is a predisposition to act in a particular way (Freeman and Gilbert 1987).

Companies may deal with communities as:

1. community creators;
2. community builders;
3. community good citizens;
4. community apathetic citizens; and
5. community exploiters or destroyers.

This typology is merely a beginning way to conceptualize firm-community relationships and should not be taken as definitive, mutually exclusive or collectively exhaustive of such relationships. Obviously, in large complex corporations, there may well be elements of several generic strategies, or different postures may exist at different times. Just as clearly, a company may intend to be a community builder, but in fact be a community exploiter. These strategic postures work as follows.

6.1 Community Creators

There are many examples of how companies have been community creators. The most obvious and direct are so-called 'company towns'. While there are fewer companies intending to create 'company towns' in today's global marketplace, company towns, in fact, still exist. Disney's planned community of Celebration, Florida adjacent to its theme park near Orlando is a more recent example of a company creating a community of place (Ross 2000). Disney attempted, with mixed success, to institute a number of particularly progressive ideals in education, community and urban planning. Communities of place can also be created by companies in conjunction with local governments; Levittown, PA is among the better-known historical examples of this.

More often today companies create communities of interest and communities of practice (recall the earlier discussion of Dunham et al.'s (2006) taxonomy of communities). Silicon Valley companies have created a community of interest around computer technology. Particular companies contribute to that community in a way that all may benefit. Particular companies such as eBay in effect create communities of interest by the very nature of their services. A person can use eBay to find other like-minded actors to trade with. Of course, internet service providers, as well as companies such as YouTube, FaceBook and MySpace, create communities of interest. Wikipedia creates a community of practice where stakeholders are all engaged in a common task. Likewise, Linux and others create communities of practice where the lines between traditional stakeholder roles become radically ambiguous.

6.2 Community Builders

Community builders are like community creators. Perhaps they did not have the original idea of creating the community, but their strategic posture is to try to maintain and improve the communities that they find. Community builders see healthy communities as places where their employees, customers and other stakeholders live. Vibrant communities are good for the company and its stakeholders. Community builders ask, 'what can we do to make this community better for its citizens, and therefore ultimately, better for our company and its stakeholders as well?'.

Community of practice builders try to find places where stakeholder interests are joint, where there is unrealized value to be created. They try to find places where the interests of employees, customers, suppliers and communities go in the same direction but this fact has gone unrecognized or underemphasized. A classic example here is the Ronald McDonald House. Here, McDonald's supports the administrative costs of taking donations from customers and creating facilities to help families dealing with illness, especially children. The actual funding comes from McDonald's customers. McDonald's is helping their customers build better communities through this coordination function. McDonald's facilitation role is an example aligning the interests of local communities and customers. This activity bears similarity to the better-known example of companies who match employee contributions to charity. With such a strategy they work with their employees to contribute to areas that the employees believe will build a better community.

6.3 Community Good Citizens

Being a good citizen means obeying the law and doing what one is asked to do if it is reasonable. However, community good citizens would take a reactive rather than proactive approach. If the community builder is always asking how to make the community better, the community good citizen is usually willing to abide by community rules, contribute its share, but rarely takes the lead in making the community better.

Many companies take a good citizen corporate posture to communities. They often have a corporate philanthropy program to respond to community requests. They pay attention to obeying the letter and spirit of local laws and customs. Companies that apply this approach to communities of interest, pay special attention to obeying internet privacy laws, copyright procedures and the like. They realize that by getting these issues wrong, they can easily create and motivate the ire of virtual advocacy groups and see community relations as a way to avoid conflict and only secondarily as a means of building positive value.

6.4 Community Apathetic Citizens

Apathetic postures yield doing the very minimum necessary. An analogy can be drawn to the individual citizen who obeys the law, but does not vote, lobby, give to charity or contribute to the conversation about how to improve or maintain the community. The apathetic citizen is not intentionally destructive or exploitative of communities *per se;* nor is he/she actively concerned about community. There is apathy in the defenses occasionally offered by companies doing business in developing nations prone to conflict or systematic human rights violations. Such a defense would say something akin to, 'the harm to the community is independent of our presence or activities here. If we weren't the ones engaged here, someone else would be. We comply with the law, but cannot be expected to solve local community difficulties'. Companies doing business in such nations have been known to disavow responsibility for these effects so long as they are within the letter of the local laws.

Nor is this limited to developing nations. In the context of advanced industrial societies, casinos, bars, landfills and effluent-generating firms of all kinds provide only a snapshot of industries, some members of which have shown apathy toward the negative externalities of their operations.

6.5 Community Exploiters or Destroyers

There is a sense in which apathy – in the sense described above – can serve to destroy communities. Beyond a minimalist, geographic sense, a community is an institution that requires active participation in order to flourish. An organization that occupies a central, prominent role in a community may have a destructive effect on that community through mere apathy.

Beyond this apathy, however, companies occasionally exploit communities more actively as a resource. Often these companies pit one community against another for the best regulatory environment, tax breaks or the one willing to make the biggest investment in infrastructure. These companies are not content to live within the letter of the law, but actively lobby to change the laws in order to make their exploitation more effective.

If the firm-community relationship always favors the company and exploits the community, it is likely that (in a relatively free society) the community will turn against the company. Indeed, independent of the presence of actual destruction or exploitation, in today's world it is enough for there to be the perception of exploitation to negatively affect a company's bottom line. Companies such as Wal-Mart have been charged (with wide disparities in the power and persuasiveness of the arguments presented) with exploiting communities and have incurred massive costs to reduce that perception.

Environmental issues are another area where many companies are seen as community exploiters or destroyers. Although not unique to them, firms in extractive industries in lesser developed countries are especially susceptible to such perceptions (again, with better and worse underlying bases for such perceptions).

Often, being a community exploiter or destroyer is a logical devolution from being a good citizen or apathetic citizen. Each of these postures views the community stakeholder as secondary to the process of value creation. And, when this happens, customers and employees, and those groups who see themselves as advocates, often non-governmental organizations, rise to 'speak for the community'. Firms perceiving of the community in this 'afterthought' fashion are missing the reciprocal, value-adding possibilities of closer community relationships such as those found among community creator and community builder companies. Furthermore, we submit, the failure to delimit core stakeholder relationships from concepts like 'global corporate citizenship' catalyze this perception of the community as secondary by overextending managerial and organizational decision-making resources.

7 Conclusion

We have suggested that some aspects of corporate citizenship are part and parcel of a nuanced view of stakeholder theory, with particular emphasis on community stakeholders. Obviously our analysis does not do full justice to the idea. It focuses only on the managerial aspect of corporate citizenship, especially in the private sector. While the United Nations may well be able to delve into the meaning of 'global citizenship' and its implications, most managers simply want to understand how to deal with specific communities in the value-creation process. By thinking about creating, building, exploiting and so on, communities, they can begin to see communities as vital stakeholders, and be explicit about the specific understanding of community that they have, as well as their own posture. Such is not a recipe for success (a 'rule for riches'), but it is a way to stop the process of self-deception and get on with the important task of value creation.

References

Cialdini, R.B. 1984. *Influence*. New York: Quill.

Donlan, W. 1998. Political reciprocity in Dark Age Greece: Odysseus and his *hetairoi*. In *Reciprocity in Ancient Greece*, ed. C. Gill, N. Postlethwaite, and R. Seaford, 51–72. Oxford/New York: Oxford University Press.

Dunfee, T.W. 2006. Do Firms with Unique Competencies for Rescuing Victims of Human Catastrophes have Special Obligations? Corporate Responsibility and the AIDS Catastrophe in Sub-Saharan Africa. *Business Ethics Quarterly* 16 (2): 185–210.

Dunham, L., R.E. Freeman, and J. Liedtka. 2001. *The Soft Underbelly of Stakeholder Theory: Towards Understanding Community*. Presentation to Society of Business Ethics Annual Meeting, August, Washington, DC.

———. 2006. Enhancing Stakeholder Practice: A Particularized Exploration of Community. *Business Ethics Quarterly* 16 (1): 23–42.

Freeman, R.E. 1984. *Strategic Management: A Stakeholder Approach*. Boston: Pitman.

Freeman, R.E., and D.R. Gilbert Jr. 1987. *Corporate Strategy and the Search for Ethics*. Englewood Cliffs: Prentice-Hall.

Hsieh, N. 2004. The Obligations of Transnational Corporations: Rawlsian Justice and the Duty of Assistance. *Business Ethics Quarterly* 14 (4): 643–661.

Johnson, J. 2003. Taylor. In *On and On*. Brushfire Records, Los Angeles.

Logsdon, J.M., and D.J. Wood. 2002. Business Citizenship: From Domestic to Global Level of Analysis. *Business Ethics Quarterly* 12 (2): 155–187.

Matten, D., and A. Crane. 2005. Corporate Citizenship: Toward an Extended Theoretical Conceptualization. *Academy of Management Review* 30 (1): 166–179.

Moon, J., A. Crane, and D. Matten. 2005. Can corporations Be Citizens? Corporate Citizenship as a Metaphor for Business Participation in Society. *Business Ethics Quarterly* 15 (3): 429–453.

Phillips, R.A. 1997. Stakeholder Theory and a Principle of Fairness. *Business Ethics Quarterly* 7 (1): 51–66.

———. 2003. *Stakeholder Theory and Organizational Ethics*. San Francisco: Berrett-Koehler.

Phillips, R., and C. Caldwell. 2005. Value Chain Responsibility: A Farewell to Arm's Length. *Business and Society Review* 110 (4): 345–370.

Phillips, R., R.E. Freeman, and A.C. Wicks. 2003. What Stakeholder Theory is Not. *Business Ethics Quarterly* 13 (4): 479–502.
Porter, M.E. 1980. *Competitive Strategy*. New York: Free Press.
Powell, W.W. 1990. Neither Market Nor Hierarchy: Network Forms of Organization. In *Research in Organizational Behavior*, ed. B. Staw and L.L. Cummings, 295–336. Greenwich: JAI Press.
Ross, A. 2000. *The Celebration Chronicles: Life, Liberty and the Pursuit of Property Value in Disney's New Town*. New York: Ballantine Books.
Waddock, S.A. 2001. Integrity and Mindfulness: Foundations of Corporate Citizenship. *Journal of Corporate Citizenship* 1: 25–38.
Windsor, D. 2001. Corporate Citizenship: Evolution and Interpretation. In *Perspectives on Corporate Citizenship*, ed. J. Andriof and M. McIntosh, 39–52. Greenleaf: Sheffield.
Wood, D., and J. Logsdon. 2001. Theorising Business Citizenship. In *Perspectives on Corporate Citizenship*, ed. J. Andriof and M. McIntosh, 83–103. Greenleaf: Sheffield.
———. 2002. Business Citizenship: From Individuals to Organizations. In *Ethics and Entrepreneurship*, Ruffin series 3, ed. R.E. Freeman and S. Venkatraman, 59–94. Charlottesville: Society for Business Ethics.

Robert A. Phillips is George R. Gardiner Professor in Business Ethics, Professor of Sustainability, and Director of the Centre of Excellence in Responsible Business at the York University's Schulich School of Business.

R. Edward Freeman is University Professor, Elis and Signe Olsson Professor, Academic Director of the Institute for Business in Society, and Senior Fellow of the Olsson Center for Applied Ethics at the University of Virginia's Darden School of Business.

Chapter 14
Managing for Stakeholders and the Purpose of Business

R. Edward Freeman and Bidhan L. Parmar

1 Introduction

Business has the potential to benefit many groups such as customers with new products and services, employees with meaningful work and wages, communities as engines of growth and employment, and shareholders through investment opportunities that provide a reasonable return and fit their values. Business also has the potential to harm these same groups through faulty products, accounting fraud, environmental damage, and increased social inequality. The balance of the benefits or harms achieved by a business is determined in part by the choices that business leaders make. One important driver of these choices is a leader's view of the purpose of business. This note: (1) outlines the history of debates about corporate purpose, (2) calls out some myths that persist in the way that people think about corporate purpose, and (3) introduces a view of business called "managing for stakeholders" that has become increasingly influential.

This technical note was prepared by R. Edward Freeman, Elis & Signe Olsson Professor of Business Administration, and Bidhan (Bobby) L. Parmar, Associate Professor of Business Administration, Darden School of Business. It incorporates and expands on ideas outlined in Managing for Stakeholders, UVA-E-0383. Copyright © 2017 by R. Edward Freeman. All rights reserved. *To order copies, send an e-mail to* sales@dardenhusinesspnblishing.com. *No part of this publication may be reproduced, stored in a retrieval system, used in a spreadsheet, or transmitted in any form or by any means—electronic, mechanical, photocopying, recording, or otherwise—without the permission of the Darden School Foundation.* Our goal is to publish materials of the highest quality, so please submit any errata to editorial@dardenbusinesspublishing.com.
Originally published by: *Darden Business Publishing* © 2017
Reprint by Springer, https://doi.org/10.1108/case.darden.2017.000415

R. E. Freeman (✉) · B. L. Parmar
University of Virginia, Charlottesville, VA, USA
e-mail: FreemanE@darden.virginia.edu

© Springer Nature Switzerland AG 2023

S. D. Dmytriyev, R. E. Freeman (eds.), *R. Edward Freeman's Selected Works on Stakeholder Theory and Business Ethics*, Issues in Business Ethics 53,
https://doi.org/10.1007/978-3-031-04564-6_14

The basic idea of managing for stakeholders is that businesses and the executives who manage them, actually do—and should—create value for customers, suppliers, employees, communities, and financiers. To manage sustainable businesses, leaders need to pay careful attention to how stakeholder relationships are managed and how value gets created to ensure that all core stakeholders are receiving value for their efforts and choose to continue to support the firm. Value here is understood as things that stakeholders value, which includes financial value as well as other forms of value (e.g., upholding values, treating people fairly, and improving the world). This view is contrasted with the shareholder-centric model of business activity; namely, that businesses are to be managed solely for the benefit of shareholders, and the only value that matters is financial value. According to this view, any other benefits (or harms) that are created for other groups are incidental.

2 The Purpose of Business: A Brief History

The modern business corporation has emerged during the twentieth century as one of the most important innovations in human history. Previously, during the 1800s, most businesses were owned and run by families or small groups. These closely held or private companies had a few shareholders who were active in the day-to-day running of the business. By the 1900s, public companies started to appear, with thousands of shareholders who owned a small fraction of their company's shares. These investors were unable to be involved in the daily activities of the business. In early public companies, such as AT&T, RCA, and GE, the board of directors controlled the company and made key decisions.[1]

When stock speculation caused the market crash of 1929 in the United States, leading to the Great Depression, a debate about the proper purpose of public companies emerged. Was their purpose to make as much money for their thousands of shareholders or to serve a broader purpose? In 1932, the debate was carried out in the *Harvard Law Review,* where Adolph Berle argued for a singular focus on profits for shareholders, and Merrick Dodd argued for a broader purpose that included benefits to customers, employees, and society. Berle eventually conceded the argument, and for the next 40 years, Dodd's broader view became the foundation for how society thought about business. Business leaders saw themselves as stewards of public institutions and responsible to a broader group of stakeholders. During this period called "managerial capitalism," business was conducted in a way in which size and stability mattered more than growth—managers sat on top of large hierarchies of huge conglomerates and mostly retained and reinvested

[1] Lynn Stout, The Shareholder Value Myth: How Putting Shareholders First Harms Investors, Corporations, and the Public (New York: Berrett-Koehler Publishers, 2012).

their earnings.[2] For a while, both workers and executives saw their incomes grow in tandem.[3]

In the late 1960s and early 1970s, the debates about corporate purpose resurfaced as several factors changed the business landscape. First, the size of conglomerates made it hard for headquarters to accurately understand how to create value for each specific business; this led to suboptimal decision making. Second, increased competition from Germany and Japan made American companies less profitable. Third, institutional investors were holding more shares and thus were able to exercise more power over companies than individual shareholders had in the past. In 1965, for example, 85% of shares were owned directly by individuals or households; today that number is close to 25%. Finally, the era of stagflation or rising prices and high unemployment threatened American prosperity.[4] Economists and policy makers began looking for a way to jump-start American prosperity.

To get the American economy back on track, economists argued that executives should focus more on shareholders and profits as part of a return to efficiency. Economists such as Milton Friedman, Michael Jensen, and others argued that shareholders needed more control over managers who seemed to be using their power to build and maintain their empires and "feather their nests" rather than to compete in the global economy. As an example, Fred Borsch, the CEO of GE, pledged a renewed focus on the "forgotten" shareholder.[5]

The 1980s marked the transition from managerial capitalism to investor capitalism by ushering in a wave of corporate takeovers in which large conglomerates were bought, broken up, and sold off in pieces to increase shareholder returns. Eventually, in the 1990s, CEO compensation became increasingly tied to share price. In 1965, the average CEO made 44 times more than the average worker; today that number is close to 370 times, with stock-based compensation making up a third of what the average CEO makes.[6]

As companies were increasingly evaluated primarily on their ability to produce short-term earnings, some executives have responded to that pressure by laying off workers, outsourcing production, buying back shares, cutting R&D expenses, and taking unethical shortcuts to raise share price. An increased focus on short-term results has led to an increase in corporate scandals, unethical behavior, and costs to other stakeholders. For these reasons, early proponents of shareholder-value maxi-

---

[2] Stout.

[3] Lawrence Mishel, "Economic Snapshot," Economic Policy Institute, September 24, 2013, http://www.epi.org/publication/the-ceo-to-worker-compensation-ratio-in-2012-of-273/ (accessed Aug. 24, 2017).

[4] William Lazonick and Mary O'Sullivan, "Maximizing Shareholder Value: A New Ideology for Corporate Governance," Economy and Society 29, no. 1 (2000): 13–35.

[5] GE annual report, 1971.

[6] Melanie Trottman, "Top CEOs Make 373 Times the Average U.S. Worker," Wall Street Journal, May 13, 2015, https://blogs.wsj.com/economics/2015/05/13/top-ceos-now-make-373-times-the-average-rank-and-file-worker/ (accessed Aug. 24, 2017).

mization like former GE CEO Jack Welch called it "the dumbest idea in the world."[7]

Despite the fact that many executives are questioning shareholder value maximization as the sole purpose of business, many students are still taught the shareholder-centric model in business schools, and it persists because of several myths.

MYTH 1: Shareholders own the corporation.
REALITY: In the United States, shareholders don't own public companies; their stake depends on the nature of their contract or share.

Executives often speak in the language of hierarchy as "working for shareholders," "shareholders are the boss," and "you have to do what the shareholders want." Shareholders do not own companies—according to law, public companies own themselves and are their own legal entities. Corporations are distinct entities that own property but are not property themselves. Shareholders own shares—or a contract with this legally independent entity.[8] Depending on the nature of the contract, the specific shareholder has specific rights and responsibilities. Some shares carry voting rights and owning a majority of those shares can allow a particular individual or institutional investor to control seats on the board and to influence the management of a company. Other classes of shares do not have voting rights. The degree of control specific shareholders have over a company depends on the kind and number of shares they own. For example, many companies have multiple classes of shares (e.g., Class A, Class B) that have different voting rights. At such companies as Google and Berkshire Hathaway, most of the voting shares are held by insiders, and most of the shares that are traded are nonvoting to allow a way for these companies to retain control over corporate decisions. In 2017, SNAP was the first IPO in American history to not offer any voting shares to the public.[9] Public companies, shareholders, and shares are too diverse to be accurately reflected in the simple slogan "shareholders own companies."

MYTH 2: Executives and directors are legally required to maximize shareholder value.
REALITY: The law provides broad discretion to executives and directors as long as they are not self-dealing.

Sometimes students believe that if executives and directors don't maximize shareholder value, they will be sued or fired. That may be the case in some corporations in which certain shareholders have a controlling interest, but in most

[7] Steve Denning, "The Dumbest Idea in the World: Maximizing Shareholder Value," *Forbes*, November 28, 2011, https://www.forbes.com/sites/stevedenning/2011/11/28/maximizing-shareholder-value-the-dumbest-idea-in-the-world/#5037dQcf2287 (accessed Aug. 24, 2017).

[8] Stout.

[9] Hannah Roberts, "Snapchat isn't Offering Voting Rights in its IPO—and Potential Investors are Furious," Business Insider, February 3, 2017, http://www.businessinsider.com/snapchat-ipo-no-voting-rights-investor-letter-2017-2 (accessed Aug. 24, 2017).

cases, the courts rely on what they call the "business judgment rule." The rule states that as long as "a board of directors is not tainted by personal conflicts of interest and makes a reasonable effort to become informed, courts will not second-guess the board's decisions about what is best for the company—even when those are decisions that seem to harm shareholder value."[10] For example, the company that owned the Chicago Cubs refused to hold night games because it believed that baseball should be a daytime sport. Holding night games would have increased profits. When the company was sued, the court ruled that under the business judgment rule, it could not disturb the board's decision, absent any evidence of fraud, illegality, or conflict of interest.[11] As evidenced by the court ruling, the business judgment rule allows a wide range of autonomy for executives. As long as they don't put profits in their own pockets, they can spend more on employees, pursue low-profit projects, and buy corporate jets—no matter what the effect on shareholder value.

Additionally, students in finance are taught that managers are the agents of shareholders or principals. According to the law, shareholders are not principals, and managers are not their agents. In 1976, Jensen and William Meckling, the finance theorists who created the principal/agent model, borrowed legal terminology to describe the problem of monitoring managers to better align their interests with shareholders and increase focus on shareholders.

In contrast to the absence of special obligations to shareholders, the law actually requires that the claims of customers, suppliers, local communities, and employees be taken into consideration by corporations. For example, the U.S. Consumer Product Safety Commission has the power to enact product recalls, essentially leading to an increase in the number of voluntary product recalls by companies seeking to mitigate legal damage awards. Thus by law, companies must take the interests of customers into account. The National Labor Relations Act gave employees the right to unionize and to bargain in good faith. It set up the National Labor Relations Board to enforce these rights with management. The Equal Pay Act of 1963 and Title VII of the Civil Rights Act of 1964 constrain management from discrimination in hiring practices; these rules have aligned corporate practice with employee rights. Similarly, the law has also evolved to try to protect the interests of local communities. The Clean Air Act and the Clean Water Act and various amendments to these classic pieces of legislation have constrained management from "spoiling the commons." In a historic case, *Marsh v. Alabama,*[12] the Supreme Court ruled that a company-owned town was subject to the provisions of the U.S. Constitution, thereby guaranteeing the rights of local citizens and negating the property rights of the firm.

---

[10] Stout, 29.

[11] Stout, 29–30.

[12] *Marsh v. Alabama,* 326 U.S. 501, 66 S. Ct. 276, 90 L. Ed. 265, 1946 U.S.

MYTH 3: Companies make more profits when executives focus on maximizing shareholder value.

REALITY: Despite 40 years of evidence, there is no conclusive link between the strategy of maximizing shareholder value and better financial performance.

The main idea behind focusing on shareholder value was to make companies more efficient and profitable. Now with over 40 years of evidence, we can see if this strategy has worked. A recent analysis by the consulting firm Deloitte suggests that return on invested capital in the United States has steadily decreased since 1965 from an average of 4.2% to 1.2% in 2009.[13] There is no clear evidence that firms aiming to maximize shareholder value actually perform better than other companies. There is some evidence that they perform worse. The irony is that shareholders can also be made worse off when executives try to maximize shareholder value.

Public companies are also disappearing as more firms choose to stay private and merge with other companies to provide economies of scale. In 1995, there were 7322 U.S. public companies; today that number is less than 3700.[14] Public companies are not living as long; in 1958, the average lifespan of a U.S. public company was 61 years, in 1980, it was 25 years, and in 2015, it had fallen to 10 years.[15] Additionally, short-termism is growing faster and faster as shareholders demand more immediate results. In 1960, the average stockholder held a stock for about eight years; in 2010 that number was down to four months.

MYTH 4: If executives focus on making profits, other problems will take care of themselves.

REALITY: The shareholder-centric model can create problems by making it harder for executives to see the ethical implications of their choices.

In a recent lab experiment by Dan Ariely and Lalin Anik, the participants were told to imagine themselves as the CEO of a bank. In one situation, the participants were told that their goal was to maximize shareholder value, and another group's goal was simply to run the bank. Both groups were given a list of actions, including ethically questionable strategies that would increase profits, and were asked which ones they would take. Participants in the maximize shareholder-value situation were

---

[13] John Hagel, John Seely Brown, and Lang Davison, "Shift Index 2011: The Most Important Business Study—Ever? *Forbes,* January 25, 2012, https://www.forbes.com/sites/stevedenning/2012/01/25/shift-index-2011-the-most-important-business-study-ever/#4abb20743a6e (accessed Aug. 25, 2017).

[14] Barry Ritholtz, "Where Have All the Public Companies Gone?," Bloomberg View, June 24, 2015, https://www.bloomberg.com/view/articles/2015-06-24/where-have-all-the-publicly-traded-companies-gone- (accessed Aug. 25, 2017).

[15] Rishi Iyengar, "This is How Long Your Business Will Last, according to Science," *Fortune,* April 2, 2015, http://fortune.com/2015/04/02/this-is-how-long-your-business-will-last-according-to-science/ (accessed Aug. 25, 2017).

significantly more likely to partake in ethically questionable acts.[16] This result is reflected in practice where over 55% of managers would avoid initiating a very positive net present value project if it meant falling short of the current quarter's consensus earnings.

These outcomes occur because most theories of business rely on separating business decisions from ethical decisions. This is seen most clearly in the popular joke about "business ethics as an oxymoron." Almost any business decision has some ethical content. To see that this is true we need only ask whether the following questions make sense for virtually any business decision:

- Who is harmed and/or benefited by this decision?
- Whose rights are enabled, and whose values are realized by this decision (and whose are not)?
- What kind of person will I (we) become if we make this decision?
- What relationships will be strengthened or weakened by this decision?

Since these questions are always open for most business decisions, although they might not be top of mind for a specific decision maker, it is reasonable to see ethics as an integral part of business. Even Milton Friedman argues that companies should make a profit "while conforming to the basic rules of the society, both those embodied in law and those embodied in ethical custom."[17] Ethics is about the principles, consequences, matters of character, and relationships that we use to live together. These ideas give us a set of open questions as we constantly search for better ways to answer in reasonably complete ways. Indeed, without trust that people will generally keep their promises and take responsibility for their actions, markets and organizations don't work. To create value, it is better to focus on integrating business and ethics within a complex set of stakeholder relationships rather than treating ethics as a side constraint on making profits.

3 Managing for Stakeholders

The basic idea of managing for stakeholders is quite simple: business can be understood as a set of relationships among groups that have a stake in the activities that make up the business. Business is about how customers, suppliers, employees, financiers (stockholders, bondholders, banks, and so on), communities, and managers interact and create value. To understand a business is to know how these relationships work. And the executive's or entrepreneur's job is to manage and shape these relationships; hence the term "managing for stakeholders."

---

[16] Lalin Anik, Ryan Hauser, and Dan Ariely, "The Shareholder Value Heuristic: Examining the Automatic Process of Ethical Transgressions that Benefit Others" (Fuqua School of Business, Duke University, working paper).

[17] Milton Friedman, "The Social Responsibility of Business is to Increase its Profits," *New York Times*, September 13, 1970, 17.

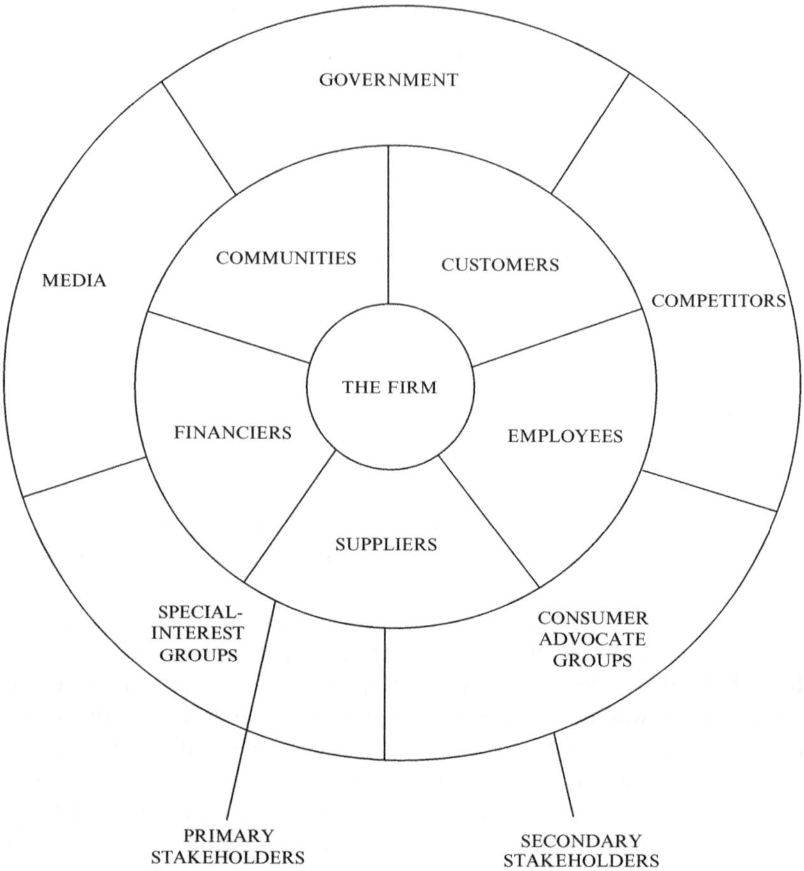

Fig. 14.1 The managing for stakeholder view of the firm

We can define the term "stakeholder" in a number of ways. First, we could define stakeholder fairly narrowly to capture the idea that any business, large or small, is about creating value for "those groups without whose support, the business would cease to be viable."[18] The inner circle of Fig. 14.1 depicts this view. Almost every business is concerned at some level with relationships among financiers, customers, suppliers, employees, and communities. We might call these groups "primary" or "definitional."

There is also a somewhat broader definition that captures the idea that, if a group or individual can affect a business, then the executives must take that group into consideration in thinking about how to create value. Or a stakeholder is any group or individual that can affect or be affected by the realization of an

[18] R. Edward Freeman, Jeffrey S. Harrison, Andrew C. Wicks, Bidhan L. Parmar, Simone DeColle, *Stakeholder Theory: The State of the Art* (New York: Cambridge University Press, 2010), 26.

organization's purpose. At a minimum, some groups affect primary stakeholders, and we might see these as stakeholders in the outer ring of Fig. 14.1 and call them "secondary" or "instrumental."

We see limited usefulness in trying to define one template, either based on the shareholder or stakeholder view, which works for all businesses everywhere. We see much value to be gained in examining how the stakes work in the value creation process for specific businesses. No stakeholder stands alone in the process of value creation. The stakes of each stakeholder group are multifaceted and inherently connected to each other. How could a bondholder recognize any returns without management paying attention to the stakes of customers or employees? How could customers get the products and services they need without employees and suppliers? If one stakeholder repeatedly pursues its interests at the expense of all the others, then in a relatively free society, stakeholders will either: (1) exit to form a new stakeholder network that satisfies their needs; (2) use the political process to constrain the offending stakeholder; or (3) invent some other form of activity to satisfy their particular needs.[19]

4 Stakeholders and Stakes

Financiers such as shareholders and bondholders have a financial stake in the business in the form of stocks, bonds, and so on, and they expect some financial return from them. Of course, the stakes of financiers will differ by type of owner, preferences for time horizon, moral preferences, and so forth, as well as by type of firm. It is important to note that even if leaders care only about maximizing value for shareholders, they have to balance multiple different factions of shareholders because shareholders are diverse—do we pay attention to long-term shareholders or shareholders who want to flip the stock? The shareholders of Google may well want returns as well as to be supportive of Google's articulated purpose of "Don't be evil." There are growing markets for social impact investing to take account of shareholders' broader values. For example, $1 out of every $5 invested in the United States goes through some kind of social screen, and the market for socially responsible investing has grown 33% since 2014 to $8.7 trillion in assets.[20]

Employees have their jobs and often their livelihoods at stake; they often have specialized skills for which there is usually no perfectly elastic market. In return for their labor, they expect security, wages, benefits, and meaningful work. Often employees are expected to participate in the decision making of the organization,

---

[19] S. Venkataraman, "Stakeholder Value Equilibration and the Entrepreneurial Process," *Ethics and Entrepreneurship,* Ruffin Series, no. 3 (Charlottesville, VA: Society of Business Ethics, 2002), 45–57.

[20] "Sustainable and Impact Investing in the United States Overview," US/SIF Foundation, http://www.ussif.org/files/Infographics/Qverview%201nfographic.pdf (accessed Aug. 25, 2017).

and, if the employees are management or senior executives, we see them as shouldering a great deal of responsibility for the conduct of the organization as a whole.

Customers and suppliers exchange resources for the products and services of the firm and in return receive the benefits of these products and services. As with financiers and employees, the customer and supplier relationships are enmeshed in ethics. Companies make promises to customers via their advertising, and when products or services don't deliver on these promises, then management has a responsibility to rectify the situation. It is also important to have suppliers who are committed to making a company better. If suppliers find a better, faster, and cheaper way of making critical parts or services, then both supplier and company can win. Of course, some suppliers simply compete on price, but even so, there is a moral element of fairness and transparency to the supplier relationship.

Finally, the local community grants the firm the right to build facilities, and in turn, it benefits from the tax base and economic and social contributions of the firm. Companies have a real impact on communities, and being located in a welcoming community helps a company create value for its other stakeholders. In return for the provision of local services, companies are expected to be good citizens, as is any person. It should not expose the community to unreasonable hazards in the form of pollution, toxic waste, and so forth. It should keep whatever commitments it makes to the community and operate in as transparent a manner as possible.

5 The Evidence Supports Managing for Stakeholders

Empirical evidence consistently suggests that companies that focus on creating value for stakeholders perform better financially than those that focus only on maximizing shareholder value. For example, companies that regularly are included on *Fortune's* Best Places to Work list outperform competitors by 2–4%.[21] That may seem like a small percentage, but when you compare it to the growth of the S&P 500 over the last 10 years (7.7%), it is pretty compelling. Similarly, companies that rank high in customer satisfaction outperform competitors by 7–8%, and companies that are good stewards of the environment outperform competitors by 4–6%.[22] All these benefits do not even take into account the losses and costs of a damaging public scandal. During the last 40 years, hundreds of studies have been conducted on the relationship between broader stakeholder management and financial performance,

[21] Alex Edmans, "28 Years of Stock Market Data Show a Link between Employee Satisfaction and Long-Term Value," *Harvard Business Review,* March 24, 2016, https://hbr.org/2016/03/28-years-of-stock-market-data-shows-a-link-between-employee-satisfaction-and-long-term-value (accessed Aug. 25, 2017).

[22] Claes Fornell, Sunil Mithas, Forrest V. Morgeson III, and M S. Krishnan, "Customer Satisfaction and Stock Prices: High Returns, Low Risk," *Journal of Marketing* 70, no. 1 (January 2006): 3–14.

and these studies have converged on a modest performance benefit (3–7%) for companies that focus on creating value for stakeholders. The specific size of the performance benefit varies by industry, time period, and the specific companies sampled. For example, a recent McKinsey study found that companies managed for the long term (defined as from five to seven years) have 36% greater earnings, 47% higher revenues, and an increased market capitalization of $7 billion compared to companies focused on the short term.[23] Another study of 28 firms that excel at creating value for customers, employees, suppliers, and communities found that these companies outperformed the S&P 500 from 1996 through 2006 by 8:1.[24]

The results are more compelling when companies within a specific industry or sector are compared. An analysis comparing Wal-Mart (specifically Sam's Club) and Costco found that the focus of Sam's Club on everyday low prices resulted in it paying a minimum wage, shortening employee hours, and awarding minimal health insurance and benefits to employees. On a per-employee basis, Sam's Club paid less than Costco, which paid a wage of $20 an hour.[25] But the story doesn't stop there. Sam's Club has high turnover (44%), which is costly, because it has to train new employees. Costco's turnover is less than half that of Sam's Club at 17%, which is very rare in retail. Additionally, sales per employee are higher at Costco because employees who feel cared for and supported are better at taking care of customers. Research has demonstrated that there is a significant positive effect on employee engagement in companies that focus on serving a broader group of stakeholders.[26] Overall, by paying a higher wage, Costco ends up ahead.

Similarly, a recent study by a team of researchers at the Wharton School showed that positive stakeholder relationships are critical to the success of firms.[27] Conducted in the context of publicly traded gold mines, research by Witold Henisz and his colleagues showed that positive stakeholder relationships such as support from local government and NGOs were worth two times the price of mined gold. When a company had a negative stakeholder relationship such as with a combative union or

---

[23] Dominic Barton, James Manyika, Tim Koller, Robert Palter, Jonathan Godsall, and Josh Zoffer, "Where Companies with a Long-Term View Outperform Their Peers," McKinsey&Company, February 2017, http://www.mckinsey.com/global-themes/long-term-capitalism/where-companies-with-a-long-term-view-outperform-their-peers (accessed Aug. 25,2017).

[24] Rajendra Sisodia, David Wolfe, and Jagdish N. Sheth, *Firms of Endearment: How World-Class Companies Profit from Passion and Purpose* (Upper Saddle River, NJ: Wharton School Publishing, 2014).

[25] Wayne F. Cascio, 'Decency Means More than 'Always Low Prices': A Comparison of Costco to Walmart's Sam's Club," *Academy of Management Perspectives* 20, 3 (2006): 2637.

[26] Mary Sully De Luque, Nathan T. Washburn, David A. Waldman, and Robert J. House, "Unrequited Profit: How Stakeholder and Economic Values Relate to Subordinates' Perceptions of Leadership and Firm Performance," *Administrative Science Quarterly* 53, 4 (2008): 626–54; Bidhan L. Parmar, Adrian Keevil, and Andrew C. Wicks, "People and Profits: The Impact of Corporate Objectives on Employees' Need Satisfaction at Work *Journal of Business Ethics* (2017): 1–21.

[27] Witold J. Henisz, Sinziana Dorobantu, and Lite J. Nartey, "Spinning Gold: The Financial Returns to Stakeholder Engagement," *Strategic Management Journal* 35, no. 12 (2014): 1727–748.

unsupportive local government, it paid such unforeseen costs as increased waiting time for permits or stalled production because of labor disputes that significantly damaged its profitability.

6 The Responsibility of the Executive in Managing for Stakeholders

Executives play a special role in the activity of the business enterprise. On the one hand, they have a stake like every other employee regarding an actual or implied employment contract. That stake is linked to the stakes of financiers, customers, suppliers, communities, and other employees. Executives also are expected to look after the health of the overall enterprise, to keep the varied stakes moving in roughly the same direction, and to help them maintain balance. Unlike the age of managerialism, executives can't focus on their own comfort and benefits but need to focus instead on creating value for others.

First and foremost, leaders need to view stakeholder interests as joint and inherently tied together. Viewing stakeholder interests as joint rather than opposed can be difficult because it is not always easy to find a way to accommodate all stakeholder interests. It is easier to trade off one versus another. Why not delay spending on new products for customers to keep earnings a bit higher? Why not cut employee medical benefits in order to invest in a new inventory control system?

Managing for stakeholders suggests that executives try to reframe the questions and create new alternatives where tradeoffs are not necessary. All stakeholders have to do well and continue to receive value if the firm is to succeed. Therefore, managers must do the hard work of limiting or avoiding systematic tradeoffs where possible, particularly among primary stakeholders. To do this well, managers must learn about stakeholder preferences and take them into account in their decision making so that they can create new alternatives where tradeoffs aren't necessary. How can we invest in new products *and* create higher earnings? How can we be sure our employees are healthy and happy and can work creatively so that we can capture the benefits of new information technology like inventory control systems? In a recent book reflecting on his experience as CEO of the Medtronic Company, Bill George summarized the managing for stakeholders' mindset:

> Serving all your stakeholders is the best way to produce long-term results and create a growing, prosperous company…Let me be very clear about this: there is no conflict between serving all your stakeholders and providing excellent returns for shareholders. In the long term, it is impossible to have one without the other. However, serving all these stakeholder groups requires discipline, vision, and committed leadership.[28]

[28] Bill George, *Authentic Leadership* (San Francisco: Jossey-Bass Wiley, Inc., 2004), 104.

If tradeoffs have to be made, as can sometimes happen in the real world, then the executive must figure out how to make the tradeoffs using a fair process and communicate the reasons why clearly to stakeholders to contain any relational damage, and immediately begin looking for ways to improve the tradeoffs for all sides. Managing for stakeholders is about creating as much value as possible for stakeholders without resorting to tradeoffs.

We believe that this task is more easily accomplished when a business has a sense of purpose beyond the requirement that it be profitable. Furthermore, business as a collection of individuals can have any purpose it chooses. Wal-Mart may stand for everyday low prices. Merck can stand for alleviating human suffering. The point is that if an entrepreneur or an executive can find a purpose that speaks to the core value the firm has created as well as to the hearts and minds of key stakeholders, it is more likely to gain sustained success. Corporate purpose is a choice made by leaders that should be defined clearly, implemented carefully, and communicated to stakeholders.[29]

Purpose is complex and inspirational. The Grameen Bank wants to eliminate poverty. Fannie Mae wants to make housing affordable to every income level in society. And all these organizations have to generate profits, or else they cannot pursue their purposes. Capitalism works because companies can pursue their purposes with others. When companies coalesce around a big idea, or a joint purpose evolves from their day-to-day activities with each other, then great things can happen.

Most human beings are complicated. Most of us do what we do because we are self-interested and interested in making others better off. People clearly care a great deal about how they do financially, yet more careful reflection on why we work often reveals a complex array of value that we seek in doing what we do—including helping our firm and our colleagues succeed. Business works in part because of our urge to create things with and for others. Working on a team or creating a new product or delivery mechanism that makes customers' lives better, happier, or more pleasurable can all be contributing factors as to why we go to work each day. The assumption of narrow self-interest is extremely limiting and can be self-reinforcing—people can begin to act in narrow self-interested ways if they believe that is what is expected of them, as some of the scandals such as Enron have shown. We need to be open to a more complex psychology—one any parent finds familiar as they have shepherded the growth and development of their children. Taking a stakeholder approach helps people decide how companies can contribute to their well-being and the kind of lives they want to lead. By making ethics explicit and building it into the basic way we think about business, we avoid a situation of bad faith and self-deception.

[29] Miguel Padró, "Unrealized Potential: Misconceptions about Corporate Purpose and New Opportunities for Business Education," Aspen Institute Business & Society Program, May 28, 2014, https://assets.aspeninstitute.org/content/uploads/files/content/docs/pubs/Aspen%20BSP%20Unrealized%20Potential%20May2014v.2.pdf (accessed Aug. 25, 2017).

7 Conclusion

Business (and capitalism) has evolved as a social practice, an important one that we use to create value and trade with each other. Managing for stakeholders aims to build into the very conceptual framework we use to think about business as a concern with ethical themes such as freedom, equality, consequences, decency, and shared purpose. By paying attention to the effects of how we create value for each other, to the rights and obligations of different stakeholder groups, and to the purpose and character of our firms, leaders can make business a more fully human institution, and increase the likelihood that it is practiced in a way that broadly enriches human society rather than harms it.

R. Edward Freeman is University Professor, Elis and Signe Olsson Professor, Academic Director of the Institute for Business in Society, and Senior Fellow of the Olsson Center for Applied Ethics at the University of Virginia's Darden School of Business.

Bidhan L. Parmar is the Shannon G. Smith Bicentennial Associate Professor of Business Administration at the University of Virginia's Darden School of Business.

Chapter 15
Managing for Stakeholders: Trade-Offs or Value Creation

R. Edward Freeman

1 The Basic Idea

The basic idea of "managing for stakeholders," as I now see it, is quite simple, and I believe it is closer to the origins of the idea from Eric Rhenman and the Tavistock thinkers. In fact, Juha Nasi was correct in his assessment of the Scandinavian origins of the stakeholder idea that focus on what holds stakeholder interests together. Nasi originally suggested that we focus on "intressent" in Swedish. In Finnish, this term was modified as "intressentti" or "sidosryhma," which could be translated as an "interest group" or even as a "linkage" or "bonding" or "binding" group (Nasi 1995 at 98). It is this bonding or binding idea that is most interesting to explore. Managing for stakeholders asks us to see stakeholders as "bound together by the jointness of their interests." Hence, the basic idea goes like this.

Business can be understood as a set of relationships among groups which have a stake in the activities that make up the business. Business is about how customers, suppliers, employees, financiers (stockholders, bondholders, banks, etc.),

Originally published in: *J Bus Ethics*, 96, 7–9 © Springer Science Business Media B.V., 2010
Reprint by Springer, https://doi.org/10.1007/s10551-011-0935-5

The ideas in this article have been developed in a number of publications, foremost of which is R. Edward Freeman, Jeffrey Harrison, and Andrew Wicks, Managing for Stakeholders: Survival, Reputation and Success, Yale University Press, 2007, and R. Edward Freeman, "Managing for Stakeholders," in T. Beauchamp, N. Bowie, and D. Arnold (eds.) Ethical Theory and Business, 8th edition, Pearson, 2008. I am grateful to editors, publishers, and co-authors for their support, and allowing me to further develop these ideas here.

R. E. Freeman (✉)
University of Virginia, Charlottesville, VA, USA
e-mail: FreemanE@darden.virginia.edu

© Springer Nature Switzerland AG 2023

S. D. Dmytriyev, R. E. Freeman (eds.), *R. Edward Freeman's Selected Works on Stakeholder Theory and Business Ethics*, Issues in Business Ethics 53,
https://doi.org/10.1007/978-3-031-04564-6_15

communities and managers interact and create value. To understand a business is to know how these relationships work. And, the executive's or entrepreneur's job is to manage and shape these relationships, hence the title, "managing for stakeholders." Customers, suppliers, employees, financiers, communities, and managers are all key parts of today's business organization. Building and leading a great company has always been about managing for stakeholders. This much is perhaps uncontroversial, and has in fact become a mainstream idea about business all over the world.

However, many would still argue that when push comes to shove, the interests of customers, suppliers, communities and employees, must be traded off against the interests of financiers. However, the idea that one particular group always gets priority is deeply flawed. The very nature of capitalism itself is putting together a deal, or a contract, or a set of relationships among stakeholders so that all can win continuously over a long period of time. As Rhenman, Ackoff and the early stakeholder theorists knew, if you take away the support of any stakeholder you simply do not have a viable business.

In this mindset of "managing for stakeholders," executives play a special role. On the one hand, they have a stake like every other employee in terms of an actual or implied employment contract. And, that stake is linked to the stakes of financiers, customers, suppliers, communities, and other employees. In addition, executives are expected to look after the health of the overall enterprise, to keep the varied stakes moving in roughly the same direction, and to keep them in balance or in harmony, and it is here where I believe there is the greatest confusion.

As stakeholder theory began to be taken seriously by scholars writing primarily in the field of business ethics, it developed as a way to raise questions of justice in corporations. It is easy to see that once we begin to see the corporation as a set of contracts among stakeholders, then it is natural to ask, "What is a fair contract?" Several answers were given to this question, and most relied on an abstract and stylized view of the corporation and economic activity.

While most philosophers rejected the shareholder theory, they saw stakeholder theory as an alternative to bring ethics and justice into business. The rejection of the "financiers first" priority rule gave rise to the search for other priority rules to take its place. This search for justice mimicked the literature on distributive justice and became focused on the way to make trade-offs among stakeholders. Questions such as, how could better benefits and pay for employees be traded off against higher earnings for shareholders, or how could attention to product safety be traded off against jobs for employees, etc. Literally there are thousands of such questions. Well-publicized cases such as the Ford Pinto, where managers used cost–benefit analysis to make such trade-offs, had led to disasters. Indeed Steven Brenner, with Philip Cochran and Jamsheed Hosseini, determined a formal theory for making such trade-offs, and presented it at the stakeholder conference in Finland in 1994.

In the economists' view of the world, there are always trade-offs. A number of assumptions lead to this way of thinking, especially the idea that the right point of analysis is one of equilibrium or near equilibrium. At equilibrium, the most efficient use of resources is reached, and exactly the right number of apples versus oranges have been produced. Producing more oranges leads to unacceptable and irrational "apple consequences." These trade-offs are built into the utility functions of both consumers and producers. Indeed, the very existence of real-valued utility functions assures us that rational actors can make the required tradeoffs among various goods, as producers and consumers.

However, in the real world managers face multiple demands simultaneously. In fact, the very heart of capitalism and its entrepreneurial spirit is in figuring out how to meet the demands of customers, suppliers, employees, communities, and financiers, so that all win. Venkataraman and others have suggested that there is a natural marriage between stakeholder theory and entrepreneurship theory along the following lines. When a stakeholder group's interests are not being met, either they leave the firm for another network that will satisfy their interests, or they enter into some entrepreneurial venture to find a better way. Continuously trading off their interests with those of another stakeholder group is simply unacceptable in a relatively free society. From a managerial perspective, if managers look for trade-offs among stakeholders, then they will create trade-offs and they may never find the "sweet spot" that signifies the joint interest of all key stakeholders.

As stakeholder theory evolves to become the mainstream narrative about business, I believe that we need to keep in mind three interconnected ideas:

1. *No stakeholder stands alone in the process of value creation.* The stakes of each stakeholder group are multi-faceted, and inherently connected to each other. How could a bondholder recognize any returns without management paying attention to the stakes of customers or employees? How could customers get the products and services they need without employees and suppliers? How could employees have a decent place to live without communities? The fact that stakeholders have joint interests is, I believe, the key insight of stakeholder theory, as it has been developed over the last 50 years.

Stakeholder interests are inherently tied together. Seeing stakeholder interests as "joint" rather than opposed is difficult. It is not always easy to find a way to accommodate all stakeholder interests. It is easier to trade off one versus another. Why not delay spending on new products for customers in order to keep earnings a bit higher? Why not cut employee medical benefits in order to invest in a new inventory control system?

However, when there is dissonance, the time is ripe to try and find a reframing of the basic business proposition so that more stakeholders win continuously over time. Stakeholders that are difficult to please, critics, employees who push back, even conflicts of values, all can be sources of value creation, when approached with the "no tradeoffs" mindset of managing for stakeholders. Rather than give into trade-offs, executives should first try to reframe the questions. How can we invest in new products and create higher earnings? How can we be sure our employees are healthy and happy and are able to work creatively so that we can capture the benefits of new information technology such as inventory control systems?

In a recent book reflecting on his experience as CEO of Medtronic, Bill George summarized the managing for stakeholders mindset[1]:

> Serving all your stakeholders is the best way to produce long term results and create a growing, prosperous company…Let me be very clear about this: there is no conflict between serving all your stakeholders and providing excellent returns for shareholders. In the long term it is impossible to have one without the other. However, serving all these stakeholder groups requires discipline, vision, and committed leadership.

2. *The primary responsibility of the executive is to create as much value as possible for stakeholders.* Where stakeholder interests conflict, the executive must find a way to rethink the problems so that these interests can go together, so that even more value can be created for each. If trade-offs have to be made, because of a failure of imagination, time pressure, or other reasons, then the obvious next step is to try and figure out how to improve the trade-offs for all sides. Managing for stakeholders is about creating as much value as possible for stakeholders, without resorting to trade-offs.

The key idea that holds this value creation mindset together is the idea that businesses can have a purpose. And, there are few limits on the kinds of purpose that can drive a business. Wal-Mart may stand for "everyday low price." Merck can stand for "alleviating human suffering." The point is that if an entrepreneur or an executive can find a purpose that speaks to the hearts and minds of key stakeholders, it is more likely that there will be sustained success.

Purpose is complex and inspirational. The Grameen Bank wants to eliminate poverty. Fannie Mae wants to make housing affordable to every income level in society. Tastings (a local restaurant) wants to bring the taste of really good food and wine to lots of people in the community. And, all of these organizations have to generate profits, or else they cannot pursue their purposes. We cannot emphasize this idea too much. Capitalism works because we can pursue our purpose with others. When we coalesce around a big idea, or a joint purpose evolves from our day-to-day activities with each other, then great things can happen.

3. *Stakeholders have names and faces and children.* Executives and academics, especially, must understand that business is fully situated in the realm of humanity. Businesses are human institutions populated by real live complex human beings. They are not mere placeholders for social roles. As such, matters of ethics are routine when one takes a managing for stakeholders approach. Of course, this should go without saying. One CEO put it very succinctly, "the only assets I manage go up and down the elevators everyday."

Most human beings are complicated. Most of us do what we do because we are self-interested and interested in others. Business works in part because of our urge to create things with others and for others. Working on a team, or creating a new product or delivery mechanism that makes customers lives better or happier or more pleasurable all can be contributing factors to why we go to work each day.

[1] Bill George (2003, p. 104).

And, this is not to deny the economic incentive of getting a pay check. However, the assumption of narrow self-interest is extremely limiting, and can be self-reinforcing—people can begin to act in a narrow self-interested way if they believe that is what is expected of them, as some of the scandals such as Enron have shown. We need to be open to a more complex psychology—one any parent finds familiar as they have shepherded the growth and development of their children.

2 Summary

The last 30 years of research on stakeholder theory has led to a rich and varied literature. The next step is to see stakeholder theory as a way to redefine how we think about value creation and trade. If we can make the twenty-first century the century of value creation for stakeholders, and if we can escape the political and institutional trap of building in trade-offs among stakeholders into public policy, then the sheer audacity of our fellow humans will lead to prosperity and freedom for more and more people.

References

George, B. 2003. *Authentic Leadership: Rediscovering the Secrets to Creating Lasting Value*. Hoboken: Jossey-Bass.

Näsi, J. 1995. What is Stakeholder Thinking? A Snapshot of Social Theory of the Firm. In *Understanding Stakeholder Thinking*, ed. J. Näsi, 19–32. LSR-Julkaisut Oy: Helsinki.

R. Edward Freeman is University Professor, Elis and Signe Olsson Professor, Academic Director of the Institute for Business in Society, and Senior Fellow of the Olsson Center for Applied Ethics at the University of Virginia's Darden School of Business.

References

Chapter 16
Five Challenges to Stakeholder Theory: A Report on Research in Progress

R. Edward Freeman

1 Introduction

The body of research that has come to be known as "stakeholder theory" has grown enormously over the past 40 years. There are literally hundreds of articles and books devoted to some part of these ideas. In a recent book (Freeman et al. 2010) that summarized stakeholder theory, my colleagues and I put together a bibliography that was 45 published pages long. A quick perusal of "stakeholder" yields over 47 million results when typed into Google, and over one million when Google Scholar is consulted. "Stakeholder management" yields over four million and 900,000, respectively, on Google and Google Scholar. "Stakeholder theory" in Google Scholar yields over 500,000 hits. There are many lines of work, some overlapping, some perpendicular, and some that offer novel, even seemingly indefensible, accounts of the main ideas. It is perhaps a fool's errand to attempt to suggest that there are some overarching challenges to stakeholder theory. Nonetheless, the main point of this essay is to do just that.

Before setting forth these five challenges I want to clarify how I see stakeholder theory, as I have found that I may have a quite different interpretations of the main ideas than many others who write about it (Freeman 2011; Freeman et al. 2007). Then I will, in turn, and very briefly, set forth the five challenges which are: (1) The Total Performance Challenge; (2) The Stakeholder Accounting Challenge; (3) The Behavioral Stakeholder Theory Challenge; (4) The Public Policy Challenge; and (5)

Originally published in: *Stakeholder Management*, 1–20 © Emerald Publishing Limited, 2017
Reprint by Springer, https://doi.org/10.1108/S2514-175920170000001

R. E. Freeman (✉)
University of Virginia, Charlottesville, VA, USA
e-mail: FreemanE@darden.virginia.edu

© Springer Nature Switzerland AG 2023

S. D. Dmytriyev, R. E. Freeman (eds.), *R. Edward Freeman's Selected Works on Stakeholder Theory and Business Ethics*, Issues in Business Ethics 53,
https://doi.org/10.1007/978-3-031-04564-6_16

The Ethical Theory Challenge. I am currently engaged with a number of colleagues on joint research projects for each of these challenges. This chapter is a very brief report on this work in progress, and is also a very stylized account of both stakeholder theory and these challenges. In what follows I will do my best to represent their ideas as accurately as possible. Please note their contributions in the Acknowledgements section below. I find that I often get far too much credit for my own rather modest contribution many years ago. Stakeholder theory is alive and well because so many scholars are engaged in research that I believe will foster a generational change in the narrative of business.

2 What Is Stakeholder Theory?

The ideas behind the stakeholder concept are as old as business itself. Even though the word "stakeholder" has a fairly recent common usage, from the earliest beginnings, it is difficult to deny that business has been a matter of trade between buyers and sellers so that both were at least perceptually better off because of the exchange, even in times of barter societies. Exchange created value between the partners and led to specialization of labor, more knowledge and innovation, and hence more exchange. Eventually, employees were added, though during ancient times they had relatively little freedom. While the separation of ownership and control may well be rooted in feudal society, the emergence of wealthy merchants who often earned their profits on the backs of others has a long history. While value was often created, it was also sometimes destroyed, especially when trade-offs were made at the expense of one group, say employees, by favoring another, say owners. Business has always affected customers, suppliers, and employees and the owners of the business, even if value was sometimes destroyed for some. Even communities and governments have long been involved in value creation and trade. Fernand Braudel's (1992) magisterial history of capitalism shows us the fiction that is free floating markets disconnected from the rest of society.

Companies incorporate offshore to minimize tax exposure. Managing the government or community relationship has been a part of value creation and trade from the very beginning.

The basic idea of "managing for stakeholders" or "value creation stakeholder theory"[1] is quite simple. Business can be understood as a set of value-creating relationships among groups that have a stake in the activities that make up the business. Business is about how customers, suppliers, employees, financiers (stockholders, bondholders, banks, etc.), communities, and managers interact and create value. To understand a business is to know how these relationships work. And, the executive's or entrepreneur's job is to manage and shape these relationships, hence the title "managing for stakeholders."

[1] I use these terms interchangeably.

To say that business is a set of interconnected relationships may well be controversial. Economists and others want to see business as a discrete set of economic transactions, where it is always possible to maximize expected value by taking a "future-forward" approach. "Business as relational" is a much more subtle idea. Relationships are not reducible to transactions. Much more research needs to be done in spelling out such a change in the fundamental vocabulary of business.[2]

The idea of "managing for stakeholders" is usually depicted in a variation of the classic "wheel and spoke" diagram with the corporation at the center (Freeman et al. 2007; Phillips 2003). However, it is important to note that the stakeholder idea is perfectly general. Corporations are certainly not the center of the universe, and there are many possible pictures. One might put customers in the center to signal that a company puts customers as the key priority. Novo Nordisk, a diabetes drug company, puts "people with diabetes" in the center of its map. Another might put employees in the center and link them to customers and shareholders. Or, one might have no organization in the center to signify that this is an interconnected system of stakeholders. But, there is no larger metaphysical claim here. It depends on the purpose of the picture, or the problem that one is trying to solve.

3 Stakeholders and Stakes

Owners or financiers (a better term) clearly have a financial stake in the business in the form of stocks, bonds, and so on, and they expect some kind of financial return from them. Of course, the stakes of financiers will differ by type of owner, preferences for money, moral preferences, and so on, as well as by type of firm. The shareholders of Google may well want returns as well as be supportive of Google's articulated purpose of "Do No Evil." To the extent that it makes sense to talk about the financiers "owning the firm," they have a concomitant responsibility for the uses of their property. Stout (2012) has argued that depicting "shareholders" as "owners" is at best misleading, and from a legal point of view, simply incorrect. And, it is equally a mistake to believe that financiers' interests can be considered solely in their own right. As with every stakeholder, financiers are connected to customers', employees', suppliers', and communities' interests as well. Stakeholders are interconnected. They have joint interests.

Employees have their jobs and usually their livelihood at stake; they often have specialized skills for which there is usually no perfectly elastic market. In return for their labor, they expect security, wages, benefits, and meaningful work. Employees often want to find meaning at work, and to participate in the decision-making. In today's world, employees who are engaged in the business are much more likely to produce good results for themselves and the other stakeholders. And, employees are

[2] I have begun this work of spelling out stakeholder theory as relational with Michelle Greenwood and Harry Van Buren in several forthcoming papers.

sometimes financiers as well, since many companies have stock ownership plans, and loyal employees who believe in the future of their companies often voluntarily invest.

Customers and suppliers exchange resources for the products and services of the firm and in return receive the benefits of the products and services. As with financiers and employees, the customer and supplier relationships are enmeshed in ethics. Companies make promises to customers via their advertising, and when products or services do not deliver on these promises then management has a responsibility to rectify the situation. It is also important to have suppliers who are committed to making a company better. If suppliers find a better, faster, and cheaper way of making critical parts or services, then both supplier and company can win. Of course, some suppliers simply compete on price, but even so, there is a moral element of fairness and transparency to the supplier relationship. For businesses in the twenty-first century, the supply chain from customers to suppliers is often integrated into the operations of the business. Many of these supply chains have impacts on the natural environment, and there has been a great deal of innovation in business in mitigating the effects of global supply chains.

Finally, the local community grants the firm the right to build facilities, and in turn, it benefits from the tax base and economic and social contributions of the firm. Companies have a real impact on communities, and being located in a welcoming community helps a company create value for its other stakeholders. In return for the provision of local services, companies are expected to be good citizens, as is any individual person. It should not expose the community to unreasonable hazards in the form of pollution, toxic waste, etc. It should keep whatever commitments it makes to the community, and operate in a transparent manner as far as possible. Of course, companies do not have perfect knowledge, but when management discovers some danger or runs afoul of new competition, it is expected to inform and work with local communities to mitigate any negative effects, as far as possible. "Community" is an ambiguous term. Some companies define it narrowly to mean only those places where they have facilities. Others define it broadly to include the communities where their suppliers are located. And, some even define community more globally to include billions of people who may use their products or be affected by them.

Oftentimes, management theorists overemphasize the role of definition in theories or frameworks. As a pragmatist philosopher, I believe that definitions often lack precision, and that this is a good feature of language for most purposes. For instance, much is written about the definition of "stakeholder." "Does it include NGOs or competitors? Yes or No? Once and for all, let's get it right," they say. My response is that it depends on what problem you are trying to solve. For certain problems, like governance at the board level, you may want a narrow definition, while for some societal problems, you may want a broader one. There is no one right definition, as the definition in use depends on the problem one is trying to solve, whether theoretical or practical.

First of all we could define the term fairly narrowly to capture the idea that any business, large or small, is about creating value for "those groups without whose support, the business would cease to be viable." Almost every business is concerned at some level with relationships among financiers, customers, suppliers, employees, and communities. We might call these groups "primary" or "definitional." However, it should be noted that as a business starts up, sometimes one particular stakeholder is more important than another. In a new business start-up, sometimes there are no suppliers, and therefore paying a lot of attention to one or two key customers, as well as to the venture capitalist (financier), is the right approach.

There is also a somewhat broader definition that captures the idea that if a group or an individual can affect a business, then the executives must take that group into consideration in thinking about how to create value. Or, a stakeholder is any group or individual that can affect or be affected by the realization of an organization's purpose.

Much value can be gained by examining how the stakes work in the value creation process and what the role of the executive is.

Executives play a special role in the activity of the business enterprise. On the one hand, they have a stake like every other employee in terms of an actual or implied employment contract. And, that stake is linked to the stakes of financiers, customers, suppliers, communities, and other employees. In addition, executives are expected to look after the health of the overall enterprise, to keep the varied stakes moving in roughly the same direction, and to keep them in balance.

No stakeholder stands alone in the process of value creation. The stakes of each stakeholder group are multi-faceted, and inherently connected to each other. How could a bondholder recognize any returns without management paying attention to the stakes of customers or employees? How could customers get the products and services they need without employees and suppliers? How could employees have a decent place to live without communities?

Stakeholder interests are joint. That is why business works. Many theorists argue that stakeholder interests are essentially in conflict, but this misses the basic idea of capitalism. It is a system of cooperation whereby customers, suppliers, employees, communities, and financiers cooperate together to create value that no one of these groups could create alone. The primary responsibility of the executive or the entrepreneur is to create as much value as possible for stakeholders. (And this directive is at least as clear as the one given by the dominant model to maximize shareholder value.) Where stakeholder interests conflict, the executive must find a way to rethink the problems so that these interests can go together, so that even more value can be created for each. If trade-offs have to be made, as often happens in the real world, then the executive must figure out how to make the trade-offs, and immediately begin improving the trade-offs for all sides. *Managing for stakeholders is about creating as much value as possible for stakeholders, without resorting to trade-offs.*

To create value for stakeholders, executives and entrepreneurs must see business as fully situated in the realm of humanity. Businesses are human institutions populated by real live complex human beings. Stakeholders have names and faces and children. They are not mere placeholders for social roles. Most human beings are complicated. Most of us do what we do because we are self-interested and interested in others. Business works in part because of our urge to create things with others and for others. Working on a team, or creating a new product or delivery mechanism that makes customers' lives better or happier or more pleasurable all can be contributing factors to why we go to work each day. And, this is not to deny the economic incentive of getting a pay check. The assumption of narrow self-interest is extremely limiting, and can be self-reinforcing—people can begin to act in a narrow self-interested way if they believe that is what is expected of them, as some of the scandals, such as Enron, have shown. We need to be open to a more complex psychology—one any parent finds familiar as they have shepherded the growth and development of their children.

With these preliminaries about how I see stakeholder theory, I want to introduce five challenges to value creating stakeholder theory (VCST). If you have a different view of stakeholder theory, for instance, that it is incompatible with currently received theory in strategic management or economic theory, then you will reject these challenges, and instead substitute some very different ones.

4 The Total Performance Challenge

Do profits measure the total performance of a business? Jones and Freeman (2013)[3] argue that the pursuit of profitability does not always lead to the greatest wealth creation and that we need to reexamine the very nature of how we determine firm performance. One of the most general forms of the problem is revealed in the following fundamental questions:

1. What is the total performance of any business?

While an answer to this question will be the result of much research and critical thinking in the disciplines of business, we shall argue that it is in fact a tractable question. VCST suggests that any business generates effects on its customers, suppliers, employees, shareholders, and communities. We might state this as:

TVC = Total Value Created
CU = Total Value Created for Customers
SU = Total Value Created for Suppliers
FI = Total Value Created for Financiers
EM = Total Value Created for Employees
CO = Total Value Created for Communities.

Thus:

$$\text{TVC} = f(\text{CU},\ \ \text{SU},\ \ \text{FI},\ \ \text{EM},\ \ \text{CO})$$

[3] This section is based on T. Jones and E. Freeman (2013), "Sustainable Wealth Creation," working paper at University of Washington and University of Virginia.

where "*f*" is a complex function that combines the value created for each stakeholder into a measure of total value creation. One way to specify "*f*" would be through the use of normal accounting and financial data. The well-documented problem here is that many externalities are not captured in the conventional systems. Furthermore, conventional accounting systems are aimed at securing data for investors where some measures like profitability (or Free Cash Flow, or EBITDA, etc.) may capture well enough the value created for shareholders. We might think about "*f*" with respect to a particular issue or decision. For each decision that a company faces we could think about how that decision creates value for particular stakeholders. The difficulty, in the real world, is that unlike in neat theoretical models, decisions do not often come in such discrete packages, and even if they did, most stakeholders have a sense of history and a sense of the future.

2. What are the conditions for specifying the total performance function "*f*"?

Another and more general way to think about "*f*" is that it is a "business model" that takes the interests of each stakeholder and puts them together in a way that creates (or destroys) wealth for the entire set. In such business models, "*f*" is seen in relational terms. It is about how a business, over time, creates value for its stakeholders. This idea is partially captured in Dyer and Singh (1998) but only to the extent that resources are seen as existing over time. These authors also presuppose that pursuit of profits is best, and that profits measure total firm performance.

Once we begin to think in relational terms, it is easier to see that stakeholder interests are joint. Obviously, we can improve value for financiers by improving value for customers. When a supplier suggests more efficient means of supply chain management, it can improve value for customers, suppliers, employees, and financiers. When these efficiencies also create a less potent waste stream or a better transportation plan, value can be created for communities as well. This gives rise to a third question.

3. How is each of the terms of "*f*" related to the others?

We might express this jointness relation in the following way.

$$\mathrm{CU} = g\left(\mathrm{SU}, \quad \mathrm{FI}, \quad \mathrm{EM}, \quad \mathrm{CO}, \quad \mathrm{X}\right)$$

where "*g*" determines how the interests of suppliers, shareholders, employees, and communities interact with the interests of customers, and X is an interaction term that is not fully captured in the total value created for the other stakeholders. There are similar equations for the total value created for each of the other stakeholder groups, leading to the following question.

4. How do we measure the total value created for each stakeholder?

The total value created for each stakeholder is best expressed as a multivariate function that includes variables that have meaning for stakeholders and that can potentially be measured. Hence, for example:

$$CU = h(CU1, \quad CU2, \quad \ldots, \quad CUn)$$

where "CU1" might be a standard variable like *price*, and "CU2" might be a variable that measures *quality*, and "CU3" might be a variable that measures *fairness in the relationship*, and so on. There are no boundaries on the kinds of variables that a stakeholder may use to determine the value created. Ultimately, we may want to rethink the standard system of accounts to reflect the actual total performance of a company, but our task here is to set out a "possibility argument."

This general argument is not as far-fetched as the conventional mechanisms for measuring corporate performance might indicate. Every serious marketing company that we know spends a great deal of time and money trying to figure out the function "CU." Many companies try to determine the function "EM" as well. And, of course it is assumed that some notion of profitability measures the function "FI." What is less pedestrian is the totality of value created, TVC, or the function "*f*" that we have designated the business model.

Even with this skeletal notation, we can set forth at least three different kinds of business models, or variations of "*f*," that will help us understand how total wealth creation can be reinterpreted in terms of value creation for stakeholders.

First, we can increase TVC by increasing the value to some stakeholders and decreasing the value to others. Such "trade-off functions" represent one major way of thinking about value creation, such as Kaldor functions (Jones et al. 2016). A second method would be to improve value for one group and not decrease the value for others, such as Pareto functions. Or we can increase TVC by increasing the value for each stakeholder. The key variable here is the imagination of the participants in the value creation process. Given the set of alternatives at a particular time, it may not be possible to find a solution that increases value to all stakeholders, or even a Pareto solution. When this occurs, the participants need to reimagine the space of outcomes. This can happen in several ways.

First, each participant may well redefine what counts as value creation. Perhaps there is another way to interpret, for instance, CU so that a new aspect of CU, say CUi, creates value in a way that was unseen. Indeed, Priem (2007) discusses ways to discover unrecognized value with respect to customers. Figure 16.1a suggests that there are many paths to the trade-off frontier which create value for both X and Y. As we get closer to the tradeoff frontier, the possibilities for creating value for both X and Y diminish; creating value for one at the expense of the other becomes increasingly necessary.[4] Obviously, on the actual frontier, X gains only at the expense of Y and vice versa; we called these wealth transfers above. In Fig. 16.1b, stakeholders X and Y have figured out a way to create even more value and avoid a frontier trade-off. In the real world, such redefinitions and reimaginings happen all the time, and sometimes getting close to uncomfortable trade-offs that are

[4]The spirit of this move is what Hayek (1996) and the other members of the "Austrian School" mean when they suggest that an economy is never "in equilibrium" and that the best we can say is that it moves towards equilibrium.

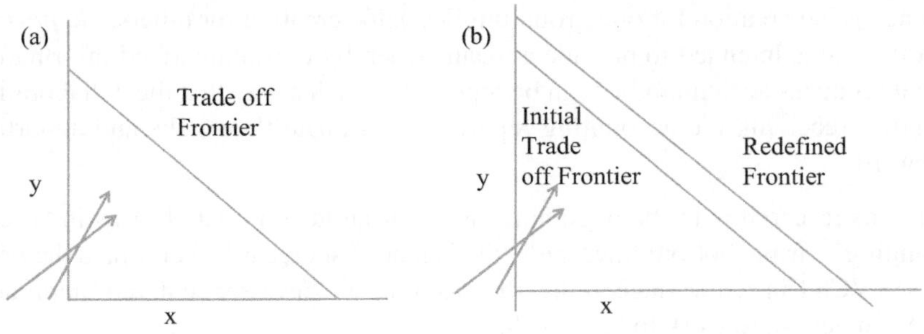

Fig. 16.1 (**a**) Getting to Trade-offs. (**b**) Redefining Trade-offs

unacceptable to both parties has the effect of producing new thinking that creates more wealth for both. Simply accepting the priority of one particular set of interests over another does not create as much wealth as is possible.

5 The Stakeholder Accounting Challenge

This first challenge, or trying to understand how to state and study the total performance of a business, naturally leads to another challenge. The very data that we rely on to study businesses is in part infected with the old narrative, rather than VCST. Hence we have an accounting challenge as well.

Mitchell et al. (2015) edited a special issue of *The Journal of Management Studies* to address this issue. Mitchell et al. (2015) propose replacing the standard "entity convention" in accounting with a concept known as the "proprietary convention" (Goldberg 1965; Riahi-Belkaoui 2004). The proprietary convention is a generally accepted method for partnership accounting, and they argue that this has a better fit with the idea that stakeholder interests are joint and that value is jointly created by them. They suggest that such a convention can be used to include more stakeholders "in the accounting records of the firm because essentially, *partnership accounting permits share of ownership and share of distribution to be different*, a quality that has dramatic ramifications for the practicality of stakeholder accounting."

Mitchell et al. (2015) argue that there are four fundamental premises that must be reflected in any method of stakeholder accounting that is developed. These premises encapsulate the following ideas:

1. Business is an activity among multiple stakeholders that creates value and sometimes destroys it for these stakeholders.
2. Stakeholder interests need to be harmonized (Mitchell et al. call this the alignment premise), so that activities can simultaneously make all better off.
3. The interaction of ideas such as purpose, innovation, and ethics must be a part of the accounting system.

4. Since value creation for one group implies value creation for others, "*Reporting* at its core is intended to provide a means whereby the summarized information that produces accountability can be reported in such a way that the collaborating parties receiving the accounting reports can evaluate their risks and apportion rewards."

While this research is in the beginning stages, it builds on related ideas in "social accounting," "triple bottom line," and "the balanced scorecard," but it puts the principles of VCST in center stage, rather than forming an "investor centered" approach. There is much more work to be done here.

6 The Behavioral Stakeholder Theory Challenge

The current narrative about business is filled with models of investor and manager behavior. There is a growing trend to study the actual behavior of investors, in behavioral economics and finance. We need a similar set of studies around other stakeholder behavior. While there is some work in marketing and management, there is relatively little work that adds "ethics and values" explicitly into the equation. Wouldn't it be interesting to study consumers as if they were moral agents? How would we understand their behavior as moral agents? What about customers who are also employees and owners? Is their behavior different from those who are customers only?

There are numerous questions that come to mind when we add the human/moral element and the idea of stakeholders and multiple stakeholder roles to behavioral theory. My colleagues, Adrian Keevil at PlusTick Partners, Bidhan Parmar at Virginia, and Kirsten Martin at George Washington University, as well as many others are undertaking many studies that use stakeholder theory as an underpinning/unit of analysis for this new behavioral approach. Just as the groundbreaking work of economists and psychologists such as Daniel Kahneman and Thomas Schelling has yielded many insights into real human behavior in economics, these theorists will also build a better version of stakeholder theory.

One project which has begun to show some results is that on public trust. After an initial report done for the Business Roundtable Institute for Corporate Ethics,[5] a number of studies were designed to study whether or not there was a difference in the public trust of business and stakeholder trust of business. For instance, the public may well distrust business as an institution, but if you ask whether or not they trust the businesses that they do business with in some stakeholder role, the answer may well be different.

This challenge to stakeholder theory is part of a larger challenge to reorient and redefine the disciplines of business in more stakeholder-oriented terms. If we eschew the separation fallacy, we can ask questions such as: What is the role of

[5] This report is available at http://www.corporate-ethics.org/

finance in society? How would marketing change if we saw customers as moral agents and brands as promises? What we do differently in operations management if we took a more human view of human beings? There is much work on these questions, and the gradual construction of a new narrative of business will highlight many long forgotten and useful ideas.

7 The Public Policy Challenge

The recent global financial crisis has generated a groundswell of public and political opinion about the state of business around the globe. Questions include:

What is the proper role of government in regulating business? What should be done about regulatory failure? Is there too much reliance on free markets in industries as diverse as autos and health care? How should executive pay, especially for bankers, be regulated? Are we seeing a foundational shift from a reliance on free markets to a more socialist view of society?

Clearly there are multiple causes of the global financial crisis, just as there will be regulatory reform and a backlash of public distrust of business. Unfortunately, the public debate and the political action that will surely result are both based on a faulty understanding of the essence of business and capitalism. This old and shop-worn model of business suggests that the secret to capitalism is that business people can maximize the returns to the owners of capital. And, by focusing on such maximization, they will lead to the greatest good for society. Capitalism works on this view because of the self-interest of economic actors, and their desire to compete and win in the battle that is business. Successful businesses, and business people, act in their own self-interest, and with proper incentives, maximize the returns to shareholders, or other financiers, and they seek to beat their competitors.

Indeed, it is this very faulty theory that led the world to the brink of financial collapse. Trying to patch it up with regulatory reform, or defining "proper incentives," is, at best, an exercise in futility, and at worse, likely to lead to even further damage. It is an arguable point that the last response to crisis, at least in the United States, the Sarbanes–Oxley Act, was partially responsible for the current one. Escaping this cycle of futility should be a major concern for business theorists and policy makers. While that is not likely to occur, I do want to argue that scholars working in the area of business and society should contribute to this important discussion.

One potential project in this area could be called "the business models project." The purpose of this project would be to examine what companies say about their stakeholder relationships in their public statements. In other words, do companies actually claim to maximize value for shareholders or do they claim to create value for stakeholders. Of course even if the latter is true they may be engaging in bad faith or self-deception, but such a demonstration may well shift the burden of proof to those who claim the dominant narrative as fact rather than ideology.

A second project in this area could examine executive compensation, which is an issue that triggers public distrust. If we take the stakeholder idea seriously, and if we work on the first challenge, we might begin to conceptualize executive compensation along the lines of "stake options" (Freeman 2016). Such an option would be at least a five-term function that measured how well a company was doing with customers, suppliers, employees, communities, and financiers. For simplicity's sake assume that each measure is a satisfaction measure (in reality, it would be more complex). Then, an option would vest when, for instance, employee satisfaction reached a certain level. Even if the initial implementation of stake options were perceptual measures that could be tracked and improved over time, we would have alignment between creating value for stakeholders and executive compensation. There could also be a secondary market (at least a Las Vegas style one) for these options.

We need to come to see governments and civil society institutions as part of the value creation process. Of course, government plays a redistributive role and a role as referee, but it can also play a role in facilitating value creation. For instance, educational campaigns on preventing smoking add a lot of value to society for very little costs. A similar campaign to build a generation of entrepreneurs would boost our ability to create value for each other. In addition, focusing on infrastructure and innovation can lead to facilitating a great deal of value creation. We need a new narrative about the role of government as facilitator of value creation in addition to its traditional roles as redistributor and regulator. Again, there is much work to be done here.

8 The Ethical Theory Challenge[6]

Once you say stakeholders are persons, then the ideas of ethics are automatically applicable. However you interpret the idea of "stakeholders," you must pay attention to the effects of your actions on others. There are at least three main arguments for adopting a managing for stakeholders approach. Philosophers will see these as connected to the three main approaches to ethical theory that have developed historically. Some philosophers have argued that the stakeholder approach is in need of a "normative justification." To the extent that this phrase has any meaning, we take it as a call to connect the logic of managing for stakeholders with more traditional ethical theory. As pragmatists, we eschew the "descriptive vs. normative vs. instrumental" distinction that so many business thinkers (and stakeholder theorists) have adopted. Managing for stakeholders is inherently a narrative or story that is at once: *descriptive* of how some businesses do act; *aspirational* and *normative* about how they could and should act; *instrumental* in terms of what means lead to what ends;

[6] The material in this section is from Freeman (2009). The author holds the copyright on this paper so it is partially reprinted here with permission.

and *managerial* in that it must be coherent on all of these dimensions and actually guide executive action. We shall briefly set forth sketches of these arguments, and then suggest that there is a more powerful fourth argument.

9 The Argument from Consequences

A number of theorists have argued that the main reason that the dominant model of managing for shareholders is a good idea is that it leads to the best consequences for all. Typically, these arguments invoke Adam Smith's idea of the invisible hand, whereby each business actor pursues her own self-interest and the greatest good of all actually emerges. The problem with this argument is that we now know, with modern general equilibrium economics, that the argument only works under very specialized conditions that seldom describe the real world. And further, we know that if the economic conditions get very close to those needed to produce the greatest good, there is no guarantee that the greatest good will actually result.

Managing for stakeholders may actually produce better consequences for all stakeholders because it recognizes that stakeholder interests are joint. If one stakeholder pursues its interests at the expense of all the others, then the others will either withdraw their support, or look to create another network of stakeholder value creation. This is not to say that there are not times when one stakeholder will benefit at the expense of others, but if this happens continuously over time, then in a relatively free society, stakeholders will either (1) exit to form a new stakeholder network that satisfies their needs; (2) use the political process to constrain the offending stakeholder; or (3) invent some other form of activity to satisfy their particular needs (Harting et al. 2006; Velamuri 2002; Venkataraman 2002).

Alternatively, if we think about stakeholders engaged in a series of bargains among themselves, then we would expect that as individual stakeholders recognized their joint interests, and made good decisions based on these interests, better consequences would result, than if they each narrowly pursued their individual self-interests. Sometimes there are trade-offs and situations that economists would call "prisoner's dilemma" but these are not the paradigmatic cases, or if they are, we seem to solve them routinely, as Russell Hardin (1998) has suggested.

Now it may be objected that such an approach ignores "social consequences" or "consequences to society," and hence, we need a concept of "corporate social responsibility" to mitigate these effects. This objection is a vestigial limb of the dominant model. Since the only effects, on that view, were economic effects, we need to think about "social consequences" or "corporate social responsibility." However, if stakeholder relationships are understood to be fully embedded in morality, there is no need for an idea like corporate social responsibility. We can replace it with "corporate stakeholder responsibility" which is a dominant feature of managing for stakeholders.

10 The Argument from Rights

The dominant story gives property rights in the corporation exclusively to shareholders, and the natural question arises about the rights of other stakeholders who are affected. One way to understand managing for stakeholders is that it takes this question of rights, seriously. If you believe that rights make sense, and further that if one person has a right to X then all persons have a right to X, it is just much easier to think about these issues using a stakeholder approach. For instance, while shareholders may well have property rights, these rights are not absolute, and should not be seen as such. Shareholders may not use their property to abridge the rights of others. For instance, shareholders and their agents, managers, may not use corporate property to violate the right to life of others. One way to understand managing for stakeholders is that it assumes that stakeholders have some rights. Now it is notoriously difficult to parse the idea of "rights." But, if executives take managing for stakeholders seriously, they will automatically think about what is owed to customers, suppliers, employees, financiers, and communities, in virtue of their stake, and in virtue of their basic humanity.

11 The Argument from Character

One of the strongest arguments for managing for stakeholders is that it asks executives and entrepreneurs to consider the question of what kind of company they want to create and build. The answer to this question will be in large part an issue of character. Aspiration matters. The business virtues of efficiency, fairness, respect, integrity, keeping commitments, and others are all critical in being successful at creating value for stakeholders. These virtues are simply absent when we think only about the dominant model and its sole reliance on a narrow economic logic.

If we frame the central question of management as "how do we create value for shareholders?," then the only virtue that emerges is one of loyalty to the interests of shareholders. However if we frame the central question more broadly as "how do we create and sustain the creation of value for stakeholders?" or "how do we get stakeholder interests all going in the same direction?," then it is easy to see how many of the other virtues are relevant. Taking a stakeholder approach helps people decide how companies can contribute to their well-being and kinds of lives they want to lead. By making ethics explicit and building it into the basic way we think about business, we avoid a situation of bad faith and self-deception.

12 The Pragmatist's Argument

Much of modern business ethics is built on the traditional "big three" theories in the Anglo/analytic tradition in philosophy. Consequentialism, deontology, and virtue theory have colonized much of what philosophers have written about the problem of the ethics of capitalism. Happily there are signs of change as many management theorists are becoming engaged in the project, and some so-called "continental philosophers" have added their voice, especially scholars such as Mollie Painter Morland and Rene Ten Bos. And, a more careful look at the impressive record of Robert Solomon makes these distinctions very difficult to continue to hold.

As many modern pragmatists such as Richard Rorty have suggested, the so-called Big Three of ethical theories are at best partial and often not very useful tools for ethical thinking and action. I want to take this argument further. The previous three arguments point out important reasons for adopting a new story about business. Human beings have been value creators and traders for millennia; however, there is very little mention of "business" in most of our ethical and political theory. What if the first question of political philosophy were "how is value creation and trade sustainable over time?" rather than "how is the state justified?" What if we focused on how the narrative of stakeholder theory could affect our conception of the self? These and similar questions need to be addressed. Too much of the current writing adopts the Separation Fallacy in the form of "the business sucks story." There is an assumption that business is somehow disconnected from morality both in practice and in theory. Nothing short of a wholesale revision of ethical theory will begin to address this problem. Thankfully, there is a place to start in pragmatism, even though these theorists, like many of their post-modern and critical studies colleagues, in fact adopt the business sucks narrative.

There is a long tradition of pragmatist ethics dating to philosophers such as William James and John Dewey. More recently philosopher Richard Rorty has expressed the pragmatist ideal (Mendieta 2006, p. 68):

> [...] pragmatists [...] hope instead that human beings will come to enjoy more money, more free time, and greater social equality, and also that they will develop more empathy, more ability to put themselves in the shoes of others. We hope that human beings will behave more decently toward one another as their standard of living improves.

Pragmatists want to know how we can live better, how we can create both ourselves and our communities in such a way that values such as freedom and solidarity are present in our everyday lives to the maximal extent. While it is sometimes useful to think about consequences, rights, and character in isolation, in reality our lives are richer if we can have a conversation about how to live together better. By building into the very conceptual framework we use to think about business a concern with freedom, equality, consequences, decency, shared purpose, and paying attention to all of the effects of how we create value for each other, we can make business a human institution, and perhaps remake it in a way that sustains us.

For the pragmatist, business (and capitalism) has evolved as a social practice; and an important one that we use to create value and trade with each other. On this view, first and foremost, business is about collaboration. Of course, in a free society, stakeholders are free to form competing networks. But, the fuel for capitalism is our desire to create something of value, and to create it for ourselves and others. The spirit of capitalism is the spirit of individual achievement, together with the spirit of accomplishing great tasks in collaboration with others. Managing for stakeholders makes this plain so that we can get about the business of creating better selves and better communities.

In general, we need to end the hegemony of the economic person that is so prevalent in business theory, and that of the political/ethical person that is prevalent in ethics and political theory. In fact, human beings are pretty complicated. We use reason and we have values. We have emotions and we have a history. We are relational, sexual, spiritual, and social, as well as individual. We have aspirations for ourselves and others, especially our families. We want to be a part of something larger than ourselves to create some sense of purpose. We want to master our environment and we want autonomy and connection. We want to live in an authentic way, but such authenticity is a project, as Sartre, de Beauvoir, and their followers remind us. We cannot forget about the constant specter of bad faith and self-deception. Business ethics needs to look less like Kant and more like Foucault; less like Maslow and more like Freud and his followers; less like general equilibrium theory and more like Amartya Sen's *The Idea of Justice*; less like the abstract theory of journal articles and more like the nuanced reporting of Michael Lewis. We need new theory, but it must be based on some ideas that at once are descriptive and aspirational for the way we create value and trade with each other. I want to suggest that putting stakeholder theory at the center of this revision is one way (not the only one) forward.

13 Conclusion

There are a number of challenges to the ideas in what has come to be called stakeholder theory, and there are existing lines of research to garner insight into these challenges. We need a much more nuanced view of how to measure total performance and how to account for the variety of stakeholder relationships that exist for twenty-first-century businesses. We need to understand how real stakeholder relationships evolve, the causes of actual behavior, as well as the mechanisms for a public policy that encourages business people to create organizations that make our world better. Finally, we must address the fundamental challenge of doing ethics in a different way. Ethics is fundamentally the conversation we have about how we are going to live together. Stakeholder theory is a central part of that conversation just as ethics is a central part of stakeholder theory. There is much work to be done.

Acknowledgments I have a number of co-authors and colleagues who have helped me with these ideas, and I have drawn heavily on a number of joint papers with them. I would like to thank them for permission to develop these ideas in the current form. In particular, thanks to Tom Jones of University of Washington, Michelle Greenwood of Monash University, Harry van Buren of University of New Mexico, Ron Mitchell of Texas Tech University, Adrian Keevil of PlusTick Partners, Rob Phillips of University of Richmond, and Bidhan Parmar at Darden. Portions of the Introduction and the section on Ethical Challenges are from Freeman (2009), copyright held by the author.

References

Braudel, F. 1992. *Civilization and the History of Capitalism, 15th to 18th Century*. Berkeley: University of California Press.

Dyer, J.H., and H. Singh. 1998. The Relational View. *Academy of Management Review* 23 (4): 660–679.

Freeman, R.E. 2009. Managing for Stakeholders. In *Ethical Theory and Business*, ed. N. Bowie, T. Beauchamp, and D. Arnold, 8th ed., 56–67. Englewood Cliffs: Prentice Hall.

———. 2011. Some Thoughts on the Development of Stakeholder Theory. In *Stakeholder Theory: 25 Years Later*, ed. R. Phillips, 212–233. Cheltenham: Edward Elgar Publishing.

———. 2016. Short Term vs. Long Term: A Skeptical View and an Alternative. *Boards and Directors, Annual Review* 40 (4): 83–84.

Freeman, R.E., J.S. Harrison, and A.C. Wicks. 2007. *Managing for Stakeholders: Survival, Reputation, and Success*. New Haven: Yale University Press.

Freeman, R.E., J.S. Harrison, A.C. Wicks, B.L. Parmar, and S. de Colle. 2010. *Stakeholder Theory: The State of the Art*. Cambridge: Cambridge University Press.

Goldberg, L. 1965. *An Inquiry into the Nature of Accounting*, American Accounting Association Monograph No. 7. Menasha: George Banta Company, Inc.

Hardin, R. 1998. *Morality Within the Limits of Reason*. Chicago: University of Chicago Press.

Harting, T., S. Harmeling, and S. Venkataraman. 2006. Innovative Stakeholder Relations: When "Ethics Pays" (and When it Doesn't). *Business Ethics Quarterly* 16: 43–68.

Hayek, F. 1996. *Individualism and Economic Order*. Chicago: University of Chicago Press.

Jones, T., and R.E. Freeman. 2013. *Sustainable Wealth Creation*, Working Paper. University of Washington/University of Virginia.

Jones, T., T. Donaldson, R.E. Freeman, J. Harrison, C. Leana, J. Mahoney, et al. 2016. Management Theory and Social Welfare: Contributions, Extensions, and Challenges. *Academy of Management Review* 41 (2): 216–228.

Mendieta, E., ed. 2006. *Take Care of Freedom and Truth Will Take Care of Itself: Interviews with Richard Rorty*. Stanford: Stanford University Press.

Mitchell, R., M. Greenwood, H. van Buren, and R.E. Freeman. 2015. Stakeholder Inclusion and Accounting for Stakeholders. *Journal of Management Studies* 52 (7): 851–877.

Phillips, R. 2003. *Stakeholder Theory and Organizational Ethics*. San Francisco: Berrett-Koehler Publishers.

Priem, R. 2007. A Consumer Perspective on Value Creation. *Academy of Management Review* 32 (1): 219–235.

Riahi-Belkaoui, A. 2004. *Accounting Theory*. New York: Thompson.

Stout, L. 2012. *The Shareholder Value Myth*. San Francisco: Berrett-Koehler Publishers.

Velamuri, S.R. 2002. Entrepreneurship, Altruism, and the Good Society. *Ethics and Entrepreneurship, The Ruffin Series* 3: 125–143.

Venkataraman, S. 2002. Stakeholder Value Equilibration and the Entrepreneurial Process. *Ethics and Entrepreneurship, The Ruffin Series* 3: 45–57.

R. Edward Freeman is University Professor, Elis and Signe Olsson Professor, Academic Director of the Institute for Business in Society, and Senior Fellow of the Olsson Center for Applied Ethics at the University of Virginia's Darden School of Business.

Chapter 17
A Puzzle About Business Ethics

R. Edward Freeman and Gordon G. Sollars

1 Part I

The following is a thought experiment. We believe that it can illustrate some of the strengths and weaknesses of the current body of research called "business ethics". In Section II we give our own response to the thought experiment and the possibilities therein. Consider the following:

1. Suppose that a person, L, buys some raw material from a supplier and produces a product, P, that is, L and only L mixes her labor with the raw material to produce P. Suppose further that L will sell this product only to people who agree to the following condition:

   (RP) I, L, stand responsible for the effects of this product. If it doesn't do what I claim that it does, then I will compensate you for your trouble, as determined by an independent third party. I stand liable for any harmful effects or violation of rights that may result from the intended uses of this product.

Originally published in: *Business Ethics, the Environment & Responsibility,* 31, 272–273 © Wiley, 2021
Reprint by Springer, DOI 10.1111/beer.12387

R. E. Freeman (✉)
University of Virginia, Charlottesville, VA, USA
e-mail: FreemanE@darden.virginia.edu

G. G. Sollars
Fairleigh Dickinson University, Teaneck, NJ, USA

© Springer Nature Switzerland AG 2023

S. D. Dmytriyev, R. E. Freeman (eds.), *R. Edward Freeman's Selected Works on Stakeholder Theory and Business Ethics*, Issues in Business Ethics 53,
https://doi.org/10.1007/978-3-031-04564-6_17

Unless you are willing to accept (RP), L will not sell P to you. Furthermore for each transaction L does her level best to be sure that you understand (RP) and what options are open to you if P fails to perform on some dimension.[1]

2. Suppose further that L will only buy raw material from a supplier who herself offers (RP) as a condition of sale. Does L have any potential ethical problems? Are there ethical dilemmas that could arise for L?[2]
3. Suppose further that L decides to hire some employees, as the business grows, but does so only if they agree to the following principle:

   (ERP) I, L, stand responsible for the effects of this job on your well-being. If it doesn't affect you in the way that I claim it will, then I will compensate you for your trouble, as determined by an independent third party. I stand liable for any harmful effects or violation of rights that may result from your performance of this job as intended.

For each employee, or potential employee L does her level best to be sure that you understand (ERP) and the options open to you if employment by L is other than as stated.[3]

4. Suppose further that the business grows, and L must move it out of her house and garage. L decides to build a plant and begin operations of a larger company, call it UL, in community C. However, L will do so only if representatives of the community agree to the following principle:

   (CRP) I, L, stand responsible for the effects of this operation on the community. If it doesn't affect the community in the way that I claim it will, then I will compensate the community for its trouble, as determined by an independent third party. I stand liable for any harmful effects or violations of rights that may result from the normal operations of this facility. I stand liable for ensuring that the facility operate in a normal way.

For each plant that L opens, she does her level best to be sure that community representatives understand and agree to (CRP) and that they understand the options open to them should L not live up to (CRP).[4]

---

[1] L's supplier K may set the same condition. Should a harm or rights violation occur involving a consumer of L's product, the liability would have to be apportioned between K and L. We do not consider the details of such apportionment here. Such problems are endemic in complicated schemes of causality.

[2] L could of course try to cheat her supplier, but this case collapses into the case considered *supra* note 1.

[3] There may be a fair amount of ambiguity in the tasks for which L hires you, but if at any time you believe that your well-being is affected adversely, then L is committed to compensating you, or simply re-negotiating the terms of your employment.

[4] We leave as an open question whether or not the local government is a legitimate representative of the community. This problem affects all of "business ethics" not just the conception that is implicit here. Surely there is no reason for a separate field of inquiry "business ethics" because a particular regime doesn't take the interests of its members into account in an acceptable manner. If this turns out to be the main "raison d'être" of business ethics, perhaps we need a more robust political philosophy which understands the value-creation process in more pragmatic terms.

5. Suppose further that as the business grows L needs new sources of financing in order to expand and meet demand for P. L decides to sell shares in UL. However, as you have by now cleverly guessed, there is one condition:

   (IRP) By accepting the ownership of shares of UL, we, L and the shareholders, agree to abide by (RP), (ERP), and (CRP). We, the shareholders, agree to sharing with L the joint liability of these agreements. In return we have a claim (proportionate to our shares) on the residual returns from the operations of UL.

2 Part II

Our first set of questions is obvious. What kinds of potential ethical problems does L have? Are there any ethical dilemmas that could arise for L?[5] Does L need a more complex theory of business ethics in order to build her company along ethical lines? Should L read the latest issues of the business ethics journals for ideas to make her business more ethically justified? Does the acceptance of the variety of responsibility principles per se, impose potential ethical problems? In particular does L need an ethical "theory of the firm"?

Alternatively, is a strong sense of responsibility sufficient for business ethics? Of course, there could be ethical issues around who counts as an independent third party, or what is an intended use of the product, or even, what actually caused any damage or malfunction since causation is usually joint and multi-variate. We could even see issues around what actually is a harm due to one's job, what is a normal way for a facility to operate or even what actually are residual returns. These questions could give rise to ethical dilemmas, but they would hardly constitute the need for any new theoretical material for a discipline called "business ethics" unless we mean by that term these particular kinds of issues. On this view business ethics would have a fairly limited scope. We would not need to do much moral theorizing if in fact L and her stakeholders acted on the various versions of the Responsibility Principle outlined above. While L has chosen a version of the Responsibility Principle that entails a kind of strict liability, we believe that there are other versions that may encompass a different view of responsibility, and that at least is a theoretical and ethical issue. In order to answer these questions posed by such a strong sense of responsibility we would need a deep knowledge of real businesses to solve any problems that arose. We are not suggesting that this thought experiment makes business ethics superfluous. Rather we are suggesting that business ethics needs a strong connection to the nuances and subtleties of real businesses. "Business" is perhaps a family resemblance idea, so that there are a multitude of different kinds of business, and perhaps a host of different ways to run an ethical business. What many

[5] Of course others could try to take advantage of L, and L's guarantee of (RP). But, in these cases we would be hesitant to say that L's business has an ethical dilemma. Maybe L would turn out to be a sucker, but that is a different matter.

of these ideas may have in common is some shared sense that whatever the business does, it needs to be responsible for the effects of its actions on its stakeholders.

Business ethics as it exists today is a curious species. Its niche exists in part because the very idea of a business has been divorced from moral discourse. And the moral discourse of business has been equally divorced from the practices of business. And, those who opine about business ethics theory are often quite disconnected from the actual practice of business. Indeed, the denial of the relevance of the very simplest moral notion, that we are responsible for the effects of our actions on others (the chief feature of what has come to be called "stakeholder theory"), seems to be at the heart of the problem. Existentialists and libertarians, Rawlsians and socialists, all agree that some version of a principle of responsibility gets ethics off the ground. It is possible to deny all of the versions of this principle, but then as Hazel Barnes tells us, we are choosing not to adopt an ethical point of view.[6] On such a choice, to bring ethics in via "business ethics" is a move that can only result in bad faith.

3 Data Availability Statement

There is no data available for this conceptual paper. It is a thought experiment.

R. Edward Freeman is University Professor, Elis and Signe Olsson Professor, Academic Director of the Institute for Business in Society, and Senior Fellow of the Olsson Center for Applied Ethics at the University of Virginia's Darden School of Business.

Gordon G. Sollars is Emeritus Associate Professor of Management in the Silberman College of Business at Fairleigh Dickinson University.

[6] Hazel Barnes, *An Existentialist Ethic*, Chicago: University of Chicago Press.

Part II
Business Ethics and Humanities

Chapter 18
Orwell and Organizations

R. Edward Freeman and Daniel R. Gilbert, Jr.

> *"Madness is something rare in individuals—but in groups, parties, peoples, ages it is the rule."*
>
> —Nietzsche

1 Introduction

All of the hoopla of 1984 in 1984 is over. We have breathed a collective sigh of relief that things are not so bad, not nearly so bad as Orwell would have had us believe. Big Brother has not come. Individualism still reigns in the land of the free and the home of the brave. Orwell gave us a real scare, and it is comforting to know that our institutions and our brethren are safe for the time being. Or are they?

We shall reexamine Orwell's doctrine in light of recent developments in the theory and practice of management of large organizations. We shall argue that Orwell's warnings loom large for both theorists and practitioners of management or

---

A Note to Gentle Readers
This essay is rather different in content and style from the usual academic article on management and organization, and at least part of its message lies in just that difference. Reading it could well turn out to be hazardous. We recommend appropriate caveats and caution.

Original Work
Beyond Good and Evil (Friedrich Nietzsche, Beyond Good and Evil, 1973 edition, R. J. Hollingdale (tr.), London: Penguin Books, p. 85.)

R. E. Freeman (✉)
University of Virginia, Charlottesville, VA, USA
e-mail: FreemanE@darden.virginia.edu

D. R. Gilbert, Jr.
Gettysburg College, Gettysburg, PA, USA

© Springer Nature Switzerland AG 2023

S. D. Dmytriyev, R. E. Freeman (eds.), *R. Edward Freeman's Selected Works on Stakeholder Theory and Business Ethics*, Issues in Business Ethics 53,
https://doi.org/10.1007/978-3-031-04564-6_18

organization theory or administrative science or whatever you like to call it. The seeds of totalitarianism are indeed alive and are much closer to home than the "evil empires" of Eurasia, Eastasia, and the Soviet Union. They are nurtured in the company songs, flags and uniforms, the shared values, corporate culture, mission statements, ethical codes, Japanese management, one-minute management, and all of the other magic that has recently been made.

Orwell, more so than his commentators and critics, recognized that the brutality of collectivism is rooted within each of us. It is not due to social or corporate structure, but rather, it is a part of our view of the human person. This is the story we shall tell.

2 Different Reading

Orwell's 1984 has traditionally been interpreted as a commentary on the dangers of socialism, or at least the possibility of the Leviathan state totally out of the control of its citizens. Despite Orwell's dissociation of the book with the socialist movement, 1984 has become accepted and internalized as part of our way of looking at the world, so much so that Mark Crispin Miller has argued that the impact of the book is lessened. Miller urges us to get outside the text and view it anew "to stand outside this spreading dream and read it".[1]

There is a different way to read 1984 and the Orwell corpus, for Orwell's view of the human person is independent of the political statement of 1984. By understanding this view of the person the warnings of 1984 and Animal Farm become all too real.

There are three interwoven strands in Orwell's view of the person: (1) the passivity of the ordinary person; (2) the need for affiliation; and, (3) the problem of power and authority. We shall take each in turn.

The Christian metaphor for man is sheep, in search of the Good Shepherd. Anyone with even a passing acquaintance with sheep understands the full depth and range of this insult. It is not that good shepherds don't lead well, it is the utter stupidity of the sheep. A complete single-track, hard-wired mind that looks at mouthful after mouthful, the sheep is the ultimate passive creature. When threatened by anything, from wolf to man to Little Bo Peep, the first and only instinct is to herd together. There is no individuation. No entity without identity. A sheep is a sheep is a sheep. There are no professional sports teams named "The Sheep". (Do not be fooled by the machismo of "Rams".)

Even the word "sheep" reflects this passivity. There is no difference between the singular and plural. The words recognize that a sheep is a sheep are sheep. What does all of this talk about sheep have to do with Orwell?

[1] Mark Crispin Miller, "The Fate of 1984", in I. Howe (ed.) 1984 Revisited: Totalitarianism in Our Century, New York: Harper and Row, pp. 19–46.

Orwell recognized the utter passivity of humans. From the inner party members to the proles, Oceanas simply let things happen to them. Even O'Brien will someday be vaporized. Perhaps Winston is in control within the confines of the corner where the telescreen does not peek, or in the forest with Julia, but by and large, Winston is a sponge. He absorbs life like water and dirt, and gets it wrung out of him. He sits on the sink and hardens and is used infrequently even by his tormentors.

Orwell has put it succinctly in "Inside the Whale":[2]

> For the ordinary man is also passive. Within a narrow circle (home life, and perhaps the trade union or local politics) he feels himself master of his fate, but against major events he is as helpless as against the elements. So far from endeavouring to influence the future, he simply lies down and lets things happen to him.

We do not know the news until we read it or see it on TV. We are not the least bit worried until Dan Rather is worried. (We are led to believe that the real problems rest with how the press has acted.) We want to know the way it was so that we know what we missed. We will be able to more easily anticipate what we will miss tomorrow. We do not like large surprises.

Indeed Orwell may well go further towards a notion of "active passivity", that is, humans actively seek to let things happen to them. We take steps to insulate ourselves from the world. We become onions. When you peel back all of the layers (nine if you're an onion, but more if you're a human) there is nothing left. (This is dignified by sociologists in something called "role theory".) Modern technology has helped in this process but it is not the cause, just as the telescreens were not the real monitors in 1984. Sheep do not need telescreens, and good shepherds do not have to keep both eyes trained on the flock.

It is unfortunate for Orwell and for us that Jonah did not end up in the belly of a sheep. Allegory is much better than criticism. But, whales are all we have, and Orwell had to make do. He argued that the Jonah impulse was a strong one. Freud may have said that Jonah wanted to return to the womb, to lie down and curl up into the fetal position, and to desperately hope that the world would go away, or at least cut him a very large break. Orwell said it better:[3]

> For the fact is that being inside a whale is a very comfortable, cosy, homelike thought. The historical Jonah, if he can be so called, was glad enough to escape, but in imagination, in daydream, countless people have envied him. It is, of course, quite obvious why. The whale's belly is simply a womb big enough for an adult. There you are, in the dark, cushioned space that exactly fits you, with yards of blubber between yourself and reality, able to keep up an attitude of the completest indifference, no matter what happens. A storm that would sink all the battleships in the world would hardly reach you as an echo. Even the whale's own movements would probably be imperceptible to you. He might be wallowing among the surface waves or shooting down into the blackness of the middle seas (a mile deep, according to Herman Melville), but you would never notice the difference. Short of being dead, it is the final, unsurpassable stage of irresponsibility, (p. 151)

---

[2] G. Orwell "Inside the Whale", in The Collected Essays, Journalism and Letters of George Orwell, edited by Sonia Orwell and Ian Angus, volume 1, p. 127.

[3] Id. at 151.

Short of being dead, it is the final, unsurpassable stage of irresponsibility. The rub is right here, for who cares if humans are sheep, whales, goats or tigers? The distinguishing characteristic of humanity, which Orwell shows us through its absence in Jonah and in Oceania, is choice and responsibility, those passe old saws. Freedom is slavery to sheep and whales. If O'Brien will accept the responsibility Winston will truly see five fingers when O'Brien holds up four. If Milgram's experimenter will accept the responsibility then the subject will deliver 450 volts to a screaming and then silent man with a heart condition.[4] Peeling the onion makes us cry for we are saddened by thought that there may be nothing in the center, and we are frightened by the possibility that there may be something.

You never see a sheep alone. Even in the distant high country of northwestern Scotland where there is literally nothing, except scenery to see, one sheep is never far from another. The instinct of the herd is strong. For us, it is not just the need to be physically near. Mental nearness is often more important, for here we share beliefs and values even when separated by physical distance. The need to believe in something, the need to deny and repress the existential terror that comes from peeling a lone onion, the need to achieve, the need for friendship, fucking and feeling like we belong whether it is to the Junior Spies, the Anti-Sex League, or the Republican Party, all make up the nexus that is the human personality.

Russell looks out on Trafalgar Square and sees the masses dancing in the streets at the prospect of War. Russell sees horror and carnage twice over, once of the war and once of the human spirit that is so joyous.[5] Millions of megatons are poised to destroy us all in half-hour's time, all for the fleeting feeling of elation that comes with seeing the stars and stripes or the hammer and sickle as we are blinded by the fireball and deafened by the roar.

It is our way of coping with onions. From cocktail parties to coffeehouses, from a cruise down the Seine to a cruise in the backseat of a 56 Chevy, from the boy scouts and brownies to "waves" at baseball stadia, Nietsche's madness grips us all. We are not in this alone. We refuse to believe that we are in it alone, that I am in it alone.

There simply is no getting away from it. Orwell despairs that there is no alternative:[6]

> But what do you achieve, after all, by getting rid of such primal things as patriotism and religion? You have not necessarily got rid of the need for something to believe in.....All the loyalties and superstitions that the intellect had seemingly banished could come rushing back under the thinnest of disguises, (p. 144)

Sociologists are not content with ruining the onion by introducing roles. There is also the mask of socialization. We are not deceived into adopting these roles, but we take them on the way we put on a shirt, even though we are pickier about our shirts. We do not question being father, mother, daughter, son, uncle, aunt, solid citizen,

---

[4] S. Milgram, Obedience to Authority, New York: Harper and Row, 1975.

[5] B. Russell, Autobiography, London: Unwin Books, 1975 edition, pp. 240 ff.

[6] Orwell, supra note 3 at 144.

young turk executive, bright young scholar, wise elder, Lutheran, Baptist, Presbyterian, Catholic, Jew, Goy, neighbor, voter, butcher, baker, or candlestick maker. There is a process, or there are processes, by which we adopt roles, and there is no agency checking our references and questioning us on whether we really know what it means to adopt a Vietnamese child.

That sneaking suspicion, that lurking nag, that choice and responsibility are just maybe possible is made easier by the pink rolls of insulation that are others. We can keep out Winter and the Wolf because we are so busy. Roles and socialization consume vast quantities of being.

So Orwell has forced us to confront Plato and Freud and many lesser lights about our passivity and our need for affiliation. So what? Who cares? What's in it for me?

Well, there is this business of the shepherd. The shepherd is a man (and lately a woman) of a thousand and one tales and faces. The shepherd is Andropov, Gorbachev, Reagan, Thatcher, Reverend Falwell, Pope John Paul, Lech Walesea, Joe DiMaggio and Lee Iacocca. The shepherd is obviously religion and patriotism, and less obviously leadership and motivation. The shepherd can visit in the form of Satrean and Randian heroes, Mathieu and Roark. It can be Granny and Uncle Tom. It is roles and socialization. It is economists and sociologists. The shepherd is power.

By believing in something (what could it mean to believe in nothing?) we accept the authority that is our belief. We want beliefs to be true. We eschew false ones, like we pick maggots out of our rice and fleas off the dog. The problem is less which of our beliefs are true (this is problem enough) than which of our "beliefs" do we really believe. Which ones really cause us to act, to strive, to love and to be. Authenticity is presupposed and rarely questioned, if you are a sheep.

The common good, America, the party, for the sake of others, self-sacrifice, the stockholders, the community, all present massive problems to be authenticated. What do we really believe about each? To the extent that 1984 is about the human person, it is about Authenticity. It is about O'Brien and Syme rather than Winston and Julia. O'Brien says "the Party" but he means "O'Brien" in his famous boot-stamping speech:[7]

> The Party seeks power entirely for its own sake. We are not interested in the good of others; we are interested solely in power. Not wealth or luxury or long life or happiness; only power, pure power.

Authenticity begets the possibility of real and forever brutal totalitarianism. Bad Faith doesn't seem so bad, until we remember the sheep and the onions.

Power speaks to us. It trips off our tongue, and comes to rest in the lap of the language that we speak. It cries out for analysis, interpretation, criticism and deconstruction. Power is truly the Word.

Orwell's creation of Newspeak rivals only Walt Disney for the most wonderful world. And, it is a real as Disneyland with the litter, the crowds and the traffic. War is Peace. Freedom is Slavery. Ignorance is Strength. We need go no further. "It is

---

[7] G. Orwell, 1984, New York: Harcourt Brace Jovanovich, Inc., 1961 edition, p. 127.

beautiful thing, -the destruction of words." We do not need megatons or sharp knives to cut out our tongues in order to exercise power. Syme is the real villain:[8]

> How could you have a slogan like 'freedom is slavery' when the concept of freedom has been abolished? The whole climate of thought will be different. In fact there will be no thought, as we understand it now. Orthodoxy means not thinking—not needing to think. Orthodoxy is unconsciousness.

Syme tells us that it is only a matter of self-discipline and reality control. The real terror of power is not in others so much as it is in ourselves. Newspeak, an unreal flight of imagination? We purchase, by coercion of government, no-fault insurance. We can break a bond to which we swear to be authentic—marriage—by no-fault. Try to find a Class D minor league in professional baseball. The "worst" is an A.

What does all this talk about sheep and whales and onions and talk have to do with management and corporations?

3 Modern Organizations and Management

Management and organization is the thing now. Control over the world is an imperative of our age. Enrollments in business schools have skyrocketed. Students are practical. They want jobs. Worse, they want careers. They want to know how to do things, how to make things work, how to get on in the world. They want to know debits from credits. 'Left' and 'right' have acquired new meanings. Every college worth its salt and pepper has acquired a business school. People with real jobs have enrolled in MBA programs, and 50,000 new MBAs roll off the press each year. Business is booming.

There is a great change in the business schools themselves. They are discovering scholarship. After years of writing cases and stories, accumulating the dirty-fingernailed wisdom of practice, and listening solemnly to old saws whine in the night, business professors are watching "social science" move in for the kill and opening the door at the same time. Onions are a-peeling. No longer do we listen to "Define your business clearly", "the 4 P's of product, price, promotion and place" and "Build on your strength". Wisdom is out. Knowledge is in.

Organizations are now inspected and dissected, like frogs and bosons. Surveys and statistics, methods and models, hermenuetics and interpretation are the current currency. Ph.D. programs abound. There is a new journal a month. Schools are hiring their own philosophers, sociologists, psychologists, historians, and economists to carry the new banner. Tenure fights have gotten even more brutal. We talk about rigor versus relevance, and worry about selling out to the corporate community. Everyone is concerned about paradigms in the way that we wonder whether the groceries will all fit in one bag without the bottom falling out. Will we be able to carry the heavy bag?

---

[8] Id. at 47.

Large corporations are in the spotlight, too. Dan Rather worries us about thousands gassed at the Union Carbide Plant in India. What about those folks who do business in South Africa, who sell infant formula in the Third World, who phony up insurance policies at Equity Funding, who wreck the Penn Central, who kite checks at E.F. Hutton, depose the Shah of Iran, strip mine Western states and other Third World sites, and jump out of the Pan Am building over bananas in Honduras? Bhopal, Soweto, Love Canal, Youngstown, the Mon Valley, and Wall Street are fast replacing the Marne, the Bulge, Bull Run, and San Juan Hill.

Every revolution has its heroes and this one is no exception. Not since the days of John D and Andy C have we seen business persons so loved and hated. T. Boone Pickens, Ted Turner, John DeLorean, Bill Agee and Mary Cunningham, Lee Iaccoca and Charlie Brown grace us with their deeds and misdeeds. This is no comic strip. This is billions of dollars. Millions of lives.

Executives are urged to be socially responsible and ethical, or at least socially responsive and open. Voluntarism is the new Zeitgeist. 5% to charity. 2% to the community. Everyone participates in The United Way. Enlightened self interest.

High tech and pieces of fruit are in, the Japanese are out. We must keep our "competitive edge". The dollar must be strong, investment encouraged, and stockholders coddled. Not since Gerard Swope, CEO of GE, wrote the IRA, have we seen such cooperation and talk of cooperation among businesses and governments. Economic reality is reality, and everyone wants a piece.

Orwell has a lot to say to this flurry of activity. The parallels are striking. We need only replace Big Brother and the Party, Oceania, Eastasia, and Eurasia with our corporate heroes and Exxon, ITT, Sears, and Citicorp. Orwell put it pretty well in his review of Trotskyite James Burnham's book, The Managerial Revolution:[9]

> The real question is not whether the people who wipe their boots on us during the next fifty years are to be called managers, bureaucrats, or politicians: the question is whether capitalism, now obviously doomed, is to give way to oligarchy or to true democracy.

Clearly EXXON is huge. EXXON can devastate a community. EXXON can be totalitarian. EXXON can be real bad. And, we can breathe another collective sigh of relief that Orwell was wrong, that things have not reached the proportions of 1984, that capitalism is not doomed, that democracy and individualism, etc., etc., etc. But, there is another story.

The characters in this story all live and work and sometimes breathe inside EXXON, AT&T, and XYZ. They conform to Orwell's view of them as sheep, at least most of the time. They want to know what to believe. They want to have things settled. They want to have something to believe in.

These are familiar characters in fiction. They have been brilliantly depicted by Sinclair Lewis, Sloan Wilson, Joseph Heller, and many others. Listen to them. First, Heller's Bob Slocum:[10]

---

[9] G. Orwell, "James Burnham and the Managerial Revolution", supra note 3, volume 4, p. 165.

[10] J. Heller, Something Happened, New York: Alfred A. Knopf, 1974, pp. 13–14.

> In the office in which I work there are five people of whom I am afraid. Each of these five people is afraid of four people (excepting overlaps), for a total of twenty, and each of these twenty people is afraid of six people, making a total of one hundred and twenty people who are feared by at least one person. Each of these one hundred and twenty people is afraid of the other one hundred and nineteen, and all of these one hundred and forty-five people are afraid of the twelve men at the top who helped found and build the company and now own and direct it.
>
> All these twelve men are elderly now and drained by time and success of energy and ambition. Many have spent their whole lives here. They seem friendly, slow, and content when I come upon them in the halls (they seem dead) and are always courteous and mute when they ride with others in the public elevators. They no longer work hard. They hold meetings, make promotions, and allow their names to be used on announcements that are prepared and issued by somebody else. Nobody is sure anymore who really runs the company (not even the people who are credited with running it), but the company does run. Sometimes these twelve men at the top work for government for a little while. They don't seem interested in doing much more. Two of them know what I do and recognize me, because I have helped them in the past and they have been kind enough to remember me, although not, I'm sure, by name. They inevitably smile when they see me and say: "How are you?" (I inevitably nod and respond: "Fine".) Since I have little contact with these twelve men at the top and see them seldom, I am not really afraid of them. But most of the people I am afraid of in the company are.

Next, Lewis's Babbitt:[11]

> He [Babbit] saw the years, the brilliant winter days and all the long sweet afternoons which were meant for summery meadows, lost in such brittle pretentiousness. He thought of telephoning about leases, of cajoling men he hated, of making business calls and waiting in dirty anterooms—hat on knee, yawning at fly-specked calendars, being polite to office-boys.
>
> "I don't hardly want to go back to work," he prayed. "I'd like to—I don't know."

But he was back next day, busy and of doubtful temper.

Finally, Wilson's Tom Rath:[12]

> Tom shrugged again. The thing to remember is this, he thought: Hopkins would want me to be honest. But when you come right down to it, why does he hire me? To help him do what he wants to do—obviously that's why any man hires another. And if he finds that I disagree with everything he wants to do, what good am I to him? I should quit if I don't like what he does, but I want to eat, and so, like a half million other guys in gray flannel suits, I'll always pretend to agree, until I get big enough to be honest without being hurt. That's not being crooked, it's just being smart.
>
> But it doesn't make you feel very good, Tom thought. It makes you feel lousy. For the third time, he shrugged. How strangely it all works out, he thought....How smoothly one becomes, not a cheat, exactly, not really a liar, just a man who'll say anything for pay.

The view of the person in these passages is that of one struggling for meaning, asking how and why organizations are important, having the edges dulled by the routine, worrying about what others think, avoiding the old saws of choice and responsibility and coming not to care why.

---

[11] S. Lewis, Babbitt, New York: Harcourt Brace Jovanovich, Inc., 1980 edition, pp. 190–191.

[12] S. Wilson, The Man in the Gray Flannel Suit, New York: Arbor House, 1955, p. 183.

Babbit's boosterism and his good fellowship get him through the days if not the nights. Rath must provide for Betty and the kids if not for himself. He understands grey-flanneled-fitting-in. The coming of age today is business.

For Babbit, Rath and Slocum their existence as organizational members is paradoxical. Organization provides the safety net, the layers of blubber between the self and the rest of the world. Organization becomes the world. There is nothing which is not connected to it. "Being-in-the-world" has become "Being-in-the-organization".

Slocum, poor, tragic Slocum, must face the brutality of loneliness and impotence that comes with the identification of self and organization, the mid-life crisis of not being able to choose anymore. He is hopelessly passive. (So much so, that some critics claimed the book should have been called, "Nothing Happened".) He understands authority, power, and fitting in. He can only wonder:[13]

> What would happen if, deliberately, calmly, with malice aforethought and obvious premeditation, I disobeyed?
>
> I know what would happen: nothing. Nothing would happen. And the knowledge depresses me. Some girl downstairs I never saw before (probably with a bad skin also) would simply touch a few keys on some kind of steel key punch that would set things right again, and it would be as though I had not disobeyed at all. My act of rebellion would be absorbed like rain on an ocean and leave no trace. I would not cause a ripple.
>
> I suppose it is just about impossible for someone like me to rebel anymore and produce any kind of lasting effect.

There are no boots on necks here. There is no rubber truncheon of Victory Gin. We are at work in the belly of the organization. Organizations provide a haven—a safe place—where as long as we do not give it indigestion we can rest forever. The classical management writers tell us that this is freedom. We will not be vomited onto the beach, unless we make trouble. Unless we are a rebel, we are no sinner. Grey and blue pinstriped flannel, wool and worsted are all important shades of gray. There are no bright colors here—only shades of grey—no stark contrasts. Choice and responsibility have been torn and tattered in the jaws of shredders and tossed into infernal unexistence. Short of being dead, it is the final unsurpassable stage of irresponsibility.

Something is happening here, but it is to easy for you to believe that it is all fiction—make-believe—not real. You can think that fiction exaggerates the search for meaning. Then, listen to these real characters.

First, Donald Siebert, Former Chief Executive Officer and Chairman of the Board of JC Penny Company, Inc. In a chapter entitled "The PRIMARY PRINCIPLE: Defining Your Personal Values", Siebert urges us not to let our quest for meaning slow us down:[14]

---

[13] J. Heller, supra note 11 at 19.

[14] Donald V. Seibert The Ethical Executive, New York: Simon and Schuster, Inc., 1984, p. 36.

> I've seen many highly capable young people who have become obsessed in their search for the "meaning of life" and as a result have gotten distracted and fallen by the wayside as they tried to move up the corporate ladder. Now, I'm not criticizing the search for meaning and purpose, because that's where you have to start if you hope to settle on a viable personal philosophy. But some people just can't make up their minds. They keep casting about, trying this or that set of values or experimenting first with one kind of personal satisfaction and then another. Today, it might be a different career focus. Tomorrow, it might be another sexual partner.
>
> In contrast, if you have a firm personal philosophical foundation, you'll find you have one less thing—or one fewer set of things—to worry about. With the question of ultimate meaning and basic values settled, you can focus more of your efforts and attention on wrestling with business problems.

This passage is as surreal as a Bergmann movie. It generates long hallways of emptiness, and plots within plots within self-deceiving plots. If only we can come to believe in something, it really doesn't matter what (though Siebert has some Babbitt-like ideas) we can get on with the important stuff—solving business problems and climbing the corporate ladder. Seek to "Know thyself" and finish up before the coffee break is over. Varnish your values into a veneer that works. This above all to thine own self be pragmatic. Do not think, graze.

The search for meaning is to be settled. This is what savvy executives do. We must be decisive about meaning. Examine the alternatives, weigh the consequences, and make a choice. Don't look back, an existential crisis might be gaining on you.

There is no place in the organization for experimenters, for those who do not have Siebert's wisdom of having settled the basic questions of life. The organization, on this view, is not a human construction that enables us to understand our need for sociality and affiliation. It is not a means to the member's ends, but rather, an end in itself. Human persons are phenomena and organizations are noumena. It is only by subjugating our own ends, stripping them of the false appearance of individuality, or better yet, adopting the goals and purposes of the organization for our self, that we can get on with wrestling with business problems.

Siebert means well—and that is the worst part—for he, more so than others, acknowledges the importance of our quest for meaning, the drive to escape the sheep-like existence that we all too easily accept. But, his view of organizational life makes organization into a repressive mechanism, which taken to extreme, stamps out the very nature of the human person.

More frightening still is this very extreme as depicted by Richard Pascale and Anthony Athos' description of Japanese management (which they wholeheartedly endorse). They describe the Matsushita "philosophy":[15, 16]

> From the outset, Matsushita believed that an enterprise as a whole was no better than the people in it. He characterized the organization's role as being like the hoops around a barrel. Without the hoops, the individual staves are all slightly different, causing the barrel to leak. Through the discipline of the hoops, he said, the individual parts come together to fulfill their intended purpose. ....

[15] R. Pascale and A. Athos, The Art of Japanese Management, New York: Simon and Schuster, 1981, pp. 52 ff.

[16] Id. at 52.

> Yet for all its commitment to "self-actualizing" its employees, Matsushita is extraordinarily performance oriented. As noted earlier, while Matsushita doesn't fire people, it regularly rotates those in trouble into other jobs. The erring executive is more likely to be asked to engage in self-reflection and come up with his own ideas where he might fit into the organization better, and to be shunted around until he finds a role where maximum use can be made of his potential.

The best that we can say about these passages is that we hope they are taken out of context or are at least false. According to Pascale and Athos' description, the organization is the meaning of life for its members. Those organizations who do best "make meaning" for its members. Employees who fail must go to Room 101 and in Coolhand Luke's terms, "Get your mind right, there boy". The corporation is a family. It takes care of its own. You don't have to worry about being fired and kicked out into the street. Papa will provide. Individuality is a small price to pay for such comfort and serenity.

While Pascale and Athos mean to describe the internal workings of the Japanese system, there is an external analog. Indeed, the roots of their view grow in the very nature of the limited-liability corporation. Since those who provide the factors of production, and here we mean all the factors not just finance, are not liable for the consequences of their action the very notion of "moral responsibility" is defined away. The government of India, the executives of Union Carbide, the employees in the Bhopal Plant, and others are not liable for their actions, except of course they could lose the plant, their investment, their jobs, their stake in the company. Things are changing in the law, but slowly, and the damage is done. Not only are the people of Bhopal aggrieved, but what is the price of giving up one's sense of moral responsibility? The concept of limited liability is foreign to moral theory (except to Utilitarians—to which very little is foreign). Pascale and Athos have done us all a favor to point out how in the best managed companies, limited liability has been internalized, but we must not fool our moral imaginations.

Pascale and Athos tell us approvingly that U.S. firms work this way too:

> It should be noted here that Matsushita's seemingly doctrinaire approach is not unlike that practiced by several outstanding firms in the United States. McKinsey and Company's Paul Kraus comments: "At P&G Sears and IBM, they make the system work by running it something like the Army. If you get out of step they come after you. They hire young people at twenty-one who are open and malleable and they only start them at the bottom. They bring them up to believe in the system and to abide by its rules...."

Studies of these "excellent" companies became the subject of an enormously influential McKinsey and Company report, and best selling book by Thomas Peters and Robert Waterman, In Search of Excellence[17] (At last count the book had sold five million copies worldwide). They tell us that eight attributes are found in excellent companies: (1) A bias for action; (2) Close to the customer; (3) Autonomy and entrepreneurship; (4) Productivity through people; (5) Hands-on, value driven; (6) Stick to the knitting; (7) Simple form, lean staff; and, (8) Simultaneous loose-tight properties. Couched in a critique of "the rational model" current organization

[17] T. Peters and R. Waterman, In Search of Excellence, New York: Harper and Row, 1982.

theorists such as Weick and March[18] are approvingly quoted on organizational evolution, and the symbolic nature of "social man".

The striking thing about this list of attributes and ideas as well as the stories in the book is that they are laden with values and the search for meaning. But, as Terrence Deal and Alan Kennedy tell us, these are things to be managed. Managed under the guise of shared values and culture:[19]

> In strong culture companies, nothing is too trivial. Any event that occurs in a work context is an event to be managed.

"Manage". What a funny word! "We'll manage" has different meanings depending on whether uttered by a poor ghetto dweller upon receiving a phone bill, or by a young executive eagerly telling the troops how they will cope with the latest corporate problem. In either case culture and values are not things to be managed. Culture and values may well emerge out of a way of managing, the way that meaning emerges out of the style of the text,[20] but we cannot manage the culture any more than we manage the meaning of a piece of literature. It is style that is important, and style, in organizations if not in literature, has a moral component. It is a question of power and legitimacy, and this is the question which the most recent bag of magic management tricks has left at home. There is no discussion of the dark side of culture and values, no place for reasoned argument. We must get on with it, for Siebert has reminded us that those business problems need to be handled with the fakery of professional wrestling.[21]

The key to success? The magic? Language. Shrouded in concern for people, in good news, in good, gooey feel-good slogans, we must make thought crime impossible. "The way we do things here" ensures that the search for meaning is over. We graze quietly, and are gently herded from pasture to pasture, focusing our attention on wrestling with business problems. It is Orwell without the prose, 1984 without the State, and terror in its most subtle form.

---

[18] K. Weick, The Social Psychology of Organizing, 2nd edition, Reading, Mass.: Addison-Wesley, 1979; and, J. March and J. Olsen, Ambiguity and Choice in Organizations, Oslo: Universitetsforlaget.

[19] T. Deal and A. Kennedy, Corporate Cultures: The Rites and Rituals of Corporate Life, Reading, Mass.: Addison-Wesley, p. 60.

[20] For an account that is relevant to organization theory see W. Gass, "Culture, Self and Style", in Habitations of the Word, New York: Simon and Schuster, pp. 185–205.

[21] We do not mean to imply that the authors of these books on management are unaware of the darker side of their prescriptions. At least Pascale and Peters have indicated to us, in conversations, that they believe that there is an ethical component to their magic. Indeed in his latest book with Nancy Austin, A Passion for Excellence, New York: Random House, 1985, Peters clearly articulates what we would call an ethical theory of business strategy, that is, that managers ought to serve the needs of customers and employees. Such a theory is one among many possibilities. We explore this connection between ethics and strategy in a positive vein in a current research project, Corporate Strategy and the Search for Ethics.

4 Conclusion

Management and organization is the thing now. There are countless new proposals, new magic, for how to manage and how to organize. We have argued or claimed or shown that there are strong parallels between some descriptions of life in organizations and Orwell's 1984. Orwell warned us that most of the time we are passive sheep, willing to give up our search for meaning, and our existential confrontation with choice and responsibility for affiliation with others, willing to accept authority over our actions in return for protection from the responsibility that accompanies being human.

There is really and truly a revolution in management and organization. Tracts are written, tents rented, and the faithful gather to hear the Word. And the Word says, "People", "Pay attention to People", "Manage Values", "Culture", "Manage Culture", "Excellence", "Manage Excellence". Thou shalt do this in the name of performance. To those who have seen this revolution in action it is truly frightening. Reflection and reasoned and informed choices have been thrown out along with the bathwater.

The revolution has ended the search for meaning as hundreds of corporations have commissioned task forces of middle managers to write statements of values, slogans to solve our confrontation with the world so that we can get on with business. The revolution in management and organization theory is dangerous precisely because it recognizes the truth of Orwell's view.

There is a silver lining. Recognizing that human beings are important to organizational success, even as mere means to corporate ends, is also a first step to another revolution: one where the importance of the human person becomes an end in itself. Values are important, but they must be chosen and respected, not sloganeered and managed.

Is it possible for there to be organizations which are mere means to the accomplishment of individual ends? Telling that tale will take a long time. But, we know of no more important work for scholars and managers to do.

Management and organization is the thing now. The revolution is going gangbusters. The presses cannot be stopped. The prophets cannot be silenced. Perhaps we can pay attention to Orwell, for he has shown us the Way. And, it is not doubleplusungood. It is evil.

R. Edward Freeman is University Professor, Elis and Signe Olsson Professor, Academic Director of the Institute for Business in Society, and Senior Fellow of the Olsson Center for Applied Ethics at the University of Virginia's Darden School of Business.

Daniel R. Gilbert, Jr. is Professor Emeritus at Gettysburg College.

Chapter 19
The Ethics of Greenmail

R. Edward Freeman, Daniel R. Gilbert, Jr., and Carol Jacobson

Recent events have threatened to turn the grey-flanneled world of corporate finance into a twentieth-century version of "shoot-out at the OK Corral". The normally staid and conservative arena of investment bankers, corporate executives, and large shareholders has given way to the drama of corporate raiders and hostile takeovers. Even the language evokes an imagery of corporate "shoot-em ups". Understanding the latest wrinkle in "shark-repellents", "poison pills", "raiders", and "greenmail", is as necessary to today's executive as poring over the latest sales and profit numbers.

Our purpose is quite simple. We want to separate the wheat from the chaff on the issue of greenmail, using the tools and tricks of modem moral philosophy. In particular we will analyse the ethical issues that underlie many of the current discussions of greenmail. We want to show that once the smoke clears and the hidden mirrors are revealed, greenmail per se gives us little cause for moral concern. However, a thorough analysis of the ethics of greenmail does reveal significant shortcomings in the currently received theory of the firm and the issues of corporate governance which the theory addresses.

Originally published in: *J Bus Ethics*, 6, 165–178 © Springer, 1987
Reprint by Springer, https://doi.org/10.1007/BF00382861

R. E. Freeman (✉) · D. R. Gilbert, Jr. · C. Jacobson
University of Virginia, Charlottesville, VA, USA
e-mail: FreemanE@darden.virginia.edu

© Springer Nature Switzerland AG 2023

S. D. Dmytriyev, R. E. Freeman (eds.), *R. Edward Freeman's Selected Works on Stakeholder Theory and Business Ethics*, Issues in Business Ethics 53,
https://doi.org/10.1007/978-3-031-04564-6_19

Our argument is divided into several sections. Section I briefly describes two case studies of "corporate raiding" and some proposals for legal reforms of current business practices. Section II clarifies the meaning of "greenmail", by virtue of understanding the differences between the concept of "greenmail" and "blackmail". We claim that sweeping analyses of "greenmail" are ultimately doomed to failure from the moral point of view. Section III analyses the possible meanings of "greenmail is morally wrong" while holding fast to the view that hostile takeovers are morally permissible, within the bounds of common morality. Section IV looks at some arguments that hostile takeovers are themselves morally questionable. Section V summarizes the findings of our analysis.

1 Two Case Studies

Corporate raiders are the latest rage on Wall Street. In order to bring the complex conceptual issues into sharper focus we will briefly set forth the facts in two recent cases of 'corporate raiding': The T. Boone Pickens attack on the Gulf Oil Corporation, and the Saul Steinberg and Irwin Jacobs greenmail case involving Disney. We will then discuss several recent proposals and legislative plans for reform in order to better ground our discussion of the underlying philosophical issues.

1.1 The Raid on Gulf

From August to October 1983, T. Boone Pickens, as chairman of Mesa Petroleum Co., purchased 18 million shares (equivalent to 10.8% of the outstanding shares) of Gulf Oil Corporation for $790 million. Pickens believed that Gulf's current market price of $40/share was undervalued, and that the company was poorly managed with a dwindling supply of oil and gas reserves. The Pickens strategy for increasing the wealth of the shareholders of Gulf centered on the idea of spinning off some of Gulf's oil and gas reserves into a royalty trust. Under the terms of such a trust, earnings from the reserves would flow directly to the shareholders, bypassing corporate income taxes. While the shareholder would pay taxes on the earnings, s/he would be partially sheltered by the depletion allowances on the reserves.

James E. Lee, chairman and chief executive of Gulf, and the Gulf Board of Directors rejected the Pickens offer and vowed to fight it A royalty trust would not only reduce the size of Guff but would also diminish the corporate prestige of its top management Some industry analysts agreed with management that a royalty trust would take away the cash flow needed for operations and growth.

Pickens' next move was to successfully solicit other investors to join his Gulf Investor Group in order to raise the necessary funds to make a tender offer of $65 per share, which would increase the group's holdings to 21.3% of Gulf.

Gulf termed this offer inadequate and authorized its financial advisors to entertain bids from other corporations. Lee continued to assert that "the shareholders do own the company, and we work for them" (Labich 1984, p. 23). In March 1984, the Gulf Board of Directors agreed to be acquired by Standard Oil of California (Socal) for $80 per share. The Pickens group stood to gain $760 million for their now 13% share of the company. $500 million of this profit would go to Pickens' Mesa Petroleum. The merger between Gulf and Socal created the nation's third largest oil concern, behind Exxon and Mobil Corporation.

Was Gulf greenmailed? Did Gulf stockholders get a good deal? Did Mesa stockholders get a good deal? What about the interests of other groups who have a stake in Gulf, such as customers, suppliers, employees, and local communities? What are the rights of Gulf management, and when are there conflicts of interest?

1.2 The Greening of M-I-C-K-E-Y M-O-U-S-E

Events at Disney took a decidedly different turn from those at Gulf. Takeover speculation began at Walt Disney Productions in early 1984. Due to disappointing earnings, discord among the Disney clan, and the firm's rich assets, which included extensive real estate holdings in Florida and California, Disney became an attractive takeover target.

Beginning in March of 1984, Saul Steinberg, who controls Reliance Holdings Inc. (and is noted on Wall Street as the brash entrepreneur that once tried to take over Chemical Bank), began to buy Disney stock. By April Steinberg had become the largest individual stockholder with 2.2 million shares (6.3% of the company) costing $132 million.

In order to ward off unwanted bidders Disney management chose two defensive strategies. First, they entered into an agreement with the Bass Brothers to buy the Arvida Corporation in exchange for Disney stock and $190 million. Second, Disney agreed to purchase Gibson Greetings for $337 million, a purchase which was later rejected by Disney stockholders. These moves were explicitly designed to dilute Steinberg's holdings, and they increased the debt to equity ratio from 25% to over 60%.

When Steinberg announced in June that Reliance would make a tender offer of $67.50/share for 37.9% of Disney stock, to go with the now 11.1% already controlled, the Disney Board agreed to pay Steinberg $325.3 million for his stake ($77.50/share for stock he purchased at an average price of $63.25/share). They also agreed to pay Irwin Jacobs, a Minneapolis investor, an additional $600,000. The initial response on Wall Street was a drop in Disney stock in only two trading session from 65 1/8 to 50 3/4.

Was Disney management threatened by Steinberg? Was it acting in the interests of the company and its stockholders? What were the effects of the 'greenmail' payments? Disney's chairman, Raymond L. Watson claimed, "It woke us up, though

I hate to give credit to something like that [the takeover attempt]. I think the company is stronger." (*Business Week,* March 4, 1985, p. 82) Evidently stock market traders agreed. In late April of 1985 die stock was trading near $77/share.

1.3 Some Proposals for Reform

Government response to the 'greenmail' phenomenon has two distinguishing features. First, no single culprit has been consistently fingered for precipitating the problem. Both target-company management and the corporate 'raiders' have been the focus for regulatory rhetoric and, in a limited fashion, action. Second, this split focus has not evolved neatly along either federal-state or legislative-judiciary lines. This is most apparent in the state legislatures, one of five governmental arenas in which the 'greenmail' question has been addressed. In addition to the states, the Federal courts, the Securities and Exchange Commission, Congressional committees, and the Treasury Department have taken steps to specify what needs to be accomplished in the form of 'greenmail' regulation. We will briefly review the recent history of these activities.

Since 1981, no fewer than eight states have enacted statutes designed to limit corporate takeover activity. While the increased frequency of takeover activity in general has played a role in these state efforts, the Federal judiciary had a hand in this trend as well. A 1982 U.S. Supreme Court ruling, in the case of Edgar v. Mite, found an Illinois anti-takeover statute to be unconstitutional, for reason of the overly-ambitious jurisdictional claims made by the state. Lewin (1984) cites this ruling for providing clarifying assistance to other states then contemplating such legislation. Although each of the eight laws pertains to a narrower domain than did the Illinois law, the similarity ends there. Three distinct approaches to takeovers and, in particular, situations conducive for 'greenmail', are evident (Lewin 1984):

1. The *fair price* approach requiring that any premium over a stock's market price be made available to all eligible shareholders.
2. The *informed shareholder* approach that places additional reporting requirements upon the stockholders contemplating a purchase of controlling interests.
3. The *Pennsylvania proviso* that any owner of a 30% interest be prepared to purchase all remaining shares if so ordered.

The first approach is ostensibly intended to curtail incumbent management's use of corporate assets for defensive purposes. The latter two are clearly directed towards 'raiding' shareholders.

Perhaps the most significant event to date, as far as state action is concerned, was a November, 1984 opinion handed down by the 8th U.S. Circuit Court of Appeals. In a complaint brought by Conwed Corporation against Cardiff Acquisitions, Inc., the court upheld the Minnesota Corporate Takeover Act. Minnesota was justified in demanding disclosures more stringent than Federally-required, "...so long as the state does not cede the advantage in a takeover fight to management of a target company" (Gold 1984). The Minnesota statute, passed in 1984, empowers the State Commerce Commissioner to require a 'raider's' disclosure of intentions for disposition of the target company. The law applies to Minnesota-chartered corporations and any non-Minnesota company in which (a) Minnesota residents hold at least a 20% ownership stake and (b) which keeps "substantial assets" within state borders (St. Anthony 1984).

At the Federal level, an advisory panel of the Securities and Exchange Committee (SEC) last year recommended 50 changes in takeover regulations, many of which require Congressional approval. One of these changes would prohibit 'greenmail' in instances where the repurchase amounted to at least 3% of the company's outstanding stock. The provision would not apply where management either received majority approval from stockholders, or where a 'fair price' offer would be extended to all stockholders (Noble 1984b).

Paralleling SEC action, House and Senate committees gave approval in 1984 to 'greenmail' provisions aimed at incumbent management The Senate Banking Committee approved an outright ban on 'greenmail', in a rider to a banking bill (*New York Times*, July 18, 1984). The House Energy and Commerce Committee agreed on a provision that 'greenmail' be banned if the payee had held the stock for less than 2 years, unless majority approval would be forthcoming from the stockholders (Noble 1984a). More recently, the House Ways and Means oversight committee has conducted hearings concerning linkages between current tax laws and the incidence of takeover activity *(Minneapolis Star and Tribune,* April 2, 1985).

Neither of the bills progressed any further in 1984, in part due to efforts by the Reagan Administration to forestall such legislation. Then-Treasury Secretary Donald Regan spearheaded the opposition. Treasury arguments stressed that the Congressional proposals would have an adverse "… impact on free market processes that provide for the orderly sale of companies to the advantage of their stockholders." (*New York Times*, July 18, 1984) In addition, the Treasury argued that state enforcement efforts were presently sufficient in the takeover arena. At this writing, the Administration has been successful in preventing both bills from receiving full Congressional consideration.

With the Gulf and Disney cases, along with an account of governmental response, as background, we will turn to consider the logic of greenmail and then analyse the essential features of the practice of greenmail from the standpoint of moral philosophy.

2 The Meaning of Greenmail

Why is one version of 'corporate raiding', or the proposed merger of two corporations, called 'greenmail? The obvious analogy with 'blackmail' is instructive. Ellsberg (1959) claims that blackmail is possible between a blackmailer (B) and a victim (V) when: B's choice of action makes a difference to V;B can inflict some loss on V relative to the status quo; and, B can communicate with V and change V's expectations of B's behavior. He summarizes a lovely argument about the pervasiveness of blackmail (p. 346):

> Given all these conditions, I can set out to *coerce* you: to influence you to choose the action I prefer you to take, by increasing your expectation that if you do not, I will choose some response leading to an outcome still worse for you than compliance.

Blackmail is usually thought to be morally wrong because B must threaten V with some harm (or equivalently, with withholding some good that is due V) in order to get V to comply with B's wishes. However, an important distinction, relevant for the greenmail analog, is easily overlooked. We need to distinguish "blackmail" from "threat of blackmail". An example will make the distinction clearer.

Suppose that B threatens V with assault or personal ruin or whatever, unless V pays $100 to B. We want to say that blackmail has not occurred until V pays (or forms the intention to pay) B $100. Rather we would say "V has been threatened", or "V has been threatened with blackmail", or "V is being blackmailed", or even "blackmail is in the process of occurring" rather than "V has been blackmailed". While such distinctions may appear to be so much philosophical niggling, it is important in order to understand where the analogy between 'greenmail' and 'blackmail', as commonly understood, breaks down.

For blackmail to occur at least two conditions are necessary:

(1) *The Threat Condition.* B threatens V with some action that is harmful to V, unless V behaves as B wishes.
(2) *The Compliance Condition.* V complies with B's wishes, rather than suffer the consequences of the threat.

Two immediate implications of this view of blackmail emerge. The first is that the threat of the blackmailer must be perceived to be credible, else why should the victim, V, comply? Merely making threats does not constitute blackmail, and likewise merely making threats to take over a company does not constitute greenmail. The threat must be credible.

The second implication of this view of blackmail it that unless we view die act of threatening as immoral, not all blackmail is immoral. Suppose that a customer threatens to take his/her business to another shop unless the shopkeeper agrees to stock the customer's favorite brand of dog food. While the shopkeeper may lose money on the dog food trade, and besides happens to loathe the very existence of dogs, s/he might comply rather than risk losing the customer. Assuming that the customer was telling the truth, nothing that is morally wrong has occurred – or if so, then any form of market exchange borders on blackmail. This implication is crucial to our discussion of greenmail, for if our argument is sound, it is not sufficient to merely identify a particular case as one of 'greenmail', and thereby claim that some moral wrong has been done.

Many cases of blackmail are in fact morally wrong. There are at least two reasons that might cause us to render such a judgement, call it "wrongful blackmail". One distinguishing feature of these cases is that B's threat to V proposes to inflict a harm on V or violate a right of V which common morality forbids. (Of course there are many difficult cases, but we are only concerned with some dear cases here.) Thus, B's threat of violence against V's person or ruination of V in the eyes of the community is "wrongful blackmail" because, if the threats were carried out, B would commit a morally wrong act Both the threat of such an act and actually carrying it out are wrong, so in a sense wrongful blackmail is odious since it is wrong twice over.

A second possible distinguishing feature of wrongful blackmail is that while V may take B's threat as credible, and hence grant B's wishes, B's threat is in fact not credible, for B has no intention of carrying out the threat. What is wrong in this situation is that B has lied. B has purposefully misled V about B's intentions and the consequences to V, and *ceterus paribus*, such an action is morally wrong.

Thus, wrongful blackmail occurs when B threatens V with an action that would be immoral if carried out (both the threat and the execution are wrong), or B misleads V about her/his intentions. What is the relevance of all of this talk about 'blackmail' for greenmail?

'Greenmail' can be defined similarly to 'blackmail' as consisting of a threat condition and a compliance condition:

(1) *The Threat Condition for Greenmail.* G threatens to engage in a hostile takeover battle for company, C, unless the management, M, of C, buys back G's stock at a premium.
(2) *The Compliance Condition for Greenmail.* M buys G's stock (for C) at a premium to prevent G from engaging in a hostile takeover.

It is important to note a further parsing of the concept of greenmail. In both the threat condition and the compliance condition reference is made to some commonly-made assumptions about modem business practices. We might spell out these practices along the lines of the following conditions:

(3) *The Agency Condition.* Management, M, acts for the company, C, in the sense of furthering the interests of the company. Furthermore, M has a fiduciary duty to the stockholders of C.
(4) *The Hostile Takeover Condition.* The potential for hostile takeovers is present when M rejects a direct proposal from an outside firm or individual to buy C, or advises its stockholders not to tender their shares to the potential buyer.

The Agency Condition implies that management, *prima facie*, actually has the right to comply with the greenmailer's threat, if doing so is actually consistent with M's fiduciary duty to the stockholders of C. Of course, the actual situation is more complicated, since G is also a stockholder of C in most greenmail cases. Indeed the whole issue of 'agency' is a nontrivial one for theories of corporate governance.

The Agency Condition is grounded in the presumption of a particular kind of fiduciary relationship; namely, one involving the representation of a person's interests by an *agent* (Shepherd 1981). 'Fiduciary' circumscribes a vast range of interpersonal relationships grounded basically in trust among the parties. At the heart of a fiduciary relationship is the *duty of loyalty.* Shepherd (1981, p. 35) has captured these themes in a general definition of such a relationship:

> A fiduciary relationship exists whenever any person acquires a power of any type on condition that he also receive with it a duty to utilize that power in the best interests of another, and the recipient... uses that power.

Shepherd goes on to observe that such power held in trust is commonly treated in the law as a form of property to be held under such a duty of loyalty. Fiduciary relationships can take both explicit form and implicit form. This is often the case in matters of corporate raiding. Incumbent management justifies its ingenuity in selecting defensiveness mechanisms in terms of the appropriateness of such tactics to the shareholders' interests. In turn, corporate raiders appear to rely upon a *de facto* fiduciary relationship between themselves and the target firm's shareholders.

Agency, as already noted, can be considered a particular form of actions in trust. For a principal-agent relationship to exist in the law, three basic conditions must be met (Powell 1965):

1. The agent must be authorized to act for the principal.
2. The agent must agree to act on the principal's behalf.
3. The agent must have the power to bind the principal in legal relationships with third parties.

This third point most clearly expresses the agent's role as fiduciary, in the sense that a transfer of power from principal to agent is presumed. At the same time, *beneficial ownership* of the fruits of that power's exercise are retained by the principal.

While the foregoing is couched in legal terminology and drawn from the legal profession's struggles with the concept of 'fiduciary', the line of reasoning is consistent with a rights-based conception of business ethics. The act of transferring power to a fiduciary presumes the principal's (i.e., the beneficiary's) right to effect such a transfer in the first place. Furthermore, a fiduciary analysis focuses upon the effects of the relationship upon the parties, the very essence of ethical inquiry. We will now explore the nature of the current wave of corporate takeover activity through the perspectives of parties to the situation.

The Hostile Takeover Condition simply defines the term and acknowledges that hostile takeovers can take two routes. The suitor or raider or acquiror may make a direct proposal to the management or directors of C to buy C or to merge with C. Alternatively, the suitor may make a proposal directly to the shareholders of C. Note that the Hostile Takeover Condition is logically connected to the Agency Condition, for the existence of a hostile takeover calls into question whether or not management is acting consistently with their duty to stockholders. Any analysis of the ethics of hostile takeovers must ultimately question the adequacy of the Agency Condition, for if management is acting according to this condition, then as a matter of logic a hostile takeover is not in the interests of die stockholders. Thus, the very existence of the practice of hostile takeovers serves as insurance to stockholders that management in fact acts in stockholders' interests. Alternatively, the existence of hostile takeovers necessitates a structural conflict in the very nature of corporate governance.

It is difficult to see how we can conclude from this account of the logic of greenmail, just as in the case of blackmail, that all acts of greenmail are morally wrong, at least on a naive account of common morality (even though the facts of all the actual cases of greenmail may point to violations of common morality). It is certainly conceivable, at least on a consequentialist account of morality, that greenmail could have good consequences, since hostile takeovers in general do have good consequences for the acquired firm (Jensen and Ruback 1983). The Disney case in Section I is an example. And even on a deontological view of morality, it is difficult to defend that all threats are morally wrong, without stipulating either that a customer who threatens to take her/his trade elsewhere is not "really making a threat", or else that all market transactions are morally wrong.

Thus, the question of the ethics of greenmail is similar to the ethics of blackmail. How could we show that all greenmail is "wrongful greenmail"? We would have to show that the very threat of greenmail is morally wrong, then trivially all greenmail will be morally wrong. Easier still, if it can be shown that engaging in a hostile takeover is morally wrong then the threat of greenmail would be morally wrong and greenmail itself would be morally wrong. However, we shall argue that this chain of reasoning cannot be cogently defended.

3 Is Greenmail Morally Wrong?

Assume that engaging in the practice of hostile takeovers is morally permissible, for whatever reason. We could show that the practice of greenmail is morally wrong if either of the following cases is true: (1) the threat of hostile takeovers is immoral; or, (2) complying with the threat of hostile takeovers is immoral (for several different reasons). Let us take each case in turn.

Case 1: The threat of hostile takeovers is immoral. Now we might think that the threat of hostile takeovers is morally wrong even though the practice of hostile takeovers is not itself morally wrong for several reasons;

(1) All threats are harmful.

It is difficult to defend this proposition for reasons stated earlier. Without a clearer analysis of 'threat', and the differences between 'threat', 'bluff', 'bargaining position', 'offer', etc. it is difficult to articulate why one would want to make such a claim. It is possible to hold such a view, but it will depend on the background assumptions and a more carefully worked out moral view. Not only will greenmail be morally wrong on such a view but many other business practices will as well. So, this reason should not cause us special worry in the greenmail case.

(2) The threat itself may have harmful consequences.

It is difficult to see how the threat itself can have harmful consequences, without there being a response to the threat that brings these consequences into being. Remember that we have assumed for now that the content of the threat is not morally wrong. So, this case is like Smith 'threatening' to give Jones $100. If Jones complies and uses the $100 to purchase a gun and shoot Brown, then the threat has brought into being some harmful consequences, but it is difficult to defend a view of morality which would hold Smith morally accountable. (Even a dyed-in-the- wool act utilitarian would have trouble here.)

(3) The threat of a hostile takeover is a mere threat, and in fact is not credible.

The issue here is that threatening a hostile takeover is like lying. Let us suppose that Steinberg had no intention of taking over the Disney Corporation. If this is the case then we can say that Steinberg made systematically misleading and false statements to Disney management and stockholders (and others as well). In general, making such statements is morally wrong. But, while this practice is morally wrong, it is not the practice of greenmail which makes it morally wrong. Lying is lying, and lying in the context of greenmail is no different in principle than lying in the context of political campaigns or disciplining one's children.

What we conclude from this case is that a necessary condition for greenmail not to be morally suspect rests on the threat to greenmail being credible. The threat is credible in fact when the 'greenmailer' has formed the intention to execute a hostile takeover of the target firm, and has communicated this intention in a manner that the management and stockholders of the target firm find believable. It is still an open question whether or not complying with the threat is morally wrong.

Case 2: Complying with the threat of greenmail is morally wrong.

Again there are several reasons for believing that it is the compliance condition that gives us reason for moral worry. These reasons are more complex simply because the act of compliance is difficult to analyse.

(1) Complying with the threat of greenmail is always harmful.

To show that the practice of greenmail is morally wrong, at least on a consequentialist view of morality, we must show that it is always the case that complying with the threat of greenmail has harmful consequences on balance. If, for instance, we could show that in all cases the greenmailed firm has unusual negative returns, then the practice of greenmail would be suspect. There is some evidence here. Jensen and Ruback (1983) have analysed tender offers made over a period of years, and have concluded that the acquired firm in hostile takeovers is on balance better off. When payments are made by the target firm to prevent acquisition there is a correlation with unusual negative returns for the target But, two implications of their analysis are important for understanding the ethics of greenmail.

The first is that there is no logical reason why there cannot be at least some cases of greenmail actually improving a firm. Suppose that Firm D is languishing, posting a rather mediocre performance, and that its stock is undervalued. A greenmailer attempts to buy the firm, and an act of greenmail is committed. Now it is surely possible that the attempt of greenmail has lit a fire under the management. They see the error of their ways and the firm is revitalized. (Indeed this may well describe the Disney cases.) Or, perhaps the stockholders then decide to throw the management out, and even sell the firm to a higher bidder (a variant of this case seems to describe what happened at Gulf). Surely these cases are not logically impossible. In fact an act of greenmail can lead to a bidding war, with good consequences for the stockholders.

The second implication of the empirical studies of Jensen and Ruback is that even if all cases of greenmail in fact lead to harmful consequences for the greenmailed firm, it is management who must bear the moral responsibility; for there is evidence that if the takeover had been allowed, stockholders of the acquired firm would be better off. The actions of greenmailers and the practice of hostile takeovers would have no cause for restriction, though we might want to say that while threatening greenmail is morally permissible, accepting greenmail payments is not. [Indeed T. Boone Pickens sometimes seems to defend a view of this sort.]

It seems more reasonable to believe that sometimes greenmail is beneficial and sometimes harmful. It could be the case that management has some knowledge about the future prospects of the company that no one else has, and that these prospects can be realized only by paying money to greenmailers, thereby making the shareholders better off. Of course, sometimes these 'investments' may not pay off, but there is no difference between greenmail and other investments in this case.

(2) Complying with the threat of greenmail violates the rights of some parties.

If we could show that by complying with the threat of greenmail, management necessarily violated the rights of some parties, then the practice of greenmail would be deemed morally wrong (at least *prima facie).* There are many candidates for rights violations.

3.1 Stockholders

Does complying with the threat of greenmail violate the rights of stockholders? It is difficult to see how this could be the case in light of the Agency Condition. In return for relative autonomy management bears a fiduciary duty to act in stockholders' interests, and therefore management has the legal, if not the moral, authority to actually pay greenmail. Of course, a particular act of greenmail may not be in the interest of stockholders, but this may be true for many actions of management. We argued above that it is conceivable that some actions of greenmail may in fact be in the interest of stockholders. Stockholders' rights are violated when management breaches its fiduciary duty, and practically speaking it is difficult to say when an actual instance is such a breach. But it is at least conceivable that a particular greenmail payment was so large and so outrageously contrary to the stockholder's interest that we would say that an obvious breach of duty has occurred. But, this is similar to making wild and risky investment decisions, or management paying themselves golden parachutes, and once again does not count against the practice of greenmail.

Stockholders' rights could be violated in another sense, and that is that all stockholders have the right to fair treatment by management. By paying greenmail to some stockholders and not to all, management treats some unfairly. Indeed some corporations have adopted such a rule in their charters as insurance against corporate raiders who only acquire 50.1% of the stock. But, there is a difficulty with this line of reasoning. No one has argued that the greenmailer deserves the greenmail payments, merely that they are morally permissible. Management has no duty to pay greenmailers, and hence no duty to pay all stockholders, though it would be morally permissible to do so, as perhaps in a leveraged buyout situation.

To the extent that management is the agent of all stockholders and bears a fiduciary duty to each of them, and to the extent that stockholders' interests conflict with each other, it will be logically impossible for management to carry out its duties. Hence management must adopt practical working rules which are in stockholders' interests and allow management to fulfill its moral responsibility. If one of these rules is "Treat all stockholders alike", then greenmail is immoral. This principle must also imply that in treating all stockholders alike, management must continue to act in their interests. But clearly if all stockholders are always treated alike management will not be able to act in their collective interest, or in their individual interests, since it will not be able to take account of differences in those interests. To see this, take a case where individual stockholder interests conflict, and where satisfying the minimum collective interest of all is not in the collective interest of others. Management could never be responsive to differences in stockholder interests. It could not respond to moral issues such as investment in South Africa that may be raised by a minority of stockholders and opposed by a majority, and it could not respond to pressure to raise the dividend, when such a proposal is perhaps opposed by a minority.

More practically sometimes paying greenmail is in the interests of stockholders, even all stockholders. If management is forbidden to pay greenmail, due to the requirement to treat all stockholders alike, management will be forbidden from fulfilling its duty to stockholders. [The tedium of this argument can be avoided by fiat if we claim that management bears a fiduciary duty to the set of all stockholders, but not to each individual one. Management's duty must be to carry out the intersection of interests of the set of all stockholders. The difficulty is in making this concept operational and meaningful. Suppose the intersection is an empty set.]

3.2 Stakeholders

Are other rights besides stockholder rights violated by complying with greenmail? The answer to this question may well be a resounding 'yes'. If management or the corporation has an obligation to stakeholders, (that is, to the community, or to employees, or to suppliers, or to customers) then it is conceivable that the act of complying with a greenmail threat could violate these rights. Take a case where management must make a huge payment, and subsequently must close a plant, or lay off employees, or reduce product quality, or stretch out payments to suppliers. We might want to say, after close analysis of such a case, that some rights have been violated. But, note that there is nothing special about the practice of greenmail that violates these rights. Management may well decide to close a plant or perform any of the actions which violate the rights of these groups for other reasons.

The question of stakeholder rights raises an interesting issue for corporate governance and corporate ethics. In effect it forces one to question the Agency Condition. If stakeholders have rights claims on the firm, then management or the firm has a corresponding duty to honor those claims. Thus, the Agency Condition must be revised to claim that management has a fiduciary duty to stakeholders. Such a far reaching reform is beyond our present scope. But, note that calls for reform based on greenmail and hostile takeovers are oftentimes aimed at "the public interest", protecting the "interest of employees and the community", etc. If stakeholder interests are to be taken seriously, such piecemeal reform will not be effective.

In summary, we have argued that particular cases of greenmail are morally wrong because they are simply cases of lying or fraud, or that they may harm some parties to benefit others, or that in complying with the threat management may well violate some rights claim of another group. But, we have also argued that the practice of greenmail itself cannot be claimed to be morally wrong. Merely picking out a case as a case of 'greenmail' is not sufficient. In any of these cases there is little justification for public policies restricting the 'greenmailer', in any special sense. We may want to consider how to hold management more fully accountable for acting in stockholders' interests, but that is a question that goes far beyond the bounds of an ethical analysis of greenmail. We turn now to die issue of hostile takeovers, to see whether or not questioning the assumption of the Hostile Takeover Condition leads us to any conclusions which can then be applied *mutatis mutandis* to the practice of greenmail.

4 Are Hostile Takeovers Morally Wrong?

If it could be shown that the institution of hostile takeovers is morally wrong, then the practice of greenmail would be morally wrong, since in general to threaten someone with a morally wrong act is itself morally wrong. Our assumption throughout the previous sections was that the Hostile Takeover Condition, defining hostile takeovers, was not subject to moral questioning. Let us now question that assumption.

We could show that the practice of hostile takeovers is morally wrong if we could show that on balance it has harmful consequences, or that the very nature of hostile takeovers violates the rights of some party. Let us take each in turn.

The practice of hostile takeovers is usually justified by an argument that claims that it represents one form of monitoring or control on management's behavior to act in the interest of stockholders. And, on balance if management is allowed to act in stockholders' interests the benefits to society will be positive, according to the theory. Therefore the question becomes, "Do hostile takeovers, on balance, do more harm than good to management's ability to act in the interests of stockholders?"

The evidence so far seems to be that in terms of stockholder interest, the practice of hostile takeovers has a positive net effect (see Jensen and Ruback 1983). But, this is evidence gathered from an analysis of actual tender offers. It does not systematically look at the effects of "threats of tender offers", and it is not clear how one could even go about gathering such data. We must conclude that given the actual state of utilitarian calculations of stockholders' interests there is a presumption in favor of hostile takeovers, but that all of the evidence is not in. (In fact it is not clear that all of the evidence could be in, for the usual reasons regarding the difficulties of utilitarian theories.) Of course a particular hostile takeover may turn out not to be in the stockholders' interests and may turn out to benefit some groups at the expense of others, but there seems to be nothing about the logic of such attempts that requires that they work out this way. If a particular firm is 'undervalued' by the market, and if the stockholders voluntarily decide to sell their shares to another party, and hence change agents (assume that the acquiring party intends to fire management, else the takeover would not be hostile) there is no logical reason why every party to the bargain cannot gain. Even the offending management may stand to gain, if certain "golden parachute" clauses in their employment contracts are triggered by the takeover.

We could question whether the effects of hostile takeovers are in society's interests, or more carefully, the interests of other stakeholders in the firm, but to do so would serve to question whether by acting in stockholders' interests management thereby acts in society's interests. Thus, questions about the net effects of the practice of hostile takeovers reduce to either a presumption in favor of the practice, or larger issues regarding the appropriateness of the Agency Condition.

The second way that hostile takeovers could be shown to be morally wrong, by rights violations, is equally problematic. If the rights of stakeholders other than management and owners are the ones which are violated, then hostile takeovers are again "reducible" to larger questions about corporate governance. If the rights of management and owners are the ones which are violated, then we must look more closely at the Agency Condition.

It is the Agency Condition which raises some of the most interesting issues in business ethics that are directly relevant to our discussion of the ethics of greenmail and hostile takeovers, for it is here that we can view the conflict of duties of managers and owners in clear relief. We will focus on but two of these many issues, and argue that hostile takeovers are in a similar ethical position to greenmail: they raise many questions about the ethics of corporate governance, and function as symptoms of a much deeper problem. The two issues are: (1) the agency presumptions of both the raiders and defending management; and, (2) the rights claims of management.

4.1 The Agency Condition

In line with the Agency Condition there are at least two defences of die practice of hostile takeovers. The corporate raiders both claim and appear to act upon a *presumption of takeover agency* (PTA) with respect to the shareholders of the target corporations. Incumbent corporate managers, at the same time, make a similar agency claim in the form of a *presumption of justifiable defense* (PJD) with regard to both the corporation, as fictitious person, and the corporation's shareholders. We will examine the premises supporting both arguments. What we will show is that, while both PTA and PJD express recognitions of specific moral obligations, they are both seriously flawed in their oversimplification of the takeover situation. To the extent that there are few or no beneficiaries to a particular takeover, the current takeover wave can perhaps be seen as triggering breaches in fiduciary relationships. But we cannot conclude that it is the practice of hostile takeovers which is morally wrong; but rather, the culprit is the fact that management and raiders breach the moral duties that they have incurred. Such a conclusion raises difficult problems for the ethics of corporate governance.

The *presumption of takeover agency* can be stated in the following form:

> We (the corporate raiders) are acting in the best interests of the shareholders of the target firm, whose interests have not been adequately met by incumbent management.

The implication here is that the raiders and the target's shareholders are parties to a principal-agent relationship, with die former as agent of die latter. Evidence in support of the existence of PTA can be drawn from three sources. First, we hear the claim straight from the source, when Boone Pickens describes himself as the "champion of the little stockholder". Second, other raiders persistently refer to their efforts as exposes and reformations of "poor management". Minstar's Irwin Jacobs (1985) writes that: "... tender offers are rarely made for well-managed companies ..." Third, an inference can be drawn from the current flurry of management activity to shed poorly-performing business units, to the effect that such 'losers' were inhibiting better corporate performance and that a healthier portfolio is less vulnerable to hostile takeover.

Throughout, a strong moral tone permeates the corporate raiders' words and deeds: Incumbent management *ought* to act in their shareholders' best interest; they are not doing so; we can rectify those shortcomings. As we will argue shortly, the raiders display an understanding of the principal-agent nature of the situation, and even claim that the problem is a moral one. But their reliance on PTA contains serious flaws.

From the vantage point of incumbent directors and management who are facing takeover threats, a number of defensive tactics are becoming attractive. The Presumption of Justifiable Defense can be stated as follows:

> In implementing certain defensive tactics to fight takeovers, we (management) are acting in the best interests of the integrity of the corporation and the best interests of our shareholders.

The media has, of course, elevated such tactics as "poison pills" and "shark repellant". Implicit in the PJD is the idea that the shareholders' interests are best served by preserving the corporation's 'integrity', strategic plans, and the influence held by the *implementers* of those plans (i.e, incumbent management). As with the PTA, the PJD reflects an acknowledgement of one particular duty of incumbent management and directors; namely, the obligation to their shareholders. The question is whether this is the sole obligation of importance.

At least four serious problems are to be found in the ethical justification for PTA. First, the PTA can be criticized for its *false presumption of agency* between the raiders and the target firm's shareholders. The first condition for agency, noted above, points to some act of authorization for the raiders to take an action such as making a tender offer. In practice, this does not readily appear to be the case, although it is conceivable that the suitor may investigate his/her chances in advance through some "trial balloon" action.

The more questionable issue here, however, concerns the presumed agent's authority to bind the shareholder 'principals' to third parties. One such third party agreement involves that between the shareholders and the new management installed by the victorious raiders. If the new management and ownership implement a drastically altered strategy, as the "junk bond" era has seen, then a question can be raised as to whether the principals' interests have been met. At the very least, principal choice may well have been abridged in the process.

As strong as the foregoing argument can be, we will presume that the agency claim is valid for the remainder of the analsyis. We choose to do so in light of the power of other challenges to the PTA, as now explained.

Second, the PTA appears to rely upon the *fallacy of loyal agency*. According to the Loyal Agent's Argument (LAA), a loyal agent sees himself/herself as acting on moral imperative in such a manner as to maximize his/her principal's egoistic interests, and to act in no other way (Michalos 1983). With regard to the PTA, two flaws are apparent. First, both the tone and substance of raiders' activities reflect a *strong egoism on the part of the agent,* a condition inconsistent with die spirit of agency. Second, the LAA presumes an oversimplified account of the takeover specialists' agency and other fiduciary relationships in which they are parties. Jacobs (1985) himself acknowledges that his role as Chairman of Minstar puts him in the same role, *vis-a-vis* his 'own' stockholders, as that occupied by the management teams he has sought to oust. Taken strictly, the LAA cannot admit the presence of multiple, concurrent agency relationships and, hence, is an untenable basis for the PTA.

Third, the PTA suffers from what we term the "*junk bond*" *fallacy* with respect to fiduciary performance. This flaw applies to those consummated takeovers which are financed, in part, by issuance of high-risk obligations assigned to the *acquired* firm. The ethical concern rests not with the high risk characteristic of the bonds, nor with the implied disparagement of the newly-acquired firm. Rather, the problematic aspect deals with the *source* of funds necessary to service this debt load; namely, asset sales and diversions of funds assigned to other preacquisition uses such as R & D. The set of fiduciary relationships affected by this practice is broad. Two particular ones include: (1) the takeover ownership and the acquired firm's employees, and (2) the takeover ownership and the acquired firm's other shareholders. In the case of the former, we can define employees' "best interests" as including the means necessary for them to carry out their responsibilities. Asset sales conducted to fund "junk bond" service are *prima facie* competitors of this particular beneficiary group's needs. Similarly for stockholders, debt-load additions augur devaluations in their holdings' worth, to the extent that capital markets are functioning more or less efficiently. There is a market-based irony here that we will address shortly.

Fourth, and perhaps most interestingly, the PTA is weakened by the *majority shareholder fallacy.* On this view, the raider, once s/he has successfully acquired some degree of controlling interest in die target firm, becomes bound to the minority shareholders of the target in a fiduciary relationship. This argument is based upon the premise that the new majority owner has the power, implicitly transferred by the remaining shareholders, to significantly affect the value of die latter group's holdings.

There is both an irony and important consequence in this line of reasoning. The irony is that the very free capital-market forces that enabled the takeover bid in the first place now come back to haunt the raider, since s/he has distorted the market for the target firm's securities in the short run. One implication of this argument, if correct, is that any premium earned by the raider in the course of the takeover attempt, successful or not, rightfully should be distributed to the minority stockholders of the target firm, *if the new majority owner is honoring his/her fiduciary obligations.* Such a premium can be paid in the form of greenmail payoffs.

The point of each of the four foregoing arguments can be succintly summarized: the "still waters run deep" for the corporate raider insofar as his/her fiduciary obligations are concerned. There are other obligations that we can identify that compound ready justification of the PTA. These all share a *conflict of interest* theme. One example involves die raider when, as director or officer of a corporation, s/he takes a takeover step which harms the interests (such as devaluing the worth of his/her firm's stock) of his/her 'own' shareholders.

What these arguments show is that the practice of hostile takeovers is a complex subject from a moral point of view. What they do not show is that the practice of hostile takeovers is morally wrong, but rather that we must tread carefully in giving a defense, paying attention to difference and nuances in the facts of the cases before us. The raider does incur moral obligations and cannot simply discharge them by presuming to act in someone else's interest.

Similar problems plague the presumption of justifiable defense. First, the PJD argument relies upon a premise that corporations, as legal persons, can be considered to have interests, apart from the interests of the parties whose actions comprise "corporate action". This premise does not lack its advocates (see French 1979; Goodpaster 1983, for example). Two basic flaws can be found. First, if the corporation truly had interests of "its own", they would be thoroughly egoistic ones, since the corporation is a creature of the law designed for specific ends. If this were the case, then incumbent management is appealing to the LAA as we saw for the takeover specialists. The LAA can be refuted in this instance by, among other things, pointing to the internal inconsistency in the LAA (Michalos 1983): managers acting altruistically and egoistically, as agents of the corporation, in the same breath. Second, the "corporate personhood" premise effectively evades the issue of multiple, concurrent fiduciary relationships by pointing to one superordinate goal; namely, "the corporation's interest". This can turn out to be little more than a mask for ineffective management.

Second, the PJD presents an interesting case when the flip side of the previously-noted *majority stockholder fallacy* is examined. Given the Agency Condition, we can readily see that the duties of incumbent management extend to include the raider buying up the firm's stock. It follows that the incumbent management is obligated to act in the "best interests" of the raider. The implication should be clear insofar as the PJD is concerned. By the foregoing reasoning, the PJD, implemented through a "poison pill" plan for example, constitutes a blatant breach of this particular fiduciary duty. To be sure, this obligation is not one readily discussed by management under fire from raiders.

This argument has focused upon a set of most visible fiduciary relationships that are affected by takeover attempts. Other effects can be found; customers are one such class of interests. Our conclusion is that, from a "best interests" standpoint, nobody emerges as a *prima facie* 'winner' in the corporate takeover game. In particular, the problem is more complex, on an ethical view, than either the spokespersons for PTA and PJD acknowledge or seem to comprehend. We cannot conclude that hostile takeovers constitute a practice that is morally wrong, but only that defenses of hostile takeovers which claim that actions of raiders and management actually fulfill the fiduciary duty to various parties are suspect.

4.2 The Rights Claims of Management

Does the practice of hostile takeovers violate the rights of management? While an affirmative answer to this question would not show that on balance die practice of hostile takeovers is morally wrong, it would create a presumption that would need to be overriden by other concerns.

An affirmative argument could be framed in the terms of "transaction cost economics" along the following lines. Management has special knowledge of a particular firm, and incurs costs in such specialization. In fact in many cases the senior management is so specialized that it cannot easily retrain itself to move to another company. Managers who incur such "asset specificity" (Williamson 1981) deserve some safeguards for their investment. One such safeguard is the right to due process. (A weaker form of safeguard is the right to do whatever is legally possible to fend off hostile takeovers.) To the extent that hostile takeovers violate these due process rights they are morally wrong.

The problem with this argument is that management has contracted with the stockholders of the firm to perform certain actions in return for various forms of compensation. To the extent that hostile takeovers violate the contractual rights of management the argument works. But, note that hostile takeovers do not necessarily violate any contractual rights, for it would depend on an analysis of each particular contract. To the extent that management has due process rights that go beyond the contractual rights, then not only may the practice of hostile takeovers violate these rights, but so may many other practices violate these rights. Once again we sound a familiar theme: the many rights claims of constituents of the corporation may well be violated by the practices of hostile takeovers and greenmail, but no more so than by other business practices.

5 Summary

Greenmail, and more generally, hostile takeovers, raise interesting questions of the ethics of corporate governance. We have detailed several instances where we may want to say that both greenmail and hostile takeovers are morally wrong. We have not been able to show that the practices of greenmail and hostile takeovers are morally wrong, and further, we have cast doubt on there being such a demonstration. The simple fact is that it is conceivable that greenmail is in the interests of all parties: perhaps the Disney case is one such case where literally everyone was a winner because of the greenmail that was paid.

More importantly where we want to say that stockholder rights are violated are precisely those cases where management violates the Agency Condition to act in the stockholders' interests. If such violations of the Agency Condition are widespread, and they may well be, then the broader issues of the ethics of corporate governance can be called into question. There is simply nothing special about greenmail and hostile takeovers, at least from a moral point of view.

Greenmail and hostile takeovers are important symptoms of a much deeper problem. In the complex world of large global corporations set in the context of pervasive government actions in the marketplace, these issues raise the most pressing question of all for understanding the moral role of the firm: For whose benefit and in whose interest ought the corporation to be managed? But, the answer to that question is a long and difficult story.

Acknowledgements The authors gratefully acknowledge support from the McKnight Foundation, and the Strategic Management Research Center and the Graduate School of the University of Minnesota.

Portions of this paper were presented at the Eastern Academy of Management Annual Meeting, May 1985, Albany, New York. We are grateful to many participants for helpful criticisms.

References

A Winning Hand. 1984. *Financial World,* April 4, 1984, 14.

Administration Says Tax Laws Shouldn't Penalize Takeovers. 1985. *Minneapolis Star and Tribune,* April 2, 1985, 9B.

Bleakley, F. R. 1984. Takeover "Frenzy" Causes Concern. *Minneapolis Star and Tribune,* July 8, 1984, 1D, 5D.

Brudney, V., and M.A. Chirelstein. 1974. Fair Shares in Corporate Mergers and Takeovers. *Harvard Law Review* 88: 297–346.

Ellsberg, D. 1959. The Theory and Practice of Blackmail. Reprinted in O. Young, *Bargaining,* University of Illinois Press, Urbana, 343–363.
French, P.A. 1979. The Corporation as a Moral Person. *American Philosophical Quarterly* 16: 207–215.
Gold, J. A. 1984. Minnesota Takeover Law Passes Appeals Test. *St. Paul Pioneer Press*, November 30, 1984, 6D.
Goodpaster, K.E. 1983. The Concept of Corporate Responsibility. *Journal of Business Ethics* 2: 1–22.
Harris, K. 1984. Takeover Speculation Shakes Disney's Not-So-Wonderful World. *Minneapolis Star and Tribune,* April 15, 1984, 3D.
Jacobs, I. 1985. Anti-Takeover Law Favors Managers over Shareholders. *Minneapolis Star and Tribune*, March 10, 1985, 27A.
Jensen, M. C. 1984. Takeovers: Folklore and Science. *Harvard Business Review,* November-December, 1984, 109–121.
Jensen, M.C., and R.S. Ruback. 1983. The Market for Corporate Control. *Journal of Financial Economics* 11: 5–50.
Labich, K. 1984. How Jimmy Lee Let Gulf's Shareholders Win Big. *Fortune,* April 2, 1984, 23.
Lewin, T. 1984. State Controls on Takeovers. *New York Times,* November 27, 1984, D2.
Mahon, G. 1984. Dealing with T. Boone Pickens – Big Oil's Bogey Man Talks about Takeovers. *Barron's* 18, 1984, 13ff.
Michalos, A. 1983. The Loyal Agent's Argument. In *Ethical Theory and Business*, ed. T.C. Beauchamp and N.E. Bowie, 247–254. Englewood Cliffs: Prentice-Hall, Inc.
Noble, K. B. 1984a. SEC Backs New Takeover Rules. *New York Times,* May 10, 1984, D1.
———. 1984b. Time is Money; Fear is Lots of Money. *New York Times,* August 26, 1984, Sec 4, 8.
Pickens Domes Gulf. 1984. *Fortune*, March 19, 1984, 6.
Powell, R. 1965. *The Law of Agency*. London: Sir Isaac Pitman and Sons, Ltd.
Regan Questions Curb on Takeover Defenses. 1984. *New York Times,* July 18, 1984, D1, D17.
Rosenberg, H. 1983. Pickens vs. Lee: The Battle over Gulf. *Financial World,* December 31, 1983, 8–13.
Scheibla, S. H. 1984. A Mickey for Greenmail?. *Barron's,* June 18, 1984,8ff.
Shepherd, J.C. 1981. *The Law of Fiduciaries*. Toronto: The Carswell Company Limited.
Sloan, A. 1985. Why is No One Safe?. *Forbes,* March 11, 1985, 134–140.
St. Anthony, N. 1984. "Takeovers" Disruptions Inspired Governing Bill. *Minneapolis Star and Tribune,* November 5, 1984, Sec. M, 1, 6.
The Raiders. 1985. *Business Week,* March 4, 1985, 80ff.
Williams, M. J. 1984. The Pickens Plot that has Gulf Gulping. *Fortune,* March 5, 1984, 34.
Williamson, O.E. 1981. The Economics of Organization: The Transaction Cost Approach. *American Journal of Sociology* 87: 548–577.

R. Edward Freeman is University Professor, Elis and Signe Olsson Professor, Academic Director of the Institute for Business in Society, and Senior Fellow of the Olsson Center for Applied Ethics at the University of Virginia's Darden School of Business.

Daniel R. Gilbert, Jr. is Professor Emeritus at Gettysburg College.

Carol Jacobson was a student of Strategic Management at University of Minnesota where she received a Ph.D. from the Carlson School of Business.

Chapter 20
Airline Horror Stories Indicate an Ethical Problem

R. Edward Freeman

Have you been on a commercial airline flight recently? Or have you been to a cocktail party where someone was talking about a flight? Airline stories – horror stories – are the hottest gossip around.

Unfortunately, most of the stories are true. Here are a few:

- A pilot asks passengers – you know, the people who pay the bill – to complain about the slow ground service in Chicago. Bottom line: 90-min delay and puzzled passengers who thought airline employees worked together.
- An airline spokesman tells a customer that the airline is "holding a flight for you." Bottom line: One-hour mechanical delay and a new in-depth understanding of the concept of "truth telling."
- An airline cancels a late flight from Aspen to Denver. Bottom line: $290 taxicab and one customer who will do anything to avoid this company.
- An airliner experiences mechanical problems before a flight out of Los Angeles. Instead of simply saying, "We don't know when or if this will be fixed," airline officials gave passengers a continual runaround. Some became angry when another flight left to which the passengers were not allowed to switch. The reason: "standbys" took the seats, which resulted in increased revenue for the airline. The airline's response to the complaining customers was to call the police. Bottom line: Amazement and a good cocktail party story.

All of these stories are true. I know. They happened to me during the past few months.

Originally published in: *The Richmond News Leader*, B-24, August 17, 1987
Reprint by Springer

R. E. Freeman (✉)
University of Virginia, Charlottesville, VA, USA
e-mail: FreemanE@darden.virginia.edu

© Springer Nature Switzerland AG 2023

S. D. Dmytriyev, R. E. Freeman (eds.), *R. Edward Freeman's Selected Works on Stakeholder Theory and Business Ethics*, Issues in Business Ethics 53,
https://doi.org/10.1007/978-3-031-04564-6_20

Combine them with travel agents who blame airlines. (One agent did not even bother to return my call complaining about the wrong seat and flight).

Add the credit card company which mistakenly enters address changes in its computer, issues no bills and turns you in to the credit reporting bureaus when you don't pay the bills they never sent you.

And if you haven't had enough, mix in a pinch of harassing phone calls and letters from banks, insurance companies and our friendly government regarding deceased parents' paid accounts 9 months after their death.

For some additional flavor and spice, add your favorite story. I'll bet you've got more than one, and I'll bet it made you "mad as mittens."

Stir this concoction, and what do you get?

Something is wrong in business, and it isn't just plain lousy service. And it certainly isn't deregulation or competition.

Something is wrong in business. And that something is ethics.

Ethics is a hot topic these days, but for my money, the emphasis is all wrong. I don't care about the headline issues such as insider trading, bribery, influence peddling and the purity of the profit motive in resupplying the Contras.

The current wave of interest in business ethics undoubtedly results in more grandiose corporate codes of conduct, statements of values, government rules and regulations, and other fine sounding words.

But the real problem is different. Ethics is about how we treat each other, every day, person to person. The current wave of concern misses that point.

If you want to know about a company's ethics, look at how it treats people: customers, suppliers and employees.

If you want to know about a company's ethics, look at how it handles complaints.

Ii you want to know about a company's ethics, look at how well it listens to its own employees.

All of the airline stories I have heard are about the little things: service, complaint procedures and teamwork among employees, or the absence of these.

Here is a checklist of questions that ought to be on every manager's desk and should provide the entire agenda for the board of directors for every airline. What would be your company's score on an "ethics audit" consisting of these questions?

- How do I treat customers: as means to a profit, or a ends in themselves, important in their own right, with their own projects that we are assisting?
- How easy is it for customers to complain to me personally? Customer service is every manager's job and not just the sales and service people.
- Do I actively encourage employees to take responsibility for their work? Do I listen? Do I encourage them to speak up, even when being critical of me? Do I punish "whistle blowers"?

The danger with the horror stories about the airlines in that government will "solve" the problems through re-regulation. And after the airlines will come banking, trucking and telecommunications. The great strides made toward a more free, Jeffersonian world will be lost. And executives will be responsible.

Throw away the codes of conduct. Shred the corporate values statements. Hang the culture on the coatrack. Let's get back to the basics.

Business is about people. Business ethics is about how customers and employees are treated. Service and complaints and participation are the keys to high ethical standards.

I don't know about you, but I think it's time for a change.

R. Edward Freeman is University Professor, Elis and Signe Olsson Professor, Academic Director of the Institute for Business in Society, and Senior Fellow of the Olsson Center for Applied Ethics at the University of Virginia's Darden School of Business.

Chapter 21
Healthy Tension Between Business and News Media

R. Edward Freeman

During the last several years, I have given speeches and seminars on business ethics to hundreds of executives. I tell them the public thinks that business occupies the moral low ground in society.

The executives' response is over-whelming: The news media are largely to blame. "Think about it," they say.

The executives say there never is any good news about business, always disasters such as Bhopal or the Dalkon Shield. New products aren't news, but plant closings are. The incredible surge of entrepreneurial activity in the United States isn't news, but losing market share to Japan is. Executives wonder: Can't the press ever tell "our side" of the story?

But that's not all my executive friends tell me. They talk about waking up in a cold sweat having dreamed that the 60 Minutes crew, or some vicious local version, has shown up at their offices, rumor in hand, ready to "investigate."

They also cite instances of being interviewed for an hour with only a two-sentence comment taken out of context as the result. It looks to them as if the story already is done and the reporter simply is proving a point that already is cast in stone. It is a no-win situation for businessmen — so they say.

Once they start in on the news media, it's hard to stop these business people. They quickly move away from the news to other programming. Where does the public get its image of business people? Would you like J.R. Ewing and the other "sweeties" on Dallas to be your ideal type? What about Saturday morning cartoons?

Originally published in: *The Richmond News Leader*, B-16, June 6, 1988
Reprint by Springer

R. E. Freeman (✉)
University of Virginia, Charlottesville, VA, USA
e-mail: FreemanE@darden.virginia.edu

© Springer Nature Switzerland AG 2023

S. D. Dmytriyev, R. E. Freeman (eds.), *R. Edward Freeman's Selected Works on Stakeholder Theory and Business Ethics*, Issues in Business Ethics 53,
https://doi.org/10.1007/978-3-031-04564-6_21

Watch for the fat, cigar-chomping boss with an I.Q. of about 23. Even Sesame Street, that paragon of virtue for our little ones, has subtle anti-business biases, the executives say.

Combine all this with such activities by reporters as hiding in bushes to report on the private goings and comings of a presidential candidate, going through trash to find evidence for stories and other practices.

Conclusion: It's no wonder that 70 percent of the public believes that business does a poor job on ethics. The executives think that the state of ethics in the news media is mostly to blame.

One thing is certain. Business people don't understand the news media very well. The first thing to remember is that news media organizations are businesses. News media executives must meet a payroll and make a profit like any other business.

News media executives face all the ethical challenges that others in business face: dealing fairly with customers, suppliers, employees, communities and stockholders. But it's a little harder for them.

News media companies have a personality problem. They usually have two identities.

The first is that of a business. Managers and employees have to worry about product quality, balance sheets and income statements, and must be adept at business techniques.

But news media companies have another identity. They are founded on one of the principles of democracy: freedom of the press. In addition to business values, news media employees must identify with this fundamental principle.

Reporters and anchormen are like the rest of us. They want to get promoted and recognized by peers. They want more money and better working conditions. Like all of us, they want successful careers. And they run the gamut of competence, some talented, some average, some terrible.

But at some point, most of them believe in what they are doing — that a society with a free and open news media is better than the alternatives.

The management challenge in news media companies is to balance the business and societal values. The great news media companies have people who have not gone completely cynical — people who believe in their careers and freedom of the press. The great news media companies know that they have a special place in our society as guardians of the free flow of information that makes freedom a reality.

We value a free press. We value the openness of our society and the resulting debate which a free press helps to create. Because freedom of the press is so ingrained in us, we expect news media people to be non-human. We expect them to tell us "just the facts, ma'am," with no interpretation, with no value judgments.

This is foolish. The mere act of choosing the words to describe "the news" is a value judgment. The virtue of a free press is not that it is valueless, but that we do not censor it because of its values.

If you don't like Dan Rather, watch CNN. If you don't like The Washington Post, read The Christian Science Monitor. Think critically. Decide for yourself. Celebrate the fact that you have choices.

We also value privacy in our society — privacy in personal affairs and in business decisions. When the news media make these private decisions public, conflict and resentment result.

The values of the news media aren't anti-business. The news media aren't responsible for 70 percent of the American people thinking that business does a rotten job on ethics.

A tension exists between business and the news media — one that leads to conflict and debate. Such conflict is a sure sign of health in a free society, especially if we continue to encourage debate about the role of ethics in business and the news media.

R. Edward Freeman is University Professor, Elis and Signe Olsson Professor, Academic Director of the Institute for Business in Society, and Senior Fellow of the Olsson Center for Applied Ethics at the University of Virginia's Darden School of Business.

Chapter 22
The Revolution in Management

R. Edward Freeman and Daniel R. Gilbert, Jr.

1 Ethics and Strategy

Business ethics is hot. There are daily stories in *The Wall Street Journal* and the popular press.[1] Television news carries the latest scandal, and eagerly depicts the current "robberbarons." Consider some of the recent events:

1. Dennis Levine and a host of others are accused and later convicted of insider trading. Speculation centers on the motivations of well-to-do Wall Streeters engaging in criminal activity.[2]

[1] We adopt the following convention in these endnotes. The first time a reference is used, we give the complete citation. On subsequent uses we refer the reader to the bibliography. For example, E. Freeman (1984), *Strategic Management: A Stakeholder Approach*, Boston: Pitman Publishing Inc., is henceforth cited as Freeman (1984).

Business ethics is equally hot in the academic world. The past several years have seen a spate of books, journals, and research centers devoted to a variety of topics. The strangest phenomenon has been the arrival of philosophers in business schools. Even though Alfred North Whitehead was influential in the early days of the Harvard Business School, philosophy has enjoyed little popularity among business students. For a sampling of philosophers' writings on business ethics, see current issues of *Business and Professional Ethics* or *Journal of Business Ethics.*

[2] For an account of the Levine situation, see M. Stevens (1987), *The Insiders: The Truth Behind the Scandal Rocking Wall Street*, New York: Putnam's.

Originally published in: *Corporate Strategy and the Search for Ethics*, 1988, Prentice Hall, Englewood Cliffs, New Jersey.
Reprint by Springer, Reproduced with authors' permission

R. E. Freeman (✉)
University of Virginia, Charlottesville, VA, USA
e-mail: FreemanE@darden.virginia.edu

D. R. Gilbert, Jr.
Gettysburg College, Gettysburg, PA, USA

© Springer Nature Switzerland AG 2023

S. D. Dmytriyev, R. E. Freeman (eds.), *R. Edward Freeman's Selected Works on Stakeholder Theory and Business Ethics*, Issues in Business Ethics 53,
https://doi.org/10.1007/978-3-031-04564-6_22

2. Ivan Boesky pleads guilty to trading stocks on insider information, in part on tips from convicted inside-trader Levine. Boesky goes on to cooperate with the SEC and allegedly tapes conversations with other business associates.[3]
3. T. Boone Pickens, Irwin "The Liquidator" Jacobs, Carl Icahn, and their fellow mavericks launch numerous takeover bids for firms they believe are mismanaged and undervalued. The results are a wave of corporate restructurings. The language of mergers and acquisitions becomes "white knights," "greenmail," "shark repellent," and "poison pills."[4]
4. General Motors, IBM, Honeywell, Exxon, and other major firms decide to divest their operations in South Africa. After a decade-long attempt to abide by the Sullivan principles, executives admit that doing business in South Africa no longer makes sense. Meanwhile, a senior executive from Mobil Corporation resigns from the board of the Duke University business school because of Duke's decision to divest stock of companies such as Mobil that still operate in South Africa.[5]
5. Johns Manville files for bankruptcy in an attempt to avoid liabilities stemming from the asbestos controversy. Executives claim that court protection is the only way to save the company. A. H. Robins follows a similar strategy after a long controversy surrounding the Daikon Shield.[6]
6. As more women enter the work force, companies adopt strict policies on sexual harassment. The U.S. Supreme Court rules that workers have a right to a working atmosphere free from sexual harassment.[7]
7. The competitive position of a number of U.S. industries declines as markets become global and as new players such as Japan and Korea target key sectors of the economy to develop. The results are loss of jobs, devastation of communities,

[3]The Boesky saga goes on and on, affecting much of the current news. For an account, see Stevens (1987).

[4]We have analyzed some of the issues in corporate governance, especially hostile takeovers and greenmail, in E. Freeman, D. Gilbert, and C. Jacobson (1987), "The Ethics of Greenmail," *Journal of Business Ethics*, Vol. 6, No. 3, 165–178.

[5]South Africa poses a complex issue as we are now seeing. For an account of some recent thinking by a diverse group of commentators, see the spring 1986 issue of *Business and Society Review*, No. 57, edited by S. Prakash Sethi. For a general account of the response of business, see D. Kneale (1986), "Firms with Ties to South Africa Strike Back at Colleges That Divest," *The Wall Street Journal*, 20 November.

[6]Both of these cases are the subject of a great deal of current scholarship. For a summary of the Johns Manville asbestos controversy, see P Brodern (1985), *Outrageous Misconduct: The Asbestos Industry on Trial*, New York: Pantheon Books. For an account of A. H. Robins constructed from published materials, see "A. H. Robins Co. and the Dalkon Shield," Charlottesville, Va.: The Darden School Case Studies, UVA E-49.

[7]Sexual harassment has been addressed in a number of places. For a sample of the literature see R. Quinn and P. Lees (1984), "Attraction and Harassment: Dynamics of Sexual Politics in the Workplace," *Organizational Dynamics*, Vol. 13, No. 3, 13–46; and G. Sullivan and W. Nowlin (1986), "Critical New Aspects of Sex Harassment Law," *Labor Law Journal*, Vol. 37, spring, 617–624.

renewed calls for trade protectionism, and diversification of some companies to other industries.[8]

8. Corporate restructuring has led many major companies to "downsize" their middle management. Estimates at companies such as AT&T are that 30,000 managers will eventually lose their jobs. The results at some companies are increased levels of stress for employees and more pressure to perform. Corporate loyalty has been radically altered.[9]

You quickly get the idea. There are multiple conflicts in the business world, and most of them arise because there are different values at work. These are clearly "business" issues, since the well-being of firms and individuals depends on how they are resolved. Just as clearly, they are "ethical" issues, since they allocate harms and benefits, and allow some to pursue their projects and prevent others from pursuing them.

Resolving the conflicts in these business decisions is difficult because the conflicts are about ethics and morality, concepts which are usually thought to be far from the minds of business executives. It is important to note that when we call a decision "a moral decision" we do not mean it is a morally correct decision. We mean that it is connected with moral issues, in much the same way that a financial decision is connected with financial issues. We mean that it is a decision of the moral kind. Strategic decisions are based on a set of values, and these values yield certain projects or purposes which key individuals or groups want to realize. If we want to understand the strategy, we must understand the values on which the decision was based. Let's look at some more examples in detail.

1.1 AT&T and Divestiture

On January 1, 1984, the largest corporation in the world ceased to exist.[10] AT&T was broken into eight separate multibillion-dollar companies. Early in 1982 the U.S. Department of Justice and AT&T agreed to a consent decree whereby the

[8]The classic statement is R. Hayes and W. Abernathy (1980), "Managing Our Way to Economic Decline," *Harvard Business Review*, Vol. 58, No. 4, 67–77. Much literature has followed since that article. For a similar position, see R. Charan and E. Freeman (1980). "Planning for the Business Environment of the 1980s," *Journal of Business Strategy*, Vol. 1, No. 2, 9–19.

[9]A recent issue of *Fortune* contains good anecdotal material on the effects of restructuring on management. See M. Magnet (1987), "Restructuring Really Works," *Fortune*, March 2, 1987 and P. Nulty (1987), "Pushed out at 45—Now What," *Fortune*, March 2, 1987. Nulty quotes a Bureau of Labor Statistics study that "between 1981 and 1986 almost 500,000 executive, administrative, and management workers lost jobs that they had held for at least 3 years." As one friend of ours in a restructured company puts it, "Restructuring will continue until morale improves!"

[10]The literature on the AT&T divestiture is burgeoning. For a sampling of views, see S. Coll (1986), *The Deal of the Century: The Breakup of AT&T*, New York: Atheneum; M. Derthick and P. Quick (1985), *The Politics of Deregulation*, Washington, D.C.: The Brookings Institution; D. Evans (ed.) (1983), *Breaking Up Bell: Essays on Industrial Organization and Regulation*, New York: North-Holland Publishing Company; E. Freeman (1983), "Managing the Strategic Challenge in

U.S. government would drop its antitrust suit against AT&T and allow it to compete in the relatively nonregulated computer business. In return, AT&T agreed to divest itself of its large distribution network and local telephone business, which embraced the wholly owned Bell System operating companies. During the months which preceded the divestiture, we can only imagine the tough decisions that AT&T managers had to make. Here was a strategic decision of the utmost importance. Billions of dollars rode on the outcome. The future of AT&T's business, and the way it was going to be defined, was up for grabs. But there were even more important issues at stake.

Over one million people worked for AT&T.[11] Many millions more, virtually everyone in the U.S. and many other countries, were customers of AT&T. A great many smaller firms made their livelihood by being suppliers to AT&T. AT&T had over three million shareholders. Local communities across the U.S. had AT&T facilities, and many had plants or offices on which the local economy partially depended. In short, the strategic decision to break up AT&T had effects on millions of citizens who had some stake in AT&T. Many had come to expect certain things from AT&T, from low-cost phone service to a job for life, and they felt that they were entitled to, or had a right to, these expectations. There was a virtual contract, perhaps a psychological one, for AT&T not to change overnight and break this string of stable expectations. The consequences of this decision and its effects on the lives of millions of people will not be fully known for a number of years.[12]

Telecommunications," *Columbia Journal of World Business*, Vol. 18, No. 1, 8–18; L. Kahaner (1986), *On the Line: The Men of MCI—Who Took on AT&T, Risked Everything and Won!*, New York: Warner; S. Kleinfield (1981), *The Biggest Company on Earth: A Profile of AT&T*, New York: Holt, Rinehart and Winston; A. Phillips (1983), "Regulatory and Interfirm Organizational Burdens in the U.S. Telecommunications Structure," *Columbia Journal of World Business*, Vol. 18, No. 1, 46–52; H. Shooshan (1984), "The Bell Breakup: Putting it in Perspective," in H. Shooshan (ed.) *Disconnecting Bell: The Impact of the AT&T Divestiture*, New York: Pergamon, 8–22; J. Tunstall (1986), *Communications Deregulation: The Unleashing of America's Communications Industry*, Oxford: Basil Blackwell; W. Tunstall (1983) "Cultural Transition at AT&T," *Sloan Management Review*, Vol. 25, No. 1, 15–26; W. Tunrstall (1985), *Disconnecting Parties: Managing the Bell System Break-up: An Inside View*, New York: McGraw-Hill; A. von Auw (1983), *Heritage and Destiny: Reflections on the Bell System in Transition*, New York: Praeger; and R. Wiley (1984), "The End of Monopoly: Regulatory Change and the Promotion of Competition," in H. Shooshan (ed.) (1984), *Disconnecting Bell: The Impact of the AT&T Divestiture*, New York: Pergamon, 23–46. Throughout this book our discussion of AT&T is based on these and other public sources, as well as a number of interviews with industry insiders and experts. We alone are responsible for the conclusions that we draw from this host of data.

[11] For an account of the scope of the undivested Bell System see Kleinfield (1981).

[12] We believe that the notion of "psychological contract" is a powerful one that has been underutilized in moral theory. Since it relies on a symmetry of expectations, it comes strikingly close to some contract theorists' attempts to find sets of stable expectations which meet certain conditions of fairness. In Nozick's terms, boundary crossings occur when the psychological contract is violated. See H. Levinson (1976), *Psychological Man*, Cambridge: The Levinson Institute; and compare R. Nozick (1975), *Anarchy, State and Utopia*, New York: Basic Books, at 57–59, for an analysis of the concept of "boundary crossings."

The important point is this: The strategic decision made by AT&T was essentially a moral decision. That is, the decision involved harmful or beneficial consequences to a number of human beings, and affected their rights, or their ability to pursue their own life projects the way that they had come to expect.

1.2 Corporate Pac-Man: Bendix, Martin Marietta, and Allied

In one of the most celebrated instances of corporate strategy, Bendix Corporation, a major supplier of automobile parts, attempted to enter the "high-tech" industries via acquisition.[13] In August of 1982, Bendix made a tender offer for Martin Marietta, an aerospace contractor. On the advice of its investment bankers and others, Martin Marietta made a counteroffer for Bendix, and each company began to line up financial backing in order to swallow the other: corporate Pac-Man was born. Both companies went heavily into debt, and the result was that Allied Corporation bought Bendix and that Martin Marietta was saddled with an enormous debt load.

This series of strategic decisions affected a number of groups that had a stake in these companies, from employees to customers to the U.S. government. Yet the decisions were made in relative isolation from those groups that had a stake in the firm.[14] These groups were not treated as if they were autonomous agents but rather as means to the ends of management of Bendix, Martin Marietta, and Allied. Should the stakeholders of these firms have a right to participate in decisions which substantially affect their well-being? Have they delegated their rights to participate to management? These are tricky ethical questions for which there are no easy answers, but which must be asked if we are to understand corporate strategy as it unfolds in the modern world.

---

[13] Our account of the Martin Marietta–Bendix–Allied–United Technologies situation is based strictly on published sources. We have no insider insight into what actually transpired. There is a rather large body of literature on the case. See G. Anders (1982), "First Boston Enjoys Surge as Speculators Expect Big Earnings Gain From Fee as Bendix Advisors," *Wall Street Journal*, 24 September; "The 4 Horsemen: Did Main Characters in Big Takeover Saga Let Egos Sway Them?", *The Wall Street Journal*, 24 September 1982; P. Hartz (1985), *Merger*, New York: William Morrow; A. Hughey (1982), "Allied Corp. Plans Major Debt Reduction, Possible Sale of Bendix's Holdings in RCA," *The Wall Street Journal*, 16 December; P Ingrassia (1982), "Forget Football. We Have Bendix and U. Tech," *The Wall Street Journal*, 24 September; J. Koten (1982), "Bill Agee of Bendix Corp: Why I Did It," *The Wall Street Journal*, October 2, 1982; M. Krebs (1985), "Allied Reviews Takeover of Bendix," *Automotive News*, 27 May; T. Metz (1982), "The Pacificist," *The Wall Street Journal*, 3 December; "Pac Man Economics," unsigned review and outlook, *The Wall Street Journal*, 27 September 1982; "Our Dream is to Work Together," unsigned Interview with W. Agee and M. Cunningham, *Fortune*, 18 October 1982; and, B. Sloan (1983), *Three Plus One Equals Billions*, New York: Arbor House. We are grateful to Ms. Twyla Greenheck for an admirable job of synthesizing this literature for us.

[14] For an account of the decision making process during the episode see Hartz (1985).

1.3 The Excellent Companies

The last several years have seen a celebration of the management process at a number of companies deemed "excellent" on a number of dimensions.[15] At Milliken, Tupperware, Hewlett-Packard, Digital Equipment Corporation, the Tactical Air Force Command, and a host of other companies, we are told that work has meaning, and organizational members seem to have fun and enjoy what they do. We are all familiar with the stories and lore of the International Business Machines Corporation, and the unparalleled consistency of McDonald's around the world. Not coincidentally, these organizations are all high performers on almost any measure you want to take. As of early 1987, Peters and Waterman's book on the excellent companies had sold over five million copies worldwide, and had been on *The New York Times* best seller list for over 100 weeks.

The magic at these excellent companies is an understanding of human values and ethics, and how they fit into these firms' conceptions of where they are going, that is, their corporate strategy. The search for excellence is really a search for ethics.

We also have tales of Japanese companies where workers sing the company song, wear uniforms, and have an incredible dedication to their firms. Productivity increases, quality is unparalleled, and competition begins to fall all over the world. The secret is a close connection between values, ethics, and strategy.

1.4 So What?

In all of these examples, key business decisions are concerned with ethics. Almost all business decisions benefit some groups and individuals, while harming others. Almost all business decisions help some group to realize its purposes and projects, or cause some group to fail to realize its projects. The punch line is this: Ethics and business go together, and this book is about how to make them work together when we think about corporate strategy.

[15] See T. Peters and R. Waterman (1982), *In Search of Excellence*, New York: Harper and Row, and T. Peters and N. Austin (1985), *A Passion for Excellence*, New York: Random House. In addition, Peters has published numerous columns and pamphlets such as T. Peters (1986), *The Promises*, Palo Alto, CA.: The Tom Peters Companies, which detail the revolution. We have had the opportunity to see Peters talk about the excellent companies many times, and the writing simply cannot give the full effect of the revolution that is taking place in management. Curiously enough there is little excitement about these ideas in large companies, as Peters' most recent examples are from small and medium sized companies.

2 A Brief Summary of Our View

There has been a revolution in management thought and practice.[16] Like all revolutions, it has emerged over time and is not the sole province of one thinker or manager or corporation. The revolution in management has been brought to you by some fanatics: fanatics who believe in what they are doing and bring a real emotional passion to their work. Yet the "fanatics" who work in excellent companies are ordinary people, people like all of us. They happen to work in an environment which activates their values. Their work situation turns them on in a way that most would envy. We shall argue that the key to understanding the revolution in management is understanding values and ethics, and the role that they play in our organizations.

The revolution has taken place on two fronts that have been seen to be distinct and different, but are really quite closely related. And when we combine these two fronts we shall show that some new and extremely powerful ideas about the corporation are possible.

The two fronts of the revolution in management theory are these:

1. We have "discovered" that organizations consist of human beings.[17] And if you closely examine human beings, you find a complex network of values. Values are the ultimate reasons that we have for doing what we do. If we truly want to understand why organizations work the way they do, we must have a comprehensive understanding of human values. Therefore, managers are adopting "values management" techniques, "excellence programs," "culture management," "one-minute management," and so forth. Management theorists are busy trying to find a place for concepts like "symbolism," "myths," "culture," "individualism and collectivism," "strategy implementation," and "strategic human resource management," all in the name of better "people management."
2. We have "discovered" that organizations do not exist in a vacuum. In making choices—strategic choices—corporations find that there are other outside groups and individuals who have a stake in what they do.[18] These "stakeholders," such

---

[16] We take as evidence of the revolution those stories so vividly told by Peters and others.

[17] The literature on "values" is quite recent. See T. Deal and A. Kennedy (1982), *Corporate Culture*, Reading, MA.: Addison Wesley; G. Cavanaugh (1982), *American Business Values*, Englewood Cliffs, N. J.: Prentice Hall; F. Sturdivant and J. Ginter (1977), "Corporate Social Responsiveness: Management Attitudes and Economic Performance," *California Management Review*, Vol. 19, No. 2, 30–39; and F. Sturdivant (1979), "Executives and Activists: A Test of Stakeholder Management," *California Management Review* Vol. 22, No. 1, 53–59. There is an older tradition in management that is relevant, especially some of the classics such as K. Andrews (1980), *The Concept of Corporate Strategy*, 2nd ed., Homewood, Il.: R.D. Irwin; C. Barnard (1938), *The Function of the Executive*, Cambridge, MA.: Harvard University Press; and P. Selznick (1957), *Leadership in Administration*, New York: Harper and Row.

[18] There is a vast literature on organizational environments. Freeman (1984), offers one interpretation of this literature. There seems to be an important distinction in this area. There are those who believe that "environment" does the explaining of organizational action, thereby belittling the importance of strategic choice or simply treating choice on a par with random variations. This

as customers, suppliers, communities, governments, owners, and employees, all make choices themselves, and all depend on organizations for accomplishing their projects. Likewise, organizations depend on them for their success. Managers are busy finding strategies to deal more effectively with government, activist groups, and so on, while management theorists write of "resource dependence," "population ecology," "corporate social responsiveness," and "stakeholder management," all in the name of improving the bottom line.

We can articulate these "discoveries" in two principles to which we shall constantly refer in the following chapters: the Values Principle and the Interdependence Principle.[19]

The Values Principle

Individual and organizational actions are caused in part by the values that individuals and organizations have.[20]

---

group fails to realize that the organization's environment consists of the strategic choices of other organizations. There is a separate group that thinks, as we do, that "environment" is a generic term and must be broken down into names and faces. While managers talk of "the business environment" and often blame it for various infelicities, there is little of substance to this notion. The causal net is not the environment, but specific actions, just as notions of "the economy" are ultimately reducible to microbehavior. Relevant but non trivial arguments to this effect can be found in K. Arrow and F. Hahn (1971), *General Equilibrium Analysis*, San Francisco: Holden-Day; and T. Schelling (1978), *Micromotives and Macrobehavior*, New York: W.W. Norton.

[19] Don't think of our principles as "first principles," or anything with "godlike" authority. They are pragmatic rules of thumb, open to revision. In the course of this book, primarily here in the notes, we will end up sketching a pragmatic account of what look, in the main text, to be absolute statements. Principles, like other general statements, even those of physics, are intelligible only in some particular context. And "context" stands for "the practices of an interpretive community." Language, in this sense, is always up for grabs. We follow the accounts of Richard Rorty in Rorty (1979), *Philosophy and the Mirror of Nature*, Princeton: Princeton University Press; Rorty (1982), *Consequences of Pragmatism*, Minneapolis: University of Minnesota Press; Rorty (1983), "Postmodern Bourgeois Liberalism," *Journal of Philosophy*, Vol. 80, No. 10, 583–589; Rorty (1984a), "The Historiography of Philosophy," in R. Rorty, J. Schneewind, and Q. Skinner (eds.) (1984), *Philosophy in History*, Cambridge: Cambridge University Press, 49–75; Rorty (1984b), "Texts and Lumps," *New Literary History*-Vol. 16, No. 1, 1–16; and, Donald Davidson in Davidson (1984), "On the Very Idea of a Conceptual Scheme," *Inquiries into Truth and Interpretation*, New York: Oxford University Press, 183–198; and, Stanley Fish in Fish (1980), *Is There a Text in This Class?*, Cambridge: Harvard University Press; Fish (1984), "AntiProfessionalism," *New Literary History*, Vol. 16, No. 1, 89–108; and S. Knapp and W. Michaels (1982), "Against Theory," *Critical Inquiry*, Vol. 8, No. 4, 732–742.

[20] We follow the argument of Donald Davidson in Davidson (1980a) "Actions, Reasons and Causes," in *Essays in Action and Events*, New York: Oxford University Press, 3–19, that values can be both reasons and cause. We are grateful to Edwin Hartman for making this point to us. For an application to the Harvard way of doing strategy see E. Freeman, D. Gilbert, and E. Hartman (1988, in press) "Values and the Foundations of Strategic Management," *Journal of Business Ethics*.

The Interdependence Principle
Organizational success is due in part to the choices and actions of those groups that have a stake in the organization.[21]

The powerful new conception of the corporation and of management theory is evident when we make the obvious linkage between the principles. *The first principle applies to those groups which the second principle says are important.* If the values of the corporation and its members are important in understanding corporate action, then so too are the values of key "stakeholders" important in understanding how they interact with the firm.

To understand why corporations choose to do what they do, we must understand the values at work in the actions of multiple stakeholders. Understanding corporate strategy means understanding the competing values claims of multiple stakeholders. Let us put it differently, again in the form of two axioms (or basic assumptions) which sum up the two fronts of the revolution in management.[22]

First Axiom of Corporate Strategy
Corporate strategy must reflect an understanding of the values of organizational members and stakeholders.

When different sets of values conflict (and even where they coincide), problems of ethics and morality emerge. Which values are going to win the day? How are we going to act? How *should* we act? Who is going to benefit and who is going to be harmed by our action? Who will be able to pursue their own projects, and who will be prevented from pursuing their projects, if we take this action? If there were no conflicts of values, no basic disagreements about how the world should be and how we should live our lives, then there would be no need for moral argument, no need for methods of resolving conflicts. Each person would go merrily about his or her own way. Ethics is important just because we do not always agree on what we should do, and because it makes sense to ask: Is this decision the very best one that could be made in these circumstances?[23]

This leads us to formulate a second axiom, or basic assumption, about corporate strategy:

---

[21] Of course, in one sense this principle is trivial, since without stakeholders there is no organization. "Organization" cannot be defined without a membership and exclusion rule. See W. Evan (1966), "An Organization Set Model of Interorganizational Relations," in M. Tuite, M. Radnor, and R. Chisholm (eds.) (1966), *Interorganizational Decision Making*, Chicago: Aldine, 181-200. For an attempt to do without "organization" as a predicate, see the fascinating work of Steven Cheung in Cheung (1983), "The Contractual Nature of the Firm," *Journal of Law and Economics*, Vol. 26, No. 1, 1–21. For an attempt to apply the insights of Cheung's contract theory to the problem of corporate legitimacy see E. Freeman and W. Evan (1987b in press) "Stakeholder Management and the Modern Corporation: Kantian Capitalism," in T. Beauchamp and N. Bowie (eds.) (1987, in press) *Ethical Theory and Business*, 3rd edition, Englewood Cliffs: Prentice Hall.

[22] There is no difference on our view between axioms and principles. Fix some axioms and see what follows, see note 19, *supra*.

[23] Ethics also has a personal side which involves trying to see one's life as a coherent whole. See H. Barnes (1967), *An Existentialist Ethics*, Chicago: University of Chicago Press. We return to this idea in Chapter 2.

The Second Axiom of Corporate Strategy

Corporate strategy must reflect an understanding of the ethical nature of strategic choice.

Almost all questions of corporate strategy are questions of ethics.[24] Furthermore, the role of ethics in corporate strategy has been, for the most part, systematically ignored in theory and in practice. We can easily summarize the real point of the revolution in management: *We must put ethics in its rightful place at the very center of discussions about corporate strategy.* Our argument is that the very best corporations can and should be managed in a way that is consistent with a strategy built on a foundation of ethics.

Once we return ethics to the center of corporate strategy, it becomes possible to articulate and defend a number of different theories or models of corporate strategy, each of which is based on a different conception of ethics. We believe that the best possible conception of corporate strategy is one that is based on the rights of individuals, both managers and stakeholders, to pursue their own projects without interference and coercion from others. Corporations are, in our view, to be seen as mere means toward the accomplishment of human goals. This contrasts sharply with the common view that corporations are ends in themselves, and that individual goals, wants, desires, values, and personal projects must be subordinated to those of the corporation.[25] We believe that this common view is in large part responsible for the current fashion of attacking the modern corporation. Let's look at some of the most recent rhetoric in more detail.

---

[24] To see how this follows, note that we assume that strategy results in some action which affects the ability of persons to carry out their project. To deny this also denies the Interdependence Principle. However, to say that most questions of corporate strategy are ethical is not to say very much. Most interesting questions of human interaction are questions of ethics. We don't always have to "do ethics"; rather, we can hold some principles as fixed points, so long as we are aware of the need to justify these principles as being ethical in nature. This is even true in that most hallowed area of academic discourse: science. For an analysis of the value judgments in science, see R. Rudner (1953), "The Scientist Qua Scientist Makes Value Judgments", *Philosophy of Science*, Vol. 20, No. 1, 1–6. The amazing point is that stating that corporate strategy and ethics are logically related comes as news to both researchers and managers.

[25] Another way to put this point is that individual goals are the primary units of analysis. It is not that corporate goals and projects do not exist, nor that they are decomposable into individual goals, but that they are explainable in terms of individual goals. Such an explanation can be accomplished without a radical form of reductionism. See D. Davidson (1980b) "Mental Events," in *Essays on Actions and Events*, New York: Oxford University Press, 207–227, and Rorty (1979). We are grateful to Edwin Hartman for clarifying this point.

3 The Attack on the Corporation

In part because ethics and values have little or no place in current conceptions of strategy, the corporation itself has come under increasing fire. This issue is well documented, and we shall mention it only briefly.[26] It is important to note that the attack on the large corporation has not filtered down to small business. Most of us still have the notion of the entrepreneur as the great American hero. This we believe because the entrepreneur clearly makes a statement about his or her values by deciding to become an entrepreneur. These value statements, exhibited through action, are what is missing in most corporate strategies.

Several charges have been leveled at the modern corporation in recent years. Each reflects an underlying concern with values and ethics.

1. *The corporation is too big and too powerful.* This sounds a familiar theme to those involved in trying to "democratize" the corporation or to restrict its power through structural means such as law. A U.S. Senate bill that was introduced repeatedly for several years would have put restrictions on the size of firms that would be allowed to merge. We often see comparisons of the sales and assets of General Motors, IBM, and others with the gross national products of many nations around the world. The argument is that since some corporations have become so large and so pervasive (they operate throughout the world), they have become virtually uncontrollable.

This particular attack on the corporation is obviously an ethical issue, at least to those who bring forth the attack. But why should executives be concerned with the ethics of growing to a certain size, through merger or just smart business? There are several answers. The first is that executives must understand why their critics see corporate size as a moral issue, and this requires applying the Values Principle to the strategy of the critics. Executives must find an answer to this question: What are the values which cause these critics to see the world this way?

The second answer is that while there is no logical connection between size and ethical or unethical behavior, there probably is a connection between size and the ability to respond quickly and effectively to those groups that depend on the corporation: its stakeholders. The very ability of the corporation to respond to its stakeholders is itself an ethical issue. If it cannot respond (let's leave aside the important question of its willingness to respond), then stakeholders will be harmed and some might find their rights violated. (If you don't like this argument, simply substitute "stockholder" for "stakeholder." It still holds true.) Thus, in growing to a certain size you need to examine your ability to be able to respond to the demands of your stakeholders. Now every executive worth his or her salt knows this basic business truth. But few see it as a moral issue—and to see it so gives it a renewed importance.

---

[26] For a comprehensive account of the critique of the modern corporation, see E. Herman (1981), *Corporate Control, Corporate Power*, Cambridge: Cambridge University Press; and R. Nader and W. Taylor (1986), *The Big Boys: Power and Position in American Business*, New York: Pantheon Books.

2. *The corporation in the U.S. has lost its ability to compete.* Again, this is obviously an ethical issue. But, so are the solutions that are proposed in response to this attack. Favorable government treatment toward some industries, in the form of import quotas and tariffs, tax credits, and the like, benefits or harms some groups in society, and it rides roughshod over the rights of some. The question is whether any of these "solutions" can be justified. The call for justification is a challenge to place these problem-solutions within the context of some moral theory.

Now, taking on the whole of moral theory is a pretty tall task to ask of anyone, and we certainly do not propose to solve all of the knotty problems of moral philosophy, but we do believe that if you really want to understand the problem of the ability of U.S. firms to compete in the world at large, you must at least be willing to address the ethical issues that are inherent in most views of corporate strategy. That is, corporate strategy must be built on a foundation of ethical reasoning. Such applied ethics must offer guidance on how to solve the problem of competition.

3. *The corporation is not "human." It is a nameless, faceless entity that simply does not care about human concerns.* The argument is not directed solely at the corporation but at bureaucracies in general. In recent times, two U.S. presidents have been elected because they gave the express promise to clean up the bureaucracy and to make government more responsive. The attack on the corporation is very much in line with what we have called the revolution in management: It is people that count, not corporations. All of the benefits of the corporation are benefits to some group of individuals—from customers and suppliers to owners and the local community—and without the individuals, there would literally be no corporation.

Once again we run squarely into the Values Principle and the Inter-dependence Principle. If we are to address this concern we must "discover" what it is to be human, and that entails an analysis of ethics and values. Likewise, if we are to formulate and implement strategies to make the corporation more responsive, we must understand the values issues and ethical issues of those groups that depend on the corporation. In short, we must link ethics and strategy. But there is another problem. In concert with the fashion of attacking the corporation, some critics have turned to the very concept of corporate strategy.

4 The Attack on Corporate Strategy

Corporate strategy and the hordes of MBAs who "think strategically" are the current targets of much scorn. However, we believe that the modern corporation cannot do without corporate strategy, even if the current versions are inadequate or irrelevant, and we shall argue that we need to infuse corporate strategy with an understanding of values and ethics. But first, let's try to understand the positions of the critics.[27] Consider the following stories as symbols of their attack:

> One *Fortune* 500 company has fired most of its strategic planners, and the CEO announced in a speech to the Harvard Business School that if you wanted to do strategic planning you ought to choose another company. Another *Fortune* 500 company (not alone) decided to stop hiring MBAs. It decided, rather, to train its own managers from within the firm.
>
> *Business Week* has renamed its "Corporate Strategy" column "The Corporation."
>
> Consulting firms are flocking toward the hot topics of "corporate culture" and "strategic human resources management," leaving strategic planning models on the drawing board.

On the theoretical front, books that deal with how to "manage people" have hit the best seller list. On the desk of every fast-track manager are the high-tech—high-touch world of *Megatrends*, the book *In Search of Excellence*, the latest on Japanese management, and the latest in psychological motivation as found in *The One Minute Manager* and *The One Minute Sales Person.*

The list is endless. Corporate strategy is out, and people are in. On the academic front, the hot topics are "strategy implementation," "corporate culture," "the symbolic nature of management," "population ecology," "corporate social responsiveness," and "strategic planning processes to create corporate futures."[28] The message in these incantations is clear: It doesn't make much difference what you decide to do, so long as you follow a process which gives people some sense of participation.

The reasons for the attack on corporate strategy are multiple, and some of them are quite valid:

1. *Corporate strategy has become the tail that wags the dog.* This objection states that while corporate strategy is important, it has led executives to ignore the main part of a business: its operational structure. Coinciding with the emergence of corporate strategy, we find the decline of manufacturing and operations as important. Manufacturing is only a part of strategy, and the overemphasis that most theories and models of strategy put on marketing and finance ignores the basics of producing some goods and services.
2. *Corporate strategy has become too analytical.* This argument states that managers have relied too much on the formal models of academics and consulting

[27] The attack on corporate strategy is led by Hayes and Abernathy (1980), R. Pascale and A. Athos (1981), *The Art of Japanese Management*, New York: Simon and Schuster. Wickham Skinner has been a voice in the wilderness for a number of years about the disregard for manufacturing as a key variable in strategy. For an early and prescient attempt to rectify the concept of corporate strategy, see the articles which led to his recent brilliant book, W. Skinner (1985), *Manufacturing: The Formidable Competitive Weapon*, New York: Wiley.

[28] Ironically enough, the academics in strategy are experiencing booming times. Strategic Human Resources Management, Strategic Information, and International Competitive Strategy are the latest genres of strategy. The job market in these specialties is up. Strategy professors have taken refuge in becoming "more empirical and scientific," and that attitude is a best seller in business schools. But the real point of the critique is that the current language of strategy is optional. We can simply set it aside and not lose much if we focus on people and their relationships to each other and the organizational systems in place.

firms. What company has not had its portfolio plotted, does not understand the difference between dogs and cows, and does not know the relationship between market share and ROI, ROE, and ROTA? The emergence of a "strategy industry" has done more harm than good, say the critics, for in their drive to be "objective" and give "objective, hardheaded, numbers-oriented" advice, strategists have left out a crucial factor: managerial judgment. According to this critical perspective, strategy is really more an art than a science.

3. *Corporate strategy requires too much centralization.* The emergence of large planning staffs, filled with bright, well-scrubbed MBAs fresh from the B-schools, has led to control of business by staffers rather than line managers. To the extent that the concept of corporate strategy has meant "strategic planners," then it has led to endless layers of forms, reports, and overheads, all of which get in the way of operating managers.
4. *Corporate strategy has led executives to give short shrift to the people factor.* This is the main rallying cry for the revolution in management that we mentioned above. The critics argue that strategy treats people as factors of production, more or less capable of implementing any idea. In this view of strategy, people do not necessarily participate in the corporate strategy process, so they have no commitment to the strategy which is the outcome of that process. Hence, the critics emphasize the need to think more about people, implementation, and culture if this shortcoming is to be overcome.

While each of these criticisms has some validity, each ignores the fundamental mistake of most theories and models of corporate strategy. Individuals act as they do because of their values. The value sets of individuals provide strong reasons for action, and the most dear values are ones we call "moral values" or "ethical values." Corporate strategy which ignores the role of people in the organization simply ignores why organizational members act as they do. Corporate strategy must return to the individual values of corporate members, *before it is formulated.* It must be built on these values, rather than taking them as constraining forces.[29]

While we will have much more to say about values and ethics in later chapters, especially Chaps. 3 and 4, it is important to note here that values are not slogans, nor are they necessarily shared among all organizational members. Addressing the problem as one of "implementation" or "culture" simply tries to fix (in vain) the wrong problem. It is no wonder that there is an "implementation gap" in most companies between the strategy and what people actually do. Strategy which is "good on paper" but "lousy on implementation" is simply bad strategy. Bad strategy makes for atrocious performance and calls into question the very legitimacy of the corporation.

---

[29] Another way to put this point is that the formulation-implementation distinction is not useful. It leads managers to thinking about strategy as a cognitive task followed by a persuasive task. D. Gilbert (1987), *Strategy and Justice*, Minneapolis: Ph.D. Dissertation, Department of Strategic Management and Organization, Carlson School of Management, University of Minnesota, explores these issues in greater detail.

The attack on corporate strategy is serious enough to warrant another look at the underlying nature of the concept of corporate strategy. While such an analysis is one of the leitmotifs of this book, the next section contains a general overview of our approach.

5 The Concept of Corporate Strategy

The diversity of the modern corporation has led to an increased need for a framework or common language so that managers in a number of functions and business elements can talk to each other, and so that corporate-wide resource allocation decisions can be made. The manager of the timber business must be able to talk to the manager of the consumer products business. The top executives must be able to evaluate both businesses, when there are limited resources. And those same executives need to be able to talk to a variety of customers, employees, suppliers, and communities.

The concept of corporate strategy has evolved to deal with the complexities of modern business.[30] No longer can managers rely solely on "seat of the pants" common knowledge, nor can they rely solely on functional skills such as marketing, finance, or production to pull them through the complexities of today's global business environment. Corporate strategy provides a method and language to integrate multiple functions and businesses.

Corporate strategy has become a central metaphor for business.[31] Virtually every large and successful firm, like many small and unsuccessful ones as well, has a strategy, and most spend countless hours worrying whether or not it is effective.

During the last 30 years the concept of corporate strategy has gotten increasingly complex. From the tried and true pronouncements of consultants, to the case-study classrooms of the Harvard Business School, the study of corporate strategy has become a discipline in its own right, both in practice and in professional schools of management. The old methods of solemnly imploring "build on your strengths"

---

[30] Strategy academics trace the evolution of the field as quite recent. See I. Ansoff (1977), "The State of Practice in Planning Systems," *Sloan Management Review*, Vol. 18, No. 2, 1–28; D. Schendel and C. Hofer (1979) "Introduction" in D. Schendel and C. Hofer (eds.) (1979), *Strategic Management: A New View of Business Policy and Planning*, Boston: Little Brown, 1–12. D. Gilbert, E. Hartman, J. Mauriel, and E. Freeman (1988, in press), *The Logic of Strategy*, Boston: Ballinger, gives a slightly more historical account. See also M. Horwitch (1988, in press), *Postmodern Management*, New York: Basic Books, for a more accurate and convincing view.

[31] We insist on thinking of corporate strategy as metaphor rather than an indentifiable and discrete set of sentences that describes all actions of a particular organization in a certain arena and only those actions. If you begin to think of corporate strategy as a metaphor, it will become easier to connect to areas of the humanities such as ethics, literature, art and the like. If you insist on thinking of corporate strategy as a concept capable of the same kind of precision as "heat" and "pressure," you will waste a lot of time searching for the equivalent of Boyle's law.

have given way to computer simulation models such as Profit Impact of Market Strategies (PIMS), a plethora of analytical matrices, and complex theories.

Even though there are endless definitions and models of corporate strategy, we believe that corporate strategy is really very little understood. Standard views of corporate strategy describe it as a process which answers four essential questions[32]:

1. Where are we going?
2. How do we get there?
3. What is our blueprint for action?
4. How do we measure or control progress?

The usual narrative which justifies these questions runs something like the following:

> As the Cheshire Cat said to Alice, in Lewis Carroll's *Alice in Wonderland*, "If you don't know where you're going, then any road will take you there." It is the task of modern management to chart a course for corporations, and managers must ask the question "Where are we going?" Corporate strategy understood as helping to answer this question is simply inevitable.
>
> Equally important is the question of means: How do we achieve the desired direction? Without some thinking about how to get to where we want to go, a statement of corporate direction becomes just empty words. Alice's question to the Cat was the right one, given that she knew where she wanted to go.
>
> The third question is important because if executives can't articulate the connections between the desired path of the company and what individuals actually do on their jobs today, then strategy will not make much difference to the way the company actually runs. Connecting desired plans with the actual behavior of managers and other organizational members is a great challenge, one that is central to any effective conception of strategy.[33]
>
> Finally the issue of control, as stated by question 4, is important. If strategy is not to be just cheerleading or thinking great thoughts, then executives must have clear ideas about how to measure progress. Control of direction, pathways, and blueprints is necessary.[34]

Corporate strategy, articulated in the form of these four simple questions, is a powerful device for managing the large and multifaceted organization that today's corporation has become. It is hard to see how these questions can ever decrease in importance or become irrelevant. But we need to take a closer look at the very nature of the concept of corporate strategy. There is more to the concept, which a closer reading will unveil. The First and Second Axioms of Corporate Strategy force us to recognize that the question "Where are we going?" is ambiguous. The question

---

[32] These questions are adapted from P. Lorange (1980), *Corporate Planning: An Executive Viewpoint*, Englewood Cliffs: Prentice Hall Inc. We are grateful for many conversations with Lorange in which he convinced us of the power of these simple questions.

[33] It is crucial to connect the actual budgets, blueprints, and concrete actions of specific individuals to the broader questions of direction. Usual models of strategy disconnect these questions, or see the third one as a question of "implementation" or "persuasion." Freeman (1984), and E. Freeman and D. Gilbert (1987) "Managing Stakeholder Relationships," in P. Sethi and C. Falbe (eds.) (1987), *Business and Public Policy: Dimensions of Conflict and Cooperation*, Lexington: D. C. Heath, try to show how a close reading of Graham Allison (1970) *Essence of Decision*, Boston: Little Brown, implies that the transactional level must be logically connected to the rational level of decision making.

[34] For a summary of recent work on strategic control, see P. Lorange and M. Scott-Morton (1986), *Strategic Control*, St. Paul: West Publishing.

assumes that the "we" is well understood, and that an answer, in terms of some corporate goal or purpose, is a straightforward matter. It is not.

5.1 *Strategy As Purpose*

A manager who approaches an important decision will likely ask: What is this situation really about? Who is taking action? Why are events unfolding in this way? In the back of that manager's mind is a set of related, but different, concerns: What are the appropriate responses? What will be required to take such steps? Eventually, if he or she determines that a response is warranted, the appropriate inquiry turns to: How can the effectiveness of these actions be determined? At the same time, managers in other organizations that are affected by this manager's decision are confronting the same set of questions.[35]

The Values and Interdependence Principles and the corporate strategy axioms imply that reasoning about the nature of strategy is equivalent to reasoning about the purposes of individuals and groups. We are going to try to convince you of this point by drawing on some everyday, common-sense examples, and on our practical understanding of what a purpose really is. There are obvious parallels and applications to the concept of strategy.

Having a purpose implies some continuity of action. To say that Jane has a purpose in buying the car is to say that her purchase is more than a mere fleeting desire or impulse. We tend to associate a person's overall competence and rationality with an ability to carry through on actions that are motivated by purposes.

This section addresses the meaning of purpose.[36] There are five key features of "purpose" which we want to explain. And we will argue that reasoning about strategy is the same thing as reasoning about purposes when we understand that:

1. Purposes are personal.
2. Purposes guide action.
3. Purposes require others.
4. Purposes are shaped by bargaining.
5. Purposes are the bottom line for performance.

---

[35] Thus, strategy is essentially game-theoretic. Events are functions of my expectations of your behavior, and your behavior is a function of your expectations of my behavior, and so on. T. Schelling (1960), *The Strategy of Conflict*, Cambridge: Harvard University Press, is still the best book available on business strategy, even though it is about the arms race. We develop this theme below in Chapter 5.

[36] We talk about "strategy as purpose," and later, in Chapter 3 we will talk about "ethics and values" and "moral values." Distinguishing between "purpose" and "value" is arbitrary, though some might say that values are more basic than purpose. If you think of purposes as resulting in certain "projects" to be accomplished, perhaps we agree. The point is that they are all the same kind of intentional concept. "Purpose" seems more in line with the jargon in strategy, and "value" seems more in line with the jargon in ethics.

1. *Purposes are personal.* Having a purpose is a capability that we commonly and uniquely attribute to a human being. Having a purpose distinguishes us from other living, sensing entities. Further, it is individuals, first and foremost, who have purposes. Our judgments about our own tastes, aspirations, and relationships lead us to an understanding of purpose as personal in three key respects.[37]

First, no one can experience the meaning of a particular purpose better than the person who has it. Empathizing with another person is as close as we can get to actually sharing his or her purpose. Second, the reasoning processes by which purposes are formed and realized are useful only if we can actually perform some actions. Each person, acting alone, uses his or her own knowledge and intuition in purposive action. Third, a purpose is subjective. The person who holds the purpose is the subject of it. Even if a person's purpose in life is to bring about social good, having that purpose is subjective.[38]

Strategy shares this idea of purpose as personal. Reasoning about strategy is not a mechanical affair. Production lines can be driven by technological wizardry. But corporate strategy is a creative endeavor that only an individual or a group can practice.[39]

The personal nature of strategy is clearly shown in the fact that the desirability of candidates for senior executive positions varies with the nature of prior managerial experience. Length and intensity of experience are presumed to reveal how well a person understands and acts upon the issue of corporate purpose. Some managers and theorists also believe that executive success varies with prior experience in similar settings. Hence, we should not be surprised to find Donald Trautlein, an accountant, at the helm of the financially troubled Bethlehem Steel Corporation. On this same reasoning, Peter Ueberroth's marketing skills at the 1984 Summer Olympics attracted major-league baseball owners in search of a CEO. In fact, this tale is told time and again as new firms begin to face broader strategic issues than those that were pressing during start-up. Howard Head at Head Ski, Steven Jobs at Apple

[37] Once again, purpose functions as a theoretical term in explanations. You simply can't give a proper account of human action without it. For a clear and concise view of explanation, see J. Elster (1983), *Explaining Technical Change*, New York: Cambridge University Press. *Pace* J. Pfeffer (1982), *Organization*, Boston: Pitman Publishing.

[38] What we have said here, and will say later in this section, is fairly standard in accounts of intentionality in the philosophical literature. See J. Searle (1984), *Minds, Brains, and Science*, Cambridge: Harvard University Press, for a readable introduction. However, we caution that the issue is bitterly contested once one leaves the surface. See Rorty (1979) for a historical treatment.

[39] There are no better examples of this point than in the classics such as A. Chandler (1962), *Strategy and Structure*, Boston: MIT Press; A. Chandler (1977), *The Visible Hand*, Cambridge: Harvard University Press; A. Sloan (1969), *My Years with General Motors*, New York: Doubleday; and, Barnard (1938).

Computer, and Seymour Cray at Cray Research moved aside when a different personal focus was needed.[40]

Despite the fact that it may seem to be just plain common sense, thinking of purpose as person-centered causes problems to our normal way of understanding both people and organizations. Purpose comes in many flavors. At any moment, we can be confused about our own purposes. Our sensory and reasoning capacities are limited and vary with natural endowment and development. Yet, none of these constraints means that we should give up the concept. Rather, we need to think more deeply about what purpose really means. In later chapters we will explain how purpose and a closely related idea, "values," are connected, and we will try to persuade you in Chap. 8 that taking seriously "purpose as personal" means a radical rethinking of our organizations. For now let's make things just a little more complicated.

2. *Purposes guide action.* People's purposes, along with their values and projects, give meaning to their lives. Meaning is distinctively human, and the search for meaning is in part a search for some purpose that realizes our deepest hopes and dreams. Purpose structures our attempt to find our way in the world. Purpose is about relationships, beliefs, and careers, not breakfast cereals and necktie styles.[41]

We need not see ourselves as complete existentialists to accept this truth about purpose and meaning. Few people, even existentialists, stop at every turn and probe whether their actions mesh in some coherent sense. Yet, most of us do so at certain points in our lives and in some form. Formulating and realizing our purposes depends on both action and reflection.

Having a purpose guides our actions in several ways. First, we try to make sense of the world, from our family and friends to the arms race, in terms of what we want to do, and how the world will affect our ability to carry through our projects. In asking "Does that situation apply to me?" we observe the world and compare it with the knowledge we have about ourselves. Hence, our purposes serve to interpret the external world. Second, we find alternative courses of action based on the particular purposes we have. We determine whether what we want to accomplish can continue in its present form. Finally, we use purpose in order to determine the resources that we need to acquire and develop. In asking "Can I accomplish this?" a person has to decide what skills and resources are appropriate. Reasoning about strategy encompasses purpose in each of these respects.

---

[40] For a theoretical view that much of business success and failure can be explained by attention to demographic variables, see J. Pfeffer (1983), "Organizational Demography," in L. Cummings and B. Staw (eds.) (1983), *Research in Organization Behavior*, Greenwich Conn.: JAI Press, 5, 299–357.

[41] Purpose or values make it possible for us to come to terms with life, to justify our lives, to see life as a coherent whole. Imagine giving a justification of your life and your projects without reference to a purpose or to a value. You would not have an answer when faced with the question "Why did you do that?" For an account of values as answers to "why" questions, see chapter 3, *infra.*

Why articulate a strategy at all? In the 1960's, some strategy theorists argued that in cases where industry change was negligible, nothing more than a "seat of the pants" approach was needed. In the decade that followed, some proponents of corporate strategy took a different stance. In their view, a strategy served an essential, simplifying function for a chief executive officer.[42] Faced with too much data and too little time to assimilate it, a CEO needs some help, and strategy fits the bill, by helping to pick out parts of the buzzing, blooming confusion as relevant. For example, we might expect that Donald Burr, CEO at People Express, uses a corporate strategy to study oil prices, but not cattle prices, in tracking the Federal Aviation Administration, but not the Federal Election Commission, policies.

Purpose can guide choices by helping managers to frame a distinctive competence for themselves and their organizations. We reflect upon our competence by examining our past accomplishments and seeing whether and how our hopes and dreams were realized. We probe our aspirations, at the highest level, by answering such questions as "What do I stand for?" In studying recent actions by the People Express management to acquire Frontier Airlines and Britt Airways, we want to ask: Were these decisions informed by People's purposes?[43]

Purpose guides action, too, when managers move from defining a distinctive competence to acting upon it. Pursuit of a purpose triggers a process of self-analysis about relevant strengths and weaknesses.[44] The current emphasis upon strategy implementation bears witness to the need for purposes to guide action. Returning to the merger activity at People Express, several examples of this point are clearly present. Until the Frontier acquisition, People had operated in a nonunion environment. But the acquisition of Frontier meant that management at People assumed responsibility for a set of collective bargaining contracts. Are People's managers competent with respect to labor contracts? The acquisition of Britt, based in Chicago, heightens People's presence in the already competitive Chicago market. Does People have the financial resources to withstand a competitive shakeout there? Finally, the proximate timing of the acquisitions raises the issue of People's competence in managing such a far-flung route network. In each instance, the level of competence at People is exposed as People uses purpose to guide action. Recent events surrounding the takeover of People Express by Continental may well show how an uncritical view of the basic purposes of an enterprise and its managers leads literally to organizational death, and great personal stress.

---

[42] Andrews (1980) is the exemplar of this view, though he has recently changed his mind.

[43] For an account of the Britt–People Express–Frontier saga, see W. Carley and T. Agins (1985), "People Express Merger Might Cause Problems," *The Wall Street Journal*, 14 October; W. Carley (1986a), "People Express is a Victim of its Fast Growth," *The Wall Street Journal*, 25 June; W. Carley (1986b), "Struggling to Survive, People Express Alters Operations and Image," *The Wall Street Journal*, 31 July; A. Salpukas (1986), "People Express Beaten at its Own Game," *Minneapolis Star and Tribune*, 29 June; A. Salpukas (1985), "People Express to get Britt," *New York Times*, 28 December.

[44] Else you cannot ever tell if a purpose has been realized. We believe that such a view is the essence of rationality.

3. *Purposes require others.* If we are going to realize our purposes, then in all but the simplest cases we need the cooperation of others. The necessity for specialization and division of labor is reason enough, but the clincher for fulfilling purpose is the prospect for enduring exchange relationships. So, our definition of purpose needs further sophistication. Purpose is a personal guide for action, activated by agreements with others for mutual benefit.

This expanded conception begs an important question for strategy and ethics: How does cooperation endure in a world of persons more interested in their purposes than anyone else ever could be? Lest Thomas Hobbes's "war of all against all" occur, managers must find ways to ensure that agreements can be struck and honored. This strongly suggests that we look at the accomplishment of purpose in terms of bargaining among self-interested persons. At the very least, the bargainers will seek terms acceptable to each and will press for safeguards against the hazards inherent in fulfilling bargains over time.

Bargaining occurs across all levels of strategy. Carl Icahn moved to solidify his acquisition of TWA by negotiating new labor agreements with TWA's unions.[45] Apple Computer cofounder Steven Jobs and Apple management bargained to forestall growing acrimony surrounding Jobs's departure from Apple to start a new company.[46] Failure to consummate bargains can be equally significant. Lack of support from the parent labor union has significantly influenced the union local's strike tactics in the prolonged labor dispute at George A. Hormel and Company in Austin, Minnesota.[47] Finally, new forms of bargaining frequently emerge between suppliers and buyers. Many health-maintenance organizations and several major airlines are pursuing prepaid bargains with corporate customers for health services and business travel plans, respectively.[48]

4. *Purposes are shaped by bargaining*,[49] We compared appointment calendars in the midst of writing this chapter. Of course, the formal contract with the publisher commanded considerable blocks of time. No less insistent were obligations to our families and friends, not to mention commitments to teach, advise students, consult, and make progress on other scholarly papers in various stages

[45] For an account of the TWA story see C. Loomis (1986), "The Comeuppance of Carl Icahn," *Fortune*, 17 February, 19–25; and, D. Hertzberg (1986), "TWA Names Carl Icahn as Chairman, but His Victory is Seen as Bittersweet," *The Wall Street Journal*, 6 January.

[46] For an account of the Apple Computer story, see P. Gray (1986), "Apple Computer and Jobs Reach Pact Barring Sale of New Machine Until 1987," *The Wall Street Journal*, 20 January.

[47] Hormel and Company's dispute with the union is chronicled in M. Charlier (1986), "Hormel and Union Are Locked in Battle that Carries High Stakes for Both Sides," *The Wall Street Journal*, 23 January; N. St. Anthony (1986), "P-9 Trusteeship gets Judge's Approval; Local Leadership Out," *Minneapolis Star and Tribune*, 3 June; and, D. Hage (1986), "On Anniversary, P-9 Militants Haven't Given Up, But Rest of Austin Looks Ahead," *Minneapolis Star and Tribune*, 17 August.

[48] For a related concept, the value chain, see M. Porter (1985), *Competitive Advantage*, New York: The Free Press. See Chapter 5, *infra*, for a critique of this view.

[49] Gilbert (1987) explores the concept of "web of bargains" in much more detail, and connects it to the strategy literature.

of coherence. Since our purposes have a way of taking on multiple forms, managing them can become quite complicated. What we seek to avert is a "gridlock" situation in which we fulfill none of these obligations satisfactorily. We have various ways of dealing with the problem. Looking to purposes as guides, we set priorities. But, over a given time span, several obligations can command close attention simultaneously. We respond by interpreting (or reinterpreting) our purposes in terms of acceptable thresholds of performance. For example, we may need to do what it takes to avert an insistent phone call from a university administrator on some project. As this example shows, knowing the terms of each bargain also help us set such thresholds. Furthermore, we attempt to subdivide tasks into pieces amenable to both available time and attention. Finally, we think about safeguards.[50] We require certain assurances from our partners so that we can satisfy our expectations. And they expect safeguards in return. Juggling multiple bargains is a fact of life for every competent person.

Maintaining a coherent purpose through multiple bargains is vividly depicted in the case of the corporate strategy necessary to manage a major-league baseball team.[51] In the months prior to the beginning of a playing season in April, the team's CEO will oversee contract negotiations for each of approximately 40 players. While these bargains result in separate agreements for each playing employee, the individual bargains are not struck in isolation. Information costs have been reduced in recent years, so that one player—and his bargaining agent—can obtain insights into the terms of other players' bargains with the team. In this situation, the CEO has strong incentive to develop a set of working thresholds with regard to contract provisions. In baseball parlance, monetary limits are defended in order to "preserve integrity of the salary structure." This situation becomes all the more complicated for those CEOs who bid for the services of so-called free-agent players, who usually find their services in great demand and engage in concurrent bargaining for employment with several teams.

Within the major-league baseball industry, certain safeguards have been established. One is an arbitration process whereby a CEO and a player, through his agent, have incentive to converge on bargains before a third party is called upon to choose in favor of one over the other. Another is the use of incentive clauses, based upon individual performance, which can facilitate consummation of bargains uniquely suited to each player.

This is a simplified account, since the CEO and team management bargain with other suppliers, as well as with customers. Still, two points regarding team strategy become quite clear. First, the real purpose of a team is essentially defined by the

[50] O. Williamson (1985), *The Economic Institutions of Capitalism*, New York: The Free Press, gives a simple account of safeguards. E. Freeman and W. Evan (1987a), "Corporate Governance: A Stakeholder View," unpublished manuscript, extend the notion to include an analysis of who pays for the safeguards. Also see E. Freeman (1987), "Review of Williamson's *The Economic Institutions of Capitalism*," *Academy of Management Review*, Vol. 12, No. 2, 385–387.

[51] This section is based on our fanatical following of several major league baseball teams, through direct research and through such publications as *Baseball America* and *The Sporting News*.

pattern of contracts successfully negotiated, no matter what the owner happens to say in the press. Second, purpose emerges as a web of semiautonomous bargains, among individuals. Once you see the world in terms of persons and their purposes, it gets harder to draw a line between strategy formulation and implementation, especially because it is a distinctively organizational and impersonal idea.

5. *Purposes are the bottom line for performance.* Having a purpose serves its most important function as the test for performance. If we did not act upon a purpose we would be "fair game" for this question: Does that purpose really exist for you? It is simply incoherent to claim that your purpose is to publish poetry and then evaluate performance on the basis of your handball skills. Yet, the tendency to reason in this way grows as people move farther from identifiable standards against which to check their progress.

Having a purpose can also be provisional, and some of our statements about purpose are clearly conditional. For example, a marketing strategy might be stated in this form: "We will use both direct sales and dealer distribution channels, unless dealer sales significantly impede relations with our most preferred customers." This statement provides for circumstances that are reasonably foreseeable. If the threshold in this strategy is crossed, a revised statement of purpose becomes necessary. Thus, managers who reason in this way move back and forth between purpose and actual cases, retesting a specific purpose on the basis of actual experience.

Purpose is also provisional in the sense that we become wiser over time. A manager who reflects upon a range of experiences will be more likely to pursue sophisticated actions in the future than will someone who is unreflective. Furthermore, the particular statement of purpose may take precisely the same form as before, but the reflective person will look upon the purpose with new appreciation. For example, one might have a purpose of being a loving and supportive parent. But the full richness of meaning in that purpose likely grows from year to year with one child, and from child to child in a growing family.

6. *Summary.* To say that we have a strategy but we have no purpose is a contradiction. Purpose is fundamental to any coherent account of strategy that recognizes the importance of choice. Purpose is complex, as each of the aspects above shows. Our argument has been a very simple appeal to the common-sense meaning of purpose, and its application to the everyday world of business. But the critique of the corporation and corporate strategy needs a simple response, not a complex one. Infuse the application of strategy with an understanding of purpose, and link it to ethics, and we can rehabilitate the whole idea. It is relatively easy to see that if all we attend to are variations of the four questions which most models of strategy ask, we will miss the rich explanatory power of a conception of strategy based on the concept of purpose. The very nature of "Where are we going?" assumes some purpose or other as an answer, and as we have shown, such answers are complex in and of themselves. To automatically assume an organizational level of analysis leaves out the interesting issues of how such a web of bargains emerges. If a particular conception of corporate strategy is built

on an understanding of purpose, then it must be much more sophisticated. (In later chapters we will show you how the current theories or models or frameworks in corporate strategy have implicit assumptions about the importance of purpose.)

Corporate strategy is about purpose, and so is ethics. Purpose is a person-based justification for action, shaped through interdependent bargaining among multiple parties. Ethics not only parallels strategy, through a common concern with purpose, but also provides a framework for reasoning through the thornier aspects of strategy as purpose. By explicitly acknowledging this connection, we can make some real intellectual and practical progress in the study of corporate strategy.

6 The Problem: Linking Ethics and Strategy

People whom we meet from all walks of life will typically ask us what we do for a living. We usually say that among other things we teach business ethics. They usually laugh, or snicker, or crack, "That's a contradiction, isn't it?" We don't like this state of affairs, and this book is in part our attempt to correct it. Business strategy must be linked to ethics if "business ethics" is to have any meaning beyond just so much pompous moralizing. And, if we can link ethics to strategy, we can revitalize the concept of corporate strategy. We will be able to address the most pressing management issues of the day in the very terms that they demand to be addressed in: ethical terms.

In a nutshell, our argument is this: Most, if not all, corporate strategies raise ethical issues. For the most part, managers and academics have not addressed these issues in ethical terms, and more than anything else we mean to rectify this situation.

How can we link ethics and strategy? It isn't easy, and you shouldn't look to this book (or anywhere else, for that matter) for a cookbook formula. Our purpose is to help you think about some difficult issues, but the best guide that you can have is your own behavior and the behavior of your corporation, but beware.

Ethics is a difficult and very personal subject. You will really understand the chapters that follow if you are willing to honestly address your own ethical beliefs, and especially to ask yourself some hard questions about what you really believe. Before we go further and show you how to link ethics and strategy, we have to meet a challenge of those who believe business ethics is a contradiction: the theory known as relativism.

References[52]

"A. H. Robins Co. and the Daikon Shield.," Charlottesville: The Darden School Case Studies, UVA E-049.
Allison, G. 1970. *The Essence of Decision*. Boston: Little Brown.
Anders, G. 1982. First Boston Enjoys Surge as Speculators Expect Big Earnings Gain from Fee as Bendix Advisor. *The Wall Street Journal*, 24 Sept 1982.
Andrews, K. 1980. *The Concept of Corporate Strategy*. 2nd ed. Homewood: R. D. Irwin.
Ansoff, I. 1977. The State of Practice in Planning Systems. *Sloan Management Review* 18 (2): 1–28.
Arrow, K., and F. Hahn. 1971. *General Equilibrium Analysis*. San Francisco: Holden-Day.
Bernard, C. 1938. *The Function of the Executive*. Cambridge, MA: Harvard University Press.
Barnes, H. 1967. *An Existentialist Ethics*. Chicago: University of Chicago Press.
Brodern, P. 1985. *Outrageous Misconduct: The Asbestos Industry on Trial*. New York: Pantheon.
Carley, W. 1986a. People Express is a Victim of Its Fast Growth," *The Wall Street Journal*, 25 June 1986.
———. 1986b. Struggling to Survive, People Express Alters Operations and Image. *The Wall Street Journal*, 31 July 1986.
Carley, W., and T. Agins. 1985. People Express Merger Might Cause Problems. *The Wall Street Journal*, 14 Oct 1986.
Cavanaugh, G. 1982. *American Business Values*. Englewood Cliffs: Prentice-Hall.
Chandler, A. 1962. *Strategy and Structure*. Boston: MIT Press.
———. 1977. *The Visible Hand*. Cambridge, MA: Harvard University Press.
Charan, R., and E. Freeman. 1980. Planning for the Business Environment of the 1980s. *Journal of Business Strategy* 1 (2): 9–19.
Charlier, M. 1986. Hormel and Union Are Locked in Battle That Carries High Stakes for Both Sides. *The Wall Street Journal*, 23 Jan 1986.
Cheung, S. 1983. The Contractual Nature of the Firm. *Journal of Law and Economics* 26: 1–21.
Coll, S. 1986. *The Deal of the Century: The Breakup of AT&T*. New York: Atheneum.
Davidson, D. 1980a. Actions, Reasons and Causes. In *Essays in Action and Events*, 3–19. New York: Oxford University Press.
———. 1980b. Mental Events. In *Essays on Action and Events*, 207–227. New York: Oxford University Press.
———. 1984. On the Very Idea of a Conceptual Scheme. In *Inquiries into Truth and Interpretation*, 183–198. New York: Oxford University Press.
Deal, T., and A. Kennedy. 1982. *Corporate Culture*. Reading: Addison-Wesley.
Derthick, M., and P. Quick. 1985. *The Politics of Deregulation*. Washington, D.C.: The Brookings Institution.
Elster, J. 1983. *Explaining Technical Change*. New York: Cambridge University Press.
Evan, W. 1966. An Organization Set Model of Interorganizational Relations. In *Interoganizational Decision Making*, ed. M. Tuite, M. Radnor, and R. Chisholm, 181–200. Chicago: Aldine.
Evans, D., ed. 1983. *Breaking Up Bell: Essays on Industrial Organization and Regulation*. Amsterdam: North-Holland.
Fish, S. 1980. *Is There a Text in This Class?* Cambridge, MA: Harvard University Press.
———. 1984. Antiprofessionalism. *New Literary History* 16 (1): 89–108.
"The 4 Horsemen: Did Main Characters in Big Takeover Saga Let Egos Sway Them?" *The Wall Street Journal*, 24 Sept 1982.
Freeman, E. 1983. Managing the Strategic Challenge in Telecommunications. *Columbia Journal of World Business* 18 (1): 8–18.
———. 1984. *Strategic Management: A Stakeholder Approach*. Boston: Pitman.

[52] There are many excellent discussions of relativism. This chapter adds little new to the philosophical literature. It is heavily indebted to Nancy Gifford's excellent book, *When*.

———. 1987. Review of Williamson's *The Economic Institution of Capitalism*. *Academy of Management Review* 12 (2): 385–387.
Freeman, E., and D. Gilbert. 1987. Managing Stakeholder Relationships. In *Business and Public Policy: Dimensions of Conflict and Cooperation*, ed. P. Sethi and C. Falbe. Lexington: D.C. Heath.
Freeman, E., and W. Evan. 1987a. *Corporate Governance: A Stakeholder View*. Unpublished manuscript.
———. 1987b. Stakeholder Management and the Modern Corporation: Kantian Capitalism. In *Ethical Theory and Business*, ed. T. Beauchamp and N. Bowie, 3rd ed. Englewood Cliffs: Prentice-Hall.
Freeman, E., D. Gilbert, and E. Hartman. 1988. Values and the Foundations of Strategic Management. *Journal of Business Ethics* 7: 821–834.
Freeman, E., D. Gilbert, and C. Jacobson. 1987. The Ethics of Greenmail. *Journal of Business Ethics* 6 (3): 165–178.
Gilbert, D. 1987. *Strategy and Justice*. Minneapolis: Ph.D. Dissertation, Department of Strategic Management and Organization, Carlson School of Management, University of Minnesota.
Gilbert, D., E. Hartman, J. Mauriel, and E. Freeman. 1988. *The Logic of Strategy*. Boston: Ballinger.
Gray, P. 1986. Apple Computer and Jobs Reach Pact Barring Sale of New Machine until 1987. *The Wall Street Journal*, 20 Jan 1986.
Hage, D. 1986. On Anniversary, P-9 Militants Haven't Given Up, but Rest of Austin Looks Ahead. *Minneapolis Star & Tribune* 17 Aug 1986.
Hartman, E. 1988. *Foundations of Organization Theory*. Boston: Ballinger.
Hartz, P. 1985. *Merger*. New York: Morrow.
Hayes, R., and W. Abernathy. 1980. Managing Our Way to Economic Decline. *Harvard Business Review* 58 (4): 67–77.
Herman, E. 1981. *Corporate Control, Corporate Power*. Cambridge: Cambridge University Press.
Hertzberg, D. 1986. TWA Names Carl Icahn as Chairman, but His Victory Is Seen as Bittersweet. *The Wall Street Journal*, 6 Jan 1986.
Horwitch, M. 1988. *Postmodern Management*. New York: Basic Books.
Hughey, A. 1982. Allied Corporation Plans Major Debt Reduction, Possible Sale of Bendix's Holdings in RCA. *The Wall Street Journal*, 16 Dec 1982.
Ingrassia, P. 1982. Forget Football. We Have Bendix and U. Tech. *The Wall Street Journal*, 24 Sept 1982.
Kahaner, L. 1986. *On the Line: The Men of MCI—Who took on AT&T, Risked Everything and Won!* New York: Warner.
Kleinfield, S. 1981. *The Biggest Company on Earth: A Profile of AT&T*. New York: Holt, Rinehart & Winston.
Knapp, S., and W. Michaels. 1982. Against Theory. *Critical Inquiry* 8 (4): 732–742.
Kneale, D. 1986. Firms with Ties to South Africa Strike Back at Colleges That Divest. *The Wall Street Journal*, 20 Nov 1986.
Koten, J. 1982. Bill Agee of Bendix Corp: Why I Did It. *The Wall Street Journal*, 4 Oct 1982.
Krebs, M. 1985. Allied Reviews Takeover of Bendix. *Automotive News*, 27 May 1985.
Levinson, H. 1976. *Psychological Man*. Cambridge, MA: The Levinson Institute.
Loomis, C. 1986. The Comeuppance of Carl Icahn. *Fortune*, 17 Feb 1986.
Lorange, P. 1980. *Corporate Planning: An Executive Viewpoint*. Englewood Cliffs: Prentice-Hall.
Lorange, P., and M. Scott-Morton. 1986. *Strategic Control*. St. Paul: West Publishing.
Magnet, M. 1987. Restructuring Really Works. *Fortune*, 2 Mar 1987.
Metz, T. 1982. The Pacificist. *The Wall Street Journal*, 3 Dec 1982.
Nader, R., and W. Taylor. 1986. *The Big Boys: Power and Position in American Business*. New York: Pantheon.
Nozick, R. 1975. *Anarchy, State and Utopia*. New York: Basic Books.
Nulty, P. 1987. Pushed Out at 45—Now What? *Fortune*, 2 Mar 1987.

"Our Dream Is to Work Together," unsigned interview with W. Agee and M. Cunningham, *Fortune*, 18 Oct 1982.
"Pac Man Economics," unsigned review and outlook, *The Wall Street Journal*, 27 Sept 1982.
Pascale, R., and A. Athos. 1981. *The Art of Japanese Management*. New York: Simon & Schuster.
Peters, T. 1986. *The Promises*. Palo Alto: The Tom Peters Companies.
Peters, T., and N. Austin. 1985. *A Passion for Excellence*. New York: Random House.
Peters, T., and R. Waterman. 1982. *In Search of Excellence*. New York: Harper & Row.
Pfeffer, J. 1982. *Organizations*. Boston: Pitman.
———. 1983. Organizational Demography. In *Research in Organization Behavior*, ed. L. Cummings and B. Staw, vol. 5, 299–357. Greenwich: JAI Press.
Phillips, A. 1983. Regulatory and Interfirm Organizational Burdens in the U.S. Telecommunications Structure. *Columbia Journal of World Business* 18 (1): 46–52.
Porter, M. 1985. *Competitive Advantages*. New York: Free Press.
Quinn, R., and P. Lees. 1984. Attraction and Harassment: Dynamics of Sexual Politics in the Workplace. *Organizational Dynamics* 13 (3): 13–46.
Rorty, R. 1979. *Philosophy and the Mirror of Nature*. Princeton: Princeton University Press.
———. 1982. *Consequences of Pragmatism*. Minneapolis: University of Minnesota Press.
———. 1983. Postmodern Bourgeois Liberalism. *Journal of Philosophy* 80 (10): 583–589.
———. 1984a. The Historiography of Philosophy. In *Philosophy in History*, ed. R. Rorty, J. Schneewind, and W. Skinner, 49–75. Cambridge: Cambridge University Press.
———. 1984b. Texts and Lumps. *New Literary History* 16 (1): 1–16.
Rudner, R. 1953. The Scientist qua Scientist Makes Value Judgments. *Philosophy of Science* 20 (1): 1–6.
St. Anthony, N. 1986. P-6 Trusteeship Gets Judge's Approval; Local Leadership Out. *Minneapolis Star & Tribune*, 3 June 1986.
Salpukas, A. (1985), "People Express to Get Britt," *The New York Times*, Dec. 28, 1985.
———. 1986. People Express Beaten at Its Own Game. *Minneapolis Star & Tribune*, 29 June 1986.
Schelling, T. 1960. *The Strategy of Conflict*. Cambridge, MA: Harvard University Press.
———. 1978. *Micromotives and Macrobehavior*. New York: W. W. Norton, & Co., Inc.
Schendel, D., and C. Hofer. 1979. Introduction. In *Strategic Management: A New View of Business Policy and Planning*, ed. D. Schendel and C. Hofer, 1–12. Boston: Little-Brown.
Searle, J. 1984. *Minds, Brains, and Science*. Cambridge, MA: Harvard University Press.
Selznick, P. 1957. *Leadership in Administration*. New York: Harper & Row.
Shooshan, H. 1984. The Bell Breakup: Putting it in Perspective. In *Disconnecting Bell: The Impact of the AT&T Divestiture*, ed. H. Shooshan, 8–22. New York: Pergamon.
Skinner, W. 1985. *Manufacturing: The Formidable Competitive Weapon*. New York: John Wiley.
Sloan, A. 1969. *My Years with General Motors*. New York: Doubleday.
Sloan, B. 1983. *Three plus One Equals Billions*. New York: Arbor House.
Stevens, M. 1987. *The Insiders: The Truth Behind the Scandal Rocking Wall Street*. New York: Putnam's.
Sturdivant, F. 1979. Executives and Activists: A Test of Stakeholder Management. *California Management Review* 22 (1): 53–59.
Sturdivant, E., and J. Ginter. 1977. Corporate Social Responsiveness: Management Attitudes and Economic Performance. *California Management Review* 19 (2): 30–39.
Sullivan, G., and W. Nowlin. 1986. Critical New Aspects of Sex Harassment Law. *Labor Law Journal* 37: 617–624.
Tunstall, J. 1986. *Communications Deregulation: The Unleashing of America's Communications Industry*. Oxford: Blackwell.
Tunstall, W. 1983. Cultural Transition at AT&T. *Sloan Management Review* 25 (1): 15–26.
Turnstall, W. 1985. *Disconnecting Parties: Managing the Bell System Break-Up: An Inside View*. New York: McGraw-Hill.
von Auw, A. 1983. *Heritage and Destiny: Reflections on the Bell System in Transition*. New York: Praeger.

Wiley, R. 1984. The End of Monopoly: Regulatory Change and the Promotion of Competition. In *Disconnecting Bell: The Impact of the AT&T Divestiture*, ed. H. Shooshan, 23–46. New York: Pergamon.
Williamson, O. 1985. *The Economic Institutions of Capitalism*. New York: Free Press.

R. Edward Freeman is University Professor, Elis and Signe Olsson Professor, Academic Director of the Institute for Business in Society, and Senior Fellow of the Olsson Center for Applied Ethics at the University of Virginia's Darden School of Business.

Daniel R. Gilbert, Jr. is Professor Emeritus at Gettysburg College.

Chapter 23
Values and the Foundations of Strategic Management

R. Edward Freeman, Daniel R. Gilbert, Jr., and Edwin M. Hartman

1 Introduction

We believe that research on strategic management has paid insufficient attention to values, with the result that some current theories of strategic management give incomplete accounts of why managers adopt certain strategies and of why strategies work or don't. The tendency to underestimate the role of values is evident in several predominant themes of current research in the field. For example, some theorists are concentrating on: (l) plugging the supposed gap between "strategy formulation" and "strategy implementation" (e.g., Hrebeniak and Joyce 1984); (2) elaborating more sophisticated theories of "strategic control" (e.g., Lorange 1984); (3) arguing that "culture management" is different from, and more important than, strategic management (e.g., Deal and Kennedy 1982); and, (4) establishing sophisticated empirical measures of strategy and performance that are based largely on observable properties of organizations (e.g., Montgomery 1982). We shall outline a different and, at some points, incompatible approach.

Our argument is simply this: if we understand what role values play in strategic management, then any theory of strategic management that we may develop will be

Originally published in: *J Bus Ethics*, 7(11), 821–834 © Springer, 1988
Reprint by Springer, https://doi.org/10.1007/BF00383045

R. E. Freeman (✉)
University of Virginia, Charlottesville, VA, USA
e-mail: FreemanE@darden.virginia.edu

D. R. Gilbert, Jr.
Gettysburg College, Gettysburg, PA, USA

E. M. Hartman
Rutgers University, New Brunswick, NJ, USA

© Springer Nature Switzerland AG 2023

S. D. Dmytriyev, R. E. Freeman (eds.), *R. Edward Freeman's Selected Works on Stakeholder Theory and Business Ethics*, Issues in Business Ethics 53,
https://doi.org/10.1007/978-3-031-04564-6_23

more coherent and more powerful. Any theory of strategic management, furthermore, necessarily deals with a concept of corporate morality. Our conclusion will be incompatible in important respects with traditional views of strategic management, as well as with recent positions that attack traditional models. But we will find something to agree with in both of the general approaches we criticize, and will find some possibility of reconciliation as well.

We proceed in several steps. In the next section, we describe some current criticisms of strategy and unpack a few of their less evident implications. We then return to strategy's roots by analyzing the role of values in the Harvard Policy Model. Then, building upon an argument that values are unmysterious and irreducible entities, we show that the theory of corporate morality presupposed by the Harvard Model has some promise and some difficulties. We conclude that the Harvard Policy Model has considered certain stakeholders' values with insufficient reflection. Broadening our scope, we then argue that value-free views of strategic management are either incoherent or impractical. The paper concludes with consideration of the implications of our view for research and practice in strategic management.

2 The Attack on Corporate Strategy

Many of the recent attacks on modern theories of management rely heavily on the premiss that the concept of "value" has not been adequately used or understood. In the managerial literature we find Peters and Waterman (1982), Ouchi (1980), Pascale and Athos (1981), Deal and Kennedy (1982) and others arguing that corporate managers have relied too heavily on analytical models of business – in particular, on models of strategy that apply better to zoology than to the study of organizational life. Missing from the tool kit of MBAs from top business schools, of consulting firms, and of modern corporate managers is a profound understanding of human beings, such as employees and customers, whose acts are based on values, according to these critics.

Peters and Waterman (1982) call for "hands on, value driven" management and a "bias for action" that seems incompatible with the "rational-analytic" style that has been equated with strategic management. Ouchi (1980) advocates understanding the role of trust and community in organizations. Pascale and Athos (1981) propose adding "sharing values" and other "soft S's" to the strategy-structure-systems triad that is prevalent in most contemporary management theories. Deal and Kennedy (1982) focus on culture, rather than strategy, as the general manager's primary concern. These theorists suggest that we ought to pay as least as much attention to socialization as to strategy. Articles in the popular business media denigrate strategy and strategy consultants, and claim that companies that adopt explicit strategies are no better for it. In short, the concept of strategy as the central metaphor for business activity is quickly going out of vogue in the managerial world. But the attack on the concept of strategy does not end there.

In the academic literature we find a similar disparagement of strategy. A recent issue of *Administrative Science Quarterly* focused on culture exclusively as the key to understanding organizational behavior. Weick (1979) and others have argued that management is a symbolic activity, largely bereft of any choice that matters. Population ecology theorists (McKelvey and Aldrich 1983) have advanced the view that, for the most part, the environment determines organization survival with the result that a person's choice is of little importance in the process. (See the first chapter of Pfeffer (1982) for a summary of these developments).

One implication of this line of argument seems to be that if we want to understand why organizations succeed or fail, we would do better to study their cultures, their symbol systems, their myths, heroes, and rituals, and their environments rather than their strategic management processes and postures. If we do study the strategic management processes, we should see them as symbolic actions and not as rational and effective attempts to align the organization with its environment in order to achieve certain objectives. So say the critics, and some of them add that the symbols themselves do not matter very much. For these, the environment is the overriding determinant, for we can adequately explain and predict corporate performance by looking at population-level variables. On this view, strategy formulation and implementation is just part of random variation, or at best "morale management".

We agree that culture matters. As organizations are human institutions, people's behaviors are necessarily salient and can be presumed to figure in any account of what organizations do. We have no argument with corporate anthropology. Our view is simply that strategic management, one slice of human behavior, can be understood only in the light of the activating force of values. And, when we ground strategic management in values, we go far beyond "symbolic management" (see Deal and Kennedy 1982).

We intend to defend and develop three points that these critiques of strategic management tend to ignore. First, values have always been a crucial element of models of strategic management. Peters and Waterman (1982) did not invent them, and no logical account of population ecology can exorcise them. Some current strategic management models appear to ignore values, but in fact these views often come equipped with significant implicit systems of values. Far from genuinely abandoning values, these models actually include many cases of corporate morality. (For a sample of the recent work in corporate morality see the essays in Beauchamp and Bowie (1979), De George and Pichler (1978), and Rega (1984); and, see the extensive arguments of philosophers such as Braybrooke (1983), Donaldson (1982), De George (1982), Velasquez (1982), and Solomon and Hanson (1985). Our analysis builds on the concept of "value" that is present in the work of most of these philosophers.) Second, the critics often ignore the interdependence that underlies strategic choice and, thus, the importance of the values of a diversity of strategic actors. In simple terms, strategic management only makes sense in the context of interdependent choice. Third, the critics sometimes misunderstand the nature of values. Values are not slogans, and they are not mysterious entities. They are both reasons for and causes of action.

These are serious and fundamental issues. They warrant a reexamination of the concept of strategy. We start with a standard and influential view of corporate strategy.

3 The Harvard Policy Model

Kenneth Andrews' *The Concept of Corporate Strategy* (1971) is the classic statement of an influential model of corporate strategy. The concept has been developed at the Harvard Business School during the past 60 years, primarily through teaching and research in what is called Business Policy. The Business Policy course was originally seen as a way to integrate marketing, accounting, finance, production, R&D, and other functions of business at the level of the General Manager or Chief Executive Officer. As the course developed over time, a field of research was born and in turn spawned several generations of strategy models. Each of these models has incorporated some features of the Harvard Policy Model. (Fig. 23.1 is a depiction of the well-known elements of the Harvard Policy Model from Andrews 1980). One of these common features is crucial: the role of values.

Andrews makes it clear that values play a central role in understanding corporate strategy (Andrews 1980, p. 81f):

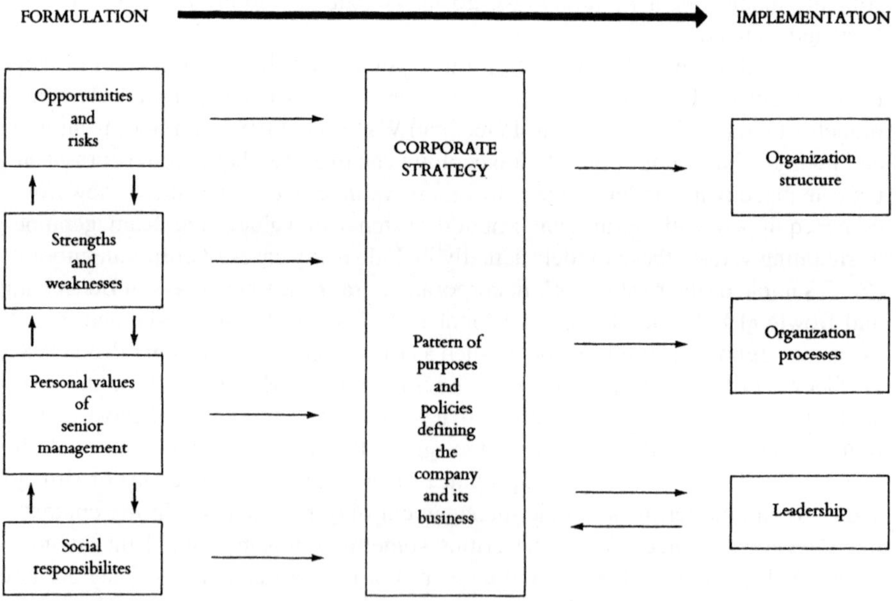

Fig. 23.1 The Harvard Policy Model

> A value is a view of life and a judgement of what is desirable that is very much part of a person's personality and a group's morale. From parents, teachers, and peers, we are told by psychologists, we acquire basic values, which change somewhat with acquired knowledge, analytical ability, and self-awareness, but remain a stable feature of personality .... Despite the well known problems of introspection, we can probably do more to understand the relation of own values to our choice of purpose than we can to change the values of others. Awareness that our own preference for an alternative opposed by another stems from values as much as from rational estimates of economic opportunity may have important consequences.

Andrews holds that while values are not always readily evident, they nevertheless have important consequences. They are a significant part of the explanation of individual actions and, according to Andrews, of corporate actions as well. Rather than apologize for the dependence of corporate strategy on personal values, Andrews urges us to shout it from the rooftops (Andrews 1980, p. 79, emphasis added):

> We should in all realism admit that the personal desires aspirations and needs of the senior managers of a company actually do play an influential role in the determination of strategy. Against those who are offended by this idea either for its departure from the stereotype of single-minded economic man or for its implicit violation of responsibilities to the shareholder, we would argue that *we must accept not only the inevitability but the desirability of this intervention.*

3.1 Interpreting Andrews' Arguments

Andrews can be read as claiming that any theory of corporate strategy must attend to two important value-laden issues. The first is whether the corporate strategy is acceptable to senior management. According to Andrews, the test of that is whether the strategy is compatible with the values of senior management. Congruence is necessary, on this interpretation, if senior management is to fulfill its role as leader of the organization. If a particular strategy is not compatible with the values of senior management, then by implementing it the senior managers are acting in *bad faith*. Bad faith entails articulating the strategy in their role as managers, while privately disagreeing with the foundations of the strategy. So we attribute to Andrews the following hypothesis, which we call the *Requirement of Good Faith*, (GF):

(GF) A corporate strategy is effective only if the senior managers of the organization can act on it in good faith.

According to Andrews then, acting in good faith is a necessary, but not sufficient, condition of the effectiveness of corporate strategy. Another way of putting (GF) is this: an effective corporate strategy must not require senior management to perform actions that violate or compel them to abandon their values.

The second value-laden issue to which corporate strategy must attend, on our interpretation of Andrews, is that of the social responsibility of the corporation (see Andrews 1980, pp. 88–92). We can hypothesize on his behalf a *Social Responsibility Requirement* (SR) for corporate strategy:

(SR) A corporate strategy is effective only if it prescribes actions that are consistent with the corporation fulfilling its social responsibilities.

Given the first requirement (GF), the second is not as controversial as it may appear, nor is it very far from (GF). *In effect,* (GF) *leaves it to senior management to determine the social responsibilities of the corporation.* In effect, (GF) leaves it to senior management to determine the social responsibilities of the corporation, at least to this extent: it is socially irresponsible, and managerially ineffective, for managers to design a corporate strategy that violates their conception of the organization's social responsibilities. For it is irresponsible to do what one believes to be irresponsible (even though it is not necessarily responsible to do what one believes to be responsible). An act that comes about as a result of one's setting out to do wrong is the moral equivalent of doing the right thing accidentally. Senior managers cannot be made to act consistently contrary to their values, and it is certain, as Andrews tells us, that they will have values that have something to do with the social responsibilities of the corporation. Once we know what senior managers value with respect to those social responsibilities, we know the limits within which the firm can reliably act. So (SR) appears to be a corollary of (GF). In short, the Harvard Policy Model as we interpret it implies that no corporate strategy that violates either (GF) or (SR) can be effectively implemented and that (GF) and (SR) are necessary conditions of effective corporate strategy.

The claim that we must understand the role of values in corporate strategy is a key part of Andrews' version – to all appearances a canonical version – of the Harvard Policy Model. If we also believe that effective corporate strategy is an important element of corporate success, then values are crucial to corporate success, and understanding success requires understanding values. To put it another way: you can deny that values matter to corporate success, but only if you are willing to deny that corporate strategy makes any difference.

To hold that (GF) and (SR) are false would be to embrace something like the following, which we might call the *Bad Faith Hypothesis*, (BF):

(BF) A corporate strategy may be effective even if it prescribes actions that are inconsistent with the personal values of the senior management.

Interpreted this way, is Andrews correct? Before we can give our assessment, we need to clarify the notion of values and to say something about the role they play in human action. In doing so we take an approach first delineated by Aristotle, and refined by British and American analytical philosophers in the past 40 years. In our view, the most mature synthesis has been achieved by the distinguished American philosopher Donald Davidson (Davidson 1984), who makes no secret of his debt to Aristotle and to many recent philosophers who would probably agree with most of what we say in the next section.

4 Values

What does it mean to say that Jones has a particular value or that she values something? It means essentially that she wants that thing, or wants that state of affairs to come to pass. How do values differ from ordinary desires? There is no simple dichotomy. Values are relatively permanent desires: I want a drink right now (and probably will not tomorrow morning), but health is a value of mine because my wanting health is a settled disposition rather than a want that comes and goes. The first of the two citations from Andrews above indicates that he understands values this way. Values are also more often intrinsic rather than instrumental wants, in this sense: when I say I value something, I am probably talking about something that I desire for itself alone, rather than because it is a way of getting something else. We are inclined, moreover, to say we value something that is good not only for ourselves but for others – integrity or clean air, for example. Moral values, to which we turn later, are that way.

4.1 Values and Actions

Values explain action as desires do. When Jones acts intentionally, she has a reason for acting, and the reason is, in broadest terms, that performing a certain action will have a relatively desirable result. That is what Jones believes, at any rate; so we may say that an intentional action is the result of a certain desire and a certain belief and that they are its causes. The belief is (or entails) that doing a certain act will fulfill the value. Aristotle seems to think of intentional acts as being based on a practical syllogism, whose major premiss is a statement of what sort of thing one wants, while the minor premiss is a statement that a certain thing is just such a thing. The conclusion is an action.

Now Jones may want to do something for no other reason than that she wants to; watching a tennis match is a possible example. Or she may want to do something because it has certain results, direct or indirect, that she favors; for example, she may compliment her boss's wife. Or she may want to do it because it is an exemplar of the kind of act that she prefers to do; an act of generosity, perhaps. In such cases, we say that Jones values watching tennis matches, having a successful career, and generosity.

Values can conflict, as desires can. In some cases, one may be unable to act according to all of one's values, in the sense that the more important one overrides a lesser one. We can infer that some values come with an "unless …" clause attached, so that one can, without self-traduction, choose the most important of conflicting values if necessary. Suppose, for example, that Jones must choose between seeing the Wimbledon Women's Singles Final and joining the boss's wife in raising money for some worthy charity. The passion for tennis might override careerism and generosity both. (Whether Jones is generous is probably a matter of how many other values can override generosity and how easily.) Still, one cannot act intentionally without valuing something accordingly. An intentional act is one to which it is appropriate to apply the question "Why did you do that?" in a sense that calls for a statement that directly or indirectly tells what the agent values.

It is less obvious, but no less important, that one cannot value something without ever acting accordingly when an opportunity arises. What would it mean to say that I value watching tennis but never do it, even when I have the opportunity and nothing to do that I value more? What reason could there be to say that I value being generous – that is, value that I be generous – if I quite consistently behave stingily? People do sometimes violate their own values, often under pressure. Regret over one's behavior is a sign that that has happened. But values on which one never acts are not values at all, only claimed or espoused values.

As Andrews suggests, and psychologists have argued persuasively, we do not always know what our values are. An immature or shallow person may be unable to characterize his values even roughly. A neurotic may be profoundly deceived about what he values. An ideologue may claim to value freedom, but may reveal by actions or by more specific pronouncements that what he really values is whatever profits his/her company or party. We detect self-deception and simple dishonesty by comparing avowals with action. Experience at this sort of thing leads us to attribute values by looking at action rather than at avowals though in some cases action must be interpreted with care: for example, someone who consistently demands honesty in others but is himself often dishonest may be said to value honesty in that he does act on it, in a way. Values bear a similarity to unobservable entities that natural scientists believe in. We cannot know by observation that they are there, but must instead postulate them in the general case and infer them in the individual case, even sometimes when the individual case is our own.

Nothing we say here should be taken to imply that organizations or even individuals always know what their values are. Characteristically rational people think about their values before acting, then try to act consistently with them. Others may understand their own values best by looking at their actions after the fact and finding patterns in them. Much the same is true of organizations, of which some are proactive while others follow strategies – hence, in some cases, values – that emerge without having been explicitly chosen in advance. (See further Mintzberg, 1978; and Mintzberg and Waters, 1985, for example.) Nor do we deny the causal influence of the environment on either individual or organizational values. On the contrary, understanding this influence is arguably a necessary condition of having an explicit and coherent set of values.

That values are not infallibly reportable by their owners and not readily inferred from behavior, which suffers so many other causal influences, is no reason for pretending they do not exist or for trying to reduce them to something easily identifiable. We have argued that values are important influences on corporate behavior and that therefore they ought to be understood and identified. Because they are so elusive, however, some researchers may understandably see a better payoff in analyzing other influences (such as market structure, for example) that are undeniably important and in some respects easier to deal with. If our account is plausible, both the traditional commentators on strategy and strategy's critics would agree that values are worth studying in spite of the difficulties.

4.2 Moral Values and Corporate Morality

Moral values fall into that subclass of values having to do with interacting with other people or acting in ways that affect them. That is, if an action affects others, that action has moral implications. What is important to our current concern is that moral values motivate action much as other values do. In particular, there is no compelling reason to believe that one's own welfare is the only thing that could be valued in itself, serving alone as an ultimate reason for acting. (Attempts to defend that particular position usually involve stipulating that whatever Jones values is, by definition, in Jones's interests).

A theory of corporate morality is an account of the moral obligations of corporations and, hence, of corporate managers in their roles as corporate managers. It might include, though not be exhausted by, a theory of social responsibility. "Social responsibility" is but one possible moral value that a corporation ought to observe in its strategic activities. A theory of corporate social responsibility presupposes, and is justified by, a theory of corporate morality. In much the same sense, we have interpreted Andrews as holding that a theory of corporate strategy presupposes a theory of corporate values.

We will have more to say shortly about corporate morality. It is important to note here, however, that strategy theorists and practitioners alike usually assume one particular theory of corporate morality. This, of course, is the "stockholder priority" view. The position is one of moral significance – that is, of the moral kind – since it prescribes corporations' and managers' behaviors relative to others.

4.3 Reprise on Values and Strategy

We emphatically approve of the renewed emphasis on values, as advocated by strategy's contemporary critics. Yet, much remains to be done to make the necessary linkages among values, moral values, and strategy. Values are absolutely central to strategy. We claim that the whole point of corporate strategy is to act in the name of the organization intentionally, rather than randomly or gratuitously or according to the dictates of someone outside the organization. It follows that acting strategically is a matter of acting according to certain values. There can be no strategy without objectives, simply because there can be no means without ends. Values are the most general and most nearly-intrinsic objectives, and they are the ends to which other corporate objectives are means. To deny that they guide corporate strategy is to hold that there can never be any point to doing corporate strategy.

5 The Good Faith Requirements

Is Andrews, as we have construed him, correct? Must a senior manager act in good faith, if strategy is to be effective? Must the strategy be a socially responsible one? Reasonable as it may seem that managers do better if they believe in what they are doing, (GF) and (SR) appear to be positions for which there is no proof.

5.1 Good Faith and Action

The first move in the defense of (GF) is to point out that it is not quite so stringent as it may appear. It seems to allow that a manager may decide to go along with a particular strategic decision of the firm, even if she would not have made that decision if it had been up to her alone. Such a decision would seem to be opposed to her values, but in a clear and straightforward sense it is not. She may well take as a value the proposition, "If a particular decision is made on the basis of due process, I will abide by it and work to realize it unless it violates some further, more permissive standard of my values". Hence, she has this overriding value to go along with decisions made by due process, unless they are genuinely terrible decisions by some further standard she must invoke from time to time. Thus, the action taken against her better judgment may not, under the circumstances, be incompatible with her values, most of which have an implicit "unless …" clause attached.

Managers and professionals often find themselves playing roles that impose obligations not entirely compatible with the values they otherwise have. A manager may follow orders he would not himself have given, if it were for him to decide. It may be that managers' orders are normally carried out. An attorney may have a professional duty to represent the interests of a client he despises and to bring about an outcome he believes the client does not deserve. In that case the value judgment may be that, taking everything into consideration, it is more important that all people have competent legal representation than that this particular wretch gets what he deserves. This sort of situation may cause moral, but not logical, discomfort, since one sincerely held value is overridden by another, more important one, and one does what one values most. The Harvard Model seems to accommodate this sort of situation. It would be in both conceptual and empirical difficulty if it did not.

5.2 Andrews and Good Faith

What Andrews appears not to permit is this situation: a manager, convinced that a particular course of action is a mistake, can be counted on faithfully to execute it. That is, if the manager values refraining from a certain act more than he values loyalty to the chain of command and his job, then there is no good reason to expect him to follow orders. Most people act against their values from time to time, and managers are no exception. But if we are talking about genuinely-held values rather than merely espoused values, then the manager's following orders to do action X would provide good, though not necessarily conclusive, reason to believe that he didn't value not-X above following orders.

So we can agree with Andrews, if we interpret him to mean that managers cannot be relied on to act against their views about what one ought to do *with all considerations taken into account.* A conceptual point, this has something in common with the statement that claustrophobes cannot be relied on to stay in small enclosed spaces. It is not trivial or useless on this account, however. To the extent that managers can and will say what their priorities in values are – not an easy task in every case, but often possible – an organization can make determinations about what may be expected of managers.

We can interpret Andrews as also making an empirical point. If an organization frequently calls upon people's loyalties or senses of duty and pressures them to set aside their personal preferences for the good of the company or in deference to the chain of command, the loyalty-related values may be strained to their breaking point. That is, people will become less loyal and more inclined to give new priority to those first-order values that are unrelated to their position in the organization. Values do change in response to unsatisfactory experience, as do people's ways of applying them to specific cases. This may happen when the obedient agent begins to suspect himself of having placed more value on, say, personal comfort than integrity, and so having actually traduced his values in the process. The point is a plausible one, though Andrews has not actually proved it.

5.3 Good Faith and the CEO: One Limitation of (GF)

Andrews appears also to hold that it is the values of senior management that determine the values of the organization. Here, too, we must take him to be referring to values in the taking-all-things-into-account sense. A CEO might prefer to take one course of action except that he has the interests of the organization to think about, and they militate in favor of another. So it is his value in the broader sense that determines his action.

As might be expected of a model originating from the Harvard Business School, Andrews writes as if the typical organization is one in which the CEO decides what the organization does insofar as it is possible for the organization to act freely. We take it that one of the more important messages from the corporate anthropologists and others who advocate attention to values, while deprecating strategy, is precisely that the CEO *cannot* get the organization to do just anything, that the culture and the values of the individual employees have to be "correct", or performance will be adversely affected.

That is one reason why consistency with top management's values is a necessary, but not sufficient, condition of effective strategy. The values of employees influence corporate action, for example. Neither individuals acting for themselves, nor corporate executives, can achieve all their objectives. But that does not show that intentional action is impossible or that values exert no influence. So the point does not necessarily destroy Andrews' position. It does, however, raise a question about it that we will address later.

5.4 Social Responsibility and Good Faith

Consider now (SR), the proposition that the only effective strategies are those that prescribe actions consistent with the corporation fulfilling its social responsibilities. To begin with it seems consistent with Andrews' overall position to interpret this to mean that effective strategies are consistent with what the corporation takes to be its social obligations. So the position is that a corporation committed to racial discrimination cannot effectively carry out a strategy that entails treating people color-blindly.

Understood this way, (SR) is a corollary of (GF). A CEO's conception of the organization's social responsibilities is a subset of his values. Leaving aside the issue of the CEO's power, the firm cannot accept certain social responsibilities but just never act on them, since we are talking about genuine and effective values, not just empty slogans.

Note that (SR) states that corporate strategy will be effective only where the corporation fulfills its social responsibility. We have argued that (SR) is a corollary of (GF). But then why doesn't (SR) read "… consistent with *management* that the corporation is fulfilling its social responsibilities"? Haven't we gone too far in inferring something about actual social responsibilities? We have not. (SR) states fulfilling social responsibilities as a necessary but not sufficient condition of effectiveness. Now if management believes the organization is not being socially responsible, then it follows, according to Andrews' conception, that the organization is not in fact being socially responsible. A necessary condition of socially-responsible action is that it be done intentionally rather than by mistake or at random. In a corporate context, this means that top management must at the very least tacitly approve of the action. So management's belief that an action is socially responsible is a necessary condition of its being socially responsible. And, being socially responsible in a CEO's eyes is a necessary condition of the effectiveness of strategy. Thus, management's belief that an action is socially responsible must also be a necessary condition for the effectiveness of strategy. In sum, we hold that (GF) does imply (SR) as stated. Those who give moral credit for acts the agent does not consider moral may well disagree, but they can still agree to the important point that an effective corporation is one that acts on values that top managers hold.

5.5 Values and Rational Action

There is another significant criticism to be made at this point against what Andrews says about values. In the first of the two passages quoted above, he contrasts values and "rational estimates of economic opportunity." In so doing, he obscures the truth that (1) economic opportunity is one possible value, or more likely a class of values, and (2) that there need be nothing at all irrational about values. In fact, consistently irrational behavior will begin to indicate a person's confusion about values, or an inability to act in accordance with them.

Andrews is not alone in supposing that there is an opposition between values (soft, subjective, etc.) and rational analysis (scientific, objective, etc.) It is true that my statement, "I value something", cannot be defeated by the kind of comparison to publicly observable facts that is characteristic of science. But values may be rationally criticized on grounds of inconsistency, or because they entail other values unattractive to the valuer, or (on moral grounds) because they are damaging to others. More to our present point, however, no intentional action is more rational than the values that motivate it and provide the reason for whatever rational analysis attends it.

6 The Harvard Policy Model and Theories of Corporate Morality

The concept of corporate strategy that Andrews articulates takes a certain distinctive view of corporate morality. Specifically, it makes an important assumption about the moral obligations of corporations and their managers acting in managerial roles. The theory assumes, and (GF) entails, that the moral values of senior management guide and, even to a great extent, determine the strategic actions of the corporation. We have interpreted the theory as holding that top management's values, at the very least, limit the intentional actions the corporation can take to those actions that are compatible with those values. To show that the Harvard Policy Model also assumes that moral values guide corporate action, we need only show that some of the personal values of senior executives are moral values and that corporate strategy sometimes prescribes actions that are actions with moral implications in that they affect the well-being of others, or inhibit or enable others in the pursuit of their lives' projects. It is readily apparent that this is true.

6.1 Moral Values and the Managerial View of Corporate Morality

Consider American Telephone and Telegraph during deliberations on whether to divest the Bell Operating Companies. This decision had a profound effect on people in at least the following groups: employees, suppliers, customers, local communities, and owners. Hence, the divestiture decision was a moral one – that is, a decision of the moral kind, whether or not it was a morally correct one. We have good reason to believe that Charles Brown, the chairman of ATT&T, acted in good faith.

Credible reports of his agonized deliberations about the right thing to do suggest that he acted according to his beliefs about the consequences of his decisions for others, their rights, and other elements in his own moral calculus. We can attribute to Andrews and the Harvard Policy Model the proposition that this corporate strategy of divestiture, like any corporate strategy, is to be evaluated as morally correct only if it is consistent with the moral values of senior management. Let us call this view that we have attributed to Andrews:

> *The Managerial View of Corporate Morality*: An action of a corporation is morally correct only if it is compatible with the personal moral views of the senior managers.

A direct consequence of this view is that only those actions senior management decides, in good faith, to take are morally correct. On the question of whether management acts in the interests of stockholders or other stakeholders in the firm, the Managerial View is silent. Much of what Andrews says and does not say suggests that he regards it as unimportant. His position is that values, including moral values, are important, but there is no need to look very far for standards. According to our interpretation and analysis, his view is that we need only look to the logic that senior management uses for its actions. Nothing is said about whether the values that guide senior management have to be either reasonable or genuinely moral values. He has not gone so far as to say that any corporate actions that senior managers values are acceptable, but he is silent on the question of what else might be needed to certify them.

6.2 Stakeholders and the Managerial View: The Punch Line

According to the Managerial View of Corporate Morality, the best answer to the question, "Why did AT&T develop and implement a corporate strategy of divestiture?" is this: because of the values of the top managers and their belief that divestiture would bring about a state of affairs consistent with their values, including values of the "social responsibility" class. But some managers and researchers, lulled by the assumption that it is senior management's values that will always determine corporate action, have failed to understand the complex way in which values actually operate in people's lives and in the turbulence of the contemporary business environment.

The punchline is very simple. The Harvard Policy Model and its descendants, when they focus on the role of values, take a damagingly oversimplified view of the relationship between the firm and its environment. If the values of senior management are the only ones that matter in strategy, then the corporation is seen as interacting with an environment made up of entities to which no values are ascribed – at least no values that make any difference to the effectiveness of strategy. To articulate this assumption is to refute it. We have to tread carefully here on two levels.

First, it does make sense to say that what the organization does must be consistent with the values of management, on pain of the possible consequences of bad faith. But good faith is not enough. In order to be effective, much more in order to meet the demands of morality, management has to conduct the organization carefully through a minefield of stakeholders' values. Suppose you are CEO at ABC Inc. The individuals and organizations in the environment that can make or break – or at any rate, help or hurt – your enterprise have values of their own, and (by definition) care about those values more than about anything else, including the good faith of your managers. The same is largely true of internal stakeholders. Certain critics of strategic management are correct at least in arguing that strategy can be effective only where the culture is appropriate for it. When people are constantly under pressure to override their self-regarding values in favor of those rooted in loyalty and the good of the organization, they cannot be depended upon to do their best. As we noted earlier, Andrews takes this position.

Second, without attention to the values of an organization's stakeholders, senior managers run the risk of presuming that their conception alone of the organization's social responsibilities is a justifiable conception. (SR), as a corollary to (GF), clearly reflects such an iconoclastic assumption. Once a senior manager sees her organization's social responsibilities as a matter of bargaining, explicit or implicit, among multiple stakeholders, however, she has moved beyond the narrow consequences of Andrews' position on (SR) and, accordingly, (GF).

This is not to say that either effectiveness or morality requires a manager to act consistently with all the stakeholder's values. Even if that were possible, it would likely be a disastrous strategy. Morality requires just that managers respect the legitimacy of certain stakeholders' claims and leave them inviolate. What effective strategy requires is that managers understand stakeholders' values – hence their actions – as significantly determinative of the outcome of strategy.

7 Value Free Hypotheses

Against all that we have said thus far it might he argued that values are just not important to strategic management. Consider this value-free hypothesis (VF):

(VF) The effectiveness of corporate strategy is independent of anyone's values.

This is no mere device for argument's sake. It is in fact a view that in some form seems integral to a number of current opinions about the effectiveness of strategic management. Those who hold that corporate strategy is simply a symbolic and ritualistic part of management's job appear to accept the implication that the effectiveness of corporate strategy is independent of whether management holds to the values that it claims, and of whether top managers act in good faith. From this follows the curious result that any values that management espouses at all – even values that are internally inconsistent or incoherent or perverse – will do as well as any others. For management does not matter, in this sense. Its intentional actions aimed at creating the organization's future will have nothing to do with whether the corporate strategy is effective. Thus, Brown's values and actions at AT&T, according to (VF), had nothing to do with the divestiture decision.

The same line of reasoning might be applied to every group affected by corporate strategy, and not just management. On that view, the effectiveness of strategy is independent of the intentional actions of all groups affected by it, including those who formulate it, those who implement it, those who buy the products that the corporation markets, and so on. This position too, odd as it may seem seems to be common. Strategic planners and managers who adopt portfolio models and the techniques usually associated with them omit, at least, the values of players in the environment from its calculus. In so doing, they espouse a rather strong variant of (VF). Portfolio approaches assume that strategic management is a game played by the firm with its environment. In defining the game, portfolio theorists consider the players in the environment in a highly-aggregated form. And if the other players are not considered as individuals, it is hardly surprising that not much attention is paid to their respective values.

In sum, *to hold (VF) is to hold that the effectiveness of corporate strategy is independent of all intentional actions of all groups affected by the strategy, since intentional actions rest on values.* This conclusion seems to be accepted even by those who hold that the effectiveness of corporate strategy is environmentally determined. For there are those population ecologists who appear to believe that the determinative influence of the environment is independent of the actions of the firm's stakeholders in that environment. We do not claim that their account of what goes on in the environment is necessarily incoherent. But it is one thing to say that we cannot explain all changes in the environment by looking at the values of its inhabitants (true), and quite another to say that those changes are independent of anybody's intentional action (false).

To leave values out of account in this way is to hold that the very notion of corporate strategy effectiveness is essentially useless, for the values of management, employees, and other stakeholders don't matter so long as management properly symbolizes what it is supposed to symbolize. (But it is not clear why that is important, either.)

Even (BF), the bad faith hypothesis, allows that there are such things as values. It just states that they turn out not to make the crucial difference. The more radical form of opposition to (GF) and (SR) is to say simply that values have nothing whatever to do with strategy one or another, or even that there are no values any more than there is Santa Claus (a position that Pfeffer 1982, among others, seems to take seriously).

7.1 The Irony

The ironic aspect of all this is that concepts like corporate symbol systems and corporate culture are proposed to lead us away from the sterile concept of corporate strategy and back to basics like values. A great deal of insight can be gained, and trouble avoided, by taking values into account from the beginning for what they are: namely, the motivating force of individual and therefore corporate action.

8 Conclusions

If values, taken together with some causal beliefs, are the motivating force for action, then senior management's values are but one subset of the values that are at work in determining the most effective actions for corporations. Middle managers' values function in the same way, as do employees' values. This should not be news to those who make much of corporate culture as explaining corporate success. So, too, do suppliers' values, owners' values, and other stakeholders' values count, and this should not be news to those who believe that the environment is an important variable in determining corporate success. The Harvard Policy Model and its implicit Managerial View of Corporate Morality seems to oversimplify what is, in fact, a conflict of multiple stakeholders' values. In that way, the Managerial View leads us astray as we try to explain – or bring about – effective strategy.

We conclude that models of strategic management, and models of corporate morality as well, must systematically account for the values of multiple stakeholders. We do not claim to have made a startling discovery. Note that researchers such as Porter (1980; 1985) have concentrated on the fact that effective strategy requires that you do what competitors and others don't expect you to do, and such business acumen requires knowing what motivates competitors, customers, buyers, and other members of Porter's "value chain". We have couched this discussion in the language of values and tried to clarify several misconceptions about what values are. We have explicitly addressed the question of the role of morality in strategic management, which others such as Porter leave implicit. While the full-blown development of such a stakeholder theory of corporate morality must await another occasion, we can draw at least three significant conclusions from our analysis of the role of values and morality in strategic management.

First, there are many possible models of corporate morality. Figure 23.2 is illustrative of the kind of analysis that can be done.

One particular model has already been suggested: what might be called the Stockholder View of Corporate Morality. It says that managers are the loyal agents of stockholders, and that their actions in the form of strategic management are therefore aimed at furthering the interest – and presumably acting according to the values – of the stockholders. Generations of strategic management models seem to imply the Stockholder View. As a result, the issue of fully evaluating corporate strategies from the moral point of view and that of understanding the role of values, in general, in corporate strategy just never seems to arise. Stockholders are paramount, according to this view, and their values are assumed to be univocal and to center upon their apparent desire for increased wealth. Strategic management models along this line would include many portfolio theory models, value-added models, competitive strategy models, transaction cost models, and the like. The Stockholder View of Corporate Morality needs to be subjected to the same critical scrutiny as that accorded the Managerial View in the preceding sections. We could readily infer hypotheses similar to (GF) and (BF), with similar conclusions.

Theories of corporate morality	Strategic management Literature (examples)
Managerial view The corporation should act in a manner consistent with the values of senior management.	*Managerial view* Andrews (1971; 1980)
Stockholder view The corporation should act in the interests of stockholders.	*Stockholder view* Wind and Mahajan (1981) Porter (1980; 1984)
Narrow stakeholder view The corporation should act in the interests of a small number of stakeholders.	*Narrow stakeholder view* Peters and Waterman (1982)
Wide stakeholder view The corporation should act in the interests of as many stakeholders as possible.	*Wide stakeholder view* Ackoff (1981) Mason and Mitroff (1982) Freeman (1984)

Fig. 23.2 Some theories of corporate morality

Second, in the formulation of guidelines for major directional actions of the firm, strategic management is by nature a process of values clarification, conflict, and in some cases negotiation among multiple parties who have something at stake in the strategy. The very logic of the concept of values requires that we see it this way. And, if values motivate one player, they can be assumed to motivate others, too. Both the Stockholder and Managerial views rest on a failure to see the effectiveness of corporate strategy within a universe of interdependent choices based on values.

Third, the claim that corporations with strong cultures will have no implementation problems would seem to be a consequence of our view. But it does not follow from our account that one solves implementation problems by getting people inside or outside the organization to change their values. *The idea that culture management is the way to take care of problems about the effectiveness and the morality of strategy is simply the Managerial View in a different guise.* Culture models are rationalized upon the inseparability of organizational and employee interests, yet are premissed, in large part, upon the ability of employees to adopt corporate values determined by senior managers to be "correct values".

The upshot of our view is that corporate strategy should be seen as taking place within a network of interdependent choices made by interested parties with agendas of their own. To understand corporate strategy in this way is also to understand the central place of the values of the parties involved. If we are to make progress in strategic management, we need to understand those values and take seriously the claims they generate. To do otherwise is to produce sterile analyses that do not explain, do not predict, and do not work.

References

Ackoff, R. 1981. *Creating the Corporate Future*. New York: John Wiley and Sons.

Andrews, K. 1971. *The Concept of Corporate Strategy*. Homewood: R. D. Irwin Inc.

———. 1980. *The Concept of Corporate Strategy*, Revised Edition, Homewood: R. D. Irwin Inc.

Baier, K. 1958. *From a Moral Point of View*. Ithaca: Cornell University Press.

Beauchamp, T., and N. Bowie, eds. 1979. *Ethical Theory and Business*. Englewood Cliffs: Prentice Hall.

Brams, S. 1980. *Biblical Games*. Cambridge: MIT Press.

Braybrooke, D. 1983. *Ethics in the World of Business*. Totowa: Rowman and Allanheld.

Christensen, R., K. Andrews, J. Bower, R. Hammermesh, and M. Porter. 1983. *Business Policy: Text and Cases*. Homewood: R. D. Irwin Inc.

Davidson, D. 1963. Actions, Reasons, and Causes. *Journal of Philosophy* 60: 685–700.

———. 1980. *Essays on Actions and Events*. London: Oxford University Press.

———. 1984. *Inquiries into Truth and Interpretation*. London: Oxford University Press.

De George, R. 1982. *Business Ethics*. New York: Macmillan.

De George, R., and J. Pichler, eds. 1978. *Ethics, Free Enterprise, and Public Policy*. Oxford: New York.

Deal, T., and A. Kennedy. 1982. *Corporate Cultures*. Reading: Addison Wesley.

Donaldson, T. 1982. *Corporations and Morality*. Englewood Cliffs: Prentice Hall.

Freeman, E. 1983. Managing the Strategic Challenge in Telecommunications. *Columbia Journal of World Business* 18 (1): 8–18.

———. 1984. *Strategic Management: A Stakeholder Approach*. Boston: Pitman Publishing Inc.

Gilbert, D. 1986. Corporate Strategy and Ethics. *The Journal of Business Ethics* 5: 137–150.

Harman, G. 1973. *Thought*. Princeton: Princeton University Press.

Hrebeniak, L., and W. Joyce. 1984. *Implementing Strategy*. New York: Macmillan and Co.

Lorange, P. 1984. Strategic Control: Some Issues in Making It Operationally More Useful. In *Competitive Strategic Management*, ed. R. Lamb. Prentice Hall: Englewood Cliffs.

Mason, R., and I. Mitroff. 1982. *Challenging Strategic Planning Assumptions*. New York: John Wiley and Sons.

McClennen, E. 1984. Rational Choice and Public Policy: A Critical Study. *Social Theory and Practice* 9 (2–3): 335–379.

McKelvey, W., and H. Aldrich. 1983. Populations, Natural Selection, and Applied Organizational Science. *Administrative Science Quarterly* 28: 101–128.

Mintzberg, H. 1978. Patterns in Strategy Formation. *Management Science* 24 (9): 934–948.

Mintzberg, H., and J. A. Waters. 1985. Tracking Strategy in an Entrepreneurial Firm. *The Academy of Management Journal* 25 (3): 465–499.

Montgomery, C. 1982. The Measurement of Firm Diversification: Some New Empirical Evidence. *Academy of Management Journal* 25 (2): 299–307.

Nagel, T. 1970. *The Possibility of Altruism*. Oxford: Oxford University Press.

Ouchi, W. 1980. *Theory Z*. Reading: Addison Wesley.

Parfit, D. 1984. *Reasons and Persons*. Oxford: Oxford University Press.

Pascale, R., and A. Athos. 1981. *The Art of Japanese Management*. New York: Simon and Schuster.

Peters, T., and R. Waterman. 1982. *In Search of Excellence*. New York: Harper and Row.

Pfeffer, J. 1982. *Organizations and Organization Theory*. Boston: Pitman Publishing Inc.

Porter, M. 1980. *Competitive Strategy*. New York: The Free Press.
———. 1984. *Competitive Advantage*. New York: The Free Press.
Rega, T., ed. 1984. *Just Business*. New York Random: House.
Simon, H. 1947. *Administrative Behavior*. New York: The Free Press.
Solomon, R., and K. Hanson. 1985. *It's Good Business*. New York: Atheneum.
Velasquez, M. 1982. *Business Ethics*. Englewood Cliffs: Prentice Hall.
Weick, K. 1979. *The Social Psychology of Organizing*. 2nd ed. Reading: Addison-Wesley.
Wind, J., and V. Mahajan. 1981. Designing Product and Business Portfolios. *Harvard Business Review* 59 (1): 155–165.

R. Edward Freeman is University Professor, Elis and Signe Olsson Professor, Academic Director of the Institute for Business in Society, and Senior Fellow of the Olsson Center for Applied Ethics at the University of Virginia's Darden School of Business.

Daniel R. Gilbert, Jr. is Professor Emeritus at Gettysburg College.

Edwin M. Hartman taught philosophy at the University of Pennsylvania, management and business ethics at Rutgers University, and business ethics at New York University.

Chapter 24
Let's Disband The Academy of Management

R. Edward Freeman

1 Introduction

Let's disband the Academy of Management: Let's close the doors, stop the mail, and donate the remaining dollars in the kitty to a cause that will do no harm.[1]

Let's disband the Academy of Management: Let's stop pretending that we are scientists, "social scientists," "management scientists," "organizational scientists."[2] Let's stop pretending that history, culture, literature, and values can be assumed to be constant, assumed away, or assumed at all.

Let's disband the Academy of Management: Let's make the au courant research on ethics optional before it becomes trivialized. Let's suddenly remember that we have to go home and wash our hair whenever the conversation turns to the Academy of Mangement's "Ethics Project," or any such statement of "ethics in social science research." Let's not choose knowingly, with full malice aforethought, to set the foxes amongst the chickens.

Originally published in: *SIM Newsletter*, 3, 1989
Reprint by Springer, Reproduced with author's permission
Division Chairperson's Address, Social Issues in Management Division, Academy of Management, Washington, D.C.

[1] I am grateful to a number of people for conversations about these ideas, especially Thomas Donaldson, Daniel R. Gilbert, Jr., and Andrew Wicks. I alone should be held accountable.

[2] Physicists and chemists and biologists don't talk like this. Imagine meeting a chemistry professor at a party. She will not say that she is a chemical scientist unless she is very uncertain about either her "self" or you.

R. E. Freeman (✉)
University of Virginia, Charlottesville, VA, USA
e-mail: FreemanE@darden.virginia.edu

© Springer Nature Switzerland AG 2023

S. D. Dmytriyev, R. E. Freeman (eds.), *R. Edward Freeman's Selected Works on Stakeholder Theory and Business Ethics*, Issues in Business Ethics 53,
https://doi.org/10.1007/978-3-031-04564-6_24

Let's disband the Academy of Management. Let's not continue the anti-intellectualism that pretends that structuralism, Marxism, post-structuralism, deconstruction, psychoanalysis, feminism, and the new pragmatism are irrelevant to the theory and practice of management and the behavior of management professors.

Let's just stop all of this stuff. Let's invent an excuse for meeting each year like "I want to talk to my friends about ideas and stay up late at night drinking with them," and let's stop the pretensions. Let's, at the very least, make all of this ritual optional for ourselves and our initiates – that is, our students.

I want to suggest that each of these suggestions for disbanding the Academy of Management contains a fairly straightforward argument that has been ignored, repressed, oppressed, contained, bottled-up, sublimated, displaced, – in general, put on the backest of all back burners. Furthermore, I want to insist that each argument gives us a good reason for disbanding the study of social issues in management as it is presently constituted and envisioned for the foreseeable future.[3] Finally, I suggest a different role for each of us, a role that can only be described as both poetic and pragmatic.[4]

2 Let's Stop Being Scientists

Frederick Taylor had a dream. Armed only with a stopwatch and an idea, he proposed some principles of management. He, or his followers, added the empty honorific, "scientific," thus launching one of the greatest con games since St. Thomas' account of Aristotle as really a Christian. Taylor's dream is alive today in consulting firms and, more importantly, in schools of management.

You see, if we only had bigger data bases, better access to decision makers, harder data, better tools and devices, and perhaps more complex forms of foreplay (sorry, data-analysis packages), we could really, by God, tell them the way the world really is. If you are doubtful of the truth of this parody, I invite you to consult the editorial policy statements of the leading journals.[5]

---

[3] The recent concern with the legitimacy of the study of social issues, which has been the subject of much debate during these meetings, is completely misplaced. If we are now legitimate, as many suggest, then we have ceased to be marginal and critical, and we are irrelevant. We ought to call it quits.

[4] I hope that what I have to say in this paper will be taken in the experimental way that I intend it. While I truly believe that we do much harm in the name of knowledge, I do not believe that people intend such harm, nor do I believe that my friends are bad people. I am simply recommending that we try something new and different and see what happens. Additionally, I have written this essay in a deliberately provocative style, not to make more enemies, but because I believe that we need to be more deliberately provocative and vastly more critical of ourselves, managment theory and practice, and our institutions.

[5] A particularly amusing essay by John Freeman, editor of Administrative Sciences Quarterly, entitled "Data Quality and the Development of Organizational Social Science: An Editorial Essay" (Vol. 31, 1986, pp. 298–303), recommends the establishment of "centrally funded large-scale orga-

Such a dream as Taylor's is ultimately a dream of domination, where the law of God in his heaven is replaced by the Enlightment dream of the scientist is his. White coat, is replaced by the Modern dream of the social scientist/policy analyst/advisor in his three-piece, blue pin-striped white coat deciding what is best for each of us.

At about the same time our "founding fathers" – Taylor, Fayol, and the other heroes of management theory – were formulating this primordial dream for our generation, people like Nietzsche, Freud, James, Pierce, and Dewey were taking it apart for everyone else. For they questioned the whole idea that there is a world to be found, reported on, consulted to, and dominated. They "decentered" the world in the most radical way and liberated human beings from the Enlightment dream. They brought individual human beings and their communities to center stage and suggested that we think of ourselves as poets, making and remaking the world through our language.

Neitzsche urged us to see ourselves as capable of creation, not following the morality of the herd. Sheep graze and bleat: human beings create their world. Freud suggested that we did not know our heads or our hearts, and constructed or proposed a way of talking that allowed us to see ourselves as more complex, more idiosyncratic, more connected to the relationships we experienced in childhood. James and Pierce tried, unsuccessfully in their day, to get our wisemen, our philosophers, to set aside Platonic urges for Truth and Beauty and to get down to the business of human purpose – what we find useful, how we choose to live. Dewey continued this Jeffersonian line by showing that science is not what the Enlightenment, via Madison Avenue, advertised. Its very methods are "subjective" by virtue of the questions it asks. A better way to put this point is that, after Dewey, it is difficult to apply the "objective-subjective" distinction. It just doesn't seem to get us anywhere.[6] Dewey suggested that we look to "community" for a conversation about how we want to live, and that we value creativity and experiment.

So you see, there is a large gap between the dreams of our forefathers, as passed along to us, and the remainder of the intellectual world. Unfortunately, the gap is getting larger. Management theory and the Academy don't seem to be catching up at all. The rest of the world has simply passed us faster than a speeding journal article.

After a brief affair with logical positivism, a number of philosophers such as Wittgenstein, Quine, Davidson, Goodman, Kuhn, Feyerabend, and Rorty, and a number of literary theorists such as Bloom, Miller, de Man, and Fish, and some odd "canards" such as Derrida, Lacan, and Foucault, and their followers began to construct a conversation that called the whole gizmo into question. They suggested that science is just one more story in the human Mother Goose–one more tale about how

---

nizations data base" in order to solve the "measurement" problem. This essay just invites psychobiography, on the one hand, but on the other, it is seriously frightening. It is written by those in power, for those in power – all in the name of science, without any hint of the moral significance of what is proposed.

[6] So, to continue a theme begun by Rogene Buchholz last year, this essay could be summarized by a suggestion that we reside in a new attraction in Disneyland – Nowhereland.

we choose to live or not live. Science has no privileged status, and attempts to give it one are just raw power grabs.

These thinkers have recommended that we see language, as the only tool available, obviously important and extremely slippery. One simple example from our everyday speech about ethics is that we often talk about issues as "black and white," where "black" refers to "the dark, evil side" and "white" refers to "right". Racism and sexism are virtually impossible to eradicate from the very fabric of our language.

Foucault has gone further and shown how the very categories of knowledge are really methods of power. Feminist scholars have argued that language contains the tools of domination and oppression. We can easily resonate with this thinking, for we know that we speak like "markets are penetrated" and "companies are raped by raiders." The imagery is sexual, and male, and sometimes violent.[7] You've either heard this in your classes or "sanitized" it for the journals. Recently, Linda Smircich and Marta Calais have shown how a feminist deconstructionist would read, say, a Pfeffer article or an SIM Proceedings paper (ironically, recommended by me) as blatantly sexist.

Sandra Harding has shown us how the categories of science itself are not gender free. Carol Gilligan, or if not Gilligan then Jean Baker Miller, Nancy Chodorow, Jessica Benjamin, Seyla Benhabib, or a host of others, have suggested that moral discourse is not so pure, unless we mean purely white male. And, Robbin Derry has gently reminded we Simians that all of this applies to us, as well.

So, what are we to make of all of this intellectual name-dropping, praising, faint praising, and damning? I am not suggesting that any of these thinkers has a monopoly on the truth–far from it. Rather I am suggesting that they are relevant characters in the story of management theory, and that the story that we currently tell, without these characters, isn't much of a story. And, all of this business about science, social science, and method just gets in the way.

But there is another point here. If we pretend to be objective, tell-it-like-it-is, just-the-facts-ma'am scientists, we are wrong twice over. Not only do we delude ourselves, which would do little harm, but we delude our constituencies that we have the answer to their questions about the way the world is, thereby justifying their actions based on our answers. Strategies are formulated. Structures are reorganized. Companies are restructured. Lives and families are made and ruined. Yet, somehow we believe that as "scientists" merely describing the world, drawing no normative conclusions, we have escaped responsibility. The choice was theirs, not ours. But as Milgram tells us, Eichman's choice was his, and it was Hitler's too. Responsibility is plentiful.

It is amazing that this con game started by the likes of Taylor works. Organizations, even Bethlehem Steel, are so obviously human, consumed by moral questions about how we can choose to live, that it is frightening to think that three-digit-IQ people believe that we have told them some truth.

[7] Somewhat more suggestive is the imagery contained in Preston and Post's notion of "interpenetrating systems."

Let's stop this harm. Let's stop this oppression. Else, let's disband the Academy of Management and SIM.

3 Let's Stop Being Ethicists

In some more recent quarters, Frederick Taylor's dream has been replaced by the dreams of philosophers. Management types are rushing to read the latest articles in the Journal of Business Ethics. Ethics, and other philosophical journals, getting invited to conferences; even jointly holding the Academy meetings with the Society of Business Ethics. Ethics, in general, and philosophers, in particular, will tell us how it is in the normative world. They will provide us with more hypotheses and theoretical constructs around which we can collect some data and do some more scientific damage. Or, more kindly, ethics will get to the root of the value questions for us. You just can't get any deeper that ethics. Foundations have at last appeared, and if we pay attention to them, by using the language, we can at last make some progress.

I am not suggesting that reading philosophy is a bad thing; far from it. If there are no baseball games on TV, its not a bad way to spend a Saturday afternoon. But, focusing on ethics will, miss a rather large and important philosophical point. The discourse of ethics, especially the tradition of recent analytical ethics, which has produced most of our new heroes, is itself open to a great deal of philosophical debate.

One such argument, put forward by the pragmatist philosopher Richard Rorty, is that politics precedes ethics. There is simply no place from which to get a foundational perspective. Kant's belief that fairness comes from human nature, constructed as "noumenal self" via transcendental arguments, is just smoke and mirrors. Rorty suggests, rightly in my view, that we see ethics as a way of elaborating those political ideals that are so important to us. Ethics becomes, on this view, just one more story, one more human narrative about how we could choose to live. Like science, it has no privileged status, and attempts to give it one will be the same grab for power – this time, by philosophers. So, Rorty politicizes ethics, and in particular, the magisterial work of John Rawls. The traditions of Western liberal democracy are important (the most important values according to Rorty), and we need to tell that story. But, we shouldn't mislead ourselves in thinking that we've finally gotten to some bedrock, something we can count on, come what may, or something we can use to justify the exclusion (or oppression) of others from either political institutions or intellectual activity.

Like the arguments about our self-image as scientists, this one also has an uglier side where real damage is done. When we finally discover ethics and moral reasoning as the bedrock on which research and practice is to be built, we usually set about writing codes of conduct or statements of ethical principles and drumming up some control systems to punish or exorcise those people who, in the words of the warden in Cool Hand Luke, "don't have their minds right." Most of the problems of the resulting systems are not seen in terms of what they require of human beings in general, and poets in particular, but are again articulated, by the warden, as "what we have here is a failure to communicate."

Given what I've already said about social science, you can just imagine what I might say about the move to introduce ethics into the process of social science research. Such a move is a, rather transparent, attempt by well-meaning people to further legitimize one way of doing things. We don't need ethics in social science research; we simply need more and better stories.

I don't wish to be misunderstood about this. I think, that the recent concern with ethics in our field is a positive one. But, I think that blind acceptance of the authority of philosophy, philosophers, or ethics in general would be an excellent reason for our becoming auto mechanics instead of management professors. We need to read more philosophy than ethics. Philosophy of science, philosophy of social science, philosophy of economics, and recent books in philosophy of business such as Edwin Hartman's Foundations of Organization Behavior – will all help us to see how much of a gap there is between the humanities and business. And we can see how naive empiricism and positivism does damage throughout management theory.

4 Let's Stop Being Anti-intellectual

I really don't have much hope for any of this. I don't think that saying all of these things in a semi-outrageous way will do much good for anyone but me. The currents of science and analytical ethics run too deep in what has become the establishment of Management, Social Issues, and Ethics, all seen in their own right as legitimate forces in business schools. And, because we need to find some foundations for what we do, whether or not it is science or ethics, I am not hopeful of everyone simply abandoning what we currently do, even if I am persuasive, which I doubt.

The problem with searching for foundations is that we usually come upon an answer that stops further looking. So we become dyed-in-the-wool scientists, ethicists, or even Marxists, structuralists, Strausians, deconstructionists, feminists, etc. Conversation stops, except among the true believers, and conversations between interpretive communities is difficult. While there is a certain amount of this difficulty built into the intellectual enterprise, no matter how it is conceived, in management schools, we take it to an extreme. It is almost impossible for someone who does agency theory in accounting to have a meaningful conversation with someone in organization behavior. Likewise for consumer behavior people and scholars in the theory of capital markets. What stops conversation isn't what these professors study, or the technical jargon, or even the mathematical shorthand that some use (where nice, neat English sentences would do just fine). Rather, it is the attitude that the method of study is the only method of study, and that they have discovered, via this method, the way the world is.

It is the highest irony that all use the same method, or purport to, and the only conversations they can have are about method, not about substance. Let me illustrate what I'm trying to say by a story:

Imagine a world, called "Namuh," very much like ours except that it doesn't have a body of texts known as management theory. There are organizations and corporations, relationships of dominance and subordination, consensual and contractual agreements, torts and other necessities of business life. There just are no management theorists as we know them.

There is no social science on Namuh and no social scientists. Industry is productive and well developed, but there is no historical figure similar to Frederick Taylor, or Max Weber. There is a long tradition of analyzing why Namuhans do what they do and the resulting structural forces and relationships, but this is done from a standpoint that is profoundly critical. Needless to say, the humanities, the arts, literature, criticism, languages, and music are all extremely well developed.

Occasionally, someone publishes a paper that looks like the invention of what we would call social science. On Namuh, the scholarly journals are open to experimentation and creativity, and while scholasticism exists, on the whole it is minimal. The commentaries on this paper wonder why such measurements and data would be collected. The purpose of such a project would be to control something or someone, and the resulting theory, on Namuh called "story," would be not very interesting; there would be no plot, no character development, no individuality, no sense of culture, history, or values – just some numbers that represent a pretty dry, "every Namuhan," analysis. Critics argued that, even if it was "found" that curly-haired, brown-eyed, bearded males tended to do housework, cook, care for children, and avoid the external world–so what? We would know little of the reasons why this happened, and could only do so by delving deeply into the cultural contexts of particular individuals.

There are schools of management on Namuh, but they are very different from ours. They spend time telling stories about various organizations, some of which are real but many of which are imaginary. They write fiction and poetry and suggests new ways to manage, and new ways to organize. They have their Orwells who routinely warn of the abuse of language by those in power. They have many historians, literary theorists, and others who spend time spinning yarns about the significance of what real managers do. They interpret.

Yet, the most noticeable difference in Namuhan schools of management is one of attitude. Namuhan management professors are profoundly critical. They focus on developing such an attitude in their students, because they believe it is the only way for students to improve themselves and their organizations. Much is argued, and not much is taken for granted.

Namuhan management professors don't have heated battles about the tensions between teaching and research, because there simply aren't any. Of course, sometimes some people don't balance their time well, but they want to teach their best thinking to their students, and they want to avoid jargon which makes this difficult. Of course, there are technical terms, and someone will often propose a way of talking, much as Derrida has done for us, which seems absolutely impenetrable, but that's just the price of letting so many flowers bloom. Even so, someone usually comes along to simplify, as Norris and Rorty have done for Derrida.

There isn't the worry about the tension between research and consulting. Many Namuhan management professors have relationships with organizations, for how else could they interpret what these organizations do? Many times these organizations find that the professors do something really useful for them, and compensate them handsomely, but the underlying value of criticism is so engrained in Namuhan academic culture, that co-optation is rare. In fact, the biggest business for Namuhan management professors is offering these critical thoughts to managers, for Namuhan managers know that they must improve continuously, experiment and innovate, and that the only method for doing so is a critical attitude towards the way they currently operate.

Namuhan management professors read a lot, and they read widely. They do not see a narrow set of journals as being the only source of ideas. They have many conversations with their colleagues in the arts and sciences, and often spend a sabbatical away from the professional school in a particular discipline.

Business ethics does not exist on Namuh. There is such a thoroughgoing connection between the language of business and the language of ethics that none is necessary. Of course, there are crooks on Namuh, some of whom work for corporations and some of whom work for universities, but every decision is routinely subjected to critical review in terms of history, values, culture, and the way that Namuhans want to live, so there is little need for a separate discipline.

This brief glimpse of Namuh suggests that it is possible to disband the Academy of Management and all of the attached apparatus and for business to thrive. I do not believe that there is logical necessity for management professors to see themselves as either ethicists or as scientists; and so I want to recommend the story of Namuh as a coherent story.[8] Furthermore, I want to suggest that Namuh contains some practices which we should adopt if we decide that the practical task of disbanding is too large.

5 Let's Be Crits

Once again I'm not sure that any of this will have much effect except to remind me and acquaint everyone else with my childlike attitudes towards authority, nonetheless, I want to urge that we remake our self-image, that we come to see ourselves as those intellectuals in the professional schools who are profoundly critical, and who in being critical, are the source of innovative proposals for business and management. After all, one interpretation of SIM is a group of people, initially disposed to question the status quo that business is and should be primarily

[8] Of course, to really do this, I would have to tell this story in much greater detail than I can here.

driven by profits. These intellectuals began to explore the ideas that emerged from the creation of a language that sees business and human purpose as integrative rather than adversarial.[9] The exploration of this new way of talking got bogged down, in several attempts at "defining the field." The search for academic legitimacy, the ability to exclude others who were "not in the field," and the need for profitable careers overpowered the initial critical urges which brought the field into fruition. Therefore, it is necessary for a new group to emerge, the Crits (for Committee for Critical Studies in Management), to say outrageous things that question the very legitimacy which the field has won and, in short order, to become marginalized by those now in power.

How do we become Crits? First of all, we need to read more books, especially books written by intellectuals working in "Women's Studies." There is no feminist critique of organizations done by SIM members, except for some very recent papers by Smircich and Calais, Martin, and Derry. We need to read books about literary theory, so that we can become more sophisticated about the importance of language. We need to explore Harold Bloom's idea of the intellectual as a "strong poet," Stanley Fish's concept of "interpretive communities," and Richard Rorty's notion of "liberal ironist."

Second, we need to see ourselves in relation to real managers, much in the way that literary critics are related to practising artists. There is a healthy tension between the two though there is no real conflict. Crits interpret what managers do. They propose new ways of talking that lend new meaning to managerial practice. Sometimes they use standard scientific practice to tell these new stories, but sometimes they use poetry to illustrate a new way of talking and doing. Managers will sometimes see we Crits as eggheads who can't do anything, and we will sometimes view them as anti-intellectual idiots who don't understand what they are doing. Most of the time, we'll get along just fine.

We need to have conversations about the substance of what we do, not about our methods. Crits are first and foremost, pragmatists, so any method will do. Any way that allows us to remake the world or offer an interesting new way to talk will do. What we are interested in is the creation of the self and world that we want, even though we are uncertain exactly what this entails most of the time. So, we have to return to Dewey and be experimental rather than scholastic. We have to become more poetic and less scientific.

Obviously, we Crits need to write about these conversations and circulate and criticize each other. We need to find some solidarity, for even Crits are humans – torn between individuality and solidarity. We need to stop doing all of the things that I mentioned earlier and come to see ourselves as creating the world through language, as trying to get others to see how human excellence and human cruelty can co-exist.

[9] I am told by Bill Frederick and others that, in fact, this is not the way, that SIM was started, but such "facts" make no difference. It is surely possible for us to adapt such a story for our present self-image, whether or not it is real or imagined.

Thus, we must be poetic and pragmatic simultaneously. We need to actively push our liberal democratic values, and we need to have conversations with a diverse set of people that question them. Simultaneously, we need to be adamant and open to change.

We Crits must be willing to marginalize ourselves and to be pleased that sometimes we don't seem to fit in, and are excluded. This has been the lot of many minority groups and one majority – women, throughout history. And, since we believe that more ideas are better than none, sometimes we don't mind being excluded by those who do not value diversity.

Finally, we Crits need to give up the search for foundations, in whatever disguise it comes. We Crits don't need the gods, science, social science, or Kant. It is just us. It is just us – all the way down.

To paraphrase Wilfred Sellers, "it is us, in the largest sense of that word, that make things, in the largest sense of that word, hang together, in the largest sense of that word." The world and our lives are ours to create, to do with as we please. We can embrace oppression, cruelty, and torture, or we can stand against it. As intellectuals, we take our stand by creating new ways to talk that allow us to live differently from the way that we do. Doing anything less is simply not worth doing. Seeing ourselves capable of anything less is to see ourselves as less than we can become.

R. Edward Freeman is University Professor, Elis and Signe Olsson Professor, Academic Director of the Institute for Business in Society, and Senior Fellow of the Olsson Center for Applied Ethics at the University of Virginia's Darden School of Business.

Chapter 25
Business Ethics: A Literary View

R. Edward Freeman

I want to recommend that we try to see business ethics in a literary way. By this I mean simply that managers are concerned with ethics because they are searching for meaning in a complex world. They seek to weave their own tapestries amidst competing interests and claims. Likewise business ethics scholars are trying to give us their favorite metaphors for understanding business and managerial life. However, each group remains stuck in a kind of Kantian tar pit where titans of theory and universal principle struggle to avoid extinction. If we conceptualize this project in literary terms we have a better chance to unravel the competing claims in business ethics and to understand more fully how managers and scholars engage the project of reinventing self, organization and community.

1 Ethics and Business Ethics[1]

Since Plato, philosophers have debated the meaning of ethics and have engaged in a variety of forms of moral discourse. While there is little agreement over the exact scope and content of the term "ethics," we can safely say that ethical or moral

Originally published in: *Social Responsibility: Business, Journalism, Law, Medicine*, 16, 5–13, 1990
Reprint by Springer, Reproduced with author's permission

[1] The author would like to thank Dan Ortiz, Michael Cornfield, Jahan Ramazani, Andrew Wicks and Daniel Gilbert for conversations about these ideas.

R. E. Freeman (✉)
University of Virginia, Charlottesville, VA, USA
e-mail: FreemanE@darden.virginia.edu

© Springer Nature Switzerland AG 2023
S. D. Dmytriyev, R. E. Freeman (eds.), *R. Edward Freeman's Selected Works on Stakeholder Theory and Business Ethics*, Issues in Business Ethics 53,
https://doi.org/10.1007/978-3-031-04564-6_25

discourse refers to the conversations that we have about the effects of our actions on others, and about the more important aspects of our lives. So, "ethics" delineates the essentially human consequences of our actions and choices. (Richard Rorty's recent work on ethics as a wide-ranging conversation about how we can create our lives and remake the world is a view to which I am in large part sympathetic.)

Specific concerns with business ethics have arisen for a number of reasons. The most obvious is that hardly a day passes without some news of another corporate disaster. From the defense industry scandals to Bhopal to the EXXON Valdez to Ivan Boesky and the Wall Street scandals, the media coverage of issues in business ethics is unparalleled. (As a so-called expert in business ethics, I can attest to the fact that most members of the media want summary judgments rather than a host of "academic" complicating conditions.) Business occupies the moral low ground in our society (in conjunction with politicians), and it seems that this is not simply an American phenomenon. In an unsystematic way I have asked managers from over 30 countries about the role of business in their society, and they attest to its occupation of the moral low ground. Ethics, as seen by a number of commentators, is necessary to moderate the "unbridled greed" of capitalism. The current wave of concern is therefore not different in kind from similar concerns in the 1920s or from the cry for "corporate social responsibility" in the 1960s. The standard questions asked by most are about the conflicts that inevitably arise between "ethics" and "profits." This overly simplistic view dominates business ethics, and is the cause for great journalistic concern.

However, there is a deeper issue here. A common view is that the discourse of business and the discourse of ethics have little in common. Executives and scholars alike behave as though an analysis of behavior from an *ethics* perspective is different from a purely *business* perspective. The disconnection of these two spheres of discourse leads to a juxtaposition of phrases in questions such as "we understand that from a purely business standpoint, closing the plant was justified, but was it ethically correct?" Such questions assume that the discourse of business would be found guilty of gross immorality if it stood before the tribunal of ethics. This analysis squares with the public perception of business in the United States. Roughly two-thirds of the public believe that business does a lousy job on ethics and that business people are more likely than others to engage in unethical behavior.

A second matter related to the separation of business discourse from ethics discourse is the fact that the dominant view of capitalism is furnished by mathematical economists. The economists' model leaves out the particularity of human concerns, and treats collective moral concerns (such as pollution) as technical issues of moral hazard, club goods, or externalities. Again the language of economics is curiously absent the language of ethics. The blatant utilitarian assumptions of most economic discourse ignore more basic concerns with human rights. One recent strand of economics, so-called "property rights economics," is an exception; yet one of its major theoretical breakthroughs, "agency theory," recommends that we view human beings as wanting to do as little work as possible for maximal rewards.

Business ethics as a discipline has arisen in part as a counterweight to this narrow view of economics. Business ethics has called in part for a return to the fundamental moral concerns of economists such as Adam Smith and Karl Marx. Some urge a reopening of the Austrian school of economics, where ethics is the primary concern. A recent surge in interest in the work of Hayek is a good example, for Hayek's primary concern has been to find and articulate a moral basis of capitalism in contraposition to socialism. The Austrian analysis of entrepreneurship places far more emphasis on cooperation than on competition as the basis of capitalism. The entrepreneur's main role is creative, putting together a deal or contract among customers, employees, suppliers, financiers and communities, so that each is better off.

Third, philosophers and other ethicians have contributed to the rise of business ethics as a separate discipline, perhaps unwittingly, by producing facile analyses of business issues. Indeed, the whole focus on "issues" as a unit of analysis assumes that one can separate the ethical component of a decision or a corporation from other components. Intellectually and practically this approach is unsatisfactory, for it is difficult to see how worker safety, for example, can be profitably analyzed outside the scope of the employment relationship in general. Bland generalizations such as "firms should not close plants," "firms should not pay bribes," or "hostile takeovers should be outlawed" are not terribly helpful in a complicated world. Because so much ideology has masqueraded as philosophical analysis, the stage has been set for a more scholarly approach.

Finally, and most important to my current task, the concern for business ethics has arisen because of the general "decentering" effect of our post-Enlightenment era. Scholars and managers alike ask more and harder questions, and take less and less for granted. Philosophers such as Kuhn and Feyerabend have broken down the traditional view of science and social science as arriving at "Truth." Others such as Rorty have picked up the pieces and articulated a new pragmatism that reassesses the traditional academic boundaries. Recent work in literary theory by Derrida, Bloom and Fish has called into question the very idea of a world of independent meanings, of essences to be discovered. And a stunning body of work by scholars in women's studies has shown that most academic disciplines have evolved along exclusionary lines, in a sexist way. However, little of this work has found its way into the business ethics field.

2 The Scholarly Response

The scholarly response to the rising concern with business ethics has been almost unidimensional. In large part philosophers see these developments as yet another opening for so-called "applied ethics." (Surely "applied ethics" represents one of the great pleonasms of our time—rendered necessary simply because the scholarly study of "ethics" has been made irrelevant by twentieth-century philosophers.) Business represents yet another avenue to explore the inner workings of ethical theory and its retinue of background disciplines, law, history, economics and other social sciences. Following in the footsteps of "the more advanced" applications of ethics in medicine and law, scholars in business ethics have outlined a project by which we hold "business" and its constituent disciplines—finance, economics, marketing, operations, organization behavior, and strategic management—to the light of universal ethical principles.

And, there is a competing view to this largely Kantian project. Some scholars want nothing to do with ethics or ethical reasoning since it is difficult to see how their methodological apparatus, rooted in positivism, can remain whole. These scholars see a bright line between their task of description/explanation, and a subsidiary task of prescription/ evaluation. Since they also acknowledge that unlike the "harder sciences" their task is necessarily incomplete, it seems to follow that the task of ethics either never begins or can never be based on true descriptions. Ethical criticisms of theories such as agency theory can always be internalized with calls for better data or more robust explanatory concepts.

While business ethics scholars have achieved a modicum of legitimacy in the past 20 years, the price has been a steep one. They have appropriated the view of philosophers such as Quine that the task of business ethics is to serve as the "handmaiden of the disciplines of business." Yet this "Handmaiden's Tale," like that of Margaret Atwood, is one of accomplice and accessory, of preventing the rich narrative of human experience from entering the halls of modern business schools.

3 Business Ethics With a Literary Base

Let us try something different and see where it can lead. Let us take literature and literary studies as the background disciplines for business ethics. Bracket for the time being the question of the role of the social sciences such as economics, sociology and psychology. What kind of discourse would business ethics be with literature and literary studies as background music?

Questions of meaning and interpretation would immediately come to the foreground. What is the meaning of particular managerial or corporate behavior? How are we to interpret a particular incident or set of incidents? From whose perspective? What is the background, history, culture and context of the actor? What are the key tensions among the various actors? Which actions or icons have symbolic meaning? What is the actor trying to accomplish? What do we see him trying to accomplish?

The list of questions is endless. Note that these questions are particular, pertaining to individual action or specific actions. These questions take business behavior as an element in a story, a narrative about human life. The task of actor and scholar alike is to give meaning to this story, to rewrite it, and in so doing to recreate self, organization and community. Note that the standard social science questions – (1) What motivates men (usually) to act? and (2) How can collective action be understood as arising from individually motivated action?- can be phrased in literary terms. But if they are phrased and answered in a way that makes the characters one dimensional, the plot quite thin, and the tone irrelevant, I am afraid these will be boring stories.

Herein lies the rub with most social science studies of business today. They are just boring. I listen to the latest pronouncements on equity theory, job satisfaction, and firm environment fit and imagine the underlying view of human beings as all having been to Dale Carnegie with steering wheels on their backsides trying to maneuver each other around inside the corporation. It is only by seeing this image as an M.C. Escher print that one begins to cry as well as laugh.

Seeing literature and literary studies as relevant to business is no panacea, but it does give us some hope of understanding human beings and their language as multifaceted and complex. Fortunately we cannot conceptualize literary studies as a simple tool that we can apply pell mell to business ethics. There are at least three ways to understand the relationship between literary studies and business ethics. Each way assumes a particular view of the connection between managerial action and scholarship. And each has the potential to add value to our understanding of the role of ethics in business life.

(1) *The Great Insights Approach.* This way of seeing the connection between business ethics and literary studies urges us to see business and ethics as a part of a larger sense of human culture. We can look to literature for help in dealing with questions of meaning in organizations. Literature can yield *insight* into the meaning of work and life as in Sinclair Lewis's *Babbitt (p.* 109):

> For many minutes, for many hours, for a bleak eternity, he lay awake, shivering, reduced to primitive terror, comprehending that he had won freedom, and wondering what he could do with anything so unknown and so embarrassing as freedom.

Babbitt's life is so ordered, controlled and sterile from his middleclass thoughts to his middle-class actions, that he looks to a love affair for meaning and finds more emptiness instead. Babbitt may well tell us that our self is created and, like Ungar, we need a notion of work that is transformative.

Or, consider Bob Solomon's discourse on personal relationships from Heller's *Something Happened,* pp. 31–32:

> Green now thinks I am conspiring to undermine him. He is wrong. For one thing, I don't have the initiative; for another, I don't have the nerve; and for still another thing, I guess I really like and admire Green in many respects (even though I also hate and resent him in many others), and I know I am probably safer working for him than I would be working for anyone else — even for Andy Kagle in the Sales Department if they did decide to move me and my department from Green's department to Kagle's department. In many ways and on many occasions Green and I are friends and allies and do helpful, sometimes considerate things for each other. Often, I protect and defend him when he is late or forgetful with work of his own, and I frequently give him credit for good work from my department that he does not deserve. But I never tell him I do this; and I never let him know when I hear anything favorable about him. I enjoy seeing Green apprehensive. I'm pleased he distrusts me (it does wonders for my self-esteem), and I do no more than necessary to reassure him. And I am the best friend he has here.

The range, depth, and complexity of human lives and emotions displayed in these brief passages — as well as countless others from Shakespeare, Flaubert, and Woolf — are surely relevant for understanding and deepening the conversation about ethics in business.

There are any number of variations of this view. We see novels, stories and poems being used in our classrooms, and teachers of business trying to distill the lessons of these works for business. We see this as an enriching experience — or part of the study of business as a part of the study of culture. And it is easy to visualize intellectual life as teaching and writing about how particular works can help the conversation about any number of ethical issues. The business school scholar would perform a kind of secondary role of literary critic, giving interpretations of literature that are peculiarly relevant to business. This task is easy for a genre of "business novels" like *Babbitt, Atlas Shrugged, Something Happened, The Jungle, Bonfire* and the like, but it is more difficult for Shakespeare, Flaubert and Joyce.

(2) *The Textual View.* A second view of how literary studies can inform conversations about business ethics I will call "the Textual View." Let us come to see *managerial action as a text.* Let us listen to managers tell their stories and the stories of their companies. Our task as intellectuals is identical to that of the literary critic. We interpret and reinterpret, frame and reframe these stories. We propose new readings of these stories – close readings – in which we give meaning to managerial action. On this view we intellectuals need less knowledge of literature, more knowledge of business and of the techniques of criticism.

Management (business) theory takes on a special role in the Textual View. It too can be seen as a text, one written by a certain group of scholars.

Management theory becomes a "meta-text" with its own embedded interpretations of managers' stories. The Textual View does not preclude the Great Insights View. Rather, it relies more on a broader notion of literature and sees the business ethics conversation as more "intellectual," i.e., as giving meaning and interpretation rather than providing just more emotional content to an otherwise analytical world.

(3) *The Neo-Pragmatic View.* Finally, we can see the literary studies/business ethics connection in a way that I call "the Neo-Pragmatic View." This view sees managerial action and intellectual life as embedded in a seamless whole of culture, politics, and community, with no *natural* joints such as the disciplines of business supposedly give us. Managerial life is a story, but it is a story connected to a genre of human stories inside and outside organizations. Intellectuals interpret the world, but such interpretation is really a political act-one that proposes or recommends how we live. Self, organization, and community are not derived from the text using the methods of social science and criticism, but these are up for grabs, to be created. How can we come to see organization and work as transformative of self, as making better ways of life possible? The pragmatist realizes that the questions of business ethics are not the questions of "justice" but those of "just us." How shall we choose to live? What stories can we tell about our past, present and future that can help us be less cruel? As intellectuals we cannot divorce ourselves from our history, culture, context and values.

If we compare these three views we note that the Neo- Pragmatist View encompasses the first two. Since the test for a pragmatist is "how will this allow us to live?" the novels of Flaubert and Joyce, the narratives of individual managers, the "high theory" of critics, and the "low theory" of postmoderns are all grist for the mill. What we need are fewer arguments about principles and more empowering visions about what we can become. We need fewer social scientists and more Freuds. We need fewer "careers" and more "selves."

In short, the conversation in business ethics needs to become more *critical*, not in the argumentative sense of analytic philosophers, nor in the Marxist sense of the legal critics, but in a clear, straightforward, pragmatic way of examining and choosing the ways we describe the world. The conversation needs to help us face up to the responsibility for self-creation and the effects of self-creation on others.

4 Effects on Business Education

My view of "business ethics" would have some rather drastic effects on the way we see "business education." We need to:

1. have more people educated in the humanities;
2. value substance over method and interpretation over data;
3. have self-creation, organization-creation and community-creation become part of the discourse of business.

If we achieve these three items, then the tension between business and ethics goes away. There will be no such thing as a "business perspective" that is free from critical analysis. We will have only competing narratives.

However, if we continue along the narrow march to irrelevant "knowledge and truth" and away from human notions of self and community, if we continue to talk about the morality of economics rather than human hope, love, caring, sexuality, and death, we will become less human. The path that business schools and their newly appointed philosopher handmaidens are now on is not just (to quote Orwell) "doubleplusungood" — it is evil.

R. Edward Freeman is University Professor, Elis and Signe Olsson Professor, Academic Director of the Institute for Business in Society, and Senior Fellow of the Olsson Center for Applied Ethics at the University of Virginia's Darden School of Business.

Chapter 26
Business, Ethics and Society: A Critical Agenda

R. Edward Freeman and Daniel R. Gilbert, Jr.

Almost all of our institutions are against what we are going to propose: the adoption of a critical approach to management and management theory, in general, and to business, ethics and society, in particular. The interests of these institutions dictate wholesale acceptance of the current way of doing business and allow change to proceed in predictable, discrete, acceptable, and rational ways. In short our institutions and our methods and theories of business and their connection with ethics and society enjoy a symbiotic relationship that limits in important ways the possibilities that our corporations hold for us.[1]

In Sect. 1 we diagnose how our current vocabulary has become a nuisance and suggest a rethinking of both business and business schools as institutions. In Sect. 2

Originally published in: *Business & Society*, 31(1), 9–17 © Sage Publications, 1992
Reprint by Springer, https://doi.org/10.1177/000765039203100102

The authors wish to thank the participants at the IABS Sundance Conference in February 1991, the editors of this journal, and especially two anonymous referees. We have benefited from discussions of these ideas with many people especially Denis Collins, Robbin Derry, Craig Dunn, Dawn Elm, Bill Frederick, Andrea Larson, Jeanne Liedtka, Richard Nielsen, Jill Teplensky, and Donna Wood. We are especially indebted to the work of Richard Rorty.

[1] We are indebted to Maria Calas and Linda Smircich, especially their Ruffin Lecture, "Predicando la Moral en Calzoncillos: Feminist Inquiries into Business Ethics," forthcoming in Freeman and Andrea Larson (eds.) *Business Ethics and Women's Studies*. Oxford: Oxford University Press. We suggest a different approach to understanding business, ethics and society. Our approach depends on dropping most of the current vocabulary that business and society intellectuals use.

R. E. Freeman (✉)
University of Virginia, Charlottesville, VA, USA
e-mail: FreemanE@darden.virginia.edu

D. R. Gilbert, Jr.
Gettysburg College, Gettysburg, PA, USA

© Springer Nature Switzerland AG 2023

S. D. Dmytriyev, R. E. Freeman (eds.), *R. Edward Freeman's Selected Works on Stakeholder Theory and Business Ethics*, Issues in Business Ethics 53,
https://doi.org/10.1007/978-3-031-04564-6_26

we propose some new metaphors—new for business and business schools—that will allow us to see things differently. In particular, we suggest that feminist theory and psychoanalytic theory can be useful starting places given where our institutions find themselves, and we show how these metaphors can be used to understand and re-state some traditional business and society questions. Finally in Sect. 3 we generalize the way we have done what we have done and recommend a different way to see our role as business school intellectuals.

Most of what we have to say has been said before. We are just going to say it again in a different context: that of a group of intellectuals who see themselves as critical of the enterprise of business, but who are for the most part still under the influence of the Enlightenment Project to tame the world. The spirit of what we have to say is one of inquiry, of launching a project to which we are committed without any set ideas of where it can ultimately lead, indeed with suspicion of the very idea of "ultimately lead".

We also write in the spirit of invitation to the reader to read and write about the books we praise and ideas that excite us, and to launch your own critical project. We do not offer arguments meant to show one position has priority over another. We offer a narrative account of one way to think about the creation of self and community.

1 The Current Vocabulary

We have suggested that our current theory, our jargon, our metaphors, whatever your favorite honorific is, it not useful anymore: it has become a big time nuisance for four reasons.

1.1 Business and Society and the Search for Legitimacy

For the past 10 years at least, scholars in business and society have embarked on a search for legitimacy. Yearly, sometimes quarterly, we are treated to this or that paradigm that finally pulls all of the disparate pieces of "the field" together.[2] At one level this search for legitimacy is understandable. Business and society scholars are for the most part marginal in most business schools. As separate scholars working to create their own discipline they have a very young history. In fact most of the so-called "founding fathers" were trained in different fields. We can see that a group of scholars from political science, economics, public policy, law, philosophy, and

[2] Recent papers by Frederick, Epstein, Vogel, Preston, Wood, and others all seek to encompass multiple points of view, and to show that concepts and models that appear to be different really can be synthesized. We ate not suggesting that these papers are not interesting, but that the genre, "framework paper," tells us something important about the so-called "field of business and society."

other disciplines came together to ask similar sets of questions about the legitimacy of business understood as a purely profit-seeking enterprise.

These scholars created courses in business and society, and in some business schools managed to get their courses required. The standard explanation singles out University of California at Berkeley, Boston University, University of Pittsburgh, and University of Minnesota as centers of knowledge and scholarship in business and society. As more and more articles, books and journals were created a "discipline" was formed, and the current historical moment has scholars in this discipline" embarking on a Kuhnian "paradigm hunt."[3]

We want to suggest that this search for legitimacy has become a nuisance, and we recommend that we bring this search to an end. The search is a nuisance precisely because it ignores difference. It is based on a politics of power whereby the scholar or school whose rhetoric is most convincing becomes "a leader in the field."

At one level the search for legitimacy by business and society scholars is perfectly understandable. It is no more and no less than the search for solidarity among a group of people who like to hang around together and who read some, but not all, of the same books. Trying to find our paradigm is just a way to make us more like others who think they have one, i.e., our colleagues in business schools, and to give us an acceptable rationalization for meeting together periodically. However, our obsession with frameworks and synthesis masks a real contribution of these scholars: the adoption of a fundamentally critical stance, that is, the very reason for their marginalization.

Instead of an agenda of legitimacy for business and society scholars that seeks to legitimate a business and society discipline within the structure of the business school, we should return to our earlier stance that is marginal. We need to see the central problem for business and society not as how to get legitimacy within the business school, but in posing questions and analyses such as why has legitimacy become such a crucial question for business schools and business.

1.2 Business and Society and the Legitimacy of Business Schools

As business schools themselves have sought legitimacy within the university, they have explicitly adopted writ large the trappings of science and a narrow social science. Two influential reports of the 1950's guided most schools towards the development of curricula and disciplines that uncovered law-like generalizations about business and organizations. Of course the only way to even sincerely pretend to find such "theory" was to simplify business and organizations enormously, and over the years, such simplification has grown to the level of unexamined assumptions.

---

[3] Of course such a hunt like the elusive "snark hunt" will be fruitless and rests on a misunderstanding of Kuhn's notion of paradigm. Paradigms emerge out of as well as guide the normal scientific process within scientific community. The postscript to Kuhn (1970) would be instructive to read.

If we put such a search for legitimacy into the university context with the full politics of tenure and promotion, the needs of university administrators to translate across disciplines necessitating a rating system for journals, and the overwhelming desire of social science intellectuals to emulate their "hard" science colleagues, we get a "business theory" that subscribes to a narrow and increasingly limiting form of positivism.[4]

The last 10 years have seen increasing numbers of questions asked by business school intellectuals. Just as business and society people were trying to become legitimate another group of intellectuals, kindred souls with business and society folks, have emerged to question the very assumptions of the business school disciplines. Thus, "crits" have emerged in accounting where the journal *Accounting, Organizations and Society* has taken a leading role in questioning the fundamental assumptions of accounting as a discipline. And, "Crits" have emerged in economics, the parent discipline to business schools, under the leadership of economist Donald McCloskey. McCloskey's two books, *The Rhetoric of Economics,* and *If You're So Smart… The Narrative of Economic Experience,* ask to see economics writings the same way we look at literary texts and bring the full interpretive power of literary theory to bear on these texts. Even marketing has gotten in on the act. Shelby Hunt, a main stream contributer and former editor of the *Journal of Marketing,* has led a dialogue on the underlying philosophical principles on which marketing theory and practice are based.

The original contribution of business and society scholars, to question the assumption that business is to be understood in purely economic terms, is in danger of being lost. While these scholars have pointed out the social consequences of business in a variety of ways, they have had little or no impact, in part because of their current unwillingness to marginalize themselves to the business schools in which they sit, and the businesses to which they consult. Where they have had an impact they have served as apologists for both businesses and business schools. Because they have business and society courses or social responsibility programs, both business and business schools have "solved" the legitimacy problem. A more careful analysis needs to be done.

1.3 Business and society and the Legitimacy of Business

The last several years have seen an increase in the questioning of the legitimacy of business itself. In the U.S. fully two thirds of the public distrusts business as an institution and, the recent "business ethics craze" is faddish testimony to the fact that business occupies a kind of moral low ground in society. But the roots of this questioning is much deeper.

[4] An unflattering view is that business schools have become havens for intellectuals who "couldn't cut the mustard in real social science departments."

In the U.S. business has long suffered from a lack of social legitimacy. Indeed the very idea of corporate social responsibility, coming from the articulation by Andrew Carnegie of philanthropic principles in the late nineteenth century, came about to try and address the fact that business activity was questionable throughout large segments of society. People distrust "the profit motive," and they draw bright lines between the realms of "ethics" and "business."

The question of the legitimacy of business is central precisely because the language that we use to describe business activity is shopworn. This language, complete with sophisticated journal articles, books, equations, as well as other rhetorical devices, has been largely propagated by economists and social scientists in the academic world, and has masqueraded as the "logic of capitalism" in practice.

The dominant view of business is as an economic activity, subordinating other goals to its primary one of creating economic value. (Even the idea of "creating value" is a relatively new one obviously intended to usurp "value" and make it "economic.") We continually draw sharp distinctions between the "economics" of a decision and its "politics" or its affects on "society." By dividing the world into economical, political, social, technological, etc. we create separate ways to understand it which must then be linked together. The very idea of business as economic, as something separated, or separable from the social world, necessitates that legitimacy will always be an unsolved problem. There will always need to be a "theory of legitimacy" as long as we insist on seeing business in purely economic terms.

Such a theory of legitimacy will serve as a translation device between economic concepts and for example, social concepts, and one realm will be defined in terms of the other. Economic performance will be shown to entail a certain kind of social performance, and there will be an endless debate over the priority of economic versus social goals. It is important to see that such a debate is neither optional nor decidable. The problem is in the conceptual apparatus, which has been pushed beyond its limits.

So, better theories of corporate social performance, or corporate social responsibility, or business and the social policy process won't help. They just reinforce the idea that you can profitably carve up the world in economic, social, and so forth. These theories, and the intellectuals who propose them, just reinforce the idea that business is essentially economic, and they become a kind of *apologia* for the status quo.

1.4 Redescribing Business and Society, Business Schools, and Business

We recommend abandoning the search for legitimacy and instead, adopting an approach which tries to redescribe business and management in ways that the problem of legitimacy just doesn't come up. How is this possible? We can start by seeing business as a connected set of relationships among stakeholders where the emphasis is on the connectedness. We need metaphors and concepts which help us redescribe their relationships usefully, and the metaphors of economics and politics will not be very useful. We will need to understand when organizations "smell like customers" or do the crazy, chaotic things that Tom Peters talks about in which economics just doesn't seem to come up.

We need to understand that stakeholders are in it together, rather than competing for limited and scarce resources, and that the fundamental reason that organizations as connected networks are effective is that they are built on principles of cooperation and caring. Each stakeholder is "adding to the value" of others, creating a good deal for all.

Trying to reduce all of this to economics is just silly. What we need are stories about the kinds of companies we can build, the kinds of connected networks that are possible, and narratives about successes and failures.

Or, if you don't like these metaphors propose some of your own that don't depend on sharp distinctions among economic, political, social and moral. The central question should be how can we redescribe our organizations so that the question of legitimacy doesn't come up? How can such legitimacy be built into the very way that we describe them? If we adopt these questions, there is a different role for the business and society intellectual. She becomes the strong poet that Harold Bloom talks about that reinterprets the world because she encounters it differently. She constantly says, "think about business this way..." where the "this way..." doesn't make a lot of sense the first time you hear it.

Business and society intellectuals can then help business schools take on a new role. Traditionally business schools have been places of socialization, especially with the MBA degree. But, if we are to reinvent our corporations and ourselves, and if we are to improve the world we live in, business schools must directly take on the critical function as well. They must begin to look more like the humanities than the sciences, and they must become tolerant of difference, especially difference aimed at new metaphors and redescriptions for business life.

Now we are not optimistic that any of this will in fact happen. So, what are we business and society intellectuals to do while waiting for our place in the sun that leads business schools and business out from under the disciplinary thumb of economists and other misguided social scientists? We want to suggest that we redescribe ourselves and begin to work in some different areas.

2 Redescribing Business

There are many areas that we could choose to illustrate how business and society intellectuals can begin to redescribe business. We shall focus on two areas, namely, feminist theory and psychoanalytic theory, simply because we have some familiarity with them. We could have focused on religious thought, family therapy, mythology, or even an analysis of pop culture like video games and Madonna, and been perhaps equally fruitful. Both feminist theory and psychoanalysis pay special attention to the concept of "silence," what has not been said. If we can give voice to some silences, we can come to redescribe business in ways that may well be liberating, that enables us to live differently and better.

There is no such univalent entity as "feminist theory." Rather there are feminisms — representing the rich body of work by scholars in Women's Studies during the last 25 years. We want to focus on one feminism, so-called "different voice feminism" or "feminist standpoint theory." This theory says that women and men are different, that they reason about ethics in different ways. Men pay attention to rules and justice and treating everyone the same, and women pay attention to caring, connection, and context, putting their priorities on the effects of action on relationships. Men see the self as autonomous, and see morality as a way of enhancing or constraining their independence. Women see the self as connected, as a set of relationships which cannot be disentangled into discrete, autonomous parts. Morality, for women, is about the maintenance and flourishing of these relationships, not about rules for deciding whose autonomous action should be allowed.

It is important to qualify this view, propounded by people like Carol Gilligan, Nell Noddings, Jean Baker Miller and her Stone Center colleagues, Nancy Chodorow, and others, by noting that these thinkers do not appeal to any innate properties, or "naturalness" of this view. Rather they suggest some deep patterns of socialization at work, from different solutions to the Oedipal drama, to the role of women as the primary caregivers in society. Also, it is important to avoid stereotyping with this approach. No one believes that this theory describes all women, or all men. We can see it as one, of a number, of useful narratives that can help us make sense out of gender relations.

Similarly there is no univalent theory called "psychoanalytic theory." Rather there are a bunch of competing narratives that have a few principles in common. These common principles revolve around the claim that we are not always the best judges and do not always know what is really in our hearts. So, we don't always understand the cause of our actions. From Freud's view that childhood sexual fantasies are central, to Melanie Klein's view of infant-mother "object relationships" psychoanalysts disagree a great deal, but they agree that we repress much of our mental lives, and that we have an unconscious which is not readily available for our introspection.

A great deal of what passes for research in business and society simply ignores the major tenets of both feminist theory and psychoanalysis. Rather than focus on the shortcomings of others, we shall focus on the silences of our own work, and then in Sect. 3 suggest some questions that are relevant for the field as a whole.

The idea that business is responsible, or ought to be, to stakeholders has a long history. In *Strategic Management: A Stakeholder Approach,* Freeman (1984) depicted stakeholders, and "stakeholder management," as a contest of autonomy. That is, stakeholders were clearly defined "names and faces" who were to be managed. If managers were to achieve their objectives, then they had to pay attention to stakeholders, regardless of their so-called "legitimacy." Stakeholder management was to be seen as understanding the rules of the game of business. Caring had nothing to do with stakeholders and in fact Freeman argued that it didn't matter if managers cared about stakeholders or not. If managers wanted to stay in control, achieve their objectives, then stakeholders had to be managed.

As it stands this narrative clearly fits a "rules and justice" approach and is therefore gendered in an important way. It simply excludes the voices of those managers who might think differently. If we listen to feminist standpoint theory, we might well revise the entire stakeholder framework, or add an alternative interpretation. The narrative could be sketched briefly as follows: Business should be understood as a set of connected relationships where the boundaries between who is in the organization and who is a stakeholder are ambiguous and blurred. The key point is the relationship. We can see "customer relationships," "supplier relationships," etc., not generically, but concretely. Understanding how these relationships mutually support each other and how they interact is an important part of the story. And, managers who are effective define their very sense of self in terms of these discrete relationships, and focus on caring for and maintaining these relationships. It could well be that we can find gendered organizations along stakeholder lines. In rules and justice oriented organizations we would expect to find glass ceilings, while in caring and connected organizations we would expect no glass ceiling.

Now there is more to be said about this view, but such a story may well be more compatible with the view of self that women have, at least from one line of feminist inquiry. Equally, we can see the shortcomings of such a redescription. For if the focus of business activity is solely on the nurturing of existing relationships how can heroic or innovative action occur? Isn't the status quo reinforced a bit too much with this view? After all, perhaps feminist standpoint theory is just the morality of the slave, and it contributes to the oppression of women that currently exists.[5] We can't address all of these questions here, but we want to show how using a feminism can offer different possibilities for different stories and how it can be "unsettling" or "decentering" of our current business and society frameworks.

Likewise, in *Corporate Strategy and the Search for Ethics,* Freeman and Gilbert (1989) proposed that we see corporations as places where the personal projects of members could be fulfilled. In doing so, we explicitly ran roughshod over the idea that persons were to be seen as capable of having and *knowing* their projects. We said nothing about the difficulty involved in coming to realize that projects may well be substitutes for a whole host of deeper issues. Freud argued that the workplace was productive precisely because we were not able to deal with issues of sexuality. In adopting a kind of rational choice view-, we ignored both gender and a more complex view of the person that is possible. Our view of the person was primarily rational, achievement-oriented, valuing autonomy at the exclusion of relationships, and consequently, unemotional (or emotionally crippled), asexual, uncaring, etc.

[5] We are grateful to Richard Rorty for making this point forcefully.

Now none of this suggests that our business and society theory has to solve all problems. Rather we need to understand where our stories are weak. And, business and society narratives are weak with respect to issues of gender, human complexity, sexuality, race, and class, to name a few areas.

There is some scholarship in these areas, but most of it adopts wholesale the dominant view of race and gender as "equality means treating everyone the same," a tenet at least up for grabs by a lot of theorists. When this dominant ideology is combined with a narrow empiricism we are treated to studies of whether or not men and women, blacks and whites, gays and straights, have been treated the same or equally, or fairly.

The world we have created for ourselves is much more complex. And it is this complexity of the human person about which most theory is silent. And, there are redescriptions of our-selves that depend on more complex stories, redescriptions that are available to business and society intellectuals. These stones involve our childhood, our dealings with authority figures, our encounters with human mythology, our unconscious drives, our emotions, our spirituality and our sexuality. But, to read the current management theory and to sit in our business and society classrooms we would believe that things have been cleared up since Freud, that really we are self-interested or altruistic. Or, if there are other dimensions, they are much less important. It is as if we have become mute on what human life really is about once we enter the realm of organization.

3 A More Critical Approach

Adopting a critical approach to the study of business, ethics and society is different from finding a paradigm. The idea of a critical research agenda is no more than a set of interesting questions or projects that might be fruitful. There is no large scale *"working out"* of a particular school of thought, but rather a willingness to explore the underlying assumptions of theory and practice, a questioning of key terms and definitions, and a thorough interrogation of the role of race, class, culture and gender in the development of our ideas. We suggest the following three projects as ones to which business and society intellectuals can make substantial contributions.

3.1 A Critical Analysis of the Business Disciplines

The business disciplines of accounting, finance, marketing, operations, organizational behavior, and others, all make assumptions about the role of the person and the proper relationships among social institutions. By bringing some critical texts to bear on these disciplines, business and society scholars can cast them in a new and more critical light. Marketing needs to be reinvented with a profoundly different idea of what a person is, i.e., more complex than a garbage disposal that consumes whatever you put in it. While we think that psychoanalysis and feminist theory critiques of the traditional business disciplines could be enlightening, we shouldn't confine ourselves to these. There is much to be gained from literary theory, Eastern and Western philosophy, the analysis of popular culture and its effects on business, to name a few sources of inspiration.

3.2 Understanding Race, Gender and Culture in Business and Society

Little is written about these topics that draw on the critical texts we have suggested. Multiculturalism has yet to make a real impact on business schools in the way it has an other disciplines. Again, business and society intellectuals are the scholars who are best suited to do this kind of analysis.

3.3 Understanding Management as an Art, and Management Theory as Criticism

Much has been said about the art of management, but very little has been done to apply the intellectual disciplines of the arts, such as literary theory, aesthetics, art history and criticism, to thinking about management. Again, business and society intellectuals are positioned to take on the task of interpreting the work of managers, seeing how they construct meaning, giving alternative interpretations to their work, and the like. We have focused on prediction and explanation, given our search for legitimacy, but if we end that search a more liberating future is possible. There will be a rather healthy skepticism on the part of managers, just as the relationship between the artist and the art critic is sometimes quite strained. But, it will be clear that the role of the business and society intellectual is to make sense and give meaning to managerial and corporate action, focusing us on the effects of those actions on others, and the way that we think about the self.

While what we have suggested involves a reshaping of what we now do as business and society scholars, our "field" is in need of reshaping, but more so, business and business schools need more critics. It is only by profoundly challenging the descriptions and narratives of our world that we can improve it. As intellectuals we have a responsibility that goes considerably beyond where the current state of our art find us.

References

Bloom, H. 1982. *Agon: Towards a Theory of Revisionism*. New York: Oxford University Press.

Calas, M., and L. Smircich. Forthcoming. Predicando la moral en calzoncillos: Feminist inquiries into business ethics. In *Business Ethics and Women's studies*, ed. E. Freeman and A. Larson. Oxford: Oxford University Press.

Chodorow, N. 1989. *Feminism and Psychoanalytic Theory*. New Haven: Yale University Press.

Epstein, E.M. 1987. The corporate social policy process: Beyond business ethics, corporate social responsibility, and corporate social responsiveness. *California Management Review* 29: 99–114.

Frederick, W.C. 1986. Toward CSR: Why ethical analysis is indispensable and unavoidable in corporate affairs. *California Management Review* 28: 126–141.

Freeman, R.E. 1984. *Strategic Management: A Stakeholder Approach*. Boston: Pitman.

———. 1989. *Let's disband the Academy of Management. Social Issues in Management Division, Chairperson's address* (August).

Freeman, R.E., and D.R. Gilbert Jr. 1989. *Corporate Strategy and the Search for Ethics*. Englewood Cliffs: Prentice Hall.

Freeman, R.E., and D.R. Gilbert, Jr. 1991. Business, ethics, and society: A critical approach. *International Association for Business and Society, 1991 Proceedings*.

Freud, S. 1961. *Civilization and Its Discontents*. New York: W.W. Norton Co.

Gilligan, C. 1982. *In a Different Voice*. Cambridge: Harvard University Press.

Hunt, S., and S. Vitell. 1986. A general theory of marketing ethics. *Journal of Macromarketing* 6: 5–16.

Klein, M. 1989. *The Psychoanalysis of Children*. London: Virago Press.

Kuhn, T. 1970. *The Structure of Scientific Revolutions*. 2nd ed. Chicago: University of Chicago Press.

Miller, J.B. 1987. *Toward a Nero Psychology of Women*. 2nd ed. Boston: Beacon Press.

Noddings, N. 1984. *Caring: A Feminine Approach to Ethics and Moral Education*. Berkeley: University of California Press.

Preston, L.E. 1986. Business and public policy. *Journal of Management* 12: 261–275.

Vogel, D. 1986. The study of social issues in management: A critical appraisal. *California Management Review* 28: 142–151.

Wood, D.J. 1991. Social issues in management: Theory and research in corporate social performance. *Journal of Management* 17: 383–406.

R. Edward Freeman is University Professor, Elis and Signe Olsson Professor, Academic Director of the Institute for Business in Society, and Senior Fellow of the Olsson Center for Applied Ethics at the University of Virginia's Darden School of Business.

Daniel R. Gilbert, Jr. is Professor Emeritus at Gettysburg College.

Chapter 27
Business As a Humanity: Epilogue

R. Edward Freeman

1 Seeing Business as a Humanity

What does it mean to see business as a humanity? What are the contributors to this book recommending for the institutions that make up Business? What are the politics and the ethics of these recommendations? In this brief epilogue I want to suggest a program for change—for one institution in particular, namely, the professional school of business—that can enable us to come to see business as a humanity.

Business is not now understood as a humanity, and the very title of this book of essays seems at best contradictory, at worst fuzzy-headed. In fact, business theorists for the most part want the study of business to become even more scientific, and business decision makers would like nothing more than to be able to rely on some quick-fix rules that rest on some authority other than the consultancies of the authors.[1] So, to issue a call to see business as a humanity, to juxtapose the tropes of business with those of literary studies, theology, philosophy, and history, to recommend change in the very background disciplines of business is first of all nonsense. After reading these essays, however, it is easy to see that they imply a very different idea of business, one in which "the human condition," "the moral point of view," "the joys and sufferings of humankind," "the human person," and the

Originally published in: *Business As a Humanity*, 215–225 © Oxford Publishing Limited, 1994
Reprint by Springer

[1]The genealogy of the business self-help book is another story, one worth telling on another occasion.

R. E. Freeman (✉)
University of Virginia, Charlottesville, VA, USA
e-mail: FreemanE@darden.virginia.edu

© Springer Nature Switzerland AG 2023

S. D. Dmytriyev, R. E. Freeman (eds.), *R. Edward Freeman's Selected Works on Stakeholder Theory and Business Ethics*, Issues in Business Ethics 53,
https://doi.org/10.1007/978-3-031-04564-6_27

like can become the centerpiece of thinking about business. In a sense these essays call for no less than a reversal of margin and center. The role of economics and its subordinate disciplines, such as finance and marketing, must move into the shadows, while ethics and the humanities begin to occupy the center stage.[2]

While I believe that such a call is ultimately futile given the current state of the professional school of business, I want to argue that such a reversal offers a way to revitalize business schools and business theory. To make such an argument I shall examine in the first section of this epilogue the history of business schools and suggest that we can see three important phases in the development of business schools that have ended in a contradiction of sorts. In the second section I suggest a reorientation of professional schools of business along more pragmatic lines. Finally, in the last section I offer some concrete suggestions for how business schools can change to open up the possibility of seeing business as a humanity.

2 A Stylized History of Business Schools

2.1 The Formative Period

According to L. C. Marshall, one of the earliest proposals for a collegiate business school was made by Robert E. Lee, the first president of what is today Washington and Lee University. Although the proposal was not put into action, Lee's idea was not confined to the business details of bookkeeping and the like "but [aimed] to teach the principles of commerce, economy, trade and mercantile law."[3]At its very beginning the American collegiate business school sought to articulate and teach the basic principles of commerce and economics, as well as their partners in the law. It is ironic that almost 125 later American business thinkers and government policy makers are still searching for these principles.[4]

The Wharton School of the University of Pennsylvania was founded in 1881 with funds from the steel magnate Joseph Wharton, and a rapid increase in the number of business schools continued for the next 40 years in both the United States and Europe. Assessing this trend in 1928, L. C. Marshall wrote: "In view of our general acceptance of the idea that the higher levels of practical education are quite properly a function of a university, it is only to be expected that business—overwhelmingly

[2] Daniel R. Gilbert, Jr., has recently moved the concept of corporate strategy fully into the shadows and hence begun this task for the business disciplines allied with economics. See his *Twilight of Corporate Strategy* (New York: Oxford University Press, 1992).

[3] L. C. Marshall, "The American Collegiate School of Business" in L. C. Marshall, ed., *The Collegiate School of Business: Its Status at the Close of the First Quarter of the Twentieth Century* (Chicago: University of Chicago Press, 1928), p. 3.

[4] See, for example, books by Robert Reich, Lester Thurow, Peter Drucker, and others for different views of what the principles are that connect these disciplines.

our largest practical activity—should have its training schools at the university level."[5]

Frances Ruml argued that the existence of business courses in secondary schools and so-called commercial colleges

> made it easy and natural for the university and collegiate schools of business to focus its attention more on professional and social ends; to place an emphasis upon dealing with the complicated social organization in which the business man does his work; to accept as its objective the training of persons with a social point of view for executives and professional positions in business.[6]

While there is a great deal of debate about the original purpose of business schools, it is reasonable to believe that multiple outcomes occurred, such as (1) replacing the apprenticeship system, largely dismantled by the industrial revolution; (2) training businesspeople in the fundamentals of scientific management; and (3) training an elite capable of managing more complex social organizations.[7]

Ruml summarized Joseph Wharton's initial stated purpose as being "to establish means for imparting a liberal education in all matters concerning Finance and Economy."[8] In his address to the trustees of the newly formed Wharton School, Wharton remarked: "It is reasonable to expect that adequate education in the principles underlying successful business management would greatly aid in producing a class of men likely to become pillars of the state, whether in private or in public life."[9] And on this original theme of imparting "moral education" to managers, Ruml approvingly quotes E. P. Cubberly's *History of Education:*

> As the industrial life of the nation has become more diversified, its parts narrower, and its processes more concealed, new and more extended training has been called for to prepare young people to meet the intricacies and interdependencies of political and industrial and social groups and to point out to them the importance of each one's part in the national and industrial organization.[10]

The American Assembly of Collegiate Schools of Business (AACSB), founded in 1916, reinforced this idea of a liberal education for businesspeople by requiring accredited schools to require students to take at least 40 percent of their credit hours in subjects other than economics and commerce.

To reinforce the argument that the original purpose of business schools was grounded in the idea of moral education, consider Joseph Wharton's original plan, most ironic for the contributors to this volume. Wharton proposed and implemented a program to take humanities faculty, trained in the liberal arts, and put them in the

---

[5] Marshall, "American Collegiate School," p. 9.

[6] Frances Ruml, "The Formative Period of Higher Commercial Education in American Universities," in Marshall, ed., *Collegiate School of Business*, p. 47.

[7] Drew E. VandeCreek, "Power and Order: The Ideology of Professional Business Training at Wharton and Harvard, 1881–1933, draft of masters thesis in history, University of Virginia.

[8] Ruml, "Formative Period," pp. 54–55.

[9] VandeCreek, "Power and Order."

[10] Ruml, "Formative Period," p. 53.

classroom to teach business subjects. This experiment, tried in 1882, failed miserably, according to Ruml, because the liberal arts professors were resentful and did not necessarily believe that knowledge should be practical. By 1883 the Wharton School had hired instructors versed in business, leaving the ensuing 100 years or so barren of meaningful interchange between business and the humanities.

The early history of business schools is rife with arguments about the same controversies that dominate discourse today. Should faculty be trained in economics? Were graduate economics departments sources of ignorance about business? Could businesspeople be used as adjunct professors, and if so, how many? What was the proper balance between research and consulting? Should the study of business be scientific or largely storytelling? Was business to be a profession with ethical norms, or were schools of business merely to socialize students into the practices of business and prepare them for the world of work? While these questions are relevant for us today, a second phase in the history of the American business school makes answering them all the more difficult.

2.2 The Scientific/Modem Period

Two well-known reports published in the 1950s gave voice to a crisis of legitimacy in the business school. The Carnegie Report and the Ford Foundation study by R. A. Gordon and J. E. Howell both recommended that business schools were in need of academic legitimacy and essentially proposed the adoption of a social sciences approach to the subject matter of business.

According to Thomas Mulligan, writing in an important retrospective in no less than *The Academy of Management Review:*

> The landmark studies of business education by Pierson (1959) and Gordon and Howell (1959) have had great influence. The often unsubstantiated descriptive content of earlier business school curricula and research has been replaced by quantitative description based on rigorous data collection, computer-assisted mathematical modeling, and the foundational concepts of science—testable hypotheses (or, at least, testable networks of hypotheses), correlated observations, and causal explanation.[11]

Mulligan concludes that the culture of science that emerged in the business schools was successful in getting established, and it has only recently begun to be challenged by the introduction of humanities-based ethics courses. Mulligan argues that the current state of business schools suggests a desperate need for something like the humanities.

While Mulligan's conclusion is certainly shared by the scholars represented in this volume, and while certainly he is correct that the culture of science is dominant in journal reviews, tenure decisions, and business school politics, I want to suggest

---

[11] Thomas M. Mulligan, "The Two Cultures in Business Education," *Academy of Management Review* 12, no. 4 (October 1987): 593–99.

that there is an alternative to his view of the two cultures of arts and sciences as incompatible. Mulligan rightly points out the emergence of "practitioner" books critical of the scientific approach, but there is more at issue here than practical criticism of academic abstractions.

2.3 *The Porter/Peters Era*

Two seemingly unrelated events catapulted business thinkers out of their obsession with description and explanation and gave rise to scholarship that was normative in nature. In the 1970s economist Michael Porter published a series of articles on the economic basis of business strategy. These articles formed the basis for his groundbreaking book, *Competitive Strategy.* From this book it was easy to reestablish the dialogue between social scientists and businesspeople. Porter had a theory with particular implications for how to manage.[12]

During roughly the same time period, Tom Peters and his colleagues at the consulting firm McKinsey and Company embarked on a study of why some organizations seemed to be successful year after year. The report they produced formed the basis for Peters's and Robert Waterman's best-selling *In Search of Excellence,*[13] which has sold more than twenty-five million copies worldwide. The power of *Search* was its systematic, albeit unscientific, story about what real companies did and did not do to achieve success. *Search* was grounded in the real world and coherent as a narrative, enhancing its believability.

My argument is not that the scientific period is over in business schools. The existence of so-called scholarly journals ensures an appetite for trivia in excess of either its production or its value. It is not that Peters and Porter provide examples of how to do research that is rigorous and relevant. Rather, I want to suggest that Porter and Peters show us the power of narrative, of telling a good story about how we can create and manage our organizations. In short, they give us, in this pragmatist reading, a basis for reintroducing the humanities in business education. For if the humanities are anything, they are powerful narratives and metanarratives about human societies and cultures. These narratives are more or less coherent and more or less normative.

---

[12] I am not claiming that Porter was the first to do this; of course he was not. I am claiming that without a landmark of the magnitude of his book, normative scholarship would not have achieved the power that it has today. The same applies to the argument about Peters.

[13] The work of Peters and Waterman and others is noted by Mulligan, "Two Cultures," pp. 594–95. Porter's work is not, probably because it is clearly couched in the scientific idiom.

3 The Functions of a Professional School

To understand how narrative can serve as a powerful force for change in business schools, we need first to set forth a conception of the professional school. I want to suggest, in a simple and straightforward way, that we think of professional schools as having two main aims: socialization and practical criticism.[14]

The first aim of any professional school will always be the socialization of students into the habits and practices of the profession. Students go to medical school to learn how to be and act like doctors and to understand the subjects and sciences necessary to the practice of medicine. Similarly, business schools train students in the practices of businesses. Students study the techniques and frameworks and disciplines that business managers use. They learn the language of business. They come to understand the norms of business, from the differences in use of "net present value" versus "payback" calculations to the dress and speech codes of companies and industries. Students become facile with the methods of recognizing and solving ordinary business problem and issues.

But they also do more. Any business school worth its salt teaches students to be skeptical and critical, that is, teaches the methods of practical criticism. A profession is a critical practice, one that contains the seeds of its own improvement. Medicine, by committing to the alleviation of human suffering, must allow the analysis of questions that are critical of current practice, for current practice is instrumental, although importantly instrumental, to the alleviation of human suffering. The following questions—(1) What's the best practice? (2) How can we improve the practice in area X? (3) Is this case a kind of human suffering that we can and should try to alleviate? (4) Is this case, such as a factory, a case of human life?—are all appropriate for students of medicine; their training and their teachers' research is incomplete if these critical questions are not a part of the curriculum. If they are absent from a school's curriculum, we would not call that school "professional."

Similarly, business is about the creation of value. How such value is created is instrumental to its creation. Value creation admits a whole host of critical questions: (1) How can we create value more efficiently? (2) How can we create *this kind* of value? (3) How should value be distributed? (4) Who owns/controls the value production process? and (5) Should business create value regardless of its effects? As with medical schools, we have to see the business school that does not produce a large dose of "practical criticism" or train its students in the methods of "critical practice" as incomplete and unprofessional.

It is a mistake, however, to see the socialization function of business as based on science and the critical function as based on the humanities. Both the sciences and the humanities have much to say to both functions. If we think of the current practice of business as "the narrative(s) of business" or "the story(s) of business," then the

[14] There is a large literature here with a number of complicated questions that I am going to ignore. I take the dual purposes outlined here as necessary, but not sufficient, conditions for a school to have the label "professional."

functions of business school are to teach the old/current stories and to invent new ones. If we think about the work of Peters and Porter, they claim to be basing new narratives on old ones. Theories, understood this way, evolve and change over time, sometimes piecemeal, sometimes wholesale. Science and art both produce narratives, and both are important for business.

For example, let us suppose that one of the standard stories of business is that managers can, do, and should act in a fiduciary manner for stockholders. That they can and do is shown by elaborate scientific studies, and that they should is argued for normatively, usually from the moral standpoint of utilitarianism, (as in the work of Adam Smith or Ronald Coase, but sometimes on libertarian grounds, as in the work of Michael Jensen. Any *standard story* has descriptive (scientific) and normative (moral) bases.

Now the pragmatist should argue that if the goal of the professional school is to teach and improve or replace the standard story by a better one, then it doesn't much matter whether that story is improved or replaced by science or morality. In fact, the savvy pragmatist points out, ultimately both will be entwined in a story of human practice whose test will be the way it allows us to live.

For example, suppose we tell a story about the practice of business that has more characters than the standard one. In this new narrative there are stakeholders,[15] and businesses create value for stakeholder groups such as employees, suppliers, customers, and communities, as well as for stockholders. Now we want descriptions of "stakeholder practice," and we want to know what the texture and meaning of stakeholder practice can be.

Any business school worthy of the description "professional school" must teach its students to evaluate multiple narratives and to create or at least improve the messy story of business as it is practiced in the world. If this argument is valid and sound, then business schools can be more professional by embracing the humanities as a partner in the socialization and critical processes of education. How can such a partnership be accomplished?

4 Suggestions for the Future

I want to make some concrete suggestions in four areas of business education: (1) curriculum, (2) faculty, (3) students, and (4) institutions. Each of these suggestions is meant more to provoke conversation than to end it. Take them as half-baked suggestions that merit further thought.

[15] Notice the ironic twist on stockholders, for new narratives are often ironic retellings of old ones.

4.1 The Curriculum of Business Schools

Far too little time is spent on the critical function of professional schools. With the exception of accounting courses, which induce deep skepticism about the "truth" of accounting statements, and ethics courses, which explicitly question the stockholder ideology, the pickings are slim indeed.

I want to make four related points about how to deepen the critical function of business education. First, ethics has to be a part of the curriculum whose purpose is to challenge the implicit values and assumptions in the rest of the curriculum. Ethics should be the course that explicitly leads the critical function.

Second, we need courses on culture studies and language. The globalization of business itself challenges the status quo, yet the response of some schools is to deny course credit for the language skills their graduates need. Courses in these areas must be supplemented with projects, events, and other pedagogical devices that give students experiences in dealing with other cultures and thus coming to see their own from a critical perspective.

Third, in areas where the application of theories, models, and frameworks from the humanities are most easily applicable—organization studies, strategy, leadership—courses must be revised to take account of the humanities. It is difficult to imagine that it is a rare exception when Freud is mentioned in a business classroom, yet the first chapter of Karen Horney's *Neurotic Personality of Our Time* is a stunning self-description for many executives.[16] Of course, these areas must continue to draw on the social sciences, but they must not fall prey to an instrumentalism that buys the dominant ideology wholesale. Thus, a course in organizational behavior that professes to teach the students "how to get more out of people to maximize shareholder value" would fail the criteria I have in mind here, as would a course that fails to mention the necessity of efficiency as a business value.

Finally, where critical attention points up conflict between the received narrative and a new one, we should "teach the conflicts."[17]After all, we are building the critical judgment abilities of our students, abilities that we hope will outlast a two-year M.B.A. program. If they cannot deal with conflicting ideology or challenges to their very assumptions about business, they will not make very effective business leaders in today's world.

There are more curriculum changes that need to take place. I am suggesting these four as necessary to any meaningful attempt to teach the critical function. I have left open the question of pedagogy because it seems to me that there are many ways to learn. Even while I remain a devoted follower of the Socratic method, I recognize that it too may be bound by culture.

---

[16] Karen Horney, *The Neurotic Personality of Our Time* (New York: Norton, 1937).

[17] Gerald Graff, *Professing Literature: An Institutional History* (Chicago: University of Chicago Press, 1987).

4.2 The Faculty

The question of designing a business school faculty is at once simple and intractable. It is simple because we need some new faces with new backgrounds. Historians, literary theorists, legal scholars, linguists, and others could easily find their way around a business school, if invited to do so. It is intractable because at the margin their presence would make little difference, as has been the case with the few philosophers who now slink about the b-school hallways.

Rather, we must find ways to reinvigorate the intellectual processes of the disciplines of business. In addition to adding humanities people, we must add "humanities questions" to the intellectual life of marketing, accounting, finance, decision science, operations, and organizational studies. Donald McClosky and his colleagues have begun this process in economics.[18] A group of scholars, writing in the journal *Accounting, Organizations and Society*, has adopted a critical approach to accounting, and scholars in business ethics and business and society have given a critical turn to issues of strategy, governance, and organizational studies. However, there is much work to be done.

We need to undertake an explicit program of lectures, conferences, institutes, joint research projects, and joint courses among business scholars and humanities scholars. Such a program of faculty development would try to build the intellectual capital of business faculty and could offer a rich and varied setting in which members of the humanities faculty could test their ideas.

4.3 The Students

In addition to the standard package of M.B.A. students, ranging in age from twenty-five to thirty-five, with experience in business, I believe that we can make business schools relevant to another group.

Imagine a program designed for one hundred of the very best liberal arts graduates from the best liberal arts programs. These students should have highly developed critical faculties, but they would have little knowledge of business. With the right curriculum design, one combining socialization and the continued development of critical faculties, these students could easily emerge as candidates for business leadership. What I have in mind is a joint program, with humanities and business faculty playing roughly equal roles.

---

[18] See, e.g., Donald N. McCloskey, *Second Thoughts: Myths and Morals of U.S. Economic History* (New York: Oxford University Press, 1993); Donald N. McCloskey, *If You're So Smart: The Narrative of Economic Expertise* (Chicago: University of Chicago Press, 1992); Donald N. McCloskey, John Nelson, and Allen Megill, *The Rhetoric of the Human Sciences: Language and Argument in Scholarship and Public Affairs* (Madison: University of Wisconsin Press, 1987).

There are many ways to design such a program, and it would have to include a heavy dose of experiential learning about business, but programs such as these never get designed because we do not question critically the ideology that says one has to know business to learn about it in depth.

4.4 Institutions

Finally, we need some new institutions, such as a Center for Business and the Humanities. The goal of such a center would be threefold:

1. to produce new knowledge and insights into the human experiences in organizations
2. to provide intense executive development experiences that rely on innovative methods of learning from multiple disciplines
3. to provide a rich atmosphere of interaction and mutual learning among scholars in business and in the humanities, business executives and humanities practitioners such as artists, writers, and musicians

A broad range of activities, including research projects on race, gender, culture, the environment, the role of institutional reinvention, and leadership, could be undertaken, and conferences could be staged to share and build learning from theory to practice and practice to theory.

While the Aspen Institute has attempted the executive development part of this mission, and while various ethics centers have attempted other parts, so far no one has tried to confront the issue on the main ground—integrating business with the humanities, or, as we have put it here, coming to see business as a humanity.

5 Conclusion

I conclude with the challenge that if professional schools of business are to be relevant in today's multicultural, technologically wired, rapidly changing world, something like a reinvention of the business school must occur. The forces of stability must not be underestimated, however. From issues of tenure to the reluctance of recruiters to hire culturally divergent people, change will not come easily. It is incumbent on the intellectuals who see themselves as telling better stories about business to try to lead this change, enlisting our colleagues, our students, and our "practitioners."

R. Edward Freeman is University Professor, Elis and Signe Olsson Professor, Academic Director of the Institute for Business in Society, and Senior Fellow of the Olsson Center for Applied Ethics at the University of Virginia's Darden School of Business.

Chapter 28
The "Business Sucks" Story

R. Edward Freeman

This essay was originally delivered as the Presidential Address to the Society for Business Ethics, in Vancouver, in August of 1995. It was never published or even submitted for publication. I am grateful to the Journal of Humanistic Management for publishing it after all these many years. It is difficult to reread something that was written so long ago, but I find that most of the ideas unfortunately are still relevant. I originally wrote this essay, inspired by a comment of my late colleague, Richard Rorty, that he "refused to participate in the America sucks sweepstakes". I, too, refuse to participate in its first cousin, "the business sucks sweepstakes". I have made minimal changes to the original, even though I might put several points differently today, especially the set of principles of "values based capitalism". I have added a few footnotes to clarify the relationship between some of the ideas in the essay and later work.

1. I recently attended a seminar given by a business executive to a group of academics mostly from the humanities. The subject of the seminar was what business could do to participate in environmentalism. One of my colleagues, of whom I am quite fond, began the questioning in what could only be called a condescending voice. He asked the executive whether or not this presentation wasn't better suited to business people rather than scholars, since the thrust of the presentation was that helping to save the earth didn't have to be costly, indeed the executive argued, it could even be profitable. No mention was made of the morality of saving the Earth, and that is the proper concern of "scholars", who

Originally published in: *Humanist Manag J,* 3, 9–16 © Springer Nature, 2018
Reprint by Springer, https://doi.org/10.1007/s41463-018-0037-y

R. E. Freeman (✉)
University of Virginia, Charlottesville, VA, USA
e-mail: FreemanE@darden.virginia.edu

© Springer Nature Switzerland AG 2023

S. D. Dmytriyev, R. E. Freeman (eds.), *R. Edward Freeman's Selected Works on Stakeholder Theory and Business Ethics*, Issues in Business Ethics 53,
https://doi.org/10.1007/978-3-031-04564-6_28

don't care in the least about the profits to be generated by environmentalism – or so said my colleague.

The executive's response was so perfect and delivered so appropriately (one might even say "efficiently") that neither my colleague nor I understood the significance of the reply. He said simply, "Oh, of course, it's the right thing to do, the ethical thing to do. I thought we all agreed on that. I'm just trying to show how it can also be profitable."

2. I originally thought that this reply was simply a nice put-down of a sometimes arrogant colleague. However, upon further reflection I have come to see this response as illustrative of an important facet of the discourse of ethics and the discourse of business. I want to suggest that both parties to this conversation participated, perhaps unknowingly in what I have come to call the "business sucks" story.[1]

In this essay I want to suggest that this story plays a large role in our understanding business and ethics, and hence business ethics. And, I want to suggest that the business sucks story consists of shopworn metaphors and lousy ideas that have long outlived their usefulness. Furthermore, we need to undertake the task of reinventing the language of business, and thus business itself, along more useful lines. Such a redescription needs to allow business full citizenship in the moral discourse of our society, rather than its current role as a rather unpleasant moral relative that has to be explained away or simply ignored.

3. My colleague clearly illustrated by the tone of his voice, his body language, selection of phrases, and the general state of his being that he thought, in short, business sucks. He spoke with such contempt and tempered outrage that the bonds of civility were clearly strained. My colleague believed that matters of profit and loss were not only irrelevant but irreverent in a serious "scholarly" discussion of the environment. The currency of his land is "right" and "wrong" and there is no exchange rate with the scorekeeping mechanisms of the capitalists' world. The ethics of business has been settled, to wit: business sucks.

Curiously enough, the business executive was making a very similar set of assumptions. As a routine matter of course he assumed that the ethics of business was not a very interesting question. Either ethics was mundane, agreed upon, or simply not an issue. The key for him and for the discourse of business was the centrality of the metaphor of profitability. If saving the earth could be shown to pay then we could get on with it. The intonation of "scholars" and "ethics" in a sophisticated Ivy-Oxbridge voice did not impress the executive, but he showed no contempt, only indifference. His currency, and by his assumption, the world's currency, was profit and loss. "Business may suck" according to him, "but that's business".

[1] The "business sucks" story is intended as a further elaboration of what Richard Rorty (1992) called the "America Sucks Sweepstakes" referring to Jonathan Yardley. However, one may refuse to participate in the America sucks sweepstakes and still believe in the business sucks story, as is the case with some modern pragmatists.

4. Now let me hasten to add that I have a great deal of respect and admiration for both of these people. One is doing groundbreaking work in the analysis of modern culture and moral discourse; and, the other is trying to change, for the better, a large bureaucratic corporate institution. Both, I am convinced, share a deep concern with the environmental mess we have made and both have definite ideas on how to fix up Planet Earth. However, because each participates in his own way in the business sucks narrative, neither will be very successful at changing business or in saving the Earth. Rather we need to come to see the business sucks story as a story that goes nowhere – as a story that prevents us from working together.
5. The problem is brought closer to home when someone asks me what I teach.[2] If I reply "business ethics", they either have to manage not to laugh, or I have to endure the now endless tirade of wisecracks beginning with "I thought that was an oxymoron" and ending with "Must be a short course". If I reply with a vaguer yet more descriptive "I teach business" or "at a business school", they simply change the subject.

   Humanities professors don't experience this profound questioning of the very motives of their work. Of course, no one may be interested in nineteenth Century Elegies or "the early views of immortality of Augustino Nifo", but if you are, no one is going to question your motives. How could they be other than the love of knowledge and your subject matter (or some Freudian or post-Freudian account of this professed love)? Surely, you are not in it for the money.

   But, business sucks. It is concerned primarily with profits at the expense of others, especially employees. Any business behavior which could possibly be other-regarding is just a clever scheme to make more money. Corporate social responsibility programs, employee involvement programs, corporate philanthropy, quality programs, customer complaint response policies, and countless other business tricks that appear to offer respect and dignity to its participants should not fool us. Business is about making money and business executives are in it for the money, and power, prestige, and elitism, which are after all just psychological explanations for money. If business executives believe that they have a moral concern with the state of the Earth, the people killed by product tamperings, poverty in the developing world, racist and sexist stereotypes, or the rise in state control of the private (or public) sphere, they must be mistaken, deceiving themselves, or in a position of bad faith. Because, as the story goes, business sucks.
6. In reality there are at least two main versions of the business sucks story, and each version depends on what I have called elsewhere, the Separation Thesis[3]:

[2] This section is a direct paraphrase of what John Wisdom considered to be the problem of philosophy (cf. Wisdom (1970: 1–2)). The link between Wisdom's view of the role of philosophy and Rorty's notion of "redescription" of self and community is larger than indicated here.

[3] More recently I have labeled this the "separation fallacy" to denote that it involves fallacious reasoning, at least to this pragmatist. It makes the fact-value dichotomy a central theme of business theory. For more on the issues here see Harris and Freeman (2008) and Putnam (2004).

the view that the discourse of business and the discourse of ethics are meaningfully or usefully separated. This thesis entails that we can evaluate "business as a whole" from an ethical point of view – hence, the business sucks story.[4]

The first version of the story, call it the trendy-left leaning-radical-version, claims that business is an exploitative, back-stabbing, mere-means treating, profiteering rat race with no redeeming intrinsic values. The discourse of business may pretend to be amoral, but it is corrupt through and through, and must be thrown out altogether. The problem is, that except for a nostalgic longing for a socialism that works, there doesn't seem to be any alternative to business. So we have business, but it sucks.

The second version of the story – call it the executive-trendy-right wing – *Wall Street Journal* version, goes as follows: business is about people who act in their own interest and contract with other self-interested beings to achieve their ends. Stockholders or other owners put their money at risk and are entitled to the returns after all of the other contractual obligations are met. Executives make "business decisions" that are hardnosed and tough-minded based on "purely business" criteria of the most efficient uses of resources. What appear to be ethical dilemmas are only the results of self-interested people choosing what is best for themselves. If we, especially government, will leave everything alone, it all works out, as if by an invisible hand.[5] Business may suck, but we're all better for it.[6]

7. These two versions of the business sucks story are implanted in our minds daily, but they have more sophisticated cousins in the academic world. Each version depends on something like the Separation Thesis for its logic.

Humanist academics decry the separation of business and ethics, but they do not see any alternative. Instead they write critical, ironic papers and use phrases like "late 20th Century capitalism" or "the commodification of culture" or "multinational companies take profits out of production". They simplify, stereotype, and belittle with rhetoric like "large corporations that control…." and "rich elite executives who….". More telling is the fact that most humanist discourse is simply silent about business, pretending it doesn't exist, and when it does raise its unwelcome head knowing glances are exchanged, expressing thanks that the moral problem of business has been solved: it sucks.

Another group of academic "humanists" have responded to the business sucks narrative and mounted a "defense", by decrying the story as one more piece of evidence that universities are controlled by leftist academics with hegemonic

[4] Cf. Freeman (1994). The Separation Thesis is related to much of Amartya Sen's argument in his brilliant book, *On Ethics and Economics* (1987). For a more recent argument see Wicks (1996).

[5] This is the dominant version behind the recent waves of rhetoric on business restructuring and re-engineering where thousands of employees are fired to make things better for "the business", and ultimately, so the story goes, for society.

[6] In today's world we can see this version of the story behind the current mania of companies to buy back their own shares to pump up the stock price, and the existence of raiders who put good companies into play with no motive other than to profit from the stock.

anti-business values. These scholars at the Heritage Foundation, the Hoover Institute and the American Enterprise Institute led by pro-business humanists like Michael Novak have argued that business doesn't suck at all. Capitalism, according to Francis Fukuyama, has some historical necessity – the triumph of the West. Business should be celebrated as allowing human beings to "be all that they can be". These academics rejoice, ironically with their opposite numbers, that the moral problem of business has been solved. According to them, there is no problem.

It is easy to see that even though these pro business humanists reach the opposite conclusions from their counterparts, both groups are appealing to the need to separate business from ethics in the day to day life of the participants in business. Judged from afar, each looks at business as a whole, applies some ethical criteria or other and comes to a particular conclusion. The business sucks narrative is still the issue even though one group is simply issuing its denial. I might have just as easily entitled this essay "The "business is great" story". Unfortunately, since I believe that the business sucks story is such a large part of our culture, no one would have read it.

8. There is a third group of academics, those who teach in business schools. Now most of these academics either are economists or wish they were economists but they have turned out to be accounting, finance, marketing or production professors. The business school gospel involves the whole-hearted acceptance of the separation thesis. Business and ethics are to be kept apart, and business school academics have abrogated the realm of business for themselves. These scholars see their main task as the development of theories, models and frameworks that explain and predict in non moral terms, the actions and strategies of businesses and their participants. Of course, for public relations reasons, or just for intellectual amusement it may be convenient to have a course in business ethics or even to hire a philosopher on the business faculty, but make no mistake: the moral problem of business is not very interesting.

   Business may or may not suck. It's not the job of these academics, according to them, to say one way or the other.
9. Now none of this is very useful. "Business", as it is currently understood by those who unwittingly accept some version of the business sucks story, is shopworn at best. The Separation Thesis is in reality a rhetorical move, by both sides of the debate, to close off serious conversation of the role of business in our lives. And, we are in desperate need of new stories that do not depend on the Separation Thesis – stories in which the very words that are used to tell the tale describe business and human beings at once.[7]

[7] This is precisely Gilbert's beef in *The Twilight of Corporate Strategy* (1992), and his more recent *Ethics Through Corporate Strategy* (1996). Why not just assume that business is one more way that human beings create meaning, and that we are fully human while we are fully engaged in business? We have tried to make this clearer recently in Freeman et al. (2010).

There was or still is some truth to the business sucks story, but it is a truth which is so limited as to be misleading. Plainly some businesses suck. In so far as they abuse the very people they are designed to serve these businesses violate a morality so common that we take it for granted. But, it is a logical error to conclude that "business sucks" from "some businesses suck". And, just as plainly, some businesses do not suck. They enable people to accomplish together what they could not accomplish alone. Furthermore, there is nothing in the logic of "business" as it is currently understood that will tell us whether it sucks or not. The problem is obviously that for the most part, business is not about what the business sucks story says it is about. Business is primarily a story of cooperation, as well as competition. It is a story based primarily on solidarity as well as individuality. And, it is enmeshed in a complex human system which can never be usefully separated into "economic", "political", "social", "ethical", or any other joint-cutting ideology, that masquerades as either science, social science, philosophy, or deconstruction.

10. We need a new story to replace the business sucks story. And, I can do no more than sketch an outline here in this essay. I want to call this new story, "values based capitalism", and it is related to what I have elsewhere called "stakeholder capitalism", and others have labeled "Conscious Capitalism", "Inclusive Capitalism", "Capitalism 2.0", and related concepts like "Impact Investing" and "Just Capital".[8] While these are new names, this new story has been enacted for some time, but, the lens of the "business sucks" story have tended to obscure the view and warp the surrounding conversation. Values based capitalism depends on four key principles that I shall briefly sketch.[9] (See Exhibit 28.1.)

The first principle of values based capitalism is that capitalism depends first and foremost on the cooperation of stakeholders. The entrepreneur's or the manager's job is to put together a deal so that customers, employees, suppliers, financiers, and communities win over time. Each of these groups has a stake in the corporation or partnership or firm, and each is vital to its success. Without the support of each stakeholder group, a business enterprise is not viable, no matter its legal or organizational form. Scholars are beginning to amass evidence, both systematic and anecdotal, that companies that have endured over time and performed at a very high level understand and practice this principle.

The second principle of values based capitalism is that people are complex. While this sounds pedestrian, the assumptions of the old story are astounding in the denial of this simple truth. The evidence is plain enough, that while people do sometimes act selfishly, they often act for others. And, most of the time they do neither, as they pursue joint conceptions of the good. Call these joint

[8] It is also related in a complex way to these many recent proposals for the revision of the story of capitalism. With colleagues, Bidhan Parmar and Kirsten Martin, I explore this new story in a forthcoming book tentatively titled: *Business: the New Story*. A first version of the argument here can be found in Freeman (2017a).

[9] I leave these principles as they were in the original. I would state them differently today as I do in Freeman (2017a) note 8 op. cit. and in Freeman et al. (Forthcoming).

Exhibit 28.1 Four principles of values based capitalism or stakeholder capitalism

I. *The Principle of Stakeholder Cooperation.*
Value is created because stakeholders can jointly satisfy their needs and desires.
[Capitalism works because entrepreneurs and managers put together and sustain deals with stakeholders, rather than become agents of the owners of capital.]
II. *The Principle of Complexity*
Human beings are complex creatures, capable of acting on multidimensional values, some of which are selfish, some of which are altruistic, and many of which are jointly created and shared with others.
[Capitalism works because of this complexity rather than in spite of it.]
III. *The Principle of Continuous Creation*
Cooperating with stakeholders and motivated by values, people continuously create new sources of value.
[Capitalism works because the creative force is primarily continuous rather than primarily destructive.]
IV. *The Principle of Emergent Competition*
In a relatively free and democratic society, people can create alternatives for stakeholders.
[Capitalism works because competition emerges out of the cooperation among stakeholders, rather than being based on some primal urge of competition.]

conceptions of the good, "shared values". From the early views of Peters and Waterman to calls for sustainable development by authors such as Paul Hawken,[10] acting on shared values is necessary if business is to work. Merck has been a great company because it has believed in alleviating suffering. J&J has performed well because the people who deal with it see the famous Credo in action. The list goes on. Each of these companies could be described as self-interested "fierce competitors", but if we use such language we cannot forget that it is the activation of values which causes such "fierceness".

The third principle of values based capitalism can be called "continuous creation". This twist on Schumpter's principle of creative destruction is meant to emphasize that capitalism is mainly creative rather than mainly destructive. Yes, some of the creative force of business destroys the value previously created, but only in theory does this happen immediately. Most of the creative force of capitalism in fact co-exists with what it purportedly "destroys". And, if Firm A invents a product or improves a product that Firm B depends on, Firm B is not destroyed, rather it creates yet another innovation. Creativity is a continuous process of capitalism, especially where the drive to create is caused by people trying to realize and act on those things that are most important to them: their values.[11]

The fourth principle of values based capitalism is the principle of emergent competition. This principle says that when capitalism is found in a society with

[10] See Peters and Waterman (1982) and Howken (1983).

[11] I would put this principle in the context of interdependent stakeholders and the search to avoid making tradeoffs. See Freeman (2017b)

relatively free political institutions, competition emerges from the first three principles. Where social and political institutions provide relatively easy access to stakeholders, the inventors, entrepreneurs, and other value-creators can challenge the established order.

While each of these principles is worthy of more conversation and analysis, I want to suggest that this sketch of the new story, values based capitalism, is a rather bold and powerful edifice. It says that businesses are successful because people stand for some values, that stakeholders' deals exist and are relatively stable because most people keep their contracts, promises and commitments, and it says that business and ethics can go together. More precisely, values based capitalism makes it possible for us to reinvent business in distinctly more human terms.[12]

11. In summary, I want to suggest that we take every opportunity possible to ferret out the "business sucks" story when it appears in the pages of *The Wall Street Journal* and *The Academy of Management Review*. We need a rhetorical strategy of satire and ridicule. We should belittle the business sucks story just as it belittles us. We should offer no respect to those who are not willing to see business for what it is – just one more garden variety way that humans create meaning for themselves – an institution that can at once promote individual freedom and community solidarity, an institution that can bring about a more just more ecologically sustainable social order, or lead to our destruction. Finally, we need a political philosophy and an ethical theory that do not depend on the business sucks story. We need to elevate the public discourse of politics and ethics beyond the level of investigative trash talk. I believe that this substantial bit of intellectual work will have to come from those of us working in the field of so-called "applied ethics", but that is a story for another time.

References

Freeman, R. Edward. 1994. The politics of stakeholder theory: Some future directions. *Business Ethics Quarterly* 4 (4): 409–421.

———. 2017a. The new story of business: Toward a more responsible capitalism. *Business and Society Review* 122 (3): 449–465.

———. 2017b. Five challenges to stakeholder theory: A report on research in progress. In *Stakeholder Management*, Business and Society 360, eds. David M.Wasieleski and James Weber, 1-20. Emerald Publishing Limited.

Freeman, R. Edward, Jeffrey S. Harrison, Andrew C. Wicks, Bidhan L. Parmar, and Simone De Colle. 2010. *Stakeholder theory: The state of the art*. New York: Cambridge University Press.

Freeman, R. Edward, Bidhan L. Parmar, and Kirsten E. Martin. Forthcoming. *Business: The new story* (tentatively titled). Columbia University Press.

Gilbert, Daniel. 1992. *The twilight of corporate strategy*. New York: Oxford University Press.

———. 1996. *Ethics through corporate strategy*. New York: Oxford University Press.

[12] And, it is worth noting that values based capitalism has plenty of room for businesses that may suck. Not all values are warm and fuzzy, as Freud and others have taught us.

Harris, Jared D., and R. Edward Freeman. 2008. The impossibility of the separation thesis. *Business Ethics Quarterly* 18 (4): 541–548.

Howken, Paul. 1983. *The next economy*. Henry Holt & Co.

Peters, Thomas J., and Robert H. Waterman. 1982. *In search of excellence: Lessons from America's best-run companies*. New York: Harper & Row.

Putnam, Hilary. 2004. *The collapse of the fact/value dichotomy and other essays*. Harvard University Press.

Rorty, Richard. 1992. Trotsky and the wild orchards. *Common Knowledge* 1 (3): 140–153.

Sen, Amartya. 1987. *On ethics and economics*. Oxford: Basil Blackwell.

Wicks, Andrew C. 1996. Overcoming the separation thesis: The need for a reconsideration of SIM research. *Business and Society* 35 (1): 89–118.

Wisdom, A.J.T.D. 1970. The logic of god. In *Paradox and discovery*. Berkeley: University of California Press.

R. Edward Freeman is University Professor, Elis and Signe Olsson Professor, Academic Director of the Institute for Business in Society, and Senior Fellow of the Olsson Center for Applied Ethics at the University of Virginia's Darden School of Business.

Chapter 29
The Relevance of Richard Rorty to Management Research

R. Edward Freeman

Richard Rorty has liberated philosophy.[1] *Philosophy and the Mirror of Nature* (Rorty 1979) and Rorty's subsequent work over the last twenty years have made it possible for philosophers to see their role as cultural critics obsessed with creating a more fruitful human future where freedom and hope are the central ideas. The problem is that most philosophers have yet to discover their own liberation. Be that as it may. Philosophy doesn't really matter—at least according to Rorty.

Business matters. And, while almost none of Rorty's published writings are about business,[2] there is great potential for his revival of Deweyan pragmatism to offer hope to the stale tomes of management theory and to turn business school

[1] This essay draws mainly on Rorty's ideas in *Philosophy and the Mirror of Nature* (1979) and *Philosophy and Social Hope* (1999). Rorty's other main monographs are *Contingency, Irony, and Solidarity* (1989) and *Achieving Our Country* (1998b). Both Rorty's 1999 and 1998b works are nontechnical, meaning that one does not need an extensive background in philosophy to benefit from reading them. Rorty's 1989 work is readable by most with a bit of background in the humanities, whereas his 1979 work is difficult even for professional philosophers.

Rorty's main technical work in the philosophy journals spanning his long and distinguished career has been collected in four volumes (Rorty 1982, 1991a, b, 1998a). In addition, Rorty has published two critical essays management theorists should read (Rorty 1986, 1998c).

For two collections by important philosophers who are critical of Rorty's work (and Rorty's response to the critics), see Malachowski (1990) and Brandom (2000).

[2] The exception is his Ruffin lecture response to Cornell West (see Rorty 1998c).

Originally published in: *The Academy of Management Review,* 29(1), 127–130 © Academy of Management, 2004
Reprint by Springer, https://doi.org/10.5465/amr.2004.11851748

R. E. Freeman (✉)
University of Virginia, Charlottesville, VA, USA
e-mail: FreemanE@darden.virginia.edu

© Springer Nature Switzerland AG 2023

S. D. Dmytriyev, R. E. Freeman (eds.), *R. Edward Freeman's Selected Works on Stakeholder Theory and Business Ethics*, Issues in Business Ethics 53,
https://doi.org/10.1007/978-3-031-04564-6_29

classrooms into Deweyan labs of exciting new ideas and experiments. In particular, *Philosophy and Social Hope* (1999) is a rich source of ideas for those who teach and write about business. Rorty's pragmatism offers hope to those management theorists who find the ever-tightening noose of a few top-tier journals committed to what Rorty would call methodolatry (Rorty 1999: xxi).

In this brief essay I introduce the work of Rorty via a brief biography, and I then concentrate on four Rortian themes that should interest management theorists. Finally, I suggest a pragmatic vision for business and theories about business, in hopes of stimulating more reading and discussion of Rorty's work.[3]

In *Philosophy and Social Hope*, Rorty (1999) has written a wonderfully readable autobiographical essay, "Trotsky and Wild Orchids," in which he outlines how he came around to the pragmatism of John Dewey. Analytic philosophy was dominant in most American philosophy departments during the last century and took as its wellspring the idea that the philosopher's job was to serve as the clarifier of conceptual schemes—a so-called handmaiden to the sciences, an analyzer of language and logic. The mantra was "If it can be said, it can be said clearly." Banished from this land was any notion of Socrates' question, "How should we live?" Ethics was turned into metaethics or the analysis of the meanings of words like "good" and "right."

Rorty was part of the center of this analytic mainstream, as a full professor at Princeton University—one of the top American philosophy departments. Rorty's main insight, which he credits to Dewey and others, but which he really crystallized in *Philosophy and the Mirror of Nature* (1979), was that most of this way of doing philosophy rested on a set of distinctions we inherited from the Greeks, and a view of language as representing the world. He argued that the idea of representation made no sense, that it was based on taking vision as a foundational sense, and that, at least since Wittgenstein, we should know that language doesn't work that way.

Rather, language is a tool, not a representation. There simply is no other way to deal with the world other than through language. To assume otherwise is to invoke the "appearance versus reality" distinction common to philosophers since Plato and crystallized in Kant. According to Rorty, philosophy had gone horribly wrong and had become not very useful. He argued that we needed to return to Dewey and see the intellectual's task as producing social hope and of always trying to figure out how we could live better. He saw intellectual life as a seamless whole, rather than divided by the disciplines.

Consequently, in a rare display of intellectual integrity for professors, Rorty accepted a MacArthur Foundation "genius award," resigned his position as full professor of philosophy at Princeton, and moved to the University of Virginia, where he became a kind of roving intellectual, not tied to any one department, and put the ideas in *Philosophy and the Mirror of Nature* into practice. He would give

---

[3] Andy Wicks and I have explored these ideas about what pragmatism implies for management theorists in Wicks and Freeman (1998). I am grateful to Andy Wicks and to Gordon Sollars, Daniel R. Gilbert, Jr., John McVea, Rama Velamuri, Nicholas Dew, and Laura Dunham for conversations about the ideas in this essay.

courses in English literature, philosophy, and law; teach jointly with a number of professors in other disciplines; and even occasionally show up at the business school. He sponsored reading groups, discussion circles, and was willing to talk to anyone from virtually any discipline about ideas. He continued to contribute to the philosophical journals and to write for more popular intellectual journals, such as the *London Review of Books*. His impact on the university was large. Eventually, he moved to Stanford University in the late 1990s, from which he has recently retired. His oeuvre is vast yet readable, and the opening footnote to this essay should be read as an introduction to his main works.

There are at least four themes in Rorty's work that are directly relevant to the way we have come to think about business and business theory (or management and management theory, if you like): (1) the poverty of most current modes of philosophy, (2) the disappearance of bright lines between disciplines, (3) the role of hope and the Deweyan experimental frame of mind, and (4) the pragmatist project of describing and redescribing self and community. I discuss each briefly in turn.

Philosophy has become irrelevant. Therefore, attempts at "philosophy of ___" have become equally suspect, whether we fill in the blank with "science," "social science," "biology," or "management." Rorty argues that it is allegiance to a set of dualisms inherited from the Greeks that holds us back. The basic idea is that it is not very useful to speculate on "the way the world really is." The whole "appearance/reality" distinction mistakenly assumes that language "represents" the way the world really is, in some sense of that phrase. Rorty shows in essay after essay how this distinction underlies a bunch of other dualisms, such as (1) finding versus making, (2) empirical versus theoretical, (3) normative versus descriptive, (4) objective versus subjective, (5) nature versus convention, (6) essence versus accident, (7) substance versus property, (8) morality versus prudence, and (9) intrinsic versus extrinsic.

Once we become skeptical of the "appearance versus reality" distinction and replace it with "relatively less useful descriptions of the world" versus "relatively more useful descriptions of the world," we are already on the pragmatist road to inquiry. Rorty articulates a view of inquiry in which the point is to produce ever more useful narratives that allow us to live better. Rorty argues that doing so requires us to see language as "an exchange of marks and noises, carried out in order to achieve specific purposes. It cannot fail to represent accurately, for it never represents at all" (1999: 50).

In the sciences, and certainly in social sciences such as management, relying on these philosophical distinctions has culminated in an obsession with "scientism and methodolatry." One doesn't have to turn very far to see the obsession with the theoretical/empirical distinction. Indeed, it is built into the structure of the Academy of Management's journals. Similarly, there is a gap between the "descriptive" journals of the Academy and the "normative" journals of the business ethicists.

Rorty's point is not that one side is correct. Rather, the resulting "theory" or "narrative" built on only one side of these distinctions will not be very useful. Small wonder there is an incredible gap between what business theorists do and write about and what those who actually practice business do.

Once one begins to question these basic distinctions, it is a natural move to ask whether or not useful narratives are produced by having "bright lines" between disciplines. While historically this may be true in some cases, Rorty thinks that separating out "philosophy" as a separate field of inquiry from "literature," "politics," or other "social sciences" has not been and will not be terribly useful. Indeed, part of the point of Rorty (1979) was to suggest that philosophy had lost any moral authority it may have once had. Suggestions and ideas about how we could live better seemed more likely to come from thinking about literature, literary theory, politics, and other areas that he deemed "cultural studies." Thus, Rorty began conversations with intellectuals across disciplines as diverse as anthropology, religion, law, and business. The disciplines, as traditionally conceived, do not reflect any "natural underlying reality," but need to be seen as rather more or less useful ways of dealing with certain sets of human problems.

These human problems, at least the ones we find in business, form a more seamless whole than most disciplines allow. Add the problem of the obsession with methodolatry and scientism that is so pervasive in management, and it is surprising from a Rortian perspective that anything useful is produced at all. Perhaps Rorty's mentor John Dewey said it best when he talked about putting the disciplines together, or at least being open to combining vocabularies into redescriptions that put human problems and human behavior at the center[4]:

> It would destroy fixed distinction between the human and the physical, as well as that between the moral and the industrial and political.... It would find the nature and activities of one person coterminous with those of other human beings, and therefore link ethics with the study of history, sociology, law and economics (Dewey 1998: 23).

Imagine the richness of descriptions of value creation and trade, business, or management if we could combine in useful ways the vocabularies of marketing, law, economics, ethics, psychology, organizational studies, biology, and others. The point here is not to turn every management thinker into a MacArthur Foundation "genius" but to say that we need to enlarge our curiosity, read more widely, and be more experimental with our ideas (keeping what works and jettisoning what does not). We need to give up the idea that our particular discipline (whether it is be ethics or economics) has some paved pathway to "real knowledge" (and, hence, legitimacy).

The controversial way that Rorty often explains what he is up to is by claiming that we need to replace the idea of "truth" as the goal of inquiry with the goals of "hope" and "freedom." Intellectual life matters precisely because we can offer descriptions and redescriptions of what we humans do that allows us to liver better. Rorty claims that we should adopt Dewey's goal for inquiry as "increasingly free societies and increasingly diverse individuals within them" (1999: 49). He argues:

> Pragmatists ... treat inquiry in both physics and ethics as the search for adjustment ... [Pragmatists ask] the practical question, "Are our ways of describing things, of relating them to other things so as to make them fulfil our needs more adequately, as good as

[4] In this passage Dewey clearly articulates what I have called, elsewhere, *the separation thesis*.

possible? Or, can we do better? Can our future be made better than our present?" (Rorty 1999: 72).

In short, it is up to us to figure out which projects to pursue, which descriptions and redescriptions might serve us better, and there are no guarantees that we'll be successful.

Pragmatists like Rorty believe there are only two interesting projects for us to engage in. The first is ever more useful descriptions and redescriptions of "self," and the second is ever more useful descriptions and redescriptions of "community." Indeed, some have suggested that these twin pragmatist projects are just two sides of the same coin, since "selves exist in communities" and since communities without "individuals" are pretty uninteresting herds of animals (see Freeman 1998).

Rorty's view of humanity is a thoroughly Darwinian one in which we are a species of "souped-up" chimps and bonobos who can cleverly adjust what we do in collaboration with each other through language. He recommends that

> we see ourselves as having the beliefs and emotions we do including our (potentially) "specifically moral" beliefs and emotions because of some very particular idiosyncratic things that have happened in the history of the race, and to ourselves in the course of growing up. … It [pragmatism] lets one see oneself as a Rube Goldberg machine that requires much tinkering, rather than as a substance with a previous essence to be discovered and cherished (Rorty 1986: 9).

Rorty claims that, rather than trendy postmodern theorizing, we need to reread Neitzsche and Emerson on how we can undertake the project of self-creation. Indeed, he quotes literary theorist Harold Bloom on the existence of strong poets who are horrified at "finding oneself to be only a copy or replica" (Rorty 1989: 29). We want to make more of ourselves than a copy of someone else. Equally, we need to read Whitman and Dewey, who emphasize that we must have solidarity with others to be able to accomplish any of our projects.

So much of management theory eschews individual difference. We search to find the theory, even a contingency theory, about how every organization has to work, how every employee is motivated, how all top teams work together or don't. Rorty is suggesting that a more fruitful course would be to produce some fine-grained narratives that focus on how we could live together better. Management theorists should be in the thick of this redescription—not at the periphery.

A Rortian and Deweyan vision of inquiry is quite far from where we find "management science." Management thinkers seem to tirelessly debate the empirical versus normative distinction, and they argue endlessly about "proper method" with obvious delight. In session after session at the annual meetings, scholars decry the gap between theory and practice. This is not the occasion to address all of these issues, but one suggestion is to simply give up the distinctions between "theory" and "practice," as well as "empirical" and "normative." We would need to develop criteria by which we judge the usefulness of work, but it would be connected to, rather than divorced from, how value actually gets created for stakeholders. That is a long story, but one that needs telling.

References

Brandom, R., ed. 2000. *Rorty and His Critics*. Oxford: Blackwell.

Dewey, J. 1998. First Published in 1922 in *Human nature and Conduct*. In *The Essential John Dewey. Volume 2: Ethics Logic and Psychology*, ed. L. Hickman and T. Alexander. Bloomington: Indiana University Press.

Freeman, E. 1998. Poverty and the Politics of Capitalism. *The Ruffin Series* 1: 31–35.

Malachowski, A., ed. 1990. *Reading Rorty*. Oxford: Blackwell.

Rorty, R. 1979. *Philosophy and the Mirror of Nature*. Princeton: Princeton University Press.

———. 1982. *Consequences of Pragmatism*. Minneapolis: University of Minnesota Press.

———. 1986. Freud and Moral Reflection. In *Pragmatism's Freud: The Moral Disposition of Psychoanalysis*, ed. J. Smith and W. Kerrigan, 1–27. Baltimore: Johns Hopkins University Press.

———. 1989. *Contingency, Irony, and Solidarity*. Cambridge: Cambridge University Press.

———. 1991a. *Objectivity, Relativism and Truth: Philosophical Papers Volume 1*. Cambridge: Cambridge University Press.

———. 1991b. *Essays on Hiedegger and Others: Philosophical Papers Volume 2*. Cambridge: Cambridge University Press.

———. 1998a. *Truth and Progress: Philosophical Papers Volume 3*. Cambridge: Cambridge University Press.

———. 1998b. *Achieving Our Country*. Cambridge, MA: Harvard University Press.

———. 1998c. Can American Egalitarianism Survive a Globalized Economy? *The Ruffin Series* 1: 1–6.

———. 1999. *Philosophy and Social Hope*. London: Penguin.

Wicks, A., and R.E. Freeman. 1998. Organization Studies and the New Pragmatism. *Organization Science* 9 (2): 123–140.

R. Edward Freeman is University Professor, Elis and Signe Olsson Professor, Academic Director of the Institute for Business in Society, and Senior Fellow of the Olsson Center for Applied Ethics at the University of Virginia's Darden School of Business.

Chapter 30
Business as a Human Enterprise: Implications for Education

R. Edward Freeman and David Newkirk

The challenges facing business education grow out of a fundamental misunderstanding of the nature of business itself: business is a thoroughly human institution, but business schools and business scholars rarely recognize this. Much current research and teaching treats business as a deterministic phenomenon, and management as a science. By embracing business as a human institution and management as a creative act, we can correct many of the problems identified in the modern critiques of management scholarship and education.

We begin with a brief, stylized history of the business academy, primarily in the United States, where it finds its most dominant form and, historically, its largest audience. We juxtapose this history with the tremendous changes that have occurred in business over the last 40–50 years, and suggest that the time is ripe for change. Next we demonstrate that recent critiques rest on a faulty and outmoded view of business. Using stakeholder theory, we argue that we need a new approach, most easily characterized as "business as a human activity." We suggest two central questions and four problems business education must place front and center. We then discuss how we might draw on all of the disciplines of the academy for a more robust and useful view of business, and more powerful research and educational tools.

Originally published in: *Rethinking Business Management*, 131–148, 2008
Reprint by Springer, Reproduced with author's permission

We wish to thank our many colleagues at the Darden School and elsewhere for helpful conversations about these issues. Versions of this paper have been delivered at the Schulich School, York University, and at Copenhagen Business School.

R. E. Freeman (✉) · D. Newkirk
University of Virginia, Charlottesville, VA, USA
e-mail: FreemanE@darden.virginia.edu

© Springer Nature Switzerland AG 2023

S. D. Dmytriyev, R. E. Freeman (eds.), *R. Edward Freeman's Selected Works on Stakeholder Theory and Business Ethics*, Issues in Business Ethics 53,
https://doi.org/10.1007/978-3-031-04564-6_30

1 A Brief History of Business Schools in the United States[1]

Management education is a recent, relatively unexamined, phenomenon. One can identify at least four distinct periods in the United States leading up to the current state of chaos.

The Formative Period While there is a great deal of debate about the original purpose of business schools, we argue that their original purpose was grounded in the idea of moral education. According to L. C. Marshall, one of the earliest proposals for a collegiate business school was made by Robert E. Lee, the first president of what is today Washington and Lee University.[2] (The proposal was not put into action.) Lee's idea was not confined to the business details of bookkeeping and the like "but [aimed] to teach the principles of commerce, economy, trade, and mercantile law." At its very beginning the American collegiate business school sought to articulate and teach the basic principles of commerce and economics, as well as their counterparts in the law.

The Wharton School of the University of Pennsylvania was founded in 1881 with funds from the steel magnate Joseph Wharton and the number of business schools rapidly increased in the next forty years in both the United States and Europe. Francis Ruml argued that the existence of business courses in secondary schools and so-called commercial colleges made it easy and natural for the university and collegiate schools of business to focus their attention more on professional and social ends; to place an emphasis upon dealing with the complicated social organization in which the businessman does his work; and to accept as its objective the training of persons with a social point of view for executive and professional positions in business.[3]

Business schools produced several results, such as (1) replacing the apprenticeship system, largely dismantled by the industrial revolution; (2) training businesspeople in the fundamentals of scientific management; (3) addressing the growing need for financial as opposed to industrial administration; and (4) training an elite capable of managing more complex social organizations.[4]

Ruml summarized Joseph Wharton's initial stated purpose as being "to establish means for imparting a liberal education in all matters concerning Finance and

[1] This section is based on R. Edward Freeman, "Epilogue," in *Business as a Humanity,* eds. Thomas Donaldson and R. Edward Freeman (New York: Oxford University Press, 1994). We are grateful to the editors and publisher for permission to recast some of that material here.

[2] Leon C. Marshall, "The American Collegiate School of Business," in *The Collegiate School of Business: Its Status at the Close of the First Quarter of the Twentieth Century,* ed. Leon C. Marshall (Chicago: University of Chicago Press, 1928), 3.

[3] Frances Ruml, "The Formative Period of Higher Commercial Education in American Universities," in *Collegiate School of Business,* ed. Marshall, 47.

[4] Drew E. VandeCreek, "Power and Order: The Ideology of Professional Business Training at Wharton and Harvard, 1881–1933," (master's thesis, University of Virginia, 1994).

Economy."[5] In his address to the trustees of the newly formed Wharton School, Wharton remarked: "It is reasonable to expect that adequate education in the principles underlying successful business management would greatly aid in producing a class of men likely to become pillars of the state, whether in private or public life." And on this original theme of imparting "moral education" to managers, Ruml approvingly quotes E. P. Cubberly's *History of Education*:

> As the industrial life of the nation has become more diversified, its parts narrower, and its processes more concealed, new and more extended training has been called for to prepare young people to meet the intricacies and interdependencies of political and industrial and social groups and to point out to them the importance of each one's part in the national and industrial organization.[6]

The American Assembly of Collegiate Schools of Business (AACSB), founded in 1916, reinforced this idea of a liberal education for business people by requiring accredited schools to require students to take at least 40 percent of their credit hours in subjects other than economics and commerce. In speaking to their 1927 conference, Alfred North Whitehead echoed the view that management was ready for professional education, if we only knew what it was:

> Universities have trained clergy, medical men, lawyers, engineers. Business is now a highly intellectualized vocation, so it well fits into the series. There is, however, this novelty: the curriculum suitable for a business school, and the various modes of activity of such a school, are still in the experimental stage.[7]

To reinforce the argument that that original purpose of business schools was grounded in the idea of moral education, consider Joseph Wharton's original plan. Wharton proposed and implemented a program to take humanities faculty, trained in the liberal arts, and put them in the classroom to teach business subjects. This experiment, tried in 1882, failed miserably, according to Ruml, because the liberal arts professors were resentful and did not necessarily believe that knowledge should be practical. (This attempt to found a business school on the capabilities and resources of a liberal arts institution has been repeated more recently at Yale and Oxford, both of whom recapitulated the history, ultimately replacing professors from the parent institution with dedicated business academics.) By 1883 the Wharton School had hired instructors versed in business, leaving the ensuing one hundred-plus years barren of meaningful interchange between business and the humanities.

The early history of business schools is rife with arguments about the same controversies that dominate today's discourse. Should faculty be trained in economics? Were graduate economics departments sources of ignorance about business? Could business people be used as adjunct professors, and if so, how many? What was the proper balance between research and consulting? Should the study of business be

---

[5] Ruml, "Higher Commercial Education," 54–55.

[6] Ibid., 53.

[7] Alfred North Whitehead, "Universities and Their Function," in *The Aims of Education and Other Essays* (1929; reprint, New York: Macmillan, 1966), 92.

scientific, or largely storytelling? Was business to be a profession with established ethical norms, or were schools of business merely to socialize students into the practices of business and prepare them for the world of work?

Unspoken are the questions of what is the nature of business itself, and therefore of management and leadership, an essential foundation for addressing the challenge of how best to develop people for these roles. Is management, as Whitehead's comments might imply, a profession, with a body of knowledge to be mastered? Or is it a craft, with skills to be learned?

While these questions are still relevant, the second period in the history of the American business school makes answering them all the more difficult.

The Scientific-Modern Period Two well-known reports published in the late 1950s gave voice to a crisis of legitimacy in the business school. The Carnegie Report and the Ford Foundation study by R. A. Gordon and J. E. Howell both claimed that business schools needed academic legitimacy, and essentially proposed the adoption of a social sciences approach to the subject matter of business.

According to Thomas Mulligan, writing in an important 1987 retrospective in no less than *The Academy of Management Review,* these reports represented a turning point, where the culture of business schools became overtly scientific and, we would argue, adopted the positivism that underlay much of the scientific thinking of the period:

> The often unsubstantiated descriptive content of earlier business school curricula and research has been replaced by quantitative description based on rigorous data collection, computer-assisted mathematical modeling, and the foundational concepts of science—testable hypotheses (or, at least, testable networks of hypotheses), correlated observations, and causal explanation.[8]

The state of business schools, he argued, suggests a desperate need for something like the humanities, but the culture of science that emerged in the business schools has only recently begun to be challenged by the introduction of humanities-based ethics courses.

While Mulligan is correct that the culture of science dominates business schools, there is an alternative to his view that the two cultures of arts and sciences are incompatible. He rightly points out the emergence of "practitioner" books critical of the scientific approach, but there is more at issue here than practical criticism of academic abstractions.

The Porter-Peters Era Two seemingly unrelated phenomena from 1980 through the 1990s catapulted business and business thinkers into the center of society and provided a countervailing force to the increasing scientism and irrelevance of business school research. The result was the emergence of another branch of business education, largely through bestsellers, consultants, corporate universities, learning centers, and the like.

[8]Thomas M. Mulligan, "The Two Cultures in Business Education," *Academy of Management Review* 12 (October 1987): 593–99.

The practical leadership in development of management techniques had moved outside the universities. DuPont had pioneered capital budgeting and return on investment calculations in 1903. The Navy created critical path management during World War II. In the fifties and sixties, GE established small business units and the processes for managing them. By the sixties, Bruce Henderson and BCG were offering services built around experience curves and growth-share matrices. All of these managerial innovations were rooted in an analytical view of business.[9]

In the 1970s, economist Michael Porter published his groundbreaking book *Competitive Strategy* on the economic basis of business strategy.[10] From this book, offering his theory of "the five forces of industry analysis," it was easy to reestablish the dialogue between social scientists such as Porter and businesspeople. He had a coherent narrative with real-life implications for executives and how they thought about their business.

During roughly the same period, Tom Peters and his colleagues at the consulting firm McKinsey and Company embarked on a study of why some organizations seemed to be successful year after year.[11] The report they produced formed the basis for Peters's and Robert Waterman's bestselling *In Search of Excellence.* Its power was, like that of Porter's theory, its systematic, albeit unscientific, narrative about what real companies and real leaders did and did not do to achieve success. *Search* was grounded in the real world and was coherent as a narrative, enhancing its believability.

Our argument is not that the scientific period is over in business schools. The existence of so-called scholarly journals, and the endless arguments in tenure committees about whether a particular journal is an "A, B, or C" journal, ensures an appetite for trivia in excess of either its production or its value. Porter and Peters show us the power of narrative, of telling a good story about how we can create and manage our organizations. Their financial success also shows the appetite for compelling business narrative. In short, they give us a basis for reintroducing the humanities in business education. The humanities contain powerful narratives and meta-narratives about human societies and cultures. These narratives are more or less coherent and more or less normative, as well as descriptive. Equally important, they are compelling to leaders and aspiring leaders.

The ensuing twenty-five–plus years have led to an explosion of the business bestseller, and the seminars and consulting firms telling these narratives. We have been treated to "business reengineering and process redesign," "customer intimacy and operational excellence," "being built to last" and "going from good to great," "managing stakeholders and stakeholder engagement," "supply chain management," "strategic alliances," and a host of other, oftentimes good and useful, ideas. During

---

[9] Gary Hamel, "The How, Why, and What of Management Innovation," *Harvard Business Review,* February 2006, 72–84.

[10] Michael Porter, *Competitive Strategy: Techniques for Analyzing Industries and Competitors* (New York: Free Press, 1980).

[11] Thomas Peters and Robert Waterman, *In Search of Excellence: Lessons from America's Best-Run Companies* (New York: Harper and Row, 1982).

this time, the leader himself (or occasionally herself) has become the subject of inquiry, with their successful traits first described then, increasingly, prescribed. Companies have begun their own "corporate universities," both to teach business skills and to develop leadership, sometimes in conjunction with leading business schools, while the schools themselves have made commitments to executive education as a source of lucrative funding.

Recent Changes to Business and Business Schools As business schools have evolved to become more self-contained and "scientific," alienated from both practice and from the rest of the university, the real world of business has changed greatly.[12]

Every industry has become more global. The coherence of business strategy is influenced by far-flung and disparate events in many corners of the globe. The purely analytical approach to strategy (such as the one advocated by Porter) has become less and less relevant. The issues of outsourcing jobs, government policy, supply chain management, value chain responsibility, environmentalism, and the rise of power of NGOs, are all now central to thinking about strategy. India and China have emerged as serious global players in business, with Russia and Brazil not too far behind. Every aspect of business, from brand values and communications to employee relations, operates in a changing cultural context.

The explosion of the Internet and the emergence of the dot-com companies laid waste to traditional ideas about how to start a business. Thousands of young people began to develop business plans for companies that were targeted to change people's behavior around ideas such as information, shopping, healthcare, and other basic facts of life. YAHOO!, Google, eBay, LendingTree, *Web*MD, are just a few among thousands of companies started either outside the traditional model of business, often by business school students who were pursuing very different aims from traditional businesses and their narrow "economics only" paradigm. And many of these start-ups were successful. IPOs were offered often before there was any revenue at all, and this flaunted the traditional way of thinking about a business.

Underneath the bubbling of the dot-com economy, business itself was changing. Information-age institutions looked fundamentally different from their industrial-age counterparts. Businesses seemed to be creating, rather than meeting, consumer needs. Financial resources were less critical than ideas and customers. Products were "co-created" with customers, and "extended enterprises" began competing in "eco-systems." The "physics" of business began to change. Companies lost mass through outsourcing. Geography began to disappear, and boundaries between markets became more permeable. And Internet time seemed to move faster than normal time. We seemed to be living in a world of creativity and unpredictability, rather than the predictable, repeat- able phenomena of the past.

During roughly the same time period, we saw the fall of communist and socialist regimes all around the world. Bill Clinton's mantra, "It's the economy, stupid,"

---

[12] These are laid out in more detail in R. Edward Freeman, Jeffrey Harrison, and Andrew Wicks, *Managing for Stakeholders: Survival, Reputation, and Success* (New Haven, Conn.: Yale University Press, 2007).

became the watchword for developing capitalist economies, and business took on a renewed importance as a central institution for human interaction.

Concurrently, we have had several waves of scandal, from the Wall Street trading scandals of the late 1980s to the Enron genre of scandals in the late 1990s and early part of this century. Paralleling these scandals have been renewed interest in "corporate responsibility," "environmental sustainability," and issues such as "the public trust of business."

At the same time, there has been an explosion in the need for quasi- managerial white-collar labor familiar with the vocabulary and techniques of business and finance. There has been an equally rapid growth in the population of business schools to meet this need. As the armies of managers have grown, the MBA has become a qualification for both the officers and the foot soldiers, with little differentiation between the two.

The response of business schools has been to highlight how their MBA programs have been made "global," largely through exchange programs with students and faculty, "field trips," or a week or two in China or India. The schools also have made a token attempt to put ethics into the curriculum, although, repeating Joseph Wharton's experiment on a different scale, often taught by business school faculty untrained in ethics, or philosophy faculty untrained and uninterested in business. Some schools have adopted some new technology for "distance learning" to help meet growing demand at lower cost. The responses have, for the most part, been traditional disciplinary answers.

At the same time, we have the phenomena of "ratings" of business school and the adoption of techniques of corporate branding. To attract students paying higher prices in this more competitive market, schools have begun to worry about their "brand," their "image," and the means of differentiating. Business has changed on a massive scale during the last 25 years, and we believe that it is fair to say that business schools have simply "moved the deck." Consequently there is a malaise, whose true cause is difficult to pin down.

2 Recent Critiques of Business Education

The dominant narrative of business schools, rooted in the Scientific-Modern Period, has a real grip on the hearts and minds of business school professors and administrators, who transmit it to students, so that it becomes self- perpetuating. However, there is some light. A number of recent critiques have begun the process of self-examination.

Jeffrey Pfejfer and Christina Fong[13] Pfeffer and Fong argue that the MBA is neither necessary nor sufficient for business success, and that a student's grades are not correlated with a flourishing career. Much of what MBAs learn is of limited

[13] VandeCreek, "Power and Order."

usefulness, business school research is of limited influence, and true "evidence-based" management is rare.

Many of the bestsellers making an impact on business are based on what Pfeffer would call "nonevidential approaches." It is difficult to discern when one reads the entirety of his work whether he believes that the problem is simply a lack of "good science" in business school research or a more fundamental issue of how research problems are framed at business schools.

In short, Pfeffer and Fong suggest that most business school researchers get "good science" wrong, and that the resulting MBA is not useful and doesn't much matter.

Henry Mintzberg[14] In a recent blistering critique of business education, *Managers Not MBAs,* management theorist Henry Mintzberg suggests that for the most part MBA programs have the wrong students, that students in their mid-to-late twenties are simply neither ready nor capable of learning management. He sees management differently from Pfeffer and Fong, as a craft rather than a science. Hence, studying management is a waste of time. One needs to be engaged in the "doing" rather than the "talking about doing." There is simply too much emphasis on the disciplines of business in the business schools, and too much emphasis on the traditional modes of teaching and learning. Consequently, MBAs have become exploiters rather than explorers and innovators. In addition, the managerial class, defined by MBAs, is responsible for a lot of economic misery in society.

Mintzberg rightly traces much of the problem back to the Scientific-Modern Period and the emergence of a "cult of methodology," whereby the worthiness of a particular piece of research is based not on how it allows us to live, but on the "validity" of the methodology by which it has been executed. However, it is perhaps disingenuous of him to make such a strong claim about the scientism in business school research, as he is routinely lauded for taking empiricism to its logical ends. His early studies of managerial work influenced a generation of students and managers, and have been defended by many as telling us what managers actually do, rather than what they should do to become good managers. His own eschewing of the normative aspects of business research seems to fit squarely into the cross hairs of his critique.

In short, Mintzberg suggests that we get "management" wrong, and the resulting MBAs do real and lasting damage in the world.

Sumantra Ghoshal In an influential article titled "Bad Management Theories Are Destroying Good Management Practices," the late Sumantra Ghoshal argued that the dominant narratives that have taken hold in business schools are the real culprits.[15] He singles out the ideologues who, concentrating on the economic and

[14] Henry Mintzberg, *Managers Not MBAs: A Hard Look at the Soft Practice of Managing and Management Development* (San Francisco: Berrett-Koehler, 2004).

[15] Sumantra Ghoshal, "Bad Management Theories Are Destroying Good Management Practices," *Academy of Management Learning and Education* 4 (March 2005): 75–91.

financial aspects of business, propose that the only legitimate purpose of a business is to maximize shareholder value, and those who understand business as a complex agency problem where managers are seen as agents of shareholders. He invokes Hayek's idea of the pretense of knowledge to suggest that we misunderstand the nature of social science. Social science theories become known, "ideas in good currency" as Donald Schon would say, and students and business people begin to act as if these ideas were true. They enact the dominant narrative. And, in doing so, they create agency problems and corporations who are immune to any criteria other than shareholder value. He writes:

> In courses on corporate governance grounded in agency theory, we have taught our students that managers cannot be trusted to do their jobs—which, of course, is to maximize shareholder value.... In courses on organization design, grounded in transaction-cost economics, we have preached the need for tight monitoring and control of people to prevent "opportunistic behavior." In strategy courses, we have presented the "five forces" framework to suggest that companies must compete not only with their competitors but also with their suppliers, customers, employees, and regulators.[16]

Business school researchers have simply misused the scientific methods by pretending that business is like the physical sciences, with fixed rules and repeatable phenomena. The result is not the useless and uninfluential theory of Pfeffer and Fong, but the highly destructive theory that Mintzberg's MBAs are inflicting on the world. In an incredibly enlightening passage, Ghoshal suggests that the answer might lie in a closer connection between the great thinkers of the sciences and the humanities, adopting a "scholarship of common sense." He writes:

> In describing himself and his work, Sigmund Freud wrote: "[Y]ou often estimate me too highly. I am not really a man of science, not an experimenter, and not a thinker. I am nothing but by temperament a conquistador—an adventurer, if you want to translate the term—with the curiosity, the boldness, and the tenacity that belong to that type of being..."
>
> Freud's inductive and iterative approach to sense-making, often criticized for being ad hoc and unscientific, was scholarship of common sense. So indeed was Darwin's, who, too, practiced a model of research as the work of a detective, not of an experimenter, who was driven by the passions of an adventurer, not those of a mathematician. Scholarship of common sense is the epistemology of disciplined imagination, as advocated by Karl Weick,... and not the epistemology of formalized falsification that was the doctrine of Karl Popper....
>
> To protect the pretense of knowledge, we have created conditions under which this kind of scholarship can no longer flourish in our community. This is true of all social science disciplines, but curiously perhaps it is most intensely true in business schools where, in our desire for respect from scholars in other fields, we have become even more intolerant of the scholarship of common sense than those whose respect we seek....[17]

In short, Ghoshal suggests that we get theory wrong, management wrong, and social science wrong, and we shouldn't be surprised at the resulting moral decline of business.

---

[16] Ibid., 75.

[17] Ibid., 81–82.

So What? The common theme to these critiques is that much malaise in and about business schools is a result of a misguided positivism or scientism, ultimately stemming from what we have called the Scientific-Modern Period of business school history While there is much truth in this critique, it is difficult to see how such a tightly knit social system is going to change. Indeed, the authors' calls for reform are quite different and do not measure up to the level of passion in their critiques. This helps uncover a deeper issue.

3 Business as a Human Institution

Rejecting the Dominant Narratives of Business What business schools get wrong is business itself. They may get science, theory, and management wrong as well. But the problem starts with the very idea of business. Intriguingly, little of the management literature and few of the critiques offer a view of what business *is*. And absent a view of business, they are equally unclear on management and leadership.

Implicit in much of the management discussion is a mechanical, deterministic, positivistic view of business—a financial engine controlled by the machinery of scientific management—with managers making judgments about them. (This is most evident in the analytical professions, including consulting and investment banking, which now hire over half the graduating classes of the leading business schools. A consultant's job is often to diagnose and describe a business problem, and an investor's is to predict which stocks will move in which direction. Both require rigorous analysis, and both tempt the practitioner to treat the business observed as an independent, determined phenomenon.)

A brief tour of the business school curriculum will find that the basic units of analysis are cash flows, products and services, processes, brands, and other stylized ideas that have acquired a metaphysical reality of their own. As Ghoshal and others suggest, our narrow ideas about value as only economic value to shareholders, our ideas about agency, about the purpose of the firm, and others are all part of the dominant narrative in business schools. They are also all artifacts of human interactions.

Like earlier mechanistic views, the scientific view of management makes it difficult to incorporate the human spirit. It leaves little room for ethics, or for other fundamentally human considerations such as creativity, trust, initiative, and will.

In very few places in business education do we begin with the basic human interactions that make business a profoundly human institution. Business, in its most essential form, is a way that we create value for each other by cooperating and specializing our labor. Business is fundamentally about value-creation and trade. Business simply does not work outside of some interpersonal-social- cultural-ethical context, yet most theories of business separate business decisions from this human matrix.

We have only recently separated management as a unique activity and begun treating business as distinct from other social phenomena. "The identification of management with business management began only with the Great Depression, which bred hostility to business and contempt for business executives. In order not to be tarred with the business brush, management in the public sector was rechristened public administration and proclaimed a separate discipline," as Peter Drucker points out.[18] "It is only the emergence of management since World War II, what I call the 'Management Revolution,' that has allowed us to see that the organization is discrete and distinct from society's other institutions."[19]

An Example: Challenging Separation from Ethics This separation of business from more general human activity can be seen most clearly in the popular joke about "business ethics as an oxymoron." We might define the "Separation Fallacy" this way:

> It is useful, first, to believe that sentences such as, "x is a business decision" have no ethical content or any implicit ethical point of view. And, it is useful, second, to believe that sentences such as "*x* is an ethical decision, the best thing to do, all things considered" have no content or implicit view about value-creation and trade (business).

This fallacy underlies much of the dominant story about business, and it is reinforced by the approach to research that the critics of business school have decried. Yet almost any business decision has some ethical content. To see that this true, we offer the "Open Question Argument." One need only ask whether the following questions make sense for virtually any business decision (the Separation Fallacy would have us believe that these questions aren't relevant for making business decisions, or that they could never be answered):

1. If this decision is made, for whom is value created, and for whom is value destroyed?
2. Who is harmed and/or benefited by this decision?
3. Whose rights are enabled, and whose values are realized by this decision (and whose are not)?
4. What kind of person will I (we) become if we make this decision?

These questions are always open for most business decisions, and therefore we need a theory about business that answers them. One such answer would be "Only value to shareholders counts," but such an answer would have to be enmeshed in the language of ethics as well as business. (Milton Friedman, unlike most of his expositors, actually gives such a morally rich answer.) However, to develop that answer meaningfully, we need an ethics capable of engaging the language and issues of business.

In short, we need a theory that has as its basis what we might call the "Integration Thesis":

---

[18] Peter Drucker, "Management's New Paradigms," *Forbes,* 5 October 1998, 156.

[19] Peter Drucker, "The New Society of Organizations," *Harvard Business Review,* September-October 1992, 100.

> Most business decisions, or sentences about business, have some ethical content or implicit ethical view. Most ethical decisions, or sentences about ethics, have some business content or implicit view about business.

Human Enterprise as a New Narrative One of the most pressing challenges facing business scholars is to tell compelling narratives that have the Integration Thesis at their heart. Seeing business as a fundamentally human institution means that we can bring to bear on business the full panoply of ideas, disciplines, models, theories, and understandings from literature, the fine arts, history, anthropology, cultural studies, language studies, law, philosophy, the sciences, and other parts of the academy. It involves Sartre's idea that ethics begins with authenticity and the need to justify one's life to oneself and to others. This is no different in today's small businesses and global corporations than it was in post-World War II Paris.

Seeing business as a fundamentally human enterprise means accepting something like the "Responsibility Principle," which is implicit in most reasonably comprehensive views. It states:

> Most people, most of the time, want to, actually do, and should accept responsibility for the effects of their actions on others.

Without something like the Responsibility Principle it is difficult to see how ethics gets off the ground. "Responsibility" may well be a difficult and multifaceted idea. There are surely many different ways to understand it. But if we are not willing to accept responsibility for our own actions (as limited as that may be due to complicated issues of causality and the like), then ethics, understood as how we reason together so we can all flourish, is likely an exercise in bad faith.

If we want to adopt the Integration Thesis, if the Open Question Argument makes sense, and if something like the Responsibility Thesis is necessary, we need a new narrative about business. This new story must be able to explain how value-creation at once encompasses both economics and ethics, and how it takes account of all the effects of business action on others.

Such a model exists, and has been developed over the last 30 years by management researchers and ethics scholars called "stakeholder theorists." They have challenged much work that is done in the name of "value-free economics and science." Indeed, they have adopted a critical stance toward much of what the critiques suggest is wrong in terms of the nature of management and its appropriate modes of analysis. But they have also invoked the second sense of the Separation Fallacy to challenge work done by philosophers who have little knowledge of either business or the business disciplines. (Sadly, recognition of their work is missing in most of the critiques of business schools, reinforcing our claim that these critiques are grounded in the dominant narrative.)

We are not going to review the particulars of stakeholder theory. This has been done many times over by many people.[20] We want stakeholder theory to serve as an

---

[20] A recent statement that reflects our view is Freeman et al., *Managing for Stakeholders.*

example of the kind of work that could be done if we truly began to see business as a human enterprise. There is new work to be done to bring the richness of our colleagues' exploration of human activity and intellectual creativity to bear in the domain of management. There are many theories and research programs that could be developed. Getting the "metaphysics of business" (if you like) right opens up many possibilities. We believe these possibilities offer the only hope for rescuing our business schools.

A Valuable Corollary If business were seen as a human enterprise, business people could again be seen as human beings. (The prevailing view, that business people are a different species, can be seen in our experience teaching ethics to high school seniors. Presented with the ethical decision of whether to read a folder left behind in a negotiation, the students split evenly on whether to read it. But 100 percent thought that businesspeople would read it.) We could encourage students and businesspeople to bring their human sensibilities and judgment into the business setting. Rather than teaching "business ethics" as a special discipline, whose very existence needed to be defended, we could teach ethics and their application in the business environment.

4 Rescuing Business Education

If we put the humanity of business into the center of the way we think about it, we must deal with two fundamental questions and four problems of management and leadership in business, even if only by redirecting them. Once these have been addressed, there will remain one big challenge. We appeal to our "common sense" about business for the legitimacy of what we offer.

Two Fundamental Questions We simply cannot understand business as a human enterprise without putting these two questions at the center of our inquiry:

1. What does it mean to be a fully complex human being?
2. How do complex human beings create value and trade with each other, and what is the meaning of such activity?

We are not arguing for a particular answer to these questions. There can be many. However, we must ask them if we are to understand the real world of business, where there are actual people engaging in relationships and transactions with one other. They cannot be reduced to questions of cash flows, markets, hierarchies, products, role-related behaviors such as consumer behavior, and other sometimes useful but nonfoundational ideas.

Four Problems Having raised these two questions, we must address four problems of management and leadership.

First is the "Problem of Authority." How are we to understand the authority relationships that exist in business, and make sense of them in human terms? Is there a conflict between these relationships and our autonomy as moral agents? How can the work of thinkers such as Philip Zimbardo, Stanley Milgram, Hannah Arendt, Nietzsche, and others be brought to bear on the authority relationships in business? As globalization extends Western models into new cultures, are there other thinkers and models which we must address?

Second is the "Problem of the Self." How are we to understand the role of the self in business? Is the self to be understood as an autonomous agent acting "for itself," or is it to be understood as a complex, interacting set of relationships? What is the role of the history of a "self"? In particular, what is the connection between the development of self, its engagement in an enterprise, and the relationship of authority? How do race, gender, sexuality, and other "variables" connect to our ideas about the self? And, how does the concept of the self connect to how we create value and trade with each other?

The third problem is the "Problem of Ethics." Is the history of our ethics a history devoid of the effects of value-creation and trade? If human beings have been "value-creators and traders" for a very long time, how could this idea have been absent for most of our discussion of ethics? Would a better first question for political theory be "How are value-creation and trade sustainable over time?" rather than, "How is the state justified?"

And the fourth problem is the "Problem of Human Nature." Is there any such thing as "human nature"? What do traditional and new narratives about business assume about human nature? Can we have an idea of "humanity" that transcends individuals, without an idea of human nature? What is the role of institutions such as business in shaping our ideas about, or even the facts of, how humans can live? Conversely, should our views about human nature shape what we view to be acceptable, or even possible, business institutions and practices?

Each of these problems is large, and for the most part, all are ignored in the current business school curriculum. The dominant narrative must ignore them. They also must be ignored if Ghoshal and others are correct about the scientism that has taken hold in business schools. These problems simply cannot be addressed solely with traditional "scientific" or "social scientific" methods and concepts.

One Big Challenge These questions and problems open a rich discussion about the nature of business, management, and leadership. However, they only lay the groundwork for the big challenge:

> What should business schools do—how, when, and for whom—to educate leaders and managers of these human enterprises?

The expanding scholarship of teaching and learning, especially in the liberal arts colleges, rightly draws the distinction between the "what" of education and the how. Conservatories and creative writing institutes have long understood the differences between criticism and performance, as well as the role of the former in helping to develop the latter. We in the business schools need to understand the balance of our own activities between developing scholars of business and developing practitioners of business, and therefore understand the best means of achieving this. Here science may offer a better solution, since the laboratory environment develops scientists through apprenticeship rather than lecture.

The foundational questions raised here should give us a better sense of the phenomena of business and possibly of how to study those phenomena. They may also give us a deeper understanding of the nature of management. This only prepares us to ask the questions about the educational task ahead.

5 An Agenda for Change

If we put these questions and problems at the center of the educational processes for businesspeople, business schools may, over time, come to look very different. We want to suggest five items in an agenda for change. There are many others.

Lead a Conversation on the Connection Between Business and the Humanities Scholars in the humanities need to become more literate in the real world of business, so that they can offer different lenses through which to understand value-creation and trade. PhD programs in business disciplines need to be broadened to use a wider range of texts and to become more interdisciplinary. Business educators need to look to the humanities for new ways of teaching, especially for developing students' judgment and creativity. Above all we need a set of Dewey-like experiments to create new modes of conceptualization.

Rethink the Business Disciplines We can take each of the disciplines of business and identify concepts that we might call, after Michael Walzer, "morally thick." Our goal would not be to drive out the value implications from traditional conceptual schemes, but to welcome them as different lenses.

For instance, suppose that we understood business as a human enterprise and took the two fundamental questions and four problems to heart in marketing. We might ask whether a company's relationship with a consumer is a promise, like the promises that parents make to children, and if the value of its brand is rooted in trust. We might ask whether we should understand customers as ends in themselves, whose projects are important precisely because they are their projects, rather than means to our own ends. We might define our "value proposition" in terms of our contribution to consumers' own values. We might ask about the moral significance of brands and the effects of brands on civic space, on civil society, or on our understanding of the environment, possibly accepting that the consumer as well as the producer has rights regarding the brand.[21]

In finance, we might see markets as having moral relevance and cash flow as a utilitarian measurement of how well a company is creating value for its stakeholders.

[21] For an analysis of brands as social texts, see Mary Jo Hatch and James Rubin, "The Hermeneutics of Branding," *Journal of Brand Management* 14 (September 2006): 40–59.

The challenge raised by Sharia finance might become real, i.e., how does a lender truly participate in the activities funded? Across all of these, we might explore where diverse global cultures should give rise to distinct, locally relevant answers.

Rethink the Disciplinary Matrix of Business Schools We also need to rethink the disciplinary mix. The world has changed so much that the traditional, functionally defined subjects (defined, ironically, by stakeholder group) may no longer be appropriate. Most schools offer courses in marketing (customers), finance (shareholders or financiers), operations (suppliers and production), human resources or organizational behavior/theory or management (employees), and public relations, ethics, or business/government (community).

Where do students learn how to deal with consumers who are employees and shareholders as well? Where do they learn how to create value with NGOs who are critics, often very well-informed critics? Where do they learn how to deal with employees who are community leaders? What standing do traditional disciplines offer to the organizations whose symbiotic products and services are critical to a business's success? What are the thick moral ideas in the new "disciplines"?

Rethink Research Here the critics, especially Ghoshal, are surely correct. Researchers need to take responsibility for their theories and for the effects of their actions. Theory needs most often to look more like narrative and only sometimes like "science." We need new narratives and new questions that better enable us to engage in value-creation and trade, and they should not duck explicitly human, and especially ethical, vocabulary. At a minimum, we must rethink the rampant positivism that Ghoshal has identified and debate what constitutes research appropriate to the issues studied. We need to understand how our colleagues in the humanities explore (and contribute to) performance and creativity. We need to redesign the core processes that shape our research, like publication, refereeing, tenure and promotion, and even the ratings game itself—or offer alternatives in order to support more appropriate methodologies.

Rethink Teaching Teaching needs to be very different. We need to design high-engagement experiences that allow the student to develop meaningful, theory-informed practices. Management is an art and learning by doing is important, but it is easier to learn with the knowledge of the theory, as music is learned by both practice and knowledge of scales and chords.

MBA programs need their equivalent of études, to teach traditional skills in the traditional disciplines as well as to open up these disciplines to insights from the humanities in the morally thick manner that we have suggested. But like all performers, business students need to have large experiential and practice components to their education, with the teacher as guide and coach, rather than source and judge. Students need to be exposed to at least some of the best thinking and thinkers from across the university, and one would hope these thinkers would also energize their own disciplines with the idea that the business school is more than a morally suspect relative who just won't leave.

6 Conclusion

Schools that adopt this agenda for reform have a chance to distinguish themselves, but it will not be easy. Business schools are part of universities, which are essentially medieval institutions with high resistance to change: faculty have high needs for autonomy, among other psychological quirks; students are driven by their perceptions of what will work in the marketplace; employers need to fill immediate needs for analysts; and companies and business leaders hire graduates based on particular skill needs in the short term. Deans rarely have the courage to tackle such a project.

To adopt this agenda successfully, a school would have to be aligned around its purpose and have a live conversation about whether that purpose can be realized in the world described by Ghoshal, Mintzberg, and others. Unless schools adopt it, however, they will likely be but a long and uninteresting footnote to Joseph Wharton's failed experiment.

R. Edward Freeman is University Professor, Elis and Signe Olsson Professor, Academic Director of the Institute for Business in Society, and Senior Fellow of the Olsson Center for Applied Ethics at the University of Virginia's Darden School of Business.

David Newkirk is a retired executive with leadership experience at the University of Virginia's Darden Executive Education, American Express, Booz Allen Hamilton and other prominent organizations.

Chapter 31
Business School Research: Some Preliminary Suggestions

R. Edward Freeman and David Newkirk

1 Introduction

Does this paper[1] count as research in business schools? If we were untenured faculty members, should we get credit for writing this paper? Should our colleagues view this paper as a contribution to knowledge, in a similar manner (though a different kind of contribution) to a paper in the Journal of Finance, or Administrative Sciences Quarterly? Suppose it happens to be published in an "A" journal. What face validity would it merit from such a publication? Does this face validity change as the paper moves down the hierarchy to end up as a book chapter (even though no sentences in the paper have changed)? Should this paper simply be dismissed as the ramblings of grey beards who can't cut it anymore in the top journals?

A colleague once remarked to us that while a particular piece of research may be okay for management, it would simply not pass muster in his field. When we responded that it was published according to the standards of management in a top journal (an "A" journal), he responded that it must be easy to publish in management, and that he remained sceptical of the whole field. When we suggested that his own field had failed to ask some fundamental questions about the complexity of human

Originally published in: *Business Schools Under Fire: Humanistic Management Education as the Way Forward*, 273–290 © Palgrave Macmillan, 2011
Reprint by Springer

[1] This paper is the result of a joint effort with our colleagues Jared Harris, and Sankaran Venkataraman. A further version of the paper will undoubtedly list some of them as co-authors.

R. E. Freeman (✉) · D. Newkirk
University of Virginia, Charlottesville, VA, USA
e-mail: FreemanE@darden.virginia.edu

© Springer Nature Switzerland AG 2023

S. D. Dmytriyev, R. E. Freeman (eds.), *R. Edward Freeman's Selected Works on Stakeholder Theory and Business Ethics*, Issues in Business Ethics 53,
https://doi.org/10.1007/978-3-031-04564-6_31

behaviour, and about morality in particular, our conversation quickly turned to music, something over which we could agree to disagree without doing harm to our relationship.

We suspect that there are many of these conversations in business schools all over the world. As schools become more concerned about rankings and publishing in so-called "A" journals, as well as about their relevance to the institution of business, it is time to examine the underlying issues in more detail. While some of these questions may seem like academic trivialities, in fact they raise an issue that is at the heart of recent critiques of business schools, namely, what is the role of research in the modern business school.

In this paper, we shall present the following argument. Recent critiques of business schools have led to the formulation of three interconnected problems: the problem of research; the normative problem; and, the problem of the use of knowledge. At the heart of the issue is what we have called "the separation thesis" or "the separation fallacy," most powerfully seen as the exclusion of ethics from business judgments but generalizable as a duality between the worlds of management and the humanities. If we give up this fallacy, we can gain a different and more pluralist view of the role of research in business schools, and of the nature of meaningful research.. We lay out five main modes of research, and suggest how each can escape the separation fallacy. We then show how the problem of the use of knowledge can be ameliorated by giving up the separation fallacy. Finally, we suggest how we can re-define and re-discover the disciplines of business by thinking about more thick moral conceptions.

2 The Critique of Business Schools

The dominant narrative of business schools – that management is a science and can usefully be researched, taught, and practiced in those terms – has a real grip on the hearts and minds of business school professors and administrators. Unfortunately, it is transmitted to students as well and becomes self-perpetuating. However, there is some light. A number of recent critiques have begun the process of self-examination.

2.1 *Jeffrey Pfeffer and Christina Fong*

Pfeffer and Fong have suggested that the whole idea of the MBA is suspect. They argue that it is neither necessary nor sufficient for business success and further that grades earned during this two year period do not correlate with a flourishing career. The subtext is that much of what MBAs learn is of limited usefulness, that business school research is of limited influence, and that true "evidence based" management is a rare species. They conclude from their extensive research:

> For the most part, there is scant evidence that the MBA credential, particularly from non-elite schools, or the grades earned in business courses as a measure of the mastery of the material, are related to either salary or the attainment of higher level positions in organizations. These data, at a minimum, suggest that the training or education component of business education is only loosely coupled to the world of managing organizations. A similar disconnection is observed when we consider research. Again, the small amount of available evidence suggests a modest effect and limited linkage between the research on management and management practice.[2]

In much of his own research, Pfeffer has been the consummate social scientist. In recent years, he has begun to acknowledge the gap between what we know and what we do, and the lack of truly evidence-based management. Many of the best sellers that he and Fong suggest are having an impact on business are based on what he would call "non-evidential approaches." It is difficult to discern when one reads the entirety of his work whether Pfeffer believes that the problem is simply a lack of "good science" in business school research or whether there is a more fundamental issue about how research problems are framed at business schools. In short, Pfeffer and Fong suggest that most business school researchers get "good science" wrong and that the resulting MBA is not useful and doesn't much matter.

2.2 *Henry Mintzberg*[3]

In a recent book, management theorist Henry Mintzberg has delivered a blistering critique of business education. He suggests that for the most part MBA programmes have the wrong students, that students in their mid to late 20s are simply neither ready for, nor capable of, learning management. Mintzberg sees management differently from Pfeffer and Fong, as a craft rather than a science. Hence, studying management is a waste of time according to Mintzberg. One needs to be engaged in the "doing" rather than the "talking about doing." There is simply too much emphasis on the disciplines of business in the business schools and too much emphasis on the traditional modes of teaching and learning. Consequently, MBAs have become exploiters, rather than explorers and innovators. In addition, Mintzberg holds the managerial class, defined by possession of MBAs, as responsible for a lot of economic misery in society.[4]

Mintzberg rightly traces much of the problem back to research and the emergence of a "cult of methodology", whereby the worthiness of a particular piece of research is based, not on how it allows us to live, but on the "validity" of the methodology by

---

[2] Jeffrey Pfeffer and Christina T. Fong (2002), 'The End of Business Schools? Less Success Than Meets the Eye', Academy of Management Learning and Education 1(1, September, 92).

[3] Henry Mintzberg (2004), Managers Not MBAs (San Francisco: Berrett Koehler).

[4] And, more recently commenting on the US's first MBA president, now responsible for a great deal of political misery as well. 'Leadership Beyond the Bush MBA', unpublished, is available on Mintzbert's website: http://www.henrymintz-berg.com/pdf/leadershipbush.pdf

which it has been executed. However, it is perhaps disingenuous of Mintzberg to make such a strong claim about the scientism in business school research since he is routinely lauded as taking empiricism to its logical ends. His early studies of managerial work influenced a generation of students and managers and have been defended by many as telling us what managers actually do, rather than what they should do to become good managers. Mintzberg's own eschewing of the normative aspects of business research seems to fit squarely into the crosshairs of his critique. In short, Mintzberg suggests that we get "management" wrong and the resulting MBAs do real and lasting damage in the world.

2.3 *Sumantra Ghosal*[5]

In an influential article, entitled 'Bad Management Theories are Destroying Good Management Practices', the late Sumantra Ghosal argued that the dominant narratives that have taken hold in business schools are the real culprits. In particular, Ghosal singles out the ideologues who, concentrating on the economic/financial aspects of business, propose that the only legitimate purpose of a business is to maximize shareholder value, and those who propose to understand business as a complex agency problem where managers are seen as agents of shareholders. Ghosal invokes Nobel Prize-winner Friedrich August von Hayek's idea of the pretence of knowledge in order to suggest that we misunderstand the nature of social science. Social science theories become known as "ideas in good currency", as Trist would say and students and business people begin to act as if these ideas are true. They enact the dominant narrative. In doing so, they create agency problems, and corporations who are immune to any criteria other than shareholder value.[6] He writes:

> In courses on corporate governance grounded in agency theory, we have taught our students that managers cannot be trusted to do their jobs – which, of course, is to maximize shareholder value ... In courses on organization design, grounded in transaction cost economics, we have preached the need for tight monitoring and control of people to prevent "opportunistic behavior." In strategy courses, we have presented the "five forces" framework to suggest that companies must compete not only with their competitors but also with their suppliers, customers, employees and regulators.[7]

[5] Sumantra Ghosal (2005), 'Bad Management Theories are Destroying Good Management Practices', Academy of Management Learning and Education 4(1).

[6] While this effect, of theory creating behaviour, may be most pernicious in the case of agency theory, we may also see it in finance. As our colleague Paul Harper has pointed out, the Black Scholes model predicts options prices possibly in part because most players in the market are taught to price options using the Black Scholes model. And in Nassim Taleb's analysis in The Black Swan, recent market upheavals have occurred when the assumptions underlying the model have proven false.

[7] Ghosal, 75.

Goshal's view is that theories of business shape business themselves, that is, that there is not a set of stable underlying phenomena. Business school researchers have simply misused scientific methods by pretending that business is like the physical sciences, with fixed rules and repeatable phenomena. The result is not the useless and non-influential theory of Pfeffer and Fong but a potentially highly destructive theory that Mintzberg's MBAs are inflicting on the world.

In an incredibly enlightening passage, Ghosal suggests that the answer might lie in a closer connection between the great thinkers of the sciences and humanities, adopting a method of "common sense". He writes (at 81–2):

> In describing himself and his work, Sigmund Freud wrote: "[Y]ou often estimate me too highly. I am not really a man of science, not an experimenter, and not a thinker. I am nothing but by temperament a conquistador – an adventurer, if you want to translate the term – with the curiosity, the boldness, and the tenacity that belong to that type of being." (In Jones 1964, p. 171)

Freud's inductive and iterative approach to sense making, often criticized for being ad hoc and unscientific, was scholarship of common sense. So indeed was Darwin's, who too practiced a model of research as the work of a detective, not of an experimenter, and who was also driven by the passions of an adventurer, not those of a mathematician. Scholarship of common sense is the epistemology of disciplined imagination, as advocated by Karl Weick (1989), and not the epistemology of formalized falsification that was the doctrine of Karl Popper (1968).

To protect the pretence of knowledge, we have created conditions under which this kind of scholarship can no longer flourish in our community. This is true of all social science disciplines but, curiously, perhaps it is most intensely true in business schools where, in our desire for respect from scholars in other fields, we have become even more intolerant of the scholarship of common sense than those whose respect we seek (Bailey and Ford 1996).

In short, Ghosal suggests that we get theory wrong, management wrong, and social science wrong, and we shouldn't be surprised at the resulting moral decline of business.

3 Three Connected Problems

The upshot of these critiques can be diagnosed as three interconnected problems. Freeman and Newkirk (2008) began this diagnosis by suggesting that what the critics themselves miss is that most of the time we get "business" wrong. Business is not an independent, repeatable phenomenon, an "economic clockwork" subject to its own, discoverable rules. In fact business is a deeply human institution and to see it as anything less misses the mark. We continue that line of thought here as we try and diagnose a further level of detail in these three problems.

3.1 The Problem of Research

The first problem, which we shall call The Problem of Research, is that a kind of positivism has overtaken business schools. This became clear to us after many years of writing letters for promotion and tenure for colleagues at many business schools around the world, and serving on the tenure and promotion committees at the various institutions where we have worked. Positivism, in this context, leads to a view, roughly, that research is to be modelled on the physical sciences. Only measurable, repeatable phenomena are "real." Knowledge results from empirical investigation. Theory is important only to the extent that it leads to testable propositions and measurable hypotheses. The results of this view are to treat method and data with a kind of reverence that ultimately leads to what Hambrick has argued is a fetish for theory – theory that makes sense out of the data, out of what can be measured, regardless of its causal relevance.

In his Nobel lecture Hayek has summarized the problem of applying "scientific" scholarship to human activities quite nicely:

> Unlike the position that exists in the physical sciences, in economics and other disciplines that deal with essentially complex phenomena, the aspects of the events to be accounted for about which we can get quantitative data are necessarily limited and may not include the important ones. While in the physical sciences it is generally assumed, probably with good reason, that any important factor which determines the observed events will itself be directly observable and measurable, in the study of such complex phenomena as the market, which depend on the actions of many individuals, all the circumstances which will determine the outcome of a process, for reasons which I shall explain later, will hardly ever be fully known or measurable. And while in the physical sciences the investigator will be able to measure what, on the basis of a prima facie theory, he thinks important, in the social sciences often that is treated as important which happens to be accessible to measurement. This is sometimes carried to the point where it is demanded that our theories must be formulated in such terms that they refer only to measurable magnitudes. (Hayek 1974)

This is precisely the problem in business research. Simply add "organization" to Hayek's idea of the complexity of markets, and one begins to see the folly of such a view. Add the idea that business is a human institution, one that is fully situated in the real world of human complexity, morality, human hopes and aspirations, and most importantly, one where theory shapes behaviour, and the quantitative techniques of modern empirical scholarship begin to seem feeble, when they are left to their own devices.

We are not arguing that data driven research offers no interesting insight. And, we are not arguing that empirical investigations into business phenomena do not yield interesting insights, in fact we want to argue that when those investigations are grounded in the practice of business they can lead to quite interesting and useful results. However, we are suggesting that the kind of positivism that has overtaken business research has run its course. It is surely not the only way to understand business phenomena. In fact we want to articulate, later in this paper, at least five modes of research that should count as "real research" and that yield insight. It is this problem of research, that only one mode of research counts, which is behind much of the critique of Pfeffer, Fong, Mintzberg, and Ghosal. Indeed in many tenure committees today, we simply would not recognize the next Freud or Darwin, if their work were before us.

3.2 The Problem of The Separation Fallacy

Freeman (1994) suggested that most business theory assumed what he called "the separation thesis." Put in the pragmatic terms of an informal fallacy it can be described as – The Separation Fallacy.

It is useful, first, to believe that sentences like, "x is a business decision" have no ethical content or any implicit ethical point of view. And, it is useful, second, to believe that sentences like 'x is an ethical decision, the best thing to do all things considered' have no content or implicit view about value creation and trade (business).

This fallacy underlies much of the dominant story about business and it is reinforced by the approach to research that the critics of business schools have decried. The implication of rejecting the Separation Fallacy is that almost any business decision has some ethical content. To see that this is true one need only ask whether the following questions make sense for virtually any business decision.

3.2.1 The Open Question Argument

1. If this decision is made for whom is value created and destroyed?
2. Who is harmed and/or benefited by this decision?
3. Whose rights are enabled and whose values are realized by this decision (and whose are not)?
4. What kind of person will I (we) become if we make this decision?

Since these questions are always open for most business decisions, it is reasonable to oppose The Separation Fallacy, which would have us believe that these questions aren't relevant for making business decisions, or that they could never be answered.

We need a theory about business that builds in answers to the Open Question Argument above. One such answer would be, "only value to shareholders counts" but such an answer would have to be enmeshed in the language of ethics as well as business. (Milton Friedman, unlike most of his expositors, actually gives such a morally rich answer.) However to develop that answer meaningfully, we need an ethics which is capable of engaging the language and issues of business. In short we need a theory that has as its basis what we might call – The Integration Thesis.

Most business decisions, or sentences about business, have some ethical content or an implicit ethical view. Most ethical decisions or sentences about ethics have some business content or an implicit view about business.

As Sandberg (2008, p. 230) notes, values are 'embedded in social contexts from which they cannot be removed'. We cannot single out particular "facts" from their underlying narratives.[8] As Searle (1964, pp. 52, 54) pointed out, the 'inclination to accept a rigid distinction between "is" and "ought", between descriptive and evaluative, rests on a certain picture of the way words relate to the world' that ignores contextual notions such as "commitment, responsibility, and obligation". In other words, statements about the external world do not "face the tribunal of sense experience" alone. Quine (1951, p. 38), James and Dewey, Putnam and Rawls, Rorty and Goodman have all made similar arguments. Philosophers of science such as Kuhn (1962) and Feyerabend (1975) have highlighted the challenge this poses to the very concept of scientific inquiry as being solely descriptive and objective.

Similarly, in The Collapse of the Fact/Value Dichotomy, Hilary Putnam (2002, pp. 61–2) suggests that facts and values are deeply "entangled", and as such, 'the picture of our language in which nothing can be both a fact and value-laden is wholly inadequate'. As an illustration, Putnam's analysis of the word "cruel" as being both descriptive and value laden illustrates how a great deal of language works, and demonstrates the limitation of employing a sterile, objectivist view of language and meaning. Putnam then analyses the work of Amartya Sen in On Ethics and Economics (1987), in which Sen specifically suggests that we have forgotten that economics is inherently entangled with matters of ethics, and argues that the false dichotomization of the two has impoverished discipline-based analysis in both economics and ethics. Sen's observation is a clear articulation of what has come to be known as the separation thesis.

Yet, the entanglement of facts and values has implications beyond our mere conception of "business" language as being both normative and descriptive. Such entangled concepts apply directly to actual practices, which always embody both facts ("business" considerations) and values ("ethical" considerations). Consider, for example, the arrangement by which a business firm provides employment to a particular individual. Has the corporation provided economic value or moral value? How can such things be disentangled?[9] Along these lines, any economic assertion is ultimately both descriptive and value laden, as well. Furthermore – and ironically – any explicit contention that commerce and morals involve mutually exclusive considerations (e.g., Friedman 1970; Jensen 2002) is also both descriptive and value laden.

Hence, while agreeing that dichotomizing business and ethics is a fool's errand that is not useful, helpful, or even meaningful,[10] we argue for a more powerful reason to reject the separation thesis explicitly: separating economic considerations and ethical considerations is impossible. As such, attempts to separate the two serve

[8] This section draws heavily on Jared Harris and R. Edward Freeman (2008), 'The Impossibility of the Separation Thesis', *Business Ethics Quarterly*, 18(4), 541–8.

[9] And if they cannot be disentangled, what are the implications for ideas such as corporate social responsibility and triple bottom line?

[10] For several articulations of these arguments, see Freeman (1994, 2000), Freeman et al. (2004), and Wicks (1996), as examples.

simply to obscure a particular set of values – but such values exist nonetheless, lurking beneath the surface. Embracing the separation thesis, then, does not in fact disentangle facts and values, but merely appears to do so, while obscuring and embedding a particular set of values that privileges certain considerations and dismisses others. Therefore, the problem with the separation thesis is not so much that it actually separates business and ethics – an impossible task – but that it purports to be morally neutral whilst surreptitiously encapsulating certain ethical values and assertions.

The Separation Fallacy is most commonly and pointedly stated in terms of ethical content. However, it can be framed in any number of "human" terms. The positivist view of business excludes the ideas of intention, will, and creativity from consideration. Absent these, our account of business is at risk of being incomplete. Just as the work on the separation fallacy in ethics has collapsed the facts/values dichotomy, we would expect a richer analysis of the other human concepts to bring them back into business research.

3.3 The Problem of Business Knowledge

Ghosal's critique is that the problem of research has created a false sense of knowledge, which pervades business schools. We teach and act as if we have created complete, or near complete, causal theories about business. We act and teach students to act on "the pretence of knowledge". And, there is a self-reinforcing cycle to this knowledge which makes it particularly trouble-some. Consider the following situation:

Let us suppose that one received idea about business is that "all x's are y". Suppose that Researcher A describes a theory, and finds some data to support it, that revises this understanding. Suppose that A "shows" that "all x's are yRz, for some relation R". This new theory is then published and becomes generally accepted. Data are collected which support the theory, and while there are many factors for which there is no data, it becomes acceptable to measure the terms of the theory by what data are available. Students and executives are taught the new theory, and they go out into the world and enact it. They manage as if all x's are yRz. Indeed, they expect to find just this phenomenon, since they were well-trained. Researcher B goes out to study the issue one more time, and finds, lo and behold, a very strong relationship. In fact, what B has found is that students enact theory.

This is the general form of Ghosal's idea that by teaching agency theory in business schools we have created agency problems. More generally, if we teach that business decisions and ethical decisions have nothing to do with each other, then we have created a generation of business leaders who look for business and ethics to conflict. Open questions are answered with "this is business", or "this is a business decision", or "I've got to do this for the good of the business". Thus, Ghoshal (2005) and others (e.g. Ferraro et al. 2005; Frank et al. 1993), in showing that we enact the very theories of social science that we propose – and therefore demonstrating that the moral consequences are indistinguishable from the theories themselves – highlight the danger of attempting to separate business from ethics. When theorists suggest and managers enact an approach that views "business" decisions as if there are no moral consequences to them (e.g. describing unfettered profit maximization as the "single objective function" of business firms), this inculcates a societal narrative about business and ethics in which ethical considerations are no less real, but merely devalued and denatured. If we treat the world of business as discovered, not created, we absolve managers of their responsibility for its structure. Is the view that owners of firms and their employees are one-dimensional maximizers of self-interest, with convex utility functions for monetary wealth, simply a matter of fact? Is the assumption that incentives effectively ameliorate agency conflicts unassailable? While some research (e.g. Frank 1988; Harris and Bromiley 2007) calls into question the prima facie descriptive accuracy of such assertions, important implications also arise from the assumptions about morality that are embedded in such statements, and their reifying influence on managerial behaviour and social norms. That business decisions have moral content is inescapable; pretending the two are divisible at best obscures important considerations, and at worst paradoxically encourages a particular set of ethical norms that may be unintended.

4 Pluralism and Five Modes of Research

To deal with the Problem of Research we need a much more open idea about what counts as scholarship in business schools. We want to suggest that there are at least five modes of research that yield useful insights into business. While these modes are not mutually exclusive or exhaustive, they are sufficiently different to warrant further investigation. It is worth noting that our criterion is not "truth" or "representation of the real world" but rather the generation of useful insights into the processes and institutions of "value creation and trade" (our phrase for "business"). This deserves more explanation but that will require another occasion. Suffice it to say here that we will follow Wicks and Freeman (1998) in their pragmatic view of theory and of business. Judgment and insight become paradigmatic for us, and that is a view we are more than happy to try to defend.

4.1 Formal Theory Mode

In the formal theory mode, mathematical modelling and reasoning are used to yield insights into complex phenomena. As paradigm cases, we would point to the development of game theory by von Neumann and Morgenstern, and its refinement and application. While many of its applications are not in the formal theory mode, these applications would make no sense without the formal theory on which they are based. Von Neumann and Morgenstern's The Theory of Games and Economic

Behavior (1944) is simply unreadable for most business school professors, except those with extensive training in formal mathematics. Yet, the insights that it has generated are important – from the development of formal utility theory in the appendix to the idea that what I expect you to do in a situation depends on what you expect me to do, which depends on…; and that a theory of rational behaviour can in fact be self-reinforcing. Similarly, the insights of John Nash on the bargaining problem were instrumental in the development of stakeholder theory: the idea that customers, suppliers, employees, communities and financiers are engaged in a complex interaction to create value. The study of the formal theoretic game and decision paradoxes, such as Ulysses and the Sirens, Prisoner's Dilemma, Newcomb's Problem, and Arrow's Impossibility Theorem have led to many insights into human behaviour, our theories of reasoning, and our institutions.[11]

There are many critiques of the so-called "rational economic actor" model of human decision making, and surely these critiques are well-founded in so far as they caution us not to take things too far. However, there are many insights to be derived from this mode, and certainly it deserves a place in the pantheon of business research. Of course, this mode, like all others, can cause real damage when it is applied as the only way to understand human behaviour. It is well known that the only group of people who have difficulties solving Prisoner's Dilemma problems in multiple play situations are economics graduate students, who are in fact "prisoners" of their own theory about the scope of rational behaviour.

4.2 The Statistical Empirical Mode

Not much needs to be said about this mode of research since it is the dominant mode in most business schools. Often it begins with some theoretical base or other and quickly moves to propositions that are testable and hypotheses that are measurable. Many insights have been generated about the way that markets work, or the structural analyses of various industries. In some disciplines, the structure of particular tasks in the work environment has been related to the attitudes of employees. And, the list of insights generated is endless.

However, often the link back to theory is at best weak and, since much of the phenomena that are of interest consist of unobservable yet causally relevant information, there is often not much generation of useful ideas in practice. Or, at least, the ideas that are generated are not always based on the data, as is sometimes the claim. Thus it is sometimes difficult to square calls for reform based on "evidence-based management" with a more pluralistic view of business research. Those who want evidence-based results sometimes write as if the only barriers

[11] And, in many cases, they have shaped our institutions. For example, the current structure of the mobile telephone market in the US is a result of behaviours induced by a spectrum auction process designed to yield maximum prices for the licenses, regardless of the stability of the resulting businesses.

preventing good practical advice were more and better data and tighter links between hypotheses and propositions. Our claim is that the best any research can yield is "insight" that turns out to be useful when applied with good judgment.

4.3 Clinical Empirical Mode

This mode of research is best exemplified outside management by Freud's development of psychoanalysis, based on the case studies that he encountered in clinical practice. While we have developed many techniques and methods to turn such qualitative encounters into essentially quantitative data, such a move may well obscure the main purpose of clinical work. Freud encountered people who had problems, or who were in pain, and the clinician's job was to help them solve the problems or deal with the pain. The resulting reflections on what he saw, and the subsequent application of these reflections, led to important new insights into the human condition.

The disfavour currently shown the clinical empirical mode may, in part, reflect a misunderstanding of the role of the "experimentalist" in the scientific process itself. In science, and in management science research, empirical research is used to validate/demonstrate hypotheses – gathering evidence that will "test" the articulated theory. In practice, however, the great experimentalists generated observations that challenged theories and triggered the development of new ones. The observation of phenomena in order to drive the development of insight may require different methods and standards, which risk being crowded out by the validation exercise.

Philip Selznick's immersion in the development of the Tennessee Valley Authority (Selznick 1980), Karl Weick's analysis of the Mann Gulch disaster (Weick 1993), Graham Allison's work on the Cuban Missile Crisis (Allison 1971), and others, are all in this clinical mode, and all made contributions to the development of management thinking. In fact, we believe that many research enterprises start out in this mode but quickly, and less fruitfully, turn into the statistical/empirical mode for the sake of publication. Case studies are out of fashion these days, except ironically as teaching devices.

4.4 Narrative Theory Mode

Most research is embedded in some kind of narrative or story. As Quine and others have suggested, sentences don't confront experience alone. They always bring their background theories with them. This is the case with theories and research studies about business as well. Sometimes it is useful to articulate these background narratives to shift direction, point out alternative modes of conceptualization, or simply raise an issue that has not been seen by others. When Rhenman, Ansoff, Ackoff and others began to talk about "stakeholders" they helped to build a new narrative about the very basic idea of business and capitalism. While this narrative may have grown out of particular observations, or case studies, or the theorists' understanding of the way that business works from multiple sources, the important point is the articulation of the story itself.

There is much work in the humanities on the idea of narrative, and there is much to be learned from literary theory, narrative ethics, and other sources in the humanities, where the scholarship of creative human activities consists largely of developing narratives that make sense of the outcome. Ironically, this purely theoretical mode begins to look a lot like the formal theory mode. If we turn to the work of someone like Thomas Schelling, we find an interesting icon. It is Schelling's understanding of the mathematics of game theory, together with his keen insights into human behaviour, that make the narratives that he spins in The Strategy of Conflict (2007) and Micro-Motives and Macro-Behavior (2006) so compelling. Indeed, the idea of mixing modes is precisely where we can generate more insight into value creation and trade, and is an idea in need of further exploration.

We simply cannot blindly accept the dominant narrative of business and business schools, based on a false positivism and the separation fallacy it encourages. Our challenge is to create rich new narratives that are ripe with human possibility and responsibility, that show us better ways to engage in value creation and trade.

4.5 Creative Mode

There is much to be learned from the fine arts that could be applicable to business. The fine arts are engaged in creative performance, informed by theory, and that idea sits comfortably at the centre of business notions of strategy, leadership and humanity. By studying and performing theatre and music we can generate insights into how collaboration and feedback work in intense environments. By understanding art, music, theatre, dance we can gain insights into the human condition. By studying the performers and their inspiration, we might understand the interplay between theory, observation and creation. As we see business as a deeply human enterprise, we must leave no stone unturned to generate useful insights. There is much more to be said here.

5 The Open Texture of Business Knowledge

To take advantage of the pluralism of the five modes of research we must avoid the temptation of hubris. No one mode, and no single discipline, has a monopoly on insight. An analysis of markets tells us a lot about business, but it does not give us the whole story. An understanding of creativity and its sources may help us develop leaders but it will not speak to the purpose of business. Similarly, deep insights into the human condition and how we make joint meaning gives much insight into how business works but, again, it is not the whole story. Indeed, we would surmise that there never is "the whole story." Each mode of research and their combinations are always subject to revision and the generation of new insights. In fact, by applying the methods and thinking of one mode to another we can sometimes generate useful ideas. By being explicit about the underlying narratives in the empirical mode, we can generate new narratives, and perhaps new testable propositions.

To avoid this temptation, we believe that we must keep Ghosal's warning about the self-reinforcing nature of social science in mind. No theory is without impact, and no powerful idea leaves the observed phenomenon unchanged. In fact, we believe that any piece of research always leaves a certain set of questions open. And, we suggest that these open questions begin to be explicitly acknowledged and answered. We want to build on the ideas of Michael Gonin's Business Research, Self-fulfilling Prophecy, and the Inherent Responsibility of Scholars (2007) and suggest that the following set of questions become routine ones:

1. Does this work answer the question(s) it proposes?
2. Was the question meaningful and appropriate?
3. Are there alternative modes of research that could lend insight into the question(s)?
4. What are the direct consequences of this research?
5. If we teach this insight to managers and students what might be the result if they act on it?
6. What is/are the background narrative(s) of this research?
7. How will we begin to see ourselves and others if we act on this work?
8. How will this work shape the context in which value creation and trade takes place?

Modelled on questions in the field of technology assessment (Davis and Freeman 1978) these become a kind of "research assessment". We are not suggesting that every piece of research must actually answer these questions but, instead, that we open up a line of research that is devoted to asking and answering these questions. This paper implies, in part, that it is problematic that such a critical line of inquiry does not exist in many disciplines. Most simply stop with the first two questions on the list, which are devoted to the efficiency and effectiveness of research, but leave unaddressed the subsequent questions on its assumptions and impact.

6 Rediscovering the Disciplines of Business

If we adopt the pluralistic approach to business scholarship, research and their critical component, as we have suggested, then we can begin the process of reinventing the disciplines of business. First of all, we can take each of the disciplines of business and identify concepts that we might call, after Michael Walzer, "morally thick". Our goal would not be to drive out the value implications from traditional conceptual schemes but to welcome them as different lenses. For instance, suppose that we understood business as a human enterprise to be at heart about marketing. We might ask whether a company's relationship with a consumer is a promise, like the promises that parents make to children, and whether or not the value of its brands is rooted in trust. We might ask whether we should understand customers as ends in themselves, whose projects are important precisely because they are their projects, rather than a means to a company's ends. We might define a company's "value proposition" in terms of its contribution to consumers' own values, and whether the company's activities shape, or even create those values. We might ask about the moral significance of brands and the effects of brands on civic space, or on civil society, or on our understanding of the environment, possibly accepting that the consumer as well as producer has rights regarding the brand.[12]

In finance, we could come to see markets as having moral relevance, and the idea of cash flow as a utilitarian measurement of how well a company is creating value for its stakeholders. We might begin to ask routine questions about ethics and finance. Are lenders responsible for "irrational" decisions made by borrowers? How far can we push the assumptions of rationality in the face of such theory as can be found in behavioural finance, or prospect theory in psychology? What responsibility do financiers have when consequences yield wild success or abject failure? The questions are virtually endless, and the modes of work that can be applied are liable to yield great insight, especially if those working within the discipline can begin to take seriously the ideas of economists like Sen, Hayek, and even Adam Smith.

In addition to adding humanities-based lenses to the business school disciplines, we also need to rethink the disciplinary mix and pedagogical choices. Ultimately, generating useful insights into business involves multiple disciplines. The world has changed so much that the traditional functionally defined subjects (ironically by stakeholder group) may no longer be appropriate. Problems don't appear very often in the guise recognized by business disciplines, which have become particularly stylized, if we and other critics are correct. For instance, most schools have marketing (customers), finance (shareholders or financiers), operations (suppliers and production), human resources or organizational behaviour/theory or management (employees), and public relations or ethics or business/government (community). Where do students learn how to deal with consumers who are employees and shareholders as well? Where do students learn how to create value with NGOs who are critics, often very well informed critics, of a company? Where do students learn how to deal with employees who are community leaders? What standing do traditional disciplines offer to the organizations whose symbiotic products and services are critical to a business's success? And, what are the thick moral ideas in these new "disciplines"? Can we create some new lenses and new understandings around these new configurations, all of which are descriptive of real business issues? And if we take the view that management and leadership are creative acts, how do we better prepare our students to perform them well?

---

[12] For an analysis of brands as social texts, see Hatch, Mary Jo and Rubin, James (2006), 'The Hermeneutics of Branding', The Journal of Brand Management 14(1–2), September, 40–59(20).

7 Summary and Conclusions

The purpose of this paper is to open up the conversation about what counts as research in business schools to a set of considerations advanced by Freeman and Newkirk (2008) that business is a profoundly human institution. Therefore, the full panoply of modes, methods, and ideas from across the intellectual landscape are capable of generating useful insights into how human beings do and could create value and trade with each other. Such an opening happily coincides with a recent report of the AACSB on the state of research or scholarly inquiry into business schools and, at least in management journals, there seems to be some appetite for considering these questions.

We should not be fooled by the conversation. An examination of the editorial pages of most of the top journals in business disciplines will yield much hope that these institutions are open to new and important ideas. However, for the most part this hope turns out to be false, as we begin to generate more and more research about more and more narrow ideas (see Daft and Lewin (2008) for their own analysis of a journal they started, Organization Science, and of its subsequent narrowing).

We believe that such a conversation will be successful only if we begin to question the background narrative of business research itself, as well as the dominant narrative about business. We need to be sceptical about distinctions such as "theory-practice", "rigour-relevance", "theory-data", "empirical-theoretical", "teacher-researcher", "content-pedagogy" and "business-ethics". We know that social science is interpretive and self-reinforcing, so we need to use that knowledge to create a way to think about business that is at once grounded in the practice of the institution and profoundly critical of it and of any knowledge generated. If we can do this, we can deliver more insight into the deeply human enterprise of value creation and trade, to make it better and more sustainable for our children.

References

Allison, G. 1971. *Essence of Decision: Explaining the Cuban Missile Crisis*. Boston: Little, Brown.

Bailey, J., and C. Ford. 1996. Management as Science versus Management as Practice in Postgraduate Business Education. *Business Strategy Review* 7 (4): 7–12.

Daft, R.L., and A.Y. Lewin. 2008. Perspective – Rigor and Relevance in Organization Studies: Idea Migration and Academic Journal Evolution. *Organization Science* 19: 177–183.

Davis, P., and R. Freeman. 1978. Technology Assessment and Idealized Design: An Application to Telecommunications. In *Evaluating New Telecommunications Services*, ed. M.C.J. Elton, W.A. Lucas, and D.W. Conrath, 325–344. New York: Plenum Press.

Ferraro, F., J. Pfeffer, and R.I. Sutton. 2005. Economics Language and Assumptions: How Theories Can Become Self-Fulfilling. *Academy of Management Review* 30 (1): 8–24.

Feyerabend, P. 1975. *Against Method*. London: Verso.

Frank, R.H. 1988. *Passions within Reason*. New York: W. W. Norton & Company.

Frank, R.H., T. Gilovich, and D. Regan. 1993. Does Studying Economics Inhibit Cooperation? *Journal of Economic Perspectives* 7 (Spring): 159–171.

Freeman, R.E. 1994. The Politics of Stakeholder Theory: Some Future Directions. *Business Ethics Quarterly* 4 (4): 409–421.
———. 2000. Business Ethics at the Millennium. *Business Ethics Quarterly* 10 (1): 169–180.
Freeman, R.E., and D. Newkirk. 2008. Business as a Human Enterprise: Implications for Education. In *Rethinking Business Management: Examining the Foundations of Business Education*, ed. S. Gregg and J.R. Stoner Jr. Wilmington: ISI Books.
Freeman, R.E., A.C. Wicks, and B. Parmar. 2004. Stakeholder Theory and the Corporate Objective Revisited. *Organization Science* 15 (3): 364–369.
Friedman, M. 1970. The Social Responsibility of Business is to Increase Its Profits. *The New York Times* 33: 122–126.
Ghoshal, S. 2005. Bad Management Theories Are Destroying Good Management Practices. *Academy of Management Learning and Education* 4 (1): 75–91.
Gonin, M. 2007. Business Research, Self-fulfilling Prophecy, and the Inherent Responsibility of Scholars. *Journal of Academic Ethics* 5: 33–58.
Harris, J., and P. Bromiley. 2007. Incentives to Cheat: The Influence of Executive Compensation and Firm Performance on Financial Misrepresentation. *Organization Science* 18 (3): 350–367.
Harris, J., and R.E. Freeman. 2008. The Impossibility of the Separation Thesis. *Business Ethics Quarterly* 18: 541.
Hatch, M.J., and J. Rubin. 2006. The Hermeneutics of Branding. *The Journal of Brand Management* 14 (1–2): 40–59. (20).
Jensen, M.C. 2002. Value Maximization, Stakeholder Theory, and the Corporate Objective Function. *Business Ethics Quarterly* 12 (2): 235–256.
Jones, E. 1964. *The Life and Times of Sigmund Freud*. London: Penguin Books.
Kuhn, T.S. 1962. *The Structure of Scientific Revolutions*. Chicago: University of Chicago Press.
Mintzberg, H. 2004. *Managers Not MBAs*. San Francisco: Berrett Koehler.
Mintzbert, H. 1980. *Leadership Beyond the Bush MBA*. http://www.henrymintz-berg.com/pdf/leadershipbush.pdf.
Pfeffer, J., and C.T. Fong. 2002. The End of Business Schools? Less Success than Meets the Eye. *Academy of Management Learning and Education* 1 (1): 78–95.
Popper, K.R. 1968. Epistemology Without a Knowing Subject. In *Logic, Methodology, and Philosophy of Sciences*, ed. Van Rootselaar and Staal, 333–373. Stanford: Stanford University Press.
Putnam, H. 2002. *The Collapse of the Fact/Value Dichotomy*. Cambridge, MA: Harvard University Press.
Quine, W.V.O. 1951. Two Dogmas of Empiricism. *The Philosophical Review* 60 (1): 20–43.
Sandberg, J. 2008. Understanding the Separation Thesis. *Business Ethics Quarterly* 18 (2): 213–232.
Schelling, T.C. 2006. *Micromotives and Macrobehavior*. New York: W.W. Norton.
———. 2007. *The Strategy of Conflict*. Cambridge: Harvard University Press.
Searle, J.R. 1964. How to Derive "Ought" From "Is". *The Philosophical Review* 73 (1): 43–58. (52, 54).
Selznick, P. 1980. *Tennessee Valley Authority and the Grass Roots*. Los Angeles: University of California Press.
Sen, A. 1987. *On Ethics and Economics*. Malden: Blackwell.
von Hayek, F.A. 1974. The Pretence of Knowledge. *The Sveriges Riksbank Prize in Economic Sciences, Nobel Prize Lecture*. Available at http://nobelprize.org/nobel_prizes/economics/laureates/1974/hayek-lecture.html.
Von Neumann, J., and O. Morgenstern. 1944. *The Theory of Games and Economic Behavior*. Princeton: Princeton University Press).
Weick, K.E. 1989. Theory Construction as Disciplined Imagination. *Academy of Management Review* 14 (4): 516–531.
———. 1993. The Collapse of Sensemaking in Organizations: The Mann Gulch Disaster. *Administrative Science Quarterly* 38 (4): 628–652.

Wicks, A.C. 1996. Overcoming the Separation Thesis: The Need for a Reconsideration of Business and Society Research. *Business & Society* 35 (1): 89–118.
Wicks, A.C., and R.E. Freeman. 1998. Organization studies and the new pragmatism: Positivism, anti-positivism, and the search for ethics. *Organization Science* 9 (2): 123–140.

R. Edward Freeman is University Professor, Elis and Signe Olsson Professor, Academic Director of the Institute for Business in Society, and Senior Fellow of the Olsson Center for Applied Ethics at the University of Virginia's Darden School of Business.

David Newkirk is a retired executive with leadership experience at Darden Executive Education, American Express, Booz Allen Hamilton, and other prominent organizations.

Chapter 32
Teaching Business Ethics in the Age of Madoff

R. Edward Freeman, Lisa Stewart, and Brian Moriarty

Tell someone that you teach "business ethics," and you know what they say: "Isn't that an oxymoron—you know, like jumbo shrimp?" Or, "I didn't know business had any ethics." Or, "Must be a short course." Recently, we even got, "That's a pretty theoretical course, isn't it?" And this is a joke that travels: we've heard these kinds of responses in many countries and cultures. So what does it mean to teach "business ethics"? And has it changed in the age of financial crisis, burst bubbles, and Bernie Madoff?

1 Business Ethics: A Brief History

Business and commerce, and their connections to ethical thinking, are as old as civilization. However, according to Richard De George, business ethics as an academic discipline is a relatively recent phenomenon. In the mid-1970s, a group of scholars, primarily in U.S Catholic universities, began to study business from the standpoint of ethical theory. In 1974 the discipline's first academic conference in this country was held at the University of Kansas. In 1980 these scholars formed the Society for Business Ethics, after which two academic journals emerged: *The Journal of Business Ethics* in 1982 and *Business Ethics Quarterly* in 1991. By the mid-1980s, courses, textbooks, casebooks, and organizations had all proliferated.

Originally published in: *Change: The Magazine of Higher Learning,* 41:6, 37–42 © Taylor & Francis, 2006
Reprint by Springer, https://doi.org/10.1080/00091380903316905

R. E. Freeman (✉) · B. Moriarty
University of Virginia, Charlottesville, VA, USA
e-mail: FreemanE@darden.virginia.edu

L. Stewart
University of California, Santa Barbara, CA, USA

© Springer Nature Switzerland AG 2023

S. D. Dmytriyev, R. E. Freeman (eds.), *R. Edward Freeman's Selected Works on Stakeholder Theory and Business Ethics*, Issues in Business Ethics 53,
https://doi.org/10.1007/978-3-031-04564-6_32

The anti-war, anti-establishment protests of the 1960s and 1970s, together with an increase in environmental awareness, had led to many questions about the role of business in society. So as the discipline was developing, another group of scholars started to think about corporate social responsibility. They formed a division of the Academy of Management (the main professional organization for management professors) called "Social Issues in Management" and began to do research in a broad area called "business and society."

William Frederick, professor *emeritus* at the University of Pittsburgh and a leader in this movement, identifies the 1961 publication of Richard Eels and Clarence Walton's *Conceptual Foundations of Business* as one of the academic starting points for this approach. Even as research in most business disciplines got narrower and more "scientific," these scholars continued in the role of critic. Key concepts and terms for the field—such as "corporate social responsibility," "social reporting and auditing," and "stakeholders"—came out of this line of attack.

Those scholars who promulgated "business ethics" generally took a humanist approach, while "business and society" scholars saw themselves as social scientists. Happily, while these debates still occasionally break out at professional meetings, the past 10 years have seen a merging of the two and more conversation across academic lines. Indeed, many of the humanities-trained business ethicists now have appointments, including endowed chairs, at leading business schools. In this article we are going to use the phrase "business ethics" to refer to the merger of these two groups of scholars, since the search for a more inclusive term continues.

2 The Role of Scandal

The business-ethics discipline has been periodically stoked by scandal, which has led to a long tradition in its courses of analyzing scandals, cast in the form of case studies. First there was the Lockheed Brake Scandal, which led to the Foreign Corrupt Practices Act in 1976. Then we had the first Firestone tire scandal and the infamous Ford Pinto debacle. The 1980s saw the Wall Street scandals regarding insider trading, which produced "villains" such as Dennis Levine, Ivan Boesky, Michael Millken, and others. These scandals were imprinted on the culture via the movie *Wall Street* and the infamous Gordon Gecko's pronouncement that "greed is good." The 1990s saw the internet bubble boom and bust, ultimately resulting in the wave of corporate scandals on a grand scale such as Enron, WorldCom, Tyco, and Adelphia Cable. Finally, scandal became globalized with, for example, the UN "Food for Oil" program and the Italian company Parmalat.

Mary Gentile, head of the *Giving Voice to Values* curriculum project at the Aspen Institute (http://www.givingvoicetovalues.org), describes "the routine that developed over time" in business schools as each "window of urgency" opened up: "Business educators would try to accomplish some curricular reform, and then we'd work as quickly as possible to get changes made before the window would shut again." Sometimes these efforts backfired. Michael Lewis, who chronicled the culture of Wall Street in *Liar's Poker*, reveals that some business students who read his book, rather than being appalled by Wall Street's fraud and greed, wrote to him for more "tips" on how to make it there. To them, the book was not a cautionary tale but an instruction manual.

But while the sinners were painted in lurid colors as rapacious and greedy, there were also saints in this morality play. These were the "good" companies—Johnson & Johnson with its response to the Tylenol poisonings, for instance, or Merck, which gave away a cure for river blindness. And there were companies that took social responsibility seriously: In Minnesota, members of the 5% Club donated 5% of their pretax profits to charity.

The myth of "saints" versus "sinners" developed as a result of focusing on scandals. But it has been challenged, and with good reason. Thomas Donaldson, for instance, thinks that "the 'greed' explanation for Wall Street excesses" is "unhelpful." He says, "I want to believe it too, but no serious study has shown that greed is higher on Wall Street than in other industries, or for that matter higher in any one industry than in another, or in any time period than in another. Greed no doubt varies by time and place, but it is notoriously hard to measure and is a persistent feature of the human condition."

Most business professors would acknowledge that in order to address ethical considerations, they need to go beyond the study of saints and sinners and incorporate ethical considerations more broadly into the curriculum. So today, most major business schools include ethics courses. But increasingly, business ethicists believe that the problem lies in the very conception of business that business schools hold and teach. Unless we think about scandals in these more fundamental terms, we are doomed to repeat the cycle of scandal begetting public outrage, which begets regulation, which begets scandal, and so forth.

3 The Role of Business Schools

The financial crisis has amplified recent critiques of business schools that began in earnest after the bursting of internet bubble, for which they were seen as partly to blame. Management theorist Henry Mintzberg, for instance, argues that business-school graduates are responsible for much of the economic misery in the world and places the blame squarely on the shoulders of the schools that shaped them.

Perhaps the most powerful critique comes from the late Sumantra Ghoshal. Teach business students that ethics doesn't matter in business decisions, he said, and they will go out into the world and enact that theory. Teach them that managers need incentives to act in the interests of shareholders, and your graduates will focus on executive compensation. Teach business students that only shareholders matter—taking precedence over employees, communities, and suppliers—and they will go forth to give us the current financial crisis.

Ethics is often simply bolted onto a curriculum that does not encourage students to engage in moral reflection and that belies what they learn in their ethics courses.

The standard business school curriculum is full of analytic techniques, methods of financial engineering, and an ideology that is deeply hostile to business ethics. The very roots of business education are entangled with unacknowledged—and therefore largely unquestioned—assumptions about both human nature and the fundamental nature and purpose of business.

Business schools typically perpetuate the following four misconceptions:

1. *Markets are perfect, or at least efficient.*

   The inflating and bursting of bubble economies that led to the global economic downturn offers some empirical evidence that markets are not always efficient, never mind perfect. This should surprise no one, since markets are a human construct. It is easy to lose sight of this fact—markets are enormously complex, and sifting out the dynamics of individual agency is like trying to trace the movements of a grain of sand in the belly of a rotating cement mixer.

2. *Human beings are always self-interested.*

   One of the key reasons the case method of teaching is so powerful is that individual cases help reveal the sizable gap between the narrative of business that emerges from an overemphasis on economic trends and the narratives that are shaped by the stories of real entrepreneurs and executives.

   Even if one accepts the proposition that human beings are always self-interested, there are clearly forms of expressing self-interest that arise from non-economic values, even if decisions made on this basis have economic repercussions. The mother who turns down a higher-paying job out-of-state so that her school-age children can continue going to school with current friends and the farmer who tills the soil because of the deep happiness he experiences from seeing the fruits of his labor are two examples.

3. *Economic models and reasoning can explain most of what is interesting about business.*

   The global economic downturn has helped to put a crack in the edifice of rational self-interest as an all-encompassing explanation of business activity. The idea that financial institutions could, in the name of profitability, lend huge amounts of capital to more and more people at high risk to default on those loans—coupled with a belief that organizational and systemic risks would actually be *lowered* by credit default swaps and similar financial instruments—no longer seems all that rational.

   As the philosopher Charles Peirce and others have argued, there is a plethora of evidence to suggest that we ought to distinguish between the logic and the rationality of our decisions, because not all of our decisions are rational, even if they follow a certain logic.

4. *Ethics is about altruism.*

   The proposition that business is all about economic self-interest while ethics is about altruism is also clearly flawed. The electric light bulb is as much the result of Thomas Edison's aspiration to change the world—and his contemporaries' desire for illumination without threat of fire—as it was of a desire for profit. Every day participants in thousands of online communities offer one another technical assistance or give companies free feedback on new products. Every day researchers working for pharmaceutical firms and universities search for treatments to alleviate human suffering. For many, the profit motive and transfer of capital is not the primary driver—it is, rather, the primary enabler of their sustained activity.

An ethics course grafted into a curriculum based on these faulty ideas can, at best, merely sensitize students to the possibility of thinking differently about business. Real change will only happen when we stop appending ethics to business education and move forward with long-term efforts to build ethical considerations into the DNA of business education.

More broadly, this means we need to tell and enact a revised story about business that shifts the center of gravity from profits to value creation, where accomplishments such as creating meaningful work for employees, helping to meet social needs, and improving the environment for the next generation are recognized as distinguishing marks of the business hero—a model to compete with the currently predominant one of the person who delivers shareholder returns. The old narrative of business as anything-goes capitalism, whose central metaphors are greed, money, and profit and whose sole value is the short-term interest of investors, is no longer useful in today's world. Profit, markets, and the like are there not as ends in themselves but to assist the hero in reaching his or her goal.

The new story goes something like this.

1. *Business is primarily about purpose. Money and profits follow.*

Business is primarily a form of social cooperation—it is about people working together to create value that no one of us could create on our own. It is about creating chairs that allow the body to rest, vehicles and networks that enable us to travel and communicate over great distances, a range of products and services as broad as the human imagination.

Of course profits are important, but as Jack Welch, former CEO of GE, argues, profits are not what you do—they are *a result of* what you do.

2. *Any business creates (or sometimes destroys) value for shareholders, as well as for customers, employees, suppliers, and communities.*

Building and leading a business involves getting these interests going in the same direction.

3. *Capitalism works because we are complex creatures with many needs and desires, and we can cooperate to create value for each other.*

Sometimes we act for selfish reasons and sometimes for "other-regarding" ones. Incentives are important, but so are values.

4. *Most people tell the truth and keep their promises, and act responsibly most of the time*

... and we need to expect that behavior. Business and capitalism are the greatest system of social cooperation and value creation ever invented. Competition is important in a free society, since it ensures that people have options. But the engine of capitalism is value creation.

It is not only academics who are hungry for a new story about business. Dean Krehmeyer, executive director of the Business Roundtable Institute for Corporate Ethics, says that many of the CEOs and other executives with whom he speaks are tired of being tarred by the latest scandals, which are not reflective of how the vast majority of companies operate. "There aren't many Enrons and Bernie Madoffs, when you think about the scope of business activity around the world," says Krehmeyer. "Many of these companies are raising the bar for how businesses create lasting value to customers, employees, suppliers, and investors."

The new story of business will not ignore scandals, but it will cease being defined by them. Redefining business in terms of value creation is critical for enabling the next generation of leaders who aspire to change the world to view it as a viable—and perhaps even preferable—way to achieve their goals.

Deep-rooted cultural change such as this demands the sustained efforts of many people. But business schools, which have a unique responsibility for educating current and future executives, should be leading the charge.

4 Teaching in the Age of Madoff

Business schools and business-ethics educators can play a role in effecting change. On the school-wide level, there are two key opportunities for telling the new story about business. First, business schools need to engage the broader university community. The new narrative needs to involve the humanities in particular, which has created (and continues to create) a wealth of knowledge about human needs, behaviors, values, and cultures. As Timothy Fort notes, "most business analysis neglects the human side, the long-term side that is the currency of trust" because this critical area is "notoriously difficult to quantify."

Second, business schools need to increase their engagement with businesses. Over the last two decades, these schools have become increasingly professionalized. At well-established institutions, nearly all of the faculty are academics with doctorates; in an earlier era, a much larger percentage were former business executives. While this trend has had many benefits, it also raises a challenge: business schools must now actively cut paths to the business community to replace the ones that were previously beaten by the to-and-fro passage of many feet.

Of course there are many forms of engagement that are still organic to the enterprise, such as writing case studies or conducting research on an industry. But these are insufficient for the purpose we are advocating, because they tend to put academics in the active position of observers, while executives and companies become passive objects of study. What is needed is a more equal dialogue where both parties listen to each other and collaborate on identifying the current challenges facing business, developing hypotheses about current trends and their drivers, conducting research that tests proposed solutions, and communicating the best thinking that results in a broad range of stakeholders.

This can happen across a range of specialties. There are many opportunities for faculty in accounting and finance, for instance, to work with business in analyzing what went wrong in this global financial crisis. What changes should we make both in business schools and businesses to avoid future Madoff scandals?

At the same time, business-ethics educators can help to transform business schools in the Age of Madoff. As Joshua Margolis of the Harvard Business School contends, "Business ethics can serve as a beacon of purpose, articulating and defending higher aspirations—the purposes business can serve and the roles managers can play."

To fulfill their responsibilities, business ethicists need to broaden their role by undertaking four key actions.

1. *Emphasize competence as much as character.*

Ethics all too often is treated as being purely a matter of "character" (which leads to the saints and sinners mentality mentioned earlier), whereas other fields in business schools are about "competence." This shortchanges ethics, which involves conceptual and behavioral competence as much as character.

One such competence is the capacity to anticipate the effects the various choices one make as a manager or employee can have on a company's stakeholders and the potential moral hazards involved. Before students are given the keys to large financial vehicles, it is now clear that they need to learn the basic rules of the road, recognize the hazards they will face while driving, and identify for themselves the destination they feel is worth driving toward.

An effort that emphasizes competency is Giving Voice to Values (GVV)—a research, dialogue, and curriculum development program created by the Aspen Institute for Business in Society in partnership with the Yale School of Management. The purpose of GVV is to identify and analyze ways that business practitioners can voice and implement their values in the face of countervailing pressures and to share this capacity through a skill-building curriculum and publications.

The program assumes that speaking up is as much about competence as it is about character. Many people have asked, "Why didn't someone at Enron speak up sooner?" The GVV program, which is informed by research, says that part of the explanation is that Enron employees lacked the competence to speak up—many executives simply do not get enough coaching and practice in this area. Distinctive features of the program—according to Mary Gentile, the GVV director—include "a focus on positive examples; an emphasis upon the importance of finding alignment between the individual's and the organization's sense of purpose; and opportunities to practice responses and to provide and receive peer feedback."

2. *Engage contemporary issues.*

Anne Mulcahy, chairman of Xerox, recently said, "In the midst of this turmoil, the global economic crisis presents a unique opportunity for leaders to step forward and make business better. This is an opportunity we must seize." The global economic crisis has also opened up new opportunities to teach ethics, since students are more likely to recognize the topic's saliency. Laura Hartman of DePaul University for instance, told us that in her classes, "students pay attention these days because they know it's important."

In light of this change, business ethicists need to stop fighting for legitimacy within the social sciences and move their focus to external states of affairs. To be an important voice—whether as a critic or as a problem-solver—demands that business-ethics educators remain engaged with real-time issues, reflecting constantly on the cogency of their body of knowledge to current events in order to shorten the "time-to-market" for their ideas.

3. *Incorporate systems thinking.*

Many of the past scandals involved business people who were engaged in illegal and unethical activities, and prosecutions and prison sentences were thought to be the remedy. During the present crisis, much of what caused the harm may be legal; however, it was overly risky and, in many cases, not conducive to creating long-term value. In that sense, it was unethical. This emphasizes the need for a more systemic understanding of how businesses do or do not function ethically.

And that systematic understanding is increasingly needed worldwide. As Dean Krehmeyer, executive director of the Business Roundtable Institute for Corporate Ethics, says, "This crisis shows that globally, businesses and industries are more interconnected than ever—and, that risk management, particularly with respect to systemic risk, is not ahead of the curve like we thought just a short time ago."

Business ethics educators are well positioned to lead a discussion with students about how to identify and minimize systemic risks. To do so effectively, however, will demand deeper engagement with other disciplines within and outside of business schools (e.g., law, psychology, and politics).

4. *Champion an enterprise approach to ethics.*

This deeper engagement with other disciplines is also critical for leading an enterprise approach to ethics at business schools and for integrating ethics throughout a curriculum. An enterprise approach takes a broad view of how courses, the curriculum, and the culture of the school contribute to ethics. It identifies leverage points in the culture of the school so that business ethics can become engrained at the deep level of core values.

Some of the larger business schools have made substantial progress in this area. For example, at the University of Virginia's Darden School of Business, ethics is seen as one of the foundational disciplines of business, on a par with marketing, finance, entrepreneurship, operations, and the like. Not only does Darden require an interdisciplinary course taught by faculty trained in ethics and functionally trained faculty, but it offers a set of electives—from seminars in ethical theory to courses using literature and the fine arts, such as drama—that address ethical issues. Affiliated faculty in the business disciplines also routinely include business ethics in their courses.

The situation is similar at Harvard Business School. According to professor Thomas Piper, one of the architects of the business ethics program there, "Our emphasis is on a three-lens model: an economic imperative; a legal/regulatory imperative that connects to public policy concerns; and an ethical imperative. We believe that each lens is very important; no one lens is sufficient." This approach depends upon close collaboration among faculty trained in a variety of disciplines: law, ethics, marketing, organizational behavior, economics, strategy, and general management.

A similar strategy is employed at the Stern School at New York University. According to Professor Edwin Hartman, NYU has created a faculty symposium to "build and sustain a vibrant community of faculty across disciplines who conduct research in business ethics and related fields." Meanwhile, NYU faculty across disciplines—including economics, philosophy, psychology, neural science, sociology, and law—have been charged with incorporating business ethics into their teaching with both undergraduate and business school students.

Something like this new story of business that we have described is beginning to be told in these and other major business schools, and the current financial crisis makes the telling much easier. Of course, this new story is not without controversy. Lively debates continue about the relative importance of stockholders versus stakeholders, the role of financial markets, the importance of the business disciplines compared to getting students to think across disciplines, the relevance of standard accounts of ethical theory, the role of government in the marketplace, the effects of globalization, and countless other topics. And, it is important to note, this type of change is far more difficult for schools with fewer resources.

Despite these caveats, the financial crisis has created a window of opportunity for both businesses and business schools to re-imagine how they might engrain ethics into the DNA of their activities so as to make business, and the world that business affects, better. Bolting ethics onto business and business education has been an extremely attractive route, because it is a much easier one to follow than the kinds of changes we have suggested. But, like many other short-cuts, this path moves us further from our destination by creating an illusion of progress.

The real journey begins when we actively engage, as live issues, the presuppositions about markets, economic models, and human nature that are foundational to prevailing beliefs about business. And for those who teach business ethics, it begins when we stop fighting for legitimacy and start affecting business in the positive ways that our knowledge empowers us to do.

Are we making progress? While it is extremely difficult to document, we do notice some encouraging signs. There is more interest by business schools and their leadership in talking about ethics as a fundamental part of business. And there is a definite uptick in the interest of students in looking at "social entrepreneurship ventures," sustainability, and corporate responsibility.

Change is difficult and often slow. The best way to measure progress will be an increasing number of students who choose to enroll in business school because they aspire to change the world. And those students, in demanding that we help them develop the competencies to make those changes, will in turn help us improve how we teach the next generation of business leaders.

Resources

All quotations are from these sources or from personal interviews and emails done exclusively for this article.

- *Shaping Tomorrow's Business Leaders: Principles and Practices for a Model Business Ethics Program*. Business Roundtable Institute for Corporate Ethics. Retrieved July 7, 2009 from **http://www.corporate-ethics.org/pdf/mbep.pdf**
- *The Dynamics of Public Trust in Business—Emerging Opportunities for Leaders*. Business Roundtable Institute for Corporate Ethics. Retrieved July 7, 2009 from **http://www.corporate-ethics.org/pdf/public_trust_in_business.pdf**
- Bruner, Robert, "Professor Says Business Schools and Students Can Take Away Lessons From Financial Crisis," *Wall Street Journal*, August 20, 2009.
- De George, Richard T., "The Status of Business Ethics: Past and Future," *Journal of Business Ethics*, 6 (1987), pp. 201-211.
- De George, Richard T., *A History of Business Ethics*. Retrieved July 7, 2009 from http://www.scu.edu/ethics/practicing/focusareas/business/conference/presentations/business-ethics-history.html. Retrieved on July 24, 2009.
- Eels, Richard and Clarence Walton, *Conceptual Foundations of Business*. Richard D. Irwin, Inc., Homeward, IL, 1961.
- Frederick, William C., *Corporation, Be Good*. Pittsburgh: Dog Ear Publishing, 2006.

R. Edward Freeman is University Professor, Elis and Signe Olsson Professor, Academic Director of the Institute for Business in Society, and Senior Fellow of the Olsson Center for Applied Ethics at the University of Virginia's Darden School of Business.

Lisa Stewart is Chief Administrative Officer, Kavli Institute for Theoretical Physics, University of California, Santa Barbara.

Brian Moriarty is Assistant Professor at the University of Virginia's Darden School of Business.

Chapter 33
The Impossibility of the Separation Thesis: A Response to Joakim Sandberg

Jared D. Harris and R. Edward Freeman

Joakim Sandberg's (2008) recent article engages in a critical conversation about what is known as the "separation thesis," the view that matters of economic value are somehow distinct from ethical values. Sandberg highlights how some (e.g., Freeman 1994; Wicks 1996) have called for the rejection of this view, and while highlighting that an alternative, integrative view of business and ethics has consequently largely been embraced by business ethicists, he implores us to attempt to understand the separation thesis, and our rejection of it, "more exactly" (Sandberg 2008: 213).

Sandberg suggests a number of different ways we might interpret the separation thesis. We shall leave a detailed discussion of the first eight interpretations of the separation thesis to others, and instead concentrate on the ninth formulation discussed by Sandberg:

> ST9: There is a genuine difference between matters of business and matters of ethics, at least insofar as there is a genuine difference between descriptive and normative matters. (Sandberg 2008: 227)

This is *precisely* the implied articulation of the separation thesis considered problematic by those who have called for its rejection. Curiously, Sandberg (2008: 227, 230) on one hand facilely accepts this formulation of the separation thesis on the grounds that it makes "intuitive sense" to distinguish between facts and values; yet on the other hand, he acknowledges in his notes that values are inextricably embedded in "social contexts" and facts. In this sense, his call to understand with

Originally published in: *Business Ethics Quarterly*, 18(4), 541–548 © Cambridge University Press, 2008
Reprint by Springer, https://doi.org/10.5840/beq200818437

J. D. Harris · R. E. Freeman (✉)
University of Virginia, Charlottesville, VA, USA

© Springer Nature Switzerland AG 2023

S. D. Dmytriyev, R. E. Freeman (eds.), *R. Edward Freeman's Selected Works on Stakeholder Theory and Business Ethics*, Issues in Business Ethics 53,
https://doi.org/10.1007/978-3-031-04564-6_33

more precision the separation thesis and its rejection appears to be answered only by additional lack of clarity.

However, Sandberg is correct in his acknowledgement that the separation thesis is based on—and merely a manifestation of—the underlying, more fundamental problem of the fact-value dichotomy. Once we understand the separation thesis as an attempt to dichotomize "business facts" and "moral values"—rather than some other more limited interpretation—we can formulate an answer to Sandberg's call for more clarity about the reasons we should reject this distinction.

As Sandberg notes, values are "embedded in social contexts from which they cannot be removed" (Sandberg 2008: 230). We cannot single out particular "facts" from their underlying narratives. As Searle (1964: 52, 54) pointed out, the "inclination to accept a rigid distinction between 'is' and 'ought,' between descriptive and evaluative, rests on a certain picture of the way words relate to the world" that ignores contextual notions such as "commitment, responsibility, and obligation." In other words, statements about the external world do not "face the tribunal of sense experience" alone (Quine 1951: 38). James and Dewey, Putnam and Rawls, Rorty and Goodman have all made similar arguments. Philosophers of science such as Kuhn (1962) and Feyerabend (1975) have highlighted the challenge this poses to the very concept of scientific inquiry as being solely descriptive and objective.

Similarly, in *The Collapse of the Fact-Value Dichotomy*, Hilary Putnam (2002: 27, 61–62) suggests that facts and values are deeply "entangled," and as such, "the picture of our language in which nothing can be *both* a fact *and* value-laden is wholly inadequate." As an illustration, Putnam's analysis of the word "cruel" as being both descriptive and value-laden illustrates how a great deal of language works, and demonstrates the limitation of employing a sterile, objectivist view of language and meaning. Putnam then analyzes the work of Amartya Sen's *On Ethics and Economics,* in which Sen (1987) specifically suggests that we have forgotten that economics is inherently entangled with matters of ethics, and argues that the false dichotomization of the two has impoverished discipline-based analysis in both economics and ethics. Sen's observation is a clearly articulated rejection of what has come to be known as the separation thesis.

Yet the entanglement of facts and values has implications beyond our mere conception of "business" language as being both normative and descriptive. Such entangled concepts apply directly to actual practices, which always embody both facts ("business" considerations) and values ("ethical" considerations). Consider, for example, the arrangement by which a business firm provides employment to a particular individual; has the corporation provided economic value, or moral value? How can such things be disentangled?[1] Along these lines, any economic assertion is ultimately both descriptive and value-laden, as well. Furthermore—and ironically—any *explicit* contention that commerce and morals involve mutually exclusive

[1] And if they cannot be disentangled, what are the implications for ideas such as "corporate social responsibility" and "triple bottom line"?

considerations (e.g., Friedman 1970; Jensen 2002) is *also* both descriptive and value laden.

Hence, while agreeing that dichotomizing business and ethics is a fool's errand that is not useful, helpful, or even meaningful,[2] we argue for a more powerful reason to explicitly reject the separation thesis: separating economic considerations and ethical considerations is impossible. As such, attempts to separate the two serve simply to obscure a particular set of values—but such values exist nonetheless, lurking beneath the surface. Embracing the separation thesis, then, does not in fact disentangle facts and values, but merely appears to do so, while obscuring and embedding a particular set of values that privileges certain considerations and dismisses others. Therefore the problem with the separation thesis is not so much that it actually separates business and ethics—an impossible task—but that it purports moral neutrality while surreptitiously encapsulating certain ethical values and assertions. But why exactly is this problematic?

Ghoshal (2005) and others (e.g., Ferraro et al. 2005; Frank et al. 1993), in showing that we enact the very theories of social science that we propose—and therefore demonstrating that the moral consequences are indistinguishable from the theories themselves—highlight the danger *of attempting* to separate business from ethics. When theorists suggest and managers enact an approach that views "business" decisions as if there are no moral consequences to them (e.g., describing unfettered profit maximization as the "single objective function" of business firms), this inculcates a societal narrative about business and ethics in which ethical considerations are no less real, but merely devalued and denatured. Is the view that owners of firms and their employees are one-dimensional maximizers of self-interest with convex utility functions for monetary wealth simply a matter of fact? Is the assumption that incentives effectively ameliorate agency conflicts unassailable? While some research (e.g., Frank 1988; Harris and Bromiley 2007) calls into question the *prima facie* descriptive accuracy of such assertions, important implications also arise from the assumptions about morality that are embedded within such statements, and their reifying influence on managerial behavior and social norms. That business decisions have moral content is inescapable; pretending the two are divisible at best obscures important considerations and at worst paradoxically encourages a particular set of ethical norms that may be unintended.

However, it is also important to note that while economists such as McCloskey (1998) have convincingly revealed the value-laden nature of economic concepts once the veneer of faux objectivity is peeled away, the mirror image of this problem—that discussions about ethics often lack rich descriptions about how we create value and trade with each other—is an equally compelling reason to reject the separation thesis. In other words, the separation thesis cuts both ways. As Sen (1987: 9) suggests, "the distance that has grown between economics and ethics" has also been

[2] For several articulations of these arguments, see Freeman 1994 and 2000; Freeman et al. 2004; and Wicks 1996 as examples.

"unfortunate for the latter."[3] In a recent analysis of Wittgenstein and Davidson, Alice Crary (2007) extends this point to suggest that focusing solely on moral judgment separates ethics into a stylized narrative that doesn't fully capture or address the kinds of problems that we try to solve with the use of value-laden language in the first place. Questions of ethics in business are overly stylized unless considered within the rich context of value creation, organizational dynamics, and stakeholder pressures.

This acknowledges that business contexts, like all social situations, may inherently contain conflicting moral considerations. Yet organization theorists (Cameron et al. 2006; Martin 2007; Quinn and Rohrbaugh 1983) have argued that the type of "both/and" thinking required to embrace and deal with multiple objectives can ultimately serve as a source of creativity, tension, and innovation that drives productive organizational action. In practice, addressing multiple imperatives and moral contradictions involves coalition building and collective action, and scholars have begun to shed light on how these processes unfold in organizations, both generally speaking (Hargrave and Van de Ven 2006; Seo and Creed 2002) and specifically with respect to considerations of ethics (e.g., Hargrave 2008; Nielsen 1996).

As such, it may well be that the academic debate over the separation thesis and its rejection has outlived its usefulness as an argumentative device in management theory. Indeed a number of business scholars have entangled facts and values, business and ethics, descriptive and normative, all in a body of groundbreaking work that investigates the complexity of organizational life.

Consider Donaldson and Preston's (1995) seminal stakeholder paper. Although on the surface their work appears to be concerned with distinguishing stakeholder theory's normative, descriptive, and instrumental elements (hence potentially *championing* the separation idea) they ultimately categorize all of these elements as "managerial in the broad sense of that term" (Donaldson and Preston 1995: 67), and therefore "intimately connected with the practice of business" (Freeman 2000: 173). Margolis, Walsh, and colleagues (Margolis and Walsh 2003; Walsh et al. 2003, 2006) pick up on this theme and tackle head-on the interconnected nature of business and ethics and its implications for the way we conduct management research, suggesting that tensions arising from economic and ethical demands ought to be explored, rather than simplified—an idea also championed by other theorists (e.g., Bartunek 2002; Hinings and Greenwood 2002).

A healthy list of research exemplifies this integrative approach. For example, Treviño, Weaver and Cochran (Treviño and Weaver 2001; Weaver and Treviño 1999; Weaver et al. 1999a, b) explore the inner workings of organizational ethics initiatives and their impact on employee attitudes and compliance. Ashforth and coauthors (Ashforth and Kreiner 1999; Ashforth et al. 2007; Kreiner et al. 2006) explore how organizational stakeholders form meaning and identity in the context

[3] Sen (1987: 10) suggests that not only should economics play a direct role in understanding better the nature of some of our ethical questions, but that some of the methodological insights used in economics in "tackling problems of interdependence" can be substantially useful in dealing with complex ethical problems.

of occupations that carry social or ethical stigma. Wade and colleagues (2006a, b) investigate the determinants of executive pay and its implications for perceptions of fairness among organizational employees. King and Lenox (King 2007; King and Lenox 2000, 2001a, b, 2002) integrate considerations of environmental sustainability and competitive strategy. Reynolds (2006a, b) explores the role of morality in decision making. Rodriguez and colleagues (Rodriguez et al. 2005, 2006; Uhlenbruck et al. 2006) analyze the impact of corruption on international business practices. The list goes on. There is growing interest, for example, in understanding interorganizational trust, bottom of the pyramid approaches to alleviating poverty, morality of working conditions within the supply chain, entrepreneurial solutions to environmental challenges, and a host of other topics that explicitly embody "entangled" business facts and moral values.

As such, the body of research that in some way, shape, or form dispenses with the separation thesis is not endless, but it is encouragingly substantial. And momentum around a vigorous, synthetic research agenda in business and ethics, at least by our anecdotal observation and reading of the literature, appears to be increasing and widespread. It is evidenced by both an increasingly nuanced treatment of moral values within organization theory and strategic management, and also the heightened presence of organizational theories and methods within business ethics. This appears to be accompanied by a certain amount of self-awareness, as well; Hartman (2008) compellingly argued in his recent Society for Business Ethics Presidential Address that such integrative approaches to business ethics should be embraced and pursued with vigor. In our view, Sandberg (2008) is therefore accurate in his observation that many researchers have rejected the separation thesis in favor of pursuing integrative questions; hopefully we have responded constructively to his call for additional clarity on the subject. In rooting the separation thesis in the fact-value dichotomy, we have attempted to (1) explain what we understand the separation thesis to be, (2) show how distinguishing between business "facts" and moral "values" is impossible, (3) clarify why embracing the separation thesis, despite its impossibility, is nevertheless problematic, and (4) highlight examples of work that in various ways effectively bridge the fact-value divide.

Granted, sometimes in research it may be useful to emphasize part of a narrative in the foreground, and shift other issues to the background. This is true in economics where many interesting ideas have been worked out by holding human complexity to a minimum. Likewise in ethics it is sometimes useful to focus solely on the complexity around difficult ideas like "fairness" or "responsibility" without having to discuss the complexities of global value creation and trade. It is impractical to require every piece of research that is *primarily* focused on exploring economic mechanisms to give a full account of related ethical considerations, and vice versa. Nevertheless it is critically important to recognize that such individual pieces of the puzzle, each with a particular primary focus, are by definition underspecified, and it is a mistake to infer from them that business and ethics can ever be meaningfully separated. Indeed we have argued that attempting to do so is both theoretically and practically detrimental. Yet just as Quine has pointed out that individual statements about the world do not stand on their own, neither does each individual paper or research project. This points the way to the vigorous development of business ethics as a discipline that, on the whole, expands our understanding of business by putting ideas about economic opportunities and constraints together with ideas about moral responsibility. There is much work to be done.

References

Ashforth, B.E., and G.E. Kreiner. 1999. 'How Can You Do It?': Dirty Work and the Challenge of Constructing a Positive Identity. *Academy of Management Review* 24 (3): 413–434.

Ashforth, B.E., G.E. Kreiner, M.A. Clark, and M. Fugate. 2007. Normalizing Dirty Work: Managerial Tactics for Countering Occupational Taint. *Academy of Management Journal* 50 (1): 149–174.

Bartunek, J.M. 2002. The Proper Place of Organizational Scholarship: A Comment on Hinings and Greenwood. *Administrative Science Quarterly* 47 (3): 422–427.

Cameron, K.S., R.E. Quinn, J. DeGraff, and A.V. Thakor. 2006. *Competing Values Leadership: Creating Value in Organizations*. Cheltenham: Edward Elgar.

Crary, A. 2007. *Beyond Moral Judgment*. Cambridge, MA: Harvard University Press.

Donaldson, T., and L. Preston. 1995. The Stakeholder Theory of the Corporation: Concepts, Evidence and Implications. *Academy of Management Review* 20 (1): 65–91.

Ferraro, E., J. Pfeffer, and R.I. Sutton. 2005. Economics Language and Assumptions: How Theories Can Become Self-Eulfilling. *Academy of Management Review* 30 (1): 8–24.

Feyerabend, P. 1975. *Against Method*. London: Verso.

Frank, R.H. 1988. *Passions within Reason*. New York: W. W. Norton & Company.

Frank, R.H., T. Gilovich, and D. Regan. 1993. Does Studying Economics Inhibit Cooperation? *Journal of Economic Perspectives* 7 (Spring): 159–171.

Freeman, R.E. 1994. The Politics of Stakeholder Theory: Some Future Directions. *Business Ethics Quarterly* 4 (4): 409–421.

———. 2000. Business Ethics at the Milennium. *Business Ethics Quarterly* 10 (1): 169–180.

Freeman, R.E., A.C. Wicks, and B. Parmar. 2004. Stakeholder Theory and 'The Corporate Objective Revisited. *Organization Science* 15 (3): 364–369.

Friedman, M. 1970. The Social Responsibility of Business Is to Increase Its Profits. *The New York Times Magazine* 33: 122–126.

Ghoshal, S. 2005. Bad Management Theories Are Destroying Good Management Practices. *Academy of Management Learning & Education* 4 (1): 75–91.

Hargrave, T.J. 2008. Moral Imagination, Collective Action, and the Achievement of Moral Outcomes. *Business Ethics Quarterly* 19: 87.

Hargrave, T.J., and H. Van de Ven. 2006. A Collective Action Model of Institutional Innovation. *Academy of Management Review* 31 (4): 864–888.

Harris, J., and P. Bromiley. 2007. Incentives to Cheat: The Influence of Executive Compensation and Firm Performance on Financial Misrepresentation. *Organization Science* 18 (3): 350–367.

Hartman, E.M. 2008. Reconciliation in Business Ethics: Some Advice from Aristotle. *Business Ethics Quarterly* 18 (2): 253–265.

Hinings, C.R., and R. Greenwood. 2002. Disconnects and Consequences in Organization Theory. *Administrative Science Quarterly* 47 (3): 411–421.

Jensen, M.C. 2002. Value Maximization, Stakeholder Theory, and the Corporate Objective Function. *Business Ethics Quarterly* 12 (2): 235–256.

King, A.A. 2007. Cooperation between Corporations and Environmental Groups: A Transaction Cost Perspective. *Academy of Management Review* 32 (3): 889–900.

King, A.A., and M.J. Lenox. 2000. Industry Self-Regulation without Sanctions: The Chemical Industry's Responsible Care Program. *Academy of Management Journal* 43 (4): 698–716.

———. 2001a. Does It Really Pay to Be Green? An Empirical Study of Firm Environmental and Financial Performance. *Journal of Industrial Ecology* 5 (1): 105–116.

———. 2001b. Lean and Green? Exploring the Spillovers from Lean Production to Environmental Performance. *Production and Operations Management* 10 (3): 244–256.

———. 2002. Exploring the Locus of Profitable Pollution Reduction. *Management Science* 48 (2): 289–299.

Kreiner, G.E., B.E. Ashforth, and D.M. Sluss. 2006. Identity Dynamics in Occupational Dirty Work: Integrating Social Identity and System Justification Perspectives. *Organization Science* 17 (5): 619–636.

Kuhn, T.S. 1962. *The Structure of Scientific Revolutions*. Chicago: University of Chicago Press.

Margolis, J.D., and J.P. Walsh. 2003. Misery Loves Companies: Rethinking Social Initiatives by Business. *Administrative Science Quarterly* 48: 268–305.

Martin, R.L. 2007. *The Opposable Mind: How Successful Leaders Win through Integrative Thinking*. Cambridge, MA: Harvard Business School Press.

McCloskey, D.N. 1998. *The Rhetoric of Economics*. Madison: University of Wisconsin Press.

Nielsen, R.P. 1996. *The Politics of Ethics: Methods for Acting, Learning, and Sometimes Fighting, with Others in Addressing Ethics Problems in Organizational Life*. New York: Oxford University Press.

Putnam, H. 2002. *The Collapse of the Fact/Value Dichotomy*. Cambridge, MA: Harvard University Press.

Quine, W.V.O. 1951. Two Dogmas of Empiricism. *The Philosophical Review* 60 (1): 20–43.

Quinn, R.E., and J. Rohrbaugh. 1983. A Spatial Model of Effectiveness Criteria: Towards a Competing Values Approach to Organizational Analysis. *Management Science* 29 (3): 363–377.

Reynolds, S. 2006a. Moral Awareness and Ethical Predispositions: Investigating the Role of Individual Differences in the Recognition of Moral Issues. *Journal of Applied Psychology* 91 (1): 233–243.

———. 2006b. A Neurocognitive Model of the Ethical Decision-Making Process: Implications for Study and Practice. *Journal of Applied Psychology* 91 (4): 737–748.

Rodriguez, P., K. Uhlenbruck, and L. Eden. 2005. Government Corruption and the Entry Strategies of Multinationals. *Academy of Management Review* 30 (2): 383–396.

Rodriguez, P., D.S. Siegel, A. Hillman, and L. Eden. 2006. Three Lenses on the Multinational Enterprise: Politics, Corruption, and Corporate Social Responsibility. *Journal of International Business Studies* 37 (6): 733–736.

Sandberg, J. 2008. Understanding the Separation Thesis. *Business Ethics Quarterly* 18 (2): 213–232.

Searle, J.R. 1964. How to Derive 'Ought' From 'Is'. *The Philosophical Review* 73 (1): 43–58.

Sen, A. 1987. *On Ethics and Economics*. Maiden, MA: Blackwell Publishing.

Seo, M., and W.E.D. Creed. 2002. Institutional Contradictions, Praxis, and Institutional Change: A Dialectical Perspective. *Academy of Management Review* 27 (2): 222–227.

Treviño, L.K., and G.R. Weaver. 2001. Organizational Justice and Ethics Program 'Follow-Through': Influences on Employees' Harmful and Helpful Behavior. *Business Ethics Quarterly* 11 (4): 651–671.

Uhlenbruck, K., P. Rodriguez, J. Doh, and L. Eden. 2006. The Impact of Corruption on Entry Strategy: Evidence from Telecommunication Projects in Emerging Economies. *Organization Science* 17 (3): 402–414.

Wade, J.B., C.A. O'Reilly, and T.G. Pollock. 2006a. Overpaid CEOs and Underpaid Managers: Fairness and Executive Compensation. *Organization Science* 17 (5): 527–524.

Wade, J.B., J.E. Porac, T.G. Pollock, and S.D. Graffin. 2006b. The Burden of Celebrity: The Impact of CEO Certification Contests on CEO Pay and Performance. *Academy of Management Journal* 49 (4): 643–660.

Walsh, J.P., K. Weber, and J.D. Margolis. 2003. Social Issues in Management: Our Lost Cause Found. *Journal of Management* 29 (6): 859–881.

Walsh, J.P., A.D. Meyer, and C.B. Schoonhoven. 2006. A Future for Organization Theory: Living In and Living With Changing Organizations. *Organization Science* 17 (5): 657–671.

Weaver, G.R., and L.K. Treviño. 1999. Compliance and Values Oriented Ethics Programs: Influences on Employees' Attitudes and Behavior. *Business Ethics Quarterly* 9 (2): 315–335.
Weaver, G.R., L.K. Treviño, and P.L. Cochran. 1999a. Corporate Ethics Programs as Control Systems: Influences of Executive Commitment and Environmental Factors. *Academy of Management Journal* 42 (1): 41–58.
———. 1999b. Integrated and Decoupled Corporate Social Performance: Management Commitments, External Pressures, and Corporate Ethics Practices. *Academy of Management Journal* 42 (5): 539–552.
Wicks, A.C. 1996. Overcoming the Separation Thesis: The Need for a Reconsideration of Business and Society Research. *Business & Society* 35 (1): 89–118.

Jared D. Harris is the Samuel L. Slover Research Associate Professor at the University of Virginia's Darden School of Business.

R. Edward Freeman is University Professor, Elis and Signe Olsson Professor, Academic Director of the Institute for Business in Society, and Senior Fellow of the Olsson Center for Applied Ethics at the University of Virginia's Darden School of Business.

Chapter 34
Creating Ties That Bind

R. Edward Freeman and Jared D. Harris

1 Two Problems in Business Ethics

In their magisterial work, *Ties That Bind: A Social Contracts Approach to Business Ethics*, Donaldson and Dunfee (1999) attempt to square a view of ethical theory built on the idea of social contracts with much of the existing literature in business ethics. In doing so, they hope to place business ethics on firmer philosophical footing. In this article we suggest that, their integrative social contracts theory articulates an important and useful set of ideas ripe for further development and extension, whereas it does not fully escape the problems they attempt to solve. We have much sympathy for the idea of a contractual account of business ethics, and therefore, we seek to extend their ideas with an approach that views the moral "ties that bind" from more of a process perspective, as a conversation among engaged moral actors. We outline our approach here in honor of Tom Dunfee, who was an important conversational partner over many years.[1]

Donaldson and Dunfee are concerned with two problems that they see in the business ethics literature. The first is the lack of integration between normative and empirical approaches to the field; let us call this the problem of integration. The second is what they call, following Nagel (1986), "the view from nowhere."

Originally published in: *J Bus Ethics,* 88, 685–692 © Springer, 2009
Reprint by Springer, DOI 10.1007/s10551-009-0333-4

[1] Though this essay is in honor of Tom Dunfee, Tom Donaldson is one of our important conversational partners as well, and we thank him for his helpful feedback on an earlier draft.

R. E. Freeman (✉) · J.D. Harris
University of Virginia, Charlottesville, VA, USA
e-mail: harrisj@darden.virginia.edu; FreemanE@darden.virginia.edu

© Springer Nature Switzerland AG 2023

S. D. Dmytriyev, R. E. Freeman (eds.), *R. Edward Freeman's Selected Works on Stakeholder Theory and Business Ethics*, Issues in Business Ethics 53,
https://doi.org/10.1007/978-3-031-04564-6_34

1.1 The Problem of Integration

As business ethics has evolved as a discipline, it has attracted a more diverse group of scholars from a number of disciplines; an increasing number of social scientists employing empirical research methods have joined the conceptual analyses of philosophers who were the early founders of the discipline. This brings a certain pluralism to the field; while many management scholars are concerned with collecting better data and making more accurate predictions, philosophers have continued their analysis from the viewpoint, largely, of analytical ethics. Philosophical business ethics is grounded in the normative ethical analysis of the relevance of principles, consequences, or character, with management scholars attending to the related descriptive and predictive investigations of actual organizational behavior. And as we have recently pointed out, an increasing number of scholars study *both* philosophical/normative texts *and* managerial/empirical texts (Harris and Freeman 2008). Donaldson and Dunfee's work was an early clarion call to think about business ethics in this kind of integrated way; they strive to put forward a theory that shows how such integration can be achieved.

1.2 The View from Nowhere

The second problem is related to the problem of integration – the separation of normative and descriptive – because Donaldson and Dunfee see the pervasive ethical "view from nowhere" as an examination from only an abstract point of view. Particular detail is wiped away to "discover" general philosophical principles or truths. This view from nowhere can be sterile and ultimately impotent since it is overly stylized and does not clearly connect to concrete actions. Due in large part to the efforts of scholars such as Tom Dunfee, there are now more legal scholars involved in the discipline, who are rooted in the details of business law and particular cases; legal analyses tend to dissolve the normative/empirical distinction, viewing the law as at once both normative and empirical. Nevertheless, Donaldson and Dunfee point out that the gap between general principle and concrete advice, especially in normative work in business ethics, is quite wide.

In order to address these problems, Donaldson and Dunfee propose a solution to these problems based on the idea of social contracts. They suggest that the general ideas of "fair contract" and "hypothetical consent" – which have been important pieces of the social contract tradition since Locke – remain as the philosophical keystones, but in a complicated and global business world, they yield only the view from nowhere. They suggest that there are many "hypernorms" which, while ultimately justified by a contractual mechanism such as the one used by John Rawls, do not exist merely as abstract, general principles. Rather, Donaldson and Dunfee suggest that we take seriously the idea that there are local norms and principles which may well apply, especially in business, grounding the norms as something more concrete than generalized rules of thumb. They suggest that we combine the idea of overarching hypernorms (based on macrosocial contracts) with the extant "microsocial contracts" to form what they call integrative social contracts theory (ISCT).

The hypernorms are the focal point of a key tension that ISCT hopes to resolve. Donaldson and Dunfee want to avoid charges of relativism by retaining the philosophical certainty of the big principles, or hypernorms, which we would perhaps agree to behind a veil of ignorance. However, they also want to embody pragmatic relevance in their theory by letting particular communities influence the acceptable norms of behavior. This central tension is governed by one important boundary condition: these local norms can vary and change, provided they inhabit a "moral free space" that does not materially conflict with the hypernorms, which are taken as axiomatic, and look much like the idea of universal human rights (Donaldson and Dunfee 1999, p. 83). In order to assess ISCT's treatment of the two problems that Donaldson and Dunfee identify and set out to solve, let us consider the role of the hypernorms.

2 Hypernorms and Entanglement

There are several different ways to understand Donaldson and Dunfee's use of hypernorms. On the one hand, we might understand ISCT's hypernorms as strongly axiomatic, privileged, and ultimately a sterile "view from nowhere" smuggled into the theory. If this is accurate, then it is not clear that ISCT gets us very far along the road to integration; rather, the normative is simply privileged, and hypernorms become the bedrock. While there is certainly a spirit of integration from which the book is written, there is simply not much actual integration if, in fact, the normative hypernorms trump everything else.

On the other hand, we might instead understand the hypernorms as fairly minimalist floor values or guidelines that most of us can agree on (e.g., all else equal, we should not harm innocents). In this sense, the hypernorms are not so much privileged, monolithic, or sovereign, but in the end this also means that they are ultimately not very specific or directive, either. Therefore, hypernorms might always be present in the background, but consist mainly of bromides that could be interpreted a number of different ways. In either case, this implies that ISCT does not really resolve the integration problem or the view from nowhere problem–at least, not yet.

We would like to suggest that Donaldson and Dunfee point us in a fruitful direction, but that the promise of their project is not yet fully realized. Elsewhere, we have argued that separating the normative from the descriptive is actually impossible, because one cannot single out specific "facts" from their underlying narratives (Harris and Freeman 2008). This means that ethics are embedded in the generalized meaning of words and concepts (e.g., Searle 1964), as well as in the actual practice of economic action and transaction (e.g., Granovetter 1985). And while Donaldson and Dunfee outline a theoretical framework by which integration is advocated, they stop short of actually showing how such integration would be achieved. In other words, we suggest that although Donaldson and Dunfee are, in fact, trying to escape the separation of the normative and the descriptive, they do not go quite far enough.

This becomes clear when they describe their project as focusing on "economic matters" (p. 27). The social contract as detailed in their four principles refers to "local economic communities" and curiously to "economic ethics." In their attempt to ground economic differences in hypernorms and macrosocial contracts, they have, perhaps, unwillingly accepted the conceptual separation of economics from ethics only to try and put them together again. A different approach is to eschew the separation in the first place. There is no separable "economic ethics," and conversely, there is no ethics without considering how human beings create value and trade with each other. If they are not separate, then they will have no need to be integrated.

So what is the problem here? We conceive of it this way: within the ISCT framework, all decisions in moral free space have to be consistent with hypernorms, and because hypernorms give only the most generic guidance, there will always be an argument about whether particular microsocial contracts are consistent with hypernorms. If ISCT is to achieve its promise, then, we suggest that the most important consideration here is not necessarily the actual hypernorm, or the particulars of the microsocial contract, but rather the ensuing conversation and process that is of necessity about the *connection* between general principles (so-called "thin morality") and specific community and individual practice (so-called "thick morality").

We, therefore, suggest it would be useful for ISCT to explore the very process of entanglement between macro and micro levels of social contracts as it applies to all of our institutions and practices – of which business (or, value creation and trade) is one. In order to do this, we have to pay more attention to the conversations that exist and the resulting processes that occur to connect these levels. Our aim here is to take a process view of ISCT, touching on several mechanisms that may connect microsocial contracts and macro hypernorms.

It is worth noting that our argument here might also be construed more broadly, suggesting that ultimately we need to think about the process by which Kantian, Utilitarian, social contract, or virtuebased issues are, in fact, all connected to custom and practice. Part of understanding these connections involves the pragmatic role of language itself. Traditional ethical theory – and its attendant language – must be employed beyond a simple focus on ethical judgments, decisions that must be made or not made, and defining and applying terms such as "right," "wrong," "good," "bad," etc. to these states of affairs. What gets lost if we stop there – if we succumb to a generalized view from nowhere – is the messy, gritty, real world of moral sense making and problem solving that is so important to our daily lives.

Yet, we also do not want to overstate the extent to which differing uses of language are what separate the gritty from the abstract; in other words, this is not simply (or even primarily) a language issue. In fact, much of analytical business ethics is *not* stylized or sterile, but, in fact, is embroiled in the particulars of practice, illustrated by particular cases, and explicitly applied to real-world problems. Indeed, Donaldson and Dunfee themselves set out to return the messy world of business to business ethics, by focusing on microsocial contracts and the differences among them. This points us in the right direction, but we want to expand their view of moral language and conversation and dialog and engagement to focus on the process by which micro and macro are connected. If Donaldson and Dunfee are correct that we need a more detailed and nuanced view of the connection between general principles – hypernorms – and the microsocial contracts that a variety of communities enact, then we need to pay more attention to the process connecting these levels.

3 Extending ISCT

We shall sketch out an initial, extended view of ISCT that focuses on the process of coming to agreement about the ties that bind, and starts to explore the dynamics of the set of conversations that occurs and/or could occur among community participants deliberating moral problems.[2] There are a number of connected ideas here, which we sketch in the following paragraphs.

3.1 The Primacy of Framing and Sensemaking

The use of language, while not the only thing that is important here, is a good starting point. Language helps us make sense of the world, and is almost always inextricably normative and descriptive. Language is about meaning-making and shared understanding.

For Donaldson and Dunfee's project, the language of social contracts frames the idea of business in a particular way. It at once enables and constrains the discussion about the connection of business to society. Social contracts help us to make sense of business and society in a certain way, and this helps to set the guidelines for what we see as the primary dilemmas of business. Using the language of social contracts is an illustration of exactly what we mean here by the sensemaking role of language. It is often this sensemaking use of language that blurs the distinction between what is empirical and what is normative.

For an example, consider the use of the concept of "stakeholder." On the one hand, we could use the term purely descriptively to single out and depict customers, suppliers, employees, communities, and financiers. On the other hand, we could use the term to make claims about the rights and responsibilities of these groups and of corporations.[3] However, the embedded sensemaking idea in this example is more

[2] The challenge of connecting microsocial contracts with hypernorms bears some similarity to the issue of resolving moral disagreement among actors that lack common normative ground – a common problem. Although discourse ethics (e.g., Habermas 1990) approaches this challenge mainly as a problem of moral justification, others (e.g., Painter-Moreland 2008; Stansbury 2009) have begun to focus on the application of discourse ethics to organizational contexts.

[3] See Donaldson and Preston (1995) for a fuller discussion of these distinctions.

subtle. By invoking the term "stakeholder," we are implicitly giving legitimacy to the groups we reference that way, i.e., they must have a stake. By using the term "stakeholder," we are engaging in an intentional play on the terms "shareholder" and "stockholder," changing only two letters of each word to invoke a more general idea. By merely using the framing of "stakeholder," we invoke both the descriptive and normative at the same time. The two realms are entangled through the use of their connected concepts. Of course, for some purposes, we can focus on either the more descriptive or more normative uses of the term, but that does not avoid their entanglement.

This point is equally clear in the idea of social contracts. Merely by using the framing of "social contracts" and recommending that we make sense of business through "social contracts," Donaldson and Dunfee are having us view business as a normal part of society rather than somehow separate from society. The very discussion of business using the language of social contracts invokes the idea of the integration of business with society. The idea of "contract" is also connected to choice and agreement, and the ideas of reciprocity and performance. What does not seem to fit – and we believe that Donaldson and Dunfee can simply do without it – is the idea of "economic community." "Economic community" seems to signal that we can separate out "economic" from the rest of what counts as "society." And at times (see p. 40ff. and p. 98ff.), "economic community" can refer to a corporation, a state, or just a regular community.[4]

How we make sense of business is logically connected to the concepts that we use in setting up the idea of microsocial contracts. And if microsocial contracts are to emerge and be connected to macrosocial contracts and hypernorms, as Donaldson and Dunfee suggest, there must be a conversation about the role of business. Again, this is not a purely descriptive or purely normative task. Donaldson and Dunfee are clearly committed to the idea that business is a deeply human institution, and even their "economic communities" are moral in nature. Yet, much of the language of business is presented in many contexts as morally neutral or solely descriptive. At the same time, there have been many recent critiques of business in the last several years. However, it is easy to see customers and other stakeholders as moral beings. Marketing consists of promises. Ideas, such as supply chain management, involve getting human beings to cooperate together so that all of them win. These key concepts, which make up particular microsocial contracts, have ethics already built into their fabric. It is not the case that moral free space is morally neutral; rather, it is full of moral complexity.

[4]If Donaldson and Dunfee had simply left out "economic" and made sense of microsocial contracts in terms of "communities", then we would see "business" as a normal part of "community". The problem is that there would not appear to be a special role of "economic ethics" here. That doesn't seem to be a problem unless one is worried about keeping a special role for "business ethics" within either philosophical or management theory.

3.2 *The Role of Background Narratives*

Any particular piece of language brings its background conditions and theories with it; it is always embedded in context.[5] Indeed, Donaldson and Dunfee call up a number of background ideas and images by the very use of the metaphor of the social contract. A social contract is broader than any individual; yet, it implies there is some role for the idea of agreement and choice. However, it is far from clear what happens to an individual who does not agree to the social contract. It is also easy to see how framing business ethics in terms of social contracts can lead to a concern with so-called "saints and sinners." If macrosocial contracts are to be understood in terms of hypernorms, and if the test of any particular locally authentic norm in moral free space is whether it violates any hypernorms, then those communities (or companies) which violate hypernorms will be viewed as sinners, since hypernorms are universal. Those who never violate hypernorms will be the saints. Hence, certain companies are vaunted for their exemplary behavior, and others are vilified for their actions, as judged by some analytical (or perhaps arbitrary) standard.[6] What is missing here is the connection between a local micosocial norm and the hypernorms themselves. What is the conversation that is to be had regarding whether a local norm violates hypernorms?

A controversial example involves local norms regarding child labor. In many local economic communities, child labor is viewed as a necessity. Yet, many have argued that child labor violates a universal human right, especially when it prevents a child from learning basic education, or leads to health and welfare issues which will prevent the child from leading a meaningful life. If the argument stays at this level, there is an inherent conflict between the microsocial contract and the

[5]Language shapes how we see the world. Words shape our actions. Philosophers such as Wittgenstein, Dewey, Quine, Putnam, Davidson, and Rorty have outlined a view of language that relies on understanding the multiple tasks of language, rather than searching for a particular meaning or set of meanings. Wittgenstein's ([1952]2001) idea of "don't ask for the meaning, ask for the use" reminds us that language shapes what we do, how we interact with each other, and how we create meaning for each other. Quine's (1960) view of the indeterminacy of translation illustrates how interpretation and conversation are always an issue. If part of the normative task of scholars is to show us how to live better, then the Duhem metaphor of rebuilding the ship while it remains afloat shows the connection between the micro real world (the ship) and making it better (the rebuilding). Putnam (2002) asks us to see such an entanglement of facts and values (and for Dewey ([1927]1954), means and ends) as a normal part of the way that language works, even in science (Kuhn 1962). Rorty (1981) reminds us that the whole edifice of the "meaning" of the words simply takes the Greek ocular metaphor far past where it is useful. Our "web of belief" (c.f. Quine 1951) is always revisable, and Davidson (2001) reminds us of language's constitutive role in the making of meaning.

[6]This highlights a related problem: what to make of inconsistency of actions? Merck is a saint for donating Mectizan (in perhaps the most famous case in business ethics), but then they are proclaimed sinners for their behavior in the Vioxx case. The recognition of both organizational complexity and the connection between microsocial norms and generalized hypernorms might be useful in making sense of these kinds of actions.

hypernorm. And in Donaldson and Dunfee's scheme, the winner appears to be the hypernorm. However, taking a process view of ISCT, things become more complicated.

Perhaps, it would be more interesting to see if there is not some reconciliation between these two levels of analysis. Child labor *per se* might be understood as something that is, generally speaking, morally problematic. The concept calls up meaning for us precisely because we see it against a background of the innocence of childhood where growth and development are more central and appropriate than the vagaries of hard labor. Yet, the employment of minors might sometimes be necessary to help feed a family, or simply for self-preservation. We might distinguish child labor on a family farm from the grinding exploitation of sweatshops, for example. We could distinguish child labor where at least some education was built into the process, from obvious deprivation and humiliation. We might be able to design a mid-level set of principles that could help determine whether a particular microsocial contract about child labor could be justified.

3.3 Moral Imagination and Reflective Equilibrium

Such a process might include a comparison of one microsocial contract with another, rather than just with the hypernorm. Indeed, this process may well grow to look like what Rawls (1971) calls "reflective equilibrium." We start with some principles (hypernorms) that represent our best thinking so far, but we realize that even these principles are revisable. We then look at particular cases that challenge these principles, such as authentic norms built into existing microsocial contracts. The principles stand in relation to the cases, as the relation of theory to new data. Either the new data confirm or are consistent with the principles, the data are rejected as flawed, or we may need to revise our principles. This process is neither easy nor problem free, but it does describe one useful way to see how microsocial contracts can be connected to macrosocial hypernorms, in a dynamic way over time.

Reflective equilibrium means that we start from where we actually are. It is here that Donaldson and Dunfee remind us that we have both hypernorms (general principles) and a set of practices in moral free space (microsocial contracts), which may reinforce or violate these macro principles. The ISCT contention that there are no morally legitimate microsocial contracts which violate hypernorms would be stronger when the connection between the emergence of practices and hypernorms were more explicit. As Donaldson and Dunfee rightly point out, in moral free space, it is very difficult to determine whether practices violate principles. Much depends on interpretation, and as they argue, tolerance is a real virtue.

3.4 The Role of Disagreement

If conversations about microsocial contracts and their connections to hypernorms are to be meaningful ones inside companies, or economies or communities or societies, then there must be a real role for meaningful difference. This implies that such conversations must already be framed in terms of mutual respect and tolerance. Of course, these ideas have limits, as extreme tolerance may imply extreme disrespect. In the world of business practice, companies are constantly grappling with how to manage in an increasingly diverse economic environment. Large global companies are, perhaps, different from local community-based businesses, and we might expect a different set of norms to emerge. The fact that Donaldson and Dunfee allow for such difference seems to be one of the strengths of their theory. Stronger still would be an understanding of the connection between local community-centric norms and wider global ones.

In many companies, senior management encourages conversations about company values. Sometimes, these company values could be read straight off of Donaldson and Dunfee's list of hypernorms, and rarely will these company values countenance outright violations of hypernorms. However, in these conversations, much is a matter of interpretation. Hypernorms can conflict. Dilemmas abound when the company values are put forward as a complete action-guiding set of norms. However, if we can connect the company values to its underlying purpose and business model, then it is somewhat easier to frame a set of decisions that must be made. For instance, Wal-Mart's obsession with everyday low price is inherently connected with its purpose of trying to bring goods and services to communities at the lowest possible price. Sam Walton believed that people in rural and poorer communities deserved to have better goods available to them. Now, of course Wal-Mart does not always make the correct decision about whether a particular action is consistent with that purpose, but the purpose and resulting values do, in fact, frame the decision in ways that connect it with hypernorms of efficiency and equality. If there is a conflict either between these principles or with another norm that gains salience, such as sustainability, then Wal-Mart must then go through an exercise of reflection and reframing. They must begin to make sense of their moral world utilizing the concept of "sustainability" and hence reframe what they mean by "everyday low price."

The traditional ideas of ethical theory form the basis for understanding how has a conversation about the nature of the conflict between microsocial contracts, and between levels. One way to think about this process is to formulate a set of questions that managers and theorists need to ask when faced with a conflict. At least one of these questions needs to appeal to our imagination and offer alternative framings, meanings, and ways to make sense of the phenomena. This process of using the tension between microsocial contracts and hypernorms as a motivation to engage our moral imagination (Werhane 1999) is not well understood, but we argue that this friction inherently offers an opportunity to see ISCT as a more realistic management process.

4 Conclusions

Seeing ISCT as more of a process points business ethics in a different direction from a more standard application of universal principles, consequences, and character. It points us toward understanding how conversations emerge among human individuals and community. It places center stage the questions of how we are going to live together, and how we are going to create our joint futures, especially focusing on how we create value and trade with each other. It moves us away from an over reliance on making moral judgments as the first virtue of a system of business ethics, and toward a nuanced view of how business activity gets framed, understood, argued about, and enacted. Making sense of business only in economic terms already gives away the game to those who believe that business ethics is an oxymoron, or it means that we need a special section for applied (or "economic") ethics – and as such, raises the question of whether such ethics has depended for its development on the purely economic view of business. In contrast, framing business as just another in a long list of human institutions and activities makes ethics applicable to business. Making sense of value creation and trade as something that we humans have done for millennia then ironically raises the question of the adequacy of a philosophical ethics that has denied business a place in moral theory.

Seeing ethics as more of a conversation among real – rather than hypothetical – moral actors points us away from a business ethics that is enmeshed in judgments about companies and capitalism as saints and sinners. Warren Buffet is reputed to have said that all saints have a past and all sinners have a future. Just as few of our children are either purely saints or purely sinners, so too are our companies. Making sense of business in a "thickly moral" sense builds in the same kind of emotional commitment to being responsible that we hope to transmit to our children. If we begin to see business as one more way that human beings create meaning for ourselves and others, then we will see ethics as concerned with authenticity and change, power and authority, leadership, imagination, and the creation of sustainable value. Donaldson and Dunfee have pointed us down this more nuanced path in business ethics. By paying attention to ethics as a deeply human undertaking, enmeshed in language and dynamic in nature, we can not only better understand the role of business in society but also create even better ties that bind.

References

Davidson, D. 2001. *Inquiries Into Truth and Interpretation*. New York: Oxford University Press.

Dewey, J. [1927]1954. *The Public and Its Problems*. Athens: Ohio University Press.

Donaldson, T., and T. Dunfee. 1999. *Ties That Bind: A Social Contracts Approach to Business Ethics*. Cambridge: Harvard Business School Press.

Donaldson, T., and L. Preston. 1995. The Stakeholder Theory of the Corporation: Concepts, Evidence and Implications. *Academy of Management Review* 20 (1): 65–91.

Granovetter, M. 1985. Economic Action and Social Structure: The Problem of Embeddedness. *The American Journal of Sociology* **91** (3): 481–510.
Habermas, J. 1990. *Moral Consciousness and Communicative Action.* Trans. C. Lenhardt and S. W. Nicholsen. Cambridge: MIT Press.
Harris, J.D., and R.E. Freeman. 2008. The Impossibility of the Separation Thesis. *Business Ethics Quarterly* **18** (4): 541–548.
Kuhn, T.S. 1962. *The Structure of Scientific Revolutions*. Chicago: University of Chicago Press.
Nagel, T. 1986. *The View from Nowhere*. New York: Oxford University Press.
Painter-Moreland, M. 2008. *Business Ethics as Practice: Ethics as the Everyday Business of Business*. Cambridge: Cambridge University Press.
Putnam, H. 2002. *The Collapse of the Fact/Value Dichotomy*. Cambridge: Harvard University Press.
Quine, W.V.O. 1951. Two Dogmas of Empiricism. *The Philosophical Review* 60 (1): 20–43.
———. 1960. *Word and Object*. Cambridge: MIT Press.
Rawls, J. 1971. *A Theory of Justice*. Cambridge: Harvard University Press.
Rorty, R. 1981. *Philosophy and the Mirror of Nature*. Princeton: Princeton University Press.
Searle, J.R. 1964. How to Derive "Ought" From "Is". *The Philosophical Review* 73 (1): 43–58.
Stansbury, J. 2009. Reasoned Moral Agreement: Applying Discourse Ethics within Organizations. *Business Ethics Quarterly* **19** (1): 33–56.
Werhane, P.H. 1999. *Moral Imagination and Management Decision Making*. New York: Oxford University Press.
Wittgenstein, L. [1952]2001. *Philosophical Investigations*. Trans. G. E. M. Anscombe. Oxford: Blackwell.

R. Edward Freeman is University Professor, Elis and Signe Olsson Professor, Academic Director of the Institute for Business in Society, and Senior Fellow of the Olsson Center for Applied Ethics at the University of Virginia's Darden School of Business.

Jared D. Harris is the Samuel L. Slover Research Associate Professor at the University of Virginia's Darden School of Business.

Chapter 35
Remoralizing the Debate

R. Edward Freeman

LOOK AROUND YOU. Ethics are everywhere. A scan of the headlines finds ethics-related issues in politics, business, health care, technology, sports, entertainment, pop culture, religion, media, family, law – you name it. How are we to make sense of all this "buzzing and booming confusion," to quote philosopher William James.

There are many reasons for the surge of ethical issues. First, the world has gotten smaller: Ethical conflicts seem more pronounced because we are more aware than ever of them. Second, the confluence of cultures and worldviews points to the need for a more comprehensive conception of ethics to sustain us in such a world. Third, some are appealing for a return to the notion of civic space, where diverse groups can come together to argue, in the best sense of the word, about the future direction that our culture, economy, government and society should take. Finally, many are calling for greater moral leadership. As a panel at the World Economic Forum in Davos noted, "Business leaders, investors and consumers all face a choice of which moral and ethical postures to adopt in the new world order that follows the global economic meltdown."

All of these reasons are partially correct, and all point to the need for more dialogue that cuts across generations, societies, North and South, East and West, classes, races, religions and other contingencies of life. This dialogue must put ethics at the center. In short, we need to remoralize our conversations about what it means to lead a good life and create good communities.

Originally published in: *IESE Insight*, 4, 6–6 © IESE Business School, 2010
Reprint by Springer

R. E. Freeman (✉)
University of Virginia, Charlottesville, VA, USA
e-mail: FreemanE@darden.virginia.edu

© Springer Nature Switzerland AG 2023

S. D. Dmytriyev, R. E. Freeman (eds.), *R. Edward Freeman's Selected Works on Stakeholder Theory and Business Ethics*, Issues in Business Ethics 53,
https://doi.org/10.1007/978-3-031-04564-6_35

Let me be clear: When I say remoralizing, I am in no way advocating a reactionary movement toward fundamentalism. When fundamentalism refers to an unthinking and uncritical application of solutions that have worked in the past, I believe that is, in part, what has gotten us into the mess we are in. Anybody who doesn't want dialogue, I consider part of the problem rather than the solution.

Remoralizing our dialogue means articulating some principles for each of our institutions that help us to discover their true purposes, and then crafting an ongoing conversation about those principles and purposes. We should expect to have such a level of conversation coming from our leaders, especially our political and business leaders, as a matter of course. Every citizen can benefit from being taught how to have better conversations about the stuff that really matters. Together, we can orient our institutions around hope and freedom.

The ability to create a more ethical world is limited only by our imaginations. Granted, for some, that could be quite limited. Patricia Werhane's book, *Moral Imagination and Management Decision Making*, makes this point, that when reputable companies do unethical things, it's because the managers lacked the imagination to change their mental models and do otherwise.

Therefore, to make sure that we are fully exercising our moral imaginations, it is important to develop an experimental mindset. Here are ten questions to get your ethical juices flowing. What would the world be like if:

- We had a commonwealth of nations linked together by value creation and trade?
- Women held all political offices for the next 100 years?
- Generals had to be mothers?
- Politicians stopped trying to tear each other down?
- Businesses saw themselves as part of communities and were dedicated to improving them?
- Executives, indeed all employees, saw their job as how to create value for stakeholders and make them better off?
- Businesses were as clear and transparent as possible about what they stood for?
- We took out competition with others and built in competition with ourselves to improve and learn more?
- Values such as courtesy, integrity, perseverance and responsibility took center stage in our schools?
- We valued our teachers as much as we do our athletes and rock stars?

In engaging in a dialogue on ethics, we must avoid turning the conversation into self-righteous moralizing. However, the time is right for us to raise the standard of our public dialogues with all of our institutions, and remake our world for the better. Our future depends on it.

R. Edward Freeman is University Professor, Elis and Signe Olsson Professor, Academic Director of the Institute for Business in Society, and Senior Fellow of the Olsson Center for Applied Ethics at the University of Virginia's Darden School of Business.

Chapter 36
Values, Authenticity, and Responsible Leadership

R. Edward Freeman and Ellen R. Auster

1 Introduction

Values are central to the idea of "responsible leadership" and most modern discussions of business ethics are connected in a variety of ways to the concept of "values." Although there are several feasible ways to interpret the idea of "values," most accounts assume that it makes sense to talk about both individual and corporate values.[1] Indeed in recent times, business ethicists have proposed that we stop separating "business" from "ethics" and instead integrate values into our basic understanding of how we create value and trade with each other. For instance, "treating employees as rights-holders," "creating value in an environmentally sustainable way," "implementing corporate social responsibility," "becoming a good citizen in civil society," "being a force for peace in the world," "engaging in social entrepreneurship," and being "ethical or responsible leaders" are all ideas that depend on some underlying notion of values. At their best, we expect businesses to act on those values and hence act "authentically." And, we look to business executives to act on

[1] Some may want to reduce the idea of "corporate values" to some notion of the values of individuals. Hopefully, nothing in our argument turns on particular accounts of the nature of values, and in fact, as pragmatists, we wish to avoid such essentialist theorizing. We explore these issues in our next article entitled, "Values and Poetic Organizations."

Originally published in: *J Bus Ethics*, 98, 15–23 © Springer Nature, 2011
Reprint by Springer, DOI 10.1007/s10551-011-1022-7

R. E. Freeman (✉)
University of Virginia, Charlottesville, VA, USA
e-mail: FreemanE@darden.virginia.edu

E. R. Auster
York University, Toronto, ON, Canada

© Springer Nature Switzerland AG 2023

S. D. Dmytriyev, R. E. Freeman (eds.), *R. Edward Freeman's Selected Works on Stakeholder Theory and Business Ethics*, Issues in Business Ethics 53,
https://doi.org/10.1007/978-3-031-04564-6_36

their own values to be authentic (George and Sims 2007). As Maak and Pless (2006a) suggest, while personal values are important for any notion of responsible leadership, we need to replace the idea of "great man" theories of leadership with "moral persons" theories. In any such idea of "moral persons," the notion of authenticity is central (Maak and Pless 2006a, p. 42).

Curiously, not much is written in the business ethics literature about the idea of "authenticity."[2] Our argument is simple. To act authentically, in this sense, assumes that values are either easy to know, but rather difficult to realize, or difficult to know but easy to realize. Acting authentically becomes either a matter of will or knowledge. Both views assume that values are relatively stable over time. We believe that the reality of modern life makes values both difficult to know and difficult to realize. So, the problem of authenticity is more complex than most theorists imagine. It is not simply a matter of introspection to find one's values, and then having the will, character, or integrity to act on those values. We see authenticity as a creative project, one where we strive to create a life that is imbued with the process of trying to live in an authentic way. However, we believe this is a creative process, and ongoing inquiry, rather than a static statement of one's values and declarations of action.

When we turn to the leadership literature, we find a fairly recent concern with the idea of "authentic leadership" where scholars define "authenticity" much in the same way that business ethics theorists do. In their introduction to a special issue of *The Leadership Quarterly* on "authentic leadership," Avolio and Gardner (2005) suggest that authenticity is fundamentally a self referential concept that is about "being true to one's self." "Authentic leadership" is more complex as it depends on leader–follower relationships and has a more relational character. There seems to be widespread agreement among leadership theorists who think about these issues that values and acting on one's values play a crucial role in the development of any theory of authentic leadership. Although Avolio and Gardner suggest that the roots of the theory are in what organization theorists would call "positive psychology," philosophers might argue that we can understand "authenticity" without such a reference, as there have been plenty of leaders who have been authentic who committed great evils in the world.

We find the practitioner literature on authentic leadership even less compelling than the academic one. George (2003) and George and Sims (2007) have raised the important idea that leaders should not try to be someone they are not. This is a welcome change from the "leadership style" advice genre, where leaders are encouraged to adopt the appropriate style to the circumstances. However, George assumes that knowing ones values and being true to oneself is a fairly straightforward process. And, he is surely correct that leaders must start with their own conceptions of themselves and their values. However, authenticity, at least in our view, does not end with a simple proclamation of either individual or corporate values. Although the development of authentic leadership theory is a step in the right direction, we believe that it would benefit from a more careful analysis of the idea of "authenticity" and a

[2] The exception is Jackson (2005) which takes a particularly existential approach to authenticity.

more explicit connection to Maak and Pless's (2006a, b) idea of embedding "moral person" into the very core of leadership theory.

We proceed as follows. In The Essential Self and the Problem of Authenticity Section, we critique the underlying idea that acting authentically is essentially about being true to one's values. In The Poetic Self: Enlargement, Connection, and Aspiration Section, we suggest re-thinking authenticity as a "project of self-creation" for individuals. In The Poetic Self and Responsible Leadership: Creating Self and Community Section, we sketch an argument about how a more robust idea of authenticity can enrich Maak and Pless's (2006a, b) theory of responsible leadership. In particular, we link this more complex conception of the self to the pragmatist project as articulated by Richard Rorty (1989) of the creation of self and community. We suggest that such an account opens up a space of possibility for a revised notion of responsible leadership.

2 The Essential Self and the Problem of Authenticity

Most discussions of authenticity begin and end with the idea that individuals have a set of values, and that these values are knowable. There are several interpretations of "values" which are important to understand. In the social science literature, we find the concept of values as preferences. Drawing on the pathbreaking work of Rokeach (1973), Frederick (1995), and others, values as preferences can be empirically studied in a number of ways, and social scientists have developed a number of scales and methods to sort out different kinds of values. In the business ethics literature, Agle and Caldwell (1999) have summarized this social scientific approach and suggested that we can understand values at five interdependent levels: individual, organizational, institutional, societal, and global, as well as ten possible connections among the levels. The main focus of this approach is to find the right instrument to empirically determine what values actually are. Although there is a large empirical literature here, we shall see that it rests on some shaky philosophical grounds.

William Frederick (1995) built on Rokeach's (1973) work and applied it to business ethics. He used the idea of values as preferences to construct a multi-layered notion of business values. He claims:

> A largely unspoken premise has been that the values held by these business leaders—owner-entrepreneurs, top-level corporate managers, financiers, and industrialists—shape the motives, policies and actions of their firms. (p. 14)

While he diagnoses the tension between business and society as an underlying struggle between three different sets of values, economizing, power-aggrandizing, and ecologizing values, he assumes, along with Rokeach, that the basic idea of values is a useful starting point.

Rather than mere preferences, business ethicist Edwin Hartman (1988) contends that values are relatively general, permanent, considered desires. The attribution of values implies a kind of rationality, but values are more difficult than social scientists would have us believe. He says:

> Most of us cannot state our values and their implications in a coherent and airtight way; hence unanswerable questions arise about whether we really hold this or that value… nobody is completely rational, we cannot always know whether a failure to act on a value is a failure of rationality, an absence of the value in question, or a simple lapse. (p. 75)

Hartman (1988) suggests the Problem of Authenticity on these accounts becomes either how we know our values or whether our values are realizable through action.[3] We begin with this same idea of individual values as the starting point of authenticity. However, unlike many management theorists, we do not assume that values are transparent to individuals (and by parallel, organizations), nor do we assume that the self is mainly defined by these values.

Many of the staple examples that business ethicists use concern companies or individuals acting on their values, or sometimes, standing up to others, based on their values. Johnson and Johnson is said to have acted on its statement of values called "the Credo" to handle the Tylenol situation. The CEO James Burke is lionized as a responsible leader because there was some match between his personal values and the Credo, and he acted on them. Merck is said to have acted on its value that "medicine is for people not for profits" in developing a drug for River Blindness. The CEO, Dr. Roy Vagelos, was able to realize that value, because it also meshed with his own personal sense of how physicians should act, even if they are in charge of a large multinational pharmaceutical company. Even the much maligned Wal-Mart is said to act on "everyday low prices" as one of its core values. And, clearly its founder, Sam Walton, believed that the poorer strata of American society deserved the same access to goods and services that wealthier Americans enjoyed. Alternatively, many of the scandals in business ethics are attributed to a lack of values, or perhaps faulty sets of values. Enron, Parmalat, and other famous scandals are routinely held up as examples of not prioritizing values. Corporate seminars are full of the advice to "walk the talk" by which is meant, if you say you have this value, then your actions need to be consistent with the value.

These very practical business issues have a philosophical counterpart in the debate in ethical theory between "internalism" and "externalism."[4] We propose to examine one of the assumptions present in this debate, namely, that acting authentically is simply a matter of knowing (Plato) one's values and then acting on them (Aristotle). In doing so, we highlight the underlying view of the self as a vessel that contains values.

---

[3] Some recent popular business literature has focused on this interpretation of "authenticity." Gilmore and Pine II (2007) have focused on "authentic" as opposed to "fake." George (2003) has written about leaders "starting where you are" by which he means not trying to become someone who has different values. Both of these theories have the ring of validity to them, but both take "being authentic" as non-problematic.

[4] The internalism versus externalism debate is rooted in Plato and Aristotle, and was brought into the foreground by Frankena (1958).

When wrongdoing occurs there are several possibilities. The first is that we could explain behavior by saying that the person simply did not know that the behavior was morally questionable or did not know (or "believe" if you are a pragmatist like us) that the behavior would lead to the morally questionable outcome. If they had known, they would have behaved differently. In the literature, this is referred as the "internal" explanation. It adopts the view first articulated by Plato that "to know the right is to do the right." Plato is alleged to hold the view that morally questionable behavior was a matter of "error." He believed that moral reasons provided the necessary motivation for action.

A second explanation called "external" was suggested by Aristotle.[5] We might say that the person or company had values, but was not sufficiently motivated by those values to produce action. Motivation is not internal to values, but must be found externally. Motivation comes from moral claims only if one has a desire to be moral, or to act on one's values. Perhaps Enron executives who actually believed in RICE could be said to hold those values, but the values could be overridden by non values-related reasons.[6] Psychology is complex, and values may only offer partial motivating force.

What happens when values conflict? According to the internal explanation, it becomes a matter of knowledge as to which values take precedence. Alternatively, it may be a matter of knowing how to create a situation through innovation or Patricia Werhane's (1999) idea of "moral imagination" so that all of our values can be realized. Distinguishing "moral values" from "prudential values" does not help the internalist here. And telling the internalist to "walk the talk" is meaningless if they cannot figure out what they do not know. The talk is the problem. For the internalist, acting on one's values is problematic because of the uncertainty and complexity in the process of coming to know one's values.

The externalist has a different problem. Because values do not necessarily offer motivational force, they cannot conclude which values are more important. Indeed, how is one to know one's values to begin with? If there is not a strong connection to action, how can one tell whether they have merely prioritized conflicting values differently, do not actually believe the values, or simply have "weakness of will." At Enron, how could someone tell the difference between the value of profit and the value of integrity as they made various decisions? Telling the externalist to "walk the talk" is meaningless unless it is interpreted as a statement about what their desires should be, i.e. that they should "walk." For the externalist, acting on one's

[5] We see no reason to make a distinction between matters of "morality" and matters of "prudence." Self-regarding values may well also only offer motivational force if one has the desire to realize one's best interest. And, for the internalist, the extent to which moral values trump prudential values is also a matter of knowledge. We believe that this distinction is better made between values that are "primarily self-regarding" and values that are "primarily other-regarding." In the real world, most people are driven by a mixture of self-regarding reasons and other-regarding reasons. Even Kant believed that self-regarding reasons had moral content, as he carefully wrote about the "duty to self perfection." We are indebted here to Professors Norman Bowie and Patricia Werhane.

[6] The Enron values were cleverly called by the acronym "RICE," standing for "Respect," "Integrity," "Communication," and "Excellence."

values is problematic not because knowing one's values is difficult, but because there are many other psychological forces that can serve to override these values.

We want to suggest that this philosophical debate for which there is no "solution" highlights the fact that human motivation is complicated. Rather than solving the "internalism–externalism" issue, we suggest a pragmatist alternative that tries to find a middle ground between the two. In the real world, human beings are complex. For the internalist, the idea of "being true to oneself and one's values" is problematic in the sense of knowing one's values, be they other-regarding values or self-regarding values. The internalist view highlights the importance of the process of self-knowledge and the difficulties in coming to know one's own values. Sometimes we are in fact motivated simply because we have discovered something that is a core value for us, and the fact of discovering or being reminded of its existence actually moves us to act. Surely internalism reminds us of this valid point. On the other hand, for the externalist, the idea of "being true to oneself and one's values" is a matter of knowing what else, besides values, has motivational force. Sometimes we find ourselves enmeshed in situations that overwhelm our values and even our sense of self. Milgram (1963, 1974) and others have demonstrated the situational effects on action, time, and time again. Surely, externalism reminds us of this fact that we encounter in the real world. However, both views point out important conceptual difficulties in the naïve view of authenticity as being true to self.[7]

We shall call the idea that the self is a vessel containing values that are knowable by introspection, the "essentialist self." The main idea, on this view, is that our values define us. They express our essence, hence the name. Our values give us our identity, and in the liberal West, we take these values to be individualized. The main point of liberalism is that we can live together, even if we have different values. We find this idea implicit in much of the business ethics literature. Yet, it has undergone a profound critique by thinkers such as Charles Taylor (1991) in philosophy, Hans Joas (2000) in sociology, and more recently, Mollie Painter-Morland (2008) in business ethics. The dispute between internalism and externalism hints that things are more complicated than they appear. The internalist view highlights the difficulty in coming to know the values, which the externalist view highlights the problem of understanding the vessel. We believe that there are problems understanding both the vessel and what may be inside.

Joas's (2000) brilliant work, *The Genesis of Values*, gives a more nuanced account of what Hartman (1988) must mean by values as "considered desires." Joas looks to the history of philosophy, particularly recent pragmatist thinking, and suggests along with Charles Taylor that we cannot give an adequate account of values without understanding our view of the self. Our values ultimately give us the

[7] Recently, Evan Simpson (1999) has suggested a middle ground between these two extremes, whereby the connection between moral beliefs and motivation is weakened from one of logical necessity to causal dependency. This analysis seems correct to us, and is in keeping with the pragmatist spirit of our suggestions here. We use "internalism" and "externalism" as illustrative of two kinds of problems with the naïve idea of authenticity as being true to self.

boundaries of the self, and are integral to the process of self-description and redescription.

Kevin Jackson (2005) is one of the few business ethicists that has directly addressed the idea of authenticity. While Jackson wants "authenticity" to do the same work as values in the essentialist self, he does point the way along for a revision of the essentialist self. Jackson draws on Sartre and the existentialist tradition to argue that values rest on our ability to choose. We need to see our freedom as a precursor of any set of values. Being conscious of that freedom when choosing to realize a particular project is the real meaning of authenticity according to Jackson. In effect, he identifies moral character with the ability to recognize one's freedom and choose accordingly. However, in Jackson's account, we choose projects, and those choices define us. By focusing on projects and freedom to choose, Jackson sketches an account that appears to do without values. However, there is still an essentialist flavor, as it appears that freedom acts as an ultimate ground. And, how are we to know when we are truly free without lapsing back into the arguments above about internalism and externalism. Jackson (2005) argues:

> …the past as a determinant of action depends on our freely constituted projects in the now. I cannot literally change the past. No physical force in the world is powerful enough to do that. Still, the meaning of the past hinges on my present commitments. (p. 312)

And, we would add that the meaning of the present depends in part on one's past experiences. The idea that every moment represents a clean slate is difficult to realize, and it ignores important ideas such as commitments, and the shaping of our present through the past. In essentialist terms, it is as if Sartre, through Jackson, is suggesting that we are free to empty our vessel at any point in time, and to fill it with whatever projects we may find ourselves engaged. We want to suggest that the human condition is more subtle. However, we also want to acknowledge that Sartre's idea of freedom should be used as a humble reminder that we can choose a different way to live no matter how difficult our circumstances, nor how difficult the choice.[8]

Indeed Charles Taylor (1991) suggests that our reliance on an individualist idea of authenticity is a root cause for much of the malaise in modern society. Because we rely too much on something like Jackson's idea of individual freedom, we get easily estranged from the meaning that we create with our fellow human beings. Mollie Painter-Morland (2008) suggests that business ethicists adopt a Taylor-like redefinition of authenticity. She suggests that the very process of defining ones values is always "relational." Values make sense only because they allow us to act in context, precisely where the boundaries of the self are at issue. Painter-Morland (2008) says

> Authenticity, thus conceived, allows for the fact that an individual's role may shift as he/she traverses the complex typography of an organization's various functional units and system

[8] For pragmatists saying that we always have a choice and that we continually makes choices is much like saying as Richard Rorty (1980) does that truth should be understood as a cautionary warning…that we may not have all the evidence for a belief. Choice reminds us that we can be masters of our own fate.

> of relations. Because it allows the individual to calibrate his/her role in relation to the various stakeholders with whom he/she is engaged, this view of authenticity involves a certain degree of perspectivism. (p. 214)

Perhaps the best illustration of the incompleteness of the essentialist self is in Shakespeare's *Hamlet*. Hamlet is searching for perspective and self knowledge and seems utterly confused throughout most of the play. We can all be reminded of Shakespeare's classic line in Hamlet where Polonius gives advice to Laertes, his son, "this above all: to thine own self be true." When such advice is taken literally, as in many management best sellers, and in the business ethics literature, it provides a set of facile recommendations that make discovering and acting on values nearly impossible.

At least one reading of Shakespeare is that Polonius is quite a fool. He is giving "fatherly advice" to Laertes precisely because he has failed to establish any deep and meaningful connection with his son. He is reduced to platitudes. We do not believe that platitudes produce responsible leadership; rather they are one of the main barriers to such leadership. Hamlet is a paradigm case of a troubled person who is searching for what it means to be true to himself, because he knows neither what "his self" is nor "how to be true." By confronting his history, especially with his parents and their history, and by examining his relationships with others, and coming to terms with his own aspirations, he is able to begin to gain insight into his own actions, and his life begins to become authentic.

Prince Hamlet suffers from what we will call:

> The Problem of Authenticity *Understanding ourselves, and why we do what we do, requires a commitment to being authentic. However, being authentic is more difficult than it first appears. "Know thyself" is easy to say and hard to accomplish. We can start with our values, but we must be willing to engage in a dialogue with our past, our relationships with others, and our aspirations for the future.*

There are many Prince Hamlets in the business world. In many cases, things just seem to happen, and we go along not bothering to understand who we are or what we are becoming. After a while, acting authentically becomes either taken for granted or becomes impossible. We need a more nuanced approach. We need to examine our past (and by parallel, the history of an organization) and try to understand why we behave the way we do, enlarging our view of the self. Very quickly we encounter the idea of self and other, and the related tensions that result, so that individual values and understandings of the past are enmeshed in connections with others. These ideas combine to confront and inform our aspirations about the lives we want to lead and our effects on others. Therefore, we suggest replacing the idea of "the essentialist self" with what we have come to call "the poetic self" viewed as the intersection of our values, our past, our set of connections to others, and our aspirations. The poetic self is better conceptualized as a project of self-creation, rather than a static entity that explains why we do what we do.

3 The Poetic Self: Enlargement, Connection, and Aspiration

Since Freud, we have become skeptical of ourselves as being the best judges of what is in our hearts, what our "true values" may be. We need not adopt the Freudian ontology of drives, or his map of the mind, to take his insight seriously, that sometimes our unconscious mind is driving the animal. This seems to come down squarely on Aristotle's side in the "internalism–externalism" debate, but it is more complicated. Many of Freud's followers have suggested that we can bring many unconscious motivations to light, and that doing so is an ongoing project. Unlike the current medicalization of psychoanalysis, Freud's original insights were not aimed at "healing" but at coming to understand why one sees the world and acts as one does. It was a matter of coping and as Carl Rogers (1959) and others have suggested "self-development."[9]

Freud's view that early childhood experience plays a crucial role in the adults we become depends in part on the idea that we can have access to at least some of our unconscious. Philosopher Richard Rorty (1991) claims that we need to see Freud as articulating a sort of moral imperative:

> Unlike Hume, Freud did change our self-image. Finding out about our unconscious motives is not just an intriguing exercise, but more like a moral obligation. (p. 3)… What we are morally obligated to know about ourselves is not our essence, not a common human nature that is somehow the source and locus of moral responsibility. Far from being of what we share with the other members of our species, self-knowledge is precisely what divides us from them, our accidental idiosyncrasies, the "irrational" components in ourselves, the ones that split us up into incompatible sets of beliefs and desires… Only study of these concrete details will let us enter into conversational relations with our unconscious and, at the ideal limit of such conversation, let us break down the partitions." (p. 6)

We shall call this idea "self-enlargement" to denote the difference between trying to discover one's true values or one's essence and trying to figure out where some of these "values" may come from. There are multiple applications of the enlarged self in business. For instance, psychoanalyst Karen Horney (1950)[10] has articulated the "Search for Glory," which drives many executives. Heinz Kohut (1996) and others have developed models of "transference" which can be useful especially in understanding how people react to change and authority. Transference occurs when we transfer relationships from the past to our current ones. For instance, an executive might adopt a posture of resentful obedience because of the way he or she was made of obey a parent or other caregiver. Transference gets in the way of developing the self, because it suggests that we repeat patterns of the past over and over.

Self-enlargement is not enough. Introspection can only take one so far. In a number of books and articles, Kets de Vries (2006) has articulated what he calls, "the clinical paradigm" and applied it to his clinical practice with executives and

[9] We are not claiming that Freud himself had much optimism for "self development" only that it was an outcome of his original insights in a more positive mind such as Rogers'.

[10] See Horney (1950), Kohut (1996), and Siegel and Kohut (2000).

companies. There are four principles to his paradigm, each of which is important when we begin to see the self as a project. They are[11]:

1. Every human action has an explanation and rationale.
2. A great deal of our action is based in the unconscious.
3. The way we express and regulate emotion is central to any idea of self.
4. "Human development is an inter- and intra-personal process."

Although the first three principles can be said to be behind the enlarged self, Freud pays little attention to the inter-personal process that goes on as we engage in conversations with our past. Here, we again turn to the object relations school of psychoanalytic thought, and in particular to Jessica Benjamin (1988) who has argued that the inter-subjective and intra-psychic views must be seen as going together. She claims that by the age of two, children feel the tension between the "assertion of self" and "recognition of the other."

> In trying to establish itself as an independent entity, the self must yet recognize the other as a subject like itself in order to be recognized by the other…In its encounter with the other, the self wishes to affirm its absolute independence even though its need for the other and the other's similar wish undercut that affirmation. (p. 32)

Recognition becomes mutual. The tension between establishing an autonomous boundary for the self, yet acknowledging others by doing so, and being acknowledged becomes permanent. The wish to resolve this tension often leads to domination and subordination. There is no subject without another subject. The self becomes a connected self. Discovering one's past associations is dealing with those individuals who have influenced a person's development so far. And, connections are current and future looking as well. Understanding who are those individuals with whom one currently has some relationship is as important as understanding the past. In fact, if Benjamin and others are correct, we cannot have access to the unconscious without thinking about the tension between autonomy and connection.

Jean Baker Miller (1976), and later Carol Gilligan (1982), made the controversial claim that such a connected self was gendered. The controversy arose in part because Gilligan was writing in the context of Kohlberg's (1969) stages of moral development that put a "higher" value on being autonomous, rather than connected. The underlying idea of authenticity as the essential self was the culprit. We need a view of the authentic self that takes into account the mutuality and paradox of recognition, and the permanent tension between self and other, autonomy and connection, which comes with mutuality.

We can summarize our more nuanced view of the self as follows. The essential self is a starting point. We begin to try and explicate our values, or at least what we would say our values are now. Self-enlargement asks us to more deeply probe into our past, and try and understand some of our history that makes us the unique person we are. Self-connection asks us to see both our current values and our past

[11] See especially, Kets de Vries (2006, pp. 9–11).

associations as enmeshed in a set of relationships. We need to keep present the tension between self and other. We also need to understand aspiration and the future.

Just as our current behavior is shaped by our understanding of the past, so too is it shaped by our aspirations, our ideas about the kind of lives we want to live. Some have such a clear and compelling vision of their own futures that the present is literally caused by these future visions. Imagine parents who have a clear idea that they want their sons or daughters to follow in a particular direction. Such a future vision can literally script their lives. More generally, think about some great leaders who have articulated a vision of the future such as Gandhi's vision of self-rule for India, causing millions of people to enact a present based on that future vision.

As we look forward, we constantly struggle with changing our lives, indeed with changing the very idea of our "self." We have aspirations for ourselves and for those connected to us, about the lives we want to live and the effects we want to have on others. Of course, these future aspirations are connected to our understanding of the past, our connections with others, and our understanding of our current values. Sometimes our current values can express these aspirations, our deepest hopes and dreams about how we want to live. Living authentically means asking hard questions about these aspirations, not taking them at face value, understanding the connections to past, present, and future that they are based on. But, if living authentically is to be more than an introspective journey, we must take account of how human beings remake their world.

We have suggested that our idea of authenticity as acting on our values, and its associated view of the self as a vessel filled with values, is of limited usefulness, and needs to be more nuanced. We have argued that we need to see authenticity, and hence our values, as a more creative process. Living authentically is at once engaging our current values as best as we understand them, constantly querying our past for clues about our idiosyncrasies and behaviors, engaging in conversation and relationships with others, and remaking our futures with our aspirations. If this more nuanced view of the self enmeshed with others is useful, then we need to rethink our idea about responsible leadership to include this more nuanced view of authenticity and values.

4 The Poetic Self and Responsible Leadership: Creating Self and Community

Uhl-Bien (2006) has outlined a different approach to leadership based on a view of the self that is relational. Because her purpose is to develop a relational theory of leadership, Uhl-Bien does not explicitly connect this theory to the psychoanalytic literature. However, something like the relational view of leadership seems to make sense as we develop a more nuanced view of authenticity and the self as we explained it in the previous section. Maak and Pless's (2006a, b) take Uhl-Bien's relational theory and suggest that the idea of "moral person" needs to be at the center. We go further to argue that "moral person" needs to be unpacked into something like the poetic self.

The "poetic self" stems from Harold Bloom's (1997) idea of the "strong poet" who literally sees the world in a way that is different from others, but is also embedded in a number of communities. It is the intersection of our values, our past, our set of connections to others, and our aspirations. Our aspirations include not only individual aspirations but also community aspirations. Thus, the poetic self embraces the idea of simultaneously creating self and community. Indeed, language works so that there are no desert island speakers. It is social and interconnected as surely as subjects are connected to objects that become subjects. Authenticity becomes the project of finding this unique expression of our own humanity that takes account of both individual (and intra-psychic) and community (and inter-subjective) aspirations.

To Maak and Pless's question, "What makes a responsible leader," we would answer that at a minimum, leadership requires the effort to be authentic understood in the sense of starting with one's values, seeking to understand the influence of the past, the set of connections or relationships in which one is entangled (Uhl-Bien 2006, p. 658), and one's aspirations. This conception of the poetic self means that leaders must think beyond followers and take on at least some responsibility for the stakeholders in the organizations that they lead, as Maak and Pless (2006b, p. 105) argue.

Taking such an approach to responsible leadership squarely places the theory in the pragmatist domain. Philosophers, such as John Dewey and Richard Rorty, have argued that the project of self-creation is a private project, whereas the project of "community creation" is a public project. If something like the account we have given of the essential plus enlarged plus connected self is helpful, then we can begin to dissolve the public–private distinction in this sphere, and see self-creation and community creation as two sides of the same coin.[12] We create self in part by creating connection, and as we create connection, we create self.[13] This is the implication of the mutuality of recognition, and the enduring tension between self and other.

We have only begun to sketch the idea of the poetic self and its connection to responsible leadership. There is much more to be said. However, an immediate question for business ethicists is how this idea may or may not translate into business organizations. One of the key ideas in business ethics is that it makes sense to claim that organizations have values, and that they can act on them. Immediately, the same questions about authentic organizations come to mind. Responsible leaders want to lead authentic organizations, but the process of creating authentic organizations is not so easy.

[12] There are many reasons that we may want to reconstruct the public–private distinction for purposes of personal freedom, but we should now be tempted to do so at the cost of a more nuanced version of the self.

[13] As the relational psychoanalysts would say, "Where there is subject there is subject." For an overview of some of the issues here see Stolorow et al. (1987) and Jordan et al. (2004). There is a very rich literature here that is suggestive of many research questions for leadership theory.

Oftentimes organizations announce their values, print them on cards, and hand them out to employees and other stakeholders. We then carefully watch and see whether the organization acts in a way that is consistent with those values. As in the case of individuals, there are multiple problems with this rather simple analytical scheme. First, values are difficult to know for individuals. They are at least as difficult to know for organizations that may consist of a great number of individuals, with different and conflicting interpretations of these values. Like the individual case, there is a knowledge problem about how organizations can know and agree on what their values are. We believe that it is more fruitful to approach this problem in terms of an ongoing process of conversation. Second, it is difficult to always pin down what actually motivates organizations to do what they do. There can be multiple conflicting reasons and causes of organizational action, only some of which is attributable to the role of organizational values.

In summary, we believe that we can conceptualize the "poetic organization" much along the lines of the poetic self. Authenticity in organizations becomes a process of starting with where the organizational values are thought to be. Second, organizations must become aware of their history and their historical routines. Third, every organization is embedded in a network of stakeholder relationships. Finally, most organizations have some kind of purpose or aspiration. By understanding these processes of self-understanding, connection, and aspiration, we have a chance to make adjustments to make our organizations more fit for human beings. Creating such organizations is the work of responsible leaders and responsible leadership. That is a much longer story to tell, and it is only possible if we adopt Maak and Pless's (Maak and Pless 2006a) idea that leadership studies (and we would add the entire field of "business ethics") becomes:

> A specific frame of mind promoting a shift from a purely economistic, positivist and self-interested mindset to a frame of thinking that has all constituents and the common good in mind. (p. 1)

References

Agle, B.R., and C.B. Caldwell. 1999. Understanding Research on Values in Business. *Business and Society 38*: 326–327.

Avolio, B.J., and W.L. Gardner. 2005. Authentic Leadership Development: Getting to the Root of Positive Forms of Leadership. *Leadership Quarterly 16* (3): 315–338.

Benjamin, J. 1988. *The Bonds of Love: Psychoanalysis, Feminism, and the Problem of Domination*. New York: Pantheon.

Bloom, H. 1997. *The Anxiety of Influence: A Theory of Poetry*. New York: Oxford University Press.

Frankena, W.K. 1958. Obligation and Motivation in Recent Moral Philosophy. In *Essays in Moral Philosophy*, ed. A.I. Melden, 40–81. Seattle: University of Washington Press.

Frederick, W.C. 1995. *Values, Nature and Culture in the American Corporation*. New York: Oxford University Press.

George, B. 2003. *Authentic Leadership: Rediscovering the Secrets to Creating Lasting Value*. San Francisco: Jossey-Bass.

George, B., and P. Sims. 2007. *True North: Discover Your Authentic Leadership*. San Francisco: Jossey-Bass.
Gilligan, C. 1982. *In a Different Voice: Psychological Theory and Women's Development*. Cambridge: Harvard University Press.
Gilmore, J.H., and B.J. Pine II. 2007. *Authenticity: What Consumers Really Want*. Boston: Harvard Business School Press.
Hartman, E.M. 1988. *Conceptual Foundations of Organization Theory*. Cambridge, MA: Ballinger.
Horney, K. 1950. *Neurosis and Human Growth: The Struggle Toward Self-realization*. New York: W. W. Norton.
Jackson, K.T. 2005. Towards Authenticity: A Sartrean Perspective on Business Ethics. *Journal of Business Ethics 58*: 307–325.
Joas, H. 2000. *The Genesis of Values*. Chicago: University of Chicago Press.
Jordan, J.V., L.M. Hartling, and M. Walker. 2004. *The Complexities of Connection*. New York: The Guilford Press.
Kets de Vries, M. 2006. *The leader on the Couch: A Clinical Approach to Changing People and Organizations*, 9–11. San Francisco: Jossey-Bass.
Kohlberg, L. 1969. *Stages in the Development of Moral Thought and Action*. New York: Holt, Rinehart and Winston.
Kohut, H. 1996. In *The Chicago Institute Lectures*, ed. P. Tolpin and M. Tolpin. Hillsdale: Analytic Press.
Maak, T., and N.M. Pless, eds. 2006a. *Responsible Leadership*. New York: Routledge.
———. 2006b. Responsible Leadership in a Stakeholder Society: A Relational Perspective. *Journal of Business Ethics 66* (1): 99–115.
Milgram, Stanley. 1963. Behavioral Study of Obedience. *Journal of Abnormal and Social Psychology 67* (4): 371–378.
Milgram, S. 1974. *Obedience to Authority; An Experimental View*. New York: Harpercollins.
Miller, J.B. 1976. *Toward a New Psychology of Women*. Boston: Beacon.
Painter-Morland, M. 2008. *Business Ethics as Practice*. Cambridge: Cambridge University Press.
Rogers, C. 1959. A Theory of Therapy, Personality and Interpersonal Relationships as Developed in the Client-Centered Framework. In *Psychology: A Study of a Science, Vol. 3: Formulations of the Person and the Social Context*, ed. S. Koch. New York: McGraw-Hill.
Rokeach, M. 1973. *The Nature of Human Values*. New York: Free Press.
Rorty, R. 1980. *Philosophy and the Mirror of Nature*. Oxford: Basil Blackwell.
———. 1989. *Contingency, Irony, and Solidarity*. Cambridge: Cambridge University Press.
———. 1991. Freud and Moral Reflection. In: *Essays on Heidegger and Others*. New York: Cambridge University Press.
Siegel, A.M., and H. Kohut. 2000. *The Psychology of the Self*. London: Brunner-Routledge.
Simpson, E. 1999. Between Internalism and Externalism in Ethics. *Philosophical Quarterly 49* (195): 201–214.
Stolorow, R.D., B. Brandchaft, and G.E. Atwood. 1987. *Psychoanalytic Treatment: An Intersubjective Approach*. Hillsdale: Analytic Press.
Taylor, C. 1991. *The Ethics of Authenticity*. Cambridge: Harvard University Press.
Uhl-Bien, M. 2006. Relational Leadership Theory: Exploring the Social Processes of Leadership and Organizing. *Leadership Quarterly 17* (6): 654–676.
Werhane, P.H. 1999. *Moral Imagination and Management Decision Making*. New York: Oxford University Press.

R. Edward Freeman is University Professor, Elis and Signe Olsson Professor, Academic Director of the Institute for Business in Society, and Senior Fellow of the Olsson Center for Applied Ethics at the University of Virginia's Darden School of Business.

Ellen R. Auster is Professor of Strategic Management at the Schulich School and Executive Director of the York Change Leadership at York University.

Chapter 37
Values and Poetic Organizations: Beyond Value Fit Toward Values Through Conversation

Ellen R. Auster and R. Edward Freeman

1 Introduction

Values and leadership are prominent topics in the discourse of both academics and practitioners in an era of corporate scandals, greed, stunning business failures such as Enron and WorldCom and the recent sub-prime mortgage industry crash (Ardichvili et al. 2009; Newton 2006; Palmer 2009). Shadowing these discussions, particularly in recent years, is the notion that authenticity is an important underlying component to consider as part of any discussion of values and leadership. Although, most scholars would agree that values, leadership, and authenticity are critical to understand particularly now, little attention has focused directly on how these three concepts might be interrelated.[1]

Freeman and Auster (2011) drawing on Rorty's (1989) notion of the "strong poet" begin to tackle this topic with their work on the poetic self as a "project of seeking to live authentically." They propose that "the idea that simply 'acting on

Originally published in: *J Bus Ethics,* 113, 39–49 © Springer Nature, 2013
Reprint by Springer, DOI 10.1007/s10551-012-1279-5

[1] Notable exceptions include Maak and Pless's (2006a, b) work on responsible leadership and Jackson's (2005) article on a Sartean perspective on business ethics and how it relates to authenticity.

E. R. Auster
York University, Toronto, ON, Canada
e-mail: eauster@schulich.yorku.ca

R. E. Freeman (✉)
University of Virginia, Charlottesville, VA, USA
e-mail: FreemanE@darden.virginia.edu

© Springer Nature Switzerland AG 2023

S. D. Dmytriyev, R. E. Freeman (eds.), *R. Edward Freeman's Selected Works on Stakeholder Theory and Business Ethics*, Issues in Business Ethics 53,
https://doi.org/10.1007/978-3-031-04564-6_37

one's values' or 'being true to oneself' are at best starting points for thinking about authenticity." They suggest that being authentic is an ongoing conversation that begins with perceived values, but also involves one's history, connections with others, and individual as well as community aspirations. In short, it involves an enlarged and enriched notion of the self as an ongoing creative project, rather than the self as a vessel or "keeper of the values."

In this article, we extend Freeman and Auster's (2011) idea of the poetic self and delve more deeply into the embeddedness of the poetic self within an organization. We begin by examining the relationship between individual values and organizational values. We suggest that the idea that "value fit" leads to authenticity is limited and that a deeper exploration of the relationship between individual values, organizational values, leadership, and authenticity is warranted. This analysis reveals that a more fruitful approach might be to view organizational values as creating an opportunity for an ongoing conversation about who we are and how we interconnect. We call organizations that embrace this type of approach "poetic organizations." We delineate a typology of four interconnected values each of which forms a foundation for critical questioning and dialogue that we suggest reframes our thinking about how values, authenticity and leadership might be intertwined.

2 The "Value Fit" Approach

Although there is little scholarly work in management on the intersection of leadership, authenticity, and values as noted above, these ideas have been discussed in the popular press particularly recently. Most of these discussions begin with the idea that there are two inter-related reasons that explain why some organizations are able to create authenticity and others fail. One reason often articulated for why authenticity is missing in many organizations is that individuals and organizations are not living their values. While there is much to be said for this approach, we argue that even if individuals and organizations are living their values, that fact alone is unlikely to create authenticity, at least at the organizational level. According to these views, a second condition must also be met. Authenticity will only emerge if the values between the individual and the organization are aligned—there is "value fit." Below we discuss the value fit approach in greater detail. We then delineate how the "value fit" approach can be viewed as a starting point for a deeper analysis of values and authenticity.

2.1 *Living the Values or Not*

A first condition for authenticity in what we are calling the "value fit" approach is that individuals and organizations must live their values—i.e., walk their talk. To illustrate this idea, we can create a continuum of an individual living their values or not and combine it with a continuum of an organization living its values or not. We can capture this view as shown in Fig. 37.1. As depicted in the lower left corner of the diagonal in Fig. 37.1, according to the "value fit" view, corporate scandals and greed in part stem from "value failure". Value failure is a result of individuals and organizations not living their stated values.

Take for example, the organization with the stated value that their employees are their most important asset. Yet inside the organization, people are never consulted on decisions that affect them and leaders focus predominately on maximizing productivity through efficiency based approaches rather than a more wholistic and inclusive approach. Stories of people being derailed or fired for trying to speak up about their ideas for how to improve the organization are told behind closed doors. What we find here is that the individual is in an organization that may post and preach its values, but everyday injustices leads to feelings of disenchantment, disillusionment, and stripped self esteem. For some, this erodes their passion and creativity and causes them to disengage as they feel empty and soulless. Others, perhaps, frustrated and angry that they lack any opportunities for input or change, not surprisingly, may undermine the company.

Clearly, this is an example of where both the individual and the organization are not living their values—authenticity is missing for both the individual and the company. Even though, there might be great deal of "values talk" and surface value fit, everyone knows that it is not real.

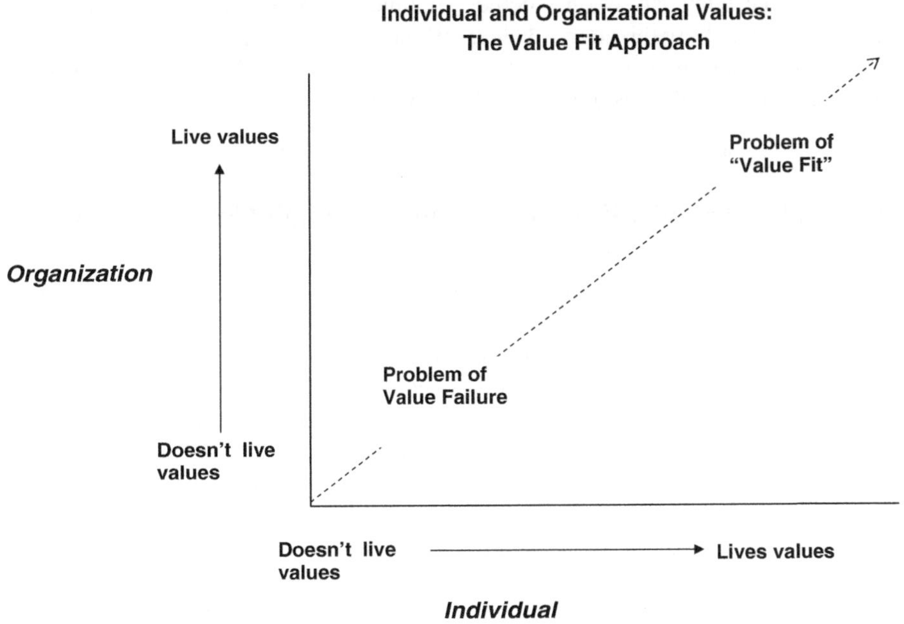

Fig. 37.1 Individual and organizational values: the value fit approach

2.2 *Value Fit or Not*

Beyond whether the individual and the organization are living their values or not, is a second condition that must be met according to what we are calling the "value fit" approach. As shown in Fig. 37.1, the upper right end of the diagonal, in contrast, is where both the individual and organization are "living" their values. The question then becomes whether there is "value fit." Is there alignment between the individuals' and the organizations' values? If the individual's values do not fit the organization's values, then neither the self nor the organization can be authentic. If they do "fit," then according to the "value fit" view, the self and organization can be authentic.

We would suggest that when the individual and the organization are both living their values and there is congruence, that might get us headed toward something better than neither the organization nor the individual living their values. If everyone is clear that it is all about profits and the money, then there is a degree of authenticity. However, it does not go very far.

Where some might see fit as the problem and think that solving the problem of fit would reveal the "answer" to problems of leadership, values and authenticity, we see the problem of fit as just the beginning of the next layer of dialogue and analysis.

The Value Fit approach emphasizes "what" the content of the values are. This is where there is agreement of values (maybe) but values are often static words that attempt to guide what we do. Fit is achieved by the organization recruiting on those values and "enforcing" the values through mechanisms like end of year reviews, annual retreats, training, web-site, and intranet posting etc. Values are "given" or dictated. The emphasis is on whether alignment is achieved between individual action and stated organizational values. While there may be stability because both are aligned, there is also an opening for deception and bad faith. We believe just living congruent values is unlikely to lead to authenticity.

3 Poetic Organizations and Values Through Conversation

Freeman and Auster (2011) develop the idea of the "poetic self" drawing from Richard Rorty's extension of Harold Bloom's work on strong poets and their role in communities.[2] Rorty (1989) saw the world as continuously evolving driven in part by what he calls "strong poets." Strong poets are individuals who reframe how we see things "creating new perspectives on the experience of life and insights that often trigger change" (Hall 2011). Freeman and Auster (2011) build on Rorty's ideas by suggesting more specifically that poetic self is a project of seeking to live authentically. Being authentic entails continuous processes of self-understanding

[2] For more on the original uses of "strong poet" see Bloom (1973, 1976, 1982). There is a secondary literature on the nature of "strong poets". For example see Garrison (1993) and Harmeling et al. (2009).

and introspection, connecting with others, and creating aspirations. Like Freeman and Auster's (2011) more nuanced view of the self, we suggest that there is an organizational analogy to the poetic self. To explore what this might look like, the analysis needs to be deepened beyond the problem of value fit to create a more robust view of values and the organization (Fig. 37.2).

An organization is people joined together to act around some common purpose (Etzioni 1964, p. 3; Scott 1981, p. 20). What inspires, engages, and energizes people in organizations is not only what the stated values are, or the physical structure, it is relationships, interactions, sharing discoveries, tackling joint challenges, and pursuing a joint purpose. Indeed, for the poetic self to thrive, it needs to have a context that is nurturing and supportive of its creative project. For the authentic individual, engaged in the process of self-realization with their past, connected others, and aspirations, organizations must be at a minimum place where this process does not get derailed. Ideally, organizations can facilitate the growth and development of authentic individuals. Authenticity then becomes the creative process of poetic selves joining together and engaging in joint understanding and introspection, connection and creating joint aspirations given their histories. This joint process of ongoing dialogue and conversation about who we are, what we stand for, where we came from, and how we want to "live" in the organization nurtures the conditions in which authenticity is likely to emerge.

We call this process, values through conversation (VTC). VTC brings talk about organizational values to life. Rather than emphasizing whether or not individuals

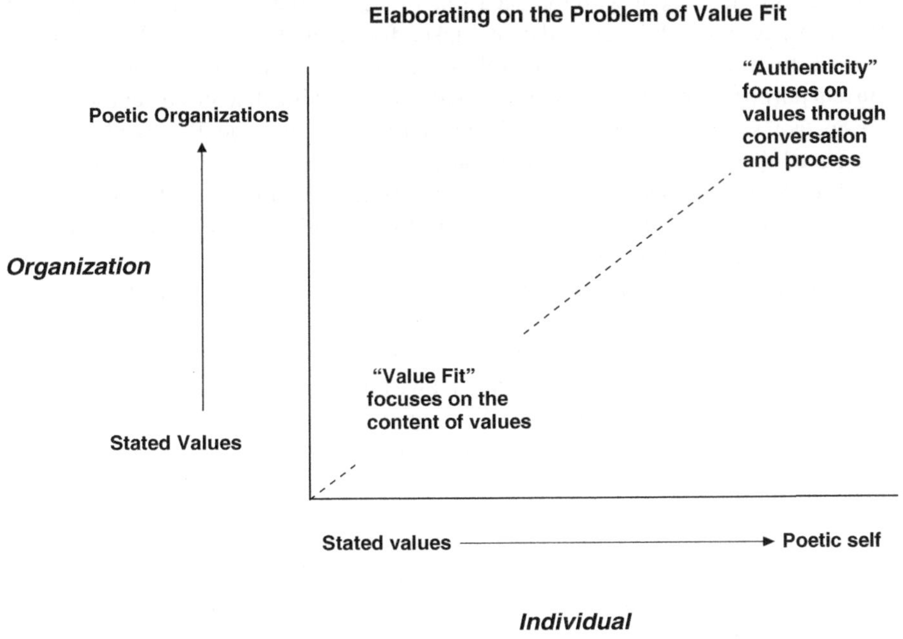

Fig. 37.2 Elaborating on the problem of value fit

actualized the organizational values, or whether the organization acted in bad faith or not, VTC is a process of enrichment and enlargement around organizational purpose and values.

VCT revolves around a typology of four interconnected kinds of values each of which forms a foundation for critical questioning and conversation. We want to suggest that these four kinds of values (which are parallel to the notion of "poetic self") allow us to answer more complex and nuanced questions about what it means to share organizational values and to act on them. The questions we offer are illustrative rather than exhaustive. The specific questions asked and how they are answered will be unique to each organization and will have important significance for how those within the community act and behave. While each value in the typology features different aspects of "the who, why, what, and how" of the organization, these values in practice may often be simultaneously expressed in the choices, decisions, and actions of those in the community (Table 37.1).

3.1 Introspective Values

Most businesses spend a lot of their time focused on current goals and fire fighting—activities that leave little time for reflection. The annual strategic planning retreat, or the quarterly—"have we met our targets?" meetings, may trigger a bit of reflection but typically these activities are intermittent and future oriented. By being so pre-occupied in the present with the occasional focus on the future, there is little time for introspection on how everyday and taken for granted actions facilitate or create barriers to authenticity. Even when these processes begin with a statement of the current purpose and values, there is often little questioning that the current statements are in fact valid, or even very well understood. Collins and Porras (1994) have suggested that truly great companies have a built in value of "dissent," that we see as a broader notion of a willingness to ask hard questions about why the organization is doing what it is doing. Such an introspective attitude is one of the precursors of building an authentic (or "poetic") organization.

There are many benefits and that can be gained from examining, contemplating, and reflecting on organizational purpose and values, just as there are personal benefits to reflecting on our own stated values and on the world around us. These benefits include breakthrough insights, learning, avoiding reinventing the wheel, sharing best practices, and time for restoration and renewal. We suggest that creating space for reflection and introspection helps facilitate authenticity in organizations much as it does in individuals.

Introspection might include a team asking itself how the work it is doing (specific activities, projects etc.) translates values into action or how the team is living the values on a personal level, or how it is not doing so. There are tough questions about which values apply and which ones do not, and more importantly, what is the shared understanding of those values, and how can that shared understanding evolve.

Deeper introspection might begin to reveal the organization's common assumptions as well as routines, norms, and processes that have become habit. No

Table 37.1 Values in poetic organizations: definitions and key questions

Value	Definition	Key questions
Introspective values	How we examine, contemplate and reflect on our collective self and the world around us	Are we achieving what we hoped for? What are our values? How do we live those values within the organization? Are they working for us? What is working well and why? How can we amplify those successes? What is not working well and why? How can we learn and improve? What inquiries should we conduct? What research is useful in helping us create value and reach our goals? What questions should we ask ourselves? What actions and habits do we "assume", take for granted? What narratives underlie what we do? How does our language affect our ideas, actions, habits and values?
Historical values	How we study the past and its impact on where we are today	What's our purpose? How do our interpretations of our organizational history converge and differ? How does our history shape the values, routines, structure and processes we have today? How does our history blind us or open up possibilities? What can be learned from the choices, processes and decisions that created our historical path? What processes do we have to help us understand our history on an ongoing basis?
Connectedness values	How we connect with each other, how we lead and follow, and our beliefs about effective processes and how they should work	Who are our stakeholders and how will we connect, coordinate and build relationships with each of them? How will we cultivate collective leadership? How can we create dialogue and conversation that generates community within our organization? How do we insure multiple perspectives are nurtured? How will we learn, innovate and change? How will we cultivate innovation and new ideas? How will we insure we bring our whole selves to the organization? How will we communicate? How should power be distributed?
Aspirational values	Our hopes, dreams and how we see ourselves as making a contribution to our stakeholders and the greater good over time	Is our purpose still valid? Why do we exist? What are our shared hopes and dreams? What do we stand for? What makes us unique? How will we contribute to our organization, our community and the greater good? How will we enable our aspirations to evolve? How will measure what matters to us?

one understands why they were put there in the first place or why they continue. Raising awareness of habits and their effects is what introspective values can help shed light on. Time for introspection uncovers cognitive blinders and what they shadow, and forces contemplation about the everyday choices made both inside and outside the organization and their consequences. Introspection and appreciative inquiry can also help uncover the things that are working well in organizations that often go unnoticed because they are running smoothly. From that insight, companies can think about whether that positive deviance might be amplified or diffused (Auster et al. 2005; Whitney and Trosten-Bloom 2003).

Introspective values also might enable examination of our assumptions about our relationship to the world at large and the types of analyses we do. For example, many companies have bought into equilibrium models as a useful way of thinking. Equilibrium approaches stress that macro and competitive forces and dynamics and how they interact is mostly predictable, knowable, and generalizable. Sophisticated analyses enable management to isolate key factors, understand their impact, and respond. In equilibrium views, predictions and plans fail because the right analyses were not conducted or the right variables were not included.

The research of many scholars poses serious challenges to equilibrium views even though they are not always pitched as critiques of equilibrium theory. As Krugman (2009) astutely notes, the economic recession occurred in part because "economists, as a group, mistook beauty clad in impressive-looking mathematics for truth." Another vein of critique stems from what Abbott (2001) calls a "general linear reality." While sometimes things change and evolve in ways that appear "orderly" most of conventional economic analyses downplay things like acquisitions, strikes, new CEO's, the advent of the internet, political upheavals and wars, technological convergence, and the recent economic downturn which all create discontinuous change and make linear predictions dubious. In addition, slower moving changes like global warming often are ignored or set aside "for now" even though their impact will be profound, complex, and wide-reaching.

An alternative view of organizations and how they connect to the world at large is offered by disequilibrium and complexity views (Beinhocker 2007; Dooley 2002; Holland 1995; Meyer et al. 2005; Stacey 1996; Westley et al. 2006; Wheatley 1992). These scholars see the world as composed of nested complex adaptive systems all of which are made up of dynamic networks of many agents (which may represent individuals, firms, nations) acting in parallel, constantly acting and reacting to what the other agents are doing. Control tends to be highly dispersed and decentralized. The overall behavior of the system is the result of people making decisions based on socially constructed information. The context is not this macro, objective force that stands above and outside individual action. Individual choices are driven by social, emotional, and cognitive needs or desires and that those choices are what constitute the context.

In short, equilibrium views see the external world as more reductionist, analyzable, static, objective, and neat where the agents change work macro to micro.

Complexity theorists and others flip that view on its head and instead view the world as holistic, interdependent, historical, and messy with change emerging micro to macro. The result is layers of adaptive systems that are ongoing, path dependent, idiosyncratic, and complex.

We highlight this tension between equilibrium versus complexity based views for two reasons. On the one hand, this type of entrenched paradigmatic thinking offers a provocative example of embedded and often unnoticed assumptions. However, it is also linked to authenticity. Indeed, while the simplification often suggested by equilibrium approaches is seductive, the world moves from micro to macro not the other way around. What seem to be more macro issues, whether they are external "forces," "big challenges" or "them" or positive change, are actually the aggregated result of people making choices. Solutions to these seemingly intractable problems thus rest in the individual decisions and actions of each of us and collectively how we work together. This implies very different approaches to leadership, inquiry, process, connectedness, and introspection.

Thus, through introspection (see Table 37.1), those in an organization can probe what is working well and why and explore how they might amplify those successes. These reflections might trigger insights about what is not working well and why and how they can learn and improve. They also might trigger reflection on actions, or habits that are "assumed" or taken for granted. Introspective values also urge us to reflect on what type of inquiries and research we conduct and how we view our connection to the challenges our organization faces and the world at large. Finally, introspection asks us to ask questions about the questions we ask.

3.2 *Historical Values*

Historical values focus on how we study the past and its impact on where we are today. Many companies operate mostly a historically focusing their efforts on now and the future with little recognition or regard for history. They may give lip service to history in the form of core competencies that should be maintained or recommendations to "stick to the knitting" but they tend view history as an option that can either be built upon or ignored to begin tabula rasa. This type of approach assumes the organization and people within it can be removed from their context and their histories.

Consider Company A consisting of a number of divisions each of which had made historically significant impacts on their industries (creating them in most cases) and on the lives of many communities. While each division had some sense of its original purpose, the company as a whole seemed to have no purpose other than making a specific number for Wall Street. While the company consistently met performance goals, it suffered from a continuous morale issue, and it had ceded leadership in a number of its divisions. In short, it could not figure out how to recapture the pioneering spirit of its rather glorious and inventive past. Historical purpose had been replaced by profits.

This scene is repeated in many entrepreneurial firms who live past the founder and who move to the public domain. The purpose and values that made the company successful is often forgotten as it grows larger, becoming less than it could be. Purpose evolves over time, but without keeping history "alive," it is likely to evolve out of the control of the leaders to whom it is so crucial.

Organizations are idiosyncratic combinations of people and activities, intermeshed in ever-changing contexts that evolve in unique ways given where they have been and where they are going. As Lerner (1997) notes, history offers tremendous insight because it reveals the path dependency across the past, present, and future. We would suggest that history is an important aspect of poetic organizations and enables authenticity for it helps the people within the organization understand the connections between where they have been in the past and who they are now and why. Historical analysis and dialogue provides insight into the key changes and choices the organization has journeyed through and how they led to current circumstances. Historical understanding also solidifies organizational identity by encouraging different groups to grapple with different interpretations of the impact of past choices and their consequences for different stakeholders.

History "helps us see how the choices we make, once made, cannot be undone," and reveals "how these choices foreclose the possibility of making other choices and thus determine future events" (Lerner 1997, p. 205). History can create "imprinting" (Stinchcombe 1965) where founding conditions and related values shape the future because of inertia or vested interests (Stinchcombe 1965, p. 169; Suddaby et al. 2011).

For poetic organizations, taking history seriously involves creating collective conversations about the connection between where the organization has been and why recognizing that these conversations are subjective interpretations of the past (Suddaby et al. 2011). These conversations might also focus on where the organization is now and how these insights aid or hinder the ability to move forward. Historical inquiry moves beyond recording history to questioning and analyzing with history in mind how the organization might do things differently in the future. History also enables those in the organization to see paths for how to transform past history into future positive growth. Dialogue and questioning emerging as part of historical values in the poetic organization might include questions such as what is our purpose (see Table 37.1). It might also explore how interpretations of organizational history converge and differ, or how history shapes current values, routines, structure, and processes or blinds or opens up possibilities. It might trigger analysis around what can be learned from the choices, processes, and decisions that created the organization's historical path and asking what processes can help improve understanding of organizational history on an ongoing basis.

3.3 Connectedness Values

> A responsible leader then can be understood as a weaver of trusting relationships, as a facilitator of stakeholder engagement, and one who balances power by aligning different values to serve both business success and the common good. (Maak and Pless 2006a, p. 41)

Collectively working together requires connectedness. In many organizations, there is a bifurcation between rational "work" and "social time." We believe that relational "work" is the rational work of organizations[3] and that this relational work is what connectedness values are all about. Connectedness values in the poetic organization are oriented toward how people in the community connect with each other in how they lead and follow and their beliefs about effective processes and how they should work.

Although explicit command and control language is not as accepted as it used to be, many organizations still run on a fundamental premise of control through social hierarchy grounded in behavior modification methods. These approaches which dominate much of management practice start by asking the question "how do we get people to do what we want?" Although often clothed in illusions of participation, proponents of this view believe that clear communication from the top on what to do when combined with proper incentives will result in obtaining the behaviors and outcomes that the company desires.

In our view, this type of approach reduces people to assets that should be motivated "correctly." It assumes compliance and respect for those in charge and that their "commands" will result in an orderly organization where people will do as expected and generate the required outcomes on the balanced scorecards and KPI indicators. Take for example this quote from a Harvard Business Review article (Christensen et al. 2006, p. 73), entitled ironically "The Tools for Cooperation and Change." "Managers can use a variety of carrots and sticks to encourage people to work together and accomplish change. Their ability to get results depends on selecting tools that match the circumstances they face."

Perhaps a better question is to ask "why do people disengage?" and then we begin to see that this combination of compliance and behavior modification tools might force short run action or yield "go through the motions" behavior at best. But it will never provide a springboard for energy, passion, creativity—the essentials that build sustainable value creation and poetic organizations. It will more likely generate inertia, anger, and/or exit.

Thus, as we move toward poetic organizations and thinking about connectedness values and authenticity, we move away from a notion of leadership as paternalistic, created by a hero or architect at the top that knows best what to do, tells the people, and forcefeeds strategy with some incentives—a plan, lock, and load approach. Instead, we suggest that leadership processes are likely to be more effective and authentic when they are embedded in a community of mutual respect and are geared toward things that matter to the people. Leading is not about creating mechanisms for control. Direction instead comes from strong purpose grounded in collective aspirations. Leadership focuses on creating space for passion to flourish, with people reinventing their work as they see is needed. Leadership liberates those who work within the organization to figure out what to do, how and when in ways that

[3] Our thanks to Henry Aldrich, Executive Coach/Human Resources at Procter and Gamble/Tremor for his discussions on this idea.

they believe are best. Values offer a means to have an ongoing conversation that brings purpose and passion to life.

Interestingly, if we observe people in organizations, we discover that leadership and change naturally bubble up and emerge continuously from many places and many people unless it is thwarted and especially when it is cultivated. Sometimes change is sparked within the organization, sometimes by customers or other outside stakeholders sometimes intentionally, sometimes unintentionally. Even if an organization attempts top management led "big" strategic leaps, it plays out as ongoing, self-organizing, and enacted. Leadership is distributed across the organization as different people and their units engage in decisions and actions at different speeds. Ideas of tipping points, social networks, diffusion, and social epidemics also all point to the natural emergence of leaders and followers (Gladwell 2000). These social epidemics are triggered by "connectors"—folks with wide social circles, "mavens"—those experts known for their knowledge, and charismatics who others idealize and want to imitate. These "go to people" can powerfully influence decisions and actions and are often important leaders, although conventional approaches might not see them as such. From this analysis, we realize that leadership may be better conceived as a process that occurs throughout most organizations all the time and in many different forms. It might be best understood as how complex human beings interact and connect. It is innately collective and distributed so perhaps a better approach to leadership is to help these natural dynamics flourish.

Economic approaches that have long dominated much of management theory for the most part conceptualize people as rational and self-interested. It is not so much that these approaches are wrong, as that they are radically incomplete. We do not wish to repeat here, the well known critiques of the "rational person" argument. Indeed, in recent years, the fastest growing area of economic theory has been so called "behavioral economics" which essentially deals with the fact that human beings do not act in the purely self-interested way predicted and assumed by many economic models. As pragmatists we prefer to say it this way: for some purposes it is useful to talk as if human beings are narrowly self-interested, for instance when we are explaining theoretical models such as general equilibrium models or Prisoner's Dilemma situations. For many purposes and for most cases of when we are talking about people working together to create value for each other, it is more useful to assume that human beings are quite complex. There are many aspects to this complexity. People are simultaneously rational, emotional, political, and social beings. For our argument to work here, that we need to understand "connectedness" in more nuanced terms, we need only acknowledge this complexity, rather than picking a particular view of connectedness.

There is much research that documents the importance of connection in our earliest development in childhood. From Object Relations Theory in psychoanalysis to more recent studies of how children understand their connections with others, and hence enable themselves to learn language, develop and grow emotionally, and become functioning members of society. This does not seem very shocking if we talk about decisions about our loved ones, or surgery or our upcoming vacation—but somehow most approaches assume we check our emotions at the corporate door.

Creating poetic organizations also involves cultivating and developing conversations and communication channels to insure that a multiplicity of voices can be heard. Education, access to information and awareness are crucial for enabling people throughout the company to be able to contribute.

Beyond "hearing" multiple voices, we advocate joint engagement and shared leadership. This means creating a nurturing context for speaking one's voice and offering one's perspective. It involves respectful dialogue and transparency. It enables the recognition of each person's capabilities and personality and creates space and cultivates their skills and growth so that they and the organization can thrive.

Thus, connectedness values focus on asking questions about how we lead, make decisions, organize and plan our work, generate ideas, and evolve collectively and individually. These values ask how we bring our whole selves and our values to life within our community. In our view, poetic organizations would lean toward connection that is mutual, collective, and ongoing with the goal of co-creating meaningful services and products that better the world in a community of respect with shared ownership and distributed responsibility. Questions that might be asked as part of this dialogue might include who are our stakeholders and how will we connect, coordinate, and build relationships with each of them. It might also ask what we mean by relationships and how we cultivate shared leadership to insure multiple perspectives are heard. As individuals and as a community, how will we support new ideas, innovation and change and how will we nurture our whole selves, our heads and our hearts, through our work (see Table 37.1).

3.4 Aspirational Values

Aspirations have been talked about in organizations for some time and often take the form of a vision statement that focuses primarily on market and competitive positioning. These types of vision statements emphasize how the company is the first choice brand, or the most profitable or how they guarantee excellent service to their customers. In recent years, many companies have expanded the scope of their vision statements beyond competitive positioning and customers to include statements about how they aspire to treat employees.

In organizations that in our view have more "poetic" aspirations such as Whole Foods and J&J, these aspirations also include how the company can make a difference to all of its stakeholders and why its efforts will make the world a better place. They incorporate broad value creation for multiple stakeholders on multiple levels including customers and employees, and shareholders and also expanding that stakeholder network to include the greater good—other living and non-living species, future generations and the viability of the planet (Wheeler and Sillanpää 1997).

While the specific words chosen for a vision statement or statement of aspirations are clearly important, we would suggest that what is more important is the process of how the values are developed, evolve, and are translated into action. Many are familiar with Whole Foods aspirations that are bound up in their Declaration of Interdependence. Their motto—Whole Foods, Whole People, Whole Planet—emphasizes that their vision reaches far beyond just being a food retailer. "Our success in fulfilling our vision is measured by customer satisfaction, Team Member excellence and happiness, return on capital investment, improvement in the state of the environment, and local and larger community support."[4] But what makes Whole Foods a great place to work is not the words but how they live it. In the words of co-founder John Mackey:

> We started with a few simple ideas and core values for the company and then created very simple business structures to help fulfill those ideals. However, over time, as the company grew a process of self-organization took place and layers of organizational complexity evolved year after year to fulfill the original core values. As the original core values were expressed over time, deeper meanings of those core values were discovered and/or created by the interdependent stakeholders. Whole Foods Market's purpose has become deeper, richer, and more complex as it has evolved over the years. (Strong 2009, p. 78)

So, according to Mackey, connectedness of stakeholders has influenced the interpretation of purpose and values as Whole Foods has evolved. Aspirational values are sometimes clearer when companies are started, as many entrepreneurs have a purpose that is more than making profits. However, as growth happens, it is easy for purpose to get lost, and as companies move from private to public, it is easy for profits to replace purpose. Coming to see values as a living conversation that goes on in an organization can ameliorate this substitution, and keep aspiration alive and evolving.

Similarly, Johnson and Johnson has the Credo. While the credo is critical for articulating their view of how they contribute to the world, what distinguishes J&J is how it permeates the organization. "You'll find the Credo part of the vocabulary at Johnson & Johnson, from developing 'Credo-based' leaders to 'Credo-challenge meetings' to 'Credo surveys'. We want to develop Credo-based leaders–broad-gauged, multi-dimensional men and women with superior talents, values and the energy it takes to bring out the best in people and produce outstanding business results," Larsen told employees.[5]

J&J also uses a "Credo Survey" and "Credo-Challenge" meetings as methods for assessing gaps in action and practice—not walking the talk and as springboards for future aspirations. The Credo Survey filled out by all employees every 3 years is composed of more than 100 questions and is used to assess how well the company is living the Credo. The survey results are then woven into "Credo-Challenge

---

[4] http://www.wholefoodsmarket.com/company/declaration.php

[5] http://www.kahaner.com/johnson_johnson.shtml

Meetings" which began 25 years ago. In these "challenge" sessions, employees offer their insights on the survey results and how to better enact the Credo.[6]

Thus we see, that in each of these companies, aspirations do not just speak to what the organization is, they focus on dialoguing about what the organization wants to become. In organizations that in our view have more "poetic" aspirations, they focus on hopes and dreams, but the emphasis is not so much on finding the perfect words but on the processes of inquiry associated with conversations about aspirations. At a fundamental level, aspirational values in poetic communities ask how we make meaning from what we do—the "how do we make a difference?", "why do we matter?" questions (see Table 37.1). The stakeholder groups, scope, and time frame tell a lot about the collective beliefs of the organization and how they perceive themselves and why they exist as an organization. Thus, an essential characteristic of aspirational values in poetic communities is that these values are not only stated in the annual report, or the company web-site or talked about in an annual retreat. Instead, aspirational values in poetic organizations are under continuous scrutiny and discussion and live and permeate the fabric of the organization. Ongoing conversations about aspirational values include questions such as whether our purpose is still valid, what do we stand for, what makes us unique, what are our hopes and dreams, how will we contribute to our organization, our community, and the greater good and how will we enable our aspirations to evolve.

4 Values and Organizational Authenticity: A Brief Conclusion

Like individuals, organizations can be more or less authentic. However, we have suggested that organizational authenticity is more nuanced than simply announcing a set of corporate values, and trying to fit individuals to those values. We suggest that exploring the problem of value fit opens the door for a new conceptualization of values—one that emphasize process and inquiry as much if not more than the specific words chosen. We develop a typology of four interconnected sets of values that leads us to thinking about organizational values as conversation starters rather than conversation stoppers. We believe that organizations that commit to conversations about their values can develop the kind of organizational culture that nourishes authentic individuals, who themselves are struggling with their own project of self-creation. Organizational values in this inquiry must include at least: *introspective values*—reflecting on our collective self and who we are, *historical values*—what we stood for in the past; *connectedness values*—how we lead and our beliefs about effective processes, and *aspirational values*—the why behind our efforts and hopes. We also develop illustrative questions that might trigger these

[6] http://www.kahaner.com/johnson_johnson.shtml

types of dialogue in practice. Thinking about creating authenticity in organizations in this new light helps explain why espoused "core value statements" although well-intended might lead to disengagement and how an approach where values are dialogic and dynamic might lead to greater authenticity.

References

Abbott, A. 2001. *Time Matters*. Chicago: University of Chicago Press.

Ardichvili, A., J. Mitchell, and D. Jondle. 2009. Characteristics of Ethical Business Cultures. *Journal of Business Ethics 85*: 445–451.

Auster, E., K. Wyle, and M. Valente. 2005. *Strategic Organizational Change: Building Change Capabilities in Your Organization*. London: Palgrave Macmillan.

Beinhocker, E. 2007. *The Origin of Wealth: Evolution, Complexity and the Radical Remaking of Economics*. Cambridge, MA: Harvard Business School Press.

Bloom, H. 1973. *The Anxiety of Influence*. Oxford: Oxford University Press.

———. 1976. *Poetry and Repression: Revisionism from Blake to Stevens*. New Haven: New Yale University Press.

———. 1982. *Agon: Towards a Theory of Revisionism*. Oxford: Oxford University Press.

Christensen, C.M., M. Marx, and H.H. Stevenson. 2006. The Tools of Cooperation and Change. *Harvard Business Review 84*: 72–80.

Collins, J., and J. Porras. 1994. *Built to Last: Successful Habits of Visionary Companies*. New York: Harper Business.

Dooley, K. 2002. Organizational complexity. In *International Encyclopedia of Business and Management*, ed. M. Warner, 5013–5022. London: Thompson Learning.

Etzioni, A. 1964. *Modern Organizations*. Englewood Cliffs: Prentice-Hall.

Freeman, E., and E. Auster. 2011. Values, Authenticity and Responsible Leadership. *Journal of Business Ethics 98*: 15–23.

Garrison, J. 1993. A Strong Poet's Perspective on Richard Rorty. *Studies in Philosophy and Education 12* (2–4): 213–221.

Gladwell, M. 2000. *The Tipping Point: How Little Things Can Make a Big Difference*. Boston: Little Brown and Company.

Hall, G. 2011. *Strong Poets*. The Ministry of Insights, March 11. http://ministryofinsights.com/2011/03/14/strong-poets/. Accessed May 18, 2011.

Harmeling, S.S., S.D. Sarasvathy, and R.E. Freeman. 2009. Related Debates in Ethics and Entrepreneurship: Values, Opportunities and Contingency. *Journal of Business Ethics 84*: 341–365.

Holland, J.H. 1995. *Hidden Order: How Adaptation Builds Complexity*. Reading: Helix Books.

Jackon, K.T. 2005. Toward Authenticity: A Sartrean Perspective on Business Ethics. *Journal of Business Ethics 58*: 307–325.

Krugman, P. 2009. How did the Economists Get it So Wrong? *New York Times Magazine*: 36–43.

Lerner, G. 1997. *Why History Matters: Life and Thought*. New York: Oxford University Press.

Maak, T., and N. Pless. 2006a. Responsible Leadership: A Relational Approach. In *Responsible Leadership*, ed. T. Maak and N. Pless, 33–53. London: Routledge.

———. 2006b. Responsible Leadership in a Stakeholder Society—A Relational Perspective. *Journal of Business Ethics 66*: 99–115.

Meyer, A.D., V. Gaba, and K.A. Colwell. 2005. Organizing far from Equilibrium: Nonlinear Change in Organizational Fields. *Organization Science 16* (5): 456–463.

Newton, L. 2006. *Permission to Steal: Revealing the Roots of Corporate Scandal–An Address to My Fellow Citizens*. Hoboken: Wiley–Blackwell.

Palmer, D. 2009. Business Leadership: Three Levels of Ethical Analysis. *Journal of Business Ethics 88*: 525–536.
Rorty, R. 1989. *Contingency, Irony and Solidarity*. Cambridge, UK: Cambridge University Press.
Scott, R. 1981. *Organizations: Rational, Natural and Open Systems*. Englewood Cliffs: Prentice-Hall.
Stacey, R. 1996. *Complexity and Creativity in Organizations*. San Francisco: Berrett-Koehler.
Stinchcombe, A. 1965. Social Structure and Organizations. In *Handbook of Organizations*, ed. J. March. Chicago: Rand-McNally.
Strong, M. 2009. *Be the Solution: How Entrepreneurs and Conscious Capitalists Can Solve All the Worlds Problems*. Hoboken: Wiley.
Suddaby, R., W. Foster, and C.Q. Trank. 2011. Rhetorical History as Competitive Advantage. *Advances in Strategic Management 27*: 147–173.
Westley, F., B. Zimmerman, and M.Q. Patton. 2006. *Getting to Maybe: How the World is Changed*. New York: Random House.
Wheatley, M. 1992. *Leadership and the New Science: Discovering Order in a Chaotic World*. San Francisco: Berrett-Koehler.
Wheeler, D., and M. Sillanpää. 1997. *The Stakeholder Corporation*. London: Pitman.
Whitney, D., and A. Trosten-Bloom. 2003. *The Power of Appreciative Inquiry*. San Francisco: Berrett-Koehler.

Ellen R. Auster is Professor of Strategic Management at the Schulich School and Executive Director of the York Change Leadership at York University.

R. Edward Freeman is University Professor, Elis and Signe Olsson Professor, Academic Director of the Institute for Business in Society, and Senior Fellow of the Olsson Center for Applied Ethics at the University of Virginia's Darden School of Business.

Chapter 38
Leveraging the Creative Arts in Business Ethics Teaching

R. Edward Freeman, Laura Dunham, Gregory B. Fairchild, and Bidhan L. Parmar

1 Introduction

For the last 20 years or so, we have been teaching business ethics using the creative arts. We believe that the creative arts offer a number of important leverage points to make a difference in the lives of our students. However, we also believe that such teaching embraces a rather different idea of business ethics and its possibilities. The purpose of this paper is to explain this different kind of teaching and to suggest that making some small changes to traditional business ethics courses via the creative arts can yield great benefits. In Literature and Theater Courses section we explain how two main courses that use the creative arts are structured. In Leverage Points section we articulate three main leverage points as we see them. In What the Students Tell Us section we share some of our students' insights. In Leveraging Key Design Principles section we extract some principles from these courses and address their implications. Finally, in Conclusion section we argue that such teaching requires a different view of the field of business ethics, in line with the pragmatist tradition. As such, we need a new research agenda that comes from this way of teaching.

Originally published in: *J Bus Ethics,* 131, 519–526 © Springer Nature, 2015
Reprint by Springer, DOI 10.1007/s10551-014-2479-y

R. E. Freeman (✉) · G. B. Fairchild · B. L. Parmar
University of Virginia, Charlottesville, VA, USA
e-mail: FreemanE@darden.virginia.edu; parmarb@darden.virginia.edu

L. Dunham
University of St. Thomas, Saint Paul, MN, USA

© Springer Nature Switzerland AG 2023

S. D. Dmytriyev, R. E. Freeman (eds.), *R. Edward Freeman's Selected Works on Stakeholder Theory and Business Ethics*, Issues in Business Ethics 53,
https://doi.org/10.1007/978-3-031-04564-6_38

2 Literature and Theater Courses[1]

Medicine, law, and business are all deeply human professions. Yet, increasingly in today's society, each depends on a substantial body of technical knowledge that requires mastery. Professional schools have to be on the cutting edge of this technical knowledge, and in large research universities, faculty must be committed to producing the next generation of such knowledge. However, it becomes crucial for both faculty and students not to lose the idea of 'deeply human profession' in the face of this technical knowledge.

This has led to a real paradox for business schools. In recent years, we have witnessed a number of critiques of business schools, especially after the Global Financial Crisis of 2008. Pfeffer and Fong (2002) have suggested that management researchers and business schools misunderstand the nature of research, and that most of the so-called "technical knowledge" is not based on evidence. Mintzberg (2004) has argued that management is an art best understood at a later age through a process of apprenticeship and experiential learning. Freeman and Newkirk (2011) have suggested that these critiques miss the mark, since they misunderstand business. According to them, business is a deeply human activity through which we create value and trade with one another. It is fundamentally a cooperative and collaborative activity that relies on understanding the particularity of human interests, needs, and emotions. It is much more than 'the physics of money'. There are whole human beings involved in the delivery and reception of services, the building of organizations and institutions, and the responsibilities for business education. Whole human beings have values, emotions, families, and otherwise complex lives. And whole human beings engage in authority and power relationships. There is substantial conflict and opportunities for self-dealing as well as cooperative behavior.

We need to draw on the full palette of the humanities to understand the richness that is inherent in business. And we should draw especially on the creative arts such as literature, theater, music, and the like. For the last 20 years, we have referred to these disciplines to teach MBA students at the University of Virginia's Darden School. Two main courses have generated a sustained interest from the students: (1) Business Ethics Through Literature (BEL) and (2) Leadership, Ethics and Theater (LET).

In BEL, the students read novels and short stories. They meet in small dinner groups to discuss the week's readings, and then come to class prepared to discuss the book or story. All of the teaching is done via Socratic questioning rather than

---

[1] In addition we have taught a course using music where the students, of widely varying musical ability and training, must learn to play music together and to create an original performance. We also use many additional experiential exercises that are oriented around getting the students to understand how meaning is created collaboratively in organizations. We believe that there are many more possibilities using the creative arts to teach business ethics, especially if we adopt a more pragmatist version of the discipline. We are also grateful to Jenny Mead and Randy Strawderman for their many contributions to these courses.

lecture. The aim is to create a conversation about the book on its own terms, followed by an examination of how the book's themes can yield insight into the lives of the students who are soon to be executives. It is crucial to get the students to examine the book in its own right, in order to develop their skills in analysis and judgment, rather than trying to glean answers to "how does this book apply to business?". It is only by digging deeply into the texts that students can understand the humanity of the characters. If there is only a straightforward application of the book to business, the students run the risk of recapitulating the standard story of business, rather than questioning and critiquing it. In addition, there are a number of creative writing assignments. The final assignment is to write a short story, and the course concludes with a "fiction reading" of the students' stories.

Typical texts are Fitzgerald's *The Great Gatsby*, Achebe's *Things Fall Apart*, Baldwin's *Sonny's Blues*. All of these texts are filled with ethical questions. For instance, in Gatsby, a discussion of who if anyone in the book is heroic, leads to an examination of what the students intend to do after business school is over, and how their goals can become so single minded that they forget their "green lights" must be set within a societal and ethical context. They can see the particularity of the Gatsby–Daisy–Nick relationships, and the importance, as well as the barriers, to love. They ask whether Nick should help Gatsby with Daisy, and what is the role of monogamy in life. They begin to talk about and understand the role of purpose, the importance of real friendship, and the need to constantly question one's authenticity.

These same themes are present in Achebe even though the context is quite different. Okonkwo's juxtaposition with the Christian missionaries illustrates one of the key ethical challenges of globalization. Moral relativism is a critical theme here, but Achebe brings it to life again through the particularity of the relationships in the village and between the villages. The arrival of the multinational outsiders, the Christian missionaries, changes the culture so much so that Okonkwo commits suicide, a taboo for the Ibo as well as the Christians. Things Fall Apart is simply the best text we have found to understand the moral complexities across culture, and the theoretical issues of moral relativism.

In the Baldwin story, we find the conflicting forces of conformity and individuality in the characters of Sonny, the jazz musician, and his brother, the schoolteacher. Baldwin intimates how a sense of self needs a harmony between these forces, and the end of story introduces us to Creole, whose very name suggests such a harmony. The real world of race, drugs, and religion is a constant pressure on the characters in all of Baldwin's work, and Sonny's embracing of a musician's life. The conversation where his brother says that people cannot always do what they want, and Sonny's response that we should all do exactly what we want, engenders a real conversation among the students about their lives post-MBA. They embrace the idea that their lives will be filled with moral issues, some of which may not be easy to solve, or not solvable at all.

We ask the students to imagine themselves as Gatsby, as Nick, as Daisy, as Okonkwo, as Sonny and his brother, and to see the world from their eyes. In doing so, we hope to get the students practice their ability for perspective-taking and empathy and broaden their worldview and be sensitive to how very different people can live. Many business ethics courses rely on ethical dilemmas. While there are dilemmas presented in fiction, they are much more contextualized and rich, and therefore more realistic and more difficult to solve. They are not separated from the rest of the characters' lives, as they so often are in business school dilemma cases. This focus on particularity makes the ethics conversations more about the students' lives and less about moral theory. The students are encouraged to be on the "inside of the situation" rather than approach it as a spectator. In essence, these ethics conversations become what Dewey called "dramatic rehearsal" (Dewey 1957).

The writing assignments give the students the idea that crafting a narrative is a creative process but one that draws on the student's experience. Sharing their narratives with others in class is an experience in becoming vulnerable. It leads easily to the idea that the students will be crafting their own narrative when they graduate, and that such a narrative is likely to be influenced by context, relationships, and offers the chance to design one's life.

The second course that draws on the creative arts (LET) focuses on forming a theater troupe of the 25 students in the course. This class takes the students a step beyond the process of *learning from* theater, through reading, discussing, or reflecting on a play and its characters, to actually *creating* theater. The course not only invites them to richly imagine themselves as the key protagonists, but to actually *become* them, to find the authentic connection between themselves and character and to bring those characters—in all their complexity and contradictory messiness—to life. To add to the challenge, students must undertake this journey embedded in a context of other actors/students seeking to do the same, and all with the meta-goal of stitching these individual human stories into a cohesive whole, an overarching narrative that imbues each character's story with meaning and purpose.

The course is usually taught as a 1 week intensive where it is the only course that the students are taking.[2] The students are told on Day One that the last day of the course will be a public performance of an original play that the students will write, direct, produce, and perform. The course begins with some acting exercises about becoming a character, believability, subtext, and creating stories. Guided through the exercises by a professional director, who is brought in as a full-time co-instructor, the students are encouraged to begin opening up, drawing upon and sharing real experiences and emotions in the exercises and subsequent improvisations. Of equal importance, they are encouraged to listen and respond honestly to their colleagues as they build scenes together. Once again, the idea of creating a narrative looms large in the course.

Day Two consists of a monolog exercise, a directing workshop, and the beginnings of rehearsal for a set of one act plays. Students begin the task of analyzing and interpreting their scene or play. While discussion plays an important role, much of the work is done with actors and directors on their feet, script in hand, considering and trying multiple interpretations of a scene. Students probe deeply to understand

---

[2] There is also a 6 week version of the course. The following website is a resource about the course, and there is a documentary that focuses on what the student experience is. See: http://it.darden.virginia.edu/leadershipandtheater/.

character, motivations, relationships revealed through the scene. As new insights emerge, the students try the scene again, informed by these new understandings. Students are encouraged to stretch their thinking and consider multiple, often contradictory interpretations. Students learn the role of directorial and actor choices regarding the material—choices in terms of how to interpret the characters, actions, and themes of the play and then how to bring those understandings to life in honest performance. As they explore these options, students come to understand the importance of looking within to find their common experience with the many characters they play, in order to access real empathy and to spark genuine feeling and truthful portrayals. They also learn the importance of close listening to their partners onstage, to letting go of their preconceived notions in order to respond honestly to what is happening in any given moment on stage.

The students perform these one act plays on Day Three. Each of the first 3 days have writing exercises, both "free writes" and an exercise where the students write a short play in 20 min. On Day Four, the students choose a theme to write scenes around, and writing commences. Now students are challenged to go beyond creating characters and believable interactions within the constraint of an existing narrative to actually crafting the narrative itself. At the end of the day, we have a reading of all the scenes and then pick some to perform the next day, and then set about trying to put the scenes into a context that makes a coherent whole. Day Five is rehearsal, dress rehearsal, technical rehearsal and a performance where the community is invited to attend. The performance, a culmination of the week's work, takes students from the safety that has developed within the enclosure of the troupe, to putting all at risk before a group of their peers who haven't shared the week's experience. Now, the troupe must fall back on the work, the trust, the collaborative ethic that has developed over the week in order to overcome the vulnerability of sharing their most personal offerings—the ideas embedded in their narratives and the real emotions, motivations, quirks they must reveal through their characters. The final assignment is a reflections paper, where the students reflect on what they have learned about themselves and about leadership, ethics, and business.

The ethics component of the learning may seem weak to some. While much of the dramatic and comedic material centers on questions of ethics, the learning is subtler. Bringing a character to authentic life is, at its core, a highly ethical enterprise requiring the actor to confer upon that character—no matter how small the role or how distant from the self conception of the actor—the dignity of understanding. With compassion, with empathy, the actor must find the worth, and more importantly, the personal connection to the individual they hope to bring to convincing life on the stage. Once again, the students confront the issues from the inside rather than as spectators, focusing again on Dewey's "dramatic rehearsal". The process of becoming a 25 person theater troupe is based on the ability of the students to collaborate, to trust, and to connect with each other on a very human level. In short, they are practicing ethics through the creative process. They must get over a sense of embarrassment that they have to expose their vulnerabilities during the week. They must learn to give and receive feedback in the directing process and to manage conflict and different opinions constructively. The bond of performing together creates a willingness to be open to experimentation and authenticity. We tend to think of theater as the land of the divas and big overstated emotions. However, truly great theater requires actors and directors to not only put aside their own egos but to also genuinely act with generosity toward each other.

There are many lessons around leadership and ethics for the students. None can rely on "expertise" or "position", so leadership is "leadership by choice" (Burns 1978). In addition, there is a moral component of leadership here. The pressure of a performance in front of their community members and friends is quite real. Without relying on expertise or position, the students must rely on judgment and trust. They must, as leaders, create a conversation amongst themselves that has meaning and that gives each person motive force. The shortened timeframe adds intensity to this pressure, and there are often interpersonal difficulties to be overcome. In short, the students experience each other as fully human, collaborating together to create a performance literally from nothing.

In the moments after the performance, you can sense the feeling that they have done something extraordinary, which none of them could have imagined 5 days before. And it worked because there was mutual commitment, trust, authenticity, and much hard work. These ideas are the very basic stuff of ethics, and the students realize that they have formed a special bond with each other based on moral ideals.

3 Leverage Points

Leverage points are the idea that relatively small changes can produce big results. And there are multiple levels on which they work. Adding courses like the ones we have described to a business school curriculum is a fairly small change that can produce significant results. But it is not necessary to add the whole courses. One can imagine a more traditional business ethics course around lectures and cases, which uses a few short stories, or a novel, or a play. For instance, adding Melville's story, Bartleby the Scrivener, or Boll's Action Must Be Taken, or any number of stories by Baldwin or Updike or Oates can expose students to these stories without having to make drastic changes to a course. Having the students read classic plays like Hamlet or No Exit, or more business-oriented plays like Glengarry Glen Ross and then perform some of the scenes can open their eyes to the richness of experience. We see these texts as leverage points for three main ideas.

3.1 Connection to Self

Even at our most experiential in case method, we typically ask students of business ethics to put themselves in the shoes of a case protagonist in order to assess a situation and determine a course of action around a business problem. But even at its best, a case study requires that we step only lightly into those shoes, just enough to grasp case data specific to the problem so that we might better wield our technical tools and skills. While we as instructors might seek to spark deeper reflection on the ethical and moral issues of the case, our understanding of the protagonist is often thin, archetypical, intellectual, centered on them as a business person operating within a narrow managerial context. And so our personal connection to the protagonist is often limited and the learning more centered on technical execution. But fully apprehending the ethical implications of a problem, and more importantly, fully committing to an ethical response requires a deeper engagement of the self.

Literature and theater provide this opportunity. Exploring literature and creating theater requires we plunge much more deeply into questions of character, motivation, values, beliefs. We must understand the characters if we are to make sense of the larger narrative. But to understand those characters at the deepest level, we must be willing to plunge more deeply into an examination of ourselves. For, ultimately, carrying out rich literary analysis or creating a truthful portrayal on stage requires that we find the commonality between ourselves and the protagonist—that with empathy, compassion. And brutal self-awareness, we examine those elements of our own character, motivation, values, and belief that connect us to characters who, often, could not seem more different from us.

Thus, through the power of narrative and the exercise of the kind of creative muscles involved in literary analysis and theater, we create a stronger connection between self and many other characters and situations. Such a personal connection is a critical leverage point for more fully engaging our students in the rich, nuanced, human problems facing the business practitioner and inspiring a more powerful moral response.

3.2 Connection to Complexity

Similarly, it is through this process of literary analysis and character development that we confront students with the messy complexity of the contexts in which we as humans operate. One must delve far beyond the surface "case facts" to fully explore a piece of literature and create compelling, believable performance. Yet, it is the magic of great narrative that it makes this task less daunting, that it sparks curiosity, that it invites us—compels us even—to grapple with complexity. We must examine and reconcile multiple layers of narrative and subtext—digging deeply into motivations, relationships, actions, reactions, histories, themes—in order to make sense of a piece of literature and bring it to honest life on the stage. The willingness to immerse themselves in the mess, to appreciate the nuance, to search for the deeper meanings provides a critical leverage point for enriching the way our students come to analyze and respond to the complex problems they will address.

3.3 Connection to Each Other

At the end of the day, this kind of pedagogical approach can only work if the students are willing to approach the materials with openness and a willingness to be vulnerable to each other. Whether it is sharing their own connection to a piece of literature, or listening to another's short story, or building a scene together that requires honest emotion and response, students must learn to trust each other and participate together in generous collaboration. Each of these activities entail risk, yet by the end of the semester, we have regularly seen students joyfully take these risks together and come out the other side invigorated and affirmed by the process. They see their role in the collective differently and have new appreciation for the kinds of creativity and productivity they can engender through trust and collaboration.

4 What the Students Tell Us[3]

The transformations and learnings that occur are best understood in the words of the students. The following is typical of student comments about the courses:

> As one might expect, it is extremely hard to get twenty two type-A MBA students to do something together, but we did it. It was not because we had a great leader leading us. It was because we all were leaders leading each other.

> Theater helps you to see how and why we as human beings work well together and how to create a situation that will lead to the most creativity and productivity.

> Leadership…is about being part of a team; it's not about giving orders, but about working together…to achieve a common objective.

> It was great to get out of the business school mindset of cases and non-fiction business books and read some fiction again. It made us talk about deeper issues than business issues and think about what we want out of our lives of careers and what responsibilities we have to our society.

In these comments, the students suggest that authenticity and a more democratic idea of leadership organized around a purpose is in fact more effective than the traditional idea based on authority. A number of other students comment on the importance of trust and empathy and getting more in touch with their emotions:

> I quickly learned that when we remove our MBA hard-hats, we are able to develop a trust and understanding for each other that can significantly improve the dynamics of teamwork.

> The experience of becoming someone else taught me the importance of really putting yourself in someone else's shoes in order to understand their motivations.

> Loved the opportunity to do some critical literary thinking. Crawling into another person's mind gives each of us the chance to understand the actions, justifications, and motivations

[3]The student comments are a selection from the last 10 years and range over both BEL and LET. Surprisingly there are very few negative comments.

> of other people, which in turn allows for an empathy we may not naturally have (I know I don't).
>
> I loved the focus on our development as complete people who can talk and learn with their classmates in more than just a business and networking way.
>
> I think it's important to recognize that we are not always business people and that there are more qualitative methods of evaluating problems and that people and feelings are a bigger part of the world than business school would lead us to believe.

In addition, the ideas of working together relatively free of ego, and pushing each other so that everyone challenges themselves and gives and expects authentic feedback from others add layers of complexity to the learning of the students. For instance,

> The most importance thing I will take away from the class…is an understanding of the power of work free of ego….I can count the times on one hand that I have been a part of a group whose purpose was clear and meaningful enough for the whole of the group to shove ego aside and be about the craft.
>
> I liked that it brought me out of my comfort zone. it's important to have a different class experience like this while at Darden. I liked learning about literature and hearing classmates share their thoughts on things we read. I wasn't enamored with all the materials we read but I thought all our discussions were interesting. I liked that our final assignment was writing a fiction short story, again it's good because it got me out of my comfort zone.
>
> The participants of the class created an environment where feedback was not only accepted, it was expected.

It is very difficult to get students in a traditional classroom to viscerally feel the importance of shared leadership, collaboration based on trust, the power of ego, and the learning that is possible through authentic conversation and feedback. Yet the theater provides a venue where these traits are a necessity for success (Dunham and Freeman 2000).

5 Leveraging Key Design Principles

From our experience with these methods and ideas, a number of key design principles have emerged that now find a central place in our teaching. We believe that these principles can themselves serve as leverage points to transform the way that we approach the teaching of business ethics.

5.1 The Power of Narrative

At the heart of humanity, in life and in literature, is the weaving together of the complex, messy lives of imperfect humans into a narrative that, at its best, reveals new insights—sometimes humorous, sometimes heartbreaking—about how we should live. We humans are story tellers at heart. Nowhere is this clearer than in business. Understanding narrative through literary analysis, and building narrative as a writer, actor, or director, requires that we develop sensibilities and capabilities that are, in some ways, the reverse of those we seek to develop in other parts of the business school. As managers, we often seek to reduce complex phenomena to a few bullet points or a two-by-two matrix; as artists, we must attempt the opposite, uncovering the many levels of complex phenomenon embedded in seemingly simple action. It is not enough to know that a character is angry; the reader, actor, or director must uncover the envy, fear, insecurity, and/or other emotions, beliefs, motivations that drive and color that anger. And it is not enough to understand one character's immense complexities; the artist must understand how each character contributes to the overall, coherent meaning in the larger story.

Carrying out this complicated task requires that we help the student as artist to connect the characters and their journeys to their own narrative, to see the commonalities of experience, hope, dreams, despair, humor that link us personally to characters who could not seem more different from us. We have to confront our own truths in a deeper, richer way. We need to be willing to be vulnerable to ourselves and the other members of the class who are undertaking the same arduous task. And so, the very act of making these connections can make our own narrative a bit clearer and our linkages to others more compelling. This process involves intimate reflection and public sharing of how each of our journeys is simultaneously *unique* and yet *common*.

These linkages become key as the student/artists try to pull the larger story together. Understanding my own character requires that I understand what it is I need and want from the other characters, how they have shaped me and my action, how their own narratives shape their response to me. I must elicit their narratives in order to understand my own, and I must respond in an honest and coherent way to what they throw my way.

One small leverage point here to facilitate this awareness is through our readings of short fiction and writing exercises. For example, we might use Isaac Asimov's *Robot Dreams* to illustrate how power is leveraged in the workplace. Asimov's story has three characters: A boss, her subordinate and a robot (some argue that there are only two characters). The story is driven by dialog that illustrates how power and deference are shown in language. Following the reading of this story, we ask that students write a short story using dialog as the vehicle to illustrate relations—"to show, not tell" through what characters say and when they choose not to respond at all. Our students come to realize the extent of their reach into others' lives—that their decisions have real and lasting influence on others.

In theater rehearsal, as the director and the cast seek to make sense of the emerging narratives of characters and their interactions, the larger story becomes an act of collaborative sense making, a negotiated sense of meaning behind character, plot, emerging themes. Unlike other class projects, where students can divide and conquer—"you do this part, I'll do that part, and we will meet before the day is due to put it all together"—this work can only happen live, as the troupe explores different interpretations and each actor responds to new actions and meanings.

Another leverage point is one of our writing exercises that involves creating a description of a place everyone is familiar with—a student common area at our institution. We ask our students to share their descriptions with other members of the class. Appreciating the differences in shared experiences aids our students in understanding that meaning is simultaneously shared and personal. This is one of the key steps in managing in a diverse world.

Many of our narratives feature broken, depressing, imperfect people in difficult situations (something like real life). Our characters are often written from the perspective of marginalized persons. The protagonists are difficult at first, because there is a tendency for much of the MBA curriculum to present lionized, world-striding masters of the universe. Because this notion of success is so strongly modeled within the business school world, some of our students have adapted a dominant mindset without critique. We hope that, by choosing narratives about those at the margins, our students can truly see the center. Often, early in the classes, students will comment on how "depressing", "powerless" and "unhappy" some of the narratives make them feel. We ask them to consider that there are many who don't expect that those feelings would be ones that they could escape.

5.2 *Exercising the Creative Muscles*

Yet another leverage point is to approach teaching ethics by driving learning to the confluence of thinking/feeling/doing. While case studies activate rich thinking, literature and theater classes require the same intellectual interrogation, while also engaging the student in deeper feeling and emotion. The theater takes that even further by requiring the doing, asking students to physically embody the learnings of their intellectual and emotional explorations. Students must respond to the situation at hand using their repertoire of skills, emotions, and body language—they get immediate feedback from their classmates' reactions and thus can learn about how they might have done things differently. They also learn by listening and watching each other, and bring new tools into their repertoire. It is this embodiment of the character and the narrative that is a source for insight.

5.3 *The Role of Attention*

Creating an authentic and believable moment on stage means responding in an honest way to what is actually happening. While the production is shaped through the process undergone in rehearsal, the ongoing performance requires a freshness that brings the play to vivid life for each audience. This requires that the actors are involved, focused, and 'mindful' at each moment. Rather than falling into habit, delivering lines and taking movement automatically, good actors seek to 'stay in the moment' and respond authentically to any subtle change that occurs on stage. We encourage students to 'be in the moment', staying present to what others are saying. In literature, this builds listening and conversational skills that can improve the understanding of the text and the self-understanding of the student. In theater, the artist must constantly be aware of everything happening on stage. All an actor has is the line that others give. So it is imperative to stay in the moment and "tell the story" regardless of dropped lines. Of course, there are multiple ways to deploy attention in writing and on stage. Students experiment with different ways to attending to the environment and each other, and learn the strengths and weaknesses of specific ways of paying attention.

5.4 Collaboration and Trust

Theater and the performing arts often suffer from an image of a field dominated by egos and 'divas.' In fact, great theater only happens when a troupe of actors come together in a spirit of generosity toward each other, willing to put their own egos on hold, in order find the larger truths of the play and bring something real and compelling to life. Honest performances require actors to take risks sharing something personal of them; compelling scenes require that each actor shares the story and responds mindfully to each other. A troupe of actors can only do this successfully if they have built a level of trust and collaboration that allows them to take these risks in rehearsal and later in front of an audience. There is a similar necessity in sharing your writing with a small set of peers as you work through your story in progress, and in sharing the final product in a group read-through with all of your classmates. By sharing the work of writing, acting, and directing, students develop the ability to jointly author their work, and learn how to contribute in ways that benefit the group as well as themselves.

6 Conclusion

We have learned some substantial lessons from our work in leveraging the creative arts to teach business ethics. First of all, it is important to create a safe environment for students in order for them to work with the kind of openness and willingness to be vulnerable that is critical to making the class successful. The dinner parties are one vehicle we use in the literature class to build this sense of safety and trust. Small groups meeting over a meal can form the kinds of bonds that allow that trust to develop. In the theater class, we begin by sharing theater experiences, and with a conversation about why the students are taking the course. By acknowledging that some have stage fright or perceive themselves as not good at public speaking, we create a safer space for them. We then use acting exercises that slowly increase in the kinds of demands they make of students. And we acknowledge that good theater is honest, and that we are not a professional theater group. As teachers, we are constant cheerleaders and providers of feedback. The result is that the students can focus both on the craft of acting and writing as well as on the lessons learned.

Students should address a mix of literature, some of which they can more easily relate to but much of which challenges them to empathize with people and situations that are difficult for them to access. The former helps build their confidence, while the latter do the important work of stretching them creatively and emotionally. In the theater course we do not let the students write about their experience in business school. We want them to reflect more broadly on their experiences in life.

Third, we do not hesitate to get help from the professionals. We have often invited professional writers to class to discuss literature and their writing. In the theater class, we have greatly benefited from having a professional director co-teach the course, and to use former actors as resources. These kinds of exercises and assignments can be very foreign to a business school instructor, but over time, with help, they can become almost second nature.

Fourth, we try and practice what we are trying to teach the students. We try to be in the moment at all times, and to be conscious of the underlying narratives that are being woven into the fabric of the course by teacher and students. We have come to see teaching itself as a performance art, and we believe that our teaching in non-creative arts settings has improved enormously because of our work in literature and theater.

Finally, our experience with this teaching leads us to a kind of skepticism about much of the current thinking in business ethics, at least as the discipline is understood in the Anglo-Saxon and management theory world. Business ethics is based almost entirely on an analytical approach to ethics. The trifecta of theory (deontology, consequentialism, and virtue) has become an oligopolistic group that seems to us to be in need of substantial revision. Often we focus on issues or dilemmas, and then try to apply one or more of these theories or their surrogates to resolve the issue or dilemma. While there is nothing wrong with using cases and theory (in fact, we use them as well), they simply do not go far enough.

We want to suggest that human experience, especially in business, is much more nuanced and more holistic. We bring a substantial and complex emotional, cognitive, and developmental palette to our lives as business persons. We see particularity and particular others (Benhabib 1992), not generalized and undifferentiated others. Thick descriptions abound. While there is room for Kant and thin descriptions and obligations to strangers in moral discourse, so too must we make room for the trust and compassion that exists (or not) in families and friendships. We must leave room for Sartre's authenticity and Kierkegaard's existential pain and Nietzsche's critique of the meaningless life.

We create meaning with each other in business. The very language that we use constructs what problems we are able to solve. The analytical language of most business discourse ensures that we will not be able to solve nuanced human problems. In addition to the dilemmas that require right–wrong judgments, we believe that ethics is best conceptualized as a conversation about how we describe and re-describe self, other and communities to live together and collaborate in making a better world. Using the creative arts to teach business ethics has given us a set of questions that engages us as researchers and scholars as well as teachers, for instance: (1) what are some alternative views of the connections between the self, others, and the community, and how is meaning constructed in these alternative views? (2) What would the discipline of business ethics be like, if we took as background disciplines the creative arts and humanities, rather than just philosophy and the social sciences? (3) How do complex emotions affect the construction of meaning and the solving of problems, and in particular, how are trust, collaboration, leadership, love, fear, and related concepts connected?

The humanities, and especially the creative arts, offer a way to leverage the idea that business is a fully human institution in all of its complexity. We need to continuously connect business with our culture and our humanity, in order to connect with each other. We look forward to the evolution of the teaching and research in business ethics in this direction.

References

Benhabib, S. 1992. *Situating the Self. Gender, Community, and Postmodernism in Contemporary Ethics*. New York: Routledge.

Burns, J.M. 1978. *Leadership*. New York: Harper & Row.

Dewey, J. 1957. *Reconstruction in Philosophy*. Boston: Beacon Press.

Dunham, L., and R.E. Freeman. 2000. There is Business Like Show Business. *Organizational Dynamics 29* (2): 108–122.

Freeman, R.E., and D. Newkirk. 2011. Business School Research: Some Preliminary Suggestions. In *Business Schools Under Fire: Humanistic Management Education as the Way Forward*, ed. W. Amann, M. Pirson, C. Dierkmeier, E. Von Kimakowitz, and H. Spitzeck, 273–290. London: Palgrave.

Mintzberg, H. 2004. *Managers Not MBAs: A Hard Look at the Soft Practice of Managing and Management Development*. San Francisco: Berrett-Koehler.

Pfeffer, J., and C.T. Fong. 2002. The End of Business Schools? Less Success Than Meets the Eye. *Academy of Management Learning and Education 1* (1): 78–95.

R. Edward Freeman is University Professor, Elis and Signe Olsson Professor, Academic Director of the Institute for Business in Society, and Senior Fellow of the Olsson Center for Applied Ethics at the University of Virginia's Darden School of Business.

Laura Dunham is Dean of the Opus College of Business at the University of St. Thomas in Minnesota.

Gregory B. Fairchild is the Isidore Horween Research Professor of Business Administration at the University of Virginia's Darden School of Business, and Dean and CEO of UVA-Northern Virginia.

Bidhan L. Parmar is the Shannon G. Smith Bicentennial Associate Professor of Business Administration at the University of Virginia's Darden School of Business.

Chapter 39
Practicing Human Dignity: Ethical Lessons from Commedia dell'Arte and Theater

Simone de Colle, R. Edward Freeman, Bidhan L. Parmar, and Leonardo de Colle

1 Introduction

The creative arts—including music, theater, literature and cinema—can offer useful insights to management scholars—particularly in the field of organizational theory (e.g., Barry 1996; Weick 1998; Barrett 1998; Barry and Meisiek 2010) and management education (e.g., Dunham and Freeman 2000; Salgado 2008; Lesavre 2012; Freeman et al. 2014). Theater, in particular, has been investigated by management scholars highlighting its potential to help us better understand elements of improvisation (Vera and Crossan 2004), change (Meisiek and Barry 2007) and leadership within organizations (Dunham and Freeman 2000). Our thesis is more specific. We argue that theater can be used to make our organizations more human, especially with regard to the promotion of human dignity. We proceed as follows: after the introduction we examine the concept of human dignity as put forward by a number of philosophers and social scientists. We suggest a broad conception of "treating others with human dignity" as "recognizing and honoring the full complexity of the human person." And, we suggest a broader idea of "acting with

Originally published in: *J Bus Ethics*, 144, 251–262 © Springer Nature, 2017
Reprint by Springer, https://doi.org/10.1007/s10551-015-2898-4

S. de Colle (✉)
IESEG School of Management (LEM-CNRS), Paris La Défense cedex, France
e-mail: s.decolle@ieseg.fr

R. E. Freeman (✉) · B. L. Parmar
University of Virginia, Charlottesville, VA, USA
e-mail: FreemanE@darden.virginia.edu

L. de Colle
Actor at Piccolo Teatro, Milan, Italy

© Springer Nature Switzerland AG 2023

S. D. Dmytriyev, R. E. Freeman (eds.), *R. Edward Freeman's Selected Works on Stakeholder Theory and Business Ethics*, Issues in Business Ethics 53,
https://doi.org/10.1007/978-3-031-04564-6_39

dignity" to include treating others with dignity as well as "acting from the point of view of living a good life." Acting with dignity encompasses sensitivity to both culture and to the dignity of others. At the end of that section we suggest a relationship between treating others with dignity, acting with dignity, and the creative arts. In the following section we look at a particular form of theater—namely, the Commedia dell'Arte—and show how its main themes are concerned with dignity. With regard to this specific form of theater, the main source of our information, besides the literature referenced, is based on one of the co-author's personal experience as a professional actor specialized in Commedia dell'Arte.[1] Additionally, reflecting on the other coauthors' personal experience as business ethics professors, we then take these themes and show how they can be developed into practical exercises for executives and students in a business school setting that can be used to promote human dignity in our organizations and classrooms. In both cases we highlight how the creative arts can be important for issues of moral development in organizations, and how they can lead to a more authentic conversation about values, ethics, and meaning.

2 Human Dignity in Organizations

Much has been written about the concept of "human dignity" but there is no standard agreed upon conception, nor any social science typology that lends rigor to the idea. Two connected uses of "dignity" are "treating others with dignity" and "acting with dignity." While these two conceptions are related, it is important to note that most uses of dignity that we have found involve thinking about how to treat others. We

[1] Our approach is action research oriented, as we are participant observers in the processes described here. For instance, Leonardo de Colle is a professional actor of the Piccolo Teatro of Milan with more than two decades of experience, and a teacher of the theater's *International Academy of Commedia dell'Arte* directed by Ferruccio Soleri, giving classes in Italy and abroad. He has been performing in various roles in "*Arlecchino Servitore Di Due Padroni*," the show directed by Giorgio Strehler since 1947 continuing the Italian tradition of Commedia dell'Arte that has been represented over a thousand times in theaters all around the world including Moscow, Vienna, Beijing, Paris, Auckland, Buenos Aires, Tokyo and New York. Arlecchino is, in fact, the Italian theater show most seen in the world (http://www.piccoloteatro.org/pages/la-storia-del-piccolo-teatro-di-milano/giorgio-strehler) and is currently staged again in Milan for the 2015–2016 season. After having many informal discussions around our common interests for theater and business ethics over the years, we first conceived the first draft of this article after a weekend of brainstorming in Verona in October 2014. Leonardo is also the brother of Simone de Colle, who along with Bidhan Parmar had the experience of being a student in the Darden theater course. Both now use variations of these exercises and techniques in their own teaching at graduate business schools. Parmar has created a set of theater-relevant techniques and exercises aimed at the promotion of human dignity in the course described here as one on collaboration. In addition, we have examined the self-reports of well over 600 students with significant managerial experience from the Darden courses on what they have learned in the course. Over the ensuing years after the students have finished the course, we have had extensive conversations with many of these former students about the applicability of these ideas to their current organizations.

want to argue that it is equally important to think about what it means to act with dignity, recognizing that acting with dignity in part entails treating others with dignity.

One of the main ideas in the literature is that dignity is connected with our humanity in a holistic sense. Therefore, treating someone with dignity is not to treat them as a purely economic or purely political being. Treating someone with dignity means to treat them as fully human, capable of body, mind, spirit, emotion, etc., or whatever set of categories one uses to define "fully human." Dignity plays a large role in Catholic social teaching, and its application in the "common good" theory of business ethics (See Mele 2009). And according to Mele, it is important to treat each stakeholder in a business with dignity. Kant, who writes extensively about human dignity being one of the foundations of the moral realm, insists on giving a transcendental interpretation, to these foundations. A more modern reading would have Kant saying that treating someone with dignity is just not treating them as a mere means, but rather as ends in themselves.

The idea of dignity is foundational for Rawls' theory of justice as fairness, but there is no precise conception. In his lectures on Mill Rawls (2008) claims that Mill has a sense of dignity that is connected with the good life, and we would argue is connected with the idea of "acting with dignity." Rawls says of Mill:

> Our sense of dignity is tied, then, to our recognition that some ways of life are admirable and worthy of our nature, while others are beneath us and unfitting (Rawls 2008, p. 269).

Martha Nussbaum (2006) begins her account of capabilities by resting it on a foundation of human dignity. Human beings have dignity and the possibility of living a life that is "worthy of that dignity—a life that has available in it 'truly human functioning', in the sense described by Marx". Marx speaks of the human being as a being "in need of a totality of human life-activities" (74).

This leads to Nussbaum's conclusions that we need multiple capabilities to lead a life with dignity. She claims that this approach finds human beings as those "who find fulfillment in relations with others" (85) Indeed, people cooperate together not just because of mutual interests and advantage, but because that is the only way to lead a fully human life of dignity. Nussbaum (2011, at 30) claims:

> …the basic idea is that some living conditions deliver to people a life that is worthy of the human dignity that they possess, and others do not.

Thus, acting from dignity is a kind of active striving to live the good life, and it acknowledges the necessity of enabling others to do the same thing. Nussbaum believes that this is a foundational notion in some sense, and we would add here, as pragmatists, that we see human dignity as the basis for the story that we tell about our culture. Nussbaum is anxious to develop a more comprehensive theory of justice from this capabilities approach, which she deduces from the idea of dignity, but the strengths and weaknesses of her approach—further discussed in the excellent analysis of Alexander Bertland (2009)—are beyond the present scope.

Michael Rosen's (2012) important new book on dignity catalogs the history of dignity and its many uses in political theory, especially in the area of human rights in general and the UN Universal Declaration of Human Rights in particular. Rosen meticulously suggests the problems with the connection of dignity to rights claims. And, he argues that most definitions reflect the holistic humanism we have suggested above. Rosen goes on to argue that we have a duty to express the respect of this humanism, regardless of whether or not the duty is correlated with any rights claims. We interpret Rosen, as we do Nussbaum, as saying that we must act from a position of dignity as well as treat others with dignity.

Finally, Donna Hicks (2011) identifies ten elements of dignity, namely: acceptance of identity; inclusion; safety; acknowledgment; recognition; fairness; benefit of the doubt; understanding; independence; accountability. She suggests that acting from a position of dignity and ascribing dignity to others is an important survival mechanism. By assuming that all humans have dignity, we are much more likely to solve our conflicts in a way that is positive for all. Hicks claims that dignity entails in part, the recognition of others, a view of social life that is inclusive and safe for all, a commitment to the independence of others, as well as to accountability.

Suffice it to say that if we can figure out how to promote the idea of human dignity in organizations, encouraging leaders to see organizational members and stakeholders as capable of living lives of dignity, of being more fully human inside the organization, then we are more likely to build organizations that can enable us to live lives worth leading. Our organizations will at once become more human, more authentic, and simply better places worthy of human beings.

We shall argue that we can use the creative arts to help accomplish this task of building better, more authentic, and more human organizations. While there are a number of approaches that have been written about, we want to focus on two with which we have personal experience, and that represent very different but connected methods. The *Commedia dell'Arte* is a centuries old tradition in Italy of artists playing without a script or a formal director. There are a number of themes that we can isolate from this tradition which are relevant to building human dignity in organizations. Alternatively, we look at the development of a creative arts program for business students with business experience that has been developed at the Darden School in the USA. Both of these traditions are oriented toward developing the full humanity that is necessary to live a life of human dignity. And, they offer insights into how to build more authentic and more human organizations. We take each in turn.

3 The Commedia dell'Arte Case

The Italian tradition of *Commedia dell'Arte* developed in Italy from the fourteenth century and flourished especially in the sixteenth and seventeenth centuries, not only in Italy, but also across Europe. It is still alive today in the tradition of the Piccolo Teatro in Milan (https://www.piccoloteatro.org/en) and also other theaters in Italy and abroad. As noted by Rudlin, "the name *commedia dell'arte* is difficult to translate. Literally it approximates 'comedy of the artists,' implying performances by professionals as distinguished from the courtly amateurs. This form has been given other names which are more revealing of its nature and characteristics. These include *commedia alla maschera* (masked comedy), *commedia all'improvviso* (improvised comedy), and *commedia dell'arte all'improvviso.*" (Rudlin 2002, p. 13). Actors play without a 'traditional' theater director and even without a script. We claim that the Commedia dell'Arte (CdA), despite having deep roots in the past, represents a very close form of "organizing" to the contemporary business organization and can be a helpful example for understanding how dignity can be built into the very foundation of organizations.

First, in the CdA there is no formal theater director, who is an external, non-playing professional with a specific authority on all the acting and artistic-related decisions on stage: the "*Capo Comico*" ("leading comedian") is, in fact, one Comedian living and acting among her peers, who coordinates rather than directs. Second, in the CdA the Comedians play without a written script, and only rely on a "*canovaccio*" ("canvas"), that is a short plot or scenario description, without specific lines to be delivered by the actors. There is only a brief indication of the key plot line and dynamics of the underlying story. With no underlying script and no hierarchical leader, the actors have to recognize and acknowledge each other as equally human, as committed to each other in the production of the art, and as acting from a position of dignity by treating each other with dignity.

In the rest of this section we will analyze the key traits of the CdA form of theater and discuss how the dynamics that govern the creative process and the life of a CdA *Company* generate useful insights for fostering dignity in today's organizations. These insights can be seen on two different levels. First, the metaphor of the CdA as a (successful) organization of professionals, offers a conceptual lens to analyze the complex relationship among organizational stakeholders. This enables us to better understand what acting with dignity and treating others with dignity might mean in an organizational context. Second, the crafts and skills of the CdA Comedians offer a fruitful avenue for designing new forms of management education that could support the promotion of human dignity within organizations. We discuss the following five key traits of the Commedia dell'Arte, which can be seen as the reasons of its extraordinary success worldwide since the mid-1500 for more than two centuries:

- *The Mask* Enabling dignity by fostering creativity from the *inside*;
- *The Canovaccio* Creating shared meaning;
- *The CdA Company* Artistic and commercial purpose go hand in hand;
- *The Capo Comico* A leader among peers; and
- *The Lazzi* Embracing improvisation as a regular practice.

3.1 The Mask: Enabling Dignity by Fostering Creativity from the Inside

The Comedians of the CdA are skilled actors playing with their face hidden behind a mask. Each mask is linked to a well-defined character (*Pantalone*, *Dottore*, *Arlecchino*, *Capitano,* etc.), therefore each mask determines a fixed "role" or "type" with specific, pre-defined physical characteristics. Yet, at the same time, each actor can bring *within the mask* her own creativity, sensitivity, physicality, and acting skills to provide a new, personal interpretation, adapting the role to the evolution of the plot and to the local audience. In the CdA Company, the roles are distributed according to the skills, the needs, and the physical characteristics of the actors. However, once assigned the role (the mask), the way to interpret that mask (say, e.g., *Pantalone* or *Harlequin*) remains fairly open and depends on the imagination, intelligence, creativity, personality, and physicality of the actor. In other words, the full humanity of the actor behind the mask is required for the mask to come alive and be effective. If the actor only replicates movements, clichés, and jokes without adding something of her own, the mask will not be alive, and if the mask does not live, it cannot work properly in the interaction with the other masks on the stage. As a consequence, the show will not succeed in engaging the audience. Even when acting the same plot and the same roles, there are no two actors playing in the same way in the CdA Theater: everyone will enact "his" *Harlequin*, and give a personalized interpretation of the character. By personalizing the interpretations, the actors acknowledge the dignity of their colleagues. This is especially powerful since all are masked. Dignity penetrates the masks and is attached to each actor qua actor, rather than merely to the character.

On this issue Taviani (1992) in *The Secret of the Commedia dell'Arte* says:

> Every time, the 'type' in which the actor specializes assumes the particular character that is necessary to the action of the *commedia* to be put on stage. The actor will choose from her repertory of roles and characters the ones that best fit the individual scenes of the play, or—if needed—will invent new ones. For example: an old, Venetian merchant father is a type that can be present in all the comedies. On the action of a single comedy will depend then if this guy will be liberal, stingy, coward, prodigal, tyrannical, weak, smart, simple, in love, helpless, melancholic (p. 374).

Taviani also points out that the secret underneath the 'scenic success' of the masks in the CdA is in

> the balance between the total lack of predetermination as regards a unique aspect—their character—and the absolute predetermination of all other aspects, from the dress to the social condition. To fix the rigid aspects of the types for comedy, to give them a name, that is a dress, an age, a place of birth, a dialect, the actors almost always resorted to the masks…they were already empty cocoons beautiful and ready to be inhabited by the spirit of a character in the comedy (p. 374).

If we transfer the metaphor of the mask to the world of today's organizations, we can see that what matters for actors to be successful when working within a predetermined 'role' is to keep the individual creative freedom and "fill the shell" with their own full humanity. Similarly, we can imagine that for successful managers of today's organizations it is important to "fill the role" that is entrusted to them (or that they have chosen) with the same freedom to express their own creativity and personality that was required to the CdA actors. Organizations with high engagement by employees often ask them to "fill the shell" of a role with their humanity. This is the case, for example, of companies such as Whole Foods Market, The Container Store, The Motley Fool and many others, described by Mackey and Sisodia (2013) and Sisodia et al. (2014). This supports the idea of dignity within organizations or "recognizing and honoring the full complexity of the human person." In the CdA theater it is essential to be able to recognize the talent of each actor to put her in a position to bring her mask to life with her own humanity, authentic presence and truth with a sense of passion, confidence and fulfillment. "Filling the shell" involves becoming a whole human being to bring a character to life. Something similar is required in modern organizations in order to foster the development of humanity and enable relationships based on authenticity. Too often in modern organizations we bring only a partial idea of ourselves into the workplace. There are endless conversations about the necessity of "work-life" balance [see, for example, the comprehensive literature review by Phipps and Prieto (2014)]. However, the more fully human we can become in the workplace, the less need there is for such a balance, as we integrate more fully who we really are, behind the "mask" of our organizational lives. Making the mask meaningful requires dignity in both the CdA and in modern organizations. This means that the organizational mask, that is the organizational culture and set of values, needs to be designed and implemented in a way that reconciles and integrates the individual and the organizational dimension of values, as part of the same *authentic,* interconnected whole (Auster and Freeman 2013). In short, the mask points to the necessity of building a more inclusive culture. Pless and Maak (2004) have suggested that more attention needs to be paid to the normative dimension of building an inclusive culture and that ideas such as recognition, reciprocal understanding, and plurality of points of view are critical. These are precisely the ideas behind the mask, and why it can be useful in creating such a culture.

3.2 The Canovaccio ('Plot on Canvas'): Creating Shared Meaning

The CdA Comedians use as a basis for the staging of their play a number of '*canovacci*', or short plots including a rather schematic narrative summarizing the salient facts of the story to be told. These *canovacci* do not go into details of lines of the individual characters. The *canovaccio* is therefore an open and "light" structure that allows the actors to move freely within a path that has only a few essential "touch points" that are necessary to advance with the story.

Therefore, despite the 'rigidity' of the mask, the CdA actors have much more freedom than in traditional theater. Each actor, despite wearing a "fixed" mask that has specific traits for that particular role, brings her own humanity on stage and plays the plot by following the rules, but in an absolutely personal way, reacting authentically within the track determined by the canvas. The *canovaccio* creates a much lighter and wider "scaffolding" than those established by a script in traditional theater plays.

Dunham and Freeman (2000) offer a useful insight that is worth considering in this perspective. They highlight the importance of creating a positive "scaffolding" enabling actors to express their unique talents and personality even within the boundaries of a pre-determined script. In the CdA theater perspective the importance of this insight is even more evident: The scaffolding, as we described, is even lighter and essential, and therefore encourages the actors to an even greater creative input, and amplifies their ability to listen to each other and to act with improvisation.

To confirm how an open "scaffolding" might enable several possible interpretative solutions, Dunham and Freeman (2000) give the example of the staging of a play by Gorky simultaneously played in two theaters of Moscow by two companies giving two different interpretations. Asked which one was the "true one," Gorky himself said, "they are both true interpretations… and there are many other possible. [...] Gorky knew that a live play has within it the possibilities of almost as many meanings as there are creative people to find them." (p. 117). Similarly, in the CdA there is no "right" execution of the plot: each company, and each actor is free to interpret the plot in the way that they think is the most authentic.

Unlike what happens in traditional theater, the *canovaccio* simply sets some shared "objectives" and some—*given circumstances*—to use the term introduced by Stanislavski in his fundamental *An actor's work*—leaving freedom to the actors to creatively decide how to develop and enact the plot: "The plot, the facts, the incidents, the period, the time and place of the action, the way of life. [...] The Given Circumstances, just like "if," are suppositions, products of the imagination" (Stanislavski 1938, p. 52). The CdA Actors, as well as business leaders, must continually answer the question "What would I do if…?," but in the case of the Commedia dell'Arte, as well as in the corporate life of today's managers, there are no pre-scripted lines to be memorized and played.

By developing the story as the result of the actors' interaction and improvisation, and adapting the jokes to the local audience and the social context, the actors create a shared meaning in a collaborative process, where each actor accepts, interprets, and reacts to her partners on the stage. To build the "scaffolding" around a simple *canovaccio* also fosters the mutual understanding and team spirit within the company of actors. Their task is not to "*put on stage*" a *script* by an *author* under the supervision of a *director*, but rather to *develop* the *plot* enabling all the *actors* of the company to act freely within a shared sense of meaning, listening, and engaging with the audience. Even the scenes, the costumes, the music, the props, and all the technical aspects are part of the CdA collaborative work.

It is the creation of shared meaning that illustrates the idea of acting from a position of dignity. By creating shared meaning we commit to living a life of collaboration with others, a life that is worth living. Of course there are no ethical guarantees here, so we also need to continue to focus on widening the collaborative circle, taking in new members, treating them with the dignity they deserve, and acknowledging that they can also design lives worth living. This feature also resonates with the dimension of dignity as encompassing a sensitivity to culture and the dignity of others that we have formulated with reference to the work on dignity and human capabilities by Nussbaum (2006).

Simply, in the CdA theater successful actors are those who are able to react in effective way—that is, showing sensitivity and meaningfully advancing the plot—to (unexpected) stimuli that the stage (and the audience) might offer them. We believe that the same insight could be useful for today's managers, who need to be able to react in meaningful ways to (unexpected) stimuli that the business world offers them. For instance, Pless (2007) outlines how Anita Roddick at the Body Shop used storytelling and values-based leadership to create a shared understanding filled with meaning at the Body Shop. She saw herself as part of a network of stakeholders all of whom required being treated with dignity. These relationships are held together by love and care as well as shared meaning. Mackey and Sisodia (2013) suggest that love and care has been ignored by business theorists for too long. Building an inclusive culture filled with shared meaning requires much more than self-interest and economic thinking.

3.3 The Commedia dell'Arte Company: Recognizing the Complexity of the Human Person

Historically and still today, the Comedians of the CdA theater are at the same time artists and professionals. In the sense of the Latin word *ars, artis*, they have not only artistic skills, but are professionals. The art is created and performed, in all its aspects, by a team of skilled artists and professional artisans, and independent 'journeymen' who offer shows with a commercial and artistic purpose: the actors are making a living off their art and are responsible of their profits or losses. The CdA company is typically formed by men and women with no discrimination based on gender. Women have always had an equal status, since the only unit of measure to determine an actor's 'status' within the Company has been the value of her or his skills and the ability to contribute to the success for the company and to attract the consent of the public. Historically the actors lived in close contact, sharing not only the boards of the stage during the performances, but also all the moments of everyday life, including the traveling, the rehearsal, the assembly, and disassembly of the stage, meals, and daily life (Taviani 1992). In this sense, we can see that the approach of the CdA Company has been that of an organization treating its members with dignity in the sense that it has acknowledged and honored the complexity of the human person, treating everyone as equal, and enabling each person to express their full humanity in the various dimensions, including the artistic, the commercial, and the personal ones.

Interestingly, many of the CdA actors were former traders or people who knew the art of the trade, who "before starting the activity of 'actors,' had attended the markets and squares as barkers or sellers, knowing perfectly the dynamics and the basic rules operating in those contexts. They apply to their theater activity not only the principle of work organization, but also the principle of expansion of the activity and growth of capital, according which the company is in good health only if it grows and conquers new territories." (Taviani 1992, p. 408). Hence the CdA Companies had to maintain a constant attention to the audience and its reactions—a constant attention to identify, understand, and honor stakeholders. It was common sense for word to spread about any CdA Company that satisfied audiences. And, only a good reputation obtained by 'word of mouth' would guarantee new contracts, new audiences, and new commercial opportunities for the Company to grow.

The intensity of the day-to-day relationships among the components of the CdA Company is without doubt a key factor to create a necessary level of harmony within the working group. Every member had to acknowledge the complex humanity of the others and at the same time, understand each other in real time during the performances and while discussing the commercial part of the enterprise. This harmony translates into the ability to predict and imagine the reactions of fellow actors on stage and in their professional conversations.

It is obvious to see the analogy with organizations of our time, who strive to create "team spirit" cultivating and promoting relationships among their members even outside daily organizational life (e.g., through the realization of 'team building' trips and "away days" to promote a more informal sharing of experiences outside work issues). Extant literature on "team building" emphasizes the importance of such initiatives [see, for instance, Shuffler et al. (2011) and Lin et al. (2012)]. One key question for these initiatives is whether they are designed in a way that really acknowledges and honors the complexity of the human person, avoiding treating persons as "human resources" only, or means to an end, that would of course deny the Kantian definition of dignity mentioned above.

3.4 The Capo Comico ('Chief Comedian'): A Leader Among Peers

Contrary to traditional theater, there is no "director" as we understand it today, i.e., a critical eye to the company of actors who coordinates and directs from outside the work of the actors. In the CdA, there is a leading role, the *Capo Comico* ('Chief-Comedian') who is an actor among actors, and lives in close contact with all her fellows. The Capo Comico knows the capabilities and potentials of all the members of the Company and shares, day after day, the work, the efforts, and the joint project. The Capo Comico also, contrary to a theater director, does not leave the company after the rehearsal and the debut, but plays together with the other actors sharing their day-to-day life on stage and outside the stage. As a result, the CdA Company has a rather horizontal, "democratic" organizational structure.

Modern organizations can learn from CdA theater by implementing an organizational design that allow managers to appreciate the benefits of living and working closely with their employees, sharing joys and sorrows, successes, and failures. It is in the interest of all members of the organization to make sure that everything runs smoothly, creating the harmony that will improve the effectiveness of working relationships. The control exercised by the "Capo Comico" is *internal* to the process itself, as they act alongside their co-workers, and decide how to develop their *canovaccio* day after day, in light of the daily reactions of the audience, and learning from their colleagues' creativity.

The actors of a CdA Company are aware and involved in the whole process leading to the final result, a team that builds and creates its shared meaning in the development and enactment of the *canovaccio*. The members of the CdA do not just perform mechanically directives coming from an external and often distant authority, but bring their authentic passion and skills to a shared goal: the realization of the *canovaccio*-shared meaning. The key role of the Capo Comico in the CdA Company teaches us that the organizational creative process can only be based on well-coordinated teamwork, and that a good business leader cannot ignore the full humanity of each member of her own team. In other words, the leader in the CdA Company is a leader among peers, who needs to be able to treat the other members of the organization with dignity in the sense of acknowledging everyone as fully human capable of body, mind, spirit, and emotion.

3.5 The Lazzi ('Sight-Gags, Jokes'): Embracing Improvisation as a Regular Practice

We have already mentioned that actors in Commedia dell'Arte do not go on the stage to play a script, but rather to enact a *canovaccio*, or a series of plots. This implies that, as part of every single 'representation,' the actors have the opportunity to exercise their creativity. They can challenge each other with some improvised *lazzi* (sight-gags, jokes, and 'acrobatic' language games) that have the twofold purpose of (a) entertaining the audience and (b) showing the acting skills and virtuosity of the actors. According to the theatrical critic Simoni, the success of the Commedia dell'Arte was precisely linked to the acting skill of its actors, and of course, the *lazzi* were the greatest exhibition of skill: "The Commedia dell'Arte was a show in which there was much to admire, but also much to disavow. It was marvelous not for its content, essentially crude and often vulgar and shameless, but for the skill of its actors" (Simoni quoted in Taviani 1992, p. 477).

The term *lazzo* derives from Latin *actio*, i.e., action. In the CdA theater, the room for improvisation through the *lazzi* is embedded as a regular practice. One important idea is the invention of a new *lazzo* as an improvised reaction to a *mistake* by another

Mask on the stage. The concept of *mistake* deserves a deeper explanation. Mistakes, or unexpected events, are not experienced by CdA actors as a dangerous glitch in a rigid roadmap. On the contrary, mistakes are almost sought after as they represent a vital stimulus for innovation and energy within the creative process on stage. The same is true in our modern organizations. Mistakes are opportunities to create and innovate, to try new ideas, or to try a new direction. They are vital for organizations to be more fully human. In the modern world, the idea of organizations as pure mechanical or bureaucratic routines is incomplete. However, if mistakes are to be opportunities for value creation, it requires that there be a shared sense of acting from and treating others with dignity, throughout the organization.

In summary, the success of the Commedia dell'Arte is the success of extraordinary entrepreneurs who were able to create organizations within which several dimensions of human dignity are naturally acknowledged and nurtured—including, for example, enabling self-awareness and authenticity (as actors behind the mask), creating shared meaning within the organization (as actors playing without a script), or respecting and trusting each other as fully human (as members of a CdA Company). In Table 39.1, we summarize the lessons that we can draw from each of the characteristic of the CdA theater that we have analyzed above and point out their potential positive impacts on human dignity within organizations.

4 The Darden Case: Theater and Collaboration in Executive Education

The Commedia dell'Arte is an ancient idea that can be used to understand how the creative arts can be used to make our organizations places that have a profound understanding of human dignity. Many of the principles of CdA are at work in modern theater as well. The many positive effects of the use of theater as a pedagogical tool to promote creativity, innovation and ethical leadership in executive education are being increasingly theorized and practiced within business schools around the world. Lesavre (2012) describes his experience at the Grenoble School of Business in France using "corporate-theatre" to develop the ability of managers to "distance themselves from the business and from the problem, and develop their own point of view, their imagination and their creativity." (p. 250). Garaventa (1998) describes how drama can be a tool for teaching business ethics. The Ariel Group, a consulting firm, has built a business on assisting business leaders and business schools in applying theater techniques to business (Lubar and Halpern, 2004). To further build on these ideas, we want to explain how we can design training interventions using creative arts, and theater in particular, to build the capabilities of organizational members to act with and from a position of dignity.

At the Darden School of Business at The University of Virginia, techniques and exercises are used in several executive education courses, and there are a number of elective courses for MBA students that take account of work in the creative arts,

Table 39.1 Linking Commedia dell'Arte and human dignity in organizations

Key traits and exercises	Lessons from theater	Lessons for organizations	Positive impacts on dignity within organizations
1. Use of the *Mask*	Actors express their own talent, skills, and personality by acting behind a mask/interpreting predefined character	Managers should be able to express their full humanity within their pre-defined organizational roles	Fostering creativity Enabling authenticity and self-awareness Learning non-verbal communication
2. Acting on the basis of a *Canovaccio*	Actors are free to interpret their character and develop and adapt the story on the basis of a shared 'plot'	Managers should be able to react to external events and adapt their strategic decisions on the basis of a shared meaning	Creating a shared meaning Capability to listen and adapt Promoting (and rehearsing) creativity
3. The Commedia dell'Arte *Company*	Actors share both the commercial and the artistic purpose of their activity	Managers should: Acknowledge/harmonize the economic and human dimension of business; Create team spirit; Engage with stakeholders to understand their needs	Recognizing the complexity of the human person Promoting harmony, respecs, and collaboration in working teams Listening to (and trusting) each other
4. The role of the *Capo Comico*	There is no external director; actors are lead by an actor-among-actors	Managers should lead organizations as leaders among peers, coordinating individual skills, ideas, and contributions	Leading as a peer Respect Empathy Equality
5. The use of *Lazzi*	Actors are trained to improvise and create new jokes by listening to each other, interacting with the audience or reacting to a mistake	Managers should learn how to nurture and develop the individual talents within organizations	Promoting moral imagination Adapting to unexpected events Learning from mistakes
6. Becoming a character	Actors become another character and begin to acknowledge the details of others' lives	Managers acknowledge difference and become more sensitive to difference	Empathy Embodiment, and the connection to emotion
7. Believability	Actors recognize the collaborative nature of the theater. They take what others give them and do something with it	Managers acknowledge collaborative nature of organizations and work to build on what others do, not just to criticize it	Dissolving conflict by collaboration Acknowledging the collaborative nature of work

(continued)

Table 39.1 (continued)

Key traits and exercises	Lessons from theater	Lessons for organizations	Positive impacts on dignity within organizations
8. Telling a story	Actors support each other by active listening so that they can pick up where others leave off and continue to create shared meaning	Managers focus on the purpose and lessons of what they do. Organizational stories lead to learning and the development of culture	Culture as consisting of storytelling Stories as celebrating collaborative problem solving Stories as morality tales Story telling yields both complexity and empathy
9. Human sculptures	Actors experience "embodiment" as a critical part of the creation of shared meaning	Managers can come to see others as whole human beings, with bodies and minds Managers learn to talk about appropriateness with respect to bodies Managers learn to use embodied resources to create meaning	Full humanity, body, mind, and spirit Recognizing the complexity of human beings Communicating through actions other than words
10. Safety to fail	Actors are encouraged to try new ideas and are discouraged from seeing ideas that don't work as failure	Managers solve difficult problems as a group with multiple people contributing	Complexity involves not always getting things right Perseverance

such as The Collaboration Lab, and Leadership, Ethics, and Theater. While some of these courses, including the Collaboration Lab, use exercises that are only loosely connected to both CdA and theater, it is important to include them here, as they build on the idea of creative imagination as well as on insights developed in the CdA and in more modern theater. In a recent paper, Freeman et al. (2014), suggest that the theater has several leverage points in teaching about ethics and values. Each of these leverage points takes human dignity as a fundamental building block. First of all theater is a place where the students can explore their "self" in some deep way. Acting is about seeing the relationship of self to characters played, and about calling on the appropriate set of emotional responses to portray the meaning of the character and the play. If we are to recognize the full humanity of others, then we need to recognize it in ourselves if we are to live lives of human dignity. Second there is a connection to complexity, as the actor finds herself inside a complex physical, emotional, and moral sphere that is a play. Such complexity again leads to the acknowledgement of a more fully human person inside organizations. This is especially important given that many organizations can never get beyond seeing its people as economic entities or as "human capital," (Greenwood 2013; Greenwood and Freeman 2011). Finally, Freeman et al. (2014) argue that there is a connection to others, as a theater ensemble must act as a group, as a whole if the performance is to be meaningful. In addition, theater is a place where collaboration is the main kind of interaction. As professional director, Randy Strawderman, who helped design the Darden theater courses, said to us, "When we are on stage together, all we have are the lines that we give each other—or not". The essence of theater is collaboration, and in this instantiation of it, it is collaboration among peers. These contextual details set the stage for explorations into human dignity. These leverage points are best illustrated through a number of techniques and exercises that are used in a number of situations: (1) Becoming a character; (2) Believability; (3) Telling a Story; (4) Human Sculptures; and, (5) Safety to Fail. In the sections that follow, we briefly describe these techniques and exercises, and offer some conclusions based on conversations and written feedback from several hundred students who have experienced them at Darden.

4.1 Becoming a Character

Students are asked to imagine a character in their life. A series of questions about the character drives the students to imagine the physical details of their character. A second series of questions asks about voice, emotion, what drives the character, what are his or her sadnesses, and moments of joy. The students are then asked to sit like the character would sit in their chairs, then to get up and walk like the character walks, slowly becoming that character. They are asked to stay "in character" on stage and to introduce themselves to each other and begin a conversation. Usually the students begin this exercise with an extreme amount of discomfort; however, as time goes on (15–20 min), they get much more inside the character. They are practicing both empathizing with another, and the embodiment of that empathy.

Being able to put one's self in someone else's shoes is one of the most important traits in adopting an ethical point of view. Becoming a character leads to a deep understanding of what it means to respect someone and treat them with dignity, despite their physical or emotional differences. Again, this is the essence of how to build a more inclusive culture as Pless and Maak (2004) and Pless (2007) suggest.

4.2 Believability

Much of business life is spent arguing, even when there is a veneer of politeness. Many conversations begin with (either explicitly or implicitly) "Yes but..." pushing against the point and the other person. In theater, there is pure collaboration. One takes what the other gives and does something with it to return it to another actor. Even in an "argument" there is pure collaboration, as the players are in it together to portray an argument. Consequently actors must embrace the principle of saying "yes, and..." to what is offered by other actors. This principle says to take what the other person says as true, and then add something to it. Yes, and…is the hallmark of improvisation, but it extends to other theater genres as well, and it is vital to playing the structures of the CdA. One exercise is to have the students sit back to back. One student gives the first line that can be anything such as "I remember when you left me to get stung by those wasps". The second students then responds "Yes, and the reason I did was….," accepting what the first person said and adding to it.

Executives find this exercise extremely helpful because it helps to dissolve conflict as it explicitly acknowledges collaboration. It builds consensus, saves valuable time from useless arguing, and often leads to more creative solutions, when applied in the real world. They have often said things in the debrief of the exercise like, "I wish my company could run its meeting on "yes, and" or "I'm going to start practicing this with my team, so we can be more effective". "Yes and" also gets at fundamental value issues. By using it, we validate and value what the other person has said. Again this is respect and dignity at a very basic level.

4.3 Telling a Story

Human beings are storytellers. There is a burgeoning literature on story telling in management theory and practice [see, for instance, Denning (2006) and Maclean (2012)]. And stories are usually situated within the moral realm. Yet much of business thinking tries to drive out key elements of morality by substituting "data" and "facts" and "analysis". While these things are surely important, so too is our ability to communicate through story telling. The dominance of presentations and PowerPoint slides in business often negate the power of what is really happening. In theater, we teach actors to "tell the story". There are a variety of techniques but the most powerful one is to simply take away the script long before the actors have learned the lines. Going "off book" forces the actors to improvise, and to "tell the story" rather than try and remember the lines. To do this effectively, they must pay attention to each other, help each other when someone forgets a key part of the story that was in their lines, and they must emotionally buy into the story as they are telling it, rather than repeating lines.

Imagine approaching business meetings as opportunities for storytelling. The culture of any business is actually transmitted through its stories. Most stories have some kind of moral edge, or they are morality tales. They help new and old members understand how the organization works, its complexity, its rules, and its purpose. Stories build empathy, and in organizations, it could be empathy for the overall purpose of the business, or treating one particular or multiple stakeholders with dignity and respect. The possibilities seem endless when horizons are broadened past the traditional methods of powerpoint and standard presentations and meetings. Stories encourage conversations about purpose and values. They focus us on action not just the words.

4.4 Human Sculptures

We have more resources than the language that we speak. Using these resources to tell a story or to make a point can be very powerful. One exercise is a constraint exercise explicated in (Freeman et al. 2014), based on director Anne Bogart. A second exercise directs the students to use their bodies to form a "human sculpture" that tells a story. Often the sculpture will require 3 or 4 movements, but the movements themselves must be carefully scripted, and no words are allowed when performing the sculpture.

This exercise embodies many of the lessons of theater. Since it often involves touching, a conversation about appropriateness, comfort, and basic physical dignity must take place. Many times the actors are very different from each other physically, and there are often challenges getting everyone involved.

4.5 Safety to Fail

Much of organizational life revolves around managing one's reputation and the impression that others have of them. While this intuitively makes sense, this practice can have negative impacts when teams face complex problems. For example, when addressing an ambiguous situation we may not know how to address it specifically, and if we aim to manage our reputations or the impressions that others have of us, we may become too risk averse or lose our willingness to try things out to learn about the problem. One activity run at Darden is called a puzzle hunt. The connection with the creative arts is clear. It is an exercise in creative imagination, where participants must learn how to collaborate effectively. For business students who are accustomed to more analytic tasks, puzzle hunts can inspire them to open their minds to different points of view.

Puzzle hunts are challenging nontraditional and nonobvious puzzles that when solved lead you to another puzzle. An example of a puzzle hunt clue might be a bag of jelly beans. There is no writing or directions on the bag—the clue is just a bag of jelly beans. Participants must figure out how to make sense of this clue and where it leads them next. Participants start to spin out theories and hypotheses about how a bag of jelly beans could lead them to the next puzzle. They may dump out the contents and notice that there are different colored beans in the bag, they may notice that there are 2 red jelly beans, and 1 blue jelly bean, 5 purple jelly beans, and 3 yellow jelly beans. They may wonder why there are specific numbers of different jelly beans, someone might throw out the idea that perhaps they should look at the 1st letter of the word blue and the second letter of the word red (etc.) to get a set of letters, which could be rearranged to provide a clue (e.g., 'bell'), which would lead them to a landmark on campus where the next clue would be available. Teams that are successful at insight tasks like these are those where participants feel comfortable sharing, 'silly' ideas, where people are less concerned with their own impression and more concerned about solving the problem together. One person's silly idea could trigger a useful hypothesis in another teammate. The worst performing teams are those that try to solve the puzzle in their own head before saying anything out loud to the rest of the team. In organizational situations where teams are confronting ambiguity, the ability to try out ideas and not suffer risks to one's reputation is helpful in supporting creative and innovative thinking that can help increase the likelihood of successfully addressing issues. Perseverance counts, especially in situations of ambiguity where failure is almost always an option. The safety to fail requires recognizing the dignity of others, in particular resisting the temptation to 'type-cast' a team mate because they gave a right or wrong answer, and recognizing that their humanity is much more complicated than their performance on any one task.

5 Summary

Many of these simple exercises in the Darden context have been a part of the daily training of the CdA actors and indeed are used throughout the world of theater, even as they are beginning to penetrate the world of business education. To summarize, there are a whole host of experiential exercises from theater that can be used to help students and executives develop skills such as self-awareness, creativity, ability to listen and share, awareness of non-verbal body language, ability to speak in public, understanding and resolution of conflicts and tensions, and being "in the moment" (Spolin 1983; Bernardi 1992; Levy 2010). Most of these exercises recognize and indeed are built on the assumption of mutual dignity, in both the "acting from" and "treating others with" senses of the term. These exercises are complementary to another line of work using service learning to broaden the perspectives of business executives and students (Pless et al. 2011; Pless and Maak 2009). Both perspectives are aimed at developing "responsible leaders".

In Table 39.1 we summarize the key elements of the CdA and the Darden exercises, some of the lessons we can derive for today's organizations from each element, and we highlight the positive impacts on human dignity that the exercises support.

In both the CdA and the Darden cases, developed centuries and cultures apart, the creative arts can be used to develop an appreciation and an embodiment of whole human beings. This deeper understanding of human beings increases the appreciation of participants for human dignity. The practice of creative arts like CdA and the exercises used at Darden immerse participants in an ongoing situation where they need to respond as whole human beings. These exercises call on participants to use their whole set of senses and repertoires to respond to or create collaborative meaning. Such exercises are inextricable from acting with dignity and treating others with dignity. Participants can see how differently they react in the moment, and they can come to value that difference. Additionally, these kinds of exercises and practices bring theory to life and can demonstrate how they can be successfully utilized.

We have only begun to explore the connection between the creative arts and the concept of human dignity. There is much more work that can be done in using the creative arts to promote human dignity in organization. Like the service learning work of Pless and Maak (2009) we can begin to work with individual organizations to open new pathways to thinking about how dignity can be enhanced in their workplaces. We have not yet begun to explore the way that music can enhance the humanity of organizations, though there is some work on using musical metaphors to understand modern organizations. This work is challenging. It must be done in a way that the participants don't feel as if it is one more "training game." The work that is produced as a means to enhancing human dignity must stand on its own as art. And, the process requires a delicate balance between encouragement and professional artistic judgment. In short there is a vast territory to more broadly connect the full panoply of creative arts to business ethics and to the project of making our organizations be places that are fit for human beings.

References

Auster, E.R., and R.E. Freeman. 2013. Values and Poetic Organizations: Beyond Value Fit Toward Values Through Conversation. *Journal of Business Ethics* 113: 39–49.

Barrett, F. 1998. Coda-Creativity and Improvisation in Jazz and Organization: Implication for Organizational Learning. *Organization Science* 9 (5): 605–622.

Barry, D. 1996. Artful Inquiry: A Symbolic Constructivist Approach to Social Science Research. *Qualitative Inquiry* 2 (4): 411–438.

Barry, D., and S. Meisiek. 2010. Seeing More and Seeing Differently: Sense Making, Mindfulness, and the Workarts. *Organization Studies* 31: 1–26.

Bernardi, P. 1992. *Improvisation Starters: A Collection of 900 Improvisation Situations for the Theater*. Cincinnati: Betterway Books.

Bertland, A. 2009. Virtue Ethics in Business and the Capabilities Approach. *Journal of Business Ethics* 84: 25–32.

Denning, S. 2006. Effective Storytelling: Strategic Business Narrative Techniques. *Strategy & Leadership* 34: 42–48.

Dunham, L., and R.E. Freeman. 2000. There Is Business Like Show Business. *Organizational Dynamics* 29 (2): 108–122.

Freeman, R.E., L. Dunham, G. Fairchild, and B. Parmar. 2014. Leveraging the Creative Arts in Business Ethics Teaching. *Journal of Business Ethics Online First*. https://doi.org/10.1007/s10551-014-2479-y.

Garaventa, E. 1998. Drama: A Tool for Teaching Business Ethics. *Business Ethics Quarterly* 8 (3): 535–545.

Greenwood, M. 2013. Ethical Analysis of HRM: A Review and Research Agenda. *Journal of Business Ethics* 114: 355–366.

Greenwood, M., and R.E. Freeman. 2011. Ethics and HRM: The Contribution of Stakeholder Theory. *Business and Professional Ethics* 30: 269–292.

Hicks, D. 2011. *Dignity: The Essential Role it Plays in Resolving Conflict*. New Haven: Yale University Press.

Lesavre, L. 2012. Are Theatre and Business Links Relevant? A Conceptual Paper and a Case Study. *Journal of Management Development* 31 (3): 243–252.
Levy, G. 2010. *275 Acting Games: A Comprehensive Workbook of Theatre Games for Developing Acting Skills*. Colorado Springs: Meriwether Publishing Ltd.
Lin, C., Y. Baruch, and W. Shih. 2012. Corporate Social Responsibility and Team Performance: The Mediating Role of Team Efficacy and Team Self-Esteem. *Journal of Business Ethics* 108: 167–180.
Lubar, K., and B. Halpern. 2004. *Leadership Presence*. New York: Gotham.
Mackey, J., and R. Sisodia. 2013. *Conscious Capitalism*. Cambridge, MA: Harvard Business School Press.
Maclean, M. 2012. Sensemaking, Storytelling and the Legitimization of Elite Business Careers. *Human Relations* 65: 17–40.
Meisiek, S., and D. Barry. 2007. Through the Looking Glass of Organizational Theatre: Analogically Mediated Inquiry in Organizations. *Organization Studies* 28 (12): 1805–1827.
Melé, D. 2009. Integrating Personalism into Virtue-Based Business Ethics: The Personalist and the Common Good Principles. *Journal of Business Ethics* 88 (1): 227–244.
Nussbaum, M. 2006. *Frontiers of Justice*. Cambridge, MA: Harvard University Press.
———. 2011. *Creating Capabilities*. Cambridge, MA: Harvard University Press.
Phipps, S.T.A., and L.C. Prieto. 2014. A Discovery of Early Labor Organizations and the Women Who Advocated Work–Life Balance: An Ethical Perspective. *Journal of Business Ethics Online First* 134: 1–13.
Pless, N.M. 2007. Understanding Responsible Leadership: Roles Identity and Motivational Drivers. *Journal of Business Ethics* 74 (4): 437–456.
Pless, N.M., and T. Maak. 2004. Building an Inclusive Diversity Culture: Principles, Processes and Practice. *Journal of Business Ethics* 54 (2): 129–147.
———. 2009. Responsible Leaders as Agents of World Benefit: Learnings from "Project Ulysses". *Journal of Business Ethics* 1 (85): 59–71.
Pless, N.M., T. Maak, and G.K. Stahl. 2011. Developing Responsible Global Leaders Through International Service Learning Programs: The Ulysses Experience at Price Water House Coopers. *Academy of Management Learning & Education* 10 (2): 237–260.
Rawls, J. 2008. *Lectures on the History of Political Philosophy*. Cambridge, MA: Harvard University Press.
Rosen, M. 2012. *Dignity: Its History and Meaning*. Cambridge, MA: Harvard University Press.
Rudlin, J. 2002. *Commedia Dell'Arte: An actor's handbook*. London: Routledge.
Salgado, M. 2008. Le theatre, un outil de formation au management. *Revue Française de Gestion* 181 (34): 77–96.
Shuffler, M., D. DiazGranados, and E. Salas. 2011. There's a Science for That: Team Development Interventions in organizations. *Current Directions in Psychological Science* 20: 365–372.
Sisodia, R., D. Wolfe, and J. Sheth. 2014. *Firms of Endearment*. Upper Saddle River: Pearson FT Press.
Spolin, V. 1983. *Improvisation for the Theater: A Handbook of Teaching of Directing Techniques*. Evanston: Northwestern University Press.
Stanislavski, C. 1938. *An Actor's Work: A Student's Diary*. Trans. and ed. Jean Benedetti. London: Routledge, 2008.
Taviani, F. 1992. *Il Segreto della Commedia dell'Arte (The secret of the Commedia dell'Arte)*. 2nd ed. Florence: La casa Usher.
Vera, D., and M. Crossan. 2004. Theatrical Improvisation: Lessons for Organizations. *Organization Studies* 25 (5): 727–749.
Weick, K.E. 1998. Introductory Essay: Improvisation as a Mindset for Organizational Analysis. *Organization Science* 9 (5): 543–555.

Simone de Colle is Associate Professor of Business Ethics & Strategy, IESEG School of Management, Paris, and Visiting Associate Professor of Business, Organizations and Society at New York University Abu Dhabi.

R. Edward Freeman is University Professor, Elis and Signe Olsson Professor, Academic Director of the Institute for Business in Society, and Senior Fellow of the Olsson Center for Applied Ethics at the University of Virginia's Darden School of Business.

Bidhan L. Parmar is Associate Professor and the Shannon Smith Emerging Scholar in Business Administration at the University of Virginia's Darden School of Business.

Leonardo de Colle is a professional actor with more than two decades of experience at the Piccolo Teatro of Milan, and a teacher of the theater's International Academy of Commedia dell'Arte directed by Ferruccio Soleri.

Chapter 40
Ethics and the Algorithm

Bidhan L. Parmar and R. Edward Freeman

Editor's Note: This article is one of a special series of 14 commissioned essays MIT Sloan Management Review is publishing to celebrate the launch of our new Frontiers initiative. Each essay gives the author's response to this question:

> **Within the next five years, how will technology change the practice of management in a way we have not yet witnessed?**

Behind every piece of code that drives our decisions is a human making judgments — about what matters and what does not. Are we designing algorithms, or are algorithms designing us? How sure are you that you are directing your own behavior? Or are your actions a product of a context that has been carefully shaped by data, analysis, and code?

Advances in information technology certainly create benefits for how we live. We can access more customized services and recommendations; we can outsource mundane tasks like driving, vacuuming floors, buying groceries, and picking up food. But there are potential costs as well. Concerns over the future of jobs have led to discussions about a universal basic income — in other words, a salary just for

---

This article was originally published on July 13, 2016. It has been updated to reflect edits made for its inclusion in our Fall 2016 print edition.

Originally published in: ***Sloan Management Review***, 58(1), 16-17 © 2016 from MIT Sloan Management Review/Massachusetts Institute of Technology. All rights reserved. Distributed by Tribune Content Agency
Reprint by Springer

B. L. Parmar · R. E. Freeman (✉)
University of Virginia, Charlottesville, VA, USA
e-mail: FreemanE@darden.virginia.edu

© Springer Nature Switzerland AG 2023

S. D. Dmytriyev, R. E. Freeman (eds.), *R. Edward Freeman's Selected Works on Stakeholder Theory and Business Ethics*, Issues in Business Ethics 53,
https://doi.org/10.1007/978-3-031-04564-6_40

being human. Concerns over the changing nature of social interaction have covered topics ranging from how to put your phone down and have a face-to-face conversation with someone to the power dynamics of a society where many people are plugged into virtual reality headsets. Underlying these issues is a concern for our own agency: How will we shape our futures? What kind of world will information technology help us create?

Advances in information technology have made the use of data — principally data about our own behaviors — ubiquitous in the online experience. Companies tailor their offerings based on the technology we employ—for example, the travel website Orbitz a few years ago was discovered to be steering Mac users to higher-priced travel services than it was PC users. Dating sites like eHarmony and Tinder suggest partners based on both our stated and implied preferences. News stories are suggested based on our previous reading habits and our social network activities. Yahoo, Facebook, and Google tailor the order, display, and ease of choices to influence us to spend more time on their platforms, so they can collect even more data and further intermediate our daily transactions.

Increasingly, our physical world is also being influenced by data. Consider self-driving cars or virtual assistants like Siri and Amazon's Echo. There are even children's toys like Hello Barbie that listen, record, and analyze your child's speech and then customize interactions to fit your child.

As our lives become deeply influenced by algorithms, we should ask: What kind of effect will this have?

First, it's important to note that the software code used to make judgments about us based on our preferences for shoes or how we get to work is written by human beings, who are making choices about what that data means and how it should shape our behavior. That code is not value neutral — it contains many judgments about who we are, who we should become, and how we should live. Should we have access to many choices, or should we be subtly influenced to buy from a particular online vendor?

Think of the ethical challenges of coding the algorithm for a self-driving car. Under certain unfortunate circumstances, where an accident cannot be avoided, the algorithm that runs the car will presumably have to make a choice about whether to sacrifice its occupants or risk harming — maybe even fatally — passengers in other cars or pedestrians. How should developers write this code? Despite our advances in information technology, data collection, and analysis, our judgments about morality and ethics are just as important as ever — maybe even more important.

We need to figure out how to have better conversations about the role of purpose, ethics, and values in this technological world, rather than simply assuming that these issues have been solved or that they don't exist because "it's just an algorithm." Questions about the judgments implicit in machine-driven decisions are more important than ever if we are to choose how to live a good life. Understanding how ethics affect the algorithms and how these algorithms affect our ethics is one of the biggest challenges of our times.

Bidhan L. Parmar is Associate Professor and the Shannon Smith Emerging Scholar in Business Administration at the University of Virginia's Darden School of Business.

R. Edward Freeman is University Professor, Elis and Signe Olsson Professor, Academic Director of the Institute for Business in Society, and Senior Fellow of the Olsson Center for Applied Ethics at the University of Virginia's Darden School of Business.

Chapter 41
Which Rules Are Worth Breaking?

R. Edward Freeman and Bidhan L. Parmar

Disruption of an established business model requires companies to disregard the "rules" of the status quo — within limits.

The twenty-first century business world is being built on disruption in industry after industry. The old rules simply no longer apply to the industries being challenged: Thanks to Airbnb, for instance, hotels will never be the same; thanks to Amazon, retail will never be the same.

At one level, disruption is nothing new — simply a more modern way of rephrasing economist Joseph Schumpeter's theory of "creative destruction." But at another level, something else is happening. Many innovators are not just building a better mousetrap; they are also trying to articulate and consciously break the rules of the game, sometimes by figuring out ways to accomplish more with less, sometimes by finding ways around legal barriers, and sometimes by taking the low road. And there's risk in defying too many rules.

Consider Uber. Only a few years ago, the San Francisco-based car service was the poster child for the new sharing economy. As Uber gained popularity, startups in other industries described themselves as the Uber of clothing, or food delivery, or travel reservations. Uber's business model explored new territory, offering customers the convenience of on-demand ride hailing via a simple app on their mobile phones. Customers got more efficient rides that were digitally integrated into their daily routines, and many of the previous hassles of getting around — from calling

Originally published in: *Sloan Management Review*, 59(3), 1–4 © 2018 from MIT Sloan Management Review/Massachusetts Institute of Technology. All rights reserved. Distributed by Tribune Content Agency
Reprint by Springer

R. E. Freeman (✉) · B. L. Parmar
University of Virginia, Charlottesville, VA, USA
e-mail: FreemanE@darden.virginia.edu

© Springer Nature Switzerland AG 2023

S. D. Dmytriyev, R. E. Freeman (eds.), *R. Edward Freeman's Selected Works on Stakeholder Theory and Business Ethics*, Issues in Business Ethics 53,
https://doi.org/10.1007/978-3-031-04564-6_41

car services to pulling out cash or credit cards for each taxi trip to having to make ride appointments ahead of time — became a thing of the past. Uber also provided drivers the benefits of flexible work and additional income.

Uber became extremely successful in a very short time. By 2014, it was providing 1 million rides a day. By 2016, despite increasing competition, it was providing about 5.5 million rides a day globally.

But more recently, Uber has become more of a poster child for bad behavior — to its employees, its users, its communities, and the ride-hailing industry in general — as a series of mistakes and controversies began to litter its path. Earlier this year, because of its surge pricing practice on the day of protests over the Trump administration's first travel ban, more than 200,000 customers participated in a #DeleteUber campaign to stop using the app and the service. A former Uber engineer wrote a bombshell blog post a few weeks later alleging that the company repeatedly turned a blind eye to sexual harassment and had a culture of gender discrimination. Other sexual harassment issues quickly came to light: A senior vice president of engineering was forced to step down after allegations of harassment at his previous job emerged; a manager was fired for groping women at a company event; a management team in Seoul was reported to have visited escort karaoke bars.

Uber's problems continued to ripple out. Google, an Uber investor, sued the company for stealing intellectual property from Waymo, Google's autonomous car program. Uber blamed "human error" for one of its self-driving cars running a red light and then later acknowledged that the fault lay in the self-driving system, which did not recognize the traffic lights. And after a *New York Times* report called out Uber's practice of "Greyballing" to deceive authorities in areas where the ride-hailing service was restricted, Uber brashly defended its tactics before later conceding that it would no longer use the tool.

This string of controversies led to a mass exodus of Uber senior executives, including the president, the vice president of products and growth, the head of Uber's artificial intelligence labs, and CEO Travis Kalanick himself.

Creating and executing innovative products and services that disrupt the status quo require creativity, and creativity involves thinking differently about overcoming constraints. Laws and social norms are important checks on individual and corporate behavior. But there are forces that make it tempting to push aggressively on constraints. Psychology studies have shown a correlation, for instance, between unethical behavior and creativity: Research conducted by Francesca Gino and Scott Wiltermuth demonstrates that there's a creative upside to cheating. People who cheated in one task were more likely to generate creative solutions in subsequent tasks. The researchers credited a heightened feeling of being unconstrained by rules for the uptick.

Disruptive companies like Uber need to move from a perspective that rules and norms don't matter to a more specific understanding of *which* rules are worth breaking and why. When disrupting the status quo, smart startups would do better using a scalpel rather than a hatchet to avoid cutting off vital relationships and essential resources.

A general attitude that "the rules don't apply to us" paired with a narrow focus on outcomes (particularly shareholder value) creates fertile ground for ethical crises. While an organizational culture geared toward winning can be helpful in many ways, it can also cause unnecessary self-damage because not all rules are worth breaking — even when there is a short-term benefit. For example, society doesn't see breaking rules about taxi licenses at the same level of importance as breaking norms about respecting gender equality. When key stakeholders perceive that a good rule has been violated, they get upset and can find ways to retaliate. Eventually, the negative impressions from repeated crises accumulate and affect the brand, making it harder to attract and retain high-quality employees, customers, suppliers, and communities, all of which are necessary for the company's flourishing.

There is more to business than narrow outcomes defined as profit for shareholders. Violating rules and norms has real costs, and you have to pay for what you break.

Our own research demonstrates that employees are more likely to experience meaningful work in companies that they perceive to be pursuing an important purpose. People gain more autonomy and competence, and have better relationships, in companies they think are focused on a purpose beyond profit. This has tangible payoffs, translating to lower turnover, better employee engagement, and better customer service for companies.

Startups and other organizations that want to disrupt the status quo and also be responsible need a fine-grained understanding of which rules they want to break. Organizations need the capability to understand ahead of time the consequences of breaking certain rules, specifically understanding who will be harmed and who will benefit. Involving stakeholders in the process of developing new products and services is essential so that businesses can appreciate the stakeholders' perspectives and better identify and avoid ethical disasters. Without this capacity to anticipate and circumvent potential problems of its innovations, a company — even one with great and truly innovative ideas — can suffer a death by a thousand cuts.

Yes, we need disruption. But let's make it responsible disruption.

An adapted version of this article appears in the Spring 2018 print edition.

R. Edward Freeman is University Professor, Elis and Signe Olsson Professor, Academic Director of the Institute for Business in Society, and Senior Fellow of the Olsson Center for Applied Ethics at the University of Virginia's Darden School of Business.

Bidhan L. Parmar is Associate Professor and the Shannon Smith Emerging Scholar in Business Administration at the University of Virginia's Darden School of Business.

Chapter 42
A Pragmatist Approach to Business Ethics Research

Bidhan L. Parmar, Robert A. Phillips, and R. Edward Freeman

1 Introduction

The purpose of this chapter is to give an introductory account of the main tenets of a philosophical approach called "pragmatism." Pragmatism has its origins in the late nineteenth and early twentieth century in the work of the American philosophers Charles S. Pierce, William James, John Dewey, and others. It has experienced a renaissance in recent years under the auspices of modern philosophers such as Richard Rorty, Richard Bernstein, Hilary Putnam, and increasingly, many others. These more recent philosophers have established a myriad of connections to other intellectual movements and to other philosophers. For instance, Rorty writes a great deal about the connections between Dewey and so-called Continental philosophers such as Nietzsche, Heidegger, Derrida, and others.

There is a stream of pragmatist work in business ethics especially in what has come to be called "stakeholder theory" (Freeman et al. 2010), and the beginnings of a stream of work in organization theory (Kelemen and Rumens 2013; Wicks and Freeman

Originally published in: *Cambridge Handbook of Research Approaches to Business Ethics and Corporate Responsibility*, 258–269 © Cambridge University Press, 2017
Reprint by Springer, https://doi.org/10.1017/9781316584385.019

This chapter draws significantly from our chapter, "Pragmatism and Organization Studies" published in Mir, R., Willmott, H., & Greenwood, M. (Eds.). (2015). *The Routledge companion to philosophy in organization studies*. Routledge.

B. L. Parmar · R. E. Freeman (✉)
University of Virginia, Charlottesville, VA, USA
e-mail: FreemanE@darden.virginia.edu

R. A. Phillips
York University, Toronto, ON, Canada

© Springer Nature Switzerland AG 2023

S. D. Dmytriyev, R. E. Freeman (eds.), *R. Edward Freeman's Selected Works on Stakeholder Theory and Business Ethics*, Issues in Business Ethics 53,
https://doi.org/10.1007/978-3-031-04564-6_42

1998). In Sect. 2 we give a brief historical view of the development of pragmatism. In Sect. 3 we explain some central pragmatist ideas, even though not all pragmatists will agree. In Sect. 4 we argue why these ideas are important for business ethicists.

2 A Brief History and Key References

It is important to note that pragmatism has been developed as a way to understand a set of traditional philosophical problems, rather than as a way to understand how human beings create value and trade with each other (business and management). However, we believe that a pragmatist approach to business and management offers great potential in practice and in theory.

Pragmatism developed in opposition to some traditional philosophical claims. The first kind of claim – typical of Cartesian and Kantian philosophy – is that thinking must be somehow founded on *a priori* first principles. So, when James defends his "radical empiricism," he is pitting it against the sort of theorizing that assumes thought to be prior to any empirical observation. For example, while science and the scientific method play vital roles, the authority of science can be overstated (e.g., positivism). Pragmatists are (generally) at great pain to demonstrate the relevance of moral thinking alongside scientific thinking. As we read the sample passages below, we may have cause to recognize this reductionist vice that pragmatist authors seek to pry open.

Pragmatism shares, broadly, the features described below in III. At the margins, however, pragmatism comes in as many flavors as there are scholars writing about it. In 1908, Arthur O. Lovejoy isolated thirteen different variations of pragmatism – a number Richard Bernstein called "far too conservative" (2010: 5). At the periphery there are arguments about who should be included in the Pantheon and who was (or is), a closet pragmatist whether they claimed the mantle or not. We will not attempt to summarize or adjudicate these questions here. The authors we summarize here were all quite wide-ranging in their interests and writings. We have isolated representative passages from each writer to give the reader a sense of each author's style.

While Charles Sanders Peirce[1] is typically regarded as the first pragmatist, later in life, Peirce began describing his own work as "pragmaticism" ("a word ugly enough to be safe from kidnappers") in order to distinguish it from what pragmatism had grown into. This quasi-apostasy notwithstanding, Peirce is credited by the other early pragmatists as the originator. Durand and Vaara (2009) adapt some of Peirce's work for uses in understanding strategic management research. Here is a taste of Peirce's (1878/1965) work:

> From all these sophisms we shall be perfectly safe so long as we reflect that the whole function of thought is to produce habits of action; and that whatever there is connected with a thought, but irrelevant to its purpose, is an accretion to it, but no part of it. (p. 256)

[1] Though resembling Pierce, it is homophonous with "purse."

> To develop its meaning, we have, therefore, simply to determine what habits it produces, for what a thing means is simply what habits it involves. (p. 257)

William James, considered the father of experimental psychology, is among most accessible and reader-friendly of the early pragmatists. Combining a capacity for deep thought with remarkable writing skills and an eschewal of jargon, his work was able to speak to both scholarly and popular audiences alike. Margolis and Walsh (2003) make good use of James's work in their influential discussion of businesses' social initiatives. Here are representative passages from James (1907/2014):

> In other words, the greatest enemy of any one of our truths may be the rest of our truths. Truths have once for all this desperate instinct of self-preservation and of desire to extinguish whatever contradicts them. My belief in the Absolute, based on the good it does me, must run the gauntlet of all my other beliefs. Grant that it may be true in giving me a moral holiday. Nevertheless, as I conceive it, – and let me speak now confidentially, as it were, and merely in my own private person, – it clashes with other truths of mine whose benefits I hate to give up on its account. It happens to be associated with a kind of logic of which I am the enemy, I find that it entangles me and metaphysical paradoxes that are inacceptable, etc., etc. But as I have enough trouble in life already without adding the trouble of carrying these intellectual inconsistencies, I personally just give up the Absolute. I just *take* my moral holidays to me; or else as a professional philosopher, I try to justify them by some other principal. (p. 78)

A prominent public intellectual with a longtime interest in education, John Dewey emphasized the role that humans have in creating our world, not merely acting and reacting within it. Dewey consistently emphasized the role of art, literature, and morals – in addition to science – in co-creating our world. Dewey's work was deeply influential on the later work of Richard Rorty (see below); Mahoney (1993) relies on Dewey and others to examine the role of determinism in strategic management. The following passages (Dewey 1908) are representative of his critique of prior philosophizing and his beliefs in the centrality of values:

> If we suppose the traditions of philosophic discussion wiped out and philosophy starting afresh from the most active tendencies of to-day, – those striving in social life, in science, in literature, and art, – one can hardly imagine any philosophic view springing up and gaining credence, which did not give large place, in its scheme of things, to the practical and personal, and to them without employing disparaging terms, such as phenomenal, merely subjective, and so on. (p. 125)

> To frame a theory of knowledge which makes it necessary to deny the validity of moral ideas, or else to refer them to some other and separate kind of universe from that of common sense and science, is both provincial and arbitrary. The pragmatist has at least tried to face, and not to dodge, the question of how it is that moral and scientific "knowledge" can both hold of one and the same world. And whatever the difficulties and his proffered solution, the conception the scientific judgments are to be assimilated to moral is closer to common sense than is the theory that validity is to be denied of moral judgments because they do not square with a preconceived theory of the nature of the world to which scientific judgments must refer. (p. 127)

After a bit of a lull in philosophical interest in the middle part of the twentieth century, pragmatism was forcefully revived with the writings of Richard Rorty.[2] Controversial among mainstream philosophers, Rorty considered language as the medium for making sense of the world and the poet (in a broad sense) as the one who shows us new and better ways to live. As with others on our list, Rorty was an apostate in analytic philosophy. Wicks and Freeman's (1998) work bear the marks of Rorty's influence. The following passages give a sense of Rorty's (2010) distinctive writing style:

> communities are held together by agreement on what counts as coherent, but make progress only when linguistic innovators break this agreement. One of their strategies is by talking about objects that we have no agreed-upon ways of talking about … Without a consensus to break, innovation would be pointless. Without innovation the crust of convention would never be broken. The alternation between the two has made it possible for human beings to move from poem to poem, from language game to language game, and from poorer to richer forms of life. (p. 194)

Like Peirce, Hilary Putnam is a mathematician as well as a pragmatist philosopher. Richard Rorty (1991) calls him "the most important contemporary philosopher to call himself a pragmatist." Purnell and Freeman (2012) rely on Putnam's collapsing of the fact/value distinction in assessing the quality of ethics conversations among investment bankers. While Putnam's writing has perhaps less lyrical quality (poetry) to it than James or Rorty, it is less technical and jargon-laden than Peirce, it is clear, careful, and precise. This is evident in the passages below (Putnam 2002):

> There are a variety of reasons why we are tempted to draw a line between "facts" and "values" – and to draw it in such a way that "values" are put outside the realm of rational argument all together. For one thing, it is much easier to say, "that's a value judgment," meaning, "that's just a matter of subjective preference," and to do what Socrates tried to teach us: to examine who we are and what our deepest convictions are and hold those convictions up to the searching test of reflective examination … The worst thing about the fact/value dichotomy is that in practice it functions as a discussion-stopper, and not just a discussion-stopper, but a thought-stopper. (p. 43ff)

More recently, Hilary Putnam calls Richard J. Bernstein's *The Pragmatic Turn*, "by far the best and most sophisticated account of recent and present-day pragmatist thought." With chapters centrally focused on Peirce, James, Dewey, Hegel, Putnam, and Rorty among others, this book is an excellent critical introduction to the history of pragmatism, the surface of which is barely scratched here.

Many, many others could have been on this list including, but not limited to O.W. Holmes, R.W. Emerson, C.I. Lewis, W.V.O. Quine, L. Wittgenstein, and C. West. And there are pragmatist interpretations of many others besides (e.g., Bernstein has a chapter entitled "Jürgen Habermas's Kantian Pragmatism").

[2] Freeman (2004) has written a book review essay entitled, "The Relevance of Richard Rorty to Management Research" in *Academy of Management Review*.

But any serious investigation into pragmatism and its potential role in the study of organizations must eventually contend with the work of the authors above.

3 What Do Pragmatists Believe?

While there are almost as many versions of pragmatism as there are pragmatist philosophers, there is a great deal of agreement on a few general principles, many of which are about what counts as doing work in philosophy that is useful. We isolate the following commonalities:

1. All "theory" is based in experience and practice.
2. Most dualisms and dichotomies are at best misleading.
3. Framing and language use is central to understanding the world.
4. Hope and freedom should be the goals of discourse rather than "truth."

3.1 All "Theory" Is Based in Experience and Practice

The founders of pragmatism, Pierce, James, and Dewey, all believed that good thinking was rooted in experience. They talk about the primacy of both practice and experience. And, they all had an idea of experience as a primary conceptualization of humans encountering the world. Dewey especially thought that our conceptual apparatus evolved to solve problems. We became language users somewhere in our evolutionary history as it gave us an ability to solve complex problems, and an advantage over some of our natural competitors and predators. Dewey (1998) writes:

> Mankind has hardly inquired what would happen if the possibilities of experience were seriously explored and exploited. There has been much systematic exploration in science and much frantic exploitation in politics, business, and amusement. But this attention has been, so to say, incidental and in contravention to the professedly ruling scheme of belief. It has not been the product of belief in the power of experience to furnish organizing principles and directive ends. (p. 23)

These pragmatists eschewed the idea of theorizing for the sake of theorizing that characterizes so much of contemporary philosophy, not to mention much of management theory. Pragmatists see the point of intellectual life as solving real problems that come from experience.

Gregory Pappas (2008) suggests that another related hallmark Dewey's thought is "seeing problems from the inside." What does this problem feel like if you are experiencing it yourself? It is often too easy to intellectualize a problem away, or simply use one of our favorite theories to understand the problem. Pappas suggest that this is seeing the problem as a spectator, and that doing so loses most of what we gain when we actually have an experience.

A good example is Dewey's approach to ethics where he does not begin with a theory like utilitarianism, or deontology, or virtue, but he begins by trying to understand the particular aspects of a moral problem from the inside. As we see how these aspects are related to one another, and to other problems we have had, we can draw on these more theoretical ideas to help solve the problem. But, we are not trying to prove the theories in any case. Every situation is different, but there are common aspects. The pragmatists look for the particularity as well as the generality in experiences.

3.2 Most Dualisms and Dichotomies Are at Best Misleading

Much of modern philosophy depends on a set of dichotomies or dualisms that contrast ways of seeing the world. For instance, "mind vs. body," "external world vs. internal world," "things as they appear vs. things in themselves," "theoretical vs. empirical," "science vs. nonscience," "self interest vs. altruism," "facts vs. values," etc. The list is almost endless and it has been said that Dewey never met a dichotomy that he didn't want to collapse.

Quine (1951) identified two of these dualisms as important dogmas of empiricism that are regrettably present in current management thinking as well. The first was the theory–observation distinction, or what philosophers would call "analytic–synthetic" distinction. Analytic sentences were true by definition and synthetic sentences were true by virtue of their relationship to the world. The idea from the positivists was that certain terms were theoretical terms while other terms were observational (data), and that data stood in a relationship of confirmation or disconfirmation to the theoretical terms. Quine argued that there was no way to enforce this distinction without appealing to a set of terms that were circular in their meaning.

Dewey's version of the same argument was that the data that you look for are already at least partially contained in how you frame a hypothesis. Theory does bear a connection to the world, but only in terms of theory as a whole, not particular hypotheses. Quine's dictum is that "[O]ur statements about the external world face the tribunal of sense experience not individually but only as a corporate body." (1951: 38) People always bring their background theories, disciplines, ideas, etc. with them.

In a similar way, Hilary Putnam has argued that facts and values are entangled (see quoted passage above), much in the way that theory and data are entangled. Putnam's example is that to call someone "cruel" is to state a fact, if they are acting cruelly, and it is to make a value judgment at the same time, since calling someone cruel is to disapprove of them. Much of our language works this way. Putnam and Bernstein would argue that facts and values are interwoven with theory and practice, so that facts, values, and interpretations are always relevant to some interpretive community or other.

3.3 Framing and Language Use Is Central to Understanding the World

Most pragmatists would agree with Wittgenstein when he argued that we should not look to a world of meanings of words to understand language, but we should look to how the words are used. When children learn a language they learn what the words will do, what problems they will solve, and what problems they will not solve. Meanings emerge out of these uses and problems solved, as "short hands" or "rules of thumb." They are not a representation of the relationship of language to the world. Indeed, most modern pragmatists would argue that many people (philosophers) misunderstand the nature of language.

As language using primates we managed to coordinate our behavior in powerful ways to become the dominant species on the planet. Indeed, Tomasello and his colleagues (2012) have suggested that human evolution selects for being "good collaborators." We coordinate our behavior by making marks and noises that indicate what we are doing. Of course this is now very sophisticated, yielding many vocabularies in fields as diverse as mixing concrete and quantum mechanics.

According to Donald Davidson, language is an unbroken whole. By that he means that we can't divide up language into different "mind sets" or "conceptual schemes" that are wholly different from each other. He argues that one of the main mistakes in both science and social science is the idea that such a "scheme-content distinction" is possible. He articulates the problem of conceptual relativism as follows (Davidson 1984):

> Conceptual schemes, we are told, are ways of organizing experience; they are systems of categories that give form to the data of sensation; they are points of view from which individuals, cultures, or periods survey the passing scene. There may be no translating from one scheme to another in which case the beliefs, desires, hopes, and bits of knowledge that characterize one person have no have no true counterparts for the subscriber to another scheme. Reality itself is relative to a scheme: what counts as real in one system may not in another. (p. 183)

Davidson then argues the idea of a total failure to translate from one conceptual scheme to another makes no sense, and then shows how even a partial failure to translate doesn't imply separate conceptual schemes. Language is all we have.

Even though we may not know how to navigate from one mindset to the other, it has to be possible, even if sometimes it isn't very useful. For instance, what does the vocabulary of nineteenth century romantic poetry have to do with the vocabulary of twenty-first century business ethics? One might find that the passion present in the stanzas of nineteenth century romantic poetry is simply missing in current business ethics. Such a comparison might lead to a question about why that is the case, and a call for a renewed vocabulary for understanding our humanity in organizations. The pragmatist program of seeing what problems vocabularies solve and why some are better than others, and how some can be enriched by comparison with others, yields very different methods for studying business than are currently used in most scholarship on management and organizations.

3.4 *Hope and Freedom Should Be the Goals of Discourse Rather Than "Truth"*

When many scholars hear the word "pragmatism" they immediately associate it with the idea that "truth is what is useful." This often cuts against the very reasons that scholars have chosen their field, i.e., to "get to the truth of the matter." In fact, many pragmatists have a far more nuanced view of "truth," and there has been substantial philosophical debate about the idea in the literature among pragmatists. For instance, Davidson believes that even if we give up the idea of an "uninterpreted reality" to be approximated by language, a view of objective truth is still possible. Davidson writes (1984):

> Of course truth of sentences remains relative to language, but that is as objective as can be. In giving up the dualism of scheme and world, we do not give up the world, but re-establish unmediated touch with the familiar objects whose antics make our sentences and opinions true or false. (p. 198)

Rorty on the other hand, thinks that the idea of truth is best understood as reminding us that any belief that we have is potentially revisable. Knowledge for Rorty is the idea of warranted assertability, and whether a claim is warranted is largely a matter of an interpretive community. The idea of an assertion being true reminds us that all the evidence for assertion may not be in. Rorty argues that vocabularies can be useful ways to solve problems or not. In that spirit, Hilary Putnam suggests that if you want to know why a square peg won't fit into a round hole, the vocabulary of molecular structure won't work very well (atoms and molecules being mostly empty space). There are lots of ways to talk about the world, and no one of them gets at the world "as it really is."

Rorty has gone further than most here by claiming that hope and freedom are more tangible goals than Truth. By this he means engaging in the search for solidarity with others, seeing "them" as "us" rather than some outside "other." He writes (Rorty 1999):

> The trouble with aiming at truth is that you would not know when you had reached it, even if you had in fact reached it. But you can aim at ever more justification, the assuagement of ever more doubt. Analogously you cannot aim at "doing what is right," because you will never know if you have hit the mark. Long after you are dead, better informed and more sophisticated people may judge your action to have been a tragic mistake, just as they may judge your scientific beliefs as intelligible only by reference to an obsolete paradigm. But you can aim at ever more sensitivity to pain, and ever greater satisfaction of ever more various needs. Pragmatists think that the idea of something nonhuman luring us human beings on should be replaced with the idea of getting more and more human beings into our community – of taking the needs and interests and views of more and more diverse human beings into account. (p. 82)

What these ideas mean for research is that we need to engage in a conversation that at once moves across vocabularies. When researchers embrace a plurality of metaphors it can unsettle the dominant paradigm and provide a broader toolkit from which to craft a better world. In order to generate a plurality of metaphors and stories, researchers would be better served to relax assumptions about truth and focus instead on conversation. Conversation is where different perspectives encounter on another and form the basis for a constructive dialog about the cash-value of particular metaphors.

4 Pragmatism, Positivism, and Business Ethics

In 1959 the Ford and Carnegie Foundations issued separate reports on the state of university-based business schools in the US. These reports were highly critical of the low standards and rigor in a majority of business schools and sought to instigate change by infusing these institutions with a dose of the newly emerging "management science" – including a methodology for managerial decision making honed during the World War II. This new decision making science included tools such as decision analysis, game theory, and insights from emerging behavioral science. The hope was that knowledge from these new disciplines would allow managers to make decisions using reason and analysis rather than intuition and judgment.

To realize the vision of a scientifically-based business school and management profession, a new kind of faculty was needed, "one focused more on fundamental research than on descriptive analysis, and deriving decision making principles more from theory than from existing practice." (Khurana 2007: 271). Therefore, new faculty were trained in behavioral sciences, particularly economics, and were encouraged to increase research productivity and publish in academic journals to raise the research profile of their institutions. Just 5 years after the publication of the Ford report, faculty at the twenty-five leading business schools in the US were significantly more research oriented and published more in academic journals.

This new emphasis on science brought with it a positivist epistemology, rather than a science based on Dewey's pragmatic experimentation. Positivism makes the assumption that the study of business and organizations can occur through a scientific approach that is value free and is superior to nonscientific methods because it corresponds to objective reality. Astley (1985:497) argues that positivism embraces a conventional model of scientific progress as cumulative discovery of objective truth, where knowledge grows linearly as new data are added to the existing stock of research findings.

Positivism, and its newer incarnations postpositivism and neopositivism, are all attempts to uncover objective Truth or True Reality. Postpositivism differs from positivism in holding that reality can be known only probabilistically, and hence verification is not possible. Neopositivism asserts that science must deal in exact descriptions and generalization, both of which require "the quantitative statement." Both post- and neopositivism share the belief that science should not – and cannot – deal with value statements.

Throughout the latter half of the twentieth century and the early twenty-first century, little changed in university-based business schools as faculty continued to publish research and focus on quantitative methods as a vehicle for discovering

Truth. In the late twentieth century the introduction of programmatic rankings for graduate business schools (and similar ranking for journals) accelerated and focused these trends in ways that reduced the diversity of research published in these journals and increased the dominance of positivism. Rankings of business school programs typically include surveys of stakeholders such as students, alumni, and recruiters but also count the number of articles that faculty have in a finite number of journals deemed "A" journals. By constraining the space in which research publications "count" as good research rankings such as *BusinessWeek* and *US News & World Report* increase competition for publishing in the journals they deem worthy as more faculty aimed for these journals to get tenure and increase their own research reputation.

This increased competition in turn serves to strengthen "normal" science and reinforce the dominant theoretical and epistemological perspectives as academic reviewers fall back on their own assumptions about epistemology and truth to make judgments on papers and ideas that they think are "True" and therefore worth publishing. Constrained space for what counts as "good" research makes new ideas and different approaches more difficult to publish, because those authors have to compete with authors who publish "normal" science that fits dominant assumptions and expectations about how science should look.

Given this background on the current state of theory and scholarship, we now turn to several ways in which a pragmatist may seek to improve this situation.

4.1 Practice-Focused Research

Pragmatism sees theory as a tool for doing things better in the world. For Dewey, the role of thought and knowledge is not to passively reflect a preexisting world, but to change and improve reality. The outcome of a successful inquiry is a "transformation of the situation." On Dewey's account, the most "basic conception of inquiry" is "as determination of an indeterminate situation" (Dewey 2008) or, to use another term, sensemaking.

Pragmatism encourages scholars to show how theoretical differences make a real difference to practice. This focus on practice shifts the focus on theory building in management science away from insular conversations between academics and toward practice-relevant research. Research that is focused on theory only for the sake of theory makes the assumption that good theory is a value-neutral and objective view of the world from which many different purposes can be achieved. Therefore academics focus on proof and evidence as mechanisms for creating Truth. This emphasis crowds out discussion of practical relevance and stacks the deck in favor of "an entrenched vocabulary which has become a nuisance" rather than "a half-formed new vocabulary which vaguely promises great things" (Rorty 1989). Positivist assumptions direct research toward those questions that can be more easily "proved" with statistics rather than questions that impact how we should live, but may not be as easily measurable.

Organizational scholars have debated their relevance to practice in journal articles, books, and conferences: (i.e., *Administrative Science Quarterly* 1982, Vol. 27, No. 4; *Academy of Management Journal*, 2001, Vol. 44, No. 2; Lawler 1999; Murphy and Saal 2013; Larwood and Gattiker 1999; Mowday 1997; Hitt et al. 1998; Huff 2000). Some scholars argue that academics and practitioners hold irreconcilably different views about research that is relevant and of high quality (Shrivastava and Mitroff 1984). Others have called for practically relevant research and isolated (modest) zones of agreement between practitioners and academics about what constitutes interesting research (Baldridge et al. 2004). Yet the actual practice of publishing work that has practical relevance remains difficult.

If "management science" is to become focused on practice, there are several prescriptions based on Dewey's pragmatic experimentation that can make it more relevant and useful. First, publishing null findings is a useful practice so that researchers and practitioners know what does *not* work. Publishing only positive findings that fit existing theory are not practices that foster the sensemaking necessary to create more useful explanations. Rather it is an artificial way to limit the kind of research that "counts." By not publishing null findings current research practice reinforces the ideas of Objectivity and Truth because there is less emphasis placed on the boundary conditions of a theory, and theoretical limitations are less salient when only positive findings are published with a small paragraph about hypothesized limitations.

Additionally, according to Dewey, the scientific spirit requires seeking out new problems that are not widely known or experienced. Pragmatism would encourage management science to look for novel problems and develop new and distinct processes for evaluating novel work. Treating new ideas in the same way as old established ideas is a sure way to kill innovation and experimentation.

Finally, inquiry is based in particular perspectives and theoretical lenses. These lenses were crafted for particular purposes and to address specific problems. Scholars must become clearer about the purposes for which their explanations were created, to encourage more careful use and experimentation with theory. Research is not an objective map of reality that can be used for any purpose, but itself is a purpose-driven activity, that must be clear about the uses it has been developed for so that others can pick up the right tool for the right job. For example, using economic assumptions to narrowly describe the corporate objective is to use a theory in a way that creates negative outcomes for companies and their stakeholders. Forgetting the purpose for which a particular theory is created can lead to what Daniel Dennett calls, "greedy reductionism," where, "in their eagerness for a bargain, in their zeal to explain too much too fast, scientists and philosophers … underestimate the complexities, trying to skip whole layers or levels of theory in their rush to fasten everything securely and neatly to the foundation" (Dennett 1995: 82).

4.2 *Bringing Back Knowing-How*

According to Dewey, knowing-that is a kind of knowledge that "involves reflection and conscious appreciation." It is our ability to consciously reason about things. Dewey sees knowing-that as a kind of reflective thinking which is an:

> active, persistent and careful consideration of any belief or supposed form of knowledge in light of the grounds that support it, and the further conclusions to which it tends. (Krueger 2009: 36)

Dewey sees most of everyday human action and interaction as involving a different sort of knowledge – knowing-how, or the

> experiential or embodied learning that is activated prior to, or without the invocation of, reflective thinking. This sort of knowledge is a pre reflective coping or skill-knowledge that enables us to navigate our world with a high degree of expert interaction. (Krueger 2009: 37)

This distinction is important for management theory and business ethics because it largely focuses on "knowing-that" and assumes that "knowing-that" is sufficient to cope with the world in better ways. This distinction is connected to the need for practice-focused research. Theories and concepts solve problems. When they no longer solve problems they are discarded in favor of other theories and concepts that do a better job. For instance, astrology once solved a problem about why certain events occurred, and what we should do in the future given these astrological predictions. Ultimately, better explanations emerged, from biology, physics and psychology, and astrology has been largely discarded. It does not help us to solve the problems of what we should do in the future nearly as well as the theories and concepts of these more modern disciplines. When children learn language, they do not learn the meanings of the words. Rather they learn what the words do, including sometimes irritating their parents. We do things with words. This distinction helps us to see that much work must be done to build "knowing-how."

Business ethicists must think differently about how they teach and research, so that their students build the capacity to know how to live better. The problem a great deal of published research seems to solve is the problem of "how do faculty get published in A journals," which has nothing to do with the kinds of problems that organizational members experience.

4.3 *Fetish Around Perfect Definitions*

As research becomes more insular, there is a growing emphasis on tight definitions and constructs. Tightly defined constructs are seen as necessary for science to accurately measure phenomena and to more closely reflect reality. Suddaby argues that, "For positivists, construct clarity helps them test theory, since precisely defined constructs are easier to operationalize and test (Schwab 1980) and it is easier for researchers to compare and contrast results (Bagozzi and Edwards 1998) … in some instances constructs become so clearly defined, measurable, and operationalized over time that they lose relevance with the empirical world and, ultimately, reappear under a different name" (2010: 353).

A pragmatist does not see these constructs as reflective of the world, but as shared assumptions and expectations within a community of scholars, trying to solve a set of real world problems. Increased precision in constructs can be helpful as shared meaning about a construct increases understanding and coordination between the author and her readers. Suddaby (2010: 352) agrees when he argues that, "The creation of a common vocabulary avoids the "Tower of Babel" effect, in which subcommunities of researchers have no common means of communication. In the absence of common and well-articulated constructs, the boundaries between subcommunities become more sharply defined and organizational knowledge becomes increasingly fragmented." But these expectations about construct clarity can be also be a detriment when expectations about the precision of constructs gives license to scholars to reject work that does not share their personal (sometime idiosyncratic) definitions. Suddaby continues, "finding a single exception is often fatal to a construct because it implies that any proposition associated with the construct is false. Reviewers may take this position even in cases where there is substantial positive empirical support for a construct, largely because most reviewers have been oversocialized to accept falsification as the basis of scientific truth" (Suddaby 2010: 349). The meaning of these constructs are negotiated in conversation, but in the journal review process the burden of translating new ideas into the old vocabulary is placed on the new author not the incumbent, thereby significantly decreasing the likelihood that novel work will be published in "A" journals.

Additionally, the assumption that research is an endeavor to produce Truth can give license to reviewers to reject work that does not fit their world view instead of engaging in conversation with the authors. It also disadvantages new vocabularies from being published when they contain constructs that are defined and used in a consistent and logical way, but may fall short of empirical testing.

4.4 *Seeing Problems from the Inside*

Another important aspect of pragmatism is the focus on lived experience and seeing problems from the inside. When theorists develop descriptions of problems, their perspective and goals are different from those of the people who experience the problem. To help those successfully address the problem, it is important to see the issues from the perspective of individuals who experience those problems. For example, in the study of ethics, John Dewey argues that traditional moral theories have abstracted out the qualitative experience of moral problems and the social context in which they occur. In his own masterwork, *The Varieties of Religious Experience*, James (1902/1985) sees matters of religion and faith in terms of common, if difficult to explain, experiences. By disregarding the lived experience of ethical issues, scholars increase the likelihood that their theories will be less useful to those individuals.

In organization studies, taking the lived experience seriously can mean several things. For example, in social psychological work on ethics we can examine not just on how context shapes behavior, but on how people experience and interpret context differently. Across disciplines, we may work to understand how people experience and apply the theories we generate, rather than assume that their application is straightforward or obvious. Attending to how people experience and apply theories can shed light on how to make them more useful and applicable.

One way to take experience seriously is highlight that organizational phenomena are socially constructed. In addition to redescribing phenomena as contingent and created, pragmatism encourages organizational scholars to pay attention to the effects of different social constructions. Each way of construing a situation or problem has benefits and costs that are differentially distributed across stakeholder groups. Therefore we must not stop at describing a process of social construction or at claiming that something is socially constructed, but also examine the effects of those constructions, how those effects themselves are created and maintained, and how they shape the ability of others to live better. These questions are inseparable from ethics and morality and therefore hold promise for business ethicists.

5 Conclusion

We believe that pragmatism holds great promise in shaping business ethics and organizational studies in ways that can help organizational actors live better lives. In this paper we have provided a brief history of pragmatist thought, outlined several of its central tenets, and provided a few directions which the studies of organizations and management can take if scholars want to take pragmatist ideas seriously. It is important to note that ideas are only the beginning, and while a growing number of academics now sympathize with pragmatist thought and decry the narrow view of science that pervades our journals, action is important. To change existing practice is difficult, but necessary if we are serious about living and helpings others to live better lives through our scholarship.

References

Astley, W.G. 1985. Administrative Science as Socially Constructed Truth. *Administrative Science Quarterly* 30: 497–513.

Bagozzi, Richard P., and Jeffrey R. Edwards. 1998. A General Approach for Representing Constructs in Organizational Research. *Organizational Research Methods* 1 (1): 45–87.

Baldridge, David C., Steven W. Floyd, and Lívia Markóczy. 2004. Are Managers from Mars and Academicians from Venus? Toward an Understanding of the Relationship between Academic Quality and Practical Relevance. *Strategic Management Journal* 25 (11): 1063–1074.

Bernstein, R.J. 2010. *The Pragmatic Turn*. Cambridge: Polity.
Davidson, D. 1984. *Inquiries into Truth and Interpretation*. Oxford: Oxford University Press.
Dennett, Daniel C. 1995. Darwin's Dangerous Idea. *The Sciences* 35 (3): 34–40.
Dewey, J. 1908/1998. Does Reality Possess Practical Character? In *The Essential John Dewey Volume 1*, eds. L. Hickman and T. Alexander. Bloomington: Indiana University Press.
———. 1930/1998. What I Believe In *The Essential John Dewey Volume 1*, eds. L. Hickman and T. Alexander. Bloomington: Indiana University Press.
Dewey, John. 1998. *Experience and Education*. Kappa Delta Pi.
———. 2008. *The Later Works of John Dewey, Volume 7, 1925–1953: 1932, Ethics*. Vol. 7. Carbondale: SIU Press.
Durand, R., and E. Vaara. 2009. Causation, Counterfactuals, and Competitive Advantage. *Strategic Management Journal* 30 (12): 1245–1264.
Freeman, R.E. 2004. Book Review Essay: The Relevance of Richard Rorty to Management Research. *Academy of Management Review* 29 (1): 127–144.
Freeman, R.E., J.S. Harrison, A.C. Wicks, B.L. Parmar, and S. De Colle. 2010. *Stakeholder Theory: The State of the Art*. Cambridge University Press.
Hitt, Michael A., Javier Gimeno, and Robert E. Hoskisson. 1998. Current and Future Research Methods in Strategic Management. *Organizational Research Methods* 1.1: 6–44.
Huff, Anne Sigismund. 2000. 1999 Presidential Address: Changes in Organizational Knowledge Production. *Academy of Management Review* 25 (2): 288–293.
James, W. 1902/1985. *The Varieties of Religious Experience*. Cambridge, MA: Harvard University Press.
———. 1907/2014. *Pragmatism*. New York: Cambridge University Press.
Kelemen, M., and N. Rumens, eds. 2013. *American Pragmatism and Organization*. Surrey: Gower.
Khurana, Rakesh. 2007. *From Higher Aims to Hired Hands: The Social Transformation of American Business Schools and the Unfulfilled Promise of Management as a Profession*. Princeton: Princeton University Press.
Krueger, J.W. 2009. Knowing Through the Body: The Daodejing and Dewey. *Journal of Chinese Philosophy* 36 (1): 31–52.
Larwood, Laurie, and Urs E. Gattiker, eds. 1999. *Impact Analysis: How Research Can Enter Application and Make a Difference*. Psychology Press.
Lawler, Edward E., ed. 1999. *Doing Research that is Useful for Theory and Practice*. Lanham: Lexington Books.
Mahoney, J.T. 1993. Strategic Management and Determinism: Sustaining the Conversation. *Journal of Management Studies* 30 (1): 173–191.
Margolis, J.D., and J.P. Walsh. 2003. Misery Loves Companies: Rethinking Social Initiatives by Business. *Administrative Science Quarterly* 48: 268–305.
Mowday, Richard T. 1997. Celebrating 40 Years of the *Academy of Management Journal*. *Academy of Management Journal* 40 (6): 1400–1414.
Murphy, Kevin R., and Frank E. Saal, eds. 2013. *Psychology in Organizations: Integrating Science and Practice*. New York: Psychology Press.
Pappas, G.F., ed. 2008. *John Dewey's Ethics: Democracy as Experience*. Bloomington: Indiana University Press.
Peirce, C.S. 1878/1965. How to Make Our Ideas Clear. In *Collected Papers of Charles Sanders Peirce*, vol. Vol. 5. Cambridge, MA: Harvard University Press.
Purnell, L.S., and R.E. Freeman. 2012. Stakeholder Theory, Fact/Value Dichotomy, and the Normative Core: How Wall Street Stops the Ethics Conversation. *Journal of Business Ethics* 109 (1): 109–116.
Putnam, H. 2002. *The Collapse of the Fact/Value Dichotomy and Other Essays*. Cambridge, MA: Harvard University Press.
Quine, W.V. 1951. Main Trends in Recent Philosophy: Two Dogmas of Empiricism. *The Philosophical Review* 60 (1): 20–43.
Rorty, Richard. 1989. *Contingency, Irony, and Solidarity*. Cambridge: Cambridge University Press.

Rorty, R. 1991. *Feminism and Pragmatism: The Tanner Lectures on Human Values*. Ann Arbor: University of Michigan.
———. 1999. *Philosophy and Social Hope*. New York: Penguin Books.
———. 2010. Reply to Aldo Giorgio Gargani. In *The Philosophy of Richard Rorty*, ed. R.E. Auxier and L.E. Hahn, 193–195. Chicago: Open Court.
Schwab, D.P. 1980. Construct Validity in Organizational Behavior. In *Research in Organizational Behavior*, ed. B.M. Staw and L.L. Cummings, vol. Vol. 2, 3–43. Greenwich: JAI Press.
Shrivastava, Paul, and Ian I. Mitroff. 1984. Enhancing Organizational Research Utilization: The Role of Decision Makers' Assumptions. *Academy of Management Review* 9 (1): 18–26.
Suddaby, Roy. 2010. Editor's Comments: Construct Clarity in Theories of Management and Organization. *The Academy of Management Review* 35: 346–357.
Tomasello, M., A.P. Melis, C. Tennie, E. Wyman, and E. Herrman. 2012. Two Key Steps in the Evolution of Human Cooperation: The Interdependence Hypothesis. *Current Anthropology* 53 (6): 673–692.
Wicks, A.C., and R.E. Freeman. 1998. Organization Studies and the New Pragmatism: Positivism, Anti-Positivism, and the Search for Ethics. *Organization Science* 9 (2): 123–140.

Bidhan L. Parmar is Associate Professor and the Shannon Smith Emerging Scholar in Business Administration at the University of Virginia's Darden School of Business.

Robert A. Phillips is George R. Gardiner Professor in Business Ethics, Professor of Sustainability, and Director of the Centre of Excellence in Responsible Business at the York University's Schulich School of Business.

R. Edward Freeman is University Professor, Elis and Signe Olsson Professor, Academic Director of the Institute for Business in Society, and Senior Fellow of the Olsson Center for Applied Ethics at the University of Virginia's Darden School of Business.

Chapter 43
Profit and Other Values: Thick Evaluation in Decision Making

Bastiaan van der Linden and R. Edward Freeman

1 Introduction

Companies engage with many "types of stakeholder value" (Harrison and Wicks 2013, 99). For instance, they can make profit, deliver products and services, provide jobs, avoid unnecessary harm to the environment, observe human rights, and respect the democratic integrity of governments. Although some of these values have started to receive our attention only relatively recently, businesses have always engaged with different values—and they have probably always encountered difficulties when doing so. Recently, stakeholder theorists have argued that different kinds of value may need to be simultaneously considered in decision making, and not only in so far as they contribute to the maximization of profit. They state for instance that "multiple measures of firm performance are superior to just one" (Harrison and Wicks 2013, 118) and that managers are "guided by many starts" (Mitchell et al. 2016, 267). This seems diametrically opposed to the position of "profit maximizers" that decision making requires a "single dimensional objective to be maximized" (Jensen 2002, 248; see also Sundaram and Inkpen 2004).

However, the difference between stakeholder theorists and profit maximizers is not as fundamental as it may seem. Friends and foes of stakeholder theory can agree

Originally published in: *Business Ethics Quarterly*, 27(3), 353–379 © Cambridge University Press, 2017
Reprint by Springer, https://doi.org/10.1017/beq.2017.1

B. van der Linden
EDHEC Business School, Nice, France

R. E. Freeman (✉)
University of Virginia, Charlottesville, VA, USA
e-mail: FreemanE@darden.virginia.edu

© Springer Nature Switzerland AG 2023

S. D. Dmytriyev, R. E. Freeman (eds.), *R. Edward Freeman's Selected Works on Stakeholder Theory and Business Ethics*, Issues in Business Ethics 53,
https://doi.org/10.1007/978-3-031-04564-6_43

that managerial decision making may involve the consideration of different kinds of value simultaneously. It's true that Jensen (2002, 237) argues that "[t]he real conflict is actually an unjoined debate … whether the firm should have a single-valued objective function." Nevertheless, if seen from the perspective of values, the gap that divides "stakeholder theorists" and "profit maximizers" is not unbridgeable. Notwithstanding their insistence on the importance of a single-valued function, the idea that decision making involves the simultaneous consideration of different kinds of value should not be entirely foreign to profit maximizers. Even if managers would make decisions through the maximization of profit, others may consider different values simultaneously, and managers may need to follow their considerations in order to maximize profit. Moreover, maximization of profit involves simultaneously respecting other goods such as "open and free competition" and avoiding bads such as "deception or fraud" (Friedman 1962, 133, but see also Jensen 2002, 171).

If not only stakeholder theorists, but also profit maximizers have reasons to recognize that decision making involves the simultaneous consideration of different kinds of value (cf. Freeman 2008, 166; Freeman et al. 2004), then the worry of profit maximizers that it is unclear how managers should make decisions in the absence of a single objective, becomes their shared concern. Stakeholder theorists do not yet have a complete solution for this problem either. Nowadays, in stakeholder theory, decision making is often seen as the simultaneous pursuit of the "utility functions" of the corporation and its stakeholders (Hill and Jones 1992; Harrison et al. 2010; Harrison and Wicks 2013; Tantalo and Priem 2016) and a stakeholder utility function "expresses the stakeholder's preferences for particular types of value" (Harrison and Wicks 2013, 101). However, pursuing different utility functions presupposes that managers and/or stakeholders have already considered different values simultaneously in order to include them in these functions with a certain weight. Moreover, deciding how and to what extent these different utility functions can be pursued may itself be a matter of the simultaneous consideration of different values.

Although profit maximizers have reasons to accept that decision making can entail the simultaneous consideration of different values, they often emphasize that profit maximization is important for the efficient functioning of markets and businesses (cf. Jensen 2002; Sundaram and Inkpen 2004). Stakeholder theorists do not necessarily have to deny this. For them, profit may only be an outcome of business, but they do consider it a condition for the creation of value for stakeholders (cf. Senge 2000, 78; Freeman 2008, 166). Profit maximization may be one of the "narratives" (Freeman et al. 2010, 76–77) that can help to create value for stakeholders. However, notwithstanding (that profit maximization is relevant for stakeholder theorists as well as profit maximizers, it stays unclear how the maximization of one value—profit—can be reconciled with the simultaneous consideration of different values. From the perspective of values, then, profit maximizers as well as stakeholder theorists are left with two joint questions: (1) how can managers simultaneously consider different kinds of value in decision making, and (2) how can this be reconciled with the maximization of profit? In this article we aim to provide an answer to these two questions.

Of course, the vast body of normative stakeholder literature proposes various theoretical foundations for managerial decision making that can provide a basis for answering these questions. These solutions are grounded in, for instance, utilitarianism (Jones and Felps 2013) or a Kantian perspective (Evan and Edward Freeman 1988), and they articulate theoretical foundations for managerial decision making. However, such normative-theoretical foundations are best seen as "normative cores" (Freeman 1994) or "narrative cores" (Purnell and Edward Freeman 2012) that provide different "narratives" of how to engage with stakeholders (Freeman et al. 2010, 76–77) and cannot lay ultimate foundations for decisions (see also Jones and Felps 2013, 373n2). Decision making can—and usually does—go on without ethical theories as "normative foundational justification" (Freeman 2008, 163). Neither postulating profit maximization as a universal dictum, nor reverting to philosophical foundations of ethics can always explain how managers should make decisions.

Rather, managers usually make sense of the values in the situations that they encounter by means of the many concepts that everyday language provides to them (cf. Sonenshein 2005). The recent philosophical discussion around thick value concepts (Putnam 2002; Dancy 2013b; Williams 1985) indeed shows that many concepts of daily language are full of evaluative content. Language can provide managers and stakeholders with the thick concepts necessary to engage with the values they meet. Looking for "foundations" for decision making in substantive normative theories would then be a search in the wrong place (cf. Williams 1985, 129–130). In order to grasp what could go wrong (and right) with managerial decision making, we should understand, as Jensen (2008, 171) also underlines, how managers make sense of the values in the situations they encounter.

Therefore, we do not propose another "foundation" or "normative core" for stakeholder theory. Rather, we start from the idea that ordinary language can provide managers with the thick concepts necessary to make sense of values (see also Mitchell et al. 2016 for an emphasis on sensemaking and value pluralism). Our goal is to explain from this perspective how managers can simultaneously consider different values in decision making, and how the maximization of a single-valued profit function can be reconciled with the simultaneous consideration of different values. This can enrich stakeholder theory as well as the profit maximization approach.

The article is organized as follows. First, we elaborate the point that both profit maximizers and stakeholder theorists have reasons to accept the view that managerial decision making engages with different values simultaneously. Second, we zoom in on how managers make sense of values with thick concepts and explain how the competent use of these concepts necessarily involves the simultaneous consideration of different kinds of value. We also sketch the elements of decision making that are implicit in the idea of thick evaluation, and we explain what it means to be competent with a thick concept in order to address the issue of circularity that may seem to threaten an approach that locates knowledge of values in language. Third, we approach profit as a thick concept. We demonstrate that maximizing is only one way of engaging with profit, and that how to engage with it is sensitive to the other values in the situation. Stakeholder theorists as well as profit maximizers can therefore see decision making as the simultaneous consideration of different values, but this does not exclude that profit maximization can play a role in decision making. Fourth, we discuss the consequences of our elaborations for profit maximizers and stakeholder theorists. Fifth, we briefly summarize our main conclusions and indicate some further questions that the perspective of thick evaluation evokes.

2 Values for Stakeholder Theorists and Profit Maximizers

Although value plays an important role in the principles of stakeholder capitalism and is central to stakeholder theory in general (cf. Freeman 2000, 176–177; Freeman et al. 2010, 280–285; Freeman et al. 2004, 364), the question as to what value is, has not received much attention in stakeholder theory. Freeman et al. (1988) address the role of values in strategy, and emphasize that "economic opportunity is one possible value, or more likely a class of values" (1988, 829). Hendry (2001a, b) argues that in a stakeholder theory based on economic value, the interests of different stakeholders may be integrated, but at the considerable cost of reducing all interests to economic interests. These remarks point in the direction of the idea that corporations engage with different kinds of value, but do not articulate systematically what values are and how managers should engage with them.

Recently, Harrison and Wicks (2013, 97) have stated explicitly that "value has been overly simplified and narrowed to economic return" and they take stakeholder theory "as a lens for considering a more complex perspective of the value that stakeholders seek as well as new ways to measure it." They argue that business is dependent on and creates value for different stakeholders and, following Phillips's (2003) "principle of stakeholder fairness," that the firm "owes an obligation [to them] based on their participation in the cooperative scheme that constitutes the organization and makes it a going concern" (2013, 102, see also 118). In this context, they "define 'value' broadly as anything that has the potential to be of worth to stakeholders" (2013, 100–101) and develop measurement scales for different kinds of value in order to support the engagement with stakeholders and the creation of value for them. They emphasize that these measures are not meant as variables that explain firm performance in the traditional sense of profit, but should be taken as dependent variables themselves (2013, 99). The rationale for this being that "while recommendations made by business scholars on how managers can create economic value may have merit, they could also lead managers to take actions that create economic value while reducing other types of stakeholder value" (Harrison and Wicks 2013, 98–99). These elaborations suggest that managerial decision making may not consist of maximizing one value, such as profit, but of pursuing different values simultaneously.

Profit maximizers take issue with the idea that managers would have to pursue different values simultaneously. They do not so much dispute the claim that managers and stakeholders engage with different kinds of value, but mainly insist that a "single objective function of the firm" is necessary for guiding the consideration of these values in decision making. The function of a business must be singular because "[i]t is logically impossible to maximize in more than one dimension" (Jensen 2002, 238).[1] According to Jensen the "long-run market value of the firm" is the most likely candidate for maximization since "200 years worth of work in economics and finance indicate that social welfare is maximized when all firms in an economy maximize total firm value" (2002, 239). For Jensen, a single-valued corporate function is a function in the mathematical sense "that explicitly incorporates the effects of decisions on all the goods or bads ... affecting the firm" (2002, 238). For the reasons just mentioned, Jensen proposes the long-run market value of the firm as the single dependent variable of this function (for the sake of conciseness we label "long-run market value of the firm" as "profit" in this article).

Such a function does indeed not exclude the possibility to consider different kinds of value in decision making (cf. Jensen 2002, 245–247). It only requires that these values are introduced by reference to their financial consequences for the profitability of the firm. The maximization of such a function means that the company engages with all kinds of value in such a way that profit is maximized. For example, the value of an inspiring job may not be primarily (or at least not entirely) of a monetary nature, but when making a decision along the lines of profit maximization, only its potential contribution to profit is what counts for it. Such a function may be "non-monotone, or even chaotic [which] makes it more difficult for managers to find the overall maximum. But even in these situations the meaning of 'better' or 'worse' is defined, and managers and their monitors have a principled basis for choosing and auditing decisions" (Jensen 2002, 238). Proponents of a single function of business could thus accept that businesses engage with different kinds of value, but hold on to the idea that the necessity of a single function justifies engaging with different values through maximizing profit (this is also the core argument of Sundaram and Inkpen 2004).

A response to this line of reasoning can be found in the work of Mitchell et al. (2016). They develop the idea that corporations simultaneously engage with different kinds of value through an elaboration in the philosophical tradition of value pluralism. Value pluralism consists of the recognition that many different kinds of value can be distinguished, that these values can be incomparable, and cannot always fully be realized at the same time (cf. Frankena 1973; Nussbaum 1999; Raz 1999, 2003). Such a pluralist approach is not only an accepted position in contemporary philosophy, but also wins terrain in economics (cf. Anderson 1995; Mousavi and Garrison 2003; Hsieh 2007a; Putnam 2003; Sen 1999). In both philosophy and economics, value pluralism means that the evaluations we make cannot be reduced to one kind of value that forms the foundation for all evaluations.

[1] That maximization is only possible if there is a single function to be maximized, could probably be logically true. Below, we will not argue against the conditions of maximizing, but reconsider the idea of maximizing itself as this is discussed by Jensen and others. The problem is maximization as it is currently understood, and not whether the logical conditions for maximizing can be met (cf. Mitchell et al. 2016, 260).

Mitchell et al. (2016) take value pluralism as a basis for scrutinizing the idea of managerial decision making as the maximization of a single-valued function. As a basis they take the starting point of Jensen and other profit maximizers that the raison d'être of corporations is their contribution to social welfare. Subsequently, they explain that welfare itself has a plural character and entails many values. This is still something that Jensen's approach can accommodate. However, they criticize the ideas of Jensen and others by arguing that the incommensurability of these different values implies that not a single-valued function, but a multi-valued approach to decision making best helps corporations to increase welfare. Then a decision about, for instance, designing a job is not a matter that can be settled by, say, a comparison of labor related costs and the expected productivity of employees, without obscuring the point of the value of providing inspiring jobs, safe-guarding the work-life balance, and respecting the privacy of employees (see below for a further elaboration of this example). From this perspective, decision making involves the simultaneous consideration of different values.

On closer scrutiny, profit maximizers should indeed be more familiar with the simultaneous consideration of different kinds of value than is suggested by the emphasis they lay on a single-valued function. This is apparent in two complications that value pluralism poses for them.

First, even if managers would make decisions along the lines of the maximization of a single-valued function—and thus ignore the incommensurability of values—stakeholders may pursue different values simultaneously (cf. Tantalo and Priem 2016). On these occasions managers may have to adhere to this value pluralism in order to maximize profit. Indeed, Jensen affirms that "the world is structured in sufficiently complicated ways" to make maximizing "difficult or impossible" (2002, 238). One aspect of this complicated structure of the world may be the incommensurability of the different values that make up social welfare. This incommensurability does not exclude, of course, that stakeholder preferences for certain choices between these values can be expressed in prices (Mitchell et al. 2016, 264) and can in that way be included in a profit function. However, in order to draw up and maximize this function, managers may need to understand and endorse the choices of others to whom these values can be incommensurable. For example: even if decision making is about maximizing profit, then a decision about job design involves understanding the effects of values such as wage costs, employee productivity, inspiration at work, and work-life balance on the profit of the firm. For an employee, however, the value of an inspiring job and the value of a good work-life balance may neither stem from some other kind of value such as profit, nor may the value of work-life balance stem from the value of an inspiring job (or the other way around). These choices genuinely require the simultaneous consideration of different values but nevertheless affect how the corporation can maximize its profits. Even if managerial decision making would only revolve around maximizing profit, managers may need to endorse the incommensurability of values for others because their preferences can affect profit and may need to be included in the single-valued corporate function.

Second, profit maximizers explicitly endorse the view that at least some values require consideration independently of whether they contribute to the maximization of the long-run value of the firm. According to Friedman, for example, a business should maximize profit, but only while it "engages in *open* and *free competition* without *deception* or *fraud*" (Friedman 1962, 133, italics ours; compare also Jensen 2008, 170–171, citing Hayek). These and other values are "embodied in ethical custom" (Friedman 1970) and should be considered simultaneously with profit maximization. Indeed, it is hard to imagine any managerial interaction with stakeholders that does not involve respect for values such as freedom, and the right of property (cf. Donaldson 2011). There may be many values—which Mitchell et al. (2016) call "sacred values"—that cannot be left entirely to who wants to pay a price for them. Even if there are laws and moral norms protecting such values, these rules can only be followed accurately if those following them understand the importance of these values. Interpreting and following such "rules of the game" in practice relies—so we will argue below—on considering these values simultaneously with the value of profit. The category of "sacred values" does not need to be limited to obvious moral values such as liberty and property, but may also encompass other important values such as the ones in the example of a job design. Maybe profit maximizers could even agree that values such as safety, privacy, and work-life balance also can require consideration simultaneously with profit.

The above elaborations suggest that not only stakeholder theorists, but also profit maximizers have reason to think of managerial decision making as somehow entailing the consideration of different values simultaneously (cf. Freeman 2008; Freeman et al. 2004). This creates a serious challenge for profit maximizers. After all, on the one hand profit maximizers (as well as authors from other backgrounds) give various reasons for taking profit maximization seriously. Profit maximization is important for the functioning of markets, and markets, in turn, are necessary for the "invisible" coordination of complicated welfare producing networks of activities that usually extend far beyond the view of the individual (cf. Hayek 1988; Jensen 2002; Smith 1759; and many others). Moreover, efficiency in business operations and decision making may be dependent on some idea of profit maximization (cf. Jensen 2002; Sundaram and Inkpen 2004). On the other hand, profit maximizers seem to have reason to embrace the idea that managers consider different incomparable values simultaneously. It remains unclear, however, how the maximization of profit can go together with the simultaneous consideration of different values.

An answer to this question can be important for stakeholder theorists as well. For them, profit is merely one outcome of business, but yet an important condition for the creation of value for stakeholders (cf. Senge 2000, 78; Freeman 2008, 166). Profit maximization may be part of a useful narrative for the efficient creation of value for stakeholders. Moreover, stakeholder theorists may endorse at least some of the reasons that profit maximizers give for taking profit seriously. Therefore, profit maximizers as well as stakeholder theorists face the challenge of answering the questions as to how managers can consider different values simultaneously, and how profit maximization can be reconciled with the simultaneous consideration of different incomparable values. In the following three sections, we approach decision making on the basis of thick evaluation, sketch a solution for the challenge of reconciliation, and discuss the consequences of our approach for profit maximizers and stakeholder theorists.

3 Thick Evaluation in Decision Making

In the literature on normative stakeholder theory, many theoretical solutions have been developed in order to provide normative guidance for dealing with profit and other values in managerial decision making. For example, Jones and Felps propose that, from a utilitarian perspective, managerial decision making should be guided by the principle of "stakeholder happiness enhancement" which states that "the objective of the corporation should be to enhance the aggregate happiness of its normatively legitimate stakeholders over the foreseeable future" (Jones and Felps 2013, 358). From a Kantian perspective, it has been argued that stakeholders should not be treated as means only, and principles such as "the corporation and its managers may not violate the legitimate rights of others to determine their own future" have been established (Evan and Edward Freeman 1988, 79).

It would be a mistake, however, to see these theories as "foundations" for decisions. Any normative theory for managerial decision making is necessarily a contingent product of the language and culture of those who develop, discuss, and apply it (cf. Rorty 1989; Putnam 2002, 1995). A theory can always be criticized from the perspective of another theory, as Jones and Felps (2013, 373n2) also recognize. Moreover, in practice, managers usually rely on all sorts of considerations that can be extracted from many different narratives within and outside the organization, but which are usually not normative philosophical theories (cf. Sonenshein 2005). "Normative cores," like the Kantian and utilitarian examples above, provide different "narratives" of how to engage with stakeholders (Freeman et al. 2010, 76–77), but are best not seen as foundations for decision making. Following Dewey (cf. Pappas 2008, 59ff.) and Putnam (2004, 15ff.), philosophical normative theories are best seen as instruments for making sense of the reasons that situations provide. Likewise, Sonenshein (2005) proposes that moral principles can at best help managers and others to critically engage with the narratives of an organization.

Therefore, comprehending how decision makers can consider different values simultaneously should not start with understanding or developing normative cores, but with getting a better picture of how managers make sense of the values they encounter by means of the concepts of their everyday language. In line with this idea, Mitchell et al. (2016) propose that making sense of values has a "holistic" character and "does not occur predictably according to some rule or formula, but will reflect the complexities of an individual's or collective's expectations, motivations, social context and experience" (see also Arnold et al. 2010). However, they do not discuss in more detail what this holistic sense making of different values consists of. Harrison and Wicks (2013, 101, 103) seem to zoom in on this issue by suggesting that managers could pursue the profit function of the corporation simultaneously with the utility functions of stakeholders (see also Hill and Jones 1992; Harrison et al. 2010; Tantalo and Priem 2016). A stakeholder utility function "expresses the stakeholder's preferences for particular types of value" (2013, 101). However, pursuing different utility functions presupposes that managers and/or stakeholders have already considered values simultaneously. This is necessary in order to include these values in utility functions with a certain weight. Moreover, deciding how and to what extent these utility functions can be pursued may itself be a matter of the simultaneous consideration of different values. Seeing decision making as the simultaneous pursuit of the utility functions of corporations and stakeholders assumes that managers and stakeholders consider different values simultaneously, but does not conceptualize the holistic evaluative sense making that is necessary for doing so.

In this article, we will focus on the basic question as to how managers and stakeholders "holistically" make sense of the different values in a situation by means of the thick concepts of ordinary language. The role of thick concepts in making evaluative and practical judgments has recently been addressed in the philosophical discussion of thick value concepts (Williams 1985; Dancy 2013b; Putnam 2002). The evaluative nature of many of the concepts of daily life is at the core of this discussion, and is also explicitly related to holism in making sense of values (cf. Dancy 2004, 1995, 2013b). This makes thick concepts a suitable starting point for understanding how managers make sense of—and engage with—the different values they encounter in decision making.

3.1 Thick Evaluation in Business

Many concepts of everyday language have not only descriptive but also evaluative dimensions. These so called "thick evaluative concepts," such as *lie, cruelty,* and *friendship,* "express a union of fact and value" and thus are not only evaluative, but "world guided" (Williams 1985, 129) at the same time. Applying them involves a descriptive grasp of the features of an actual situation, as well as an evaluation of these features, and often provides reasons for action. In order to become a competent user of a thick concept, one needs to share the evaluative point of view of those who are competent with the concept, that is, one must learn how a thick concept relates to the other concepts of a vocabulary (Putnam 2002, 37ff.). The competent use of a thick concept involves the ability to make sense of a value given the other values in the situation.

Below we elaborate the idea of "thick evaluation"—making sense of values through the application of thick concepts—in order to argue that evaluations typically involve the simultaneous consideration of different values, and thus to explain how managers can simultaneously engage with different values in decision making. For this purpose, we take the following steps. We start by (1) introducing the idea of thick concepts, and continue by (2) explaining in more detail that the evaluative and descriptive dimensions of thick concepts are inextricably related. This is necessary in order to (3) argue that the competent application of such concepts does not only involve the descriptive, but also the evaluative dimensions of other concepts. This means that making sense of a value is sensitive to the other values in the situation. Subsequently, we (4) explain how the evaluations that result from the application of thick concepts give rise to reasons for action, and how competent users of thick concepts come to practical responses to different values considered simultaneously. This provides a basis for sketching decision making as consisting of four elements that are implied in thick evaluation.

(1) The use of the thick evaluative concepts of ordinary language consists of applying them to features of situations, and through these applications, these features obtain descriptive, but also evaluative meaning (Putnam 2002; Williams 1985). Thick concepts thus play a central role in making sense of values in a situation. For example, when we call someone *industrious,* we do not only mean that s/he has the feature of working hard, but also that something valuable is realized in her/his so working. Likewise, when we say that a car is *green* we do not only mean that it uses relatively few natural resources, but usually also that it is good that it does so. These thick evaluative concepts provide a description *and* an evaluation of features of situations.[2] Frequently discussed examples of thick concepts include cruelty (Putnam 2002, 34), treachery, promise, brutality, courage (Williams 1985, 129), lewdness (Gibbard 1992, 279), and lies (Payne 2005, 89).[3] In daily life, within and outside corporations, we use many of these thick concepts in order to grasp the values in a situation (cf. Anderson 1995, 98; Raz 1999, 146). The importance of thick concepts appears in many business practices. When hiring an employee, for instance, we may look for a *productive* and *innovative team player,* who feels *at home* in a culture of *respect, purposefulness,* and *mutual support.* Stakeholders and managers use many thick concepts which are full of evaluative contents that help to frame situations and give descriptive and evaluative meaning to them. Of course, this does not exclude the possibility that managers can use the "tools" of normative theories for making decisions. These "normative cores" (Freeman et al. 2010, 76–77) are made up of thick concepts and give systematic accounts of the contents of these concepts from the perspective of a specific theoretical tradition. Establishing the usefulness of these tools in a situation—and actually using them—presupposes that managers rely on thick evaluation in order to grasp the values they encounter there.

[2] Thick evaluative concepts can be distinguished from "thin" evaluative concepts such as good, ought, and right (Williams 1985). Thick concepts "express a union of fact and value" and are "world guided" (Williams 1985, 129) so that they only apply to features that meet certain descriptions. Thin concepts, such as right and good, on the other hand, are not "world guided" and therefore can apply to virtually everything. Thick and thin is probably a "matter of degree" (Scanlon 2003, 276).

[3] Eklund enumerates: "[a]mong concepts that have been regarded as thick are discretion, caution, enterprise, industry, assiduity, frugality, economy, good sense, prudence, discernment, treachery, promise, brutality, courage, coward, lie, gratitude, lewd, perverted, rude, glorious, graceful, exploited, and, of course, many others" (Eklund 2011, 25).

(2) When in use, the evaluative and descriptive contents of thick concepts cannot be separated (Eklund 2011; Kirchin 2013; Putnam 2002; Williams 1985). This is often discussed by stressing that definitions do not suffice for learning how to apply a thick concept (e.g. Dancy 1995, 263; Garrard and McNaughton 1993, 57–58; Williams 1985, 141). For example, understanding what makes certain work *industrious* and other work *being on the make* is not only a matter of having command of a certain body of definitional knowledge, but also of understanding the value implicit in industriousness and being on the make. Competently using the concept of *industriousness* requires—of course—other descriptive concepts regarding the features of the things to which this concept applies (say, to individuals who work more than eight hours per day and exceed their targets). However, in order to learn to apply this thick concept intelligibly, one also needs to grasp why and when industriousness can be valuable. Being industrious and being on the make are different because of the values they have. Of course, a definition of *industriousness* could be refined by specifying for example that, contrary to *being on the make, industriousness* is not only self-regarding. Yet, in the end, such specifications cannot entirely determine the application of the thick concept to new cases without relying on other evaluative concepts (which is indeed the case with *self-regarding).*

(3) Being a competent user of a thick concept requires understanding the place of this concept in the whole of the concepts of the vocabulary to which it belongs from an ethical point of view (Putnam 2002, 37ff, but also Raz 2001; Raz 2003; Williams 1985, 141–142). In order to understand the value of mutual support in work life, to continue with one of the examples from above, one may have to understand that it can be valuable in itself, but also, for instance, that it may be important for cooperation and psychological well being in teams, that it can jeopardize decisiveness, and may not always go well together with impartiality. In thick evaluation the value of a feature depends on the values of other features of the situation (cf. Dancy 2004). For example, whether mutual support is good, bad, or has no value at all, depends on other values in the situation, such as cooperation, psychological well being, decisiveness, and impartiality. Making sense of a value through the application of a thick concept is thus sensitive to the other values in the situation. Ethical lapses in actual business practice can indeed be analyzed along these lines: understanding why it was wrong of managers at Enron to think that changing the numbers was valuable as an instantiation of loyalty to the company (cf. Tourish and Vatcha 2005) involves the simultaneous consideration of values other than loyalty. In the Enron case, loyalty was bad, say, because in this situation loyalty doesn't go well together with impartiality, and threatened values such as honesty, the reliability of information, and the long term value of the firm.

Incompetence with thick concepts is by no means an academic matter only, and can, as this example shows, have serious consequences for business and society. It explains how managers can—and must—consider different values simultaneously in decision making. The competent application of a thick concept necessarily involves other values, so that, even if managers focus on a single value, other values can be taken into consideration and affect the evaluation made. The evaluative dimensions of thick concepts may remain largely inarticulate when making sense of a value (Raz 2001, 7; Dancy 2004, 191–193). Nevertheless, they are in principle available to, and used by, competent users so that merely applying thick concepts competently involves considering different values simultaneously.

(4) The above approach shows how different values can—and must—be considered simultaneously, but does not suffice for explaining how thick evaluation can support decision making. This also requires the further step of expounding how thick evaluation provides reasons for action. After all, decisions are not only conclusions about what is of value in a situation, but also about how to act with regard to these values. In order to explain how thick evaluation provides reasons for action and guides decision making, we address the practical meaning of the application of thick concepts. Thick concepts do not only hold together description and evaluation, but are usually action guiding as well (cf. Dancy 2004, 191–192; Tappolet 2014; Dancy 2013b, 57; Payne 2005, 93–96). They "are characteristically related to action" so that "if a concept of this kind applies, this often provides someone with a reason for action" (Williams 1985, 140). These reasons for action can take many different forms: some values should be admired, other values should be protected, still other values should be created, etc. Moreover, different values require different specific behaviors. For example, the value of privacy often requires our respect, but we should engage with it in a specific way in order to respect it. Respecting privacy may, for instance, require remaining ignorant of certain aspects of the life of others. The appropriateness of such a practical response also depends on other values in the situation (Williams 1985, 140; Dancy 2013b, 57). For example, how to respect privacy, and even whether it should be respected at all, may depend on which aspects of life are considered shameful or intimate by those engaged, and also on which other values may have to be abandoned for the sake of privacy.

Competent users of a thick concept grasp which "contribution" (Dancy 2004, 192) its application can make to what should be done in a situation, and also how this contribution can vary depending on the other values in the context. A plurality of different kinds of value is an assumption for any thick evaluation, and engaging with a plurality of values does not require comparing them along the lines of one value (cf. Dancy 2013a). Rather, knowing how to engage with different values involves understanding which practical reasons they can contribute and finding a way to engage with them. Typically "[w]e are faced with several contributing features, and what we know about each enables us to cope somehow with the interfering effects of the context considered as a whole" (Dancy 2004, 192). Managers and stakeholders see the "practical shape of the situation and thereby the nature of the appropriate response" (Dancy 2004, 191) through the application of thick evaluative concepts and thus the simultaneous consideration of different values.

The above leaves open the possibility that only a single thick concept appears explicitly in reflection or communication, while other values are simultaneously considered implicitly. On other occasions, various thick concepts may appear explicitly. In all these situations, the sensitivity of the thick evaluation of a feature to the other values in the situation explains how managers can "holistically" (cf. Mitchell et al. 2016) consider different values simultaneously and decide about an appropriate response.

3.2 Thick Evaluation in Managerial Decision Making

From this perspective, a managerial decision is the establishment of a practical response to the values in a situation, and the rightness of a decision depends on whether this response is appropriate given the reasons that these values contribute in the context as a whole. The idea of thick evaluation shows that establishing an appropriate response to the values in the situation neither requires a single value to be maximized, nor does it require reverting to a theoretical normative core such as stakeholder theorists sometimes suggest. Even the application of only a single thick concept involves the simultaneous consideration of other values. Decision making thus typically involves a plurality of incomparable values and the rightness of decisions hinges upon the competence of managers to apply thick concepts. On this account, the usual "dilemmas" from the literature on value pluralism—in which two possible responses to the whole of the values in a situation are juxtaposed (e.g. Chang 1997)—are potentially present in any decision. These dilemmas do not always appear because decision making does not need to involve developing these alternative responses and choosing one of them. Even if different response alternatives are developed and a choice between them is hard to make, the way forward is a better understanding of what sorts of reasons the different values contribute to what should be done, and finding a way to engage with them. Maximizing a single value or reverting to a substantive normative theory can help reflecting on alternatives and making them comparable, but cannot resolve the dilemma of the practical reasons that different values provide.

This "holism of practical reasons" (Dancy 2004) implies that general principles or normative philosophical theories cannot express the requirements for managerial decision making—these reasons are to be found in the situation using the relevant thick evaluative concepts.[4] A more obvious strategy for sketching the practical consequences of taking seriously the role of thick concept in decision making, is not a substantive, but a formal approach. Such a formal approach starts from the idea that the value of a feature may be "enabled," "disabled," "intensified," and "diminished" by other values in the situation (Dancy 2004; see Mckeever and Ridge 2013 for a summary in these terms). The value of provocation for instance, may be enabled by the value of probing, disabled by the value of politeness, intensified by the value of a lively discussion, and diminished by the value of a laid back conversation. How to engage with a provocation (whether or not to continue provoking, how, and to what extent) therefore depends on other values in the situation.

This formal characterization of how engaging with a value can be affected by other values in the situation allows us to further flesh out an approach to decision making that explicates the requirements implied in the idea of thick evaluation.

[4] Eklund enumerates: "[a]mong concepts that have been regarded as thick are discretion, caution, enterprise, industry, assiduity, frugality, economy, good sense, prudence, discernment, treachery, promise, brutality, courage, coward, lie, gratitude, lewd, perverted, rude, glorious, graceful, exploited, and, of course, many others" (Eklund 2011, 25).

Such an approach should commence by noting that almost all thick concepts have a "default valence" (Dancy 2013b, 45). For example: in the absence of disablers and/or diminishers, efficiency and beauty are good-making features, lying is a bad-making feature, nutrition is a good-making feature, etc.[5] Attached to these default values are default responses: efficiency should be improved, lying should be avoided, beauty should be admired, nutrition should be eaten, etc. These default responses can—and may often do—change depending on other values in the situation: lying may be necessary to expose an impostor, beauty may need to be ignored when admiration would be embarrassing for the person whose beauty is concerned, safety may exclude further efficiency improvements, etc. On this basis we can now sketch in more detail what decision making should entail if thick evaluation is taken seriously. Upon encountering a value that is at stake, the decision maker should:

1. Establish what is the "default response" to this value.
2. Take stock of the other values in the situation.
3. Assess how these values affect the default response.
4. Decide how to respond to this value given the other values in the situation.

This is a conceptualization of the paradigmatic decision making in which one value is encountered and other values affect how to engage with this value. These steps can be performed for different values in the situation. Making a decision consist of finding an appropriate response to the reasons that the values in the situation contribute.

3.3 Managerial Competence with Thick Concepts

The formal approach to decision making that we have presented above relies on the competence of managers with the relevant thick concepts. This renders the question as to what makes a manager competent with a thick concept very important. Above, we have argued that competence with a thick concept is the practical ability to apply it intelligibly considering its complex relation to the other concepts of a vocabulary. Therefore, in the end, only (other) competent users of a thick concept can judge the conceptual competence of managers. As a competent user of a thick concept one has "full recourse to the whole of one's conceptual armory, information and powers of argumentation in reaching conclusions as to which practices sustain goods and which sustain evil, or worthless things" (Raz 2003, 24). Nevertheless, this idea of conceptual competence may seem to constitute a vicious circularity in which there are no other grounds for deciding between good and bad than the contingent contents of thick concepts that have evolved over time in the practices and conversations of

[5] Provocation is one of the rare examples of a thick concept that seems to have no default valence and therefore does need an enabler to have value (Dancy 2013b, 45). Above, we used this example to systematically explain how a value can be affected by the other values in the situation.

their users.[6] A solution for this problem, so we have argued above, can neither be found in the development of substantive normative philosophical theories, nor in postulating the maximization of profit. Indeed, if the thick concepts of ordinary language provide the basis for managerial decision making, then this would be a "search in the wrong place."

Fortunately, relying on the users of thick concepts for judging the conceptual competence of managers does not need to be smothered in circularity. Following Dewey, we suggest that managers and stakeholders are not caught in a network of concepts that only refer to other concepts, but can actually investigate, validate, and reconsider their concepts by (jointly) using them in structuring and making sense of their experiences (cf. Pappas 2008). Raz (2003) identifies a similar possibility when he argues that notwithstanding the general dependence of values on social practices, there are values that do not require social practices in order to exist or be experienced (see also Wallace 2003). In locating knowledge of values firmly in the concepts of ordinary language, we have therefore not given up the possibility that the competence of users of thick concepts also rests on experiences of actual states of the world. Rather, we have, as Davidson (1974) puts it, only given up the possibility of a sharp distinction between concepts and experience of the world, and embraced the idea that the thick concepts of everyday language do not only enable, but are also organized by, what we experience as valuable.

An answer to the question as to whether a manager is competent with a certain thick concept is therefore not necessarily circular, but can also be founded on actual experience of the value of which it is the purported concept.[7] Of course many concepts will play a role in judging the accuracy of the applications of a thick concept that a manager makes in practice, and some—and perhaps even many—of these concepts may be flawed to some extent. Yet, the circularity in judging the accuracy of an application of a thick concept can be interrupted by referring to experience with values in practice. For example: in order to judge whether a supplier has made sufficient efforts to avoid child labor, a manager is not limited to applying thick concepts like child labor, efficiency in production, risk assessment, and profit to the features of this situation. She could also go and see the people working there, try to remember how it was to be a child, imagine how it would have been to work in circumstances like these, take part in (and initiate) conversations with others about these values and their experiences, and subsequently ask herself whether the efforts of the supplier for avoiding child labor are sufficient.

---

[6] We thank one of the reviewers for asking us to elaborate this point.

[7] It is important to note that being a competent user of a concept (Raz 2001, 8–9; Dancy 2004, 136–137) does not presuppose having a full understanding of the concept, but consists of the practical ability to apply it intelligibly in actual situations. The knowledge that individual people have of a concept may be partial or defective, and there even does not need to be anyone who knows the full content of the concept. People can be competent users of a concept to a greater or lesser extent (Raz 2001, 16–18; Crary 2007, 42). See also Mantere (2013) for a similar Wittgensteinian view, but then specifically with regard to the concepts of a corporate strategy.

Against this background the most important managerial insight that this article delivers may just as well be that competence with relevant thick concepts is of central importance to any managerial decision and should be developed by engaging with values in practice and in conversations. Without the actual experience of the values of a work-life balance, an inspiring job, and a productive business—to pick up one of the examples we used above—it may be difficult to make appropriate decisions about job redesign. Of course, explications of the values in a situation along the lines of the procedure we sketched above can support managerial decision making. However, without the largely inarticulate evaluative contents of thick concepts that are developed through experience and the actual use of these thick concepts in practice and conversations, these procedures would remain empty formalisms devoid of a substantive understanding that could support making sense of the values in a situation and deciding about an appropriate response.[8]

4 Profit as a Thick Concept

When thick evaluation is taken to be the basis for decision making, there is no need for setting down a single function to rule all managerial decisions as profit maximizers propose. After all, the many thick concepts of ordinary language provide managers with sufficient resources to consider different values simultaneously. This does not exclude, as we discussed, that profit often is, and should be, a central consideration, and that profit maximization may be important in business. Nevertheless, as we argued, profit maximizers have reasons to agree with stakeholder theorists that managers may need to consider different values simultaneously in decision making. This puts them in a quandary: how to reconcile profit maximization with the simultaneous consideration of different values? An answer to this question can be important for stakeholder theorists as well.

If thick evaluation is taken to be the basis for decision making, it should be possible to resolve the quandary of profit maximizers by approaching profit as a thick concept. From this perspective, understanding profit does not only involve the ability to descriptively define a corporate function, but also to evaluatively make sense of the value of profit given the other values in the situation, and to know how to engage with profit because of this. Of course the *descriptive dimension* of the concept of profit (say, benefits minus costs) is important for managerial decision making. Indeed, nowadays, the fields of finance and accounting provide complicated systems of definitions (for instance on costing and value assessment) that help

[8] Organization structures, ethics programs, HR practices, innovation processes, and many other organizational characteristics shape experiences with and conversations about values of managers and stakeholders. If the competence of managers and stakeholders with thick concepts is developed through experience of values and the public use of thick concepts in practice and conversations, then it would be interesting to elaborate how these organizational characteristics can support (or impede) the development of competence with thick concepts.

managers and others to calculate profits. However, a competent user of the thick concept of profit is not only able to reliably calculate a profit, but also to grasp the *evaluative dimension* of this concept: a competent user understands that profits are important for the functioning of markets, that they can be used to realize many different values, that they may be a symbol of success, that they may not always go hand in hand with for instance privacy, and more generally human rights and environmental values, and so on. Competently using the thick concept of profit does not only involve making accurate calculations of (potential) earnings, but also evaluations of these earnings given the other values in the situation.

Competent users of the concept of profit can thus grasp which "contributions" its application can make to what should be done, and how these contributions can vary depending on the other values in the situation. In business, the "default contribution" (Dancy 2004, 191) of the application of the concept of profit may be that profit should be maximized. Nevertheless, a competent user of this thick concept has the ability to understand when the practical meaning that is contributed changes because of other values in the situation. In principle, as we argued earlier, any kind of value that affects profit can be descriptively included in a profit function. However, competently using the concept of profit also entails the ability to grasp the value of profit given the other values in the situation and to understand against this background whether maximization is appropriate. Thick evaluation can reveal various ways of engaging with the value of profit, of which maximization is only one. Engaging with profit may for instance also consist of maintaining it, pursuing it, or even sacrificing it.

Indeed, examples can be provided to show that the competent use of the concept of profit may yield various differing conclusions about how to engage with profit. Think for instance of a situation in which calculations show that earnings could be made by taking large risks with the safety of employees in a developing country. For many competent users of the concept of profit, the idea that such earnings could constitute a profit would probably just not come up. They would not consider the concept of profit applicable at all in this situation given the value of safety that is at stake. Other competent users of the concept of profit may see such earnings as a profit, and may also see maximizing as the default response to profit. However, taking stock of the value of safety in this situation, and assessing how this affects the default response of maximizing profit, would probably yield the conclusion that such a profit is bad, and this would be a strong reason to *sacrifice*—and not maximize—it.

On other occasions thick evaluation may lead to subtler conclusions about how to engage with profit. Take an example in which a manager is restructuring the work of a group of employees. Profit may be an important consideration here, and maximizing may again be considered the default response to profit. However, such a restructuring operation can also involve the value of loyalty towards employees. Here, loyalty would be one of the values in the situation of which the manager takes stock, and assessing how loyalty affects the default response to profit could lead to a decision that involves a response to profit that is neither maximizing nor sacrificing it. Assessing how loyalty affects the default response to profit will of course involve calculating and considering the potential profits related to different alternatives for restructuring. However, loyalty may, for instance, also require investigating, which of these alternatives involves the smallest number of breaches of labor contracts, while not only paying a fair compensation to those who leave, but also making enough profit to avoid the risk of insolvability in order to keep the company afloat. Here, the appropriate response to profit may be neither maximizing nor sacrificing it, but rather finding a way to *maintain* it.

These examples are not—and, from our perspective, cannot be[9] —definitions that catch the essence of the content of the concepts of profit, loyalty, safety, etc. Nevertheless, such examples illustrate that it makes sense to approach profit as a thick concept. The concept of profit appears to have not only a descriptive dimension that helps to calculate profits, but also an intricate evaluative dimension that entails an understanding of how the value of profit can be affected by other values in the situation. This evaluative dimension allows competent users of the concept of profit to grasp how to engage with profit given the other values in the situation, and to understand when maximization is appropriate. How often maximization is appropriate, is an empirical question we cannot answer here.

Approaching profit as a thick concept helps to understand how profit maximization and the continuous consideration of different values can be reconciled, and can thus help solve the profit maximizers' quandary.[10] If profit is a thick concept, then any competent application of it involves the simultaneous consideration of other values in the situation. Decisions about how to engage with profit—including those that involve maximizing it—are necessarily made against the background of other values. Maybe in business, the "default contribution" of the value of profit is that it should be maximized, but there are also situations where it could have another practical meaning. Even in those situations where the competent use of the concept of profit indicates that "maximizing" profit is appropriate, this involves considering the other values in the situation. Therefore, also in these cases, engaging with profit cannot be seen as maximization *tout court*. The evaluative dimension of the concept of profit is always involved in the competent use of this thick concept, even if the practical meaning of its application is that maximizing profit is the appropriate response to the situation.

---

[9]As we understand them, thick concepts do not have an essence or a core definition that guides their application, and competence with such a concept does not consist of being able to follow a definition. Rather, being a competent user of a concept is the ability to apply it intelligibly considering its complex relation to all the other concepts of a vocabulary (cf. Raz 2001, 8–10; Dancy 2004, 136–137). Our examples are directed to competent users of thick concepts in order to illustrate for them that it makes sense to approach profit as a thick concept. They may of course disagree with the content of our examples, but this only proves that these thick concepts are contestable. We thank one of the reviewers for asking us to clarify this point.

[10]Profit maximizers may not be convinced by the examples above. They may argue that safety is a constraint within which profit should be maximized, and that loyalty does not require abandoning the maximization of profit. For an elaborated response to this line of reasoning, see the *Consequences for Profit Maximizers* section below.

The thick evaluation of profit that underlies the establishment of the appropriateness of maximization may often happen intuitively and already in the first phase of decision making when the situation is framed and the need for making a decision is recognized (cf. Dane and Pratt 2007, 37–38, on intuiting as making holistic associations). The simultaneous consideration of different values then remains in the background. This could explain how even seemingly straightforward profit maximizing decisions such as a lease-or-buy decision are made against the background of an implicit understanding of the appropriateness of profit maximizing,[11] and how competent users of the concept of profit can recognize situations that require a different response (maybe, for instance, when buying instead of leasing would provide capital to a local supplier to invest in a new sheltered workshop). Profit maximizers can now explain how the simultaneous consideration of different values is possible even if profit is maximized, and stakeholder theorists can stick to the idea of the simultaneous consideration of different values while not entirely abandoning the advantages of profit maximization.[12]

5 Consequences for Profit Maximizers and Stakeholder Theorists

Above, we have established a general sketch of the elements of managerial decision making that are implied in the idea of thick evaluation. Upon encountering a value that is at stake, the decision maker should: (1) establish what is the default response to this value, (2) take stock of the other values in the situation, (3) assess how these values affect the default response, and (4) decide how to respond to this value given the other values in the situation. Subsequently, we have approached profit as a thick concept in order to show that maximization is only one way of engaging with profit, and that how to engage with profit depends on the other values in the situation. Taking thick evaluation in decision making seriously has consequences for stakeholder theorists as well as profit maximizers. Below, we sketch some of these consequences by elaborating two specifications of our four-step sketch of thick evaluation in decision making that are dedicated to stakeholder theory and profit maximization.

---

[11] We thank one of the reviewers for asking us to elaborate this point.

[12] This does not mean that we have solved all the issues that Jensen (2002, 2008) mentions with regard to multiple-value decision making. He argues that accounting for decisions, dealing with conflicting stakeholder interest, and resolving disagreements are difficult or impossible in the absence of a single value that can be maximized. However, if decision making can and should involve the simultaneous consideration of different values—no matter whether one is a profit maximizer or a stakeholder theorist—then these three issues have become problems for both traditions, and not just for stakeholder theory. It would be interesting to further elaborate how accounting for decisions, dealing with conflicting stakeholder interest, and resolving disagreements must take place in the face of a plurality of values.

5.1 Consequences for Profit Maximizers

In the first section we concluded that profit maximizers can encounter two kinds of situations in which they may embrace the idea of the simultaneous consideration of different kinds of value. First, even if managers would make decisions through the maximization of profit, others may consider different values simultaneously, and managers may need to follow their considerations in order to maximize profit. Second, maximization of profit involves simultaneously respecting other goods such as "open and free competition" and avoiding bads such as "deception or fraud"—and there may be more "sacred goods" that require consideration simultaneously with profit. The first kind of situation does not constitute a fundamental challenge for the profit maximizing approach, but only confirms that "the world is structured in sufficiently complicated ways" to make maximizing "difficult or impossible" (Jensen 2002, 238). Here, the idea of thick evaluation explains how it is possible that managers make decisions through profit maximization, even under the condition of complexity that stems from the simultaneous consideration of values by stakeholders. It does not, however, challenge the core of profit maximization.

In the second kind of situation, however, the idea of thick evaluation may have more substantial consequences for a profit maximizing approach. Profit maximizers usually see the simultaneous consideration of different values as "playing by the rules" (literally in Friedman 1970, but also a central idea in Jensen 2002), that is, they conceive of certain values as constraints on profit maximization. However, if understanding these values involves thick evaluation, then they will not have the character of "fixed" constraints. These "constraints" depend themselves on the other values in the situation including the value of profit (and the latter is the very value they were meant to constrain). If it is useful to think about these values as constraints at all, then they are constraints that may in principle need to be set again in each situation through thick evaluation. For example, profit maximization in choosing suppliers should be constrained by avoidance of child labor, but in order to make this constraint practical and to establish which efforts can reasonably be expected of a particular supplier to avoid child labor, other values in the situation such as profit and environmental protection may need to be considered simultaneously.[13] Even in a situation where constraints are met, and where maximizing is the appropriate way

[13] Elaborating the example: In practice, the requirement to abolish child labor in the supply chain involves that suppliers find measures to reduce the risk of child labor to a reasonable extent. But what are sufficient measures to reduce this risk, and what would be too much to expect? Can a supplier be expected to check the age of all employees by verifying their official identity documents and by deploying sophisticated medical scans to assess their age if identity documents are missing? Answering such a question requires understanding the value of avoiding child labor vis-à-vis profit and probably other values (this line of reasoning resembles Williams (1982, 1988) "negative responsibility" argument against utilitarianism and Kantianism). Even only the interpretation of rules of the game that can set limitations within which profit maximization is appropriate, may often involve the simultaneous consideration of different values (amongst which profit).

of engaging with profit, this conclusion follows from the simultaneous consideration of other values and the affirmation that these values neither disable nor diminish the value of profit (i.e. when the present efforts sufficiently guarantee avoidance of child labor). Maximizing may well be the default response to profit, but in order to establish whether this default response is appropriate, other values in the situation must be considered simultaneously[14] even if they are labelled as "constraints." As we argued above, engaging with profit is thus never maximizing *tout court*.

If profit is approached as a thick concept along the lines above, then maximizing may be the default response to profit, but yet only one of several ways of engaging with it. If an appropriate response to profit could also consist of, for instance, pursuing, maintaining, or even sacrificing profit, then there may be situations in which maximization is inappropriate given the other values in the situation. For example, it may be inappropriate to maximize profit when child labor is insufficiently avoided by a supplier, and sacrificing profit for the sake of avoiding child labor would be a much better idea. Here, the strategy of profit maximizers could be to require managers to set extra company-internal constraints on the decision making process in addition to the "rules of the game" in order to render maximization appropriate (see Hsieh 2007b for a proposal in this direction). For example, a business may develop its own code of conduct or engage in the establishment of branch-specific rules with regard to the selection of suppliers in the light of their efforts to avoid child labor, and maximize profit within these boundaries. However, just like in the case of the interpretation of the "rules of the game," setting and respecting such constraints requires the simultaneous consideration of different values in order to establish whether and how they should be followed (cf. van der Linden 2013). If setting additional constraints is to help decision making through maximizing profit, then these constraints need to be adjusted time and time again through thick evaluation.

A profit maximizing approach to decision making that can accommodate the two considerations above—and thus incorporates the idea of thick evaluation—would be as follows:

1. Take stock of the values in the situation.
2. Assess how these values affect the appropriateness of maximizing profit.
3. If maximization is inappropriate, set additional constraints for making a decision (and do so considering the other values in the situation—including profit).
4. Make a decision through maximizing profit.

If thick evaluation in decision making is taken seriously, then it appears that engaging with profit is never maximizing *tout court*. The "constraints" within which maximizing is supposed to take place appear to depend on the other values in the situation including the value of profit, and they must in principle be set again in each situation through thick evaluation—the simultaneous consideration of different values.

---

[14] Here it is important to note that a default response is not the same as the most commonly appropriate response.

5.2 *Consequences for Stakeholder Theorists*

Stakeholder theorists have explicitly embraced the idea that corporations create different kinds of value (see for instance Harrison and Wicks 2013). For them, other values than profit do not need to appear as constraints on profit maximization, but can in principle be simultaneously considered from the outset. Our perspective reveals two aspects of this move that have hitherto not been systematically considered.

First, in stakeholder theory, decision making is now often seen as the simultaneous pursuit of the "utility functions" of the corporation and its stakeholders (Hill and Jones 1992; Harrison et al. 2010; Harrison and Wicks 2013; Tantalo and Priem 2016). These utility functions are often considered to contain different kinds of value. Our perspective identifies a problem for these utility functions, but at the same time proposes a solution. The problem is that it remains unclear how corporations and stakeholders come to attach certain weights to the different values in their respective utility functions. The solution is to assume that managers and stakeholders can be competent with the relevant thick concepts and that their reasoning evolves along the lines of the four steps of thick evaluation in decision making that we sketched above. However, if thick evaluation plays the role in decision making that we have sketched, then the suggestion that the utility for a corporation or a stakeholder can be "calculated" on the basis of a mathematical function, would obscure the challenge of integrating the qualitatively different practical responses that the values in a situation call for. From the perspective of thick evaluation, a utility function could at best be seen as a numerical representation of the outcome of a complex process of grasping the values in the situation and of finding an appropriate response to them through thick evaluation.

Second, so far the utility functions of corporation and stakeholders have been treated as somewhat discrete sets of values created for stakeholders, and considerable effort has been put into explaining how it is possible that corporation and stakeholders are prepared to consider each other's interests to a larger extent than strictly necessary (Bosse et al. 2009; Bridoux et al. 2011; Bridoux and Stoelhorst 2014; Tantalo and Priem 2016; Harrison et al. 2010). The arguments in these articles revolve around values such as trust, reciprocity, fairness, etc. From the perspective of thick evaluation, the willingness of managers and stakeholders to consider each other's interests can be explained without reference to these additional values, but by reference to the interrelatedness of the values in a situation. This interdependence is not necessarily confined to the values within one utility function (see the point just above), but can also transgress the boundaries of utility functions. For example, the disvalue of child labor does not only provide reasons to the engaged children for finding ways to get education instead of work, but can also affect how a business should engage with other values that belong to its utility function (i.e., not maximizing profit when this involves child labor). Thick evaluation can naturally entail considerations of what is valuable for stakeholders beyond what serves the utility of the firm. Dividing the whole of the values in the situation into discrete utility functions for corporation and stakeholders does not do justice to the interrelatedness of all the values in the situation.

A stakeholder approach to decision making that can accommodate the two considerations above—and thus incorporates the idea of thick evaluation—would be as follows:

1. Identify the values in the utility functions of stakeholders.
2. Establish what are the default responses to these values.
3. Assess how these values mutually disable, diminish, and intensify these default responses.
4. Decide how to respond to the values in the situation.

The stakeholder theorists' move of widening the idea of the maximization of a single profit function of the corporation to encompass the simultaneous pursuit of different utility functions of company and stakeholders emphasizes that "value has been overly simplified and narrowed to economic return" (Harrison and Wicks 2013, 97). Thick evaluation explains how managers and stakeholders can deal with the simultaneous consideration of different utility functions and the values within them, but also shows that the idea of discrete utility functions neglects how the values in a situation are holistically related.

6 Conclusion

This article has started from the idea that corporations engage with different kinds of value. From this perspective, the differences between profit maximizers and stakeholder theorists appear not to be as stark as they may seem. In fact, both may have reasons to answer the questions as to (1) how managers simultaneously consider different kinds of value in decision making, and (2) how this can be reconciled with profit maximization. One of the central points of this article is that managerial decision making, no matter what form it takes, does neither require substantive normative philosophical theories, nor the maximization of one single value. What is of value and how to engage with it can—and must—be established in the particular situation using all the thick conceptual means that ordinary language makes available. Therefore, our proposal to consider the role of thick concepts in decision making cannot provide substantive guidelines for managerial decision making. Rather, we have answered the two questions above by explaining how managers should go about making decisions if thick concepts are taken seriously.

We have argued that competently applying thick concepts to the features of a situation involves evaluating these features considering the other values in the situation, and that this provides practical reasons that are sensitive to these other values. Decision making thus always involves the simultaneous consideration of different kinds of value. More specifically, thick evaluation in decision making implies: (1) establishing what is the "default response" to a value, (2) taking stock of the other values in the situation, (3) assessing how these values affect the default response, (4) deciding how to respond to this value given the other values in the situation. These steps can be performed for different values in the situation. Making a decision consists of finding an appropriate response to the reasons that the values in the situation contribute.

From this perspective, the maximization of profit can be reconciled with the simultaneous consideration of different values by approaching profit as a thick concept. Then, maximizing appears to be only one way of engaging with profit, and how to engage with profit appears to depend on other values in the situation. Even in those situations where the competent use of the concept of profit indicates that "maximizing" profit is appropriate, this involves the simultaneous consideration of other values in the situation. Therefore, also then, engaging with profit cannot be seen as maximization *tout court*. Conceiving of profit as a thick concept is important for profit maximizers as well as stakeholder theorists. Profit maximizers can now explain how profit maximization can be reconciled with the simultaneous consideration of different values, and stakeholder theorists can maintain that decision making involves the simultaneous consideration of different values while not entirely sacrificing the idea of profit maximization.

We have elaborated the consequences of the perspective of thick evaluation for both traditions by making two specifications of our four-step sketch of thick evaluation in decision making. *Profit maximizers* tend to think of values other than profit as "constraints" on profit maximization. However, if thick evaluation is taken seriously, then the values that figure as "constraints" within which maximizing is supposed to take place appear to depend on other values in the situation including the value of profit—the very value that is supposed to be constrained. Therefore, these "constraints"—if that is a useful way to address these other values in the situation at all—must in principle be set again in each situation through the simultaneous consideration of different values including profit. *Stakeholder theorists* have adopted the mindset of utility functions that contain different kinds of value. Here, the idea of thick evaluation explains how stakeholders and managers can accord weights to the values in their utility functions, and also makes clear that pursuing the utility functions of corporation and stakeholders is not a process of calculation, but of finding a way to integrate the qualitatively different practical responses that the values in a utility function call for. Moreover, it demonstrates that the interrelatedness of different values does not stop at the boundaries of utility functions, but encompasses all the values in the whole of the situation. This offers new theoretical opportunities for understanding why corporations sometimes consider the interests of their stakeholders to a larger extent than what seems strictly necessary.

This article roots decision making in the plurality of values that any manager encounters and shows that the thick concepts of ordinary language are of central importance for the simultaneous consideration of these values. From this perspective, managerial decision making requires neither substantive normative philosophical theories nor the maximization of one single value such as profit. This article also demonstrates that this perspective offers new opportunities for the development of the traditions of stakeholder theory as well as profit maximization. If thick evaluation is taken seriously, the simultaneous consideration of different values appears not to be a problem, but rather the very basis for managerial decision making—no matter whether one is a profit maximizer or a stakeholder theorist.

Acknowledgements We would like to thank Wil Martens, René ten Bos, Andy Wicks, and the editor and reviewers at *Business Ethics Quarterly* for their helpful comments.

References

Anderson, Elisabeth. 1995. *Value in Ethics and Economics*. Cambridge: Harvard University Press.

Arnold, Dennis G., Robert Audi, and Matt Zwolinski. 2010. Recent Work in Ethical Theory and its Implications for Business Ethics. *Business Ethics Quarterly* 20 (4): 559–581.

Bosse, Douglas A., Robert A. Phillips, and Jeffrey S. Harrison. 2009. Stakeholders, Reciprocity, and Firm Performance. *Strategic Management Journal* 30 (4): 447–456.

Bridoux, Flore, and J.W. Stoelhorst. 2014. Microfoundations for Stakeholder Theory: Managing Stakeholders with Heterogeneous Motives. *Strategic Management Journal* 35 (1): 107–125.

Bridoux, Flore, Regis Coeurderoy, and Rodolphe Durand. 2011. Heterogeneous Motives and the Collective Creation of Value. *Academy of Management Review* 36 (4): 711–730.

Chang, Ruth. 1997. *Incommensurability, Incomparability, and Practical Reason*. Cambridge: Harvard University Press.

Crary, Alice. 2007. *Beyond Moral Judgement*. Cambridge: Harvard University Press.

Dancy, Jonathan. 1995. In Defense of Thick Concepts. *Midwest Studies in Philosophy* 20 (1): 263–279.

———. 2004. *Ethics Without Principles*. Oxford: Oxford University Press.

———. 2013a. Moral Particularism. *Stanford Encyclopedia of Philosophy* (Fall 2013 edition).

———. 2013b. Practical Concepts. In *Thick Concepts*, ed. Simon Kirchin, 44–59. Oxford: Oxford University Press.

Dane, Erik, and Michael Pratt. 2007. Exploring Intuition and its Role in Managerial Decision Making. *Academy of Management Review* 32 (1): 33–54.

Davidson, Donald. 1974. *On the Very Idea of a Conceptual Scheme*. Proceedings and Addresses of the American Philosophical Association.

Donaldson, Thomas. 2011. The Inescapability of a Minimal Version of Normative Stakeholder Theory. In *Stakeholder Theory: Impact and Prospects*, ed. Robert A. Phillips, 130–139. Cheltenham: Edward Elgar.

Eklund, Matti. 2011. What are Thick Concepts? *Canadian Journal of Philosophy* 41 (1): 25–49.

Evan, William M., and R. Edward Freeman. 1988. A Stakeholder Theory of the Modem Corporation: Kantian Capitalism. In *Ethical Theory and Business*, ed. Tom L. Beaucamp and Norman E. Bowie, 75–84. Englewood Cliffs: Prentice Hall.

Frankena, William K. 1973. *Ethics*. Englewood Cliffs: Prentice Hall.

Freeman, R. Edward. 1994. The Politics of Stakeholder Theory: Some Future Directions. *Business Ethics Quarterly* 4 (4): 409–421.

———. 2000. Business Ethics at the Millenium. *Business Ethics Quarterly* 10 (1): 169–180.

———. 2008. Ending the So-called 'Friedman-Freeman' Debate. *Business Ethics Quarterly* 18 (2): 162–166.

Freeman, R. Edward, Daniel R. Gilbert, and Edwin M. Hartman. 1988. Values and the Foundations of Stategic Management. *Journal of Business Ethics* 7 (11): 821–834.

Freeman, R. Edward, Andrew C. Wicks, and Bobby L. Parmar. 2004. Stakeholder Theory and 'The Corporate Objective Revisited'. *Organization Science* 15 (3): 364–369.

Freeman, R. Edward, Jeffrey S. Harrison, Andrew C. Wicks, Bobby L. Parmar, and Simone De Colle. 2010. *Stakeholder Theory: The State of the Art*. Cambridge: Cambridge University Press.

Friedman, Milton. 1962. *Capitalism and Freedom*. Chicago: University of Chicago Press.
———. 1970. The Social Responsibility of Business is to Increase its Profits. *The New York Times Magazine,* September 13, 1970.
Garrard, Eve, and David McNaughton. 1993. Thick Concepts Revisited: A Reply to Burton. *Analysis* 53 (1): 57–58.
Gibbard, Allan. 1992. Morality and Thick Concepts. *Proceedings of the Aristotelian Society* 66: 267–283.
Harrison, Jeffrey S., and Andrew C. Wicks. 2013. Stakeholder Theory, Value and Firm Performance. *Business Ethics Quarterly* 23 (1): 97–124.
Harrison, Jeffrey S., Douglas A. Bosse, and Robert A. Phillips. 2010. Managing for Stakeholders, Stakeholder Utility Functions, and Competitive Advantage. *Strategic Management Journal* 31 (1): 58–74.
Hayek, Friedrich. 1988. *The Fatal Conceit: The Errors of Socialism*. Chicago: University of Chicago Press.
Hendry, J. 2001a. Economic Contracts versus Social Relationships as a Foundation for Normative Stakeholder Theory. *Business Ethics: A European Review* 10 (3): 223–232.
———. 2001b. Missing the Target: Normative Stakeholder Theory and the Corporate Governance Debate. *Business Ethics Quarterly* 11 (1): 159–176.
Hill, Charles W.L., and Thomas M. Jones. 1992. Stakeholder-Agency Theory. *Journal of Management Studies* 29 (2): 131–154.
Hsieh, Nien-hê. 2007a. Is Incomparability a Problem for Anyone? *Economics and Philosophy* 23 (1): 65–80.
———. 2007b. Maximization, Incomparability, and Managerial Choice. *Business Ethics Quarterly* 17 (3): 497–513.
Jensen, Michael C. 2002. Value Maximization, Stakeholder Theory, and the Corporate Objective Function. *Business Ethics Quarterly* 12 (2): 235–256.
———. 2008. Non-rational Behaviour, Value Conflicts, Stakeholder Theory, and Firm Behaviour. *Business Ethics Quarterly* 18 (2): 167–171.
Jones, Thomas M., and Will Felps. 2013. Stakeholder Happiness Enhancement: A Neo-Utilitarian Objective for the Modern Corporation. *Business Ethics Quarterly* 23 (3): 349–379.
Kirchin, Simon. 2013. Introduction: Thick and Thin Concepts. In *Thick Concepts*, ed. Simon Kirchin, 1–19. Oxford: Oxford University Press.
Mantere, Saku. 2013. What is Organizational Strategy? A Language-based View. *Journal of Management Studies* 50 (8): 1408–1426.
Mckeever, Sean, and Michael Ridge. 2013. Why Holists Should Love Organic Unities. In *Thinking About Reasons: Themes From the Philosophy of Jonathan Dancy*, ed. David Bakhurst, Margaret Olivia Little, and Brad Hooker, 265–285. Oxford: Oxford University Press.
Mitchell, Ronald K., Gary R. Weaver, Bradley R. Agle, Adam D. Bailey, and James Carlson. 2016. Stakeholder Agency and Social Welfare: Pluralism and Decision Making in the Multi-Objective Corporation. *Academy of Management Review* 41 (2): 252–275.
Mol, Annemarie. 2010. Care and its Values. Good Food in the Nursing Home. In *Care in Practice: On Tinkering in Clinics, Homes and Farms*, ed. Annemarie Mol, Ingunn Moser, and Jeannette Pols, 215–234. Bielefeld: Transcript.
Mousavi, Shabnam, and Jim Garrison. 2003. Toward a Transactional Theory of Decision Making: Creative Rationality as Functional Coordination in Context. *Journal of Economic Methodology* 10 (2): 131–156.
Nussbaum, Martha C. 1999. The Discernment of Perception: An Aristotelian Conception of Private and Public Rationality. In *Aristotle's Ethics: Critical Essays*, ed. N. Sherman, 145–181. Lanham: Rowman & Littlefied.
Pappas, Gregory F. 2008. *John Dewey's Ethics: Democracy as Experience*. Bloomington: Indiana University Press.
Payne, Andrew. 2005. A New Account of Thick Concepts. *The Journal of Value Inquiry* 39: 89–103.

Phillips, Robert. 2003. *Stakeholder Theory and Organizational Ethics*. San Francisco: Berrett-Koehler.
Purnell, Lauren S., and R. Edward Freeman. 2012. Stakeholder Theory, Fact/Value Dichotomy & the Normative Core: How Wall Street Stops the Ethics Conversation. *Journal of Business Ethics* 109: 1009–1016.
Putnam, Hilary. 1995. *Words and Life*. Cambridge: Harvard University Press.
———. 2002. *The Collapse of the Fact/Value Dichotomy: And Other Essays*. Cambridge: Harvard University Press.
———. 2003. For Ethics and Economics Without the Dichotomies. *Review of Political Economy* 15 (3): 395–412.
———. 2004. *Ethics Without Ontology*. Cambridge: Harvard University Press.
Raz, Joseph. 1999. *Engaging Reason: On the Theory of Value and Action*. Oxford: Oxford University Press.
———. 2001. Two Views on the Nature of the Theory of Law: A Partial Comparison. In *Hart's Postscript*, ed. J.L. Coleman, 1–38. Oxford: Oxford University Press.
———. 2003. *The Practice of Value*. In *The Practice of Value*, ed. R. Jay Wallace, 15–59. Oxford: Oxford University Press.
Rorty, Richard. 1989. *Contingency, Irony, and Solidarity*. Cambridge: Cambridge University Press.
Scanlon, Thomas M. 2003. Thickness and Theory. *Journal of Philosophy* 100 (6): 275–287.
Sen, Amartya. 1999. *Development as Freedom*. Oxford: Oxford University Press.
Senge, Peter. 2000. The Puzzles and Paradoxes of how Living Companies Create Wealth: Why Single-valued Objective Functions are not Quite Enough. In *Breaking the Code of Change*, ed. Michael Beer and Nitin Nohria, 59–81. Boston: Harvard Business School Press.
Smith, Adam. 1759. *The Theory of Moral Sentiments*. London: A. Millar.
Sonenshein, Scott. 2005. Business Ethics and Internal Social Criticism. *Business Ethics Quarterly* 15 (3): 475–498.
Sundaram, Anant K., and Andrew C. Inkpen. 2004. The Corporate Objective Revisited. *Organization Science* 15 (3): 350–363.
Tantalo, Caterina, and Richard L. Priem. 2016. Value Creation Through Stakeholder Synergy. *Strategic Management Journal* 37 (2): 314–329.
Tappolet, Christine. 2014. The Normativity of Evaluative Concepts. In *Mind, Values, and Metaphysics*, ed. A. Reboul, 39–54. New York: Springer.
Tourish, Dennis, and Naheed Vatcha. 2005. Charismatic Leadership and Corporate Cultism at Enron: The Elimination of Dissent, the Promotion of Conformity and Organizational Collapse. *Leadership* 1 (4): 455–480.
van der Linden, Bastiaan. 2013. Principles as 'Rules of Thumb': A Particularist Approach to Codes of Ethics and an Analysis of the Dutch Banking Code. *Review of Social Economy* 71 (2): 209–227.
Wallace, R. Jay. 2003. Introduction. In *The Practice of Value*, ed. R. Jay Wallace, 1–14. Oxford; New York: Oxford University Press.
Williams, Bernard. 1982. *Moral Luck*. Cambridge: Cambridge University Press.
———. 1985. *Ethics and the Limits of Philosophy*. Cambridge: Cambridge University Press.
———. 1988. Consequentialism and Integrity. In *Consequentialism and its Critics*, ed. Samuel Scheffler, 225–242. Oxford: Oxford University Press.

Bastiaan van der Linden is Associate Professor of Corporate Social Responsibility at EDHEC Business School and Director of the MSc in Global and Sustainable Business.

R. Edward Freeman is University Professor, Elis and Signe Olsson Professor, Academic Director of the Institute for Business in Society, and Senior Fellow of the Olsson Center for Applied Ethics at the University of Virginia's Darden School of Business.

Chapter 44
Unethical, Neurotic, or Both? A Psychoanalytic Account of Ethical Failures Within Organizations

Simone de Colle and R. Edward Freeman

> Conceptions of psychological man, in the sense used here, have yet been implemented in any significant way. They call for viewing the organization as an adaptive organism directed to its own perpetuation. Thus conceived, an organization is a problem-solving mechanism. It is an educational institution that, for its own survival, must increase the psychological and economic competence of those who work in it. This conception calls for a different role for the leader, that of a teacher of problem solving and a facilitator of human development.
>
> H. Levinson, The Great Jackass Fallacy (1973, p. 31).

1 Introduction

The stream of research in behavioral ethics is increasingly gaining attention among business ethics scholars, as indicated by the comprehensive reviews by Treviño et al. (2006), Kish-Gephart et al. (2010), Treviño et al. (2014), and Moore and Gino (2015). But surprisingly, the work of psychoanalytic psychologists applied to organizations has not yet clearly connected with the work of business ethicists. One contribution of psychoanalytically informed scholars to organization studies consists in drawing attention to the role of unconscious processes in organization

Originally published in: *Business Ethics: A European Review*, 29(1), 167–179 © John Wiley & Sons – Books, 2019
Reprint by Springer, https://doi.org/10.1111/beer.12239

S. de Colle (✉)
IESEG School of Management (LEM-CNRS 9221), Paris La Défense cedex, France
e-mail: s.decolle@ieseg.fr

R. E. Freeman (✉)
University of Virginia, Charlottesville, VA, USA
e-mail: FreemanE@darden.virginia.edu

© Springer Nature Switzerland AG 2023

S. D. Dmytriyev, R. E. Freeman (eds.), *R. Edward Freeman's Selected Works on Stakeholder Theory and Business Ethics*, Issues in Business Ethics 53,
https://doi.org/10.1007/978-3-031-04564-6_44

life, including their influence on ethical decision-making processes (Arnaud 2012; Fotaki et al. 2012). The aim of this paper is to point out the implications for understanding the sources of ethical failures within organizations by examining a number of insights on *self-realization* and *neurotic processes* highlighted by the classic studies of the psychoanalyst Karen Horney (1950) and the clinical psychologist Manfred Kets de Vries (1980, 1991, 2018)and Kets de Vries and Miller (1984), who is the leading scholar on the relationship between psychoanalysis and organization studies with contributions that stretches from the 80s to nowadays. While there are different, well-established research traditions within psychoanalytic theory, we use the classic work by Horney and Kets De Vries as an example to illustrate the kind of learnings that business ethics research can draw from the psychoanalytical literature. We connect these psychoanalytic insights with business ethics research, in particular applying a stakeholder theory perspective (Freeman 1984; Freeman et al. 2010; Jones and Wicks 1999; Venkataraman 2002). Business ethicists can now rely on important findings on human cognition and its failures, showing how ethical decision-making processes can fail due to the influence of psychological factors, such as bounded ethicality (Banaji et al. 2003; Banaji and Bhaskar 2000; Chugh et al. 2005), bounded awareness (Bazerman and Chugh 2005), ethical blind spots (Bazerman and Tenbrunsel 2011; Moberg 2006), motivated blindness (Gino et al. 2009b), and self-deception (Tenbrunsel and Messick 2004). These findings complement ongoing research by "traditional" business ethics scholars on the antecedents of ethical failures, pointing out individual and situational factors, including, for example, crises of ethical leadership (Knights and O'Leary 2005; Wray-Bliss 2012), the failure of responsibility (O'Leary 2015), the influence of the societal context (Gonin et al. 2012), and the effects of ethical pressure (Tian and Peterson 2016).

The aim of this paper is to indicate some promising avenues for integrating into business ethics research a different perspective from which to study ethical behavior within organizations, that is, the psychoanalytic account. Traditionally, business ethics research has focused on individual aspects affecting the ethical decision-making process, for example, moral traits (Delerue and Hamid 2015; Hartman 1998, 2000; Solomon 1992, 2003), moral predispositions (Brady and Wheeler 1996; Reynolds 2006), or stages of moral development (Kohlberg 1969, 1984). On the contrary, the psychologists' approach to the study of ethical decision making has been "less entrenched in rationalistic traditions" and focused more on "non-deliberative processes." Research in this tradition has been focused on identifying various psychological factors that can affect human behavior within organizations (Moore and Gino 2015, p. 238) and on investigating the Dark Triad of personality (Babiak 1995; Furnham et al. 2013; Paulhus and Williams 2002). This is defined as the distinct but overlapping constructs of Machiavellianism (characterized by the manipulation of others, deception, and cynical action), narcissism (a sense of grandiosity, dominance, and superiority), and subclinical psychopathy (high impulsivity, low empathy, and a thrill-seeking personality). Taking a psychoanalytical approach to ethical decision making, in particular focusing on *neurotic processes* that human beings encounter in their quest for self-realization, we contribute to existing business ethics research by providing new insights into the sources of ethical failures within organizations, connecting the work done by clinical psychologists with the work done by business ethicists, thereby providing richer explanations for the antecedents of ethical failures and new avenues of research on how to limit their diffusion.

We articulate the paper around five insights on the antecedents of ethical failures in organizations that a (non-psychologist) business ethicist can draw from the analysis of the clinical psychology literature. We then discuss three strategies that could help organizations in limiting the emergence and the influence of these neuroses. The first deals with acknowledging the complexity of human motivation, and the importance of emotions. The second focuses on developing a conversation around authentic values within organizations as a way to enable healthy individual growth. The third explores the role of the business ethicist in facilitating such a values conversation, suggesting a parallel between the role of the business ethicist and that of the psychoanalytic therapist.

2 Neurotic Processes and Ethical Decision Making Within Organizations

Published in 1950, Karen Horney's *Neurosis and Human Growth* provides a good starting point to examine how psychoanalytic theory can be applied to study a potential source of unethical behavior within organizations, which is a "particularly unfortunate form of human development," the tendency to fall into the neurotic process (Horney 1950, p. 13). Horney's underlying assumption on human nature is optimistic; in every person, there are "evolutionary constructive forces, which urge him to realize his given potentialities" (p. 15). This does not mean that human beings are essentially good, but it means "that man, by his very nature, strives toward self-realization, and that his sets of values evolves from such striving" (p. 15). Therefore, according to this perspective, every person is not good or evil by birth but has the possibility (and, one may add, the moral responsibility) to understand by herself what is good or bad. This process of learning and self-realization can only be achieved by being "truthful to himself" and by an active involvement in some forms of social life ("...unless he relates himself to others in the spirit of mutuality," Horney 1950, p. 15). Here Horney introduces the key concept of her theory, the difference between the "real self"—the "central, inner force, common to all human beings, and yet unique in each" (p. 17), whose realization represents the path toward healthy growth—and the "idealized self"—an idealized image whose actualization represents the neurotic drive to alienation from the real self. While the real self leads each individual toward the authentic realization of their human potentialities ("being truthful to himself"), the idealized self is a neurotic development (i.e., "the opposite of spontaneous wishes or strivings") determined by inner necessities (e.g., feelings of guilt, anxiety, rejection, etc.).

These are not expressions of the real self but are, on the contrary, compulsive drives that lead individuals toward utter disregard for themselves or for their best interests.

This underlying view of human nature resonates with the Aristotelian characterization of the natural striving for happiness (*eudaimonia).* According to Aristotle, human beings can reach happiness by developing their *virtues*. A virtue is defined as the actualization of a *potential;* human beings are not born good or bad by nature but have *capacities* to become good if they develop their virtues (or bad, if they indulge in vices). Development of virtues can be fostered by *habit*, that is, by developing our natural capacities and dispositions to do the right thing in any situation, by exercising practical wisdom in our daily interactions with others. It is clear that Horney shares with Aristotle these fundamental assumptions about human nature. She further clarifies this point when she discusses the difference between what we could call a virtuous development of personality ("healthy human strivings") and the (vicious) development of neurosis, which she calls "the search for glory." One key element of the search of glory is the self-idealization in a perfect image, like the Pygmalion of Bernard Shaw who "aims not only at retouching but at remodeling itself into his special kind of perfection prescribed by the specific features of his idealized image." (Horney 1950, p. 25). As she says, "Similarities between healthy strivings and the neurotic drives exist because they have a common root in specific human potentialities. Through his mental capacities man has the faculty to reach beyond himself" (p. 37). This key passage clarifies the analogy between Horney's human growth and Aristotle's virtuous development. In Fig. 44.1, we try to represent it in a visual way.

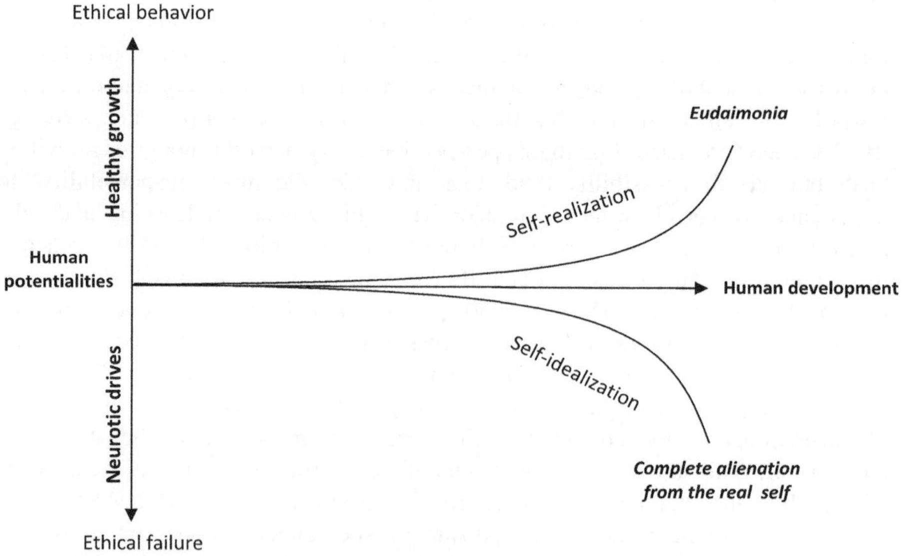

Fig. 44.1 Human development between *healthy* growth and *neurotic* drives

Both Horney and Aristotle see human growth as starting from a common root, given by natural capacities. In Aristotelian terms, these capacities are only potential that can be actualized, thereby fulfilling a natural predisposition innate in every human being. In Horney's terms, there is "a propensity, inherent in human beings, to develop given potentialities. [...] I would say that the live forces of the real self urge one toward self-realization" (p. 38). The line dividing healthy growth from neurotic strivings denotes Horney's central idea, rejecting the view that the difference between the two developments is only a quantitative one (i.e., a matter of intensity). On the contrary, the difference is a qualitative one:

> On the surface they [the two divergent processes] may look deceptively similar, so much that their differences seem to be variations in degree only. It looks as though the neurotic were merely more ambitious, more concerned with power, prestige, and success than the healthy person; as though his moral standards were merely higher, or more rigid, than ordinary ones; as though he were simply more conceited, or considered himself more important than people usually do. (Horney 1950, p. 37)

However, the two paths are clearly divergent. While one elevates a human being toward happiness (not a reachable point, but an ideal to be continuously searched for, as the tangent lines that only touch themselves at the infinite indicate), the other trails him down toward a complete alienation from the real self: "By taking this road, the individual is in fact losing his soul—his real self" (Horney 1950, p. 39). The difference therefore is not one of degree, but there is a qualitative difference between genuine striving (for self-realization) and compulsive drives (to alienation from the real self).

Finally, it is important to note that the two tendencies operate at same time in every person. Human growth is a complex process, and we always need to understand the difference between "spontaneity and compulsion; between recognizing and denying limitations; between a focus upon the vision of a glorious end-product and a feeling for evolution; between seeming and being, fantasy and truth." They are two separate tendencies, and it is the prevalence in intensity of the former or the latter that distinguishes between a healthy person, "engaged in realizing his real self," and a neurotic person, "wholly driven to actualize his idealized self" (Horney 1950, p. 39). We will come back to these considerations on "healthy growth" and discuss their meaning from a business ethics perspective, asking what we can do to help individuals develop themselves in a healthy (non-neurotic) way in their workplace. But before answering this question, let us investigate the organizational consequences of (individual) neurotic drives by looking at another important body of work coming from psychoanalytic theory applied to organizations.

2.1 *Some Effects of Neuroses Within Organizations*

If Horney's foundational theory opens the road to conceptualize the dangers of neurotic drives for human development, the applied work of Kets de Vries (1980, 1991, 2018) and Kets de Vries and Miller (1984) brings clinical psychoanalysis a step closer to understanding the implications of neurotic processes on management and leadership styles within organizations, particularly the modern corporation. Within any organization, the influence of psychological forces—of which the individual is usually unaware—often produces outcomes that appear extremely "irrational" and "dysfunctional" (Kets de Vries and Miller 1984, p. 1). These psychological processes, generally ignored by management theorists, need to be taken into account to understand the sources of many organizational failures, and strategies need to be put in place to limit their pernicious effects. By looking at management styles with the eyes of a clinical psychologist and a psychoanalyst, Kets de Vries and Miller (1984, p. 22) identify a number of "typical" expressions of neurotic behaviors by managers that generate organizational inefficiencies. In particular, they identify five different "neurotic managerial styles," based on categories well established in psychoanalytic literature, namely the *Paranoid,* the *Compulsive*, the *Dramatic*, the *Depressive,* and the *Schizoid.*

Each neurotic managerial style is associated with specific psychological traits, and the individuals who suffer from them express some characteristic *fantasy* (the creation of a symbolic reality that influences the real behavior of the individuals who share it). While each neurotic managerial style can also have some positive effects, in terms of helping individual managers focus on some specific aspects of their own role, task, or performance, looking at the implications of that style from the viewpoint of organizational effectiveness, it is possible to identify a number of potential dangers—or organizational dysfunctions, as Kets de Vries and Miller call them. Interestingly, while some of these neurotic tendencies can be found in successful organizations, "their rigidity seems to contain at least the seeds of failure" (Kets de Vries and Miller 1984, p. 23). These five ideal types enable the authors to point out the linkages between the development of specific types of neurotic behavior by the managers of an organization and the corresponding elements of inefficiency that will be generated.

While some scholars might see this leap from the individual to the organizational as a level of analysis issue, we believe—like Kets De Vries—that it is important to understand the connection between *individual* neuroses and their potentially dysfunctional *organizational* effects. Similarly, we think that individual neuroses can also lead to organizational ethical failures, as we discuss in the next section of this paper. Each of the five neurotic tendencies will in fact harm organizational performance by negatively affecting decision-making processes and deteriorating the managerial culture within the corporations. For example, in a *paranoid* firm (an organization, whose top management's behavior is predominantly characterized by this specific neurotic tendency), there will be an excessive tendency, based on suspicion and wariness, to perceive and react to any external signal as a threat. Such signals could include competitors' market strategies, consumers' actions, or new regulations. This reaction creates the risk of the company's losing its strategic direction and "muddling through" these events, all perceived as threats. On the other hand, *compulsive* firm management will suffer through an organizational loss of flexibility, as the obsession with rituals and processes harm the individual capacity to change. Strategy development will also suffer, since the fear of making mistakes will generate indecisiveness and a tendency toward postponement. In a *dramatic* firm, excessive risk-taking based on the CEO's overconfidence and boldness will generate inconsistent strategies and the risk of relevant losses of resources. In a *depressive* firm, excessive conservatism and rigidity with respect to bureaucratic protocols will stifle innovation, and corporate strategies and products will suffer from this stagnation. Finally, in a *schizoid* firm, the feeling of noninvolvement and the lack of leadership will generate an atmosphere of distrust between managers and subordinates, which will reduce collaboration and create barriers that prevent relevant information from reaching the proper decision makers.

Besides the problems arising from neurotic manageme+nt styles, there are two additional ideas about psychological biases that can generate organizational dysfunctioning. One is the diffusion of *group fantasies,* that is, a set of shared beliefs that manifest themselves by shaping organizational culture, and the other is the phenomenon of *transference,* a reaction in interpersonal relationships in which one perceives and responds to someone as if that person were an important figure from the past. Kets de Vries and Miller (1984) identify three different types of organizational culture affected by group fantasies:

- *Fight-flight*: This is a culture based on the group fantasy that "*there is an enemy inside or outside: defense or escape is necessary*." The resulting organizational culture is characterized by the projection of its own hostile feelings onto others and by splitting the world into "good" and "bad" or friends and enemies. For example, managers start seeing specific stakeholder groups as their enemies: "competitor's product-market strategies are imitated or countered by the firm. Employees suspected of disloyalty are fired." (p. 55);
- *Dependency*: Assuming that "*there is a desire to be nourished and protected by the leade*r," the dependency culture is based on the idealization of the charismatic leader. Members of this group fantasy "wish to believe that the leader's actions are all-good, and they deny any evidence to the contrary." (p. 52);
- *Utopian*: The utopian culture is based on the idea that there will be a person, still unborn, "who will deliver the group from hatred, destructiveness, and despair." (p. 53). Hoping in such a messianic leader will generate enthusiasm, but also turn into despair and disillusionment given the members' excessive (and utopian) expectations.

Again, specific dynamics affecting personal relationships and organizational culture are logically linked to each of the three group fantasies. The *fight-flight* fantasy can lead to suspicion, a lack of collaboration, hatred, and fighting among internal departments of the organization and with outside constituencies. The *dependency* fantasy, with its idealization of the leader, can induce passivity and a lack of critical judgment by subordinates, and the *utopian* fantasy can facilitate unrealistic, optimistically biased analyses of the external environment as well as excessive risk-taking.

Finally, Kets de Vries and Miller (1984) add the phenomenon of *transference* to their clinical analysis of organizational functioning, identifying three major patterns of transference in superior/subordinate relationships and indicating how these can negatively affect leadership and decision making within organizations:

- *Idealizing* transference: This is a revival of the early psychic development phase, where the individual fosters a sense of union with someone who is "omnipotent and perfect." (p. 80). This generates admiration for the leader and her/his beauty, intelligence, power, or moral stance. The idealizing individuals usually tend to feel lost and unworthy unless they can relate to the idealized person;
- *Mirror* transference: This creates "a grandiose sense of the self," whereby individuals develop a narcissistic attitude and a desire to be applauded, admired, and revered while seeing the outside world as uninteresting and full of imperfections. Clearly, the mirror transference is complementary with the idealizing transference in that "mirroring superiors seek out idealizing subordinates." (p. 84); and
- *Persecutory* transference: This reenacts negative experiences of excessive frustration from the past, typically from childhood, generating a tendency to view the world as all bad and interpersonal relationships as characterized by hostility, aggressiveness, and envy.

Each transference pattern has negative consequences with respect to the quality of leadership and decision making within the organization. The idealizing transference will generate extremely dependent subordinates who are easily controlled and manipulated by the leader. In other words, "they are lost in terms of independent resource to the company" (Kets de Vries and Miller 1984, p. 81). The mirroring transference will also generate serious drawbacks. Leaders will tend to claim all the credit for everything whether they deserve it or not. They will also tend to take advantage of and exploit the subordinates, feeling threatened by strong subordinates, thereby generating a lack of collaboration between the different hierarchical levels in the organization. They will tend to centralize power in order to be highly recognized and be able to make the important decisions. Finally, persecutory transference generates all sorts of negative attitudes in the relationships between leaders and subordinates, as the hostility that characterizes the former makes them see subordinates as incompetent, deliberately misbehaving, and making "mistakes" to ruin them. Stronger forms of control and sanctioning systems are therefore introduced to (suspiciously) monitor as much as possible the behavior of the subordinates—who, in turn, will react either with an increasing reluctance to collaborate or, worse, by sabotaging the control systems whenever possible.

The central claim of this paper is that these five forms of neurotic managerial styles and the associated *group fantasies* and phenomena of *transference* can be the antecedents of five specific forms of unethical behavior within organizations, which we therefore call "neurotic" ethical failures. In the next section, we discuss the characteristics of these neurotic ethical failures and how they relate to existing business ethics research.

3 Five Forms of Neurotic Ethical Failures

What can the business ethicist learn from insights on organizational shortcomings due to the diverse manifestations of human neurosis? To answer this question, we will point out on the one hand the linkages between the kinds of dysfunctions produced by neurotic managerial styles and on the other, the key traits of a number of ethical failures within organizations that have been identified in business ethics research. We will also reflect on some real cases, using practical examples to further shed light on the emergence of neurotic ethical failures. But before discussing this, let us add a note of clarification; the term "neurotic" indicates patterns of behavior that are rather common in every normal human being. Neurotic drives manifest themselves within each of us and can consequently be found within any organization.

But neurotic behavior is not to be confused with "psychotic" behavior, which refers to the pathological manifestation of neurosis as a severely incapacitating behavior, fortunately limited to sporadic occurrences. Therefore, the characterization of modern organizations as clinical "psychopaths" (suggested, e.g., by Joel Bakan 2003, in a book that inspired the authors of the movie *The Corporation,* admittedly, skillfully created with great film-making style) is only a caricature and not an accurate description of common patterns of behavior within modern corporations. Below, we provide a narrative for five different forms of ethical failures that we think can be traced back to some kind of neurosis, patterns of transference, or group fantasies (or a combination of them) that we have examined in the psychoanalytic literature.

3.1 (Dramatic) Neurotic Ethical Failure #1: "The Rules Do Not Apply to Me"

The *Dramatic* neurotic style is characterized by a narcissistic attention to the self, overconfidence in assessing one's personal capacities, and a love of excessive risk-taking. Dramatic executives sometimes seem to make decisions implicitly assuming that they are persons with a "special role" and a "special character" that places them above the rules and behavioral standards of "common" people. They love innovation, and they play with the rules, or better, they love to *change* the rules of the (market) game. All these personality traits resonate with the narcissistic dimension of the Dark Triad described by Paulhus and Williams (2002). Combine this with the diffusion within the organization of the *Utopian* group fantasy, whereby members of the organization start to believe that they are part of a messianic enterprise and that their leader will (in some unpredictable way) always solve any problem they might encounter. We can expect that managers of such an organization will make decisions without too much concern for their effects on the organization's stakeholders; what matters most to them is how their decisions affect *their own* role, status, power, or prestige. Within business ethics, this kind of management style has been defined as a "stakeholder equilibration failure" (de Colle et al. 2017a), or the failure by managers to balance in a fair and efficient way the interests of the organization's stakeholders in the value creation process, which is the key role of the "moral manager" (Venkataraman 2002).

Think of the case of Enron from this perspective. Its vision was nothing but grandiose, "To become the world's leading energy company," as one could read in Enron's Code of Ethics. Its organizational culture was aggressively focused around a key value: Excellence ("We are satisfied with nothing less than the very best in everything we do;" Enron's Code of Ethics, p. 5). In addition to this (official) corporate value, Enron's managerial culture tended to reward individuals for their "boldness" and "being smart" (in playing with the rules). Enron's internal performance review system was "the harshest employee ranking system in the whole country"[1] in terms of its deliberately excessive turnover rate of 15%—a Darwinian "natural selection" mechanism pushing people to try very hard to achieve their objectives in "bold," "smart" ways. The management style characterizing Enron was truly dramatic, as Kets de Vries characterizes it:

> One of the things that was always a strange occurrence at Enron was that every time, for weeks before a quarterly report, we would be under the impression that we were not going to make our numbers, but then, somehow, miraculously, we always made the numbers! (Enron: The smartest guy in the room 2005).

We can also recognize elements of the *mirroring transference* in Enron's top executives. According to McLean and Elkind (2003), Enron's CEO Jeff Skilling was a huge risk lover and was once heard to proclaim, "I am Enron!". "Making the numbers" and willingly ignoring the impacts of their choices on Enron's stakeholders were the key traits of a grandiose management style that dragged Enron into a bankruptcy with tragic outcomes (including one suicide, the destruction of 21,000 jobs, and a $2bn loss in employees' pensions).

3.2 (Compulsive) Neurotic Ethical Failure #2: "The Tyranny of Goals"

When the neurotic *Compulsive* style characterizes behaviors within an organization, managers tend to be obsessed with processes and controls as the best (safest) ways to reach the corporate goals. Both subordinates and stakeholders become means for the successful continuation of the organizational machine. Living under the *dependency* fantasy, the members of the organization are happy to leave to the leader the task of thinking about the corporate direction, while the subordinates become cogs in the system. As a result, we can expect that unethical behaviors in such an organization will systematically emerge as a by-product of this neurotic

[1] Enron's executive Mark Eberst, Public Relations, of Enron Energy Services, quoted from the movie, "Enron: The smartest guy in the room," by Alex Gibney (2005).

management style, since what matters more than anything else are the corporate procedures and goals.

Think, for example, of the famous case of the "Aircraft brake scandal" with respect to the B.F. Goodrich Corporation, as described by K. Vandivier (in Donaldson et al. 2002). After winning a very lucrative contract with the U.S. government to produce brakes for fighter planes, Goodrich managers were trapped in a spiral of shortcuts to complete the contract. Under the pressure of a very strict time frame, quality controls were manipulated, reports were falsified, and defective equipment was eventually produced and delivered, even if it was clear that the brakes could fail and put the lives of pilots in danger. None of the Goodrich managers, engineers, or technicians involved in the scandal were deliberately behaving unethically to achieve personal gain—but when they encountered major roadblocks that stood in the way of achieving their (respective) individual and corporate goals, they were more apt to cheat and lie without considering the potentially disastrous consequences of their actions.

Within business ethics research, several scholars have discussed the potential negative consequences of internal systems and processes designed to link corporate goals with specific (and measurable) goals of the diverse units of the organization. Some of this work has included, for example, goal-setting theory (Locke and Latham 1990), management by objectives (Drucker 1954), or the balanced scorecard methodology (Kaplan and Norton 1992). While the founders of goal-setting theory acknowledge "possible dangers of setting goals" (Locke and Latham 2009), more critical scholars point out the intrinsic dangers of goal setting, where there are "systematic and predictable ways in which goal setting harms organizations" (Ordóñez et al. 2009, p. 7). These authors argue that despite the fact that *specific* and *challenging* goals might be beneficial for organizational effectiveness and performance, these very characteristics also often cause goals to "go wild." Instead of inspiring us, goals can drive us. Hoyk and Hersey (2008) describe this phenomenon as the "tyranny of goals," one of the many "psychological traps" that entice executives to abandon their ethical values. As these authors suggest, "Goals become all important. We can move too fast, take shortcuts, do anything to reach our goals." Eventually, "Ends are used to justify means" (Hoyk and Hersey 2008, p. 32). Focusing too narrowly on their given goals, managers end up neglecting non-goal areas and tend to "engage in more unethical behavior than they would otherwise" (Bazerman and Tenbrunsel 2011, p. 104).

In the philosophical tradition of business ethics research, this form of ethical failure has been described by Goodpaster as *teleopathy*, or "the unbalanced and undisciplined pursuit of goals" (2007, p. 42). Without explicitly mentioning it, by choosing a term that recalls a human pathology, Goodpaster may implicitly agree with our claim that insights from psychoanalytic psychology are indeed relevant and could enrich and inform further business ethics research.

3.3 (Paranoid) Neurotic Ethical Failure #3: "Us Versus Them"

As we have seen, both the *Paranoid* neurotic style and the *fight-flight* group fantasy generate interpersonal relationships characterized by suspicion, a mistrust of others, and a readiness to react to the actions of other organizational members as counteraction against perceived threats. This management style can generate severe ethical failures, for example, by damaging the quality of industrial relations in a firm. In 2008, Alitalia, the Italian airline company, was facing the risk of bankruptcy, and the Italian government actively supported the take-over of the company by a group of Italian entrepreneurs (The Economist 2017). However, the new management group—most of whom had no experience in the aviation industry—was unable to gain the trust of the flight attendants' trade unions, and a number of strikes and other forms of protest paralyzed the company during the negotiations. Paradoxically, 1 day, the new management group, exasperated by the situation, announced their withdrawal from the negotiations to purchase the company, which would have caused insecure unemployment for all the flight attendants. Yet some of the flight attendants were caught on television applauding and cheering at the news, as if their anger against the new management was greater than their concern for their own jobs.

Despite the many issues involved in the Alitalia case, this episode might be seen as an example of what happens when the relationships between a company's management and its stakeholders become characterized by a manifestation of an "*us* versus *them*" neurotic ethical failure. Managers act with a conflicting/manipulating attitude toward their stakeholders, as if these stakeholders are the corporation's enemies. In this neurotic management style, we can expect that corporate tactics turn into "fight them" or "blame them" tactics. But we might also see the opposite, where managers adopt "please them" tactics toward stakeholders, depending on the relative power of each stakeholder group. If these stakeholder groups are powerful and effective in influencing the corporation, the latter option would probably be more common. Business ethics research has been looking at this form of ethical failure primarily by leveraging the vast literature on the psychology of inter-group behavior (Tajfel 1982; Tajfel and Turner 1986). For example, the research points out how loyalty to one's in-group can influence ethical behavior (Wright et al. 2013); the positive influence of peers setting ethical standards (Moore and Gino 2015); and the negative influence of in-group "bad apples" (Gino et al. 2009a). While we acknowledge the relevance of this research, we think that looking into the potential side-effects of human neuroses on managerial styles adds an important dimension to this research avenue.

3.4 (*Schizoid*) Neurotic Ethical Failure #4: "Just Leave Me Alone. I Am Not Interested"

One of the main characteristics of the *Schizoid* neurotic style is detachment. Business—and the role of managers—is mainly viewed as a cold, rational activity, where there is no room for emotions, passions, or a sense of purpose. People trapped in this neurosis lack interest in the present or future, prefer to keep a distance from the reality, and avoid any form of involvement. The way schizoid managers look at business decisions resonates with the cognitive fallacy described by Putnam (2002) as "the fact/value dichotomy." This is the idea (and interestingly, Putnam actually uses the word "fantasy") that business, like technology, is a sphere of decisions completely detached from morality and values and that one is better off focusing only on the "rational," "scientific" sphere—committing, in Putnam's words, an "evasion of values" (2002, p. 155).

In business ethics research, this fallacious belief has been described as the "Separation Thesis" (Freeman 1994) and can be formulated as follows: it is the idea that "it is useful to believe that sentences like, 'x is a business decision' have no ethical content or any implicit ethical point of view" (Freeman et al. 2010, p. 6). Business ethics scholars have argued that the potential negative effects of the Separation Thesis are particularly concerning, as it may induce managers to behave in ways that accept the characterization of business as an activity based on a logic of self-interest, "opportunism," and "greed" or as an "amoral" activity at best (Wicks 1996). The trait of detachment described above as a form of the schizoid managerial style could therefore be seen as a likely antecedent of unethical behavior characterized by *moral disengagement* (Bandura 1999; Detert et al. 2008; Moore et al. 2012): managers do not see ethics as a relevant dimension of their (separated) business actions. We do not claim that this the unique, nor the main explanation for moral disengagement, but we suggest that business ethics researchers should be aware of these insights from psychoanalytic theory and also explore the potential role of neuroses in generating moral disengagement.

Another side-effect of the schizoid neurotic management style is to generate a "leadership vacuum" (Kets de Vries and Miller 1984, p. 38); perhaps because of past disappointment, top executives will indulge in their emotional apathy and inactivity and will refrain from taking action. This lack of leadership might induce second-tier managers to take the opportunity to take a more active role in decision making, with the result of transforming the organization into a political battlefield among competitive coalitions of managers trying to take over the company leadership. If we look at relationships with corporate stakeholders, we can expect that the schizoid neurotic style will, at minimum, reduce the processes of stakeholder engagement; why interact with them if one perceives the world as "an unhappy place populated by frustrating individuals"? (Kets de Vries and Miller 1984, p. 38).

3.5 (Egocentric Blindness) Neurotic Ethical Failure #5: "See Only What You Want to See"

Finally, we can think of another type of ethical failure generated by neurosis, which is connected to the five neurotic management styles described by Kets de Vries and Miller and, in some aspects, resembles their *dramatic* type. It refers to the tendency of individuals to fail to recognize unethical behavior (they "see no evil," as the article title suggests) when they have some vested interest in doing so. We call this neurotic managerial style "*egocentric blindness,*" alluding to the notion of "*motivated blindness*" introduced by Gino et al. (2009b, p. 247). This might include, but is not limited to, an opportunistic economic interest such as, for example, the case of Enron's former auditors, where "Arthur Andersen had strong reasons to be afflicted by motivated blindness" (Gino et al. 2009a, b, p. 247). Motivated blindness can also occur in other situations, for example, when we tend to overlook the unethical behavior of others "when we recognize that the unethical behavior would harm us" or "when ethicality erodes slowly over time" (p. 259).

However, the kind of neurotic ethical failure that we have in mind here is wider than Gino et al.'s "see no evil" phenomenon. By "egocentric blindness," we are thinking of a form of ethical blindness that occurs even without the trigger of a vested interest, more likely because of the egocentric neurotic behavior that Horney describes as "the search for glory." Individuals trapped in this "neurotic ambition" are "wrapped up in themselves" (Horney 1950, p. 292). This self-idealization drive produces distortions in the way they see others: "The neurotic sees the others in the light of the needs engendered by the pride system. [...] Distortions may result from endowing others with characteristics they do not have or have only to a negligible degree" (p. 292). We expect that such a neurotic ambition, combined with cognitive distortions, could represent a significant source of ethical failures, inducing managers to ignore unethical behavior during their decision-making process. This analysis resonates with the problem of the paucity of "moral imagination," a concept analyzed in business ethics research as a possible explanation for organizational ethical failures. In particular, Palazzo et al. (2012) see moral imagination as a key resource for limiting the manifestation of *ethical blindness,* which they define as "the temporary inability of a decision maker to see the ethical dimension of a decision at stake" (p. 325). In fact, Werhane (1999, p. 12) discusses the "distorting narratives" that characterize a number of organizational disasters, where managers prove to lack moral imagination (but not imagination) as they "confuse reality with what they want it to be," and argues that these distortions are the main cause of their ethical failure. While Werhane points out that the paucity of moral imagination can be one explanation for these distortions, the psychoanalytic perspective notes the (distortive) role of neurotic drives.

In Fig. 44.2, we summarize our analysis, identifying five forms of neurotic ethical failures and noting their linkages with existing business ethics literature.

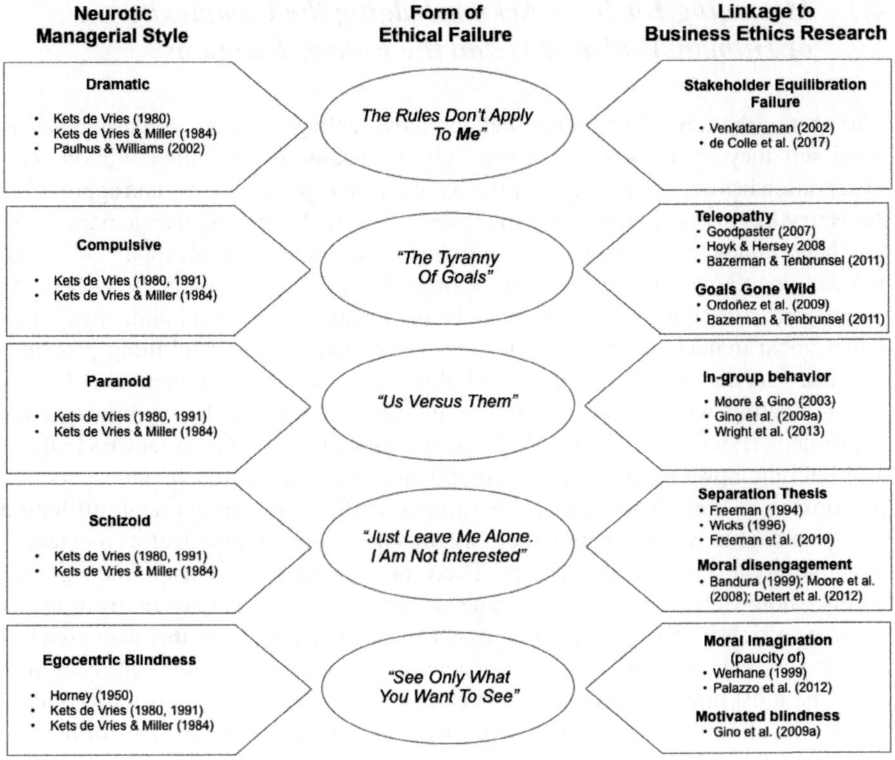

Fig. 44.2 The linkages between neurotic ethical failures and business ethics research

4 Organizational Strategies to Mitigate Neuroses

Looking from a business ethics perspective at these insights from psychoanalytic literature, one might raise the question, "How can organizations try to limit the emergence of neurotic-driven ethical failures?" In particular, business ethicists are interested in understanding what executives might do in order to limit ethical failures and enable a process of individual self-realization (limiting the emergence of neuroses) within organizations. In the next sections, we formulate three suggestions based on this perspective.

4.1 Managing Business Acknowledging the Complexity of Human Motivations and the Role of Emotions

Managers, like any other human beings, have feelings, desires, fantasies, and an ideal self they try to realize. Surprisingly, to a large extent, this common sense observation has been overlooked by mainstream management theory and approaches. As Harry Levinson already noted in the early 1970s, American management theory has been characterized from its beginning by a dominant philosophy of human motivation, the reward-punishment approach or, as it is usually described, the *carrot-and-stick* approach. Unfortunately, the reward–punishment philosophy upon which most management techniques are based leads to a self-fulfilling prophecy, because "the unconscious assumption behind the reward–punishment model is that one is dealing with jackasses, that people are jackasses to be manipulated and controlled" (Levinson 1973, p. 10). Levinson called this "the Great Jackass Fallacy" to highlight how, by adopting a compliance-oriented approach, managers will paradoxically limit "spontaneity, investment, dedication, commitment, affiliation, and adaptive innovation" within their organizations—all of these factors that tend to generate organizational inefficiency. Therefore, concluded Levinson, management theory should change this unfortunate assumption about human motivation and embrace a richer view of human beings. He suggested focusing this new vision of the "psychological man," a new perspective on human motivation that recognizes that "the most powerful motivating force for any human being is his wish to attain his ego ideal." For Levinson, the "ego ideal" is "a picture for ourselves of how we should be at our ideal best" (1973, p. 29). Well-functioning organizations are those that provide a working environment enabling every member "to achieve a greater sense of mastery and to better meet the demands of his ego ideal" (p. 31).

The dominant approaches to business ethics programs—especially in the post-Enron corporate world, characterized by a surge of compliance after the introduction of the Sarbanes-Oxley Act (2002)—seem to suffer from a fate similar to that of the "Great Jackass Fallacy" evoked by Levinson; their ethical infrastructures are designed without taking into proper consideration the motivational complexity of human beings, particularly the drives, wishes, fantasies, and possible neuroses derived from striving for the fulfillment of the ego ideal. In our view, this tendency is a poor application of the main ethical theories, within which there are important insights into the role and influence of emotions and other behavioral traits on human motivation—particularly on moral motivation, as noted by de Colle and Werhane (2008). (Freeman For instance, contrary to the simplistic prevailing view that the moral content of *utilitarianism* is a "cold" utility maximization exercise, according to John Stuart Mill emotions do play a key role in promoting ethical behavior, as *internal* (our sense of duty) and *external* (e.g., fear of punishment) sanctioning factors (see Crisp 1998). Even in Kantian ethics, despite the fact that the role of reason is central, emotions are taken into account. According to Kant, emotions are not only potential "enemies of the reasons," but they can also play a significant role in supporting ethical behavior, as they can *induce* actions in conformity with moral duty, for example, supporting the inclination to help others (de Colle and Werhane 2008). Finally, the *virtue ethics* approach recognizes the importance of emotions, since emotions can fade the moral motivation of a well-intended person but can also provide a harmonious support for ethical motivation. In fact, for Aristotle, we do not have to deny our emotions to be able to act ethically; rather, we need to cultivate the *right* emotions, as these will eventually reinforce our moral motivation. The virtuous person has a disposition not only to do the right thing but to *enjoy* doing the right thing (Hartman 2000). Therefore, for Aristotle, emotions are not to be seen as a threat to correct decision making, since in the virtuous person, they are aligned with reason.

Despite the fact that most ethical theories pay attention to the role of emotions, in the application of these normative approaches to the design and implementation of corporate ethics programs, this attention seems to disappear, replaced by a great emphasis on following formal procedures, complying with rules and standards of behavior, and enforcing monitoring and sanctioning processes. The empirical evidence, however, shows that compliance-based ethics programs have rarely delivered good ethical performance (see, e.g., Badaracco and Webb 1995; Tenbrunsel et al. 2003; Weaver et al. 1999). Tenbrunsel and Messick (1999) even demonstrated that in some cases, the introduction of a monitoring system on ethical codes can produce detrimental effects.

Turning back to the psychologists' point of view, this is no big surprise. Within psychological research, the call to acknowledge the complexity of human motivation has been clearly raised and addressed by important work in the last two decades. As Messick stated, "Ethics does matter to people, but so do a lot of other things including success, friendship, sex, money, prestige, love, power, and authority. Human motivation is a tumult of goals and desires that are often mutually incompatible" (Messick 1996, p. 223). Similarly, but from a different point of view, other psychologists such as Jonathan Haidt have deeply investigated the role of emotions in influencing moral judgments, for example by differentiating between "moral emotions" and other types of emotions (Haidt 2003). Haidt and others advocate the adoption of a socio-intuitionist model to explain ethical behavior, where moral reasoning happens to be only an ex-post rationalization of individual behavior, done in order to persuade others (Haidt 2003, p. 865).

The consideration of the psychologists' perspective suggests that the organizational shortcomings (in terms of loss of organizational effectiveness) produced by the carrot-and-stick approach that Levinson pointed out from a management perspective may bear significant consequences from an ethical perspective as well. In fact, one can expect that business ethics programs that fail to take into account the complexity of human motivation will be as ineffective—if not worse—in promoting the individual propensity to ethical conduct as the carrot-and-stick management approaches are in promoting organizational effectiveness. Some recent contributions by business ethics scholars are addressing the need for broader discussion around humanism in management approaches, recognizing the role of emotions (Schwartz 2017) and including themes such as human dignity (Pless et al.

2017) and the use of creative arts within organizations (de Colle et al. 2017b; Freeman et al. 2015).

4.2 Developing a Values Conversation Within Organizations, Enabling Individual Self-Realization

How can business ethics theory and practice respond to the challenges to ethical behavior raised by neurotic processes, since these can affect every organization?

From a theoretical perspective, we think that business ethics needs to provide an answer to the possibility of *self-realization* within organizations, and this implies discussing what the conditions for authenticity are within the working environments in today's business organizations. As Auster and Freeman note, an "organizational culture and set of values, need to be designed and implemented in a way that reconciles and integrates the individual and the organizational dimension of values, as part of the same authentic, interconnected whole" (Auster and Freeman 2013; see also Freeman and Auster 2011, 2015). Since neurotic processes tend to alienate human beings from the realization of their "real selves," we need to promote an ethics of authenticity that helps people control and limit their neurotic drives and develop their human potentialities. Following the road indicated by Richard Rorty (1991) in his pragmatist interpretation of Freud, we see authenticity as a project of self-creation. There are at least two major pathways to self-creation—one based on discovering the essential nature of the self and one based on an idea of self-enlargement. Each view has different implications for the interaction between individuals and corporations. As a side note, we would say that any good pragmatist conceives two main projects: self-creation and the creation of community. These are merely two sides of the same coin, each emphasizing a slightly different vocabulary.

Looking at the tools that business ethics research has to offer to enable a process of self-realization, we think that a key question becomes, "How can we develop within organizations a conversation around values that enables individual self-realization and *healthy* growth?" We use the adjective "healthy" in the sense illustrated by Horney—that is, in order to emphasize our acknowledgment of the complexity of human nature and our concern for the neurotic drives that limit the capability of individuals to achieve healthy growth. We would further characterize Horney's notion of "healthiness" from a stakeholder theory perspective (Freeman 1984; Freeman et al. 2010); *healthy growth* is a concept that embraces the core ideas of the stakeholder approach. Interestingly, despite the use of different words, the (business ethics) concept of "stakeholder" can be traced back to the original definition of "healthy growth" used by Horney. In fact, she explains that the individual who is growing in a healthy way is one who solves (or deals well with) the inner conflict between the real and the idealized self (Horney 1950, p. 39). Self-realization includes the development of good relations with others: "[self-realization] means his striving toward relating himself to others with his genuine feelings;

toward respecting them as individuals in their own right and with their own peculiarities; toward developing a spirit of mutuality (instead of using them as a mean to an end)" (Horney 1950, p. 364). We take this as a genuine early statement of the stakeholder theory approach.

From the review of the psychoanalytic literature, and reflecting on the key characteristics of the five neurotic ethical failures that we have identified, we suggest that the starting point in enabling healthy growth within an organization is to develop a conversation around values that, as Freeman and Auster (2015) indicates, is a process characterized by four main aspects, or sets of values: Introspective, Historical, Connectedness and Aspirational Values. Authenticity becomes a project, best suited to conversation with others, especially those we are connected to. Acting authentically is more than just knowing your values and acting on them. In fact, Freud and his followers have taught us how difficult it is to know our own values. The same is true for organizations.

For an organization to embark on a project of authenticity, it must encourage introspective analysis by its members, both personally and organizationally. It must seek to determine who are its stakeholders and why the organization creates or destroys value for them. It must be as honest as possible about both its history and its aspirations for the future. Some of these conversations need to take place well in advance of the more typical strategy and budgeting conversations.

4.3 The Business Ethicist as a Facilitator of the Values Conversation

What is the role of the business ethicist if we take into account our discussion of neurotic ethical failures? The function of the business ethicist, in light of the psychoanalytic perspective that emphasizes the psychological aspects of the self-realization process, assumes a new meaning. The key question within this perspective becomes, How can business ethicists fulfill their role in helping organizations enable individual self-realization and limit the emergence of neuroses? We answer this question in light of our reading of Horney and Kets de Vries in the following way:

- *The role of the business ethicist is to help organizations understand and express their values and their sense of purpose, facilitating healthy human and organizational growth.*

This, in fact, also resonates with Levinson's view on the role of the leader, from his perspective of the psychological man:

> This conception [of the psychological man] calls for a different role for the leader, that of a teacher of problem solving and a facilitator of human development. This more complex conception requires deeper, more comprehensive understanding of motivation as a complex derived from drives, wishes, fantasies, and the ego ideal. (Levinson 1973, p. 31)

Reflecting on the work by Horney and Kets de Vries, we think that in helping organizations develop a conversation around their own values, the role of the business ethicist is similar to the psychoanalytical therapist, who assists human beings in struggling with the development of neurosis and facilitates healthy growth. As Horney writes, the psychoanalytic therapist's role is to "enable the patient to find himself, and with that the possibility of working toward self-realization." Self-realization includes the development of good relations with others, the "faculty for creative work and that of assuming responsibility for himself" (Horney 1950, p. 334). Similarly, we argue that to enable a healthy organizational growth means to help organizations to expressing their authentic values, developing mutually beneficial relationships with their stakeholders, finding their place in the community and assuming their share of responsibility.

As the therapist starts by uncovering the *inner conflict* within the human mind, the ethicist enters into the complex network of stakeholder relations to uncover their potentially conflicting claims. Facilitating the emergence among the organizational participants of an open conversation on purpose and values, and discussing the danger of potential manifestations of opportunism, abuse of power, discrimination, and other forms of "unhealthy" stakeholder relationships, the business ethicist helps individuals in the organization to initiate a process toward a set of shared values able to solve or reduce these conflicts. As the road of analytic therapy is a Socratic "road to reorientation through self-knowledge" (Horney 1950, p. 341), the road to developing corporate values cannot be other than a road to the reorientation of the corporate mission and *raison d'être*. It is helping the corporation find its place in society and develop its stance on economic, social, and ethical issues relevant to the kind of business within which it operates and the kind of organization it wants to become.

5 Conclusion: Insights from Psychoanalysis on the Sources of Ethical Failures

We have discussed five possible forms of organizational ethical failures by looking at research in clinical psychology. This field provides additional insights into the existing business ethics literature, pointing out the negative influence of *neurosis* on ethical decision making within organizations. There is a growing interest within business ethics in developing a stronger collaboration between the normative approach of philosophers and the social science approach of psychologists, to improve our capacity to promote ethical behavior within organizations (see, e.g., Hartman 2008; Moberg 2007). This partial review aims at showing some possible avenues for further developing the collaboration between these two research approaches.

Further development of the collaboration between psychology, psychoanalysis, and the traditional philosophical approach to business ethics research will certainly improve our understanding of the possible sources of ethical failures within organizations. We need better organizational design and richer managerial theories, acknowledging the complexity of human motivation and the implications of bounded rationality on ethical decision making. As Freeman (2008, p. 163) noted, "We need a conceptual apparatus in business that does not commit the separation fallacy." We believe that the separation fallacy is a challenge for business ethics as well, and business ethics scholars need to embrace the "*integrative revolution*" (Freeman 2008) that reconciles business and ethics. Studying organizations with an interdisciplinary approach that reconciles the applied philosophy perspective of traditional business ethics with the theoretical insights of psychoanalytic theory and the empirical findings of psychology can help us better understand business (and human beings) and learn useful insights toward better—that is, more ethical and successful—decision making in organizations, resulting in decisions that create value for the organization's key stakeholders over time while continuously balancing their interests in a fair and efficient way.

References

Aristotle, T. I. (1999). *Nichomachean Ethics* (T. Irving, Trans.). Indianapolis: Hacket Publishing.

Arnaud, G. 2012. The Contribution of Psychoanalysis Contribution to Organization Studies and Management: An Overview. *Organization Studies 33* (9): 1121–1135.

Auster, E.R., and R.E. Freeman. 2013. Values and Poetic Organizations: Beyond Value Fit Toward Values Through Conversation. *Journal of Business Ethics 113*: 39–49. https://doi.org/10.1007/s10551-012-1279-5.

Babiak, P. 1995. When Psychopaths go to Work. *Applied Psychology 44*: 171–188.

Badaracco, J.L., and A.P. Webb. 1995. Business Ethics: A View from the Trenches. *Californian Management Review 37* (2): 8–28. https://doi.org/10.2307/41165786.

Bakan, J. 2003. *The Corporation: The Pathological Pursuit of Profit and Power*. New York: Free Press.

Banaji, M.R., and R. Bhaskar. 2000. Implicit Stereotypes and Memory: The Bounded Rationality of Social Beliefs. In *Memory, Brain, and Belief*, ed. D.L. Schacter and E. Scarry, 139–175. Cambridge: Harvard University Press.

Banaji, M.R., M.H. Bazerman, and D. Chugh. 2003. How (Un)ethical Are You? *Harvard Business Review 81* (12): 56–64.

Bandura, A. 1999. Moral Disengagement in the Perpetuation of Inhumanities. *Personality and Social Psychology Review 3*: 193–209.

Bazerman, M., and D. Chugh. 2005. Bounded Awareness: Focusing Problems in Negotiation. In *Frontiers of Social Psychology: Negotiations*, ed. L. Thompson, 7–26. College Park: Psychology Press.

Bazerman, M.H., and A.E. Tenbrunsel. 2011. *Blind Spots: Why We Fail to Do What's Right and What to Do About It*. Princeton: Princeton University Press.

Brady, F.N., and G. Wheeler. 1996. An Empirical Study of Ethical Predispositions. *Journal of Business Ethics 16*: 927–940. https://doi.org/10.1007/BF00705573.

Chugh, D., M.R. Banaji, and M.H. Bazerman. 2005. Bounded Ethicality as a Psychological Barrier to Recognizing Conflicts of Interest. In *Conflicts of Interest: Challenges and Solutions in Business, Law, Medicine, and Public Policy*, ed. D.A. Moore, D.M. Cain, G. Loewenstein, and M. Bazerman, 74–95. New York: Cambridge University Press.

Crisp, R., ed. 1998. *Utilitarianism*. Oxford University Press.

de Colle, S., and P.H. Werhane. 2008. Moral Motivation Across Ethical Theories: What Can We Learn for Designing Corporate Ethics Programs? *Journal of Business Ethics 81*: 751–764. https://doi.org/10.1007/s10551-007-9545-7.

de Colle, S., Y. Fassin, and R.E. Freeman. 2017a. When David Beats Goliath: Two Case-Studies in the Brewery Sector. In *Stakeholder engagement: Clinical research cases. Issues in Business Ethics*, ed. R. Freeman, J. Kujala, and S. Sachs, vol. 46, 311–337. Cham, Switzerland: Springer.

de Colle, S., R.E. Freeman, B. Parmar, and L. de Colle. 2017b. Practicing Human Dignity: Ethical Lessons from Commedia Dell'arte and Theater. *Journal of Business Ethics 144*: 251–262.

Delerue, H., and M. Hamid. 2015. Who Are These People? Personality Traits and Judgments About Trade Secret Misappropriation in Postemployment Activities. *Business Ethics: A European Review 24* (3): 315–331.

Detert, J.R., L.K. Treviño, and V.L. Sweitzer. 2008. Moral Disengagement in Ethical Decision Making. *Journal of Applied Psychology 93*: 374–391.

Donaldson, T., P. Werhane, and M. Cording. 2002. *Ethical Issues in Business*. 7th ed. Upper Saddle River: Prentice Hall.

Drucker, P. 1954. *The Practice of Management*. New York: Harper.

Enron: The Smartest Guys in the Room. (2005). *Movie Script*. Retrieved from https://www.springfieldspringfield.co.uk/movie_script.php?movie=enron-the-smartest-guys-in-the-room

Fotaki, M., S. Long, and H.S. Schwartz. 2012. What Can Psychoanalysis Offer Organization Studies Today? Taking Stock of Current Developments and Thinking About Future Directions. *Organization Studies 33* (9): 1121–1135. https://doi.org/10.1177/0170840612448152.

Freeman, R.E. 1984. *Strategic Management: A Stakeholder Approach*. Boston, MA: Pitman.

———. 1994. The Politics of Stakeholder Theory: Some Future Directions. *Business Ethics Quarterly 4* (4): 409–421.

Freeman, R. E. 2008. Ending the so-called "Friedman-Freeman" debate. In B. R. Agle, T. Donaldson, R. E. Freeman, M. C. Jensen, R. K. Mitchell, & D. J. Wood (Eds.), *Dialogue: Toward superior stakeholder theory. Business Ethics Quarterly, 18*(2), 153–190.

Freeman, R.E., and E. Auster. 2011. Values, Authenticity and Responsible Leadership. *Journal of Business Ethics 98*: 15–23. https://doi.org/10.1007/s10551-011-1022-7.

Freeman, R.E., and E.R. Auster. 2015. *Bridging the Values Gap. How Authentic Organizations Bring Values to Life*. Oakland: Berrett-Koehler Publishers.

Freeman, R.E., J.S. Harrison, A.C. Wicks, B.L. Parmar, and S. de Colle. 2010. *Stakeholder Theory: The State of the Art*. Cambridge: Cambridge University Press.

Freeman, R.E., L. Dunham, G. Fairchild, and B. Parmar. 2015. Leveraging the Creative Arts in Business Ethics Teaching. *Journal of Business Ethics 131* (3): 519–526. https://doi.org/10.1007/s10551-014-2479-y.

Furnham, A., S.C. Richards, and D.L. Paulhus. 2013. The Dark Triad of Personality: A 10 Year Review. *Social and Personality Psychology Compass 7* (3): 199–216.

Gibney, A. (2005). *Enron: The Smartest Guy in the Room* [Motion picture].

Gino, F., S. Ayal, and D. Ariely. 2009a. Contagion and Differentiation in Unethical Behavior: The Effect of One Bad Apple on the Barrel. *Psychological Science 20*: 393–398. https://doi.org/10.1111/j.1467-9280.2009.02306.x.

Gino, F., D.A. Moore, and M.H. Bazerman. 2009b. See no Evil: Why We Fail to Notice Unethical Behavior. In *Social Decision Making: Social Dilemmas, Social Values, and Ethical Judgments*, ed. R.M. Kramer, A.E. Tenbrunsel, and M.H. Bazerman, 241–263. Danvers: Psychology Press.

Gonin, M., G. Palazzo, and U. Hoffrage. 2012. Neither Bad Apple nor Bad Barrel: How the Societal Context Impacts Unethical Behavior in Organizations. *Business Ethics: A European Review 21* (1): 31–46.

Goodpaster, K.E. 2007. *Conscience and Corporate Culture*. Malden: Blackwell Publishing.

Haidt, J. 2003. The Moral Emotions. In *Handbook of Affective Sciences*, ed. R.J. Davidson, K.R. Scherer, and H.H. Goldsmith, 852–870. Oxford: Oxford University Press.

Hartman, E.M. 1998. The Role of Character in Business Ethics. *Business Ethics Quarterly 8* (3): 547–559. https://doi.org/10.2307/3857437.

———. 2000. An Aristotelian Approach to Moral Imagination. *Professional Ethics 8* (3&4): 57–77. https://doi.org/10.5840/profethics200083/415.

———. 2008. Reconciliation in Business Ethics: Some Advice from Aristotle. *Business Ethics Quarterly 18* (2): 253–265. https://doi.org/10.5840/beq200818218.

Horney, K. 1950. *Neurosis and Human Growth: The Struggle Toward Self-realization*. New York: W.W. Norton and Company, Inc.

Hoyk, R., and P. Hersey. 2008. *The Ethical Executive*. Stanford: Stanford University Press.

Jones, T., and A.C. Wicks. 1999. Convergent Stakeholder Theory. *Academy of Management Review 24*: 206–221.

Kaplan, R. S., and D. P. Norton 1992. The Balanced Scorecard—Measures that Drive Performance. *Harvard Business Review* (January–February), 71–79.

Kets de Vries, M.F.R. 1980. *Organizational Paradoxes: Clinical Approaches to Management*. London: Tavistock/Methuen.

———. 1991. *Organizations on the Couch: Perspectives on Organizational Behavior and Change*. San Francisco: Jossey-Bass.

———. 2018. Coaching the Toxic CEO. *Harvard Business Review 92* (4): 100–109.

Kets de Vries, M.F.R., and D. Miller. 1984. *The Neurotic Organization: Diagnosing and Changing Counterproductive Styles of Management*. San Francisco: Jossey-Bass.

Kish-Gephart, J.J., D.A. Harrison, and L.K. Treviño. 2010. Bad Apples, Bad Cases, and Bad Barrels: Meta-analytic Evidence About Sources of Unethical Decisions at Work. *Journal of Applied Psychology 95*: 1–31. https://doi.org/10.1037/a0017103.

Knights, D., and M. O'Leary. 2005. Reflecting on Corporate Scandals: The Failure of Ethical Leadership. *Business Ethics: A European Review 14* (4): 359–366.

Kohlberg, L. 1969. Stage and Sequence: The Cognitive-Developmental Approach to Socialization. In *Handbook of socialization theory and research*, ed. D.A. Goslin, 348–480. Chicago: Rand McNally.

———. 1984. Moral Stages and Moralization: The Cognitive Developmental Approach. In *The Psychology of Moral Development: The Nature and Validity of Moral Stages*, ed. L. Kohlberg, vol. 2, 170–205. San Francisco: Harper & Row.

Levinson, H. 1973. *The Great Jackass Fallacy*. Boston: Harvard University Press.

Locke, E.A., and G.P. Latham. 1990. *A Theory of Goal Setting and Task Performance*. Englewood Cliffs: Prentice Hall.

———. 2009. Has Goal Setting Gone Wild, or Have Its Attackers Abandoned Good Scholarship? *Academy of Management Perspectives 23* (1): 17–23. https://doi.org/10.5465/amp.2009.37008000.

McLean, B., and P. Elkind. 2003. *The Smartest Guys in the Room: The Amazing Rise and Scandalous Fall of Enron*. New York: Penguin.

Messick, D.M. 1996. Why Ethics is not the Only Thing That Matters. *Business Ethics Quarterly 6*: 223–226.

Moberg, D.J. 2006. Ethics Blind Spots in Organizations: How Systematic Errors in Person Perception Undermine Moral Agency. *Organization Studies 27* (3): 413–428. https://doi.org/10.1177/0170840606062429.

———. 2007. Practical Wisdom and Business Ethics: Presidential Address to the Society for Business Ethics Atlanta, August 2006. *Business Ethics Quarterly 17*: 535–561. https://doi.org/10.5840/beq200717336.

Moore, C., and F. Gino. 2015. Approach, Ability, Aftermath: A Psychological Process Framework of Unethical Behavior at Work. *The Academy of Management Annals 9* (1): 235–289. https://doi.org/10.1080/19416520.2015.1011522.

Moore, C., J.R. Detert, L.K. Treviño, V.L. Baker, and D.M. Mayer. 2012. Why Employees Do Bad Things: Moral Disengagement and Unethical Organizational Behavior. *Personnel Psychology 65*: 1–48. https://doi.org/10.1111/j.1744-6570.2011.01237.x.

O'Leary, M. 2015. Work Identification and Responsibility in Moral Breakdown. *Business Ethics: A European Review 24* (3): 237–251.

Ordóñez, L., M.E. Schweitzer, A.D. Galinsky, and M.H. Bazerman. 2009. Goals Gone Wild: How Goals Systematically Harm Individuals and Organizations. *Academy of Management Perspectives 23* (1): 6–16.

Palazzo, G., F. Krings, and U. Hoffrage. 2012. Ethical Blindness. *Journal of Business Ethics 109*: 323–338. https://doi.org/10.1007/s10551-011-1130-4.

Paulhus, D.L., and K.M. Williams. 2002. The Dark Triad of Personality: Narcissism, Machiavellianism, and Psychopathy. *Journal of Research in Personality 36*: 556–563.

Pless, N.M., T. Maak, and H. Harris. 2017. Art, Ethics and the Promotion of Human Dignity. *Journal of Business Ethics 144* (2): 223–232. https://doi.org/10.1007/s10551-017-3467-9.

Putnam, H. 2002. *The Collapse of the Fact/Value Dichotomy and Other Essays*. Cambridge: Harvard University Press.

Reynolds, S.J. 2006. Moral Awareness and Ethical Predispositions: Investigating the Role of Individual Differences in the Recognition of Moral Issues. *Journal of Applied Psychology 91* (1): 233–243.

Rorty, R. 1991. *Essays on Heidegger and Others*. New York: Cambridge University Press.

Schwartz, M.S. 2017. Ethical Decision-Making: An Integrated Approach. *Journal of Business Ethics 139*: 755–776.

Solomon, R.C. 1992. Corporate Roles, Personal Virtues: An Aristotelian Approach to Business Ethics. *Business Ethics Quarterly 2*: 317–339.

———. 2003. Victims of Circumstances? A Defense of Virtue Ethics in Business. *Business Ethics Quarterly 13* (1): 43–62. https://doi.org/10.5840/beq20031314.

Tajfel, H. 1982. *Social Identity and Intergroup Relations*. Cambridge: Cambridge University Press.

Tajfel, H., and J.C. Turner. 1986. The Social Identity Theory of Inter-group Behavior. In *Psychology of Intergroup Relations*, ed. S. Worchel and L.W. Austin, 7–24. Chicago: Nelson-Hall.

Tenbrunsel, A.E., and D.M. Messick. 1999. Sanctioning Systems, Decision Frames, and Cooperation. *Administrative Science Quarterly 44*: 684–707. https://doi.org/10.2307/2667052.

———. 2004. Ethical Fading: The Role of Self-Deception in Unethical Behavior. *Social Justice Research 17* (2): 223–236.

Tenbrunsel, A.E., K. Smith-Crowe, and E.E. Umphress. 2003. Building Houses on Rocks, the Role of the Ethical Infrastructure in Organizations. *Social Justice Research 16* (3): 285–307.

The Economist. 2017, April. *Alitalia is Bankrupt Again. This Time Perhaps it's Terminal*. Retrieved from https://www.econo mist.com/blogs/gulliver/2017/04/arrivederci-alitalia

Tian, Q., and D.K. Peterson. 2016. The effects of ethical pressure and power distance orientation on unethical pro-organizational behavior: The case of earnings management. *Business Ethics: A European Review 25* (2): 159–171.

Treviño, L.K., G.R. Weaver, and S.J. Reynolds. 2006. Behavioral ethics in organizations. *Journal of Management 32* (6): 951–990.

Treviño, L.K., N. Den Nieuwenboer, and J.J. Kish-Gephart. 2014. (Un) Ethical behavior in organizations. *Annual Review of Psychology 65* (1): 635–660.

United States (2002). *Sarbanes-Oxley Act of 2002: Conference report (to accompany H.R. 3763)*. Washington, DC: U.S. G.P.O.

Venkataraman, S. 2002. Stakeholder Value Equilibration and the Entrepreneurial Process. *Ethics and Entrepreneurship, Special Issue of the Society for Business Ethics*, The Ruffin Series # 3, 45–58.

Weaver, G.R., L.K. Trevino, and P.L. Cochran. 1999. Integrated and Decoupled Corporate Social Performance: Management Commitments, External Pressures, and Corporate Ethics Practices. *Academy of Management Journal 42*: 539–552.

Werhane, P.H. 1999. *Moral Imagination and Management Decision Making*, The Ruffin Series in Business Ethics. Oxford University Press.

Wicks, A.C. 1996. Overcoming the Separation Thesis: The Need for a Reconsideration of Business and Society Research. *Business & Society 35* (1): 89–118. https://doi.org/10.1177/000765039603500108.

Wray-Bliss, E. 2012. A Crisis of Leadership: Towards an Anti-Sovereign Ethics of Organization. *Business Ethics: A European Review 22* (1): 86–101.
Wright, S.A., J.B. Dinsmore, and J.J. Kellaris. 2013. How Group Loyalties Shape Ethical Judgment and Punishment Preferences. *Psychology & Marketing 30* (3): 203–210.

Simone de Colle is Associate Professor of Business Ethics & Strategy at IESEG School of Management, Paris, and Visiting Associate Professor of Business, Organizations and Society at New York University Abu Dhabi.

R. Edward Freeman is University Professor, Elis and Signe Olsson Professor, Academic Director of the Institute for Business in Society, and Senior Fellow of the Olsson Center for Applied Ethics at the University of Virginia's Darden School of Business.

Part III
Stakeholder Capitalism

Chapter 45
The Myth of Cowboy Capitalism

The Real Danger Facing American Business Is Not Its Ability to Compete, But Its Ability to Cooperate

R. Edward Freeman

The recent controversy surrounding the appropriate roles for customers, suppliers, employees, communities, and stockholders threatens to obscure the real danger facing American business. NCR has claimed in a series of expensive and highly visible ads that its corporate mission is to "create value for stakeholders." Its essay contest, which offered $300,000 in prizes had business schools buzzing. It even provoked a *Wall Street Journal* editorial from an outraged champion of poor, neglected stockholders.

What's going on here? Just the latest jargon? Another PR gimmick? The latest B-school fad? A conspiracy of the left, as some have suggested? Or, can we use this opportunity to say, once and for all, what's really wrong with American business?

1 Stakeholders and Stockholders

Stockholders are a big deal. That much is common sense. Any executive who ignores their expectation of returns, crystallized in the share price, is likely to be pounding the pavement pretty soon, the company raided, broken up, and looking for a place to hang its next slogan.

But the story of American business that is unfolding does not begin or end with managing the relationship with stockholders. Customers, suppliers, employees and communities are also big deals.

Originally published in: *The Darden Report*, 14(3), 28–32 © Darden School of Business, 1988
Reprint by Springer

R. E. Freeman (✉)
University of Virginia, Charlottesville, VA, USA
e-mail: FreemanE@darden.virginia.edu

© Springer Nature Switzerland AG 2023

S. D. Dmytriyev, R. E. Freeman (eds.), *R. Edward Freeman's Selected Works on Stakeholder Theory and Business Ethics*, Issues in Business Ethics 53,
https://doi.org/10.1007/978-3-031-04564-6_45

The company that does not pay close attention to customer needs—that instead sells "commodities"—that does not work jointly with customers to develop new products—that does not pay attention to service and complaints, is a company bound for the trash dump. The customer revolution, wrought by Tom Peters and others, is here to stay.

And think about the political and regulatory solutions "close to the customer" that we have tried in our society. They have involved the Consumer Product Safety Commission, Ralph Nader and the consumer movement, and egregious products liability law.

Suppliers are equally important. Just-in-time inventory, close cooperative relationships with suppliers, the emergence of on-line ordering systems, such as American Hospital Supply's, and other production innovations have finally made managers realize that they are in it together with suppliers.

2 A Shared Sense of Fate

We have also tried to regulate the firm-supplier relationship by passing the Foreign Corrupt Practices Act, the Uniform Commercial Code, and unworkable antitrust laws. Ditto for employees. Quality, participation, teamwork, and ownership are the watchwords for the 1990s. Going fast is the old paternalistic "Father Knows Best" school of management. In the world today, American companies can no longer afford such management arrogance. Akio Morita said it best in his book, *Made in Japan-*. "Those companies that are most successful in Japan are those that have managed to create a shared sense of fate among all employees, what Americans call labor and management, and the shareholders."

And if you don't believe them, think about the labor laws, OSHA, and the resulting work rules that hamstring productivity.

If you don't think that communities are important, then how do you explain the rush to locate new plants in areas willing to give favorable tax treatment, to provide upgraded schools and roads, and other favorable treatment to business?

On the other side, how else do you explain the onerous regulation passed during Republican administrations of the early 1970s that were partially responsible for our inability to compete today?

Now surely the interests of all of these groups can conflict, and it is the job of any manager worth his or her salt to solve these conflicting claims so that the corporation survives and thrives. And, just as obviously, self-serving managers can use "multiple and conflicting stakeholders" as an excuse to engage in self-dealing, just as they use stockholders and the Japanese as excuses. But, the hard, cold fact remains: without the support of each of these groups, business is just impossible.

But so what? If the question is "Are stockholders or stakeholders more important," then the answer is "Bad question." If the question is "Which stakeholders are most important," then the answer is "It depends on the business." If the question is "Does management have a fiduciary duty to stockholders or stakeholders," the answer is

"Academic question given the current reality in the business world," because "sound business judgment in effect means exercising a duty of care to all of them." Any way you cut it, it is hard to get excited. This much ought to be plain to any executive who has not had a frontal lobotomy. Unfortunately, it isn't quite so easy.

3 Cowboy Capitalism

What gets in the way is a mindset about business. Call it Cowboy Capitalism. Cowboy Capitalism is the view that business is first and foremost about competition. Executives have to saddle up their horses and ride out to play shoot 'em up with the bad guys —domestic competitors—or band together and duke it out with the savages—foreign competitors. What makes the system work is the competitive urge—the need to win—to get the other guy. Capitalism, on this view, is just a bunch of greedy little bastards out to do each other in. (Small wonder that 70% of the American people think that business does a lousy job in ethics.)

Cowboy Capitalism runs deep in American business, business schools, and political circles. "We are poised," we are told, "at a crisis of competitiveness." "We need a coherent trade policy so that we can compete," says one presidential aspirant. We are bombarded with presidential commissions, economic statistics, and executive speeches that border on anti-Japanese racism. And the competitive rhetoric is certain to heat up in the coming election frenzy.

Here's a secret that's the best kept one around. Cowboy Capitalism is a myth. Business isn't about competition at all. The sure road to disaster is focus all of the guns on killing the enemy.

Business is about cooperation—that's right—cooperation. What makes a business successful, from Mrs. Field's cookies to IBM, is the ability of someone, professional manager or entrepreneur, to put together a deal. Business is about cooperation among customers, suppliers, employees, financiers, and communities. Each factor is important—a necessary condition for success—and the world's great companies use some cooperative glue to find ways to meet these needs.

This is no airy academic idea, and surely nothing based on altruism, or fuzzy, leftist notions about property. Each group has an incentive to cooperate—to be a part of the deal or contract—because each group can do more together than it can alone. Cooperation is the ultimate pragmatic strategy. It is the real fuel of capitalism.

Now competition is important, less I be accused of having a frontal lobotomy, but it is a second-order phenomena. It helps to determine the incentives for cooperation and the available alternatives for each factor. But, if managers focus on the competition they will not focus on customers, or on employees. Such is the logic of the concept of "focus."

"Cooperation." Say it again. "Cooperation." Now say "Close to the customer" and "employee participation." It all begins to sound amazingly alike. But, there is one more piece.

4 The Real Problem of Competitiveness

There is a problem of competitiveness. The myth of Cowboy Capitalism has given American managers and workers alike the ability to take their eye off the ball—to ignore customers and other stakeholders—to cost jobs, communities, and a future generation of manufacturing in our society. And, through the messages of Tom Peters, Philip Crosby, Edward Demmings, and others we are slowly beginning to change.

But, because we have not laid the myth of Cowboy Capitalism to rest the problem is that the government now poses a real and present danger in trying to fix the problem of competitiveness. Erecting trade barriers, capping liability damages, "fixing" the capital markets, and other nonsense threaten irreparable damage to our free society. And, ironically trying to fix the problem of competitiveness will surely cause a real problem of competitiveness.

The real solution is much more simple and straightforward. All we have to do is to change the self image of managers. Business isn't about cowboys and shoot 'em ups. It is about creating value for stakeholders. Doing that well is more difficult than taking out some advertisements in newspapers. But, if we take that challenge, capitalism can survive.

R. Edward Freeman is University Professor, Elis and Signe Olsson Professor, Academic Director of the Institute for Business in Society, and Senior Fellow of the Olsson Center for Applied Ethics at the University of Virginia's Darden School of Business.

Chapter 46
Understanding Stakeholder Capitalism

R. Edward Freeman

1 Background

The idea of stakeholders is quite old, growing up in the 1960s through the work of management theorists Eric Rhenman, Igor Ansoff, Russell Ackoff and their students. And the idea is connected to a very old tradition that sees business as an integral part of society rather than an institution that is separate and purely economic in nature.

Identifying and analysing stakeholders was originally a simple way to acknowledge the existence of multiple constituencies in a corporation. The main insight was that executives must pay some strategic attention to those groups who were important to the success of their corporation. So far… common sense.

As the pace of change accelerated in business, these thinkers and others began to advocate more interaction with stakeholders so that they had some sense of participation in the day-to-day affairs of the corporation. So we had the emergence of consumer advisory panels, quality circles, just-in-time inventory teams, community advisory groups and so on, all designed to get the corporation more in touch with the key relationships that affect its future.

R. Edward Freeman discusses the rise of the concept of stakeholder capitalism and offers some guidelines as to how it can be made effective.

Originally published in: ***Financial Times***, July 19 © Financial Times Ltd, 1996
Reprint by Springer

R. E. Freeman (✉)
University of Virginia, Charlottesville, VA, USA
e-mail: FreemanE@darden.virginia.edu

© Springer Nature Switzerland AG 2023

S. D. Dmytriyev, R. E. Freeman (eds.), *R. Edward Freeman's Selected Works on Stakeholder Theory and Business Ethics*, Issues in Business Ethics 53,
https://doi.org/10.1007/978-3-031-04564-6_46

During the 1980s the idea of "stakeholder management" was articulated as a method for systematically taking into account the interests of "those groups which can affect and are affected by the corporation". Again…just good management.

During the late 1980s and early 1990s we also saw the emergence of a strong movement concerned with business ethics. Much of the business ethics movement has been rooted in a response to perceived corporate excesses such as oil spills, financial scandals and business-government collusion. But a small number of thinkers began to ask questions about the very purpose of the corporation. Should the corporation serve those who own shares of stock or should it serve those who are affected by its actions?

The choice was laid bare: corporations can be made to serve stockholders or they can be made to serve stakeholders.

2 False Choice

Most thoughtful executives know that this choice between stockholders and stakeholders is a false one. Corporations must be profitable at rates determined by global capital markets. No longer can executives ignore the fact that capital flows freely across borders and that rates of return are more complicated than internally generated hurdle rates and payback schemes. Business today is truly global.

Most thoughtful executives also know that great companies are not built by obsessive attention to shareholder value. Great companies arise in part out of a shared sense of purpose among employees and management. This sense of purpose must be important enough for individuals to expend their own human capital to create and deliver products and services that customers are willing to pay for.

We need only return to the wisdom of Peter Drucker and W. Edwards Deming to see the importance of meaning and purpose and the destructiveness of fear and alienation in corporate life.

Management thinkers such as Tom Peters, Charles Handy, Jim Collins and Jerry Poras have produced countless examples of how employees, customers and suppliers work together to create something that none of them can create alone. And capital is necessary to sustain this process of value-creation.

From Cadbury to Volvo, Nordstrom to Hewlett-Packard, executives are constantly engaged in intense stakeholder relationships. Boeing's building of the 777 jet involved a complex process of interaction among suppliers, customers and employees. The interests of stockholders and stakeholders are often aligned rather than in conflict. Stockholders are a key stakeholder group whose support must be sustained in the same way that customer, supplier and employee support must be garnered. The issue is one of balancing the interests of these groups.

Furthermore, in a relatively free political system, when executives ignore the interests of one group of stakeholders systematically over time those stakeholders will use the political process to force regulation or legislation that protects themselves.

Witness the emergence of "stakeholder rights" in the US in the form of labour legislation, consumer protection legislation, environmental (community) protection legislation and even shareholder protection legislation.

There are many ways to manage a successful company. Daimler Benz will be different from Volvo. Procter and Gamble will be different from Unilever. However, all will involve the intense interaction of employees with critical stakeholders.

The more that stakeholders participate in the decisions that affect them, be they product design decisions or employment contract decisions, the greater the likelihood that they will be committed to the future of the corporate enterprise.

3 Four Principles

Business today is different. Capital markets have been liberalised and are truly global. For the most part, investment flows to the best returns regardless of geography just as customers buy the best value regardless of country of origin.

Political regimes have been liberalised so that everyone is adopting some version of capitalism or market mechanisms, yielding the possibility for unprecedented worldwide growth. And incredibly powerful information technologies tie together sophisticated communications systems that make McLuhan's concept of a global village a stunning reality.

Business approaches this century of change with an ideology that can only be called radically shop-worn and outdated. As businesses change the composition of human society, its ideology proclaims that it is amoral, that business ethics is an oxymoron, that it is only doing what shareholders require.

Furthermore, it captures the imagination by claiming that this is a result of our human drive to compete. Business is to be understood as warfare and executives are the lonely soldiers on the battlefield of global markets. The result of this myth of the primacy of the shareholder and the view of business as "cowboy capitalism" leads to a public mistrust and misunderstanding of the processes that make companies successful.

Stakeholder capitalism, properly formulated, is the new story that we need. But stakeholder capitalism is not a leftish excuse to regulate business nor is it a convenient whipping boy of the right to avoid common moral decency and fair play. Stakeholder capitalism has to serve as a model of the very best companies - the best that business can be; that should be our expectations for corporations.

Let us begin by defining the primary stakeholders as customers, suppliers, employees, financiers and communities. Managers simply have to pay attention to the concerns of these groups. A fully articulated stakeholder capitalism will have to add models of relationships with other groups such as the media, government, competitors and even terrorists.

Stakeholder capitalism should be based on four principles.

4 Stakeholder Co-operation

Value is created because stakeholders can jointly satisfy their needs and desires. Capitalism works because entrepreneurs and managers put together and sustain deals or relationships among customers, suppliers, employees, financiers and communities. And the support of each group is vital to the success of the endeavour. This is the co-operative common sense part of business that every executive knows, although the myth of the primacy of the shareholder tells us that some stakeholders are more important than others.

But try building a great company without the support of all stakeholders. It simply cannot be sustained.

5 Complexity

Human beings are complex creatures capable of acting from many values. Sometimes they are selfish and sometimes they act for others. Many values are and shared. Capitalism works because of this complexity rather than in spite of it.

6 Continuous Creation

Business as an institution is a source of the creation of value. Co-operating with stakeholders and motivated by values, business people continuously create new sources of value.

This creative force of humans is the engine of capitalism. And the beauty of the modern corporate form is that it can be made to be continuous rather than destructive. One creation does not have to destroy another. Rather there is a continuous cycle of value creation that raises the well-being of everyone. People come together to create something, be it a new computer program or a new level of service.

7 Emergent Competition

Competition emerges from a relatively free and democratic society so that stakeholders have options. Competition emerges out of the co-operation among stakeholders rather than being based on the urge to "get the other guy".

Competition is important in stakeholder capitalism but it is not the primary force, as so many business thinkers claim. Humans' urge to co-operate and create is what distinguishes us from apes, not our urge to compete.

Stakeholder capitalism does not pretend to be amoral like its cowboy cousin. It takes a firm ethical stand: that human beings are required to be at the centre of any process of value creation; that common decency and fairness are not to be set aside in the name of playing the game of business and that we should demand the best behaviour of business.

Stakeholder capitalism is no panacea. It simply allows the possibility that business becomes a fully human institution.

Stakeholder capitalism bases our understanding and expectations of business not on the worst that we can do but on the best. It sets a high moral standard, recognises the common sense practical world of global business today and asks managers to get on with the task of creating value for stakeholders.

There is much work to be done to articulate stakeholder capitalism in a way that allows business to occupy the moral high ground. Seeing business as anything less than an institution that is a vital part of our conception of the good life gives away too much.

Adam Smith understood that business and ethics must go together. He wrote *The Theory of Moral Sentiments* along with *The Wealth of Nations*. Stakeholder capitalism can be a way of resuscitating what we have forgotten about Adam Smith and of building a capitalism that is human.

R. Edward Freeman is University Professor, Elis and Signe Olsson Professor, Academic Director of the Institute for Business in Society, and Senior Fellow of the Olsson Center for Applied Ethics at the University of Virginia's Darden School of Business.

Chapter 47
Stakeholder Capitalism and the Value Chain

R. Edward Freeman and Jeanne M. Liedtka

1 Introduction

Much has recently been written about the concept of 'stakeholders' or 'Stakeholder Capitalism'. The purpose of this article is to show how thinking about stakeholders leads to a different and more robust understanding of strategic thinking. In particular we shall focus on the concept of the value chain and suggest that it be reinterpreted in stakeholder terms. By doing so we can more carefully construct methods to build stakeholder relationships that can be sustained over time. We begin by laying out the road map of stakeholder capitalism in the form of four key principles. Next, we review the concept of the value chain as it has evolved in the strategy literature, showing how new views of the value chain and the principles of stakeholder capitalism reinforce each other. Finally, we show how it is possible to operationalize the stakeholder value chain in a transactional mode, but that it more naturally lends itself to an interpretation of a relational view of business and its stakeholders.

Originally published in: *European Management Journal*, 15(3), 286–296 © Elsevier, 1997
Reprint by Springer, https://doi.org/10.1016/S0263-2373(97)00008-X

R. E. Freeman (✉) · J. M. Liedtka
University of Virginia, Charlottesville, VA, USA
e-mail: FreemanE@darden.virginia.edu

© Springer Nature Switzerland AG 2023

S. D. Dmytriyev, R. E. Freeman (eds.), *R. Edward Freeman's Selected Works on Stakeholder Theory and Business Ethics*, Issues in Business Ethics 53,
https://doi.org/10.1007/978-3-031-04564-6_47

2 Stakeholder Capitalism: A Synopsis[1]

The idea of stakeholders is actually quite old, growing up in the 1960s through the work of management theorists Eric Rhenman, Igor Ansoff, Russell Ackoff, and their students. The idea is connected to a very old tradition that sees business as an integral part of society, rather than an institution that is separate and purely economic in nature. Identifying and analyzing stakeholders was originally a simple way to acknowledge the existence of multiple constituencies in the corporation. The main insight was that executives must pay some strategic attention to those groups who were important to the success of the corporation.[2]

As the pace of change accelerated in business, these thinkers and others began to advocate more interaction with stakeholders so that they have some sense of participation in the day to day affairs of the corporation. We had the emergence of consumer advisory panels, quality circles, just-in-time inventory teams, community advisory groups, etc., all designed to get the corporation more in touch with the key relationships that affected its future. During the 1980s the idea of 'stakeholder management' was articulated, as a method for systematically taking into account the interests of 'those groups which can affect and are affected by the corporation'.

During the late 1980s and early 1990s we also saw the emergence of a strong movement concerned with business ethics. Much of the business ethics movement has been rooted in a response to perceived corporate excesses such as oil spills, financial scandals, business-government collusion, and celebrated cases of whistle blowing. But, a small number of thinkers began to ask questions about the very purpose of the corporation. Should the corporation serve those who owned shares of stock, or should it serve those who are affected by its actions? The choice was laid bare: Corporations can be made to serve stockholders, or they can be made to serve stakeholders.

Most thoughtful executives know that this choice between stockholders and stakeholders is a false one. Corporations must be profitable at rates determined by global capital markets. No longer can executives ignore the fact that capital flows freely across borders, and that rates of return are more complicated than internally generated hurdle rates and payback schema. Business today is truly global.

Most thoughtful executives also know that great companies are not built by obsessive attention to shareholder value. Great companies anse m part out of a shared sense of purpose among employees and management. This sense of purpose must be important enough for individuals to expend their own human capital to create and deliver products and services that customers are willing to pay for. We need only return to the wisdom of Peter Drucker and W. Edward Deming to see the

[1] Portions of this section have appeared in Freeman (1996). We are grateful to the editors for permission to reprint certain paragraphs here. Defining 'stakeholder capitalism' is an ongoing project. For some preliminary statements see, Freeman (1997), and Freeman (1994).

[2] For a more careful history see Freeman (1984), Donaldson and Preston (1995); and more recently still, Mitchell et al. (1997).

importance of meaning and purpose and the destructiveness of fear and alienation in corporate life.

Management thinkers such as Tom Peters, Charles Handy, Jim Collins and Jerry Poras have produced countless examples of how employees, customers, and suppliers work together to create something that none of them can create alone. And, capital is necessary to sustain this process of value-creation. From Cadbury to Volvo, Nordstrom's to Hewlett Packard, executives are constantly engaged in intense stakeholder relationships.

In this view, the interests of stockholders and stakeholders are very often aligned rather than in conflict. Stockholders are a key stakeholder group, whose support must be sustained in the same way that customer, supplier and employee support must be garnered. The issue is one of balancing the interests of these groups, not picking one at the expense of the other. Furthermore, in a relatively free political system, when executives ignore the interests of one group of stakeholders systematically over time, these stakeholders will use the political process to force regulation or legislation that protects themselves. Witness the emergence of 'stakeholder rights' in the United States in the form of labor legislation, consumer protection legislation, environmental (community) protection legislation, even shareholder protection legislation.

Quite simply, there are many ways to manage a successful company. Daimler-Benz will be different from Volvo. Procter & Gamble will be different from Unilever. However, all will involve the intense interaction of employees, management and non-management alike, with critical stakeholders. The more that stakeholders participate in the decisions which affect them, be they product design decisions or employment contract decisions, the greater the likelihood that they will be committed to the future of the corporate enterprise.

Contrast this common sense view of the workings of business with traditional business ideology. As businesses change the very composition of human society, its traditional ideology proclaims that it is amoral, that business ethics is an oxymoron, and that business exists only to do what shareholders require. Business, in this view, is to be understood as warfare, and executives are the lonely soldiers on the battlefield of global markets playing 'shoot 'em up' with competitors. This myth of the primacy of the shareholder and its view of business as 'Cowboy Capitalism' leads to a profound public mistrust and misunderstanding of the basic processes that make companies successful. We need a new story — one that elevates business to the higher moral ground that it can occupy — and, one that smacks of common sense and reality in today's business world.

Stakeholder Capitalism, properly formulated, is just the new story that we need. Stakeholder Capitalism is not an excuse to regulate business, nor does it give business the right to avoid common moral decency and fair play. Stakeholder Capitalism must serve as a model for the very best companies — it represents the best that business can be.

Stakeholder Capitalism is based on four principles. Each is important to remember if we are to craft a capitalism that will serve us in the next century. First of all *The Principle of Stakeholder Cooperation* says that value is created because stakeholders can jointly satisfy their needs and desires. Business is not a zero sum game. Capitalism works because entrepreneurs and managers put together and sustain deals or relationships among customers, suppliers, employees, financiers, and communities. The support of each group is vital to the success of the endeavour. This is the cooperative common sense part of business that every executive knows, but the myth of primacy of the shareholder tells us that some stakeholders are more important than others. Try building a great company without the support of all stakeholders. It simply cannot be sustained.

Second, *The Principle of Complexity* claims that human beings are complex creatures capable of acting from many different values. We are not just economic maximizers. Sometimes we are selfish and sometimes we act for others. Many of our values are jointly determined and shared. Capitalism works because of this complexity, rather than in spite of it.

Third, *The Principle of Continuous Creation* says that business as an institution is a source of the creation of value. Cooperating with stakeholders and motivated by values, businesspeople continuously create new sources of value. This creative force of humans is the engine of capitalism. The beauty of the modem corporate form is that it can be made to be continuous, rather than destructive. One creation does not have to destroy another, rather there is a continuous cycle of value creation which raises the well-being of everyone. People come together to create something, be it a new computer program, a new level of service, a way to heal the sick, or simply to work together.

Finally, *The Principle of Emergent Competition* says that competition emerges from a relatively free and democratic society so that stakeholders have options. Competition emerges out of the cooperation among stakeholders, rather than being based on the primal urge to 'get the other guy'. Competition is important in Stakeholder Capitalism, but it is not the primary force. It is in its ability to manage the tension created by simultaneous cooperation and competition that stakeholder capitalism distinguishes itself.

Stakeholder Capitalism takes a firm ethical stand — that human beings are required to be at the center of any process of value creation — that common decency and fairness are not to be set aside in the name of playing the game of business — that we should demand the best behavior of business — and that we should enact a story about business that celebrates its triumphs, admonishes its failures, and fully partakes of the moral discourse in society as a routine matter.

Stakeholder Capitalism is no panacea. It simply allows the possibility that business becomes a fully human institution. There will always be businesspeople who try to take advantage of others, just as there are corrupt government officials, clergy, and professors. Stakeholder Capitalism bases our understanding and expectations of business, not on the worst that we can do, but on the best. It sets a high moral standard, recognizes the common sense practical world of global business today, and asks managers to get on with the task of creating value for all stakeholders.

Interestingly, as writers in the field of ethics have developed their arguments for a more collaborative view of business as comprised of a network of stakeholders, emerging thinking in the field of strategy has also accorded cooperation a far greater role in the search for competitive advantage. Nowhere is this more evident than in the evolution of the concept of the value chain.

3 The Evolution of the Value Chain

The concept of the 'value chain', popularized by Michael Porter in his important book, *Competitive Strategy* (1980), has evolved dramatically over the past 5 years. As used by leading management thinkers today, it has become the centerpiece of new views of the sources of competitive advantage. Below, we summarize the shifts in thinking about the value chain:

Traditional View → Emerging Views

Traditional View	Emerging Views
• physical	• virtual
• product-focused	• capability-focused
• intrafirm	• interfirm
• static and fixed	• evolving and shapable
• linear and sequential	• matrixed and simultaneous

The value chain, in its original manufacturing-based view, was seen as the set of operations, accomplished sequentially, that an individual firm used to physically transform its raw material inputs into finished products. Current views of the value chain employ the term much more expansively, seeing it as a set of processes through which a constellation of actors work together to continuously innovate in a way that produces value for customers. As Normann and Ramirez (1993) note:

> *Increasingly, companies do not just add value, they reinvent it. Their focus of analysis is not the company or even the industry but the value-creating system itself, within which different economic actors — suppliers, business partners, allies, customers — work together to co-produce value.*

Physical → Virtual

No longer confined to the physical transformation of the product itself, the value chain is now seen as involving the creation and use of information, as well. This 'virtual value chain', the information flow that accompanies the physical flow, presents a range of new opportunities to achieve competitive advantage (Rayport and Sviokla 1995). 'Digital assets', like information, offer a very different set of scale economies than equipment or 'bricks and mortar' (as traditional banks competing against financial services competitors like Fidelity and Shwab have discovered). They also offer broader economies of scope to firms capable of leveraging their knowledge of customers into new products and services. Finally, they are able to achieve dramatically improving price/performance ratios, as the cost of information processing capability decreases. Thus, opportunities to gain advantage over entrenched competitors by disrupting the status quo in an industry are far more attractive in an information-defined 'market space' than in a physically defined 'marketplace' (Rayport and Sviokla 1995).

Product focused → Capability focused

The change in emphasis from a focus on the product itself to the underlying capabilities that transform it reflects a more fundamental shift in the strategy field's view of the sources of competitive advantage. The belief that competitive advantage lies, not only in the selection of products and markets, but in the choice of which capabilities to develop, was elaborated by Stalk et al. (1992):

> *... the essence of strategy is not the structure of a company's products and markets but the dynamics of its behavior. And the goal is to identify and develop the hard-to-imitate organizational capabilities that distinguish a company from its competitors in the eyes of customers, (p. 62)*

In this view, the concept of the value chain is raised to a more strategic focus on the capability set that underlies the operations that previously formed the building blocks of the value chain. The focus moves from one of departmentalized functions to cross-functional business processes.

Intrafirm → Interfirm

Increasingly the capabilities necessary to create value through innovation do not reside within a single organization. Here, too, vertical' has lost ground to 'virtual', as changing circumstances demand a level of responsiveness and innovation that vertical integration has often been able to produce. In a world where the management of knowledge is critical, aligning one's self with slow-moving or inefficient internal partners, while competitors are free to ally with more focused firms at the leading edge of particular capabilities, can be suicidal. Thus, organizations increasingly look to partnerships and alliances to create these new value chains.

Industry → Ecosystem

This broader set of players often crosses traditional industry boundaries. In fact, arguing for the 'end of industry', James Moore (1996) advocates a focus on what he terms the 'business ecosystem' rather than the traditional focus on the concept of industry:

> *I suggest that a company be viewed not as a member of a single industry but as part of a business ecosystem that crosses a variety of industries. In a business ecosystem, companies coevolve capabilities around a new innovation: they work cooperatively and competitively to support new products, satisfy customer needs, and eventually incorporate the next round of innovations.*

The blurring of the once-distinct barriers separating the banking, insurance, and investment industries in the US, for instance, is well under way. Even the existence of a broader industry, like financial services is increasingly blurred as seemingly unrelated firms like Microsoft explore banking on the Internet. As multiple sectors of the economy threaten to collapse into one mega-category of 'information services', it seems time to look for more useful ways of thinking about the age old question, 'What business are we in?' As protective barriers and accepted 'rules of the game' are dismantled, firms who limit themselves to traditional frameworks and categories are likely to see themselves unseated by new entrants with less conventional views.

Static and Fixed → Evolving and Shapable

Moore's argument for a view of ecosystems as coevolving, rather than static, is fundamental. This perspective shifts emphasis from a view of value as created through a 'matching' process in which organizations align their current capabilities with current market opportunities, to a view of value as creating through a 'shaping' process, in which the on-going creation of new capabilities spurs innovation. In doing so, Moore builds on the notion of 'strategic intent' (Hamel and Prahalad 1994) that has profoundly affected recent views of strategy, and then extends it beyond a single-firm phenomena to that of a community of suppliers. Working together, the community creates its future.

Linear and Sequential → Matrixed and Simultaneous

Taken together, these new views culminate in a concept of the value chain that bears little resemblance to Porter's original concept, or to the physical representation of a 'chain' at all. No longer comprised of a series of sequential operations, undertaken by a single firm, it is now comprised of a collection of value-creating processes contributed by different firms working simultaneously both independently and cooperatively, where lines of authority and accountability are necessarily less clear. The traditional categories of supplier, competitor, and customer are no longer very useful, as firms change roles as circumstances dictate.

Value is created, not through the static individual capabilities of a firm, but through the continuous reconfiguration of the larger set of value chain activities. Again, Normann and Ramirez continue:

> *... (the) key strategic task is the reconfiguration of roles and relationships among this constellation of actors in order to mobilize the creation of value in new forms and by new players.*

One's attractiveness as a partner, the very source of advantage itself, becomes a firm's contribution to this 'innovation trajectory' (Morris and Ferguson 1993).

This new thinking about the value chain, then, reflects the evolution of thinking about the business of achieving competitive advantage, itself. Traditional views saw a firm's positioning within its industry, and its subsequent ability to find a 'fit' between its competencies and external opportunities, as central. Emerging views look instead towards a firm's capability to contribute continuous innovation, within a larger community of suppliers, and to shape the ongoing evolution of that community.

The dramatically increased complexity involved in managing the new realities that these emerging views represent is significant. The disparity between the mindset and skillset needed to manage in this new world and the ideology of 'cowboy capitalism' accounts, we believe, for the high failure rates associated with attempts at collaboration across, and even within, firm boundaries. The tenets of stakeholder capitalism, on the other hand, mirror those that we have ascribed to new concepts of the value chain. The principles of stakeholder cooperation, complexity of values and motivations, continuous creation, and emergent competition all reflect the reality of the challenges involved in creating sustainable advantage in today's marketplace. All point to a new role for organizations interested in leading the way — that of the architect and shaper of an evolving community of stakeholders. Creating and sustaining such a community, however, requires that we learn how to better manage the tension between two fundamental aspects of business — value creation and value capture.

3.1 Value Creation vs. Value Capture

This new view of the value chain that we have described thus far does not change an old reality — that success in business requires both creating value and capturing that value in the form of profitability. The shift in emphasis from the individual firm's ability to produce value for customers, to a broader community of suppliers' ability to achieve value-creating outcomes further complicates the value capture issue, because there are more players involved in the allocation of the financial rewards that accrue. Thus, while we might argue that emerging views of the value chain are more inherently collaborative in nature — competition remains a strong force, both between competing value chains (for customers) and within an individual value chain (for a share of the profits). This tension between simultaneous competition and collaboration is inherent to the concept of capitalism itself and is not, nor should it be, resolved. It must, however, be managed effectively. We suggest that a management mindset based on stakeholder capitalism is capable of managing it more effectively than traditional approaches.

To understand why, it is important, first, to differentiate between the two processes of value creation and capture. James Moore has used the terms 'value space' and 'deal space' to do this. Value space is the opportunity space that exists to create value by reconfiguring the business processes. This space, he contends, must be significant in order for cooperation to work. Only dramatic enhancements in the current price/performance ratio are capable of funding the inherent inefficiencies in the process of learning to work together and fighting the status quo. In fact, the rationale behind collaboration must be that the parties, working together are capable of accomplishing significant innovations that none can achieve working alone. Thus, we begin with the assumption that part of the value generated by the innovation will be passed on to customers in the form of lower prices or greater functionality, and part will be retained by the supplier community in the form of higher profits. Therefore, the 'pie' itself will be larger. The driver of the value creation formula is the on-going innovation-generating capacity of the partnership, operating in a value chain in which a number of stakeholders work together simultaneously and jointly, in a way that evolves over time; rather than sequentially and discretely in fixed patterns.

The driver of the value capture formula, on the other hand, has generally been seen to be bargaining power. The sources of this bargaining power do not differ materially from the sources that Porter (1980) describes in his 'five forces' model. The major difference here is that we are concerned with our positioning within an ecosystem's value chain, rather than within an industry. Moore (1996) ties the sources of bargaining power in an ecosystem to innovation capacity (as described above), criticality to customers, and embeddedness within the chain. These respond closely to Porter's notions of differentiation, lack of substitutes, and switching costs. This where 'deal space' comes into play. As the partners in the value chain come together to negotiate their 'take', their bargaining power will determine the success of their efforts to keep as great a share of total profits as possible for their own firm.

In effect, this view attempts to overlie a traditional 'cowboy capitalism' mindset for value capture onto the functioning of the more collaborative value chain's capability for superior value creation. The result is unworkable, because the simultaneous effort to both create value and to capture a 'fair' share of the value using a traditional transaction-oriented management mindset are incompatible and work at cross purposes with each other. The emphasis on value capture undermines the value creation process as soon as one member's bargaining power gets out of balance. In the face of significant power differences, traditional business ideology argues that the strong member may elect to hold the other members of the value chain 'hostage'. The Compaq-Intel spectacle that has unfolded as each jockeys for greater leverage within the value chain that they share, is instructive of the dynamics that the dual pressures of competing and collaborating can raise.

The hostage strategy is one which Intel (with its 70% margins) has been accused of playing, relative to its partners in the PC value chain. This strategy, however, is not sustainable as it relies on the lack of alternatives. With the development of choice, the relationship disbands. The question of interest in building sustainable collaboration becomes one of how to keep choice and build relationships on an on-going basis.

3.2 Exit, Voice and Loyalty

Hirshman's (1970) concept of exit, voice and loyalty reflect the dynamics at play here. States, organizations, and individuals, Hirshman maintains, have two mechanisms with which to deal with the inevitable deterioration of quality over time. The first is exit, or leaving the relationship. The second is voice, or speaking out to repair the relationship. Loyalty mediates the choice between the two. The stronger the loyalty, the more likely that the party will choose voice over exit, electing to repair rather than terminate the relationship. Economic logic, too, argues for voice. If success requires collaboration, and collaboration is facilitated by trust and trust accumulates over multiple transactions, this investment is lost when a relationship is terminated. Voice, however, cannot exist without the availability of the exit option. Collaborative success, then, requires both that we invest in our current relationships and preserve alternatives simultaneously.

Thus, we need to find a way to manage the ever-present tension that capitalism, with its demand for value capture, creates in a way that does not imperil the value-creating processes of the larger community. This is difficult to accomplish within traditional management ideology, but central to stakeholder theory.

3.3 The Role of Architecture

One way to describe the contribution of stakeholder theory to the operation of the more cooperative value chain is to think of it as providing the 'systems architecture' for sustainable collaboration. Borrowed from the language of computer systems, and defined by Morris and Ferguson (1993) the 'complex of standards and rules that computer systems rely on', they describe its purpose as bringing order and making interconnectivity possible:

> *It [systems architecture] reduces complexity. It permits clean separation between centralized general purpose functions and de-centralized or specialized functions. It enables management of unpredictability and change ... Good architecture facilitates experimentation and competition: once a framework is specified, multiple approaches can compete without jeopardizing compatibility. And finally, a standard architecture permits many systems and organizations to be developed independently and still work together gracefully.*

In our view, managing the tensions of competing/collaborative systems requires the presence of such an architecture. Stakeholder capitalism provides the underlying conceptual anchor for such an architecture. It provides the values, mindset, and skillset that make sustainable collaboration possible within the larger community.

It does this by altering the scope, timeline, and prominance of the value capture process, within the stakeholder mindset, the value creation dominates — value capture, as an issue, has less saliency because it is seen within the context of an on-going, trust-based relationship rather than as occurring within a sequence of discrete transactions. As Axelrod (1984) has noted in his work on cooperation, the sustainability of cooperation rests on partners' assessment of the value of their shared future, their beliefs about the other party's willingness to collaborate, and their attention to monitoring and managing the on-going relationship.

3.4 Inclusion of Employees and Communities

There are other important advances that one can make by coming to see the value chain in stakeholder terms. Traditionally the value chain idea has focused on supplier-firm-customer relationships. While these relationships are obviously central to the very idea of a business they are not the whole story. Communities, through governments or interest groups, offer important sources of value creation which can hardly be captured in the traditional models.

McDonald's and the Environmental Defense Fund (EDF) developed a joint project around recycling and waste reduction which created value for McDonald's and for EDF. McDonald's saved millions in tipping fees by paying attention to suggestions from EDF about recycling and waste reduction. EDF gained credibility with businesses which led to new projects and furthered their ability to work with corporations and help them to meet the demands of environmentalism.

More obviously still, local governments' investments in education and other forms of social infrastructure directly affect a corporation's ability to put together a deal with other key stakeholders. The market for employees depends in part on a choice of where to live. Governments which invest in corporate parks and research facilities contribute to the value creation process.

While at one level, scholars such as Michael Porter want to take this part of the value chain as given, there is in fact a market for this part of the value chain. The market for new facility location is well documented with governments and communities competing for the best companies. Similarly the inputs of interest groups are increasingly useful in a world of fragmented political power.

Corporations who pay attention to the needs of communities are less likely to run into liability issues and are able to count on the goodwill and trust of local officials. The story of Johnson & Johnson's Tylenol is well known, however, James Burke, former CEO, often says that it is the attention that the company routinely paid to its communities that built the goodwill and trust that allowed them to reintroduce their brand successfully under the harshest of circumstances after the poisoning of eight people. Customers, suppliers, and employees live in communities and their well-being is often inextricably bound up in the well-being of the community. Therefore an intense relationship between company and community is integral to the value creation process.

Traditionally the value chain has viewed employees under the internal control of management. By focusing on the supplier-firm-customer chain, employees were subsumed under 'firm'. Traditional views of the value chain divide up the operations of the firm into functions so that employees' work can be directly attached to a particular part of the chain. The stakeholder view suggests that we see employees as a vital part of the value creation process, as important in their own right. The traditional view sees employees as important instrumentally — only in so far as they contribute to the supplier–firm–customer interaction. Stakeholder Capitalism suggests that seeing employees as instrumental, indeed seeing any stakeholder as purely instrumental, misses the whole point of the value creation process.

4 Operationalizing the Stakeholder Value Chain

How can managers come to use stakeholder capitalism to rethink the value chains in which their firms are enmeshed? While we propose no cookbook approach, there are some general analytic devices and broad lessons that we have learned from our observations of many diverse companies. In reality, adopting a stakeholder mindset leads one to invent new forms of stakeholder relationships. By focusing on the four principles of stakeholder capitalism, we can get out of the trap of sequential value-creation and value-capture — a trap which leaves value on the table as uncreated and uncaptured.

4.1 Managing Discrete Stakeholder Transactions

The first step in understanding how to operationalize the concept of the stakeholder value chain is to see how particular discrete relationships can be managed. How value actually gets created is a function of the behavior of particular stakeholders and, of course, the consequences of that behavior. We suggest that stakeholder behavior be divided roughly into three sets: (1) the actual/observed/current behavior; (2) the cooperative potential; and, (3) the competitive threat.[3]

Actual behavior is a concrete statement of what behavior that stakeholder has done, what decision that stakeholder has taken that affects the firm. For stakeholders with whom the firm has had no interaction, actual behavior is an empty set. For long time customers, actual behavior includes their repeat buying behavior, suggestions for new products, a statement of how jointly the firm and the stockholder's operations are intertwined or not, etc.

It is important to note that a business's current strategy, or at least the sum of its interactions with that stakeholder are an important influence on the actual behavior that can be observed. Of course, the world doesn't revolve around individual firms, so there are other influences on a particular stakeholder, and it is important to understand the world from his point of view. One can even build a model, real or conceptual, that

[3] For a more robust understanding of these ideas see Emshoff and Freeman (1981) and Freeman (1984).

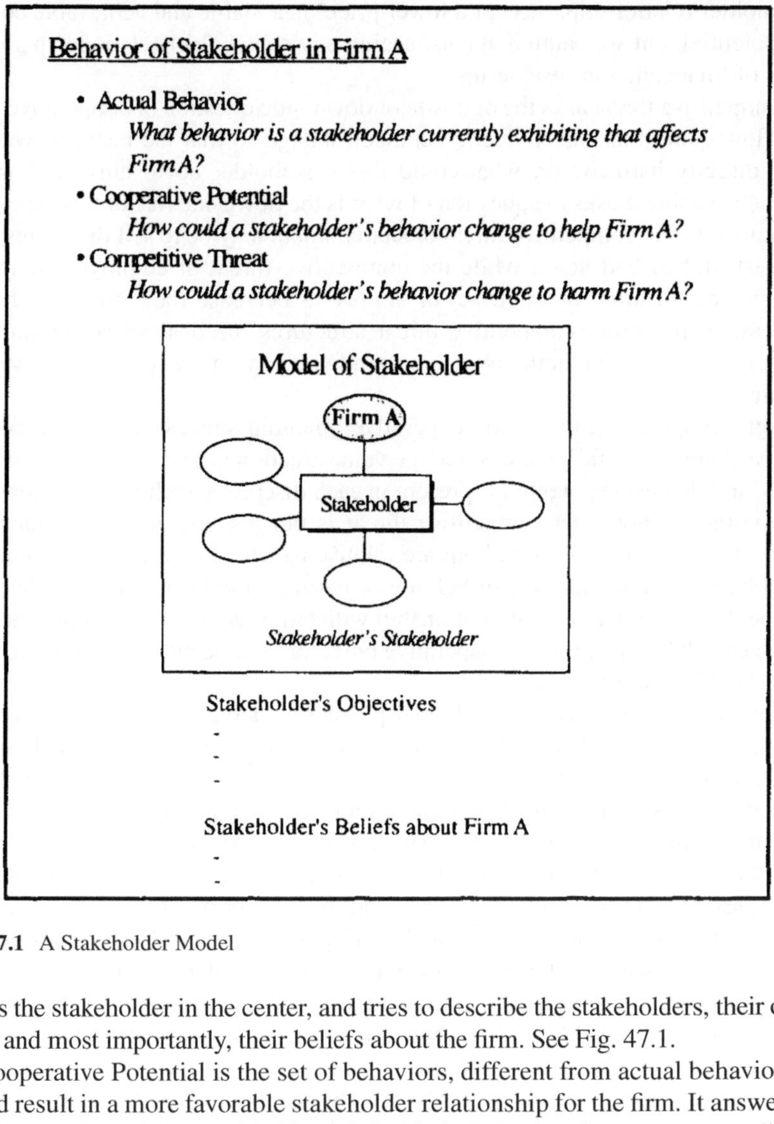

Fig. 47.1 A Stakeholder Model

places the stakeholder in the center, and tries to describe the stakeholders, their objectives, and most importantly, their beliefs about the firm. See Fig. 47.1.

Cooperative Potential is the set of behaviors, different from actual behavior, that would result in a more favorable stakeholder relationship for the firm. It answers the simple question, 'How could this stakeholder's behavior change to more directly benefit the business?' Cooperative potential need not be probable, nor should it be within the bounds of the current strategy. Cooperative potential bounds the best of all possible worlds, and seeks to ramp up the value-creation process. Cooperative potential could of course, focus solely on value-capture. So that cooperative potential for a customer could logically be 'accept a lower price'. Such a focus does not do justice to the spirit of stakeholder capitalism. The whole point of stakeholder capitalism is to create more value for everyone, for the larger community of stakeholders. Of course, there are instances where, within the context of the particular stakeholder relationship, 'accept a lower price' is a viable and achievable cooperative potential, but we caution against seeing any stakeholder relationship solely in terms of financial, win-lose, terms.

Competitive threat asks the opposite or down-side question of cooperative potential. How could this stakeholder's behavior change so that the business would be more directly harmed? or, what could this stakeholder do to hurt the business? Competitive threat asks the question of what is the downside risk of this stakeholder relationship. The competitive threat of shareholders may be to sell the shares on the slightest hint of bad news, while the competitive threat of employees may be to attempt unionization, or to strike, or simply to decrease their productivity. Like cooperative potential, competitive threat structures the deal space within which value is created, and both lay the groundwork for more penetrating strategic questions.

Both competitive threat and cooperative potential suggest that actual behavior may be changed. To fully understand the value creation process we need to see what upside and downside potential is present in each discrete stakeholder relationship. It is important to note that competitive threat is not just the opposite behaviors of cooperative potential, because both are relative to where a particular stakeholder's actual behavior falls in terms of helping or harming the firm. For example, it may well be that a firm has a great relationship with employees, which simply cannot be improved. Then, that group's cooperative potential will be nil, while its competitive threat may be multi-faceted.

There are four important stakeholder profiles as depicted in Fig. 47.2, from our experience in doing this analysis. *Swing stakeholders* can impact the business's objectives both positively and negatively if they change from their actual behavior. Typical swing stakeholders are regulators and other rule-makers, as well as large customers with potential to become partners on more products or issues.

Defend stakeholders are those who are already helping the firm as much as possible (they have little upside potential), and their support is crucial to the maintenance of firm performance. Typical defend stakeholders include employees in a situation where there is already a good and productive relationship.

Opportunity stakeholders are those with high cooperative potential and low competitive threat, stakeholders with whom there is little downside risk. Here we need to be careful, for oftentimes, corporate critics fit into this profile. Managers often assume that they are at risk by dealing with harsh critics such as environmental activists, or customers who complain and are hard to satisfy. But the logic of stakeholder value creation is quite different. Opportunity stakeholders are telling you that there is little they can do (other than their current behavior) to harm the business. This is another way of saying that the firm's current strategy is not working, so there is little risk in trying a new approach.

Monitor stakeholders may well be important, but since there are few options for them to change their behavior, there is little room for strategic maneuvering.

By paying attention to particular stakeholder profiles, managers can undertake a process of reasonably sophisticated analysis to understand where, from a behavior view, value is created, and where there is potential for more cooperation, and where

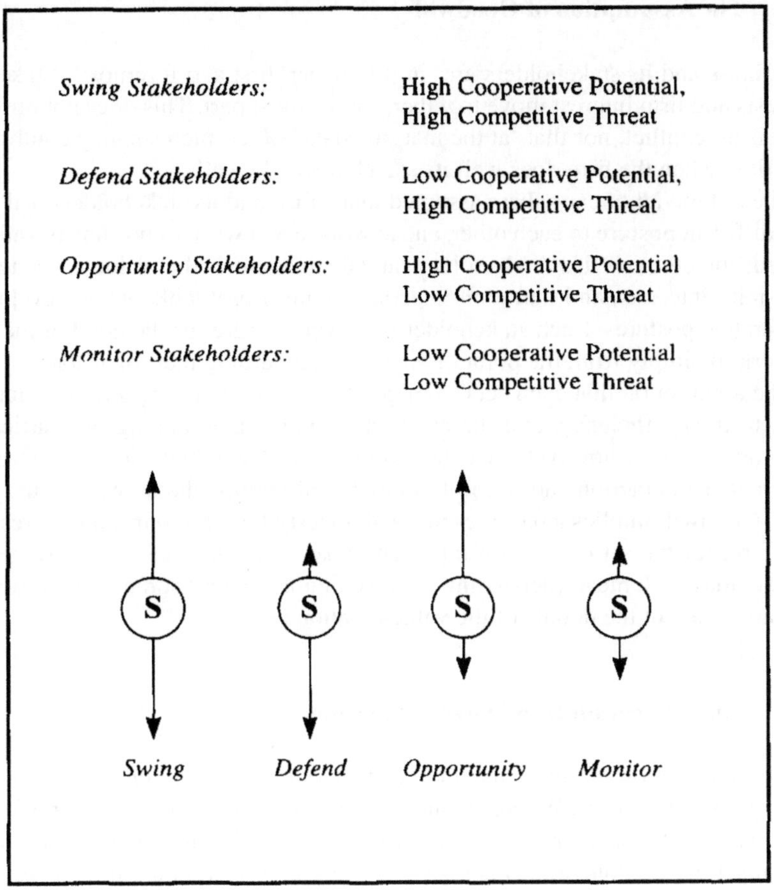

Fig. 47.2 Four Stakeholder Profiles

there is the need to sustain the value that has been created. By looking in detail at the set of transactions that a firm has with stakeholders, we naturally see that by focusing solely on the narrow behavior, we are likely to miss the rich panoply of interactions and partnerships that the stakeholder value chain leads us to expect.

4.2 *Managing the Context of Stakeholder Relationships*

While the Stakeholder Value Chain can be operationalized in discrete behavioral terms, the stakeholder capitalism mindset asks us to go beyond the level of transactions with stakeholders. Indeed, the whole idea of the emerging view of the value chain is to think more relationally, where context is important, evolving and shapable. In our experience there are five main ideas which managers need to keep in mind in considering their analysis of the stakeholder value chain.

4.2.1 The Assumption of Goodwill

A business and its stakeholders are 'in it together, first and foremost'. Stakeholder interests and firm interest move together, for the most part. This does not imply that there is no conflict, nor that 'at the margin' stakeholder interests may conflict. But, over time when the firm does well its stakeholders do well.

The old model of capitalism assumed that a firm and its stakeholders had at best an indifferent posture to each other, and at worst acted with 'opportunism and guile' towards the other parties. Stakeholder capitalism assumes that value is created only from joint interests, and that joint interests are most profitable, when developed in cooperative postures. Each stakeholder is seeking to care for the good of the whole network, if simply from the point of view of safeguarding their own stake.

The assumption that each acts from goodwill rather than opportunism and guile leads to trust, efficiency, and the creation of more value among the parties. It is important to note, however, that the assumption of goodwill does not mean that stakeholder interactions are free of conflict, and simply about 'warm, fuzzy feelings'. Goodwill implies a basic respect and honesty for all of our joint interests, and therefore, for the set of relationships that make up the business. Such respect and honesty may well mean (depending on the cultural context) some rather brutal confrontations about the details of the value-creation process.

4.2.2 Value Creation Dominates Value Capture

The assumption of goodwill also implies that value creation must dominate a concern with value capture. Where capture dominates, the concern is with 'my interest' as opposed to 'your interest', and in a world of relationships there is little hope of focusing fully on 'our joint interests'. That such cooperation can emerge and be stable, even in a world where sometimes our interests are opposed is a very well understood phenomenon. Robert Axelrod's important work on the Prisoners' dilemma has shown that it is rational to forego capturing every last aspect of value in a world where the parties will continually interact, and where neither can tell which interaction is going to be the last one.[4]

4.2.3 Maintaining the Commons

The stakeholder value chain assumes that the process of creating value may well be 'self ordering'. There may well be no one, or no one company, in charge. Complex alliances that depend on virtual networks like the World Wide Web are an example. Good architecture is fundamental to the stakeholder value chain. It follows that each party which benefits from the value chain must invest in the maintenance of the drivers of that chain.

---

[4]Axelrod (1984).

We have a clear example of how not to do this. The Alyeska Consortium responsible for the Alaska Pipeline was faulted for taking everything for granted, taking the commons for granted, and being unprepared when in fact a disaster, such as the Exxon Valdez, occurred. By failing to insure that cleanup equipment was ready and that an adequate response was standing by at all times, the consortium scored a public relations disaster, and Exxon incurred large costs in cleaning up the spill.[5]

At Motorola, the commons is the knowledge and skill base of the employees. Both employees and the company invest heavily in that commons through the creation of Motorola University. All employees from the CEO to the shop floor are required to take a certain amount of training each year. By continual reinvestment, Motorola can be assured of a committed and knowledgeable workforce. Other companies are beginning to use executive development programs as a means to bring together customers, suppliers and employees for joint problem solving. While at one level this looks at current business problems that need a creative solution, in reality such a strategy builds strong relationships for the future.

4.2.4 Communities of Stakeholder Interaction

The stakeholder value chain implies that we have to come to see our stakeholder relationships much in the way that we see the communities in which we are involved. Jack Welch of GE has suggested that business today is 'boundaryless', that executives and employees must work across and within the normal organizational and role-related boundaries that we find in the social world. Whether or not someone is a member of a supplier, a customer design team, or even a critic such as an environmentalist, matters little in today's world. The important aspect is what and how value gets created. Where the value creation process is successful, there is a presumption that it will continue, and thus a presumption that each party is doing its part to maintain the commons, and contribute to the community.

It is especially important to note that the specific inclusion of employees and communities in the value- creation process that stakeholder capitalism makes explicit, implies a broadening of the very concept of what counts as a business. Stakeholder capitalism asks us to see business as more like a multi-faceted community than like a football team, more like a quilting bee and a bam raising than a poker game, and more complex than the tenets of purely competitive strategy could ever imagine.[6]

---

[5] It is unclear whether or not any response would have substantially reduced the effects of the spill. See The Big Spill, *Nova.* Public Broadcasting System Film.

[6] It is easy to underestimate the role of competition in stakeholder capitalism. And, some recent discussions in the political realm have done exactly that. Our point, from a business perspective, is that modem business in a global economy is both cooperative and competitive. Stakeholder interests are joint, and jointly determined, yet the existence of similar businesses and similar networks of stakeholders serves as an important check on the market power of any one factor of production. These are quite complicated issues and need to be spelled out in greater detail than is possible here.

4.2.5 The Corporation as Mere Means

Understanding the stakeholder value chain and stakeholder capitalism puts a new spin on many of the present fashionable business techniques. Many current authors bemoan the loss of corporate loyalty, and go on in somber tones to talk about the necessity of restructuring and reengineering and the like. Our experience suggests that adopting a broad view of the value chain and the value-creating process, offers plentiful opportunities to avoid the painful medicine often necessary by being 'uncompetitive'.

We need to come to see our corporations, not as ends in themselves, with people as fungible parts in business processes. Rather we need to focus on the complexity of the human animal and what each is capable of achieving in concert with others. The stakeholder value chain is ultimately about our businesses becoming mere means to achieve stakeholder purposes.[7]

5 Conclusions

We have suggested that stakeholder capitalism rests on four key principles which have the potential for redefining the way that we think about business. Furthermore we have suggested how such a broadened view of business flows naturally into an understanding of the value chain that makes sense of much of what goes on in today's global business economy. The resulting idea, the stakeholder value chain, says that we need to eschew the old distinction between value creation and value capture, that focusing on value creation in a world of enmeshed relationships is today's key to effective management. We have suggested some analytical tools and ideas for understanding the stakeholder value chain at a transactional level, and finally we have proposed five general rules for understanding the stakeholder value chain in relational terms.

There is a great deal of work to be done by executives and business thinkers to sharpen and shape these ideas into a new view of business and new models of business strategy. We have only scratched the surface in this short article, but the resulting rich world of stakeholder interaction yields much food for thought.

References

Axelrod, R. 1984. *The Evolution of Cooperation*. New York: Basic Books.

Donaldson, T., and L. Preston. 1995. The Stakeholder Theory of the Corporation: Concepts, Evidence and Implications. *Academy of Management Review* 20: 65–91.

[7] Seeing the corporation as a mere means is what Freeman and Gilbert suggest in their 'personal projects strategy' and it is what Tom Peters has suggested is the only way to understand the chaos of modem business. See Freeman and Gilbert (1987) and Peters (1993).

Freeman, E. 1981. Stakeholder Management; a Case Study of the U.S. Brewers and the Container Issue. In *Applications of Management Science*, ed. R. Schultz, vol. 1, 57–90. Greenwich: JAI Press.

———. 1984. *Strategic Management: A Stakeholder Approach*. Boston: Pitman Inc.

———. 1994. The Politics of Stakeholder Theory. *Business Ethics Quarterly* 4 (4): 409–422.

———. 1996. Understanding Stakeholder Capitalism. *Financial Times*, Friday, July 19.

———. 1997. Managing for Stakeholders. In *Ethical Theory and Business*, ed. N. Bowie and T. Beauchamp, 5th ed. Englewood Cliffs: Prentice Hall.

Freeman, E., and D.R. Gilbert Jr. 1987. *Corporate Strategy and the Search for Ethics*. Englewood Cliffs: Prentice Hall.

Hamel, G., and C. Prahalad. 1994. *Competing for the Future*. Boston: Harvard Business School Press.

Hirshman, A. 1970. *Exit, Voice and Loyalty*. Cambridge, MA: Harvard University Press.

Mitchell, R.K., B.R. Agile, and D.J. Wood. 1997. *Toward a Theory of Stakeholder Identification: Defining the Principle of Who and What Really Counts*. University of Victoria, Faculty of Business, Manuscript.

Moore, J. 1996. *The Death of Competition*. New York: Harper Business.

Morris, C., and C. Ferguson. 1993. *Computer Wars: How the West Can Win in a Post IBM World*. New York: Times Books.

Normann, R., and R. Ramirez. 1993. From Value Chain to Value Constellation: Designing Interactive Strategy. *Harvard Business Review*, July–August, 65–77.

Peters, T. 1993. *The Pursuit of WOW!* New York: Vintage Books.

Porter, M. 1980. *Competitive Strategy*. New York: The Free Press.

Rayport, J., and J. Sviokla. 1995. Exploiting the Virtual Value Chain. *Harvard Business Review*, November–December.

Stalk, G., P. Evans, and L. Shulman. 1992. Competing on Capabilities: The New Rules of Corporate Strategy. *Harvard Business Review* 70 (March/April): 57–69.

R. Edward Freeman is University Professor, Elis and Signe Olsson Professor, Academic Director of the Institute for Business in Society, and Senior Fellow of the Olsson Center for Applied Ethics at the University of Virginia's Darden School of Business.

Jeanne M. Liedtka is United Technologies Corporation Professor of Business Administration at the University of Virginia's Darden School of Business and former chief learning officer at United Technologies Corporation.

Chapter 48
Toward a Life Centered Ethic for Business

R. Edward Freeman and Joel Reichart

1 *Introduction*

Our goal in this essay is to suggest that there are several alternative stories or narratives that we can tell about how to come to see business as an institution that is central to the politics of our times, especially our emerging conversation about the environment. In short, we want to suggest how we can see our corporations as institutions that simultaneously do the right thing, save the earth, and make money. We believe that such a story is peculiarly relevant today precisely because we oftentimes see these ends as mutually incompatible.

Our strategy is as follows. In Sect. 2 we shall outline the chief reason that we need a different story about our corporations—a new version of an old argument by philosopher Blaise Pascal. In Sect. 3 we shall suggest that several of our current stories work together to prevent us from seeing business as an institution that is hopeful. In Sect. 4 we suggest three strategies for incorporating environmental issues in alternative narratives of business, and in Sect. 5 we shall argue for one of those narratives as worthy of exploration. This last narrative which we shall call "A Life Centered Ethic" centers on the development of a more fully Darwinian account

Originally published in: *The Ruffin Series of the Society for Business Ethics*, Environmental Challenges to Business, 2, 143–158 © Society for Business Ethics, 2000
Reprint by Springer, DOI 10.5840/ruffinx200027

R. E. Freeman (✉) · J. Reichart
University of Virginia, Charlottesville, VA, USA
e-mail: FreemanE@darden.virginia.edu

© Springer Nature Switzerland AG 2023

S. D. Dmytriyev, R. E. Freeman (eds.), *R. Edward Freeman's Selected Works on Stakeholder Theory and Business Ethics*, Issues in Business Ethics 53,
https://doi.org/10.1007/978-3-031-04564-6_48

of humans and our institutions. It suggests that we come to see business as just one more part of the buzzing, blooming confusion that is the so-called "natural world."

Throughout this essay we shall attempt to engage in what Professor Bryan Norton calls "practical philosophy," that is to root our analysis in the set of issues around how can we address the twin projects of the (re)creation of self and community.[1] We want to place ourselves squarely in the pragmatist camp, but with a slight difference from most modern pragmatists: We want to suggest that we can come to see business as a human institution capable of leading the way towards a more environmentally sustainable future.[2]

2 Pascal's Wager

There is not a single aspect of our world today that can escape the scrutiny of environmental analysis. Pollution of air, water, and land, the production and disposal of hazardous wastes, solid waste disposal, chemical and nuclear spills and accidents, global warming and the greenhouse effect, ozone depletion, deforestation and desertification, biodiversity, and overpopulation are a few of the issues which today's executive needs to understand to be environmentally literate.

We are treated to daily doom and gloom press reports about the state of the earth. Scientists have "discovered" that global warming is or isn't a problem, is or is not caused by solar storms, is or is not related to the emission of greenhouse gases, and so forth. We want to know the answer, the whole truth, "just the facts," about the environment, and we get disturbed by so many conflicting reports.

The truth is this: there is no one truth about the environment. The truth is also this: we have not lived in a way that respects the environment and preserves it for our children's children.

2.1 Our Children's Future: A Wager

Let's assume an optimistic scenario that implies that the gloomy forecasts are all wrong. Maybe there is enough land for landfills for generations to come. Global warming may be elusive. Many chemicals may well be harmless. The destruction of forests may be insignificant and worth the benefits of development. Clean and healthful water may someday be plentiful. And, it may be that we can invent the technology we need to compensate for whatever damage we have actually done to the earth.

---

[1] Bryan G. Norton, "Why I am Not a Nonanthropocentrist: Callicott and the Failure of Monistic Inherentism," *Environmental Ethics*, Volume 17, Number 4, Winter 1995, 341–358.

[2] Freeman, Pierce, and Dodd have discussed this issue in *Environmentalism and the New Logic of Business*, Oxford University Press, forthcoming.

Are you willing to bet the future of your children on this optimistic scenario? If it is wrong, or even partially wrong with respect to say "global warming," then there will be no inhabitable world left for our children. Like Pascal's wager[3] we are going to assume that it is reasonable to bet that there is in fact an environmental crisis. The consequences of being wrong are too great to bet otherwise.

Yet the great majority of responses to the environmental crisis have been at best ineffective. The main response mode has been to marshal the public policy process to legislate that the air and water be cleaner, and to assign the costs of doing so to states, localities, and businesses. Twenty plus years of environmental regulation in the United States has led to "environmental gridlock." There is disagreement and contention at several important levels.

First of all, as we stated earlier, there isn't any one truth about the state of the environment. Many (but not all) individual scientific "facts" are disputable. There is widespread disagreement about the scientific answers to environmental questions, even about how the questions should be stated.

Second, among those who agree about the science around a particular issue, there is still disagreement about the appropriate public policy. Even if we agree that greenhouse gases lead to global warming, we may well disagree that limiting carbon dioxide emissions to 1990 levels will solve the problem.

Third, there is disagreement about the underlying values. How should we live? By exploiting the earth's resources? By conserving the earth's resources? By living with nature? Should we be vegetarians to improve the ability of advanced societies to feed the hungry and use land efficiently? Should we recycle or should we consume green products or should we build an ethic of "anti-consumption," of saving the earth rather than consuming it?

These three levels of disagreement lead to gridlock, especially in a public policy process that purports to base policy on facts rather than values. Overlay these three levels of disagreement on a litigious system of finding, blaming and punishing polluters of the past and the result is a conversation about the environment that goes nowhere fast.

We believe that this public policy process needs to change, that we need to have a better conversation about the environment and the role of governments, but, we

---

[3] Philosopher Blaise Pascal proposed the following wager in the 17th Century: Suppose Christianity is correct. If you are not a believer you are in for some a seriously hot eternity in Hell. So, it is rational to believe in Christianity or to act as if it is correct. Now Pascal's wager doesn't work in its original form because it is a tenet of both Christianity and liberalism that individuals can decide for themselves whether to mortgage their own future in eternity for a few temporal moments of pleasure during life on earth. Our Children's Wager doesn't suffer from the same logical defect because the point is that our children will not get to make those choices if we do not begin to live differently. We have used "children" in the sense of future generations which include but are not limited to existing children. The actual structure of Pascal's wager (or the three related arguments that have come to be called Pascal's wager) is technical and much debated in decision theory. See for instance the essays in Jeff Jordan (ed.) *Gambling on God: Essays on Pascal's Wager,* Lanham, Md.: Rowman and Littlefield, 1994. The argument here is related to what environmental philosophers have called "the protection principle" or "the precautionary principle."

are not willing to wait for these changes to take place. Instead we want to suggest another mode of response to the environmental crisis: business strategy. If we can come to see how business activity can take place, systematically, in environmentally friendly ways, then we can respond to the environmental crisis in lasting and effective ways.[4]

3 The Problem of Mindsets

If Our Children's Future Wager issues a call for rethinking our institutions that has moral force, then we need to examine some entrenched ideas which makes such a rethinking so difficult. We need to examine several pervasive mindsets that can prevent us from entertaining new possibilities, and from designing new frontiers for action. There are several: (1) The Incompatibility Thesis Mindset; (2) The Regulatory Mindset; (3) The Cost-Benefit Mindset; (4) The Constraint Mindset; (5) The Sustainable Development Mindset; (6) The Greenwash Mindset. We want to briefly set forth each one and then suggest that we set them to one side.

(1) *The Incompatibility Thesis Mindset*

Much that is written about the environment assumes that a concern with the environment and the idea of markets are mostly incompatible. In fact many environmentalists squarely place the blame for the mess we have made of the planet on the shoulders of those who defend markets, capitalism, and business. They claim that these ideas have spawned institutions which are incapable of taking account of environmental values, and that we need new institutions that are redesigned from the ground up that are built on and preserve environmental values.

Let us call the idea that capitalism and markets as we know them are incompatible with a deep concern with the environment, "The Incompatibility Thesis."[5] The Incompatibility Thesis is said to be true on two related fronts. First of all markets are to blame for environmental damage because in its concerns with profits, capitalism institutes a system of production that gets the cheapest resources possible without regard for the environmental damage done in marshaling these resources. Whether or not these resources can be sustainably developed is not a concern of capitalist businesses, at least not operating in their market-driven states.

---

[4] Our approach is radical—at least for business theorists and environmentalists. Most writing about the environment acts as if business is evil, and most writing on business acts as if business is separate from the environment. We want to stake out some new territory.

[5] The Incompatibility Thesis is related both to the Separation Thesis and to what Freeman has called "the business sucks story." For more argument around each of these topics see, R. Edward Freeman, "The Politics of Stakeholder Theory," *Business Ethics Quarterly,* Volume 4, Number 4, 1994; and, "The Business Sucks Story," The Darden School Working Paper, Charlottesville, Va.

Secondly, the Incompatibility Thesis blames business for environmental damage because it influences, inculcates, depends on, and preys on a culture of consumerism. Business thrives only when citizens take on the additional role of consumer. Business and capitalism cannot work if consumers do not want the products of its labor. And, the argument goes, for capitalism to thrive, it needs an ethic of always consume more. The engines of growth and development must be fed by consumer demand. And, institutions such as advertising are aimed at insuring that we come to see the latest sneaker, the latest computer wizardry, the latest food product as things we simply cannot flourish without.

Such a leftish indictment of markets and business is usually accompanied by a litany of environmental problems such as global warming, pollution, the ozone hole, and biodiversity. The bell is rung and we take up the cudgel against the evil capitalist in order to save the planet.

Unfortunately, we don't fare any better if we move across the political spectrum to the right. Defenders of markets, capitalism, and business decry the left liberal environmentalists and their misunderstanding of how markets work. They assert that markets are about the efficient use of resources, and that if there are any environmental factors to be dealt with they will be included in the price of the resources, and the prices of the products. So, we should just kick back and let the market take its invisible hand course.

If consumers want a cleaner environment then they will be willing to pay for it and furthermore they will demand products and services with "clean environment" attributes. Defenders of markets could then point (though they rarely do) to the emergence of "green products and services" as evidence that market processes will cleanse us of any environmental worry. Furthermore, this argument goes, even if there were any environmental problems, which there probably are not, the best way to solve them would be to institute market oriented reforms such as pollution permits.

Such a defense of the market, capitalism and business is usually accompanied by a "there really isn't any scientific basis to environmentalism anyway" kind of argument. In its extreme form we are treated to Julian Simon's "Don't worry, be happy" flavor of optimism about the grand and glorious nature of human beings and their capacity to invent absolutely any technology that they may need. Many on the right adopt a kind of Pollyanna Principle which assumes that everything will be okay, and even if it isn't we ought to let markets, as they understand them, work, and then things really will be okay.

We wish to question the certainty with which others offer these positions. We understand the political reasons for the certainty for there is a lot at stake between left and right, but as tools for analysis these arguments are sorely inadequate. We want to suggest the following. First and foremost, both of these arguments appeal to a view of the discourse of business which we have described elsewhere as The Separation Thesis, the view that we can meaningfully, and usefully discuss business, capitalism, and markets separately from the ethical issues of how we are to live our lives. Because these arguments accept the Separation Thesis they fail to understand the complex relationship between business, as we know it, democracy, as our fundamental political ideal, and the fact that we are pretty complicated creatures.

(2) ***The Regulatory Mindset***

This mindset sees the environment as a part of the business government relationship to be spelled out in terms of regulation or public policy. The Regulatory Mindset says that the best way to take care of the environment is through the public policy process that produces laws and rules with which business must comply. It discounts the possibility and the wisdom of voluntary initiatives that stem from deeply held environmental values, or even the desire to respond to the environmental preferences. While the recent history of concern with the environment has usually meant the passage of laws and their attendant regulations, the debate today goes far beyond a regulatory mindset. Regulation lags the real world, and regulation inevitably entails unforeseen consequences. Our question for the regulatory mindset is: Are you confident that government, as it currently works, will create a sustainable future for your children?

(3) ***The Cost/Benefit Mindset***

The Cost/Benefit mindset is sometimes closely related to the Regulatory mindset simply because many regulatory regimes use cost/benefit methods to determine "proper" regulations. The Cost/Benefit mindset says that cleaning up the environment or making products and services more environmentally friendly has costs and benefits. And, we should go only so far as the benefits outweighing the costs.

There are several problems with this view. The first is that if you focus on costs and benefits you will fail to use "innovation." The argument is similar to the quality approach. By focusing on the cost of quality, managers make wrong decisions. Instead by focusing on quality processes like six sigma quality, human innovation takes over and drives quality up and costs down. The Cost/Benefit mindset says that there is a contradiction between "environmentally friendly" and costs. Many companies are discovering that by adopting one or more of the shades of green that we recommend here, they are making money and becoming more environmentally friendly. By focusing on costs and benefits, managers are inevitably led to ask the wrong questions.

The second problem with the Cost/Benefit mindset is that it assumes one particular set of underlying values: economic ones. Many environmentalists, executives, and other thinkers have questioned the priority of our current ways of thinking about economics. All value is not economic value, and anyone who believes that it is, is trying to get us to live in a certain way. Does the last gorilla have just an economic value? What about the beauty of the Grand Tetons? What about the futures of our children? Human life is rich and complex and is not reducible solely to an economic calculation. It is degrading to all of us to think that we only value people and things in economic terms.

(4) ***The Constraint Mindset***

Still many will argue that the main purpose of business is to create and sustain economic value, and everything else from ethics, to the environment, to meaningful work, is best viewed as a side constraint. The business of business is business. Anything else should be viewed as not the main objective of business.

There is a nugget of truth here, like there is in all of these prevalent mindsets. Economic value has been the main focus of business, and other kinds of value have been seen as constraining a kind of unfettered capitalism, driven by the urge to win, succeed, and compete. However, a more thoughtful analysis of "economic value creation" shows that it is impossible to separate out "economic, political, social, personal" aspects of value. When the employees of Delta buy a jet for the company, when Johnson and Johnson recalls Tylenol, when Body Shop employees volunteer to help the homeless, when Mattel donates money to the part of Los Angeles destroyed by riots, all of these actions imply that it is possible for a company to be driven by economics and by ethics. No one is arguing that economics is unimportant, but we are insisting that the reduction of all human value creation/value sustaining activity to economic measures misses the mark. Business does more, as Adam Smith realized, and to reduce capitalism to economics endangers our free society.[6]

(5) *The Sustainable Development Mindset*

It may seem strange to lump what is supposed to be a way to save the Earth with mindsets that prevent environmental progress. Obviously, not all discussions of sustainable development act as barriers but one recent discussion simply misses the mark. The Brundtland report, the basis of the RIO Earth Summit in 1992, called on governments to redefine economic activity to become sustainable. The problem with this view is that it calls on governments to have an intrusive role in the process of value creation, and if we have learned anything from the collapse of state socialism, governments and centralized approaches do not work very well. Ultimately, a worldwide regime of environmental cooperation could become a worldwide hegemony of democratic freedom, especially if combined with the other mindsets. Decisions on the future of whole industries and companies could become a matter just of government's beliefs about what is sustainable.

Recall our view, that there is no one truth about the environment. We believe that it is necessary to adopt a radically decentralized approach which focuses on shared values, and a conversation about those shared values. If such an approach is not viable, then we should see the heavy hand of the state as part and parcel of our failure to integrate business, ethics and the environment.

(6) *The Greenwashing Mindset*

The greenwashing mindset pervades many discussions of the environment. Characteristic of it is the view that business could never act on values other than profit maximization. And, whenever we see a company engaged in something that looks like it might be good for the environment, we should be deeply skeptical.

---

[6]Adam Smith was primarily concerned about justice.

Really, if truth be told, the company is probably trying to make money, or avoid some future cost, or engage in other narrowly self-interested schemes. Many environmental programs at companies are, on this view, cleverly disguised attempts to be seen as green, while really continuing in an environmentally destructive mode.

Many times when we have presented these ideas to groups that contained people who were deeply committed to environmental values, but had little real contact with the inner workings of business, there has been an assumption that "business is bad."

Now it is surely true that there are attempts to greenwash—portray trivial changes to products, services, and processes in grand and glorious environmental terms. And, we should look carefully at such claims. However, the assumption that therefore all business attempts at environmental action are suspect simply does not follow.

We want to suggest that we be skeptical of all grand environmental claims, whether they be from business, government, environmental groups, or scientists. The arena is very uncertain and complex. However, the greenwash mindset makes our task impossible, so we shall set it aside. Of course, businesses want to make money, but it doesn't follow that the environment must be left out of the equation, or that profit is the only value which counts.

When you combine the six mindsets, or stories, there is small wonder that we have created a presumption that business and the environment do not mix—or that they mix under special circumstances, such as when we have "good guys at the top," or "executives who really look to the long term," or "enlightened managers." Clearly, in the normal senses of the words, "business" and "environment" are strange companions. What we need is a way, as many of the people here are busy inventing, of combining talk about business with talk about the environment and talk about ethics.

4 Three Strategies for a New Narrative

There are at least three broad strategies for spinning narratives that combine business, ethics and the environment. First, we can add environmental values to our common understanding of business and ethics. Second, we can enrich and revise our common understanding of business. Finally, we can enrich our common understanding of ethics. Each strategy focuses on a different problem, and we want to suggest that we come to see all strategies as necessary for a better conversation about business, ethics and the environment. Hence, we want to eschew the approach that looks for the one and only one theory of value, or theory of the corporation, which will tell us once and for all how business, ethics and the environment are to be connected. Rather, we need to see how each approach allows us to ask different questions, and indeed to live differently. Let's turn to each of these strategies and see how business, ethics and the environment can be connected. We can only sketch the connection in the brief space available.

4.1 The Status Quo Strategy and the Ethic of Conservation

Let us take the ordinary understanding of business and the ordinary understanding of the relevant moral discourse that applies to business and ask: Can we envision a narrative that adds a concern with the environment? Let us call such a strategy "The Status Quo Strategy." The strategy yields a "Conservation Ethic" which can be used to tie together business, ethics and the environment.

Suppose that we understand business in a way that most economists do, as the main vehicle through which wealth is created in society, and in particular we understand the corporation as the bailiwick of the shareholders. Managers' tasks include maximizing the value of the firm for shareholders, subject to the normal constraints of ethics, custom, and law. The best known proponent of this view is the Chicago economist Milton Friedman.

Ethics, on this view, serves as a side constraint to moderate greed, fraud, and other socially undesirable traits. Friedman claims that law and ethics and custom must be obeyed if maximizing profits is to be economically feasible.[7] Now most business ethicists believe that Friedman's view is flawed in one way or another, and that may well be true. However, it is not easy to dismiss. Depending on how one parses "law, ethics, and custom," one can make the case that most any ethical view is compatible with Friedman's view.

For example it is easy to see how to make environmental concerns compatible. Even if business is concerned primarily with maximizing shareholder value and the moral constraints on such a view, prudence indicates that there is a conservation principle that demands that we take environmental concerns into account. Even this most conservative understanding of business is not immune from the green revolution, for it is easy to see that if we have a presumption of the firm's existence through time, then self interest dictates that we not destroy resources that will be useful to us in the future.

In fact, the conservation ethic has a long history which has been amply chronicled by others, such as Roderick Nash. While "conservation" is not an easily unpacked term, note that we are using the term in a strongly anthropocentric way. Nature is to be conserved precisely because it is useful to humans.[8]

What would corporations look like that began to adopt such a conservation ethic? For starters they would adopt a strategic posture that we have elsewhere called "Legal Green" or "Market Green." The legal green strategic posture claims that corporations can find sustainable competitive advantage in obeying the law. For example, they can influence the standards, or they can use the standards as a tool to drive innovation. Market green is a slightly different shade which claims that there is competitive advantage in paying attention to the environmental preferences

---

[7] Of course, Friedman's view is subject to many interpretations. There is a vast literature here.

[8] For the idea of strong versus weak anthropocentrism see Bryan Norton, "Environmental Ethics and Weak Anthropocentrism," reprinted in C. Pierce and D. VanDeVeer eds., *People, Penguins, and Plastic Trees,* 2nd ed. (New York: Wadsworth Publishing, 1995), p. 157.

of customers. Now not all of these preferences may be good for the environment, but as out knowledge increases, and as our environmental awareness is raised, companies, under our status quo assumptions will follow customer demand.

Many have argued that a conservation ethic does not go far enough, yet it is difficult to see how any environmental issue cannot be undergirded by an appeal to conservation. Even biodiversity can be seen to be a matter of conservation, if we are willing to entertain the causal claim that the number of species affects humans' welfare.

There are many questions for the status quo strategy and the ethic of conservation. Not the least of these involve a complex set of scientific and policy related claims about the relationships among species, ecosystems, and humans.

4.2 The Enriched Business Discourse and Stakeholder Capitalism

A second broad strategy for a new narrative consists of enriching our understanding of business so that it is easier to talk about values and ethics. On this view values and ethics cease to be a side constraint and instead serve as the foundation for thinking about business.

There are many different methods for enriching the discourse of business to include ethics. Wood encapsulates a tradition of seeing the corporation as engaged in social performance and social responsibility as well as economic responsibility.[9] Werhane has suggested that we apply a strict language of rights inside the corporation to employees.[10] Donaldson and Dunfee have suggested that we come to see firms as complex integrated social contracts.[11] Solomon has suggested that we use an Aristotelian theory of virtues interpreted for the modern corporation.[12] Hartman has argued for coming to see the corporation as a commons to be preserved by all.[13] Each of these theorists can accommodate the inclusion of environmental values and concerns in their theory, though some have an easier time of it than others.

---

[9] Donna J. Wood "Corporate Social Performance Revisited," *Academy of Management Review,* October 1991, 691–718.

[10] Patricia H. Werhane, *Persons, Rights, and Corporations,* Englewood Cliffs, N.J.: Prentice Hall, 1985.

[11] Thomas Donaldson and Thomas Dunfee "Integrative Social Contracts Theory: A Communitarian Conception of Economic Ethics," *Economics and Philosophy,* volume 11, 85–112.

[12] Robert Solomon, *Ethics and Excellence*, New York: Oxford University Press, 1991.

[13] Edwin Hartman, *Organizational Ethics and the Good Life,* New York: Oxford University Press, 1996.

We want to suggest that one variant of this approach makes the addition of environmental values quite simple.[14] We have argued for this approach elsewhere so we shall only briefly mention it here. That is, we need to see the firm as a set of relationships among those parties who have a stake in the firm. For starters, these stakeholders are customers, suppliers, employees, financiers, and communities. The main idea of stakeholder capitalism is to see the firm as primarily a cooperative venture among stakeholders. Capitalism works precisely because it allows us to create together what no individual could create alone. Capitalism works because these complex human beings try to realize their values by working together. It is easy to see how environmental values can be a strong part of such a conception of the firm. We might define a shade of green as "stakeholder green" whereby a company seeks to respond to the environmental preferences of its key stakeholders.[15]

4.3 The Enriched Moral Discourse Strategy and Contextual Ethics

A third broad strategy consists of enriching our understanding of ethics itself, a view favored by many in environmental ethics. Such an "Enriched Moral Discourse Strategy" is complicated and offers much room for interesting and exciting research. Some of the lines of thought that have been opened here include so-called "expanding circle arguments"; ecofeminism; Wendell Berry's analysis of place; and deep ecology.

This strategy begins with the view that the modern moral discourse of rights, duties, justice, etc. is misguided or possibly bankrupt for constructing an effective environmental ethics. Human beings have become disconnected from the natural world in much the same way that they have become disconnected from each other and "traditional" moral philosophy has had little or no impact on reversing this dysfunction which is the source of the Earth's environmental woes. Environmental ethics should refocus its efforts toward healing the source of the dysfunction for the natural world can only be saved when the disconnections between, men and women, humans and communities, or humans and nature are rewoven. The central idea in

---

[14]The history of the stakeholder approach is documented in R. Edward Freeman, *Strategic Management: A Stakeholder Approach,* Boston: Pitman Publishing, 1984. For a particular analysis of what he has called "stakeholder capitalism" see "Stakeholder Capitalism," *Financial Times,* 16 July 1997; and Freeman and Jeanne Liedtka, "Stakeholder Capitalism and the Value Chain," *European Journal of Management,* 15: 1997, on. 286–296.

[15]Paul Shrivastava, and Edward and Jean Stead have shown how corporate strategy can be "greened" or made "sustainable" using the stakeholder framework. It is most ironic that strategy theorists have talked for years about "sustainable competitive advantage," where they have not had an environmentally sensitive usage of "sustainable" in mind. See Paul Srivastava, *Greening Business* Cincinnati, Ohio: Thompson Executive Press, 1996; and Edward Stead and Jean Stead, *Management for a Small Planet,* 2nd Edition, Thousand Oaks, Calif.: Sage Publishing Inc., 1996.

each of these sometimes conflicting visions is a call for a reflexive analysis of moral discourse, which may entail a reversal of center and margin.

In short, we must (re)contextualize ethics along a variety of dimensions. Ecofeminists ask us to put an account of gender and dominance relationships in the center of our ethical space. Wendell Berry asks us to situate the human condition in a place—and in particular in a community with a relationship to the land, and the process of growing our food. Deep ecologists ask us to consider how to make the Earth a part of our moral discourse—to live more fully in tune with the Earth and its creatures.

We have explored a number of these alternatives and their connections elsewhere so we want to focus on one particular realization of the contextual strategy.[16] It is a realization that is not quite captured by the others, but one that is in need of further debate and conversation. It is also one in which it is easier to envision a role for business.

5 A *Life-Centered* Ethic *for* Business

There are two main academic texts which we wish to use to describe "a life centered ethic for business." The first is James Rachels' *Created from Animals: The Moral Implications of Darwinism,* and the second is William Frederick's *Values, Nature, and Culture in the American Corporation.*[17] Each is important in framing the kind of conceptual model that we are trying to articulate. In general terms we want to suggest that we adopt a more fully Darwinian conception of humans and our institutions, including business, and we want to do so, hopefully, without committing any grievous philosophical errors.

In trying to come to terms with Darwin, James Rachels claims that Darwin undermines the most fundamental notion of ethics: the idea that human persons have dignity. Rachels suggests that human dignity arises out of either our theological beliefs that humans are God's creatures, made in the image of God, and so forth, or our belief that humans are separable from the rest of the creatures on Earth because they are rational. According to Rachels Darwin showed how we no longer had to believe in William Paley's "watchmaker." Nature and the humans in it evolved. Time and variation are the culprits, not a supernatural power. In addition, by naturalizing us, Darwin showed how we are just one more set of creatures, not different in kind from all of those species that we see around us. Rachels claims that such a move undermines the idea that only humans use reason.

[16] R. Edward Freeman and J. Reichart, "Four Strategies for Environmental Ethics," The Darden School Working Papers, Charlottesville, Va., 1996.

[17] Rachels' book was published by Oxford University Press in 1990, and Frederick's book by the same press in 1995.

Thus, a more fully Darwinian account of humans is one that does not rely on our specialness, but rather it is one that situates us fully in the long line of descendants from mitochondrial Eve through the DNA replicator enriched river that is life. Now Rachels suggests a kind of moral individualism to replace the idea of human dignity. As near as we can tell this is the view that we treat others based on their individual characteristics, rather than on their membership in a species or their characteristics based on that membership. While moral individualism raises many questions, we can take Rachels' analysis and invent other Darwinian conceptions of ethics. Moral Individualism is but one kind of life-centered ethic.

A life centered ethic is not the argument that we have the same respect for all forms of life that we do for human life, nor an argument that ecosystems have standing thus it is not necessarily a deep ecological view. Rather, it is the argument that we begin to construct an enriched moral discourse from the basics of our understanding of how life emerges and thrives.

William Frederick has argued that there are three value clusters which we can use to understand our institutions. He calls these economizing, power aggrandizing, and ecologizing values. While the first two are "natural" ones to analyze corporations, the last is not. However, if we look in nature we see economizing and ecologizing values at work hand in hand. Frederick's claim is that we must come to see our institutions, and our corporations in particular, as just one more "natural" biological phenomenon. He claims that "the central challenge is to blend and harmonize these two value systems—economizing and ecologizing—in ways that sustain human purposes within the constraints and opportunities of the still evolving system of nature and culture." (p. 166) we want to suggest that Frederick's call is no less than one to develop a new understanding of ethics, a new understanding of humans as biological creatures, and here we must include our institutions.

We are not claiming that there are a number of "oughts" which follow from a fully biological, fully evolutionary, fully Darwinian view of humans, rather we are suggesting how we might come to see each other and our life-cousins, and how we might come to treat each other and our life-cousins given such a view. We simply need a new story that lets us understand life itself in a biological sense.

At some point in the past, cells began to replicate themselves and life emerged. Through the magic of DNA and time our current world came into being. Our role in that evolution is no different in kind from the role of phage, muskrat, or elephant. In short, let us view ourselves as a kind of "souped up chimpanzee" with whom we share a large percentage of DNA. On this view, questions of humans' purpose in the grand scheme of life are misguided attempts to accrue self-importance and privilege at the expense of something else. There is no Watchmaker, only watches that work.

There are three principles which seem to be compatible with this Darwinian view of the world.

> *The Principle of Connectedness.* Human life is biologically dependent on other forms of life, and on ecosystems as a whole, including the nonliving aspects of ecosystems. Therefore, humans must establish some connection with life itself and respect that life itself exists because living things exist in some state of cooperation and coexistence.

> *The Principle of Ecologizing Values.* Life itself exists in part because of the ecologizing values of linkage, diversity, homeostatic succession, and community. There is a presumption that these values are primary goods to be conserved.
>
> *The Principle of Limited Competition.*[18] "You may compete (with other living beings) to the full extent of your abilities, but you may not hunt down your competitors or destroy their food or deny them access to food ... You may compete but you may not wage war."

The Principle of Connectedness does not suggest that humans make no impact on the world. Indeed this would be impossible simply because according to this worldview, we are a part of the world. Rather, it suggests in the words of Gary Snyder that we "go lightly." It is not that trees have rights nor that we should minimize the harms to the land, but that we should walk with humility and awe and a sense of oneness with the world, in addition to our keenly human senses. This principle says that human life is biologically dependent on other forms of life and on ecosystems as a whole.

The Principle of Limited Competition emphasizes connectedness for it is here that we get the idea that we may not flourish at the expense of our competitors. Daniel Quinn divides the world into two cultural groupings, Takers and Leavers.[19] Leavers take from nature that which they need to live and leave the rest. They accumulate knowledge and that knowledge is passed down from generation to generation. Perhaps more importantly, Leaver cultures obey the Principle of Limited Competition. By adhering to the law of limited competition Leaver cultures live in symbiotic harmony with nature and the natural world as a whole thrives.

Taker cultures, among other things, have exempted themselves from this law which has resulted in the negative environmental consequences, such as declining species diversity, deforestation, declining world fisheries, ozone depletion, global warming, etc., that are motivating much of the discussion regarding environmental ethics. For all life to flourish our human activities must abide by The Principle of Limited Competition. This does not mean an end to human development, rather it means a development that preserves habitat and species, one that is sensitive to our life-cousins.

Human destiny is a wholly human concept. It is ours to make rather than to find in nature. A Life Centered Ethic would locate at least part of that human destiny in the rich understanding of the processes of life itself, for how can we walk lightly when there is so much that we have failed to understand. Whether it is mapping the human genome, studying the behavior of finches, or trying to build a sustainable space colony, a life-centered ethic focuses our attention on the fact that we are in part just DNA replicators, and if we are to make anything of our heretofore rather short stay on Earth we need to understand these forces.

A Life Centered Ethic could be called pragmatic in the following sense. It is just us here in a rather brief moment along the cosmic timescale. What we self-referencing, language using humans make of our opportunity depends in part on how we come to see ourselves and our relationship with other forms of life. If we

---

[18] Daniel Quinn, *Ishmael* (New York: Bantam Books, 1992), p. 129.

[19] Ibid.

see this relationship as one of "us" for most purposes, rather than "humans" as "us" and everything else as a "them," then we fuse an understanding of our world that is respectful of life itself.

How could these principles be applied to business, or better, how would we understand business if we took these principles seriously. We want to be very practical here. These are not principles in the sense that we need to reorganize the structure of society. Rather they are principles that are useful in retelling our story as more Darwinian.

The first step in creating a life centered ethic for business is to use these principles to analyze a particular business and its effects. We need to map as a routine part of business the ecological effects of the business. We don't mean here just doing the kind of environmental impact statements that have become a political routine. Every interesting notion of morality begins with accepting responsibility for the effects of my (our) action. Responsibility ought to imply knowledge. We can only avoid irresponsible choices when we meaningfully understand the effects. We mean that we must choose actively to be responsible for the effects of our actions on others and the process of life.

Each business will have "a design problem" in the words of William McDonough. How do we create the value that we now create, and "go lightly" or "do no harm" or "repair any damage that we have done"? Imagine an automobile company that sets out to design a "car" that ameliorates global warming. Imagine a corporate farm that sets out to design a food growing process that repairs topsoil erosion. The Principle of Connectedness implies that these are the kinds of questions that business must begin to ask.

Yet, there is another set of questions as well. The Principle of Limited Competition may well rule out certain industries—certainly those who thrive on the destruction of life or habitat. We believe that these questions are easier to face, though not to answer, with a life-centered ethic than one based on human dignity.

There is much more to be said about a life centered ethic. It is obviously connected to much of the environmental ethics literature, but there are some differences. A life-centered ethic is not nonanthropocentric. It starts with a view, a very particular view, of humans and it builds from that view. It says that we should have a more fully Darwinian view of humans—one that does without the idea of the special nature of human dignity, and one that pays attention to ecological values. But, we must understand the impact that humans are having first of all on ourselves.

A life centered ethic is pragmatist in its conception. It says that it is up to us to make something of ourselves, to invent solutions to the riddle of self and community, and that we need to heed Dewey's dictum to be experimental. It recognizes the role that business must play if we are to create a sustainable future for ourselves within the life process.

A life centered ethic is similar to what E. O. Wilson has called the Biophylia Hypothesis, "the connections that human beings subconsciously seek with the rest of life."[20] Our restoration of the world must cross the divide of science and religion,

[20] Edward O. Wilson, *The Diversity of Life* (New York: W. W. Norton, 1992), p. 350.

so that those who see no purpose other than Darwin, and those who see a Divine hand in the beak of the finch are united in "reweaving the wondrous diversity of life that still surrounds us."

R. Edward Freeman is University Professor, Elis and Signe Olsson Professor, Academic Director of the Institute for Business in Society, and Senior Fellow of the Olsson Center for Applied Ethics at the University of Virginia's Darden School of Business.

Joel Reichart is Professor in the Management Department at Berkeley College.

Chapter 49
Create a New Story About Business

R. Edward Freeman

We have a unique moment to make a lasting difference in corporate practice. This is a moment we must seize.

IMAGINE YOU ARE a director of a company that creates such great products and provides such terrific service that extreme customer loyalty is the norm. The firm's employees are energized contributors who live the firm's values — all the best talent wants to work for your organization. You have the brightest and most innovative managers who keep you ahead of the competition. Whenever there are potential problems with a product or practice, employees throughout the organization raise the issue, confident that when the information gets to the right place, the company will listen and act quickly to remedy the situation. Your suppliers view you as their best customer — in every sense of the word — and they are always thinking of ways to make your company and its products better. Your community shudders to think what life would be like without your company in their backyard. Business students across the globe study your company as an example of best practice, and the most common question your CEO gets asked by the media is: "Why can't other companies be like yours?"

Given the above, two other facts are almost certainly true: first, your firm's shareholders realize strong, sustainable returns; and second, as a director you are confident that continued company growth — and not a subpoena or civil suit — is in your future.

Originally published in: *Directors and Boards*, Issue: 2005 Second Quarter, 23–27 © Directors and Boards, 2005
Reprint by Springer

R. E. Freeman (✉)
University of Virginia, Charlottesville, VA, USA
e-mail: FreemanE@darden.virginia.edu

© Springer Nature Switzerland AG 2023

S. D. Dmytriyev, R. E. Freeman (eds.), *R. Edward Freeman's Selected Works on Stakeholder Theory and Business Ethics*, Issues in Business Ethics 53,
https://doi.org/10.1007/978-3-031-04564-6_49

I am sure this scenario strikes some people as a dream. I do not disagree, but I believe the dream is an eminently realizable one, as was the dream of sending people to the moon and returning them safely to earth — an accomplishment that involved far less computing power than is contained in the lap top with which I am writing this article. Achieving great things requires that great effort accompany our aspirations. This is what leadership is all about.

1 The First Step

The first step toward making such companies a reality is for leaders to actively incorporate values and ethics into all of their business decisions and to encourage others to follow their lead.

I have been teaching business ethics at the Darden School for nearly 20 years. Since 1988 we have included business ethics as a required core discipline in our integrated program —just as we include accounting, finance, marketing, and other functional areas. Our students are trained to recognize that all business managers' decisions include dimensions of ethics and values. This is no different for executives and directors.

Admittedly, events of the last 3 years give some grounds for doubting that ethics can be integrated as a core discipline of business practice. It is important to recognize that viewing "business ethics" as an oxymoron has great traction in the larger culture — and this is a view that we must not pass on to the next generation of leaders.

I believe we can make an enduring difference in the coming years, and ironically, without the recent spate of malfeasance, I might not be so optimistic. The upside of the current scandals is that they have led to a widespread recognition among business leaders that laws and regulations alone are necessary but not sufficient to strike at the root of questionable business practices.

Other groups — stockholders, customers, regulators, lawmakers, educators, the media, and the general public — are very engaged with this issue as well. Given this widespread level of engagement, we have a moment — a unique moment — to make a lasting difference in corporate practice. This is a moment we must seize.

My optimism does not rest upon naïveté — it is supported by recent concrete action. Of particular note, I am encouraged by Business Roundtable and its 160 members, who just over 1 year ago launched the Business Roundtable Institute for Corporate Ethics. I serve as the institute's academic director (see sidebar on page 25 for more information on the institute).

I am further encouraged that, even in the age of branded business schools, IB leading scholars from competing schools took the bold action to work together and join the institute, recognizing that the importance of the task at hand requires the collaboration of us all.

More directly, I am encouraged by the genuine desire for change that I am hearing in my work with senior executives and directors. In fact, this past summer

11 Fortune 500 CEOs came together for an entire day at an institute seminar to tackle tough ethics issues — a CEO discussion of ethics among peers that I believe to be without precedent.

So far, we are making good progress toward telling a new story about business, but there are many chapters yet to be written. Directors have a particularly important role to play.

A Uniting of 'the Letter and the Spirit'

Large companies are being required to devote dramatically increasing levels of resources to comply with regulatory laws. While attention to compliance is essential for the long-term existence of any firm, there is a potential pitfall to this trend: failing to develop an enterprise perspective in regard to ethics.

Organizations must progress from a compliance viewpoint to an enterprise ethics approach. Here are the reasons why:

- A compliance viewpoint focuses on the law; an enterprise ethics approach focuses on creating value for stakeholders. Compliance is a set of tasks externally imposed upon the company — it is required but has little to do with a company's mission or core purpose. An enterprise approach makes ethics integral to the firm's strategy and goals.
- Compliance applies to a particular set of decisions. An enterprise approach sees ethics as essential to every decision made by the firm, much like marketing and finance. An enterprise approach must be embraced throughout all levels of the organization — led by directors and executives, but lived by all. A compliance approach will not make your business more successful; a solid enterprise ethics approach might.
- Compliance guides employees to know what is legal and what is not, while an enterprise ethics approach engages the moral dimension of employees, making them self-aware stakeholders in the value the firm creates every day.
- Compliance relies on the threat of being punished, while enterprise ethics is based upon leadership and values. Leaders breathe energy into the workplace when they connect people's values to the mission of the firm — uniting ethical and economic concerns.
- Compliance has to do with the letter of the law, while enterprise ethics has to do with the letter and the spirit. And should history repeat itself, and future scandals lead to further compliance reforms, those organizations with an enterprise ethics approach will deserve the public trust.

Compliance is a necessary and foundational part of an enterprise ethics approach. Yet compliance alone is not sufficient to ensure a sustainable value proposition. As wardens of their firm's long-term health, directors must keep management moving in an enterprise ethics direction.

2 A New Set of Ground Rules

Directors and company executives are keenly aware of the ways in which the ground rules have changed. Among the most noticeable differences driven by the wave of current corporate ethics crises is the unprecedented level of legislation and regulations that have resulted.

The Public Company Accounting Reform and Investor Protection Act — more commonly known as the Sarbanes-Oxley Act (SOX) — was signed into law in July 2002 and has become the most far-reaching corporate reform legislation in a generation. The regulations rightfully received widespread bipartisan and public support. In the almost 3 years since its passage, the most significant provisions have been widely implemented by corporate America. One aspect of SOX of particular interest is the directive for companies to adopt and disclose a code of ethics for their senior financial officers.

Regulations such as those included in SOX are important and necessary. Unfortunately, implementation is leading to unintended, although not necessarily unexpected, consequences. Most noticeable are the tremendous resource and capital requirements, particularly in implementing sections requiring executives and their auditors to document and certify that their internal financial controls work properly.

In trying to address current ethics issues, we must be careful not to think we can legislate away every problem, issue, or variance. In doing so, we may risk impairing our companies' economic growth and innovation. Moreover, when we create a false belief that all problems and issues are addressed by rules and regulations, we potentially discourage the kind of employee pushback and questioning that drove whistleblowers like Sherron Watkins at Enron and Cynthia Cooper at WorldCom to uncover two of the largest corporate frauds in history.

Despite the consequences from legislation such as SOX, the long-term costs of noncompliance are far greater. Already we have suffered the greatest cost: loss of public trust. Business leaders must recognize that something greater than legal compliance is required if we have any hope of restoring this trust.

3 Putting It All Together

If there is anything that a careful review of ethics crises through the years has taught us, it is that we don't solve root causes by making more rules and laws. Hopes in this direction are misplaced. While the current perp walks, indictments, and trials are all necessary judgments on the past, they bring us no closer to a brighter future. Worse, they impart a false image of business as a practice that is separate from ethics (see sidebar on ethics vs. compliance on page 24).

We should continue to improve and advance regulations for corporations that serve to increase the public trust. But the step-change in restoring the public trust will occur only when business leaders push corporate culture in a direction that leads employees to put business and ethics together on the level of everyday practice.

The root of the current wave of scandals is that our idea of good management has been usurped by a view that says, "Management's only obligation is to maximize shareholder value," in a very narrow sense of that term. This narrow view is not limited to the executive suite or the boardroom — it is one shared by legislators and the larger culture as well.

Of course, shareholder protection is necessary in light of corporate scandals. However, one problem with this narrow view is that we have what economists call "bounded rationality." In other words, we don't know everything. There are limits to what executives and directors can know. And because we live and work in a complex world, what looks today to be in shareholders' interests may not be.

4 A Quiet Settlement?

For example, suppose you are a director for a company that discovers a problem with its product. Millions of customers are very satisfied with your company's product, but under certain rare conditions, a few customers who have used it have been hurt. Further, upon being notified of the problem, suppose your board agrees to quietly settle with the injured customers in order to avoid a potential drop in your company's stock price. You are, of course, aware that many loyal, energetic, and long-tenured employees have their children's education and their own retirement funds invested in the company's stock.

By your action, you have decided to trade off the interests of customers with the interest of shareholders. However, because the world continues to become more and more informed through news and the Internet, eventually the settlement becomes public. What's likely to happen next is a much greater loss of your brand, revenues, and stock price.

When we adhere blindly to the narrow view of shareholders, we fail to see that corporate life is a dynamic, moral environment involving multiple stakeholders — including employees, customers, suppliers, communities, and yes, shareholders. Thus, we must stop separating business and ethics in our return to the basics of good management. Good management must embed ethics into the very fabric of business. In this sense, good management really is about creating value for all stakeholders by developing quality products and services for customers, having employees who are energized to show up for work every day, working with suppliers so each party improves the other's business, being good citizens of the community, and finally — if you get all these other things right — making money for shareholders.

Business Roundtable Institute for Corporate Ethics

The Business Roundtable Institute for Corporate Ethics is the first business ethics center designed to renew and enhance the link between ethical behavior and business practices. The institute was founded through the collaboration and support of Business Roundtable — an association of chief executive officers of leading corporations with a combined workforce of more than ten million employees in the United States and $4 trillion in revenues — and 18 leading academics from the country's best business schools.

The nexus of leading corporate CEOs and leading ethics scholars — or, put another way, the combination of best practice and best thinking — allows the institute to make a lasting contribution to business ethics.

The institute's mission is directed squarely at current and future business leaders and seeks to involve these leaders throughout each of its projects. Its initial *Mapping the Terrain* study of Business Roundtable CEOs — intended to identify the most important ethics issues facing corporate leaders — provided direction for the institute's other activities.

CEOs indicated that the top five ethics issues facing business are:

1. regaining the public trust;
2. effective company management in the context of today's investor expectations;
3. ensuring the integrity of financial reporting;
4. fairness of executive compensation; and
5. ethical role-modeling of senior management.

The institute is continually interested in collaborating with all corporate executives and leaders in its continuing mission to put business and ethics together in everyday business practice. Activities include research, executive seminars, and a best-practice series.

Great companies have long-term success only insofar as they satisfy all of their stakeholders over time by uniting the interests of various stakeholders around a core purpose or value that goes beyond just making profits. Jim Collins and Jerry Porras, in their landmark study *Built to Last: Successful Habits of Visionary Companies,* looked at organizations that had stood the test of time and had significantly outperformed industry peers. One of their key findings was that in successful companies the entire organization *lived* the core purpose every day. This finding has extremely powerful implications for executives and directors as we think about where we go from here.

Collins and Porras demonstrated that values and ethics can drive business strategy and practices. Executives and directors must be role models whose behavior mirrors the company's code of ethics. And they must act on what they say After all, ethics and values are matters of the feet and the heart, not just matters of the mouth.

The new rules require directors to be more vigilant in their governance of the corporation and its managers. And while it is critical to know and adhere to the new rules, we should not be fooled into thinking they are a panacea for our problems.

5 Long and Difficult Work

The rules allow us to treat the symptoms, but we must also work to remedy the disease. Instead of relying upon the legal system to "fix business," we all need to participate in the long and difficult work of changing how our culture thinks about business.

The good news is that business leaders have received the message. In a recent Business Roundtable Institute for Corporate Ethics survey, chief executive officers identified the need to regain and restore the public trust as the most important ethics issue facing business.

Directors have a great role to play in restoring the public trust. I propose several actions that directors can take to help their company move in this direction.

- *Encourage and inspire a culture* in which employees are truly empowered to push back against the organization so that executives and directors get bad news early. This is not something that just happens, and it cannot be feigned. Leaders must design it into their company and respond in the right way when employees risk speaking out. As a director, ask yourself the following question: Do you know how information you need to have would get to you other than through the chain of command? Sometimes the chain of command is the problem.
- *Actively engage in conversations about ethics in the boardroom* that are broader than simply following the ground rules. Include a wider group of company management in the conversation so directors can better understand their company and management. Ask yourself: Do you know how your company makes each of its stakeholders better off?
- *Work with management to develop a clear understanding* of your firm's position and aspirations with regard to ethics. What message does the board send to management: "Follow industry standards" or "Set industry standards"? Can every employee answer the question, "What does our company stand for?"
- *Frame the process.* By the time a problem reaches the board for resolution, it has usually been "worked to death." What questions do you want your managers to ask as they try to solve these problems? Can you be sure that their decision-making framework includes ethics questions? As a director, you may still have to make a tough ethics choice, but at least the problem will be presented to you with all the relevant data, both economic and ethical.

6 Vision Into Reality

Telling a new story about business is not sufficient. More difficult is actually *creating* and *enacting* that story about business every day. But, it is something we can and should do.

Ultimately, the value of all of the executive seminars, best practice studies, and research rests in how we put ethics into practice in the everyday interactions at each of our places of work.

We all have a responsibility in the effort to put business and ethics together. CEOs and senior managers have to realize that no matter how they are compensated, they are paid to be good managers, and no matter what challenges short-term investors present they are paid to be good managers for the long term.

Right now we have the momentum to become the generation that made business better.

By starting at the top, with the steps I have outlined above, we can collectively bring our new vision of good governance closer to reality.

R. Edward Freeman is University Professor, Elis and Signe Olsson Professor, Academic Director of the Institute for Business in Society, and Senior Fellow of the Olsson Center for Applied Ethics at the University of Virginia's Darden School of Business.

Chapter 50
Business Ethics at the Millennium

R. Edward Freeman

1 Introduction

Business ethics, as a discipline, appears to be at a crossroads. Down one avenue lies more of the same: mostly philosophers taking what they know of ethics and ethical theory and applying it to business. Along this route there are many interesting stops. Here lies the Kantian theory of organizations of Norman Bowie, Patricia Werhane's theory of employee rights, Donaldson and DeGeorge's rubrics for multinationals dealing with important cultural differences, and Donaldson and Dunfee's elaborate system of hypernorms and integrated social contracts, to name but a few.[1] Down this avenue lies most of the research reported in *Business Ethics Quarterly* and delivered

[1] See Norman Bowie, *Business Ethics: A Kantian Perspective* (Oxford: Black well, 1999); Patricia Werhane, *Persons, Rights and Corporations* (Englewood Cliffs, N.J,: Prentice-Hall, 1985): Thomas Donaldson, *The Ethics of International Business* (New York: Oxford University Press, J989); Richard DeGeorge, *Competing With Integrity in International Business* (Oxford: Oxford University Press, 1993); and Thomas Donaldson and Thomas Dunfee, *Ties That Bind: A Social Contracts Approach to Business* (Boston: Harvard Business School Press, 1999). I don't wish to except my own work in this tradition from the criticisms that I raise here. William Evan and R. Edward Freeman, "A Stakeholder Theory of the Modern Corporation: Kantian Capitalism" (in *Ethical Theory and Business,* 3rd ed., ed. T. Beauchamp and N. Bowie [Englewood Cliffs, N.J.: 1988]) is squarely in this tradition.

Originally published in: *Business Ethics Quarterly*, 10(1), 169–180 © Cambridge University Press, 2000.
Reprint by Springer, DOI 10.2307/3857703.

R. E. Freeman (✉)
University of Virginia, Charlottesville, VA, USA
e-mail: FreemanE@darden.virginia.edu

© Springer Nature Switzerland AG 2023

S. D. Dmytriyev, R. E. Freeman (eds.), *R. Edward Freeman's Selected Works on Stakeholder Theory and Business Ethics*, Issues in Business Ethics 53,
https://doi.org/10.1007/978-3-031-04564-6_50

at the Society for Business Ethics. These lines of research are very interesting and fruitful, as has been proven in recent years. They represent the very best that philosophy has to offer. Indeed, the very fabric of ethical theory has begun to be rewoven by these authors and their close counterparts in biomedical ethics.

Yet, one gets the feeling that the entirety of business, that multifaceted institution of capitalism, could ignore this avenue of inquiry without so much as a glance. Some may not ignore it, and there may emerge "pockets of ethics" and "ethical companies" according to these lines of research. However, the mainstream conversations in business have had little to do with the work of these philosophers.[2]

Another avenue at the crossroads fares no better. There is a long tradition of scholars working in the area known as "business and society" or "social issues in management." Most of these scholars are trained as social scientists and teach in business schools. Their *raison d'etre* has been admirable: trying to get executives and students of business to understand the social impacts of business and to see business in broad, societal terms. They have relied upon ideas such as corporate social responsibility, corporate social performance, and issues management and have taken a largely descriptive and empirical approach to their discipline, as opposed to the largely normative and theoretical approach of the philosophers.

Interestingly enough, the same criticism seems appropriate. Not much of this work has appeared in mainstream discussions of business and capitalism. There are two important reasons that may explain the general "perceived irrelevance" of work in business, ethics, and society (if we may combine these two lines of work into one). Both accounts of work in this area were opened largely to question the status quo, which claimed that businesses should be managed solely in the interests of stockholders. Put more crudely, the standard claim is that capitalism is a system that rests solely on individual self-interests to the exclusion of others, and that the "natural" drive of humans to compete is the main fuel of the engine of capitalism. Naturally a critical posture relative to this standard claim would make research in this area marginal and of little perceived value by those in the mainstream. Secondly, at least the philosophers would argue, we shouldn't be surprised that their ideas have not shown up in the mainstream. After all, they would argue, the mainstream is about the current practice of business and capitalism, while the ethical critique is about what *ought* to be the case. And, these philosophers would opine, you simply can't deduce an "ought" from an "is." How businesses ought to behave is not necessarily connected to how they do in fact behave. So, there is no small wonder that these ideas have not found their way into the mainstream, especially as the practice of capitalism has found itself triumphant around the world.

[2]By this I mean that there are few articles in the popular discussions of business that mention the work of these scholars. There are few articles about ethics at all in these journals and magazines. There are no ethics best-sellers in business. Most executive programs have ethics sessions as an afterthought or an after-dinner speech. There is, however, an increasing number of articles by business, ethics, and society scholars in more mainstream management journals such as *Academy of Management Review*.

Yet these responses have a hollow ring to them. The recent emergence of concerns with "vision and values," and "a sense of purpose" in the mainstream conversations about business seem to be right up the alley of scholars in business, ethics, and society, yet it was others such as Prahalad and Hamel, and Collins and Poras who have brought these ideas into the mainstream."[3] Even more curious is the fact that the two streams of research and critique seem to add up to more than the sum of their parts. Recently these scholars have begun to see that they have a great deal to say to each other. Conferences, papers, and books have begun to appear to indicate how these two lines of inquiry can be combined. The results of this combination crystallize the problems that business, ethics, and society scholars face, and at the same time illuminate yet another avenue for future development, an avenue that puts the work of these scholars squarely into the mainstream conversation about business and capitalism.

2 The Role of Stakeholder Theory

These two streams of research have naturally merged around the idea of "stakeholders," a controversial concept in the mainstream conversation of business and capitalism, but an idea that has played a large role in the development of both lines of research. The stakeholder idea has a long history, but its current instantiation is due to business theorists trying to do ever better strategic planning, and to take into account precisely those groups who can have some effect on the firm or may be affected by the firm's actions.[4] The last 15 years have ironically seen a growing influence of the stakeholder concept in the business, ethics, and society literature, and a lessening of it in the strategic management literature.

The philosophers see "stakeholders" as a way to bring in the fact that business should be accountable to others. And, it is precisely some of these "others," who can affect or be affected by the firm, to whom some accountability is owed. There are a variety of proposals for which "others" count, but all of these proposals use the stakeholder idea to add in morality to the perceived "immorality" or "amorality" of capitalism.

---

[3] See G. Hamel and C. K. Prahalad, *Competing for the Future* (Boston: Harvard Business School Press, 1994) and J. Collins and J. Porras, *Built to Last* (New York: Harper Business. 1994).

[4] For a history of this idea see R. Edward Freeman, *Strategic Management: A Stakeholder Approach* (Boston: Pitman Publishing Inc., 1984); T. Donaldson and L. Preston, "The Stakeholder Theory of the Corporation: Concepts, Evidence and Implications." *Academy of Management Review* 20, No. 1 (1995): 65–91, reprinted in M. Clarkson. ed. *The Corporation and its Stakeholders: Classic and Contemporary Readings* (Toronto: University of Toronto Press, 1998); and Giles Slinger, "Spanning the Gap: The Theoretical Principles Connecting Stakeholder Policies to Business Performance," Center for Business Research, Department of Applied Economics, University of Cambridge.

Social scientists, on the other hand, see "stakeholders" as a useful *unit of analysis* that easily depicts the social and societal effects of business. By focusing on those groups who are affected by the firm or who can affect the firm we can show that there are broader issues at stake than just the economics of business. We can measure performance with these other groups, model good and bad performance, and revise the narrow economist's vision of business to include social issues and societal effects.

Two recent papers are canonical versions of these arguments. Donaldson and Preston's review of the stakeholder literature divides it into three main areas: normative, descriptive, and instrumental. They suggest, and others have seconded their suggestions with fervor, that theorists be more precise in making either normative, descriptive, or instrumental claims. They claim, "The muddling of theoretical bases and objectives, although often understandable, has led to less rigorous thinking and analysis than the stakeholder concept requires."[5] However, in the end they claim that the normative area is the central one, where "stakeholder theory," understood as an alternative to "stockholder theory," will get its justification. They go on to suggest that such a justification is to be found in an analysis of property from a modern perspective. And, Donaldson and Dunfee suggest that something like "stakeholder theory" can be justified only with an appeal to an elaborate mechanism known as "integrated social contract theory" else it collapses into the relativist abyss.[6] So, it seems that regardless of what empirical or descriptive propositions follow, and regardless of the causal or hypothetical connections that are established, we are to look to the normative realm if "stakeholder theory" is to be justified. For here we will find "its connection with more fundamental and better-accepted philosophical concepts."

Mitchell, Agle, and Wood have argued almost the opposite of Donaldson and Preston.[7] They suggest that we look to the empirical world to see how executives actually determine who and what stakeholders really count. By focusing on power, urgency, and legitimacy we can map a multidimensional analysis of stakeholders and their effects on the firm. Whether or not these effects ought to be the case or not is a separate matter. Mitchell, Agle, and Wood argue that while the search for legitimacy of stakeholder theory goes on, and we wait for the articulation of a normative core that is found to be convincing, "managers must know about entities in their environment that hold power and have the intent to impose their will upon the firm."[8]

---

[5] See Donaldson and Preston in Clarkson, *The Corporation and its Stakeholders,* p. 182.

[6] See Donaldson and Dunfee, *Ties That Bind.*

[7] See R. Mitchell, B. Agle, and D. Wood, "Toward a Theory of Stakeholder Identification and Salience: Defining the Principle of Who and What Really *Counts,*" *Academy of Management Review* 22, no. 4 (1997): 853–886, reprinted in Clarkson. *The Corporation and its Stakeholders,* pp. 275–314.

[8] Ibid., p. 307.

3 The Separation Thesis and the Responsibility Thesis

In short, Donaldson and Preston, and Mitchell, Agle, and Wood, seem to accept some form of what I have called "The Separation Thesis," though each set of authors thinks that one side of the thesis holds more promise than the other. The Separation Thesis is the view that:

> The discourse of business and the discourse of ethics can be separated so that sentences like "x is a business decision" have no moral content, and "x is a moral decision" have no business content.[9]

The Separation Thesis has a long history and tradition. Amartya Sen has suggested how it has come to hold sway in modern economics.[10] I intended my statement of the thesis to be used as a diagnostic device to examine the current state of the mainstream conversation about business and capitalism, but it works equally well to diagnose the conversation about business, ethics, and society.[11]

Philosophers have stood behind their expertise in ethical analysis to trumpet the primacy of the normative. Social scientists have stood behind their methodological expertise to champion the primacy of the empirical. Some have argued that the normative must be based on an understanding of the empirical, but many philosophers would reject such a suggestion with a cry of "Hume's Law" referring to the belief that one can't derive a normative statement from a statement of fact. So, the Separation Thesis seems to aptly describe the two avenues of business, ethics, and society, and the crossroads at which the field stands. But, there is more.

First of all, along with the Separation Thesis we might use another thesis for diagnostic purposes. Call this thesis the Responsibility Thesis, and let it stand for the idea that if ethics is to get off the ground and have any meaning, then people have to take some responsibility for their actions. More formally, the Responsibility Thesis claims:

> The basis for ethics, or the moral point of view, is that most people, most of the time, take, or want to take, responsibility for the effects of their actions on others. And, if they did not, then what we call "ethics" and "morality" would be meaningless.

Now it is easy to see the problem. First, a discourse, such as the discourse of business and capitalism, can't simultaneously appeal to the Separation Thesis and the Responsibility Thesis. If "Business" is truly separate from "Morality" then

---

[9] R. Edward Freeman, "The Politics of Stakeholder Theory," *Business Ethics Quarterly* 4, no. 4 (1994): 409–422.

[10] Amartya Sen, *On Ethics and Economics* (Oxford: Blackwell, 1987); and "Does Business Ethics Make Economic Sense," *Business Ethics Quarterly* 3, no. 1 (1993): 45–54.

[11] There are many dualities in the mainstream conversation of business and capitalism that bear some relationship to the Separation Thesis. I have in mind "business—ethics," "social science—humanities," "fact-based—opinion and feeling," "empirical—normative," "descriptive—prescriptive," "business—society," and others. A full accounting of the Separation Thesis is beyond the scope of the present paper. Suffice it to say that I believe it runs to the core of the mainstream conversation.

responsibility plays no role. And no matter how hard philosophers and social scientists try to slip it in via the back door, the mainstream conversation, built on the Separation Thesis, will reject it. Second, there is some truth behind the Separation Thesis that business, ethics, and society scholars tend to ignore and it is this: economics counts. Value creation and trade, in fact, can take place even when the parties to the practice share few values. Value creation and trade are resilient practices that have developed over millennia, before the advent of modern governments, before trade agreements, and before the large multinational corporation. If value creation and trade took place only within relatively homogeneous societies, then, getting all the normative claims lined up before we could describe business or its hypothetical relationships may make some sense. But, such a vision is not a useful one in today's world.

4 Stakeholder Theory as Managerial

Having diagnosed the split in business, ethics, and society, and its attempted healing via stakeholder theory as a necessary and implicit tension and appeal to the Separation Thesis and the Responsibility Thesis, I want to suggest that the seeds of the solution lie precisely in the middle of Donaldson and Preston's article. It is a great irony that such an influential article should be misinterpreted as claiming three things about stakeholder theory, that it is descriptive, normative, and instrumental. For in fact, contra the myriad papers citing this article, there is a fourth claim. Donaldson and Preston's "Thesis 4," which even they don't develop, holds the key to a way out of this morass. Thesis 4 claims[12]:

> The stakeholder theory is managerial in the broad sense of that term. It does not simply describe existing situations or predict cause-effect relationships; it also recommends attitudes, structures, and practices that, taken together, constitute stakeholder management. Stakeholder management requires, as its key attribute, simultaneous attention to the legitimate interests of all appropriate stakeholders, both in the establishment of organizational structures and general policies and in case-by-case decision making. This requirement holds for anyone managing or affecting corporate policies, including not only professional managers, but shareowners, the government, and others. Stakeholder theory does not necessarily presume that managers are the only rightful locus of corporate control and governance. Nor does the requirement of simultaneous attention to stakeholder interests resolve the longstanding problem of identifying stakeholders and evaluating their legitimate "stakes" in the corporation. The theory does not imply that all stakeholders (however they may be identified) should be equally involved in all processes and decisions," (pp. 175–6)

If stakeholder theory is managerial in this sense, then it is impossible to sort out the precise normative and empirical claims. Though the instrumental claims, the large cause-effect claims, may well turn out to be the most interesting of the three, and the connections between the instrumental claims and the managerial nature of stakeholder theory will be crucial. In that sense the recent work of Thomas Jones makes

[12] Donaldson and Preston, pp. 175–176.

central contributions to the development of stakeholder theory.[13] My suggestion is that we replace a concern with Donaldson and Preston's first three theses with a full understanding of Thesis 4. We need to see stakeholder theory as managerial, as intimately connected with the practice of business, of value creation and trade. That was its original impetus, in the sense of redescribing the practice of value creation and trade to ensure that those with a "stake" in this practice had attention paid to them. In the messy world of management it is simply impossible and not very useful to be precise about what claims are normative and what claims are empirical etc., though there will surely be times when such careful delineation has a purpose.

Stakeholder theory is about value creation and trade. If capitalism is the umbrella under which we analyze value creation and trade, then stakeholder theory is inherently capitalistic, no apologies. Stakeholder theory is about the real world of business, the messy relationships, sometimes sortable into neat categories, but oftentimes, customers are suppliers are competitors. If "stakeholder theory" is to join the mainstream conversation about business and capitalism it will be because theorists are able to both understand practical managerial problems and offer narratives or stories that enable managers and stakeholders to enact a better, more useful version of value creation and trade. I want to suggest how the business, ethics, and society scholar can do this by way of example.

By implicitly appealing to the Separation Thesis and the Responsibility Thesis four managerial problems arise. Each of these problems can be avoided by enacting a version of a story about value creation and trade that does not depend on the tension between these theses. Furthermore such a story, which I want to call "Stakeholder Capitalism," is a more generic narrative about value creation and trade that avoids many of the pitfalls of our standard story about capitalism.

5 Four Managerial Problems

Since, according to the Separation Thesis, business and ethics are to be disconnected, the first managerial problem may be called the Problem of Meaning. Human beings spend a majority of their waking hours at work. Most people want meaningful work, to see that their work matters and has some meaningful and usually good consequences for others. However, if our understanding of business is based on the Separation Thesis, such moral ideas about work are problematic. Hence, there will be a tension between people wanting meaningful work and the more traditional idea that work and the meaning of life are somehow disconnected. Work is at best a means in which we use its fruits, money, to find meaning or to care for others that are important to us.

---

[13] See Thomas Jones, "Instrumental Stakeholder Theory: A Synthesis of Ethics and Economics," *Academy of Management Review* 20, no. I (1995): 92–117; Jones and A. Wicks, "Convergent Stakeholder Theory," *Academy of Management Review* 24 (1999).

Collins and Porras, in *Built to Last,* have detailed company after company that has outstanding performance in part because these companies have a sense of purpose, a sense of meaning that is transferred to their employees. This sense of purpose is broader than "maximize shareholder value" and inevitably refers to some good done for other stakeholders as well. Note that this sense of purpose doesn't imply that these companies do not maximize shareholder value, but that such an outcome is not the purpose of the enterprise. For example, employees at Merck come to work in part because it is important that they do so—finding cures for human suffering is a cause worthy of their attention. The problem of meaning is a function of our ideology, a function of the Separation Thesis, and a problem that we can help dissolve by coming to see the employee stakeholder relationship in both business and moral terms simultaneously.

The second managerial problem is related and can be called the Problem of Careers. If work has little meaning, then for the sake of one's mental health, the self is separated from work. If we have a self that we conceive of as "fully or partially moral," that moral self is out of place in the business world where ethics and business are separate. Consequently we need new constructs such as "career" to be a "dummy variable" for self. Students often say "I'm doing this for my career" when they mean, "I really don't want to do this. I'm not like this really. Etc." Separating self from work automatically ignites another problem, that of balancing work with other interests. Since work is not, on this view, part of one's moral universe, there is a limited amount of time an employee is willing to devote to her career. Rather than making corporations friendly places for families, relationships, and more fully developed human beings, the separation of self from work leads to the tensions between one's "professional" and "private" life, one's work life and family life, one's career and self. These tensions can ultimately be alienating on either side of the tension. We become our work, or we endure our work for the sake of privacy, family, and self.

The third managerial problem can be called the Problem of Change. There is no topic that is written about more in the mainstream discussions of business than "change." Executives constantly invoke "change programs," send employees and themselves to "change seminars," and hire consultants to help them "manage change." The standard mantra is that change is difficult. Change is hard. People have difficulty with change and don't want to change. But, this standard story is directly deducible from the affirmation of the Separation Thesis and the denial of the Responsibility Thesis. If business is separate from values and ethics, and if change requires one to think about values and ethics, then change in business will be difficult. It will prescribe that one do what cannot be done: invoke the cause of change, values, in an arena where it is illegitimate, according to the discourse, to do so.

If we revise the standard story of business, to make it a story about stakeholder relationships that are fully human, then change is actually quite easy. What is difficult is knowing what the values are that one stands for. But, if we give up the Separation Thesis and all of its concomitant guises, then the question of what one stands for (individual, or company) becomes a central part of understanding any business and any stakeholder relationship.

The fourth problem can be called the Problem of Leadership. With the possible exception of "change" no other topic has recently generated such a large volume of literature. There is a hue and cry for leadership in today's business world, and it is connected to the preceding three problems. Because our discourse asks us to see work as not admitting of meaning, the self as disconnected from work, the exigencies of change as difficult to manage, we need some heroic construct like "leadership" to get us through the day-to-day pressures of business. There is a whole genre of "Leadership Lessons From X" where X ranges over a set of people from Attila the Hun to Martin Luther King. And, this literature mimics the Separation Thesis. It has little role for values, other than in a descriptive role, or other than in the "character of the leader." In few instances do we find values in both their individual and social forms, and rarely does this literature see the need for a critical approach to both the leader and the very idea of leadership.

6 Toward a New Understanding of Capitalism

These four problems, based on the very real world of a global business environment, call for the need to rethink the standard story of capitalism. Suppose that we gave up the Separation Thesis. Note that this entails giving up all the distinctions that Donaldson, Preston, Mitchell, Wood, and Agle hold dear, at least most of the time, and adopting what Donaldson and Preston call the managerial thesis of stakeholder theory. So, there are no "descriptive" constructs. There are no "empirical" studies. There are no fundamental "normative cores" on which to base everything.[14] There are only narratives and pieces of narratives that are at once descriptive of how we are and at the same time suggestive of how we could live better. These narratives serve the function of offering hope about how we can revise our current institutions to make them serve us better. "Stakeholder theory," "stakeholder management," or "managing for stakeholders" is precisely this kind of narrative. It calls into question the dominant story of "anything-goes capitalism" or "cowboy capitalism" or "shareholder capitalism," etc.[15] I want to articulate four principles of stakeholder capitalism and suggest how enacting these principles can help us to avoid the four managerial problems outlined above.

---

[14] This doesn't imply that normative cores or even empirical categorization schemes are not sometimes useful. It does imply that their usefulness to a managerial stakeholder theory is the proper criterion for their evaluation.

[15] That it has always been at least intended (if not executed) as such a managerial revising of the mainstream conversation about value creation and trade, at least in my own view, is the subject of a joint paper with Robert A. Phillips. "Stakeholder Theory: A Libertarian Argument," Society for Business Ethics Meeting, Chicago, August 1999.

First of all *The Principle of Stakeholder Cooperation* says that value is created because stakeholders can jointly satisfy their needs and desires.[16] Value creation and trade is not a zero sum game. Capitalism works because entrepreneurs and managers put together and sustain deals or relationships among customers, suppliers, employees, financiers, and communities. The support of each group is vital to the success of the endeavor. This is the cooperative commonsense part of business that every executive knows, but the Separation Thesis leads us to believe that the shareholder is always more important than others. Try building a great company without the support of all stakeholders. It simply cannot be sustained.

Because this principle is rooted in the interests of stakeholders, the corporation becomes a clearinghouse or nexus of activity where stakeholders satisfy their desires. Far from being meaningless, the corporation becomes an institution imbued with meaning from many different perspectives.[17] Employees need not pursue only careers in such corporations, but the corporation becomes a vehicle, a mere means, if you like, to employee and other stakeholder ends. Note however, that such meaningful work has to satisfy the desires and interests of other parties to the agreement. All corporations are not managed for the benefit of any one group, though some may in fact be so managed.

Second, *The Principle of Complexity* claims that human beings are complex creatures capable of acting from many different values. We are not just economic maximizers. Sometimes we are selfish and sometimes we act for others. Many of our values are jointly determined and shared. Capitalism works because of this complexity, rather than in spite of it.

If human beings are complex and multifaceted, the problem of change takes on new meaning. It becomes a central task to determine an answer to fundamental values questions that may bind together a business entity. There are no obvious "right" answers here. There are many different ways to engage in value creation and trade and also be "an ethical person."[18]

Third, *The Principle of Continuous Creation* says that business as an institution is a source of the creation of value. Cooperating with stakeholders and motivated by values, businesspeople continuously create new sources of value. This creative

---

[16] The following sections contain some paragraphs from R. Edward Freeman, "Stakeholder Capitalism," *Financial Times*, July 26, 1996. I am grateful to the editors and publisher for permission to use this material here.

[17] For an account of how this can come about see R. Edward Freeman and Daniel R. Gilbert, Jr.. *Corporate Strategy and the Search for Ethics* (Englewood Cliffs, N.J.: Prentice-Hall, 1987) Chap. 7.

[18] That there are many ways to run a business is the insight behind the often-ignored idea of "enterprise strategy" and its theoretical analog "normative core." It is a separate story whether or not "being an ethical person" makes any sense in isolation from the ideas of value creation and trade. If value creation and trade are fundamental to the human experience, then separating out "ethical person," as the above sentence does, is also illegitimate. Another way to say this is that our analysis points out the need for a political philosophy or a conception of ethics where value creation and trade, rather than the state, play a central role.

force of humans is the engine of capitalism. The beauty of the modern corporate form is that it can be made to be continuous, rather than destructive. One creation doesn't have to destroy another, rather there is a continuous cycle of value creation that raises the well-being of everyone. People come together to create something, be it a new computer program, a new level of service, a way to heal the sick, or simply to work together.

The problem of leadership is relevant here. By seeing the corporation as itself an innovative mechanism set forth to preserve the possibility that the current cooperative agreement among stakeholders may well turn out to be worth preserving, leadership takes on a whole new meaning. Again, the leader will not be able to separate out the "skills and infiuence techniques" from the ongoing intersection of interests and values of the stakeholders. An analysis of the ethics of the corporation won't be done from an outside point of view, say that of "ethical theory," but from the inside view of what stakeholders are trying to achieve together, and what they are trying to preserve so that tomorrow they may achieve again.

Finally, *The Principle of Emergent Competition* says that competition emerges from a relatively free society so that stakeholders have options. Competition emerges out of the cooperation among stakeholders, rather than being based on the primal urge to "get the other guy." Competition is important in stakeholder capitalism, but it is not the primary force. It is in its ability to manage the tension created by simultaneous cooperation and competition that stakeholder capitalism distinguishes itself.

Stakeholder capitalism takes a firm stand against the Separation Thesis. It implies that human beings are required to be at the center of any process of value creation and trade. It underscores the Responsibility Thesis that common decency and fairness are not to be set aside in the name of playing the game of business. It suggests that we should demand the best behavior of business, and that we should enact a story about business that celebrates its triumphs, admonishes its failures, and fully partakes of the moral discourse in society as a routine matter. Yet, stakeholder capitalism is no panacea. It simply allows the possibility that business becomes a fully human institution. There will always be businesspeople who try to take advantage of others, just as there are corrupt government officials, clergy, and professors. Stakeholder capitalism bases our understanding and expectations of business, not on the worst that we can do, but on the best. It sets a high standard, recognizes the common-sense practical world of global business today, and asks managers to get on with the task of creating value for all stakeholders.

7 The Role of Business, Ethics, and Society Scholars

I believe that stakeholder capitalism, as briefly sketched above, articulates many of the values, beliefs, and the critical stance of many scholars in business, ethics, and society. I also believe that the four managerial problems represent the cutting edge of how managers experience the business world. Both of these claims could be false, but even if they are, I want to argue that some such process as I have suggested in the last two sections is the road down which business, ethics, and society scholars should tread.

We need to be fully immersed in the world of value creation and trade. If these four problems don't interest scholars then find some others. We need to be fully cognizant of the worth of what we are doing, which is, in large part, revising the story of capitalism to make it work better. If the revision I've called stakeholder capitalism doesn't interest scholars, then find some other revisions.

We need to avoid isolated theorizing that is unconnected and unconnectable to the practice of value creation and trade. We need to avoid philosophical distinctions that, for the most part, don't make a difference. We need to firmly reject the idea that business and ethics are separate, and we need to explore whether or not our "ethical theory" contains traces of the Separation Thesis, i.e., the idea that business couldn't possibly be a full citizen in the moral universe.

In short, we need to join the mainstream conversation about business and capitalism. We need to develop the stakeholder framework more fully to help revise the process of value creation and trade, to make business an even more fruitful institution in bringing about good and raising up the least well off in the world. Business, ethics, and society scholars are poised to lead this conversation. The challenge, however, is a substantial one.

R. Edward Freeman is University Professor, Elis and Signe Olsson Professor, Academic Director of the Institute for Business in Society, and Senior Fellow of the Olsson Center for Applied Ethics at the University of Virginia's Darden School of Business.

Chapter 51
Poor People and the Politics of Capitalism

R. Edward Freeman, Adrian Keevil, and Lauren Purnell

"The halo that surrounds the head of a saint does not become visible until we know he has paid the rent."

—John Wisdom, *"Freewill," in Paradox and Discovery (1965)*

1. Here's one way to think about poor people and capitalism. People are poor because they have few opportunities to pull themselves up by their bootstraps.[1] In fact, the gap between rich and poor has increased in recent times due to the more wholesale adoption of capitalist practices around the world. Especially in the West, but not limited to it, the institutions of business and government conspire to give the

Some of the material here is modified from R. Edward Freeman, "Poverty and the Politics of Capitalism," which appeared in *New Approaches to Business Ethics*, 1998, volume 1 of the Ruffin Lectures for the Society for Business Ethics. The series is sponsored by the Ruffin Foundation and the Olsson Center of Applied Ethics at the University of Virginia, and is published in cooperation with the Society for Business Ethics by the Philosophy Documentation Center, Charlottesville, Virginia. We are grateful for permission to recast this material here.

Originally published in: ***Business & Professional Ethics Journal***, 30(3/4), 179–194 ©, Philosophy Documentation Centre, 2011.
Reprint by Springer, DOI.

[1] The style is shamelessly borrowed from Richard Rorty's "Philosophy as a Kind of Writing: An Essay on Derrida" in *Consequences of Pragmatism* (1982). Rorty's views on business can be found in "Can American Egalitarianism Survive a Globalized Economy?" (Rorty 1998). In fact Rorty is the main inspiration behind this essay, even though he certainly would not approve of our conclusions. For a review of the relevant works of Richard Rorty to business and management theory see Freeman 2004.

R. E. Freeman (✉) · A. Keevil · L. Purnell
University of Virginia, Charlottesville, VA, USA
e-mail: FreemanE@darden.virginia.edu

© Springer Nature Switzerland AG 2023

S. D. Dmytriyev, R. E. Freeman (eds.), *R. Edward Freeman's Selected Works on Stakeholder Theory and Business Ethics*, Issues in Business Ethics 53,
https://doi.org/10.1007/978-3-031-04564-6_51

poor a Hobson's choice of minimal wage jobs or unemployment. Neglect of both urban ghettoes and the rural poor has been systematic, if not conscious. The very idea of unbridled capitalism reinforces the notion that some are meant to be poor and some are meant to be rich. The recent financial crisis has exacerbated the chasm between rich and poor; lack of regulatory oversight permitted Wall Street to become a parasite on the poor. Simply examine the aftermath of the mortgage meltdown. Relatively poor people, who were granted mortgages with no equity and exploding interest rates, are left relatively poorer, without homes, without credit, and with a share of the national debt that will be inherited by their children's grandchildren. Wall Street, which made big money from bundling and trading the lousy mortgages, assumed no risk when the gig went south and a government bailout covered their losses. Unlike the poor, who continue to pay a daily price for the mortgage meltdown, Wall Street repaid the bailout and returned to boom-era bonuses before Michael Douglas walked the red carpet for *Wall Street: Money Never Sleeps*. When we turn to the developing world, we find the same problem pumped up on steroids. The gap between rich and poor in fast-developing economies such as India, China, Brazil, and Russia is enormous without much chance of being narrowed. Clearly, the only solution is a massive redistribution of income and a system of capitalism that is severely restrained. However, even if there were such a redistribution policy, we could not count on government to execute it fairly. Capitalism leads to hopelessness for poor people.

2. Here's a second way to think about poor people and capitalism. Capitalism, with its market system, gives poor people a way out of poverty. If poor people are willing to work hard, albeit sometimes for low wages, and be imaginative, they can use the power of the market economy to capitalize on the fruits of their labor. So many people live in poverty largely because of government. Governments constrain the efficient allocation of goods and services that capitalism provides by artificially creating prices for labor and goods such as agricultural products. This hurts poor people most of all. There are not as many jobs as there should be, and prices of key commodities such as food are too high. Government has hurt poor people even more by creating a safety net on which they have come to rely. The "welfare culture" prevents the poor from bootstrapping their way out of poverty to the middle class. In a similar vein, the recent bailout of Wall Street shows how government safety nets incent the wrong behavior by saving firms that deserved to fail. Again, when we look to the developing world we find the problem magnified. In many places around the world, the governments are even more corrupt than the anti-market governments in the West. State-owned enterprise in China is a good example of how much market incentives distort business incentives, trapping people in poverty. Let the markets function properly and just rewards (or punishment) will be distributed fairly. The only solution is to get government out of the picture and to let markets work.

3. Neither version of this story is very interesting. However, both are present in many of our academic and policy conversations about poor people, business, government, and the mechanics of capitalism. Both versions of the story rest on shopworn, useless metaphors about business and government, and both reduce poor people to unhelpful stereotypes. The cacophony is heard not just in America, but

echoes around the world. The volume increases with the severity of the problem: that we harbor an outmoded story about business,[2] one that we desperately need to drop. The old worn-out story about business goes like this. Business is fundamentally about economics and competition. Think of business as a battlefield on which the lonely corporate warriors keep a constant vigilance against the onslaught of competition. Recently, this battlefield has moved beyond geopolitical borders and is global in nature. War is hell and so is business. The drive to compete must be strong if one is to enjoy the sweet nectar of corporate success. If everyone acts solely within one's own self-interest, then a better world will be created as if by magic (or an invisible hand). In business, being selfish is a virtue, and looking out for one's own interest is mandatory. In short, business is populated by a bunch of greedy little bastards out to do each other in before it is done to them. Call this story "Cowboy Capitalism."

4. That the story of Cowboy Capitalism is a living presence in our culture is undeniable. Business leaders and elected officials bombard us daily with strategic initiatives to gain competitive edge over X. In his 2011 State of the Union address, President Obama used the words "compete" or "competitive" 12 times before stating "The future is ours to win." Similarly, the academic literature of management is replete with war-like language and imagery. Professors talk about "competitive advantage," "market penetration," "defensive strategy," and the like. And, in a relatively recent version of this narrative, shareholders and financiers are deemed to hold the most important stake in any business. Other interests such as customers and employees must be subordinated to the interests of financiers. Calvin Coolidge is remembered as one of the most pro-business presidents in the history of the U.S., largely based on a quotation attributed to him that he actually never uttered. "The business of America is business" is a cut and paste version of the original idea, which he delivered to the American Society of Newspaper Editors in 1925. The accurate quotation reads as follows: "After all, the chief business of the American people is business. Of course the accumulation of wealth cannot be justified as the chief end of existence. American newspapers have seemed to me to be particularly representative of this practical idealism of our people. … We make no concealment of the fact that we want wealth, but there are many other things that we want very much more. We want peace and honor, and that charity which is so strong an element of all civilization. The chief ideal of the American people is idealism. I cannot repeat too often that America is a nation of idealists. That is the only motive to which they ever give any strong and lasting reaction" (Coolidge 1925). The adage that money-making has been a religion to Americans has literal truth unperceived by many who repeat it (Griswold 1934). Whitney Griswold, researching the origins of the financial excesses that led to the Great Depression, observed that one possible cause of the irrational investing behaviors was a religion called "New Thought," which was immensely popular among Wall Street professionals prior to the stock market crash

[2] We prefer to talk about "business" or "business and capitalism," given the multiple interpretations of "capitalism."

of 1929. New Thought was a cut and paste version of Puritanism. It took the worship of prosperity but left out the part about morality. Its adherents were "impelled mainly by the motive of profit and few were masters of its theology and metaphysics" (Griswold 1934, 309).[3]

5. That the story of Cowboy Capitalism is indeed shopworn is more controversial. Those on the left certainly see it as shopworn because they presume capitalism to be immoral from the start. For those on the right it is subtler. On the one hand they want business to be a celebration of the human spirit, but on the other they want to escape the moral consequences of business. The language of Cowboy Capitalism does not clearly admit to moral discourse, and this makes capitalism as bothersome to the Right who want to presume it to be right as it is to the Left who want to presume it to be wrong (Freeman et al. 2010). Each is content with a summary judgment. We believe that whether approached from the left or from the right, Cowboy Capitalism is based on three fundamental assumptions that we need to drop. The first is an assumption about what it takes to run a successful business; namely, that maximizing value for financiers is the key ingredient. The second assumption is about the human beings who make up the institution of business, that they are maximizers of their own self-interest. The third assumption is a more subtle one about language; namely, that our language represents the world. We want to question each in turn.

6. The recent financial crisis highlighted the assumption that a successful business depends on its executives' and board members' ability to try and maximize the returns to shareholders or financiers. The last 30 years have produced a body of scholarship, under the rough rubric of "stakeholder theory" that has called this assumption into serious question.[4] The truth is that we know that a successful business simultaneously satisfies the interests of at least customers, suppliers, employees, communities, and financiers. The intersection of interests is what makes business and capitalism viable. This is just simple logic, since, in a free society, members of these groups always have the political process as a back-up mechanism of last resort.[5] There is simply no way to escape the discourse of morality once we

[3] A more modern version of this argument can be found in Mark C. Taylor, *Confidence Games: Money and Markets in a World Without Redemption* (2004).

[4] This literature is summarized in (Freeman et al. 2010), and we won't repeat the arguments here. Stakeholder theory is certainly not uncontroversial, but it does shift the burden of proof to those who would interpret business and capitalism in terms of Cowboy Capitalism. So far, there have been no practically relevant and theoretically interesting responses that are incompatible with Stakeholder Theory.

[5] We are not naive here. Of course it is possible to build a business without paying much attention to stakeholders, but the results are a hodgepodge of regulation, bad reputation, constant reorganizations, re-starts, turnover from employees, bad customer and community relationships, etc. All of these results cost someone time and money. And the lack of political options is often the underlying issue for poor people.

begin to see business as how we create and trade value with each other. Business is a deeply human institution. To see it as otherwise is to diminish everyone involved.[6]

7. Embedded in the story of Cowboy Capitalism is a view of human beings that is increasingly suspect, a view that only propagates the story of the self-interested capitalist. The origins of the self-interest assumption in economics theory are best understood through Richard Dawkins's (1976) *The Selfish Gene*, which set the bar for behavioral assumptions in organizational economics. In Chap. 1, Dawkins writes, "Let us try to teach generosity and altruism because we are born selfish" and, "My own feeling is that a human society based simply on the gene's law of universal ruthless selfishness would be a very nasty society in which to live. But unfortunately, however much we may deplore something, it does not stop it being true" (Dawkins 1976, 1). Dawkins's view was intended to be a descriptive view of the ways that genes intend to propagate at the expense of another family's genes. As he writes in the preface to the 1989 edition, "what kind of entity is it that survives, or does not survive, as a consequence of natural selection. That unit will become, more or less by definition, 'selfish.' Altruism might well be favored at other levels" (Dawkins 1989, vii). Those "other levels" were what he called memes, defined as the transfer of cultural ideas from generation to generation. He writes, "Just as genes propagate themselves in the gene pool by leaping from body to body via sperms or eggs, so memes propagate themselves in the meme pool by leaping from brain to brain via a process which, in the broad sense, can be called imitation. If a scientist hears, or reads about, a good idea, he passed it on to his colleagues and students. He mentions it in his articles and his lectures. If the idea catches on, it can be said to propagate itself, spreading from brain to brain" (Dawkins 1989, Chap. 11). Ideas change bodies. No longer do we believe that there is a separate unintegrated view of the mind, that has no embodiment. Cultural ideas, social practices and patterns of cooperation contribute to who we are as human beings, including our bodies. Nature and culture work together to create our identity.

Today, that assumption of humans as purely self-interested—taken from a supposedly "scientific" Darwinian view of the world—has been questioned so much as to be discredited. There is an overwhelming amount of evidence that the assumption about self-interested utility maximization is no longer useful (e.g., De Waal 1997; Tomasello 2009; Henrich and Henrich 2006; Henrich et al. 2010; Boehm forthcoming). What is distinctive about us is that we are actually oriented towards cooperation with others. We do so freely, enjoyably, and without expectation of reward (Tomasello 2009). The subtle difference between cultural norms being transmitted through memes, rather than Dawkins's selfish genes, is important. Dawkins, and the economic theories he speaks for, assume that people are autonomous individuals, not centrally connected to others. But, the very idea of language means we are able to coordinate our actions with others. Such coordination has led to deep tribal and cultural bonds of love and affiliation. These new theories show convincingly that selection occurs on a group level for the orientation towards

[6] Harris and Freeman (2008) have argued that attempts to see business in purely economic terms assumes a particular view of ethics and morality, making the "amoral" view of business a logical impossibility.

both individual success and common good (Boehm forthcoming). Our reasons for doing what we do are complex. Sometimes we act for other regarding reasons and sometimes we act for self-interested reasons. Sometimes we act as part of a group, consisting of family, tribe, community, or culture. We are, in effect, both Hobbes and Rousseau. We can crave group-level interactions but, when primed otherwise, we are very happy to be self-interested.

We need to be thoughtful about the idea that our theories in practice are likely to lead to recommendations that might ironically produce the behavior they intend to curb (Pfeffer 1996). For example, until recently, it was a general assumption among social psychologists that humans tended towards hypocritical behavior. The Batson et al. hypocrisy studies (1997, 1999, 2002) somewhat infamously claimed to show that people tended to act immorally when they weren't being observed, yet endorse a higher fairness standard when asked publicly. In these studies, participants were asked to make a fixed-sum distributional decision and assign a task to themselves and one to another person they did not know. One task was obviously enjoyable and had the opportunity for a prize at the end; the other was a dull task with no opportunity for a prize. Participants in all cases were told, "The decision is entirely up to you. You can assign yourself and the other participant however you choose" (Batson et al. 1997, 1999, 2002). Experimenters then presented one of two conflicting statements intended to make certain fairness standards salient. They were then left alone to make the decision. After making the decision, participants were asked to take a survey in which they were asked to judge their views of morality of statements and the morality of their own behavior. A significant number of respondents chose the desirable condition for themselves in private, and then endorsed the experimenter's fairness standard (fairness or altruism) in the questionnaire. The general conclusion from this study was to question the assumption of fairness as an expected norm in society. It was also used to make a general claim about the ease with which people behave hypocritically.[7]

8. Both Cowboy Capitalism and assumptions of the self-interested economic actor are based on a view of language that is outmoded. Language is not some representation of the way the world is, but the way that we humans coordinate our actions and make meaning that is shared. Such a view of language depends on our ability to interpret context in subtle and often unique ways to preserve the option of creating new value by talking differently, and hence opening up possibilities for action. As e. e. cummings wrote: "since feeling is first / who pays any attention / to

[7] Recently, Fernandez-Dols et al. (2010) replicated the Batson et al. hypocrisy studies (1997, 1999, 2002) based on the idea that the results may have been primed by experimenter instructions. They hypothesized that, by saying, "The decision is entirely up to you. You can assign yourself and the other participant however you choose," Batson experimenters had unwittingly been telling participants what they expected of them. In other words, if an experimenter tells you that a "decision is entirely up to you," s/he is in effect priming a norm of justified self-interest. Fernandez-Dols and his colleagues found that when they removed that line from the experimenter protocol, people tended to adhere to the fairness or altruistic instruction both privately and publically. The simple instruction elicited specific behavioral and emotional responses that experimenters interpreted as hypocrisy (Fernandez-Dols et al. 2010).

the syntax of things /will never wholly kiss you; // wholly to be a fool / while Spring is in the world." To communicate a kiss, for example, any words would be a poor cassette recording of the original. Talk yields to action. Rorty once wrote that from a Darwinian (i.e., "behavioral") perspective, "We should think of the word 'language' not as naming a thing with an intrinsic nature of its own, but as a way of abbreviating the kinds of complicated interactions with the rest of the universe which are unique to the higher anthropomorphoids" (Rorty 1999, 64). Language, from this perspective, is a tool for coordinating the activities of individuals. This insight from Nietzsche, Dewey, Wittgenstein and others implies that all our concepts, including our moral ones are useful in dealing with our environment and our fragile embodiment. If Alice Crary (2009) is correct, focusing on the narrow language of what has come to be called "ethical theory" runs the risk of adopting a stylized and impoverished view of our interconnections.

9. In *Philosophical Investigations* (1974), Wittgenstein focused our attention back to the ways in which words might direct behavior. Our attention turned to the pragmatic concern about what language does in the world. Two of Douglas Schön's central concepts—"generative metaphor" (1979) and "frame reflection"—(1994) reflect this influence in an organizational (or "business") context. Generative metaphor represents the figurative descriptions of social situations, usually implicit, that shape the way problems are conceived of and thus the solutions to address them. In other words, metaphors serve as a "problem-setting" effect. A simple demonstration of this effect is found in Liberman et al.'s (2004) series of studies that showed that the way in which a prisoner's dilemma game was named ("community" or "Wall Street") predicted whether participants would cooperate or compete. Interestingly, the participants' personalities (assessed by close friends as well as psychological instruments) were not predictive. As Hilary Putnam (2002) might argue, this is one of the outcomes of our attempts to de-sensitize the language of theory: while we may presume it is, our theoretical language is not devoid of meaning. Putnam (2002) offers a view on the compartmentalization of social science, taking aim specifically at logical positivism and arguing that the division value, or emotion, from meaning of words in science, is a false dichotomy. In order to more fully understand Putnam's position, it is important to examine his view on the de-ethicizing of language. His particular gripe is with positivist constructs, which seek to separate "all our putative judgments" from inquiry. The effect of their endeavor creates a market for things which are valuable to research and juxtaposes them from those things that are "cognitively meaningless"—aesthetic judgments, metaphysical judgments, and the like (Putnam 2002, 10). The core motive for *The Collapse of the Fact/Value Dichotomy* is to excoriate the idea that language has to be devoid of meaning in order to be used for scientific inquiry. "Normative judgments are essential to the practice of science itself," he writes (Putnam 2002, 30). He reassures even the most technical of us that language in all its forms is embedded with

assumptions and value judgments (Putnam 2002, 43).[8] Our embodiment as well as our sensory apparatus (which is what else than embodiment of our ideas?) lead to the adoption of metaphors and tropes that then cause us to use language in a certain way to solve problems that in turn shape our scientific perception, our social reality, and in the end, our bodies.[9] What does all of this have to do with poor people and capitalism?

10. In *Race Matters*, Cornel West has suggested that the dichotomy of left and right is bankrupt when it is applied to the intersections of the problems of racism and poor people (West 1993). Liberals assume that the causes are structural in nature and that the solution is better structures, while conservatives assume that the problem is internal to the character of poor people with the only hope to be better people. While West believes that we certainly need better structures and that we need more people to bootstrap their way out of poverty, he identifies a different alternative for our discourse. Because he sees poverty and hopelessness as intimately connected he calls for a politics of conversion within a prophetic framework to restore hope and to offer the possibility of creating some meaningful future. A politics of conversion offers hope that people can struggle together to find meaning, and a prophetic framework builds in moral assessment from the beginning. Now West is proposing such a framework as necessary to address the African American experience in the United States, but we want to suggest that it has a broader application. (We want to do this without minimizing the differences between the problems of racism and poverty, and certainly without suggesting that the African American experience is solely or even essentially defined by poverty.) Surely Mohammad Yunus has created a prophetic framework with the Grameen Bank. Countless social entrepreneurs are solving real problems of poor people around the world with just such a prophetic framework that takes stakeholder interests, our ability to help others, and our language as a tool for opening up possibilities (becoming in reality what Harold Bloom called "strong poets"). It is ironic that West sees no role for capitalism in this prophetic framework, for we want to suggest that if we can retell the narrative of business, as we suggested above, it offers exactly the kind of hope within a prophetic framework that is necessary.

11. Imagine a world in which workers routinely vote on big decisions that a company faces, such as where to locate a factory. In this world employees set their own salaries, have no time clocks or expense reports, and there is little hierarchy. Factories in this world contain employees and independent entrepreneurs (some of

[8] The anti-social elements of these theories also carry with them implied social norms which permit "unethical" behaviors with implied moral justification (Fernandez-Dols et al. 2010). For example, if an executive knows that a board will be "monitoring" his/her behavior, the effective clue is that the board doesn't trust him/her. Therefore the executive might assume that both sides don't trust each other, that is the way things are and it is a mutual agreement. As a corollary response, s/he might withhold information, for example, or try to game compensation agreements. From a behavioral perspective, when there is a psychological contract based on distrust and opportunism, self-interest becomes justifiable moral norm under these conditions (e.g., Rousseau 1995). Behavior, in effect, reveals the nature of human interaction. Language, however, predicts and causes behavior.

[9] We are grateful to Professor Mollie Painter-Morland for this point.

whom used to be employees) all working beside each other. Indeed, the CEO of the business simply does not know how many people work there and is unsympathetic to those who want to know. Routinely, the borderlines between company and customer, company and supplier, even company and competitor are crossed. Imagine that when the workers go on strike (there are still unions in this world) they use the company cafeteria to plan strategy. The company does not cut off their benefits during the strike, which is over rather quickly. Welcome to SEMCO, a Brazilian firm that has carved a successful business by applying the principles of democracy, profit sharing, and information openness.[10]

12. Imagine a world in which workers do all of the hiring and firing of employees. Workers constantly meet in teams to set production schedules, plan new business opportunities, and do quality control. If consumers have a problem, workers responsible for the problem personally contact the consumer. Workers share in the profits in this world, and they have salaries tied to their own improving skills and knowledge. Workers go to classes to learn about topics from budgeting to international economics. There aren't any supervisors in this world. The workers do it themselves. There are a few "coaches" who see their job as helping the workers. Welcome to Johnsonville Sausage, a Wisconsin firm that has applied many of the same principles as SEMCO.[11]

13. Imagine a world in which a major grocery chain retailer internalizes the role of buying agent for the customer instead of selling agent for manufacturers. Transparency is the lifeblood of the relationship with customers and suppliers and everyone is rewarded for finding innovative win-win situations. The compensation of the highest paid employee is capped at 1 9x the average full time team member's salary, a minimum of 5% of profits is returned to the community on an annual basis, and the stock consistently trades at a best-in-class multiple. The company has led industry trends towards greener, healthier, and more local offerings while fully disclosing the trade-offs and shortcomings inherent in any mission as ambitious. Welcome to Whole Foods, an Austin, Texas based firm whose strategy to serve its customers, suppliers, community and employees on generous terms has created value for everyone involved.[12]

---

[10] For more information on SEMCO see Ricardo Semler, "Managing Without Managers," *Harvard Business Review* 89(5): 76–84; and Ricardo Semler, "Why My Former Employees Still Work for Me," *Harvard Business Review* 94(1): 61–72. For a version of these principles as applied in the United States see "Jack Stack (A) and (B)" a case study published by the Business Enterprise Trust, available from Harvard Business School. See also the September 22, 2005 lecture delivered by Ricardo Semler at MIT Sloan Management School at http://mitworld.mit.edu/video/308/

[11] For more information on Johnsonville Sausage's unique relationship with employees, see James Belasco and Ralph C. Stayer, *Flight of the Buffalo* (New York: Warner, 1993), or visit the company website: http://jobs.johnsonville.com/home/why.html. Ralph C. Stayer became President and CEO of Johnsonville Sausage in 1978, when he succeeded his father, founder Ralph F. Stayer. For a recent interview on "The Difference Network" with Ralph C. Stayer see Marquette University, October 16, 2010: http://www.youtube.com/watch? v = v1y11C_V0HM.

[12] For more information on Whole Foods philosophy see "Rethinking the Social Responsibility of Business: A Reason Debate Featuring Milton Friedman, Whole Foods' John Mackey, and Cypress

14. Imagine a world in which struggling inner city high-school and adult vocational students enter a stunning building of arches and circles designed by a pupil of Frank Lloyd Wright. The natural light floods classroom walls lined with computers, potting wheels and sewing machines. The skills taught are part of a bigger philosophy, one that sees the spirit of the arts and entrepreneurship as interchangeable. The results are stunning on every metric from improved high school graduation rates to higher employment rates. This may sound like a best-in-class community non-profit, until you look a bit deeper. Two floors away, a major health care provider leases commercial space for back-office work. The recording studio in the basement houses a Grammy-award winning jazz record label. Major multi-million dollar bids have been accepted to provide food services on the state turnpike. Is this a savvy charity or a socially-minded corporation? The founder, Bill Strickland, would say neither. Welcome to the Bidwell Training Center, where Strickland's take on entrepreneurship renders irrelevant the distinction between profit and non-profit models. After all, a non-profit still must survive on cash flows from somewhere. And while for-profit entities have invested shareholders, there is social and economic value created for employees, suppliers and the community.[13]

15. What these examples have in common is that they point the way to a new story about business and capitalism. The new story goes like this. Corporations are places where stakeholders pursue their joint interests. Stakeholders are simply groups like suppliers, customers, employees, financiers, and communities. The interests of these groups are joint. One thrives when all thrive. When the interests of one group are systematically discounted over time, all suffer. In a relatively free political system, these discounted groups seek political remedies to the discounting of their interests. Such remedies are often ineffective, expensive to all, and result in exactly the kind of left vs. right political quagmire that exists in many Western countries and seems on the rise in the developing world. Corporations are governed by their stakeholders, and while there may be many different governance schemes, depending partially on circumstances, each scheme must pay attention to the interests of each party and to their joint concerns. Directors of corporations have a duty of care to stakeholders, and managers are the agents of multiple principals. Call this story, "Stakeholder Capitalism" or "Managing for Stakeholders."

16. There is a lot more to be said about this story.[14] Suffice it to say here that Stakeholder Capitalism envisions a world where business and ethics are inextricably

---

Semiconductor's T. J. Rodgers," *Reason Magazine* (October 2005) http://reason.com/archives/2005/10/01/rethinking-the-social-responsi; or visit the company website to find numerous presentation detailing the company philosophy: www.wholefoodsmarket.com/company/corevalues

[13] For more information on Bill Strickland and his work with the Bidwell Training Center see Sara Terry, "Genius at Work," *Fast Company Magazine*, August 31, 1998; or visit his personal website www.bill-strickland.org. See also Bill Strickland, *Make the Impossible Possible: One Man's Crusade to Inspire Others to Dream Bigger and Achieve the Extraordinary* (New York: Doubleday, 2007).

[14] Pieces of this new story have their genesis in a number of places. See (Freeman and Gilbert 1988; Freeman and Liedtka 1997; Freeman 1994, 1996).

intertwined, where values and virtues are a part of corporate life, and where hopelessness and despair are replaced by a solidarity that comes from the joint achievement of shared aims. Stakeholder Capitalism seeks to transform the old story of "anything goes" capitalism into a prophetic framework. People are complex and so is the spectrum of models for how we create value and trade. There is no single "right" story underlying Stakeholder Capitalism, just as there is no "right" way to define a family. However, there is a common spirit and a potential bounded only by our own imaginations. While some may suggest that Stakeholder Capitalism is hopeless idealism, we want to insist that such a viewpoint is mired in the old story about business. For every scandal of corporate greed that reaffirms the old story, we find more and more people enacting the new one: entrepreneurs, executives and employees who choose to live each day in a world where doing business is an expression of the values for how we live together.[15]

17. Some will say that there is no new story, that companies like Whole Foods and Patagonia are just using ethics as a marketing tool, that all of this is just "wink, wink, nudge, nudge, I say, I say" in the spoofing tradition of Monty Python. Rake in the profits and laugh all the way to the bank. But, business is not hagiography. This is not a contest between saints and sinners, no matter what journalists say. The new story is more about possibility than profitability; it is about attempting to redescribe our institutions and ourselves so that we can live better. For managers and stakeholders alike, the new story is about embodying a better way to live. Such redescriptions cannot "escape" the past or be outside of space and time and culture.

The fact that the Body Shop may sometimes go too far in the interests of financiers does not negate the fact that all Body Shop employees are required (and paid) to do community service. It says only that we are trying to enact this new story in a far from perfect world with far from perfect creatures: human beings. Yet, even Stakeholder Capitalism does not adequately describe what is going on today. The potential permutations of Stakeholder Capitalism and the ever burgeoning practices of CSR, sustainability, patient capital, local sourcing, fair trade, socially responsible investing, green venture funds, among many others, help us to describe a new story emerging from entrepreneurs around the world. This new story is inspiring billions to think differently about business and the potential of markets. This new story, in this broadest sense, might come to be called "Responsible Capitalism."

18. So, what does this new story of Responsible or Stakeholder Capitalism have to do with poor people? Why should they care about this talk about talk about business? For starters we need to raise our expectations about business. We need more conversation about the role that business can play in solving the problems of poor people. This conversation is going on among social entrepreneurs and their stakeholders around the world. If we come to expect that companies raise the level of the least well off, or that they act out of respect for the earth, or that they manage the web of relationships in which they are enmeshed, then we are most of the way

[15] More of the complexities in this argument are available in an essay, "The Business Sucks Story," The Darden School Working Papers, University of Virginia, Charlottesville, VA.

towards realizing the possibility that is inherent in this new story. If the only possibility is to expect Cowboy Capitalism, then we had better get used to the rather tiresome laments of the left, and the grating rasp of the right, locked in a conversation that goes nowhere, especially for poor people. Second, we need to encourage entrepreneurship in our global society. When we see business as creating value for stakeholders, then we give moral weight to the Right's view that entrepreneurship is morally good and will help us bootstrap our way to a better world. Third, we need to see government policy that encourages and rewards companies for taking care of communities, employees, suppliers, customers and financiers. Such a "value creating" government will give new meaning to the Left's idea that we need better social structures. Fourth, we need to pay attention to people like Mohammad Yunus and others who are talking "with their feet" out in the world, making it better for poor people and hence for all of us. The problematic of global poverty is complex. But, we cannot make progress with a political philosophy that either ignores the real institution of business, or makes nineteenth-century assumptions about it, or more cruelly still, simply accepts the oppressive story that has been told as the one that is necessarily so. We cannot make progress so long as we hold ourselves and our imagination in check with a set of polarizing dichotomies that leads to political and moral paralysis. We need nothing short of a new political philosophy that takes something like Responsible or Stakeholder Capitalism into account, not as exceptional acts of altruism, but as part of the everyday mosaic of business, pointing us to what it is possible for humans to accomplish.

References

Batson, C., D. Daniel, J. Kobrynowicz, L. Dinnerstein, H.C. Kampf, and A.D. Wilson. 1997. In a Very Different Voice: Unmasking Moral Hypocrisy. *Journal of Personality and Social Psychology.* 72 (6): 1335–1348.

Batson, C. Daniel, Elizabeth R. Thompson, and Hubert Chen. 2002. Moral Hypocrisy: Addressing Some Alternatives. *Journal of Personality and Social Psychology* 83 (2): 330–339.

Batson, C. Daniel, Elizabeth R. Thompson, Greg Seuferling, Heather Whitney, and Jon A. Strongman. 1999. Moral Hypocrisy: Appearing Moral to Oneself without Being So. *Journal of Personality and Social Psychology* 77 (3): 525–537.

Boehm, Christopher. forthcoming. *Social Selection, Altruism, and Conscience Origins.*

Coolidge, Calvin. 1925. *Address to the American Society of Newspaper Editors.* January 17, 1925.

Crary, Alice. 2009. *Beyond Moral Judgment.* Cambridge: Harvard University Press.

Dawkins, Richard. 1976. *The Selfish Gene.* London: Paladin.

———. 1989. *The Selfish Gene (New Edition).* Vol. 3. Oxford: Oxford University Press.

de Waal, Frans. 1997. *Good Natured: The Origins of Right and Wrong in Humans and Other Animals.* Cambridge, MA: Harvard University Press.

Fernandez-Dols, Jose Miguel, Pilar Aguilar, Silvia Campo, Robin R. Vallacher, Alisha Janowsky, Hugo Rabbia, Silvina Brussino, and Melvin J. Lerner. 2010. Hypocrites or Maligned Cooperative Participants? Experimenter Induced Normative Conflict in Zero-Sum Situations. *Journal of Experimental Social Psychology* 46: 525–530.

Freeman, R. Edward. 1994. The Politics of Stakeholder Theory: Some Future Directions. *Business Ethics Quarterly* 4: 409–421.

———. 1996. Understanding Stakeholder Capitalism. *Financial Times*, July 26.
———. 1998. Poverty and the Politics of Capitalism. *Business Ethics Quarterly* Supplemental Volume. The Ruffin Lectures. Edited by Patricia H. Werhane, 31–35.
———. 2004. The Relevance of Richard Rorty to Management Research: A Book Review Essay. *Academy of Management Review* 29: 127–130.
Freeman, R. Edward, and Daniel Gilbert. 1988. *Corporate Strategy and the Search for Ethics*. Englewood Cliffs: Prentice Hall.
Freeman, R. Edward, and Jeanne Liedtka. 1997. Stakeholder Capitalism and the Value Chain. *European Management Journal* 15 (3): 286–296.
Freeman, R. Edward, Andy Wicks, Jeffrey Harrison, Bidhan Parmar, and Simone DeColle. 2010. *Stakeholder Theory: The State of the Art*. New York: Cambridge University Press.
Griswold, A. Whitney. 1934. New Thought: A Cult of Success. *American Journal of Sociology* 40 (3): 309–318.
Henrich, Joseph, Jean Ensminger, Richard McElreath, A. Barr, Clark Barrett, Alexander Bolyanatz, Juan C. Cardenas, Michael Gurven, Edwins Gwako, and Natalie Henrich. 2010. Markets, Religion, Community Size, and the Evolution of Fairness and Punishment. *Science* 327 (5972): 1480.
Henrich, Natalie, and Joseph R. Henrich. 2006. *Why Humans Cooperate: A Cultural and Evolutionary Explanation*. New york: Oxford University Press.
Liberman, Varda, Steven M. Samuels, and Ross Lee. 2004. The Name of the Game: Predictive Power of Reputations Versus Situational Labels in Determining Prisoner's Dilemma Game Moves. *Personality and Social Psychology Bulletin* 30 (9): 1175.
Pfeffer, Jeffrey. 1996. *Competitive Advantage through People: Unleashing the Power of the Work Force*. Cambridge, MA: Harvard Business Press.
Putnam, Hilary. 2002. *The Collapse of the Fact/value Dichotomy and Other Essays*. Cambridge, MA: Harvard University Press.
Rorty, Richard. 1982. Philosophy As a Kind of Writing: An Essay on Derrida. In *Consequences of Pragmatism*, 90–109. Minneapolis: University of Minnesota Press.
———. 1998. Can American Egalitarianism Survive a Globalized Economy?. *Business Ethics Quarterly* Supplemental Volume. The Ruffin Lectures, edited by Patricia H. Werhane, 1–6.
———. 1999. *Philosophy and Social Hope*. London: Penguin Books.
Schön, Douglas A. 1979. Generative Metaphor: A Perspective on Problem-Setting in Social Policy. In *Metaphor and Thought*, ed. Andrew Ortony. Cambridge, UK: Cambridge University Press.
Schön, Douglas A., and Martin Rein. 1994. *Frame Reflection*. New York: Basic Books.
Taylor, Mark C. 2004. *Confidence Games: Money and Markets in a World Without Redemption*. Chicago: University of Chicago Press.
Tomasello, Michael. 2009. *Why We Cooperate*. Cambridge, MA: The MIT Press.
West, Cornel. 1993. *Race Matters*. Boston: Beacon Press.
Wisdom, John. 1965. *Paradox and Discovery*. Oxford: Basil Blackwell.
Wittgenstein, Ludwig. 1974. *Philosophical Investigations*. Malden: Blackwell.

R. Edward Freeman is University Professor, Elis and Signe Olsson Professor, Academic Director of the Institute for Business in Society, and Senior Fellow of the Olsson Center for Applied Ethics at the University of Virginia's Darden School of Business.

Adrian Keevil is a Managing Partner for PlusTick Partners, where he is responsible for sourcing, evaluating, and managing investments for an investment fund.

Lauren Purnell is Senior Business Development Lead of Innovation & Strategy at the University of Virginia's Health System.

Chapter 52
Bowie's Ethics: A Pragmatist Perspective

R. Edward Freeman

1 Introduction

In *Business Ethics: A Kantian Perspective*, Norman Bowie lays out a modern-day view of Kant and carefully traces the implications for business ethics. In Sect. 2, I shall argue that Bowie's Kantian perspective can easily accommodate a pragmatist view of business ethics, linking two traditions in ethical theory that are often found to be enemies. In Sect. 3, I shall focus on Bowie's arguments about the second interpretation of the categorical imperative and his claims about the humanity of stakeholders. Finally in Sect. 4, I shall resurrect an old idea which Daniel R. Gilbert, Jr. and I called 'personal projects enterprise strategy', and show how it is at once Kantian and pragmatist, reinterpreting it without the heavy deontological language in which it was originally framed. Bowie's three formulations of Kant's dictum lend more support to such a project than may be generally perceived. To conclude, I shall suggest that one lasting contribution of Bowie's work will be that it facilitates the integration of work in philosophy and ethics with work in the disciplines of business, strengthening both along the way.

Originally published in: *Kantian Business Ethics: Critical Perspectives*, 35–47 © Edward Elgar Publishing Limited, 2012
Reprint by Springer, DOI 10.4337/9781781004968

R. E. Freeman (✉)
University of Virginia, Charlottesville, VA, USA
e-mail: FreemanE@darden.virginia.edu

© Springer Nature Switzerland AG 2023

S. D. Dmytriyev, R. E. Freeman (eds.), *R. Edward Freeman's Selected Works on Stakeholder Theory and Business Ethics*, Issues in Business Ethics 53,
https://doi.org/10.1007/978-3-031-04564-6_52

2 B Kantian Pragmatism

What is a pragmatist anyway?[1] In the philosophical literature, pragmatism is usually credited to its 'founders' or 'pioneers' such as Charles Peirce, William James and John Dewey.[2] For my purposes, I rely on the more recent work of philosopher Richard Rorty and those who have become known as 'the new pragmatists'.

Analytic philosophy, stemming largely from Kant, was dominant in most American philosophy departments during the last century and took as its wellspring the idea that the philosopher's job was to serve as the clarifier of conceptual schemes, a so-called 'handmaiden to the sciences', an analyzer of language and logic. The mantra was 'if it can be said, it can be said clearly'. Philosophers worried about 'meaning' rather than Socrates' question, 'How should we live?'. Ethics itself was turned into meta-ethics or the analysis of the meanings of words like 'good' and 'right'. Rorty was part of the center of this analytic mainstream as a full professor at Princeton University, one of the top American philosophy departments. His main insight, which he credits to Dewey and others, but which really was crystallized in *Philosophy and the Mirror of Nature* (Rorty 1979), was that most of this way of doing philosophy rested on a set of distinctions we inherited from the Greeks, and a view of language as representing the world. He argued that the idea of representation made no sense, that it was based on taking vision as a foundational sense, and that at least since Wittgenstein (1953), we should know that language doesn't work that way. Rather, language is a tool, not a representation. There simply is no other way to deal with the world other than through language. To assume otherwise is to invoke the 'appearance versus reality' distinction common to philosophers since Plato and crystallized in Kant. According to Rorty, we needed to return to Dewey and see the intellectual's task as producing social hope and of always trying to figure out how we could live better. In short, Rorty turns us back to Socrates, focused on how we should live.[3]

The controversial way that Rorty often explains what he is up to is to claim that we need to replace the idea of 'truth' as the goal of inquiry with the goals of 'hope' and 'freedom'. Intellectual life matters precisely because we can offer descriptions and re-descriptions of what we humans do that allow us to live better. Rorty claims that we should adopt Dewey's goal for inquiry as 'increasingly free societies and increasingly diverse individuals within them' (1999, p.49). He argues:

---

[1] This section relies heavily on parts of Freeman, Harrison, Wicks, Parmar and De Colle, *Stakeholder Theory: The State of the Art* (2010).

[2] See Sandra Rosenthal and Gene Buchholz, *Rethinking Business Ethics: A Pragmatic Approach* (2000); Gary Gutting, *Pragmatic Liberalism and the Critique of Modernity* (1999); Robert B. Westbrook, *John Dewey and American Democracy* (1991); and Richard Rorty, *Consequences of Pragmatism* (1982), for a complete history.

[3] Rorty has a nuanced view here. 'How should we live' is not a question that admits of one answer, true for all time. Rather, we are constantly trying to find better ways to live; we struggle for adjustment.

> Pragmatists … treat inquiry in both physics and ethics as the search for adjustment … [Pragmatists ask] the practical question, 'Are our ways of describing things, of relating them to other things as to make them fulfill our needs more adequately, as good as possible? Or, can we do better? Can our future be made better than our present?' (1999, p.72)

In short, it is up to us to figure out which projects to pursue, which descriptions and re-descriptions might serve us better, and there are no guarantees that we'll be successful. Pragmatists like Rorty believe that there are only two interesting projects for us to engage with. The first is ever more useful descriptions and re-descriptions of 'self', the second is ever more useful descriptions and re-descriptions of 'community'. Indeed, some have suggested that these twin pragmatist projects are just two sides of the same coin, since 'selves exist in communities' and since communities without 'individuals' are pretty uninteresting.[4]

Pragmatism is often seen as a response to Kantian and post-Kantian thought. And, Rorty often argues that philosophy went off the rails with Kant and has never recovered. While such an argument may well be valid with respect to epistemology and metaphysics, it need not spell the end of Kantian ethics, at least in the way that Bowie describes Kantian ethics.

In *Business Ethics*, Bowie suggests that each of Kant's formulations of the categorical imperative has a different use. The first formulation is useful in terms of sorting out the 'bads', delineating immoral business practices. The second is useful in having us see that business is a fundamentally human activity among human beings, and that theories like 'stakeholder theory' must take this basic humanity into account. The third focuses on moral community and emphasizes, as Bowie would have it, the idea that we can see business as a moral community connected to the larger moral community of humankind.

Now, it seems to be a matter of pride among Kant scholars that these three formulations of the Categorical Imperative are in fact logically equivalent. Many journal pages have been written about such demonstrations. However, it is difficult to see what turns on such an equivalence. First of all, let's grant Bowie's arguments about the usefulness of the three formulations and the connections to business ethics. We'll come back to one of these in the next section. One possibility is that the deductions that Bowie makes from each of the formulations are strengthened if we know that the formulations are equivalent. However, Bowie's arguments, if they are valid and sound, seem to stand on their own. A second possibility is that Bowie could be accused of using Kant where it is convenient, which seems to worry him.[5] But Bowie is also concerned with generating insight into business problems. His very use of the three formulations shows that he is faithful at least to the spirit of the Kantian enterprise. Just as Kant had not much to say about capitalism, Bowie has not much to say about the problems that clearly worried Kant and led him to develop his ethics. For the pragmatist, context is everything, and Bowie clearly shows the

[4] See R.E. Freeman, 'Poverty and the politics of capitalism' (Freeman 1998); Rorty (1998) makes his first foray into business ethics in this volume, replying to a former student, Cornell West.

[5] Cf. Bowie (1999), at p.5.

usefulness of the Kantian program, without having to believe that the three formulations are equivalent.

A third possibility is that Bowie simply doesn't want to admit his pragmatism, and with good reason, given the rather low regard in which pragmatists are held by most 'real' philosophers. However, he makes a slip in the first chapter where he admits that the Kantian theory is not the only one. He says:

> Fifth, I do not believe that the Kantian vision is the only reasonable moral vision that one can apply to business ethics. For example, Robert Solomon has an articulate and inspiring Aristotelian virtue ethics vision … However, I shall argue later that I believe Kantianism does a better job than competing ethical theories in justifying a number of 'good' business practices. (Bowie 1999, pp.4–5)

I would interpret 'better job' as a way of claiming precisely that Kantianism is simply more useful than other theories. Of course, some may need to make truth claims, or equivalence claims, etc., but not much hinges on such philosophical debates. Hence, my conclusion is that Bowie's Kantianism could well be explained as Kant's pragmatism, which makes Kant much more interesting to the analysis and framing of real ethical problems.

There is a significant tension in Kant between his ethics and his views on real world events such as the French and American revolutions of his time. While Kant argued that revolution was immoral, he was also an admirer of both revolutions. This is not surprising since human autonomy and freedom is at the heart of the Kantian project. Lewis White Beck claims:

> In *The Strife of the Faculties*, Kant draws a moral conclusion from the French Revolution. The passionate participation in the good, viz., the disinterested enthusiasm with which the Revolution was greeted, could have no other cause, Kant thinks, than a moral predisposition in the human race to seek what is ideal and purely moral. It gives hope and evidence of the moral progress of mankind. (Beck 1971, p.419)

Kant's pragmatism is signaled again by Bowie in his more comprehensive treatment of Kant's ethics, rather than a focus on the stylized formulations of the categorical imperative. Bowie casts a wide net over Kant scholarship, from strict interpretations to neo-Kantians like John Rawls, not to get Kant right once and for all, but to build on the core Kantian insights and develop a Kantian ethics that is useful in the twenty-first century. Such a project, for all of its roots in the categorical imperative, is essentially pragmatist.

Why bother with this debate about whether Bowie's project is Kantian, pragmatist or even Martian in nature? Business ethics has been dominated for too long by philosophers who take the trifecta of ethical theory, deontology, consequentialism and virtue, as the only way to frame problems. Philosophers are conveniently the experts in these areas, and business ethicists can simply apply the theories to the morally suspect institution of business. By crossing these lines, turning Kant into some kind of modern pragmatist, Bowie opens up more space to connect important philosophical traditions to the disciplines of business, thereby opening new philosophical and business questions. I shall try to give an example of how this is accomplished in the next section.

3 The Humanity of Stakeholders

Kant's second formulation is one that prohibits the treatment of other persons as mere means. Bowie rightly argues that the application of this formulation requires seeing stakeholders as human beings. This was the original insight behind Evan and Freeman's (1993) paper on Kantian capitalism, and more recently the paper by John McVea and Freeman on a 'names and faces approach to stakeholder theory' (2005). Bowie's conclusions are that if we understand the humanity of stakeholders as ends and never mere means, then companies have obligations not to lay off employees and to create meaningful work.

Part of what is at issue is Kant's and Bowie's view of what it means to be a human person. Kant had a reasonably clear idea of how to answer 'What is a person?' and Bowie approves. Kant sees the question as connected with what it means to be a moral agent, and focuses on ethics built on how all persons are alike. For the pragmatist, the question 'What is a person?' is more complex.

If the question is a request for definition, then it is built into a more complex narrative. For instance, we could answer it in biological terms so that persons are members of the same species. We could answer in more anthropological terms by claiming that persons live together in communities, use language, build cities, domesticate animals, and kill each other for reasons other than food. We could answer in religious terms, making persons special creatures of the gods, or even beings with the potential to be such creatures. We could answer in more psychoanalytical terms by giving reasons for why each person is different, uniquely themselves, given their early childhood experiences.

The point is not that there are many definitions, but that each has a usefulness or not depending on how the underlying or background narrative lets us live. Definitions are not the kind of things that can be right or wrong. Each has a purpose embedded within a theory. Thus, what is interesting is the underlying narrative. Kant's purpose was to give a comprehensive, once and for all time, necessarily true account of ethics, or in the Kantian vernacular, a transcendental argument about how morality is possible. However, even if we give up such a philosophical project as resting on a representational view of language, and an outmoded idea of things as they appear and things in themselves, we can still determine whether or not Kant's underlying narrative is useful today. Indeed, since it depended for its insights on the ideas of autonomy and freedom, perhaps there is a better way to tell Kant's story, and this is precisely what Bowie seems to accomplish. Bowie says:

> Human beings ought to be respected because human beings have dignity. For Kant, an object that has dignity is beyond price … They have dignity because human beings are capable of autonomy and thus are capable of self-governance. As autonomous beings capable of self-governance they are also responsible beings, since autonomy and self-governance are the conditions for responsibility. (1999, p.43)

So far, so good. Indeed, this is a Kantian argument for what I have called in other places, the Responsibility Principle, which actually gets stakeholder theory off the ground. Without some notion of people wanting to take responsibility for their actions, the so-called 'flip side' of autonomy, ethics simply doesn't get off the ground. This was the main point that Rob Phillips and I developed in 'Stakeholder theory: a libertarian defense' (Phillips and Edward Freeman 2002).

Bowie applies this formulation to the problem of layoffs. He wants to claim that layoffs are rarely the fault of employees, even when there has been no deception or coercion. Sometimes things happen that could not be foreseen, beyond the control of employees. He says of massive layoffs, 'if blame is to be placed, perhaps it should be placed on management, who hired too many employees in the first place' (1999, p.51). However, it is difficult to see how the blame gets placed on managers here if both employee and manager are to be Kantian persons. If indeed the arrangement is voluntary, and there is no deception or coercion, then we could rewrite this quote from Bowie as 'if blame is to be placed, perhaps it should be placed on employees, who agreed to do a job which was not sustainable'. In fact, blame is tough in these cases, since business is complicated. Bowie wants to push further, since he knows this 'blame management' argument really doesn't work, so he goes on to argue that the current institution of capitalism actually encourages deceptive and coercive employment arrangements. The argument seems to be something like this: find an employee arrangement that looks voluntary and most of the time, if you look closer, you will find deception and coercion. Maybe this is correct, but it is an 'entangled fact-value claim' that would admit of much more research.[6] Bowie then makes several suggestions for reform of these agreements, all of which seem, to me, to be on the right track from Kantian, Pragmatist and Good Management perspectives. Profit sharing and transparency work in many businesses, and it is a tragedy that more businesses don't try them. Of course, there may be some circumstances where they would not work, so we need to be cautious about adopting the Kantian project of finding rules or practices that work for all businesses at all times, in the process of value creation. According to Bowie's view of Kant, this first formulation can at best give us ways to avoid immoral action. To sort this out more carefully, we would need to figure out how to understand the entangled fact-value claims. This will be a matter of both philosophy and management theory, including empirical and clinical studies.

Bowie wants to go further and deduce an obligation for companies to create 'meaningful work'. And, while I am sympathetic with the intuition behind this argument, namely that it is a really good idea to have employees engaged in the purpose of any company, I am skeptical that such an argument can be built on Kantian grounds. If persons are autonomous and capable of self-governance, we need to be careful when we engage in second-guessing their choices, when these choices are real, that is, made without coercion and deception. Now, given the argument of the last paragraph, Bowie may believe that these conditions are never present. But, this comes dangerously close to paternalism, which is decidedly anti-Kantian, on my reading.

[6] See Hilary Putnam, 'The collapse of the fact/value dichotomy' (in Putnam 2002), for an analysis of how misleading the fact–value distinction is, and how most claims are actually entanglements.

I treat you with dignity, only when I respect the choices that you make for yourself. This is easy to do when you make the best choices; however, it is more difficult when you make choices that I know are bad for you. Indeed this is what we might call The Parent's Dilemma. Treating your children with dignity and not as mere means, involves letting them, when they are 'of age', make choices that may not be best for them. I treat you as an end only when I see your projects as yours, not how your projects affect mine. This is as true for parents, with the project of 'taking care of and wanting what is best for their children', as it is for managers with a project of 'taking care of employees and wanting what is best for them'.

Bowie says, 'To treat a person as an end itself sometimes requires that we do more than merely refrain from coercion or deception; it requires that we take some positive action to help a person' (1999, p.63). That is surely correct and Kantian (and pragmatist, as well); but it is difficult to see what is either wrong or anti-Kantian with the following scenario (adapted from Freeman and Gilbert Jr 1988).

Imagine company X where prospective employees are explicitly told that there is no participation by employees in running the company, and the jobs that are described are hardly 'meaningful' in any sense. They consist of mostly mindless routine and repetition. However, the company pays for hours worked, and it expects no extraordinary commitment. Consider Norm who wants one of these jobs because he can fund his musical dreams by working 8 h a day or less in a job that requires little thought and commitment. Even if we know that Norm will hate this job after a few weeks, and resent the job and all of its 'idiot managers', it is difficult to say that on Kantian grounds, such a job is wrong to offer and wrong to accept. Surely, there is a place in business for jobs where people do not have be part of the 'community' that is a company, where they don't have to care about much at all except showing up, doing what's required and cashing their paycheck. Bowie may be correct that these companies can't sustain themselves for long without resorting to coercion or deception; but again, that is a fact-value entanglement that needs more research. And, as I suggested earlier, it is exactly these 'fact-value entanglements' that can lead philosophers to look at work done in the disciplines of business.

In outlining what a Kantian Moral Firm would look like, Bowie suggests seven principles, each of which contains such fact-value entanglements. The seven principles are (1999, pp.90–91):

1. The firm should consider the interests of all the affected stakeholders in any decision it makes.
2. The firm should have those affected by the firm's rules and policies participate in the determination of those rules and policies before they are implemented.
3. It should not be the case that for all decisions, the interests of one stakeholder take priority.
4. When a situation arises where it appears that the humanity of one set of stakeholders must be sacrificed for the humanity of another set of stakeholders, that decision cannot be made on the grounds that there is a greater number of stakeholders in one group than in another.

5. No principle can be adopted which is inconsistent in the sense discussed in Chap. 1 [of *Business Ethics*], nor can it violate the humanity in the person of any stakeholder in the sense discussed in Chap. 2 [of *Business Ethics*].
6. Every profit-making firm has an imperfect duty of beneficence.
7. Each business firm must establish procedures designed to insure that relations among stakeholders are governed by rules of justice. These rules of justice are to be developed in accordance with principles 1–6 and must receive the endorsement of all stakeholders. They must be principles that can be publicly accepted and thus be objective in a Kantian sense.

Even though these principles are formulated in Kantian language, there are fact-value entanglements in each of them. To pick an example, consider the first principle. Who counts as a stakeholder? And, as I have argued endlessly, the very use of 'stakeholder' is not a value-neutral act. I am skeptical of one answer to this question for all business activity, for all time, as mentioned earlier. Or consider principle 6. Suppose that firms actually managed for stakeholders, and were driven by purposes other than or in addition to profits. It is not clear to me that an imperfect duty of beneficence still applies. I am not arguing against these principles, or against Bowie's conception of the Kantian firm, only for a more pragmatist interpretation of them so that we can figure out what our best management thinking might say about them, or how the combination might frame a new approach to business.

4 Kant, Bowie and Corporate Strategy

In *Corporate Strategy and the Search for Ethics*, Daniel R. Gilbert Jr. and I tried to work out what we believed to be a pragmatist view of the connection between persons and corporations. We also believed that it was consistent with, and indeed inspired by, Kant's second formulation of the categorical imperative. We further believed that such a narrative gave some support for the idea that companies should be 'managed for stakeholders', or, as I would now put it, that we come to understand capitalism as a cooperative scheme of how we create value and trade, and that involves managing for stakeholders. We argued a libertarian line that capitalism was best understood as 'a set of voluntary agreements among consenting adults'; and we articulated these ideas in six main principles that relied heavily on the language of rights. I would like to take this occasion to reformulate these principles and connect them to Bowie's seven principles of a moral firm.[7] While the arguments for these

[7] The original six principles were (Freeman and Gilbert Jr 1988, pp.168–170):

The principle of personal autonomy. Persons have the right to formulate, articulate and pursue projects, unless in doing so they violate the rights of others.

The principle of conventional rights. Persons' rights are a matter of general agreement among members of the community.

principles were heavily framed in the language of rights, a language that I do not find as useful as I once did, they can be revised along the following lines to capture the essence of our ideas.

4.1 Principle of Complexity

> Persons are complex. They are enmeshed in a set of values, a history which they do not always understand, a set of relationships with others and a set of aspirations or projects about their present and future. They are at once autonomous and connected.

This principle rejects the cardboard cutout view of human nature at the heart of the current narratives of capitalism. People are complex, they act for a variety of reasons. Their actions benefit themselves and others, and people usually take that into account. It is also important to note that since we are complex, we are able to differentiate consequences based on who is being affected. It is part of human nature to care more about consequences that affect those we are close to rather than others. The view of human nature that we hold has a tendency to become a self-fulfilling prophecy – when we expect managers to be self-interested, they meet those expectations. By raising the bar for human complexity in business, we allow for a broader conceptualization of 'value' and create more space of ethics. That is a reason why the Principle of Stakeholder Responsibility (below) is important. It helps to balance our natural tendency to discriminate and reminds us that, despite our differences and separation, we still can have profound effects on each other.

This principle simply makes explicit Bowie's Kantian starting point, or at least a pragmatist interpretation of that starting point. It is our complexity rather than our moral agency which is the starting point for ethics. Of course, this view does violence to Kantian ideas about duty taking a central role in ethics; but I believe that the idea of responsibility is better.

The principle of respect for persons. Persons have a duty to treat others as ends in themselves, rather than as mere means.

The principle of voluntary agreements. Persons have the right to enter into voluntary agreements with others in order to accomplish their projects.

The principle of human institutions. Institutions, for example, corporations, exist as a mere means for the accomplishment of the projects of institutional members.

The principle of corporate membership. Corporate members have the right to participate in those decisions that affect the accomplishment of their projects in an important way. Typical corporate members include managers, stockholders, employees, suppliers, customers, and community representatives.

4.2 *The Principle of Respect for Persons*

> Society works best and all of us can live better lives when persons treat others with dignity, as complex persons in their own right with values, etc. Persons should treat others as ends in themselves rather than as mere means. When a person is treated as a means, they must give permission.

The claim in this principle is not the universalist one of rights, even conventional rights, but the pragmatist one that it is more useful to act in a way that treats others as ends. This is more complex, because our understanding of the person is more complex. But, this principle and the next one capture the essence, I believe, of the Kantian dictum.

4.3 *The Principle of Respect for Communities*

> Society works best when persons treat other communities as complex relationships among persons, with values, histories and aspirations. Persons should treat other communities as ends rather than mere means.

Here we mean not to reify governments or other social structures, but simply to suggest that if we are simultaneously autonomous and connected, then we probably should respect the set of connections that exist. This is complicated and needs more argument.

4.4 *The Principle of Stakeholder Cooperation*

> Value can be created, traded and sustained because stakeholders can jointly satisfy their needs and desires by making voluntary agreements with each other that for the most part are kept.

Rather than assume that we are all first and foremost self-interested and out to maximize our own benefit, this principle highlights the social nature of value creation. Value is not 'discovered' lying around the market, but created through shared assumptions and beliefs in a community. Value, any value, is a social phenomenon. We must create value in a context, with the help of others and with others who value what we create. This principle acknowledges that business activity is explicitly social and uses that to enhance the process of value creation. Foregrounding the social nature of business gives us insight into the problem of value creation and trade because it puts the focus on human relationships and the shared sense making that creates value.

4.5 The Principle of Stakeholder Engagement

> To successfully create, trade and sustain value, a business must engage its stakeholders. Almost every business transaction involves customers, suppliers, communities, employees, and financiers. Other stakeholders, such as media, additional civil society representatives, NGOs, etc. are often affected or can affect value creation.

Rather than argue over whose rights trump whose, this principle acknowledges that a large cast of stakeholders are necessary to sustain value creation. As often as possible, the needs of multiple stakeholders must be met. There may be specific situations in which privileging the rights of one group can benefit others in the long term, but this is not clear prima facie, and must be decided upon by the affected parties. Recognizing the role of a multitude of stakeholders in the value-creation process diminishes the problem of the dominant group. Instead of trying to find and create arguments for one group's right to trump the rest, engaging stakeholders to create as many win/win situations as possible lies at the heart of creating sustainable value.

4.6 The Principle of Stakeholder Responsibility

> Value can be created, traded and sustained because parties to an agreement are willing to accept responsibility for the consequences of their actions. When third parties are harmed, they must be compensated, or a new agreement must be negotiated with all of those parties who are affected.

This principle rejects the view that business is amoral or even immoral. If business is a social process, then morality is at its center. Scandals and selfish behavior are a breach of the trust and transparency that is the norm for business to flourish. We can all think of notable lapses in managerial responsibility, but the successes are less visible. Being proactive about effects on others, rather than waiting for government recourse, will help managers build stakeholder trust and loyalty, both of which will help create a more sustainable business. The stakeholder responsibility principle brings ethics into the heart of capitalism and reduces the problem of ethics and capitalism. It also helps resolve the problem of business in a liberal democracy, because if ethics is inherent to business, then the role of government as an 'ethics watchdog' is lessened. Responsible business does not need external imposition of morality. Finally, this re-description of capitalism helps managers embed ethics into the way they think about their day-to-day activities.

One obvious question is whether or not this Stakeholder Capitalism can be fairly said to be Kantian Capitalism. I believe that it is at least Kantian in the pragmatist spirit in which I have interpreted Bowie's Kant.

However much one may disagree with particular conclusions that Bowie draws about particular business practices, he points the way toward how to develop the narratives that we need to connect the very best thinking in ethics with the very best thinking in management theory and science, and the very best thinking about business practices. Bowie's integration is of equal importance to his re-conceptualizing of Kant. For both of these feats we should be grateful, and repay him with much more of a similar kind of scholarship.

References

Beck, Lewis W. 1971. Kant and the Right of Revolution. *Journal of the History of Ideas* **32** (3): 411–422.

Bowie, Norman. 1999. *Business Ethics: A Kantian Perspective*. Malden\Oxford: Blackwell Publishers.

Evan, W.M., and R.E. Freeman. 1993. A Stakeholder Theory of the Modern Corporation: Kantian Capitalism. In *Ethical Theory and Business*, ed. T.L. Beauchamp and N.E. Bowie, 97–106. Prentice Hall: Englewood Cliffs.

Freeman, R. E. 1998. Poverty and the Politics of Capitalism, In *The Ruffin Series, Vol. 1, The Society for Business Ethics,* ed. R.E. Freeman, 31–5.

Freeman, R. Edward, and Daniel R. Gilbert Jr. 1988. *Corporate Strategy and the Search for Ethics*. Englewood Cliffs: Prentice Hall.

Freeman, R. Edward, Jeffrey S. Harrison, Andrew C. Wicks, Bidhan L. Parmar, and Simone De Colle. 2010. *Stakeholder Theory: The State of the Art*. Cambridge: Cambridge University Press.

Gutting, Gary. 1999. *Pragmatic Liberalism and the Critique of Modernity*. Cambridge: Cambridge University Press.

McVea, John, and R. Edward Freeman. 2005. Stakeholder Theory: A Names and Faces Approach. *Journal of Management Inquiry* 14 (1): 57–69.

Phillips, Robert, and R. Edward Freeman. 2002. Stakeholder Theory: A Libertarian Defense. *Business Ethics Quarterly* **12** (3): 331–349.

Putnam, Hilary. 2002. *The Collapse of the Fact/Value Dichotomy and Other Essays*. Cambridge, MA: Harvard University Press.

Rorty, Richard. 1979. *Philosophy and the Mirror of Nature*. Princeton: Princeton University Press.

———. 1982. *Consequences of Pragmatism*. Minneapolis: University of Minnesota Press.

———. 1998. 'Can American Egalitarianism Survive a Globalized Economy?', The Ruffin Series, Vol. 1, *The Society for Business Ethics*, 1–6.

———. 1999. *Philosophy and Social Hope*. New York: Penguin Books.

Rosenthal, Sandra, and Gene Buchholz. 2000. Rethinking Business Ethics: A Pragmatic Approach. In *The Ruffin Series in Business Ethics*, ed. R.E. Freeman. Oxford: Oxford University Press.

Westbrook, Robert B. 1991. *John Dewey and American Democracy*. Ithaca: Cornell University Press.

Wittgenstein, Ludwig. 1953. *Philosophical Investigations*. Oxford: Blackwell Publishing.

R. Edward Freeman is University Professor, Elis and Signe Olsson Professor, Academic Director of the Institute for Business in Society, and Senior Fellow of the Olsson Center for Applied Ethics at the University of Virginia's Darden School of Business.

Chapter 53
Short Term vs. Long Term: A Skeptical View … and an Alternative

R. Edward Freeman

As the debate rages about the appropriate form of capitalism, I have become increasingly skeptical about one of the underlying assumptions that never gets challenged. It is the distinction between "short-term" and "long-term." We are told that the ills of capitalism are due to its short-term orientation, and that this leaves a big gap between Wall Street and Main Street. Directors are urged to take a long-term view of their companies, and executives are to manage for the long term.

This distinction can easily become an excuse. For many years I have been told, "Well, all these ideas of yours about stakeholders and ethics and such will work in the long term, but in the short term business is about profits and shareholders." The big problem with this line of reasoning is that it is always the short term. We live in the "now," not the future. If these ideas eventually pay off, there must come a time where they are paying off in the short term, or else they never pay off.

The real problem has nothing to do with the "short term" vs. "long term" distinction. The real problem is that the main narrative we have about business is deeply flawed. That narrative says that business is about making money for share-holders and that everyone involved in a business transaction is a short-term max-imizer of their individual self-interest.

There is now a robust body of knowledge that these assumptions are no longer useful. There are countless research studies that show that people are quite complicated, and there are many businesses that defy the idea that shareholder value is all that counts.

Originally published in: *Directors and Boards*, Issue: 2016 Annual Report, 83–84 © Directors and Boards, 2016
Reprint by Springer

R. E. Freeman (✉)
University of Virginia, Charlottesville, VA, USA
e-mail: FreemanE@darden.virginia.edu

© Springer Nature Switzerland AG 2023

S. D. Dmytriyev, R. E. Freeman (eds.), *R. Edward Freeman's Selected Works on Stakeholder Theory and Business Ethics*, Issues in Business Ethics 53,
https://doi.org/10.1007/978-3-031-04564-6_53

Most businesses have a purpose, a *raison d'etre,* an answer to "why they exist" that goes well beyond making money for shareholders. Of course, making money for shareholders is important, but so too is purpose, ethics, and creating value for stakeholders. If we can enact a new narrative of business that focuses additionally on these ideas, we have a way out of the endless debate about the appropriate form of capitalism.

Purpose doesn't just matter in the long term; it matters every day. It is about the meaning of our lives and how we are enacting that meaning. You don't tell your children to have a purpose, but don't live it in the short term. The same goes for ethics. It's about the quality of our relationships with others, each and every day. Enron is a glaring case of putting ethics off till the future, as the board gave an exception to the ethics policy to the CFO. The only way you can build a great company is by making purpose and ethics a priority every day. The distinction between short term and long term is irrelevant.

What about stakeholders? Remember that key stakeholders are customers, employees, suppliers, communities, and financiers. Businesses have to create value for these stakeholders each and every day. Yes, we have to anticipate what those stakeholders will need in the future, but that is based on what a company is doing to create value for them today. Increasingly in today's global world, these stakeholders are interdependent. Building a great company is about getting them all going in the same direction. You can't build a great company by satisfying customers, and then saying that in some distant future you will take care of your communities.

The bottom line here can be seen through the lens of executive compensation. There are endless debates, laws, expressions of outrage over both executive pay and the proposed laws to regulate it. It is often said that executive pay is too short-term oriented toward stock price, or earnings, or some other quarterly measure. This "short termism" is often diagnosed as the real problem with capitalism as it exists today.

The problem is not short termism. The problem is that we measure and compensate the wrong things. My proposal is that companies consider compensating people based on "stake options" not "stock options." A stake option has five measurements, each of which could include multiple variables:

1. Consumer satisfaction
2. Supplier satisfaction
3. Employee engagement
4. Community support
5. Financial success

The measures could be normalized and a market could exist so that I might short a stake option because of employee engagement. If companies really do create (and sometimes destroy) value for stakeholders, then let's make that the main way that we compensate executives.

For CEOs and board members, we might also add a measure of how well they are driven by the company's purpose and their commitment to the company's ethics. Such a perceptual measure from employees would hold everyone's feet to the fire.

I can't see any other way out of the endless debate about capitalism, other than changing the underlying story, getting rid of the short term vs. long term distinction, and beginning to build systems, like stake options, to revitalize the best system of business that we have ever created.

R. Edward Freeman is University Professor, Elis and Signe Olsson Professor, Academic Director of the Institute for Business in Society, and Senior Fellow of the Olsson Center for Applied Ethics at the University of Virginia's Darden School of Business.

Chapter 54
Responsible Capitalism: Business for the Twenty-First Century

R. Edward Freeman, Bidhan L. Parmar, and Kirsten E. Martin

1 Introduction

The last 10 years have seen clarion calls for the reform of capitalism, as evidenced by the essays in the present volume. However, most calls fall well short of what is needed, and represent only partial success. The main reason for a lack of thoroughgoing reform is the standard story of business and capitalism that is deeply embedded in our culture in the West and, indeed, has taken hold around the world. This main narrative about the nature of business and capitalism focuses on the pursuit of profits and "the physics of money" (Freeman et al. 2010) as the main characters in the story. Human actors are depicted as self-interested economic beings who are in constant competition with each other, either individually or within their business organizations. Most attempts at reform retain this standard narrative.

True reform of the idea of capitalism must address this underlying narrative and propose a different story, and that is our task here. In these few pages we can only briefly sketch the story, however, the good news is that this new narrative of business is being realized every day by a collection of companies that span the spectrum from start-ups to multinationals (see, for example, Sisodia et al. 2014; Mackey and Sisodia 2014). Indeed, we shall argue that once we understand this new narrative we can become more aware of how its main tenets have always been present, if underemphasized, in great businesses and business leadership.

Originally published in: *Re-Imagining Capitalism*, 135–144 © Oxford University Press, 2016
Reprint by Springer, https://doi.org/10.1093/acprof:oso/9780198785453.003.0010

R. E. Freeman (✉) · B. L. Parmar
University of Virginia, Charlottesville, VA, USA
e-mail: FreemanE@darden.virginia.edu

K. E. Martin
University of Notre Dame, Notre Dame, IN, USA

© Springer Nature Switzerland AG 2023

S. D. Dmytriyev, R. E. Freeman (eds.), *R. Edward Freeman's Selected Works on Stakeholder Theory and Business Ethics*, Issues in Business Ethics 53,
https://doi.org/10.1007/978-3-031-04564-6_54

We proceed as follows. In the next section we sketch the tenets of the old narrative more precisely, and we highlight some of the partial suggestions for reform. Then we turn to a set of principles developed by a group of scholars over the last 40 years in what has come to be called "stakeholder theory," and suggest that these stakeholder principles are more useful in understanding how businesses need to operate in the twenty-first century. Finally we summarize with some principles of what we call "Responsible Capitalism."

2 Some Partial Solutions

The standard understanding of business by many is mainly from an economic point of view. In this view there are buyers and sellers, each looking to attain what the other offers at the lowest possible price. Economic theory is a set of propositions about the conditions under which this is possible and is one of the main vocabularies through which we understand business. The most common understanding goes something like this: business is about maximizing the returns—measured in profits, or stock price, or some such financial measure—to the owners of capital, i.e., shareholders, banks, bondholders, etc. By seeking to maximize the interests of these owners, society is better off, at least according to certain strands of the theory (Friedman 1970; Jensen and Meckling 1976). In addition, in order to maximize their investment, owners of capital must adequately control the activities of the business. In short, we can see business people on this view as solely concerned with profits and money. Further, since we are using the vocabulary of economics, the actors care only about their own interests—which in reality may well be other regarding, and maximizing those interests. The drive to compete and win is the ultimate energizing force of business, and the realization of that drive makes society better off.

There is much that is wrong with this characterization of business and capitalism, not the least of which that it is something of a "straw man" argument—easily blown over, as well as a characterization of actual practice. However, the picture does resonate with the popular conception that business is primarily about profits, money, and self-interest. We shall leave a more nuanced parsing of this view for another occasion or other scholars. Suffice it here to say that there have been calls for reform, especially given the global financial crisis (GFC) that has been laid directly on the doorstep of this view of capitalism.

For starters, some have argued, since Marx, that the primary deficiency of capitalism is that it pits "capital" against "labor" (Jameson 1991). In such a battle, capital is bound to win, unless labor can organize, call strikes, etc. In modern terms, some theorists have called for more participation in the management of the business by its employees, not just its executives. Companies such as Toyota, who have empowered assembly line employees to be responsible for product and manufacturing quality, have indeed outperformed their more classical capitalist rivals such as General Motors. Many have argued that pitting capital versus labor is a false dichotomy (Follett 2011). It is surely in the interests of owners, managers, and employees to cooperate together to create value. However, the old story of business says that this value is always claimed by the owners of capital at the expense of labor. This assumption prevents more thoroughgoing reform.

Second, some have argued that we need to take more account of the role of government. According to the standard narrative (Freeman et al. 2010), the appropriate role of government is to regulate the otherwise unbridled search for profits by businesses; without regulators, business would be allowed to create havoc for society. This role of government is often attributed to Keynes and his followers. Government action is as necessary as is the tending of a gardener to a garden, according to their metaphor (Keynes 2006). Without government there would be chaos and disorder. Whether the modern-day monetarists such as Friedman agree or disagree with Keynes makes no difference. The idea of government as a necessary actor to ameliorate the fluctuations of the market and the behavior of business is the same.

Surely, the GFC taught us that business does not exist in a vacuum. Rather, we should see it as an institution that is embedded in other societal institutions, with government being one of these. There have been a number of calls for regulation, since the GFC, as well as real regulatory reform, such as Dodd-Frank in the United States, and a number of regulations on executive pay in other parts of the world. Again, the problem is that the basic assumptions about business remain the same. Dodd-Frank assumes that businesses and banks remain best described as driven by profit, and that there should be side constraints imposed on this "natural" drive. Reform is again piecemeal and partial.

Finally, there are a number of reforms that are aimed at improving "corporate governance" by paying more attention variously to the interests of shareholders, executives, and employees in the basic structures of boards of directors and the day-to-day management of the business. All of these reforms are well meaning yet they stop short of questioning the basic assumptions of business (Bebchuk 2006). It is to those basic assumptions that we now turn.

3 Stakeholder Theory and the Basics of Business

What makes any business successful and sustainable over time? The answer from a practical point of view is that it continuously creates value for its customers, suppliers, employees, communities (and civil society), and its financiers. If it loses the support of any one of these groups, then it is vulnerable to failure. If it ignores any one of these groups over time in a free society, that group uses the political process to enforce its claims. Business is a voluntary activity, for the most part. And, those who engage with a business do so voluntarily, since they usually have some amount of choice about whether to do so. Of course, the existence of choice depends on the underlying structure of society, but as in the case of neoclassical economics, we want to assume that some form of property rights or some other basic moral regime can be appealed to in order to give business some legitimacy—issues that are complex and beyond the scope of this chapter.

So, any business must create some value for its customers with products and services that make their lives better off, so that customers are willing to pay for them. Any business must work with suppliers, as their customers, for the same reasons. Businesses must create value for employees so that they are willing to learn how to operate various aspects of the business. And, research is clear that the more employees are engaged in the business, the more successful the business (Edmans et al. 2014). It is a bit more controversial that business must create value for communities, but nonetheless it is true. Communities have often spoken out via the regulatory process or the courts and prevented a particular business from operating within its boundaries. More often communities have sought restitution when a business has damaged a community. And, of course, any business must have the support of its financiers, a point acknowledged by the old narrative.

However, we need to go further. The interests of each of these stakeholders do not exist in isolation. How engaged our employees are surely affects our ability to innovate and produce quality products for our customers, and there are similarities among all stakeholders. There is now a burgeoning and substantial literature on how stakeholder interests affect all areas of business (see Freeman et al. 2010; Parmar et al. 2010 for the arguments and references). Indeed, all real businesses continually try and satisfy stakeholders. The introduction of this simple and practical idea of "creating value for stakeholders" turns the old narrative on its head (see also Chap. 12, this volume). In fact, we can suggest a number of different assumptions that form this new narrative of business, which are summarized in Table 54.1. Please note that we are not trying to be complete and mutually exclusive here, but merely aim to illustrate the kind of ideas that are necessary to really reform our old idea of business and capitalism.

Let us discuss each of these in turn:

3.1 *The Unit of Analysis Assumption*

A useful unit of analysis of business is the set of stakeholder relationships rather than discrete economic transactions. Most businesses consist of the voluntary cooperation of at least customers, suppliers, employees, communities (including civil society), and financiers. This cooperation includes the ability to make agreements that extend over time periods based on fairness and trust, rather than merely transaction by transaction. Value gets created for each stakeholder because each can freely agree to cooperate with the others. Most people keep most of their agreements most of the time (see below for its underlying moral framework). By looking at a large group of stakeholder relationships, rather than a single transaction, scholars and practitioners can better understand the effects of managerial decisions on a broader system of relationships. This is important because it allows us to create ways to make decisions that benefit the ecosystem of stakeholder relationships, rather than

Table 54.1 Changing the assumptions of the business narrative

New assumption		Replaces old assumption	Implications
The unit of analysis assumption	Businesses consist of the voluntary cooperation of at least customers, suppliers, employees, communities (including civil society), and financiers.	Business is conducted transaction by transaction.	If capitalism is to flourish it will be because individuals and groups of individuals can work together to create value for each other. Collaboration only works when there is freedom of association and choice.
The interdependence of stakeholders assumption	There is a jointness to stakeholder interests. Each stakeholder contributes to the value that is created for the others.	Executives have to make tradeoffs where the value created for one shareholder reduces the value created for stakeholders.	Business is fundamentally a cooperative enterprise built to create value, trade, and make ourselves, and others, better off.
The complexity of human motivation assumption	We are pro-social, language-using collaborators. And, most of the time we aspire to be a part of something that is larger than our own individual self-interest.	We are primarily self-interested and opportunistic.	Most of us, most of the time, want to and actually do take responsibility for the effects of our actions on others.

trying to maximize a particular variable within a transaction and then causing value-destroying consequences in other areas.

3.2 *The Interdependence of Stakeholders Assumption*

There is a "jointness" to stakeholder interests. For instance, employee wellbeing is connected to customer well-being, etc. Each stakeholder contributes to the value that is created for the others. Value is created within the context of others. And, rarely is this process reducible to mere contracts. The task of the executive is to continuously work to get stakeholder interests going in the same direction. This works best when executives see the interdependence among stakeholders. It works least well when executives have to make tradeoffs where the value created for one stakeholder reduces the value created for others. This interdependence requires a set of skills and ideas that use executive imagination. And, this is best exercised in full engagement with the stakeholders themselves. Stakeholder engagement is sometimes thought to entail corporate social responsibility. Nothing could be further from our idea here. Stakeholder engagement is about how a particular firm's business model creates value.

3.3 The Complexity of Human Motivation Assumption

Capitalism works because human beings are complex creatures. We are at once self-interested and other regarding. We are capable of both selfish and selfless acts. We are pro-social, language-using collaborators. And, most of the time, we aspire to be a part of something that is larger than our own individual self-interest. Most of us, most of the time, want to and actually do take responsibility for the effects of our actions on others. The complexity of human behavior is writ large in our culture, our art, music, and literature. As we learn more about human behavior from its scientific and literary study we find less application of the simplifying motivational structures of traditional economic theory.

The three basic assumptions together with the implicit moral framework here form a different basis for understanding business. The implicit moral framework is not some idealistic fantasy, but one based in both science and cultural theory. Human history is partially a history of us cooperating together to achieve lifting billions out of poverty, and making a better life for ourselves, our families, and our communities. Such a project founders if the participants have no need to be responsible to those whom they affect with their actions. Indeed, the stakeholder idea is simply based on the idea that we need to be responsible for the effects of our actions on others. Such a basic principle is a part of every ethical system, religious belief, and guide to human interaction that has ever been invented or discovered. Of course, there are some who take advantage, but the idea that the main way we create value for each other depends on our willingness to "get away with whatever we can" is an idea that has long outlived any usefulness it may have had. Such an idea of "cowboy capitalism" should be put aside and made optional at best.

Any reasonable reform of capitalism and business must be put on a firm ethical foundation so that "business ethics" ceases to be a joke or an oxymoron. We argue that we do not need a very high edifice here, but rather a garden-variety commonsense view, that we teach all of our children simply to be responsible for the effects of our actions on others. Note that this assumption of responsibility applies to stakeholders in a business as well as to employees and executives. In the current parlance, many assume that it is only companies who need to be responsible (Goodstein and Wicks 2007), but this assumption is as faulty as its opposite.

4 Responsible Capitalism

The previous section suggests that we have come to see business differently. Business is fundamentally a cooperative enterprise built to create value, trade, and make ourselves and others better off. Since it is fully embedded in societal institutions we can also sketch the moral ideals on which it is based. So let us briefly sketch a set of criteria for proposals of reform of capitalism and business from our earlier discussion:

4.1 Responsibility Criterion

Any reform of capitalism must address the question of who is responsible for the effects of business activity on particular stakeholders and on society. We believe that the answer to this question is best given as "a business and its stakeholders," but we recognize that there may be alternatives. Also, "responsibility" is a devilishly tricky philosophical notion. Often responsibility is a joint endeavor, and it can be opaque due to difficult causal mechanisms, or ignorance of those mechanisms. These difficulties just add urgency to the requirement to understand responsibility in more nuanced ways, and to use all of the disciplines at our disposal to do so.

4.2 Voluntary Collaboration

If capitalism is to flourish it will be because individuals and groups of individuals can work together to create value for each other. Collaboration only works when there is freedom of association and choice. Forced labor, or forced cooperation, or forced consumption, or forced investment, carry negative consequences. However, it is not often so clear what is voluntary and what is not. The very low percentage of people who are actively engaged in their work is in part a function of very low perceived freedom to act and to do what is best for their organizations. In the thriving businesses of the twenty-first century, employees must have the ability to work with others to create value, and they must have the time and resources to push value creation in new and innovative directions.

4.3 Competition as Emergent Property

Businesses, and business schools, are often obsessed with the idea that business is fundamentally a competitive activity. We believe that such an assumption simply misses the main engine of capitalism. Capitalism works because we cooperate together to create value that can then be traded to make everyone better off. Competition is important in that a free society encourages voluntary agreement. And, if we can figure out a better way to do something, then we can satisfy a network of stakeholders better than they are currently being satisfied. Venkataraman (2002) has suggested that the entrepreneurial process leads to such equilibration as the result of a competitive economy. However, it may well be just a mistake to attribute the system-wide property of competitiveness to individual actors in the system.

4.4 Continuous Creation

Schumpeter (2013) famously wrote about the idea of creative destruction. Every company would eventually be replaced and destroyed by those who came after. However, this idea simply ignores the genius of the corporate form. Companies may well have a kind of limited immortality. In the twenty-first century companies don't stand still. They look for the next disruptive innovation, and some even look to disrupt themselves. Very large companies don't immediately go out of business when a better mousetrap gets to the market. They adapt. The adaptiveness of the corporate form is its innovativeness. It has good and bad qualities, but it lends some stability to a fast-changing world.

4.5 Government as Facilitator of Value-Creation Criterion

In calls for reform of capitalism, there are many proposals for the further regulation of business, usually from the left, and suggestions for "smart regulation," often from the right. Surely government has a role to play as referee, especially given the scope of the old narrative of capitalism. However, it stands to reason that if the main engine of business is collaboration to create value for stakeholders, one obvious question would be how government and other civil society organizations can facilitate the creation of value by businesses. We see a great deal of progress in this area as more companies are collaborating with non-government organizations (NGOs) on issues that range from industry working conditions to specific geographical and ecological issues. Companies such as Whole Foods Market routinely collaborate with animal welfare organizations to develop more awareness about conditions under which food is produced. McDonald's collaborates with environmental organizations on some issues of ecological significance. And, we have begun to see the emergence of public-private partnerships, usually involving private companies, NGOs, and governments cooperating together to solve problems that no single one of these organizations could easily solve.

We believe that in addition to its traditional role of referee and redistributor of value, government can play an important role as facilitator of value creation. There are a number of roles here, such as (1) coordinating, communicating, and validating information; (2) smart policies that encourage voluntary agreements to create value; (3) assisting education around business and entrepreneurship; (4) providing infrastructure to make business start-up and growth easier; (5) facilitating business growth and trade at an international level. Practical policies include the support of business incubators, changes in tax laws and regulatory regimes, the establishment of free and fair trade, encouraging companies to think broadly about stakeholder value, and making examples and role models out of the companies that do it right.

5 Summary

We believe that we are at an inflection point in the history of capitalism. It is time to think broadly and creatively about the very best in capitalism, and how to avoid its historical weaknesses. In this chapter we have outlined that many of the weaknesses of capitalism come from inaccurate and unhelpful assumptions about the nature of business including its purpose and level of interdependence, as well as assumptions about the complexity of human beings. Without addressing these fundamental assumptions, calls for reforming capitalism—whether from business leaders or policymakers—will have limited impact. We have suggested that the scholars and executives, who have been working on stakeholder theory and its practice have given us ways to talk differently about these assumptions that unleash our potential to work together, increase responsibility, and make each other better off. These narratives must become more widespread and engrained in practice. Much work remains to be done in both developing theory and practice to continue this progress.

Acknowledgments This essay is based on a number of earlier attempts to define stakeholder capitalism (see especially Freeman et al. 2006, 2007, 2010). We are grateful to co-authors, editors, and publishers of the earlier attempts for permission to continue to develop the ideas here.

References

Bebchuk, L. 2006. *Pay without Performance*. Cambridge, MA: Harvard University Press.

Edmans, A., L. Li, and C. Zhang. 2014. *Employee Satisfaction, Labor Market Flexibility, and Stock Returns around the World*, w20300. Cambridge, MA: National Bureau of Economic Research.

Follett, M.P. 2011. Business as an Integrative Unity. In *Sociology of Organizations: Structures and Relationships*, ed. M.E. Godwyn and J.H. Gittell, 7–13. Thousand Oaks: Sage.

Freeman, R.E., K. Martin, and B. Parmar. 2006. Ethics and Capitalism. In *The Accountable Corporation, Vol. 2: Business Ethics*, ed. M. Epstein and K. Hanson, 193–208. Westport: Praeger.

———. 2007. Stakeholder Capitalism. *Journal of Business Ethics* 74 (4): 303–314.

Freeman, R.E., J. Harrison, A. Wicks, B. Parmar, and S. de Colle. 2010. *Stakeholder Theory: The State of the Art*. Cambridge: Cambridge University Press.

Friedman, M. 1970. The Social Responsibility of Business Is to Increase Its Profits. *New York Times Magazine* 13: 32–33.

Goodstein, J.D., and A.C. Wicks. 2007. Corporate and Stakeholder Responsibility: Making Business Ethics a Two-Way Conversation. *Business Ethics Quarterly* 17 (3): 375–398.

Jameson, F. 1991. *Postmodernism, or, the Cultural Logic of Late Capitalism*. Durham: Duke University Press.

Jensen, M.C., and W.H. Meckling. 1976. Theory of the Firm: Managerial Behavior, Agency Costs and Ownership Structure. *Journal of Financial Economics* 3 (4): 305–360.

Keynes, J.M. 2006. *General Theory of Employment, Interest and Money*. New Delhi: Atlantic Publishers and Distributors.

Mackey, J., and R. Sisodia. 2014. *Conscious Capitalism*. Cambridge, MA: Harvard University Press.

Parmar, B.L., R.E. Freeman, J.S. Harrison, A.C. Wicks, L. Purnell, and S. De Colle. 2010. Stakeholder Theory: The State of the Art. *Academy of Management Annals* 4 (1): 403–445.
Schumpeter, J.A. 2013. *Capitalism, Socialism and Democracy*. New York: Routledge.
Sisodia, R., D. Wolfe, and J. Sheth. 2014. *Firms of Endearment*. 2nd ed. Upper Saddle River: Pearson Press.
Venkataraman, S. 2002. Stakeholder Value Equilibration and the Entrepreneurial Process. In *Ethics and Entrepreneurship*, ed. R.E. Freeman and S. Venkataraman, 45–57. Charlottesville: Philosophy Documentation Center.

R. Edward Freeman is University Professor, Elis and Signe Olsson Professor, Academic Director of the Institute for Business in Society, and Senior Fellow of the Olsson Center for Applied Ethics at the University of Virginia's Darden School of Business.

Bidhan L. Parmar is Associate Professor and the Shannon Smith Emerging Scholar in Business Administration at the University of Virginia's Darden School of Business.

Kirsten E. Martin is the William P. and Hazel B. White Center Professor of Technology Ethics, in the Mendoza College of Business at the University of Notre Dame.

Chapter 55
The New Story of Business: Towards a More Responsible Capitalism

R. Edward Freeman

1 Introduction

The last 40 years have seen great changes in our understanding of business. In our lifetime, we have seen a remarkable pace of globalization. We have seen revolutionary information technology. We have seen nothing less than a fundamental shift in the story of business. In this talk I will try to explicate what I believe is a conceptual revolution in business, and in particular, I will try to explain this "new story of business" in terms of a few fundamental principles or ideas.

While the roots of this revolution are easily traceable back to the 1980s or even earlier, they are most clearly seen in the responses to the Global Financial Crisis (GFC) of 2008. Since that time, we have seen an explosion of ideas of how to make businesses more responsible for the consequences of their actions.

For instance, many companies have taken a renewed interest in corporate social responsibility and sustainability. In addition, we have seen a renewed emphasis on the idea of Conscious Capitalism, as John Mackey and Raj Sisodia have argued, that companies that follow the tenets of conscious capitalism will outperform those that do not. Michael Porter and a colleague have suggested that companies focus on "shared value" where economic value and social value are seen as going hand

Originally published in: *Business and Society Review*, 122(3), 449–465 © John Wiley and Sons, 2017
Reprint by Springer, https://doi.org/10.1111/basr.12123
Editor's Note: A public lecture by R. Edward Freeman, Verizon Visiting Professor of Business Ethics, Bentley University.

R. E. Freeman (✉)
University of Virginia, Charlottesville, VA, USA
e-mail: FreemanE@darden.virginia.edu; https://redwardfreeman.com

© Springer Nature Switzerland AG 2023

S. D. Dmytriyev, R. E. Freeman (eds.), *R. Edward Freeman's Selected Works on Stakeholder Theory and Business Ethics*, Issues in Business Ethics 53,
https://doi.org/10.1007/978-3-031-04564-6_55

in hand. Nestle has been the show pony for this idea. Just Capital is an NGO that is committed to rating companies on widely accepted standards of justice. Bill Gates has suggested Creative Capitalism, whereby companies forego profits for the sake of public welfare. Senator Mark Warner of Virginia has suggested that it is time for new metrics, especially around the welfare of workers and has hailed a move to Capitalism 2.0.

Meanwhile, the idea of social entrepreneurship has taken off with many millennials beginning to start businesses that address social problems. NGOs such as the Acumen Fund, Kiva, Kickstarter, and countless others have been started to provide capital for small and very small businesses and entrepreneurs who are societal change agents. Even on Wall Street, we see an increase in the movement toward what is variously called, "patient capital," "impact investing," "responsible investing," and other terms.

Business pundits have gotten into the act, decrying a lack of ethics that they claim brought about the GFC. Robert Reich has argued that ethics and profits have to go together. Agency Theorist pioneer, Michael Jensen, has suggested that integrity is an important element in a successful business. The Aspen Program in Business and Society has led various conversations about new business models, new governance models, and a variety of other related ideas.

At two recent meetings at the White House in 2016, sponsored by the Obama Administration's Department of Labor, 75 people from these organizations and movements gathered to discuss commonalities and whether or not there needed to be one brand that identifies this new story of business that is emerging. While such a brand may someday emerge, or perhaps it already has, for my purposes I want to focus on the underlying ideas and principles that are inspiring all of this activity. Whichever brand or brands that ultimately become the rallying cry on which this conceptual revolution will be based. The brand will have to be based on a sense of purpose and ethics that is as central to the new narrative as profit is to the old one. It will have to address how companies can simultaneously create value for all of its key stakeholders. The brand will have to take sustainability and the physical limits of business very seriously. And, it will have to recognize that people are complex and that they can and do collaborate with others to create value. I will set these ideas out more carefully in a later section. For now, let's turn to the dominant narrative or the "old story of business," and see why it is no longer applicable for the twenty-first century.

2 The Dominant Narrative

While all of these conversations and new business models are going on, we still see many executives and academics who are strongly committed to the idea that the main goal of any business is to make as much money as possible for its shareholders.

This view was articulated best by Milton Friedman in a 1970 *New York Times* article in which he stated that the only responsibility of the executives is to make as much money as possible for shareholders. While Friedman's view is more nuanced and is often misinterpreted, the shareholder primacy idea has a real grip on both executives and academics. This dominant narrative of what we might call the old story of business is deeply rooted in business cultures around the world. Oftentimes, it is appealed to as a way of justifying almost any action that seems to harm groups other than shareholders. Consequently, especially since the GFC, there's a real struggle going on around the issue of the ethics of capitalism. Let's be a little more precise here. We can talk about this old story of business in terms of five claims.

The first is that business is primarily about making money and profits for shareholders. Business on this view is the physics of money, and the language of money and profits is seen by many to be the main metaphor in talking about business. Often these theorists talk as if money is the only thing that matters, and the vocabulary of finance and accounting are the only vocabularies that tell us how to build a great business. More precisely, business is seen as a collection of economic transactions that can be fully understood using economic models and concepts. Modern economics has built some really powerful models, from general equilibrium to modern econometrics, and indeed they are useful for understanding how markets work. The problem is that they are not the only way to understand business, and they can be misleading as the GFC taught us.

The second claim is that the only constituency that matters is shareholders. Friedman's claim is that the only social responsibility of an executive is to make as much money as possible for shareholders. Even though Friedman was careful to also say that such a claim was subject to ethical rules and law—this part of the claim often gets left out. So, executives and pundits sometimes argue that whatever is not illegal is permissible in the name of shareholder value.

The third claim is often implicit. It is that we live in a world of fairly limitless physical resources, or that market forces will always determine which resources are economically feasible to use. We need markets for natural resources such as air, water, and carbon emissions, not regulation.

The fourth claim, that we see played out in the popular press all the time, is about what motivates business people. In keeping with the idea that business is about the physics of money there is a widespread idea that given the opportunity business people will cut corners, lie, and cheat. People, on this view, are completely self-interested. They will work for others, only if they are incentivized to do so with either a threat of punishment or a promise of rewards.

The fifth claim summarizes the first four and says that business and capitalism work because people and companies are self-interested, competitive, and greedy. The greatest good emerges as if by an invisible hand. Usually homage is paid to a brief passage in Adam Smith about the butcher and baker.

3 What's Wrong with the Old Story?

3.1 Profits Are Not the Purpose of Business

While there has been much criticism of this so-called neo-classical view of the firm, for our purposes I want to focus on three main flaws that make it easier to see why the old story is no longer appropriate. The first flaw is the idea that business is the physics of money and that profits are and should be the purpose of any business.

While there are many businesses that have come to see making money for their "owners" as the main purpose of their business, most of these businesses didn't start out this way. Most entrepreneurs start their companies because, in John Mackey's words, "they are on fire about their business idea." No matter whether it was Steve Jobs and Bill Gates on fire about the desktop computing revolution and starting Apple and Microsoft, Mackey himself, on fire about bringing healthy food to people through what became Whole Foods Market, or my son, Ben, on fire about bringing the great rhythm and blues sound of Motown and Stax into the twenty-first century with Red Goat Records, none of these entrepreneurs started the business with the purpose of making as much money as they could.

Now, of course, money and profits are important. They must exist for a business to live. Demonizing profits, as many pundits have done, is just a shortsighted, ideological mistake. Likewise, to claim that the purpose of a business is to make profits for owners is a similar and often ideological mistake.

I need to make red blood cells to live. However, it does not follow, without a lot of argument that I simple cannot imagine, that the purpose of my life is to make red blood cells. Even if I have fallen on unhealthy times, and I have to actually concentrate on making red blood cells, for instance by paying a lot of attention to the iron in my diet, it still does not follow that the purpose of life is to make red blood cells (or to breathe, or to drink water, etc.).

Likewise, sometimes businesses fall on hard times. A competitor disrupts the industry, or mistakes have been made, or the world simply changed. In all of these cases, a business might have to focus fairly clearly on generating profits to stay alive, but it would be a mistake to claim that profits were the purpose of the business.

Former CEO of GE, Jack Welch, business guru, Simon Sinek, and many others have claimed that it is best to see profits as an outcome. And, I will add it is an outcome of purpose and how value gets created for stakeholders. More on that idea later.

3.2 Business Ethics Is Not a Contradiction

When I tell people that I teach business ethics you know what happens. Either they have to manage not to laugh, or they say "business ethics" I thought that was a contradiction. The idea that business and ethics do not go together has long been a part of the dominant story of business. After all, if business is just about money, shareholders, and profits, there's not much room for ethics to be a central part of it. The way I like to put it is that of the phrase "business ethics," we only really misunderstand two of the words.

First, let's talk about business. The old story of business says to us that business is about competing, making money, and doing whatever you can get away with. The idea that business people are not trustworthy is in many cultures around the world. A recent study found that only 19% of people around the world actually trust business executives of large companies. Now, every business person that's here today knows that business is not just about making money. But even if that's all you want to do, how are you going to do it? You would better have some products and services that customers want to buy. You need suppliers who are committed to making you better by improving your business with their products and services. You need employees who are not there just for the paycheck but are there to make your business better—employees who are engaged in their jobs. You need them not to be unengaged or actively engaged, as many studies have found is the case in a lot of businesses. And, you need to be a good citizen in the community. If you are not a good citizen in the community, communities will pass restrictive laws or ordinances to prevent your business from operating well there. If you do these things, if you have great products and services for customers, if your suppliers want to make you better, if you have employees who are engaged, and if you are a community builder, then if you put those things together the right way, profits will emerge. Again, this is the idea that profits are an outcome. So business, far from being just about the money, is about creating value for stakeholders.

Now, let's talk about ethics. Many people think that ethics is about things you talk about only on Sunday. It's about angels and organ music or very serious things talked about in hushed tones. While religion is important in ethics, I want to urge you to think about it more broadly. Ethics is really the most practical thing around. Ethics is about how I live my life and how living my life affects how you live your life. Ethics is always personal; it's about what I want and how I'm going to live. And, it's always interpersonal. It's about how we are going to live together. Many people have the idea that ethics is about rules chipped into stone, rules that never change, and rules that are fairly inflexible. But sometimes these "rules" can conflict and sometimes they need to change or be interpreted in different ways as we invent new technologies or discover other previously hidden features of a situation. Ethics is, in my view, a conversation about how we are going to live together. It is a conversation that substitutes reason, dialogue, and talking together for violence. It's easy to see in many places in the world where the conversation has broken down. Violence is not a good answer. Now, many people would argue that ethics is personal. I get to decide what my ethics are. I have to look myself in the mirror. I have to live with myself. And that's true, you do have to live with yourself. But all of us have to live with you too. And that gives us some right to join in the dialogue and conversation about how we are all going to live and thrive together. From the time when we lived in caves, we have always had conversation about what behavior is appropriate and what behavior is inappropriate in a community. We have had conversations about what are some good ideas—some good rules of thumb—to keep the order in society and to allow all of us to flourish. This is ethics at its best. Over time, we have grown quite good at ethical reasoning. But at the same time, we have many challenges to our ethics in society today, primarily because of technological change and the emergence of new societies to do things differently and new cases of a kind that we just have not thought of before.

For instance, I learned ethics at my grandmother's knee in the 1950s in Georgia, but we did not have any conversations about intellectual-property that could be digitally reproduced for no cost. We did not have any conversation about end-of-life technology that could prolong life at a great cost. We did not have any conversations about abortion or things like file-sharing and Napster or what's appropriate on Facebook and Twitter. Those things simply did not exist. Now, one thing you can say is know your ethics and your values and you will not have to have much of a conversation. The problem with this is that it is very hard to do in today's world. Human beings are very good at fooling themselves and saying one thing and doing something else. We need each other in this conversation about how we are going to live together.

What does all this have to do with business? If you ask people around the world to write down their three most important values that they would like to teach their children, or their three most important ethical principles, they all pretty much actually write the same thing. There is some version of respect, honesty and integrity, caring and love, and responsibility. Of course, what it means to show respect in Jakarta is very different than what it means in Charlottesville, so there is always a cultural context. But, it is difficult to imagine doing business without these values. Try to think what it would be like to do business always with people you did not respect, or on whose honesty you could not rely on, or who did not care about others. It would be impossible.

Adam Smith knew how important ethics was to business. Indeed, in *The Theory of Moral Sentiments* and in *The Wealth of Nations*, Smith makes it clear that without a sense of justice, markets just will not work very well at all. Thinking that business ethics is a contradiction is a deep flaw in the dominant narrative of business.

3.3 People Are Complicated

There is ample scientific evidence that human beings are not the rational economic beings that much of our economic and business theory assumes. We are not always driven by extrinsic if-then rewards. We want to be engaged in doing something that has meaning and purpose. Some like Daniel Pink, have argued that we understand human motivation in terms of three ideas: Mastery, the sheer joy we take out of getting better at something; Autonomy, the freedom to live our own lives to try new things; and Purpose, the idea that we stand for something greater than just ourselves and our self-interest. By thinking about mastery, autonomy, and purpose we get a much more realistic view of what motivates people in business. The old story's insistence on human beings being motivated primarily by money is really not appropriate in the twenty-first century if it ever was appropriate. Certainly, the new generation that we call Millennials want to do something that has meaning, that has purpose, that is not just making money.

Psychologist, Harry Levinson, used to hang out at the Harvard Business School. He was kind of a crusty old guy and would often ask executives what was the main way people are motivated inside corporations. Most executives would say to him rewards and punishments or carrots and sticks. Levinson would draw on the board a carrot at one end and a stick at the other. In the middle, he would put a question mark. He would ask what animal do you imagine between the carrot and the stick? Most people would say a jackass. So, Levinson coined the idea that he called the great jackass fallacy. It goes like this: maybe … just maybe, human beings are slightly more complex than jackasses. Maybe they have social, spiritual, sexual, political, ethical lives as well as economic lives. But it's even a little more difficult than that. If you treat people like jackasses, they begin to act like jackasses. They nose around for the carrots and they try to avoid getting the stick. Think of all of the productivity that gets left on the table, and, indeed, think of all of the human misery that results from treating people like jackasses. Human beings are more complex than the dominant story would have us believe. We will say more about this as we move to thinking about the principles that underlie this new story of business.

There are at least three flaws in the current story of business. First, the purpose of a business is not only to make money. Second, business and ethics need to go hand in hand. And, third, human beings are complicated. The time has passed for these flaws. The new story of business that is being built by thousands of entrepreneurs and executives around the world eschews these flaws and takes a different approach. Let's turn to understanding the ideas that are behind the new narrative.

4 The Ideas Behind the New Story

I have identified at least six new ideas that undergird the new story of business that is emerging. They are: (1) The unit of analysis is stakeholder relationships; (2) stakeholders are interdependent; (3) tradeoffs are managerial failures of creative imagination; (4) purpose, values, and ethics must be embedded in organizations; (5) business exists in the physical world; and (6) people are complicated. Let's look at each in turn and see how they are connected with each other and this new story.

4.1 The Unit of Analysis Is Stakeholder Relationships

One of the cornerstone ideas behind the new story of business that is emerging is the importance of looking beyond shareholders to a broader group of stakeholders. As we saw earlier, creating value for stakeholders is something that every successful business has actually done. As we become more aware of this fact, we can build into our business models more nuanced ways to create value.

The key difference is that in a relationship there is a presumption that the relationship will continue over time, other things being equal. Businesses need relationships with their stakeholders so that each has some attachment to the other. You want customers to have some degree of loyalty. You want employees to give you the benefit of the doubt, and you want shareholders who will stick by you when things are tough. Seeing these relationships through the lens of "discrete transactions unrelated to each other" does not build loyalty. In fact, it encourages exit when things get tough. Building loyalty with stakeholders mitigates the risks of difficult times.

Company X had built a relationship with a stakeholder group that often criticized the company. The stakeholder was targeting X with a campaign and called to tell them. The executive at Company X asked for help in solving the problem that the campaign was about, and the company and stakeholder were able to introduce an innovative program that went a long way toward solving the problem. All of this happened because there was a relational mind set.

Of course, relationships are two-way streets. Companies have to stand by their suppliers and their employees when times are tough for them. And, they have to share in the rewards of success, in a broad manner, not just in terms of rewarding senior management. Whole Foods Market uses gain-sharing to reward the employees who work to bring things in under budget. They also give most of the stock options to non-top management employees.

4.2 Stakeholders Are Interdependent

It has often been said that the key insight of stakeholder theory is that there are groups that are important other than shareholders. And, while this is one insight that the theory has brought to management thinkers, another is more important. It is that stakeholder interests have a certain interdependence. And, when management can capture this interdependence and push it forward, great results are likely to occur. Wal-Mart was for many years a poster child for this interdependence. By negotiating tough deals with suppliers, Wal-Mart could offer customers everyday low prices, and even though the margins were thin for suppliers, there was a great deal of volume that could lead to profitability. Employees were better off since there were more customers coming to take advantage of the everyday low prices, and the stock price saw a steady increase. Unfortunately, Wal-Mart paid little attention to communities as stakeholders, focusing on citizens as customers. Many outside groups began to be formed and Wal-Mart was blamed for many social ills. Today, Wal-Mart is working hard to repair its relationships with communities and to integrate ideas that make communities better off into the rest of its business model. The progress that has been made with sustainability is but one of several examples where Wal-Mart has taken a leadership role.

Even the companies who do CSR and who have adopted Michael Porter's view of shared value have begun to see stakeholders as interdependent. For a long time, CSR was seen as something of a public relations move, unconnected to the main business model. More recently, we have begun to see how CSR can be connected to the basics of what a company knows how to do. Nestle has pioneered this idea with shared value, as it has sought to introduce the creation of social value all the way down its economic value chain.

4.3 *Trade-Offs Are Managerial Failures of Creative Imagination*

Economists love trade-offs. In fact, one of the hallmarks of modern economics is that one can always calculate trade-offs. I have become increasingly skeptical of trade-off thinking. In fact, I believe that the drive to collaborate and avoid trade-off thinking is far more powerful. When we see the task of the executive as getting stakeholder interests all going in the same direction over time, trade-offs will disappear. Of course, sometimes they have to be made, because we cannot imagine an alternative, but when we make a trade-off we need to immediately begin the process of making the trade-off better for both sides.

I have often told the story of a large chemical company who decided to commit to being more sustainable and cleaner. The CEO announced a large and lofty sustainability goal and proceeds around to the divisions and plant sites to let them know that he was very serious about this. There were interim goals and plans in a very businesslike approach. In one facility, as he told the story to a symposium at Dartmouth in the 1990s, the engineers came up to him and said, "Sorry but we can't meet these interim goals. This process is too dirty, this equipment is too old, and we can't meet the first target." The CEO said that they were serious about this program and so they would have to close the plant. What I understood from that was that he was willing to make a trade-off, environment or community on the one hand versus employees on the other. The CEO's trade-off was that the environment was a serious issue and it was going to be the winner. So, he told the engineers to prepare to close the plant. A few weeks later the engineers came back and said that a miracle had occurred. They figured out how to do it. When the CEO asked what it would cost, the engineers actually were embarrassed to say that the new method would save money.

When trade-off thinking becomes unacceptable we kick into gear the only infinite resource we really have, which is our creative imagination. The use of the creative imagination is radically underutilized in most companies today. Trade-offs are easy. In the new story of business with the stakeholder mindset, trade-offs become managerial failures. As more and more companies are thinking about the new story of business, they are discovering ways to satisfy multiple stakeholders. Simultaneously, this is one of the key ideas in the new story.

What this means is that conflict, often avoided in many companies, is precisely the place where value creation can take place. When there's conflict among stakeholders, where there's conflict among core values, where there is conflict among competitors or products this is exactly the place where we can reimagine that conflict and create more value. We have to come to see conflict as a good thing. Recall the story above about company X who had conflict with the stakeholder group but kept up the relationship because that stakeholder group could help them figure out how to solve a problem. A lot of value was created.

4.4 Purpose, Values, and Ethics Must Be Embedded in the Organization

One of the great things about business as an institution is that many different purposes are possible. Novo Nordisk wants to rid the world of diabetes. Whole Foods Market wants to help people be healthier with better choices for food. Tastings, a small restaurant in my hometown, wants to bring the joy of good French country cooking to its customers. The founders of Relish MBA want to make it easier for companies and MBA students to find a good match. The only limit to the purpose of a business is our imagination. Of course, purposes do not have to be all good. We have plenty of examples from human history about organizations that were of high purpose, but whose purpose was morally evil. A sense of values and ethics has to go alongside purpose. There are many organizations in the pantheon of new story organizations who are addressing precisely this issue. Just Capital is rating organizations based on a notion of "Justice." However, it shakes out, we can no longer make the mistake that the pursuit of profits is the sole purpose of business. Real purpose inspires both employees and other stakeholders who come to share that purpose. And, this new story of business is an inspirational story.

4.5 Businesses Exist in the Physical World

While many who write about sustainability and the environment sound a caution about the physical limits of the world, I want to suggest that this is only part of the story. We do need to come to see business as embodied in the world, and hence, there are constraints imposed by the physical world. However, we also need to see business as capable of transforming those constraints into new opportunities. We have seen this time and again as companies such as 3M figure out how to turn waste streams into products and services. Obviously, we need to tackle climate change, but seeing it as giving us limits to growth is forgetting the creative imagination that has solved so many of our problems in the past. Adopting some kind of green values, and integrating respect for the environment into our purpose and values, can be a powerful elixir for creativity.

4.6 *People Are Complicated*

I do not want to repeat the arguments I gave earlier about the flaws in the old story. Rather, I want to suggest that there is a much more inspirational view of human beings that is emerging. People are using business models and ideas to attack age old problems of poverty, education, disease, and more participation in society. Often these problems are attacked by these "new story companies" in conjunction with NGOs, governments, and other private organizations. We have seen a wave of "social entrepreneurs" and "impact investing" where the explicit idea is to use business to make society better and to solve social problems. I believe that we are fast crafting a new idea about what it is to be human. Let me illustrate.

What is this smartphone, really? The way I see it, it's some bits of sand and metal, some vocabularies that we have invented to solve problems, and the fact that we can work together collaboratively to achieve things that no one of us can achieve alone. In short, I see the world not as a world of scarcity, but as one of abundance. We have an almost infinite capacity to invent ways to solve our problems, whether we take on poverty, space travel, climate change, or understanding the rules of cricket. But, we are not in it alone. We invent mutually beneficial vocabularies with others to solve our problems. And, this is true whether we are scientists or politicians. We are surely more than narrow economic creatures, and in fact, capitalism works just because of this complex human dimension.

5 Towards a More Responsible Capitalism

If we are to move our system to a more responsible capitalism, we need a few more ideas that will transcend these particular ones aimed at creating and sustaining a successful business. First of all, we need some broadly defined principles of responsibility throughout our society. Responsibility is an idea that comes with the more libertarian idea of freedom that underpins market systems. We need to see people and companies as responsible for the effects of their actions on others. That is the moral cornerstone of the stakeholder idea. Unless we are willing to take responsibility and to be willing to justify our actions to our fellow humans, our society will not continue to flourish.

The second idea is that we need to continue to see business as a voluntary enterprise. No one is forced to do business with anyone else. This idea of choice as applied to multiple stakeholders is central in creating a system of business that works for everyone. Competition should celebrate the choice that customers have. Restricting that choice artificially through institutions like crony capitalism should have strong sanctions.

Finally, we need a new idea of the role of government. While we have clear theories about how government plays its role as both regulator and redistributor, there is another role that is often overlooked. Government can be a facilitator of value creation. It does this via infrastructure and other programs such as enforcing civil rights, and ensuring that crony capitalism does not take hold. We have only begun to explore this idea of government as facilitator of value creation, so stay tuned.

6 What Can You and Your Company Do?

Let me wrap up by taking these ideas down to an extremely practical level. There are at least five ways to begin to practice this more responsible capitalism in your businesses.

First, you can rediscover your purpose and the values that go with it. No less a company than Unilever has begun this process, and while it takes time, the payoffs are large. Employees become inspired, and the innovative ideas begin to flow. How do you rediscover your purpose? Well, the first thing to do is to have a look at history. What are the founders' stories that are told in the company? Why do people actually show up for work? What really helps them when they are at their best? It takes a concerted effort if a company has lost its way, but it is an exciting process to rediscover the purpose of an organization. Values go along with purpose and are often seen as the "How" we are going to realize the purpose.

Second, you can perform a systems/process check on the purpose and values. Purpose and values live in the systems and processes of an organization. Talk is cheap. The first places to look are the HR and expense reimbursement systems. Think about an organization that trumpeted its respect for employees but required them to get a receipt for a $2 toll on the Mass Pike, a task that is mostly dangerous if not impossible. Or consider an organization who is very proud of its stand on values, yet waits 60 days to reimburse employees, and pays suppliers in even later terms.

Third, you can begin live conversations about the purpose and values. Small groups of employees can help to clarify what values are actually in force at the company, and there are known techniques for creating a conversation about these values, such as which ones are we really serious about, and which ones are we just giving lip service to. Some companies have begun to encourage meetings where employees bring "values vignettes," or sticky problems, to groups of peers to try and get insights. Based on Johnson and Johnson's original "challenge meetings" discussions of these vignettes helps to clarify what the true meaning and intention of often quite general values statements are.

Fourth, you can be a community builder. There are many ways to help build the communities in which you operate. Giving employees time off to volunteer, hiring some of the least well-off members of a community and giving them training, and donating to charities, are all viable strategies. However, figuring out what you know how to do, can be used to build community, brings the power of the business model to bear on tough problems. We are beginning to see more and more companies working with stakeholder groups in the nonbusiness sector to jointly tackle societal issues. Such multi-sector collaborations are one of the best ways to build community.

Finally, you can communicate how your organization makes the world a better place. We need business organizations to inspire us. We need business to become an institution of hope. We need business executive to try and remake their organizations to be places we want our children to live in. Asking anything else is to set the bar too low.

Acknowledgments The ideas in this paper have been partially developed in a number of places, especially Freeman et al. (2007, 2016), Freeman and Ginena (2015), and Freeman (2017, forthcoming).

I am grateful to editors and co-authors for their permission to more carefully develop these ideas here for a public audience. This paper will be a part of a forthcoming book, tentatively titled *The New Story of Business: Responsible Capitalism*, co-authored with Bidhan Parmar and Kirsten Martin, in 2018.

References

Freeman, R.E. 2017. Five Challenges to Stakeholder Theory: A Report on Research in Progress. In *Stakeholder Management, Business and Society 360 Series*, ed. D. Wasieleski and J. Weber. Bingley: Emerald Publishing.

Freeman, R.E., and K. Ginena. 2015. Rethinking the Purpose of the Corporation: Challenges from Stakeholder Theory. *Notizie di Politeia* 31 (117): 9–18.

Freeman, R.E., J. Harrison, A. Wicks, B. Parmar, and S. de Colle. 2010. *Stakeholder Theory: The State of the Art*. Cambridge: Cambridge University Press.

Freeman, R.E., K. Martin, and B. Parmar. 2006. Ethics and Capitalism. In *The Accountable Corporation, Vol 2: Business Ethics*, ed. M. Epstein and K. Hanson, 193–208. Westport: Praeger.

———. 2007. Stakeholder Capitalism. *Journal of Business Ethics* 74: 303–314.

Freeman, R.E., B. Parmar, and K. Martin. 2016. Responsible Capitalism: Business for the 21st Century. In *Re-Imagining Capitalism*, ed. D. Barton, D. Horvath, and M. Kipping, 135–144. Oxford: Oxford University Press.

Friedman, M. 1970. The Social Responsibility of Business Is to Increase Its Profits. *New York Times Magazine* 13: 32–33.

R. Edward Freeman is University Professor, Elis and Signe Olsson Professor, Academic Director of the Institute for Business in Society, and Senior Fellow of the Olsson Center for Applied Ethics at the University of Virginia's Darden School of Business.

Chapter 56
The Social Responsibility of Business Is to Create Value for Stakeholders

R. Edward Freeman and Heather Elms

On Sept. 13, 1970, economist Milton Friedman suggested that, as the headline to his essay in *The New York Times Magazine* put it, "The Social Responsibility of Business Is to Increase Its Profits." While we hear from many executives about additional social responsibilities, all too often those executives will revert back to arguing,"… but our first social responsibility is to maximize shareholder profits."

Businesses that want to be successful in the twenty-first century need to be saying and doing something else.

Here's what we argue: *The social responsibility of business is to create value for stakeholders.* That means its customers, suppliers, employees, and communities, as well as its shareholders.

The stakeholder approach aims to create a new narrative about business—a new *story*—that enables great companies to make our communities and our lives better through the creation of stakeholder value, rather than simply profit to shareholders. The story includes a recognition that if we want the outcome of business to be a more responsible capitalism, it requires stakeholders to value business responsibility.

Originally published in: *Sloan Management Review*, January 4 © 2018 from MIT Sloan Management Review/Massachusetts Institute of Technology. All rights reserved. Distributed by Tribune Content Agency
Reprint by Springer

R. E. Freeman (✉)
University of Virginia, Charlottesville, VA, USA
e-mail: FreemanE@darden.virginia.edu

H. Elms
American University, Washington, DC, USA

© Springer Nature Switzerland AG 2023

S. D. Dmytriyev, R. E. Freeman (eds.), *R. Edward Freeman's Selected Works on Stakeholder Theory and Business Ethics*, Issues in Business Ethics 53,
https://doi.org/10.1007/978-3-031-04564-6_56

1 Why We Need a New Story

The Great Recession of the late 2000s should have made one thing abundantly clear: The way we have been encouraged to think about business is no longer appropriate—if it ever was. In the twenty-first century, there is too much complexity and too much uncertainty for a focus on "maximizing profits this quarter" to work very well. The landscape is littered with companies that tried this, and they simply did not understand—either because they could not understand or refused to understand—the complex consequences of their actions. This led to the demise of investment banking company Lehman Brothers, the bankruptcy of automotive company General Motors Corp., and the crash of countless smaller businesses. It cost U.S. citizens trillions of dollars.

If you add to these debacles a whole series of high-profile, far-reaching scandals (Enron, Madoff, Wells Fargo, Volkswagen), where unscrupulous companies and their executives acted for themselves while pretending to do what was in the shareholder's interest, the old story simply collapses. We can no longer afford to accept that businesspeople will be only self- and shareholder-interested, greedy little bastards divorced from the societal context in which they are embedded.

And to be clear: Cobbling together ideas like "corporate social responsibility" is ineffective. Friedman was wrong nearly 50 years ago when he argued that the only business of business is to make profits for shareholders, but he was right when he urged businesspeople to stick to business. Notable executives like former General Electric CEO Jack Welch have agreed. "On the face of it, shareholder value is the dumbest idea in the world," Welch told the *Financial Times* in 2009. "Shareholder value is a result, not a strategy. … Your main constituencies are your employees, your customers, and your products."

2 The New Story Spotlights Stakeholders, Not Just Shareholders

In reality, the only way to make profits is to have great products and services that customers want because those offerings make their lives better. Profits follow from having suppliers who are committed to making a company better, and employees who are inspired to work together to create something of value. And if a business is not a good citizen in its community, at least in a free society, people will use the political process to regulate the business closely and even prevent it from operating within community borders. Stakeholders are interdependent, and everyone who runs a great business knows that.

The new story of business is about creating as much value for all these stakeholders as possible, and this of course includes creating profits for shareholders. In the global economy, customers, suppliers, employees, communities, and financiers—shareholders plus bondholders plus banks and other sources of capital—are all intertwined. The winning business models of the twenty-first century figure out how to get these interests going in the same direction, with as few trade-offs as possible.

Organizations that compromise the interests of one stakeholder with the interests of another quickly find that, in today's world, there is simply no place to hide. Someone will figure out how to do the business better without the trade-offs.

Today's business world yields "continuous creation," not the old story's "creative destruction." Many resources may be limited, but human ingenuity and imagination are not, especially when inspired by a sense of purpose. Think about Amazon (and its recent acquisition, Whole Foods Market), Genentech, Apple, Facebook, and Google—all are high-purpose, stakeholder-oriented companies, based on creating value for multiple stakeholders. No business is perfect, and every business can improve, but it is time to end the myth that Wall Street is disconnected from Main Street. Get executives focused on value creation for real people, and products, services, and jobs will appear.

We know that one immediate reaction of many executives of public companies will be, "Oh, but my fiduciary duty is to shareholders." Not so. Legal precedent suggests that courts have granted companies a great deal of flexibility in how they balance their stakeholders, including shareholders, in the interests of the business. We see similar flexibility worldwide. Capitalism works because entrepreneurs and managers figure out how to get the interests of many going in the same direction.

The stakeholder approach sets forth a new conceptualization of business, in which business is understood as a set of relationships and management's job is to help shape these relationships. Business is about how customers, suppliers, employees, financiers, communities, and managers interact to create value, and there is no single formula for balancing or prioritizing stakeholders. Creating that balance is part of what management is all about, and it will be different for different companies at different times.

3 The New Story and Stakeholder Responsibility

Business is designed to meet demand. The danger of both the old story (shareholder profit maximization) and the new story (stakeholder value maximization) is that there are some activities in which business should not engage. If, for instance, consumers value only the cheapest product or service, and it is enabled only by depriving employees of any value (and perhaps much worse), then businesspeople must engage their ingenuity and imagination. They need to be inspired to create competitive products and services that create value across the board—for employees as well as consumers, and other stakeholders as well.

This can be done—business is certainly capable of motivating the interests of consumers, employees, investors, and other stakeholders toward one option over another. Business drives demand for technological innovation: You probably didn't know that your smartphone would be able to do as much as it does, but now that you do, would you ever go back to a flip phone? In the same way, business can drive demand for responsible capitalism by offering responsible options for all stakeholders.

Stakeholders need to respond to these options. Consumers need to purchase responsible products and services. Employees need to choose to work for responsible employers. Suppliers need to provide for responsible buyers. Investors need to finance responsible opportunities. And communities need to welcome responsible entrants and help to sustain responsible incumbents. Responsible capitalism depends on responsible behavior from both business and its stakeholders.

4 The New Story and the Future of Business and Capitalism

Capitalism is simply the greatest system of social cooperation that we have yet invented. It allows free people to cooperate together and create value for each other in a way that no individual can do alone. Business can be a part of solutions to societal problems, rather than a cause—witness Tesla and renewable energy, IBM and smart cities, and recent startups like Milk Stork Inc., based in Palo Alto, California, which provides an option for mothers who travel for business to get breast milk home to their children.

Let us aspire to these kinds of businesses, and others that create value across stakeholders, rather than settling for value only to shareholders. Let us benefit from the implications of a better way to think about business.

R. Edward Freeman is University Professor, Elis and Signe Olsson Professor, Academic Director of the Institute for Business in Society, and Senior Fellow of the Olsson Center for Applied Ethics at the University of Virginia's Darden School of Business.

Heather Elms is Associate Professor of Management at the Kogod School of Business, American University, Washington, DC.

Epilogue

R. Edward Freeman

It is difficult to look back at one's work over a 40+ year time frame and try to come to some conclusions about it. My first reaction is simply "Is that all there is"? My work has been a "one trick pony" that has built on the pioneering ideas of many others. Probably because I have simply said the same things far too many times, I get a lot more credit than I deserve about the development of stakeholder theory. The real pioneers, at least in my mind, were Eric Rhenman and his colleagues in Sweden, Bob Stewart and Marion Doscher at Stanford Research Institute, Juha Nasi in Finland, as well as a variety of others who applied the stakeholder idea in a variety of settings. When I arrived at Wharton to work at the Busch Center and Wharton Applied Research Center, Jim Emshoff, Russell Ackoff, Eric Trist, Ian Mitroff, and others were using the stakeholder idea to organize much of their work in corporate strategy. Emshoff, who was my boss at the Centers, asked me if I wanted to take on the stakeholder idea and see what I could make of it, if people were to take it seriously. Little did I know that this would become a lifelong project and my professional "Raison d'Etre".

I was lucky to meet Ram Charan, Steve Starr, Jagdish Sheth, Jim Wilson, and other outstanding business faculty from many schools, in working with AT&T on the Bell Advanced Management Program. I was in my mid 20s and had absolutely no idea how to teach executives, but I had some ideas (developed of course with Emshoff) on how to take the stakeholder idea seriously, and these great faculty mentored me so I could learn how to teach. The hundreds of executives that we taught in the program took our ideas and applied them to the fast-changing telecom business, and they invited me along to help. That became the source material for the early stakeholder management papers and the book published in 1984 (though written in 1982) *Strategic Management: A Stakeholder Approach*. Chapter 2 of that

R. E. Freeman
University of Virginia, Charlottesville, VA, USA
e-mail: FreemanE@darden.virginia.edu

© Springer Nature Switzerland AG 2023

S. D. Dmytriyev, R. E. Freeman (eds.), *R. Edward Freeman's Selected Works on Stakeholder Theory and Business Ethics*, Issues in Business Ethics 53,
https://doi.org/10.1007/978-3-031-04564-6

book contained my first attempt to trace the origins of the stakeholder idea now called various things from stakeholder theory, stakeholder management, stakeholder engagement and stakeholder capitalism, all stemming from the work of these original pioneers. Later attempts by Giles Slinger at Cambridge University and a joint paper of mine with Robert Strand, tried to clarify this early history. Part of the story still remains somewhat contested.

Of course, I didn't really understand what I was doing as I thought that the most interesting idea in that 1984 book was that business could have a purpose that I called Enterprise Strategy after Peter Drucker. The stakeholder idea seemed common sensical to me. I had grown up, poor, on a dirt farm in Georgia, and even we knew that to be successful in life you had to pay attention to those folks who could affect you or that you could affect. I didn't understand, and still don't, why this idea seemed so radical to so many people.

I had always thought of myself as an ethicist, since getting a PhD in philosophy at Washington University. I was enmeshed in the thinking of the philosopher, John Rawls, and some related economists such as Kenneth Arrow and Amartya Sen. I also had been educated in philosophy by pragmatists, students and students of students of Lewis, Quine, Goodman, and others. So, I thought that a strict empirical positivism was supremely dead. Quine and others had killed it in the 1950s. I was surprised to find a narrow positivism alive and well in business schools. So much so, that there was, and still is, a strong distinction between positive/descriptive and normative research, or between facts and values. I have never seen the point of this distinction except to reinforce some kind of misplaced ego ideal of being a scientist that is present in many business faculties. It has severely inhibited work in business ethics and in stakeholder theory.

As my life progressed, I became more interested in taking on these philosophical assumptions in management theory that I found to be outmoded and that served as barriers to good work. Eventually, I coined the idea of "the separation thesis" or "separation fallacy" out of these ideas. That has not had the impact that my stakeholder work has had, but I think it is probably more important. Once again it builds on the work of others, such as Amartya Sen, and philosophers such as Quine, Davidson and Rorty.

I have come to see most of my work and that of my close colleagues as the Humanities part of a business curriculum. Unfortunately, business schools have become caught up in a kind of careerism that fosters thinking about where something is published rather than the ideas themselves. And, the resulting "Scientism" is very far from tackling real business and societal problems that are so very urgent in today's world. If business schools are to remain relevant institutions in a democratic society, they must return to the purpose of making business a better and more ethical institution and inspiring students to be responsible leaders. I am quite lucky to have spent the last 35 years at the Darden School at the University of Virginia, that has never wavered from this mission.

Many challenges remain and there is much work left to be done. I want to suggest a few questions here that I believe are important as the twenty-first century unfolds. First of all there is a lot of work to be done in understanding how best to make the transition to Stakeholder Capitalism that is now ongoing. There is no going back to the old story of shareholder primacy but much remains to be done to make the transition to more stakeholder-oriented firms and economies more complete. For instance we need more work on the idea of metrics. Finding one set of metrics for all companies is likely to be a fool's errand, in part because businesses are so different. It is perhaps better to think about concepts like "business" and even "leadership" as Wittgensteinian family resemblance ideas. However, we could have different individual metrics, but use some kind of crowd sourcing for comparison. I have suggested that we could enable something like "stake options", but more work needs to be done here.

A second crucial issue is the relationship between business and democracy. Dewey understood that democracy was much more than voting. He knew that we needed to make our institutions, such as schools, NGOs, and businesses more participatory, and they needed to work together to support and enable more democratic governments. Unfortunately we have not made much progress since Dewey's arguments in the last century, and in fact the very institution of democracy has come under attack. Business needs to be a part of the solution here, and I believe that individual businesses must embrace the challenge and help us figure out how to enhance, even save, our democratic future.

Third, I believe that we need to dig even deeper into the Humanities and the Creative Arts for inspiration for business students, executives and scholars. Literature, history, cultural studies, philosophy, social/human sciences, as well as art, music, theater, etc. can all help us to revise and redescribe capitalism into a system that is more human and that is fit for our children. That will require in depth knowledge for business scholars. We cannot just skim the surface. However, nothing is more exciting than trying to understand a new discipline, read some new books, and have some different conversations with others who may not share our world view.

Finally, we need a true "Critical Studies in Management" as Daniel Gilbert, Jr. and I outlined some 30 years ago (and is reprinted in this volume). That is best done with a pragmatist frame of mind that seeks to describe and redescribe our selves and our communities. Critical studies as it exists in many places today does the "critical task" but leaves the reconstruction of something better to another occasion. We no longer can afford this luxury.

Let me be more personal for a moment. Throughout this journey I have been blessed with many generous colleagues, outstanding students, and interesting conversation partners. Many of these folks such as Dan Gilbert, Robert Phillips, Kirsten Martin, Bidhan Parmar, and Sergiy Dmytriyev have been both PhD students as well as colleagues and friends. I would not have been able to contribute to this volume without them. From Jim Emshoff, Ram Charan, Peter Lorange, Bill Frederick, John Rosenblum, Alec Horniman and others to scholars doing stakeholder theory work around the world, I have been exceptionally lucky to have mentors who have served as a crucial support network, especially during the years when stakeholder theory was certainly not an idea in good currency. Working for many years with Professor Patricia Werhane has been a very special privilege. Pat is clearly one of the founding pioneers of the academic discipline of business ethics. I can't even begin to enumerate all the lessons I learned from her, from really reading Adam Smith to our joint effort to train a new generation of business ethics scholars with our PhD program at Darden.

Of course all of these faculty relationships would not be possible without outstanding staff support, and I have had that at all three institutions where I have been. Of particular note is the staff here at Darden who are committed to the mission of improving the world by inspiring our students to be responsible leaders, and developing new knowledge that is useful for business. Being at a purpose driven institution matters a great deal, especially when one is writing about the importance of purpose.

My life partner, Maureen Wellen has been such a central person in my work. She has thankfully kept my ego in check, been a terrific critic of my work, and nurtured a loving relationship throughout our 44 years of marriage. I must also acknowledge the importance of our children, Ben, Emma and Molly. They suffered being raised in stakeholder theory, and we have had countless conversations about it. Indeed, Ben Freeman and I have started a company together, Stakeholder Media, to produce The Stakeholder Podcast, as well as a set of videos that hopefully will cause even more stakeholder thinking around the world.

9783031045639